AIR MONITORING FOR TOXIC EXPOSURES

AIR MONITORING FOR TOXIC EXPOSURES

Second Edition

HENRY J. MCDERMOTT
H. J. McDermott, Inc.
Moraga, California

WILEY-
INTERSCIENCE

A JOHN WILEY & SONS, INC., PUBLICATION

Library of Congress Cataloging-in-Publication Data:
McDermott, Henry J.
   Air monitoring for toxic exposures / Henry J. McDermott.—2nd ed.
      p.   cm.
Rev. ed. of: Air monitoring for toxic exposures / Shirley A. Ness. c1991.
Includes bibliographical references and index.
   ISBN 0-471-45435-4 (cloth)
   1. Air—Pollution—Measurement.   2. Biological monitoring.   3. Air sampling apparatus.   I. Ness, Shirley A. Air monitoring for toxic exposures.   II. Title.
   TD890.M38 2004
   628.5'3'0287—dc22
                                                                                    2003026039

CONTENTS

PREFACE

Shirley Ness contributed a wealth of valuable material in the first edition of *Air Monitoring for Toxic Exposures*. I want to acknowledge her contribution to occupational and environmental health in her excellent overview of the entire topic of air sampling.

When gathering information to revise the first edition, I was impressed by the advances in sampling technology that have occurred in the 15 years since Shirley researched the first edition. Today even pocket-size direct-reading instruments have data-logging capability, which allows them to collect integrated exposure measurements. We are also seeing instruments designed be used with hand-held computers in the field so that data are stored directly in the computer rather than downloaded in a separate step. Sensor technology, microprocessors, and miniaturization have increased the range of direct-reading instruments available, and they also allow sophisticated instruments such as GC/mass spectrometers and Fourier transform infrared devices to be field-portable. Colorimetric systems continue to develop: More sensitive detector tubes that measure more chemicals are on the market, and useful accessories such as battery-powered sampling pumps have been introduced.

The need air sampling to be performed during terrorism events is another reflection of the changing times since the first edition was published. Air sampling can help to determine whether an event has actually occurred and, if so, identify the agent(s) and quantify exposure levels. To address this need, a completely new chapter on air sampling during emergency response including terrorism events has been added.

I appreciate the help of equipment manufacturers and others in providing the photographs used in this edition. Of course there are many quality instruments available other than the ones highlighted in this book; it was not feasible to include all options.

There are a few people who deserve special thanks. Bob Esposito, Jonathan Rose and Lisa Van Horn of John Wiley & Sons were very helpful. Jack Chou of International Sensor Technology graciously permitted use of many diagrams and Appendix C on sensor calibration from his fine book *Hazardous Gas Monitors—A Practical Guide to Selection, Operation and Applications*. Dr. E. C. Kimmel provided the electron microscope photographs of smoke particles to illustrate Chapter 7. Michelle Filby of SKC, Inc. spent extra time providing many photographs in a format that I could use. Galson Laboratories furnished their table of guidelines for air sampling and analysis for Appendix D. Also, a special salute goes to C. W. Pots for editorial advice and general encouragement.

This edition is "dedicated" to all of the air sampling practitioners who perform air monitoring for toxic exposures. They include industrial hygienists, environmental specialists, safety and health technicians, safety professionals, environmental health specialists, chemists, laboratory technicians, firefighters, emergency responders, hazardous materials specialists, and others. Each group has unique skills and background; together they work to help protect people and the environment. I hope that this book is helpful to them.

HENRY J. McDERMOTT,
Moraga, California   CIH, CSP, PE

PART I

BACKGROUND CONCEPTS FOR AIR MONITORING

CHAPTER 1

AIR MONITORING OVERVIEW

The field of air monitoring is broad from every perspective: the reason for performing the monitoring; the types of air contaminants and the potential hazards they pose; the monitoring techniques and equipment; and the background and skills of the people who carry out the sampling. The purpose of this chapter is to give a general overview of the entire topic in order to make the remaining chapters in the book easier to understand and apply. Remember, almost everything summarized in this chapter is covered in more detail later in the book. Additionally, many of the terms used in air monitoring can have more than one meaning, so this chapter presents the nomenclature that will be used throughout the book.

The decision to perform air monitoring for toxic exposures is based on either (a) a regulatory standard that requires monitoring or (b) a hazard evaluation that indicates monitoring is needed to identify or quantify exposures. Hazard evaluations recognize that chemical, physical, and biological agents can cause injury, disease, or death. Potential hazards are evaluated by skilled practitioners based on:

- Toxicity of the material—the material's inherent capacity to cause disease or injury.
- Physical or chemical form of the contaminant—whether a gas, vapor, or particulate matter (aerosol).
- Route(s) of exposure—inhalation, skin, or ingestion.
- Physical hazards—such as the fire or explosion risk posed by the material.
- Likely dose based on an understanding of the exposures that occur as part of the industrial process, from predicable release scenarios, or other exposure situations.
- Effectiveness of exposure controls (engineering interventions, safe work practices, or personal protective equipment) in preventing harmful exposures.

Air Monitoring for Toxic Exposures, Second Edition. By Henry J. McDermott
ISBN 0-471-45435-4   © 2004 John Wiley & Sons, Inc.

An illustration of regulatory-driven monitoring is the U.S. Federal OSHA Benzene Standard. It requires monitoring to determine exposure levels for workers handling benzene-containing liquids containing >0.1% benzene, except for specific situations listed in the standard that have been shown not to cause significant exposures. As an illustration of a typical air sampling "case study," the Federal OSHA Benzene Standard will be cited throughout this chapter as it impacts air monitoring for occupational exposures. Table 1.1 lists the aspects of the standard[1] that are related to exposure monitoring.

AIR SAMPLING IN PERSPECTIVE

Air sampling (or monitoring) is one part of an overall process called *exposure assessment*, which is aimed at defining an individual's or a group's exposure to chemical, physical, and biological agents in the environment. The source of the agents may be natural, industrial operations, vehicle emissions, homes, agriculture, demolition operations, waste disposal sites, accidental releases, intentional releases from terrorism or similar events, or others. The population whose exposure is being measured may be employees of the organization

---

TABLE 1.1. Air Monitoring-Related Requirements—OSHA Benzene Standard

The following items are requirements from the U.S. Federal OSHA Benzene Standard that relate to exposure monitoring. These requirements are a good illustration of how exposure monitoring applies in the occupational setting. Also see Table 1.3, which is a summary of sampling and analytical methods for benzene from the standard.

General
- The standard establishes two permissible exposure limits (PELs):
  - Time-weighted average limit (TWA): 1 ppm as an 8-hour time-weighted average.
  - Short-term exposure limit (STEL): 5 ppm as averaged over any 15-minute period.
- The "action level," which requires some follow-up actions, is 0.5 ppm TWA. Additionally, the term "employee exposure" means exposure to airborne benzene which would occur if the employee were not using respiratory protective equipment.
- STEL compliance is determined from 15-minute employee breathing zone samples measured at operations where there is reason to believe that exposures are high. Objective data, such as measurements from brief period measuring devices, may be used to determine where STEL monitoring is needed.
- Except for "initial monitoring," in cases where one shift will consistently have higher employee exposures for an operation, monitoring is required only during the shift on which the highest exposure is expected.

Exposure Monitoring
- *Initial monitoring* is to cover each job on each work shift and be completed within 30 days of the introduction of benzene into the workplace. For TWA exposures:
  - At or above the action level but at or below the TWA, the monitoring is to be repeated annually.
  - Above the TWA, the monitoring is to be repeated at least every six months.
  - Below the action level no follow-up monitoring for that employee is required unless conditions change or during spills or releases.
- The monitoring schedule may be reduced from every six months to annually for any employee for whom two consecutive measurements taken at least 7 days apart indicate that the employee exposure has decreased to the TWA or below, but is at or above the action level.

TABLE 1.1. *Continued*

- Monitoring for the STEL shall be repeated as necessary to evaluate exposures of employees subject to short term exposures.
- Periodic monitoring may be discontinued if at least two consecutive of the required periodic monitoring exposures, taken at least 7 days apart, are below the action level, unless conditions change or during spills or releases.
- Whenever spills, leaks, ruptures or other breakdowns occur that may lead to employee exposure, monitoring is required (using area or personal sampling) after the cleanup or repair to ensure that exposures have returned to the level that existed prior to the incident.

Other Monitoring-Related Requirements
- Employees are to be notified of monitoring results within 15 working days after the results are received, either individually in writing or by posting results in an appropriate, accessible location. Whenever the PELs are exceeded, the notification must describe the corrective action being taken to reduce the employee exposure to or below the PEL.
- Exposure controls are required for exposures above the PELs.
- Respiratory protection, if used, is to be selected based on exposure levels:

Airborne Concentration or Condition of Use	Respirator Type
≤10 ppm	• Half-mask air-purifying respirator with organic cartridge.
≤50 ppm	• Full facepiece respirator with organic cartridges or chin-style canister.
≤100 ppm	• Full facepiece powered air-purifying respirator with organic cartridges
≤1000 ppm	• Supplied air respirator with full facepiece in positive-pressure mode.
>1000 ppm or unknown concentration	• Self-contained breathing apparatus with full facepiece in positive pressure mode. • Full facepiece positive-pressure supplied-air respirator with auxiliary self-contained air supply.
Escape	• Any organic vapor gas mask. • Any self-contained breathing apparatus with full facepiece.
Firefighting	• Full facepiece self-contained breathing apparatus in positive pressure mode.

- Medical surveillance program (meeting the requirements of the standard) must be made available for employees who are or may be exposed to benzene at or above the action level 30 or more days per year; for employees who are or may be exposed to benzene at or above the PELs 10 or more days per year; and for employees who have been exposed to more than 10 ppm of benzene for 30 or more days in a year prior to the effective date of the standard when employed by their current employer. Information provided to the physician is to include the employee's actual or representative exposure level.
- Employees require initial training. If exposures are above the action level, employees are to be provided with information and training at least annually thereafter.
- A record of all exposure monitoring is to be maintained for 30 years. This record is to include:
  ○ The dates, number, duration, and results of each of the samples taken, including a description of the procedure used to determine representative employee exposures;
  ○ Description of the sampling and analytical methods used;
  ○ Description of the type of respiratory protective devices worn, if any; and
  ○ Name, social security number, job classification and exposure levels of the employee monitored and all other employees whose exposure the measurement is intended to represent.
- Observation of monitoring: Employees, or their designated representatives, are to have an opportunity to observe the measuring or monitoring of employee exposure to benzene conducted pursuant to the standard.

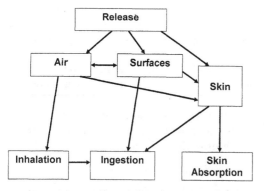

Figure 1.1. Pathways showing how occupational and environmental releases can result in human exposure to contaminants.

releasing the contaminant (e.g., occupational exposure), members of the general population, emergency responders or hazardous materials specialists, or another group. Figure 1.1 shows a general model of exposure pathways typical of airborne release of toxic contaminants. For very comprehensive exposure assessments, Figure 1.1 might have to be expanded to include other pathways such as ingestion of contaminated water.

The topic of exposure assessment is a technical specialty by itself;[2] more information on the topic is contained in Chapter 3. The focus of this book is on the air monitoring phase of the exposure assessment process.

AIR SAMPLING STRATEGY AND PLAN

Air samples are collected for a variety of purposes and under different circumstances:

- *Occupational exposures*—generally the contaminants are known and monitoring is performed to evaluate legal compliance or conformance with recommended exposure limits. Emphasis is on collecting enough samples to ade-

quately describe the *typical* average work shift and peak short-term exposure patterns and also describe the *worst-case* exposures.

- *Community environment*—longer-term sampling to assess community exposure to a wide variety of expected pollutants including industrial emissions, vehicle exhaust and evaporated fuel components, airborne dust, and so on. Short-term sampling is performed to (a) identify odors or irritants based on community complaints or (b) identify toxic components in smoke or uncontrolled releases.

- *Indoor air quality evaluations*—long- and short-term monitoring of carbon dioxide, humidity, carbon monoxide, asbestos, organic vapors, bioaerosols, and radon and its progeny.

- *Emergency response* (including terrorism events)—sampling to identify unknown agents; direct reading tests for flammable vapors and other common contaminants to evaluate immediate hazards; exposure measurements to document exposures to responders to hazardous material releases and other emergencies and to determine the appropriate level of respiratory protective gear.

The air sampling approach in each of the above applications will be different. For example, in occupational exposures there is a choice about whether to carry out a statistical sampling campaign to gather enough samples for a valid statistical analysis of results or to identify and evaluate expected "worst-case" exposure patterns to ensure compliance. For emergency responders the emphasis is on preplanning to ensure that the right monitoring equipment is available and is deployed quickly to identify/assess potential hazards. During an emergency response, sampling devices that give immediate readings at the scene are preferred (where feasible) over possibly more accurate sampling methods that require laboratory analysis. However, for emergency response, laboratory samples are often used to verify initial field readings.

To ensure that the air samples meet the needs of each situation, a *monitoring strategy* is developed to clearly answer questions such as:

- What is the reason for this monitoring?
- What jobs, tasks, or exposures should we monitor?
- What contaminants should we measure?
- What monitoring techniques should we use?
- How many samples should we collect?
- When should we monitor?
- Who will perform monitoring and laboratory analysis?
- How will the results be used and communicated?

It is also important to note that even very accurate air sampling may not reflect an individual's total exposure to a material if ingestion, skin absorption, or mucous membrane absorption contributes significantly to the overall exposure. Additionally, any periods where the individual wears respiratory protection must be accounted when evaluating the health implications of the exposure. These topics should be addressed in the *strategy*.

After the strategy is completed, a written *monitoring plan* is prepared to spell out the details of the monitoring, such as number and types of samples that will be collected. Both the strategy and plan are continually reviewed as initial samples are collected or more experience is gained, and both are modified to reflect the new information. See Chapter 3 for more details on the *monitoring strategy* and *plan*.

TYPES OF AIR MONITORING

There are three main types of air samples: personal, area, and source.

Personal Sample

These samples measure a particular individual's exposure to airborne contaminants. These measurements are performed using sampling devices that the person wears or that are otherwise positioned in the person's breathing zone as they go about their activities. Personal samples are usually collected for comparison with an exposure standard such as: a regulatory requirement; a recommendation from a consensus group, professional association, or government agency; or an internal standard adopted by the employer or other organization concerned with the exposure. Often the results of personal monitoring are considered to be representative of a larger population group than the people who were actually monitored, such as community residents or workers in the same job classification. Because the individual usually wears the sampling device, its weight and bulk are an important factor when selecting the sampling device. Typical sampling devices include either small battery-powered pumps with a suitable collection device (Figure 1.2), a passive sampler that relies on diffusion of the contaminant into the sampler where it is captured (Figure 1.3), or a small real-time direct reading instrument with a high-level alarm and data storage capability (Figure 1.4).

Area Sample

With this type of sample the contaminant levels are measured at a particular area, either continuously, periodically, or at a discrete point in time. This type of sampling is used where:

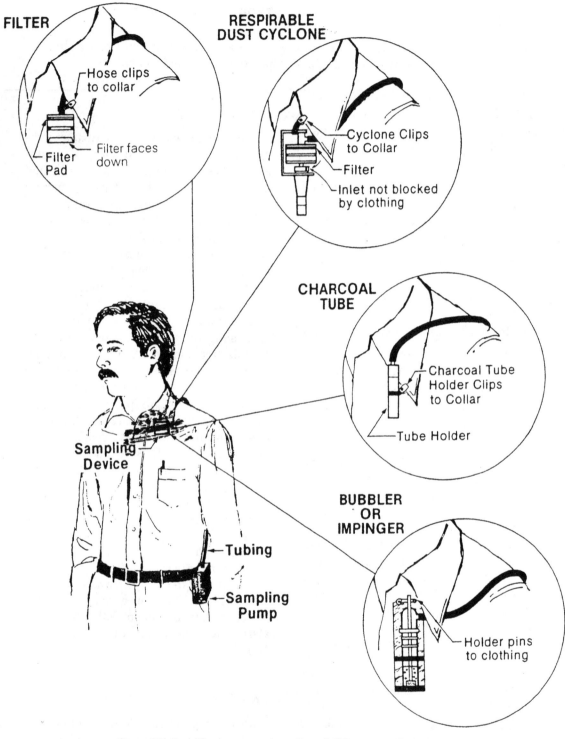

FILTER

Hose clips to collar

Filter Pad

Filter faces down

RESPIRABLE DUST CYCLONE

Cyclone Clips to Collar

Filter

Inlet not blocked by clothing

CHARCOAL TUBE

Charcoal Tube Holder Clips to Collar

Tube Holder

Sampling Device

Tubing

Sampling Pump

BUBBLER OR IMPINGER

Holder pins to clothing

Figure 1.2. Breathing zone samples collected with a personal pump.

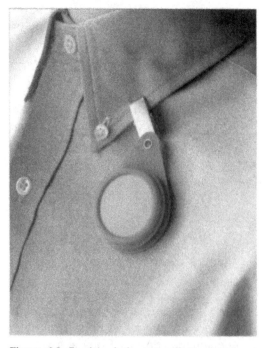

Figure 1.3. Passive dosimeter collects gases and vapors without a sampling pump. (Courtesy of SKC, Inc.)

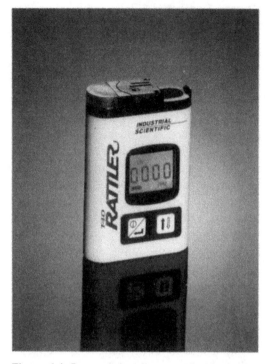

Figure 1.4. Personal-size direct reading electronic instrument for air contaminants. (Courtesy of Industrial Scientific Corporation.)

- Suitable personal sampling (portable) techniques are not available, and so area measurements are the best surrogate for personal samples.
- A fixed direct reading system (often with multiple sensors or pickup points located throughout the area) that periodically or continuously measures airborne levels is advantageous for confirming that concentrations are low, as well as for identifying leaks or other causes of higher-than-expected levels. The sampling points must be near enough to likely leak or release locations so that high airborne levels in the area trigger warning alarms before people are overexposed.
- Portable survey instruments ("sniffers") are used to measure airborne levels before people are allowed to enter or remain in the area. These are often used for emergency releases and when people work in confined or enclosed areas.

Where the area sampler is a direct reading device, it is usually equipped with audible and/or visible alarms to warn of high levels.

Area samples are generally not a good way to estimate personal exposures because people normally move around and so are not continuously exposed to the concentration that exists at the fixed measuring location. The likelihood of underestimating actual exposures is obvious if an employee works nearer to the contaminant source than the sampling point is located. However, there are also many opportunities to overestimate exposures using area readings, so caution should always be used. If area samples are used to estimate personal exposure, it should be done as part of a detailed study where the location and movement pattern of the people are recorded along with the variables (such as production rate, release rate, amount spilled, etc.) that govern airborne levels.

Source Sample

These are samples collected directly at the contaminant source either to measure normal release rates or to immediately identify any leaks, releases, or control system malfunctions. These may appear to be similar to "area" samples, but are different in that they do not purport to reflect concentrations in the general work or occupied area. For example, some air pollution regulations require periodic "sniffing" using a direct reading device of pipe flanges, seals on rotating pumps, other locations where volatile organic compounds may be released, or other piping connections. Another application is where a laboratory hood containing flammable gases is equipped with a sensor and alarm device to warn of concentration buildup due to gas leaks or a malfunctioning ventilation fan. It is important to understand when source rather than area data are being collected since source values cannot be used to estimate personal exposures.

Integrated Versus Grab Samples

Air samples can be further categorized:

- *Integrated sample*—estimates exposure over a time period such as a work shift by collecting one or more personal samples that cover the entire time period. The term "integrated" means that the measured exposure level integrates or averages all of the different concentrations and time durations during the time period of interest. Community environmental standards may refer to a 24-hour exposure, while occupational exposure standards often specify the allowable integrated exposure over the work shift or other time period, which is called a *time-weighted average* (TWA) exposure level. Some exposure standards specify an allowable *short-term*

exposure limit (STEL) measured over 15 minutes or so—in this case, one or more separate sample(s) would be collected to cover the 15-minute time period(s) of expected highest exposure. This 15-minute sample would also be referred to as an *integrated* sample since it covered the time period of interest (15 minutes).

- *Instantaneous or grab sample*—a sample collected over a very short time period, usually less than 5 minutes. These samples are used to evaluate a peak or "ceiling" (maximum) exposure. Some sampling devices can only collect this type of sample since they have a very short sampling time (such as a "hand pump and detector tube," described later). It is very difficult to estimate a full-shift personal exposure from a series of grab samples, although there are statistically based sampling strategies that can be applied. Typical uses of grab samples are: to determine maximum peak exposures; for range finding sampling to determine if an airborne contaminant is present as a precursor to more detailed monitoring; or for "go/no go" decisions following purging or ventilation of a work area.

AIR SAMPLING TECHNIQUES

There are two main categories of sampling techniques: (a) *sample collection devices* that are analyzed in a laboratory (b) and *direct reading instruments*. Both of these techniques have application with the major types of contaminants: gases, vapors, and particulate matter (aerosols):

- *Gas*—a material with very low density and viscosity and that readily and uniformly distributes itself throughout any container at normal temperature

TABLE 1.2. Air Monitoring Techniques

Sample Collection Device Sampling	Direct Reading Methods
Sample collection methods	Electronic instruments
Pump-based methods	Combustible gases
Passive methods	Specific gas or vapor
Laboratory analysis	Nonspecific gas and vapor
Chemical analysis	Particulate matter
Other analysis	Colorimetric systems
	Detector tubes
	Paper-tape or liquid devices
	Other colorimetric devices

and pressure. As used in this context, a gas is a material that exists in a gaseous state at normal ambient conditions. Concentrations of gases are expressed in mass of contaminant per unit volume of contaminated air (mg/m^3) or parts of contaminant per million parts of contaminated air (ppm) by volume.

• *Vapor*—the gaseous form of substances that are normally solid or liquid at room temperature and pressure. Vapors are generally formed when volatile liquids, such as solvents, evaporate. Concentrations of vapors are expressed in the same units as gases.

• *Particulate matter*—discrete units of fine solid or liquid matter, such as dust, fog, fume, mist, smoke, or sprays. Particulate matter suspended in air is commonly called an *aerosol*. Concentrations of particulate matter are generally expressed in mass of contaminant per unit volume of contaminated air (mg/m^3), although units such as fibers/cm^3 of air are used for asbestos and some other fibrous materials.

Table 1.2 shows how the specific monitoring techniques are categorized under these two main headings. Appendix D is a table listing the recommended sampling techniques for common contaminants. For air sampling considerations, gases and vapors often behave similarly and thus are discussed together below.

SAMPLE COLLECTION DEVICES

With this approach a sample is collected in the field and returned to a laboratory for analysis. The summary below focuses on air samples to determine an individual's personal exposure level; similar principles apply to sampling for environmental quality purposes except that the equipment is generally larger. To ensure accurate results, all aspects of sampling and analysis including type and size of sampling device, sampling rate and total sample volume, and sample storage and handling *must* follow the sampling procedure or protocol for the method and the material(s) being analyzed. Table 1.3 (from the *OSHA Benzene Standard*) is an excerpt of a typical air sampling procedure. It is especially important to discuss the sampling beforehand with the analytical laboratory so the sample will be collected and handled in a manner that permits accurate analysis once it reaches the laboratory. Factors to consider include:

TABLE 1.3. Summary of Sampling and Analytical Methods for Benzene

Methods for Benzene Monitoring
(Excerpt from OSHA Benzene Standard—Appendix D)

General:
- Measurements are best taken so that the representative average 8-hour exposure may be determined from a single 8-hour sample or two 4-hour samples.
- Short-time interval samples (or grab samples) may also be used to determine average exposure level if a minimum of five measurements are taken in a random manner over the 8-hour work shift. The arithmetic average of all such random samples taken on one work shift is an estimate of an employee's average level of exposure for that work shift.
- Air samples should be taken in the employee's breathing zone (air that would most nearly represent that inhaled by the employee). Sampling and analysis must be performed with procedures meeting the requirements of the standard.
- The employer has the obligation of selecting a monitoring method that meets the accuracy and precision requirements of the standard.
- The method of monitoring must have an accuracy, to a 95% confidence level, of not less than plus or minus 25% for concentrations of benzene greater than or equal to 0.5 ppm.

Methods:
- Collection of the benzene vapor on charcoal absorption tubes, desorption in the laboratory with carbon disulfide, and subsequent chemical analysis by gas chromatography (OSHA Method 12).
- Portable direct reading instruments, real-time continuous monitoring systems, passive dosimeters, or other suitable methods.

Overview of OSHA Method 12 for Air Samples:
- Recommended air volume and sampling rate when using standard charcoal tubes containing 100 mg of charcoal in the front section and 50 mg of charcoal in the backup section of the tube: 10-L sample collected at 0.2 L/min. Air sample size is limited by the number of milligrams that the tube will hold before overloading.
- When the sample value obtained for the backup section of the charcoal tube exceeds 25% of that found on the front section, the possibility of sample loss exists.
- The limit of detection for this analytical procedure is 1.28 ng with a coefficient of variation of 0.023 at this level. This would be equivalent to an air concentration of 0.04 ppm for a 10-L air sample.
- Field Sampling Considerations:
  - Calibration of personal pumps. Each pump must be calibrated with a representative charcoal tube in the line.
  - The charcoal tube should be placed in a vertical position during sampling to minimize channeling through the charcoal.
  - Air being sampled should not be passed through any hose or tubing before entering the charcoal tube.
  - A sample size of 10 L is recommended. Sample at a flow rate of approximately 0.2 L/min. The flow rate should be known with an accuracy of at least (+ or −) 5%.
  - The charcoal tubes should be capped with the supplied plastic caps immediately after sampling.
  - Submit at least one blank tube (a charcoal tube subjected to the same handling procedures, without having any air drawn through it) with each set of samples.
  - Take necessary shipping and packing precautions to minimize breakage of samples.
- Gas Chromatograph Parameters:
  - Use a flame ionization detector (FID).
  - Column: 10-ft × 1/8 -in stainless steel packed with 80/100 Supelcoport coated with 20% SP 2100, 0.1% CW 1500.
  - Operating Conditions: 30 mL/min (60 psig) helium carrier gas flow; 30 mL/min (40 psig) hydrogen gas flow to detector; 240 mL/min (40 psig) air flow to detector; 150°C injector temperature; 250°C detector temperature; 100°C column temperature; 1-μL injection size.

- Physical and chemical characteristics of the contaminant(s)
- Possible interferences to the collection or analysis
- Required accuracy and precision
- Type of sample (personal, area, source)
- Duration of sampling period

Pump-Based Sampling

The sampling arrangement consists of a calibrated battery-powered pump, collection device, and connecting tubing (Figure 1.2). Some procedures may also include an airflow control valve, an airflow measuring device, and a special particle size separator device (for some particulate matter sampling). The pump sampling rate is adjustable and is chosen to match the airflow requirements of the sampling procedure—often personal pumps are categorized as either "high flow" (1–10 L/min) or "low flow" (1–1000 cm^3/min). Some battery-powered sampling pumps can operate as either high- or low-flow pumps. Two different low-flow designs are available: (a) a stroke-volume (pulsating flow) pump that functions using a known-volume piston to move the air while a counter records the number of piston strokes and (b) a continuous-flow pump with an airflow rate indicator.

Before use, the pump is calibrated (see Chapter 5) so the airflow rate is known; after use, and periodically for a long-duration sample, the pump calibration is checked to ensure that it has not changed. Once the pump flow rate is determined, the volume of air sampled is calculated from either:

- Airflow rate (L/min) × sampling time (minutes) for a continuous-flow pump
- Stroke volume (cm^3/stroke) × number of pump strokes for a piston-type pump

Sampling using evacuated containers can be considered a modification of pump sampling (Figure 1.5). In this case a vacuum pump is used to produce a vacuum in a suitable container. The container may be plain stainless steel or have a specially treated interior surface for sampling reactive chemicals. Container size depends on sampling time; a 6-L volume usually is sufficient for a 24-hour sample. A valve on the container is opened to start sampling as ambient air is drawn into the container. A flow control orifice or other device is used to regulate the airflow into the container so that the flow rate is constant over the sampling period. The valve is closed at the end of the sampling period.

For all pump-driven sampling techniques, a key parameter to be decided is the sampling rate for the sample. The concepts of *limit of detection* and *limit of quantification* help determine the best sampling rate to use:

- Limit of detection (LOD) is the least amount of contaminant that can be "seen" by the analytic method even though the amount cannot be accurately determined.
- Limit of quantification (LOQ) is the minimum amount of contaminant that can be quantified with some specified degree of accuracy.

Analytical results below the LOD are reported as "<LOD (with the LOD specified)." Analytical results greater than the LOD but below the LOQ are reported as "detected, but not quantified (with the LOQ specified)." Some analytical methods (particularly older versions) may not specify a LOQ, but the LOD should always be specified in the written procedure.

The concept of LOQ is used to determine the minimum sampling volume needed for accurate analysis, which in turn determines the sampling rate. In order to perform this calculation, you must first

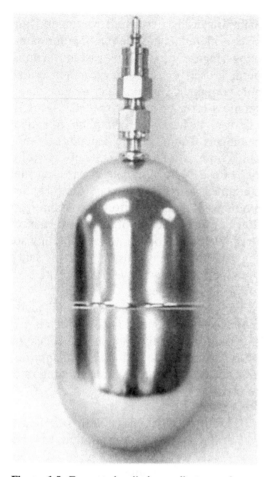

Figure 1.5. Evacuated cylinders collect samples at a controlled rate without use of a pump. They are available in a variety of sizes. (Courtesy of Galson Laboratories.)

decide the minimum airborne concentration that you need to quantify. Usually this is some percentage (often 10% or lower) of the allowable exposure limit.

$$SV = \frac{LOQ}{EL \times F}$$

where SV is the minimum sample volume (L), LOQ is the lower limit of quantification (µg), EL is the allowable exposure limit (mg/m^3), and F is the percent of EL that needs to be quantified.

Once the sample volume is determined, that value can be used to select the sampling rate and sampling time since

$$SV(L) = \text{Sampling rate (L/min)}$$
$$\times \text{Sample time (min)}$$

In some cases the sampling time is fixed (e.g., for a 15-minute short-term exposure sample), and so the above equation is used to determine the sampling rate. In other situations the maximum sampling rate is dictated by the pump capacity and so the equation determines minimum sampling time. If the sampling method you are considering will not collect enough contaminant to accurately measure the minimum airborne level you are interested in, then it is necessary to look for another method.

For many sampling/analytical methods there is also a *maximum* amount of contaminant that can be quantified in a sample either because of overloading the collection device or other problems. Generally the sampling duration for each sample can be reduced to avoid this type of problem. For example, instead of collecting one 8-hour sample during the shift, it might be necessary to collect four 2-hour samples to cover the shift to avoid overloading the sample device. The time-weighted average concentration is then calculated to yield the average for the shift.

The specific sampling collection device depends on the material being sampled:

Gases and Vapors. For gases and vapors, typical collection devices include:

- Adsorption tubes (also called *sorbent tubes*) containing charcoal (activated carbon), silica gel, or another sorbent (Figure 1.6), depending on the contaminant. The tubes contain two sections of adsorbing material separated by an inert spacer. As contaminated air is drawn through the tube, the airborne chemical is adsorbed (deposited on

Figure 1.6. Charcoal adsorption tubes collect organic vapor for later laboratory analysis. (Courtesy of SKC, Inc.)

Figure 1.7. Midget impingers collect air contaminants in a liquid for later laboratory analysis. (Courtesy of SKC, Inc.)

the surface) on the adsorbent particles in the front section. If the front section becomes saturated with the contaminant molecules, the remaining contaminant molecules "break through" the front layer and are collected on the backup section. At the laboratory, each section is analyzed separately; and if significant breakthrough has occurred, the results will be questionable and so the sampling should be repeated. A special *screening* adsorption tube is available that has several different adsorbent layers in a single tube. These are useful for initial sampling for unknown contaminants and in some indoor air quality investigations. Vendors can also provide custom-packed adsorption tubes to meet special sampling needs.

- Absorption devices such as a small vial with a bubbler fitting. The vial holds deionized water to absorb the airborne gas or vapor, or a dilute chemical reagent to react with the contaminant (Figure 1.7). These are used where adsorption tubes are not available, but the added steps of handling liquids in the field often make these techniques less desirable than an adsorption tube.

- Specially treated filters similar to those described below for particulate

sampling which are used for monitoring certain reactive chemicals. The airborne chemicals react with the compounds are on the filter to form other chemical compounds that are stable and can be measured in the laboratory.

- Sampling bags made of Tedlar or Teflon can be used when the air concentration is above the LOD (and so the ambient air can be analyzed directly back in the laboratory). A sampling pump or other means is used to fill the bag with ambient air. Bags are rarely used when another sample collection method is feasible due to the bulk of the bag, shipping limitations, and the possibility of sample degradation prior to analysis.

Particulate   Matter. For   particulate matter, high-flow pumps along with these sampling devices are most commonly used:

TABLE 1.4. Typical Air Sampling Filter Materials

Filter Material	Characteristics	Typical Applications
Glass fiber	High particulate retention and wet strength	Gravimetric analysis
Polyvinylchloride (PVC)	Low tare weight	Gravimetric analysis
Mixed cellulose ester (MCE)	Dissolves easily	Metals and fibers
Gelatin	High moisture content, can be pre-sterilized	Airborne microbes
Teflon	Strong and chemically resistant	Acids, bases, and solvents
Polycarbonate	Glass-like surface, transparent and straight-through pores	Gravimetric and microscopic analyses

Source: Reference 3.

• *Filters.* Different filter materials are available, depending on the contaminant being sampled; Table 1.4 lists a few common filter materials. Since sampling with the incorrect filter often makes analysis impossible, it is critical for the sampling practitioner to coordinate with the laboratory to ensure that the proper filter is used for each type of air sample. The usual filter is 37 mm in diameter, enclosed in a filter cassette unit attached to the pump. The cassette has a small hole for the air being sampled to enter, which avoids deposition of larger particles in the ambient air onto the filter. The filter by itself is used to collect total particulate samples, which may include particles too large to actually reach or be deposited in the human respiratory system when inhaled. Various-size separators are used to collect only those particles within the size distribution of interest. For example, a small "cyclone"-size separator can be mounted before the filter to separate and discard particulates that are larger than an established "respirable size" distribution. Asbestos samples are collected on 25-mm-diameter mixed cellulose ester (MCE) filter in an electrically conductive cassette to avoid fiber loss due to electrostatic

Figure 1.8. Six-stage inertial impactor for particulates. (Courtesy of Thermo Electron Corporation.)

effects. Asbestos sampling is performed with the full filter face open to the ambient air in order to get an even distribution of fibers across the filter since the analytic techniques involves optical counting of the fibers.

• *Inertial Impactors.* These retain particulates due to impaction as the airflow hits a "collecting" surface. The collection efficiency is determined by the mass of the particulate, the characteristics of the collecting surface, and the velocity of the air stream. Inertial impactors with several "stages" can be used to obtain a size distribution of the particulate matter (Figure 1.8). The stages can be weighed after sampling to determine the mass of contaminant retained, or they can be analyzed using chemical methods. For bioaerosol sampling, agar plates or other suitable growth media are used as the collecting surfaces.

- *Impingers.* These capture particles in a liquid (usually water) after being drawn at high velocity into a liquid-filled vial (Figure 1.7). The particles impinge on the vial bottom, then lose their velocity and are trapped in the liquid. A sample of the liquid is counted in a special cell under a microscope to determine particle count and size distribution. These devices are generally used if a filter or other suitable technique is available.

Passive Sampling

With this technique no pump or other air moving device is used—the contaminant diffuses at a predictable rate according to a scientific principle called Fick's Law. The contaminant molecules move from the area of higher concentration (the ambient atmosphere) to the zone of lower concentration (inside the collection device) where they are trapped for later analysis.

Gases and Vapors. A passive monitor for gases and vapors is typically a small badge-like device that clips to the person's collar or can be mounted for fixed area samples (Figure 1.3). Other devices resemble small glass vials. These devices consist of a protective membrane covering the opening and a collection medium pad or matrix inside the device. The collection medium (such as charcoal) for any contaminant is generally the same as used in the adsorption tubes described above, and the laboratory analysis is similar. Some devices have two collection stages: the main collector plus a backup that will indicate if the

main collector pad was saturated or overloaded during air monitoring. Passive monitors are available to monitor for over 200 substances.

The "sampling rate" for any passive device for a specific contaminant is determined by the "diffusion coefficient" of the chemical and also by the diffusion path or distance based on the internal dimensions of the device. The effective sampling rate is provided by the device's manufacturer.

Sample collection occurs as long as the monitor is exposed to the atmosphere, and it continues until the device is resealed. Depending on the design of the particular device, they are provided in a protective package or with a protective cover over the front membrane. The start and end times are recorded and used to calculate equivalent sample volume. For accurate measurements the devices need a certain minimum airflow across the membrane to ensure that the air layer at the membrane represents the ambient atmosphere. Normal worker motions are sufficient to produce the needed velocity, and for fixed badges normal air currents should suffice. If the devices are mounted in stagnant locations, a room fan or other similar method of generating air movement may be required.

The major advantage of passive sampling compared to pump-driven monitoring is the simplicity of not having to deal with a pump with the attendant pump calibration, maintenance, and power (battery charging or line power supply) issues. For short-term samples at low contaminant levels the limit of quantification may become an obstacle for some passive devices. The unit cost of passive monitors is greater than the cost of comparable adsorption tubes, but the overall cost (when the pump and people's time is included) is generally lower for passive sampling. Note that there are also direct reading passive dosimeters; these are discussed later under the "Colorimetric Systems" heading later in this chapter.

Particulate Matter. For particulate matter samples collected for laboratory analysis, passive monitors do not compete with pump-driven sampling as they do for gas and vapor sampling. Settled dust samples can be collected on an open Petri dish or similar surface for chemical identification, or on nutrient plates to culture airborne bioaerosols. However, passive measurements cannot be used to determine breathing zone concentrations of particulate matter in mg/m^3, $fibers/cm^3$, or similar concentration units.

Laboratory Analysis of Samples

When performing monitoring with the *sample collection devices* described above, the samples are sent to a laboratory for analysis of the collected material. This section provides a brief overview of the various analytical techniques. Often the person collecting the samples is not directly involved with their analysis, and they may view the laboratory analysis as a function too detailed or complex to get involved with as long as the results are timely and accurate. This is understandable, but the likelihood of valid sample results is increased if the sampling practitioner has at least a general understanding of the laboratory methods and possible interferences for their samples. In every case it is important to consult with laboratory specialists *before* the samples are collected in order to ensure that the laboratory will be able to analyze the samples for the contaminants of interest in the concentrations anticipated and with any possible interference from other airborne compounds.

Although there are a variety of analytical techniques used for air sample analysis, a key function of every laboratory is quality assurance. This is the process of following the appropriate method, performing equipment calibration, analyzing reagents for interferences, determining sample recovery from the collection device, evaluating shelf life of samples, running analytical blanks, performing duplicate analysis on split samples, and training and evaluating staff that helps ensure that the results are accurate and reproducible. "Pre-spiked" adsorbent tubes and passive dosimeters can be purchased to assist in the quality assurance effort. These are usually "loaded" with the contaminant of interest to represent half of the allowable exposure level, and they are submitted to the laboratory along with field samples. A good indication of strong quality control is when the laboratory has achieved accreditation by a recognized body. The Industrial Hygiene Laboratory Accreditation Program administered by the American Industrial Hygiene Association is a leading example of an accreditation program.[4] Similar programs are also available from other organizations.

Chemical Analytical Techniques. These techniques identify the mass or amount of a chemical compound or class of compounds in a sample. There is some overlap between the chemical analysis techniques used for gases, vapors, and particulate matter, so they are covered together:

Gas chromatography (GC) is commonly used for gases, organic vapors such as solvents or alcohols, and some compounds that are solids at room temperature but can be volatized sufficiently at 225°C. It operates on the principle that a volatilized sample is mixed with a carrier gas and injected into a column that separates the components in the sample according to the time it takes the component to travel through the column (Figure 1.9). As the molecules emerge from the column, a detector measures the amount of each material. On a chromatogram, each emerging compound is represented by a "peak" based on its elution time. Component identification is made using a data "library" developed by injecting samples of known chemicals into the column and measuring the travel time for each. The concentration

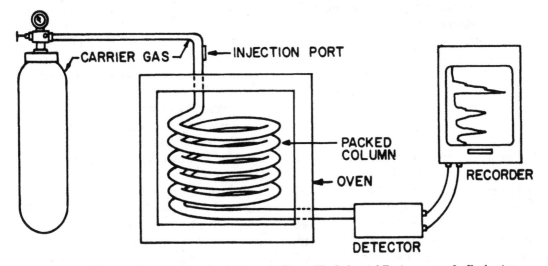

Figure 1.9. Diagram of a typical gas chromatograph. (From *The Industrial Environment—Its Evaluation and Control*, NIOSH, Washington, D.C., 1973.)

of each material is represented by the area under its peak on the chromatogram.

The GC's ability to identify many different chemicals, especially those in mixtures, is achieved through proper selection of column, detector, and temperature programming. The ideal situation is to choose the operating parameters to yield sharp and narrow peaks that are easy to identify and quantitate for the materials of interest:

- There are two types of columns: packed and capillary. The packed column, typically 1/8 inch in diameter and up to 20 feet long, contains an inert solid support that is coated with a liquid material. The capillary column is a very narrow tube (<1 mm in diameter) with a coating on the wall of the column, which may be over 100 m long. The function of the column geometry and coating is to control the movement of sample molecules so the optimum separation is achieved for the materials of interest as they emerge from the column.

- Different detectors are used to measure the materials as they emerge from the column. The most widely used is the flame ionization detector (FID) that functions by burning the sample in a hydrogen flame and measuring the current produced by the ionized material. Organic materials like hydrocarbons that burn have a much greater response than materials that do not burn. Other common detectors include (a) the electron capture (EC) detector used to measure chlorinated pesticides, PCBs, and other halogenated materials and (b) the thermal conductivity detector (TCD) that responds to gases like oxygen and nitrogen as well as organic compounds. The GC/mass spectrometer (GC/MS) is a special detector that passes the ionized molecules emerging from the column through an electric field that focuses the molecules by atomic mass. This permits identification of specific compounds using a library of data developed for the specific instrument.

- Temperature programming involves increasing the temperature of the column in a predetermined manner as the analysis proceeds. This enhances separation of many complex mixtures because the lower-molecular-weight compounds move through the column at the lower temperature, while heavier molecules begin to move more quickly as the column temperature is increased.

The most common air sampling application is use of a charcoal sorbent tube for field air sampling, such as for benzene (see Table 1.3). The GC column, detector, and temperature program are selected as specified in the analytical procedure for the compound(s) being measured. The GC instrument is set up and tested using quality assurance steps, which may include (a) checking the solvents and other reagents for interferences and (b) calibrating the instrument using known concentrations of the contaminant being measured. In the laboratory, the actual field sample tubes and one or more "field blank" tubes are handled the same way: The main and backup sections of charcoal are removed from the tube and placed in separate vials. A small quantity of solvent is added to each vial and the contents agitated to desorb the captured contaminant. A portion of the solvent/contaminant sample is injected into the GC, and the analytical cycle is initiated. The GC runs through the temperature program, and the peak area on the chromatogram for each contaminant is analyzed by computer to yield the amount of that contaminant in the sample. If the backup section of the sample contains more than 25% of the amount on the front section, the possibility of sample loss due to breakthrough in the field is noted on the analytical report. If needed, the analytical results can be adjusted to account for any contaminant that was not desorbed by the solvent based on desorption efficiency tests. Because only a portion of the solvent/contaminant sample was injected into the GC, the run can be repeated to

check reproducibility or if there were any instrument problems during the first test. Sometimes the analysis is repeated on one or more instruments using different test conditions in order to measure all contaminants of interest.

Another desorption technique is *thermal desorption*, where the adsorption tube is heated and the collected contaminants are "blown" off the tube with the carrier gas into the GC column. This technique can achieve a lower LOQ than the solvent desorption method since the entire sample (rather than a small portion) is analyzed. However, since the entire sample is analyzed, the GC run cannot be repeated with the same sample if there are any analytical problems.

High-pressure liquid chromatography (HPLC) is a technique that may be used for compounds unsuitable for GC analysis because they are thermally unstable or not volatile enough. HPLC uses a column to separate compounds, but the carrier material is a liquid rather than a gas. To enhance separation, the composition of the carrier solvent may be changed as the analysis progresses, similar to the temperature increased used with GC to improve separation. The detector for HPLC must be suitable for measuring the compounds of interest in a liquid and may be based on principles of electrochemical detection, refractive index, ultraviolet or visible light absorption, or fluorescence.

Infrared spectroscopy is a technique that uses an optical instrument that measures the absorption of "light" in the infrared spectrum by the sample. It can be used for solids, liquids, and gases. The general principles of operation are described in Chapter 12 on direct reading instruments.

Atomic absorption spectrophotometry (AA) is used for analyzing air samples (often collected on filters) for metals. The technique involves dissolving a portion of the sample filter in an acid, and then atomizing this solution in a flame and measuring the how much light at one or more specific wavelength(s) that the flame absorbs. The light source usually is made with a cathode made from the metal being analyzed for, which emits a spectrum of that element. AA is highly specific and has few interferences, and it routinely used for over 50 metals including lead, mercury, chromium, arsenic, copper, zinc, cadmium, nickel, beryllium, and others of occupational health or environmental concern.

Ultraviolet-visible spectroscopy (UV-VIS), or wet chemistry, is based on the absorption of light by a colored compound where the absorption is proportional to the concentration of the material. It can be used for molecules that absorb either UV or visible light, or where the molecule can be reacted to form another compound that does. UV-VIS is generally less sensitive and less specific than the techniques discussed above; in particular, UV-VIS is subject to more interferences. For some materials, UV-VIS is preferred because it will distinguish between different oxidation states of the material. For example, UV-VIS will identify Cr^{6+} (i.e., hexavalent chromium) from total chromium, which is useful because Cr^{6+} is a recognized human carcinogen.

X-ray diffraction (XRD) and similar tests are based on the principle that a beam of X rays is affected by crystals in compounds in such a way that the amount of the material can be measured. It is used to determine the amount of free crystalline silica (i.e., quartz, cristobalite, and tridyimite) in air, settled dust, or bulk samples.

Other (Nonchemical) Analytical Techniques. *Gravimetric analysis* means that the filter or other sample collector is weighed to determine the amount of particulate matter collected. This is a nonspecific technique: All material on the filter is included even if it is not the contaminant of

interest. While most contaminants are now determined using other analytical methods that give the quantity of the compound in the air sample, materials such as wood dust, coal dust (<5% silica), metal working fluid mist, cotton, and grain dust are still measured gravimetrically. There are two main approaches to gravimetric analysis:

- Weighing the filter before sampling (determining the *tare weight*) and then reweighing it after sampling. Generally the filter has to be desiccated (dried) in a controlled manner before each weighing to avoid error from moisture on the filter.
- Use of stacked, matched weight filters. Using this technique, the manufacturer places (or *stacks*) two filters weighing the same in a single sampling cassette. The particulates are retained on the first filter during sampling, and the weight of the second filter is unchanged. At the laboratory, each filter is weighed separately; the amount of matter collected equals the difference in weight between the two filters.

Microscopy involves counting the number of particles or fibers with some type of microscope. One technique for counting asbestos fibers is to dissolve part of the filter with a solvent and then count certain fibers using phase contrast microscopy (PCM) according to a rigorous counting methodology (Figure 1.10). Since this techniques counts all fibers, in a "mixed" fiber environment it will tend to overstate the amount of asbestos. In these cases, scanning electron microscopy (SEM) or transmission electron microscopy (TEM) are two techniques that can identify and count only the asbestos fibers. Microscopy is also used for bioaerosols as described below.

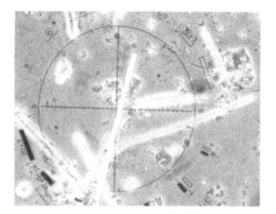

Figure 1.10. Asbestos fibers under phase contrast microscopy. (Courtesy of Forensic Analytical.)

Ionizing radiation concentration or activity is generally determined by performing standard counting techniques on filters or other collection devices. Radon (a gas) is determined by collecting the gas in a charcoal canister or by counting the alpha tracks caused by its decay on a sample of special polymer material.

Bioaerosols—the technology of identifying and quantifying bacteria, fungi, and viruses in air samples has progressed rapidly in recent years, primarily due to increased concern about indoor air quality and mold problems. Still there are few accepted standards for this type of analysis, and so close contact with the analytical laboratory is needed when first considering a bioaerosol sampling project. Often the laboratory provides the bioaerosol collection media that is suited to their analytical methods rather than the sample practitioner using off-the-shelf air sample collection devices. Common analytical methods, in addition to specific identification of toxic agents, include counting colonies that grow on nutrient plates, counting fungal spores using microscopy techniques, and determining the amount of endotoxin (a portion of the cell wall of gram-negative bacteria) using chemical or biological methods.

Understanding Laboratory Reports

With practically all of the air sampling devices and analytical techniques described above, the laboratory determines the *amount* of the contaminant(s) in the sample. This is expressed in weight or mass units such as milligrams (mg) or micrograms (µg). For optical counting techniques the report will indicate the number of fibers or colonies in the sample according to the counting protocol used.

In order to calculate the airborne concentration of the contaminants, it is necessary to divide the amount of contaminant by the volume of air that the sample represents. This is the reason for careful calibration of the sampling pump, scrupulous attention to pump start and stop times during sampling, and periodic checks of pump performance during the actual field sampling.

Airborne concentration (mg/m^3)

$$= \frac{\text{Laboratory result (mg of contaminant)}}{\text{Air sample volume } (m^3)}$$

The unit mg/m^3 applies to the concentration of gases, vapors, and particulate matter (except for fibers or colonies that are counted). However, the allowable exposure level for some gases and vapors is expressed as "parts of contaminant per million parts of contaminated air" by volume (ppm). Use this equation to convert mg/m^3 to ppm:

Concentration (ppm)

$$= \frac{\text{Concentration} (mg/m^3) \times 24.45}{\text{Molecular weight of contaminant}}$$

If only one air sample was collected to cover the entire exposure period, then the calculated air concentration for the sample represents the average concentration for the period (called the *time-weighted average*). However, if two or more sequential samples were collected (one after the other) rather than a single sample to cover the exposure period, the TWA is calculated from this equation:

$$\text{TWA} = \frac{(\text{Conc}_1)(\text{Time}_1) + (\text{Conc}_2)(\text{Time}_2) + \cdots + (\text{Conc}_{\text{last}})(\text{Time}_{\text{last}})}{\text{Exposure period}}$$

where TWA is the time-weighted average exposure for the period, Conc is the concentration in each sample, Time is the time period that each sample covers, and Exposure period is the total exposure period. In this equation the subscript ($_{1,2,\text{last}}$) refers to each sample collected sequentially to cover the time period. Also, for occupational exposures, "8 hours" (or 480 minutes) is generally used as the exposure period even if the work shift or exposure period was longer since the occupational exposure limits are based on an 8-hour workday. More details on this concept and the calculations are in Chapter 5.

DIRECT-READING DEVICES

The airborne concentration of many gases, vapors, and particulates can be measured in the field using direct-reading devices, thereby eliminating the time delay and effort of sending field samples to a laboratory for analysis. Most direct-reading devices are electronic instruments, although there are some where concentration is indicated by color change (called *colorimetric devices*). Direct-reading devices vary in purchase and operating cost, so cost comparison between direct-reading and laboratory sample approaches should be evaluated if appropriate. This chapter only gives an overview of the subject; Chapters 9–14 give details about the different types of direct-reading devices. The discussion below first covers electronic instruments and then colorimetric devices.

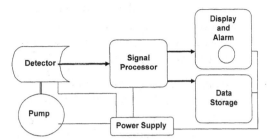

Figure 1.11. Diagram showing components of typical electronic direct reading instrument.

Electronic Direct Reading Instruments

Figure 1.11 shows a generic diagram of a typical direct-reading electronic instrument for gas, vapor, or particulate matter consisting of a detector, pump, signal processing unit, data display, power supply, and perhaps a data storage section. Of course all of these can be included in one small package—sometimes small enough to fit into a shirt pocket—but it is easier to understand the explanation of the different instrument types and capabilities in this section using this simple model. An ideal direct reading instrument is rugged, lightweight, easy to calibrate, and simple to operate, can operate over the temperature and humidity range it will be used in, functions for a long period without requiring maintenance, and, if battery-powered, has a long battery service life between charging. Typical components are:

- *Detector*—senses or measures the contaminant in the air. A pump may be used to drawn ambient air into the detector, although some devices rely on diffusion. The idea sensor is specific to the contaminant being measured, not responsive to potential interferences, not subject to fouling, is able to react quickly to changes in concentration, features a long service life before replacement is needed, and is inexpensive and easy to replace. The sensor is often the most important selection criteria when choosing between direct reading instruments for a specific application.

- *Signal processor*—the electronic circuitry that takes the signal from the sensor and converts it into the concentration reading. Advances in microchip technology have allowed manufacturers to add more features and make this part of the instrument smaller and more powerful. This has allowed some instruments, such as infrared devices, to store a large library of infrared spectra from different materials that can be used in compound identification. A possible downside of the technical advances is that some devices have a lot of flexibility but a very small keyboard and readout screen—as a result considerable scrolling through menus is required to set up and use the instrument. This added complexity of operation can be a major problem for the occasional user since they do not use the instrument enough to become proficient.

- *Data storage*—many instruments can store a large number of individual readings collected at a preset interval in one second or less to several hours and then integrate these readings to give a time-weighted average value for all or part of the sampling period. Often the peak value is stored and can be displayed at a later time. Some devices store the individual reading that are collected throughout the sampling period so the data can be downloaded to a computer for further analysis and permanent storage. This data storage feature is also referred to as data logging. Separate, stand-alone data loggers may be used with some direct-reading instruments without their own internal storage capability.

- *Data display* shows the concentration levels and other information such as sampling time. The display can be a meter device or a digital readout. Often a meter will have multiple scales, so it is important that the user know which scale is in use at any time. For complex instruments with a digital display (especially with a small display), considerable scrolling may be required in order to read all of the data values. The instrument may also have audible or visual alarms if an established concentration is exceeded or if the instrument fails.

- *Power supply*—rechargeable batteries are a requirement for portable instruments. For these devices the operating life between charges is critical, and the charging period is also important. Portable instruments may often be used for long periods with the charger plugged into line current to allow sampling periods longer than the battery life. Fixed instruments generally operate on line power.

Recent advances in microchip and sensor technology have resulted in dramatic improvements in direct reading instruments. They are more sensitive, more accurate, more specific, smaller, and lighter and exhibit longer battery life than just a few years ago. They also have extensive data-logging capability. The improvements permit devices that were once barely portable to be conveniently used in the field, and also allow manufacturers to combine several sensor devices in a single instrument for measuring several contaminants. Personal size devices with datalogging can often be used in place of sample-pump methods requiring laboratory analysis. Since the technology is constantly changing, consult manufacturers' literature when considering direct reading devices for airborne contaminants.

A method of field calibration or at least "calibration check" is an important feature for any direct reading instrument. This is a means of exposing the sensor to a known concentration of contaminant in the concentration range of interest and then confirming that the instrument is measuring the proper level. Without this assurance, it is difficult to rely on the readings from these devices. Often the calibration system is a cylinder of a compressed gas containing the appropriate level of contaminant that is used to fill a plastic bag for calibration, or a small glass vial containing the compound that can be broken inside a test chamber to generate a known concentration. Availability of a suitable calibration check system can be a limiting factor for some applications of direct reading instruments; for example, gas mixtures may not be stable over time and generally the calibration system cannot be transported by common carrier to field sites without special shipping documents and procedures. A "check" button on the instrument that tests the electrical circuitry is *not* an adequate substitute for a calibration check using the challenge contaminant at the sensor.

Types of Electronic Direct-Reading Instruments

Combustible Gas Monitors. These instruments read the "percent of lower explosive limit" (LEL) of a flammable gas in air and have been a mainstay of firefighters, gas utility company personnel, and safety inspectors for many years. They function using one of these operating principles: the change in electrical resistance or thermal conductivity of a sensor as the flammables are oxidized in a chamber within the instrument; or a change in electrical conductivity of a metallic oxide sensor when flammable compounds are adsorbed on its surface. Proper calibration and use of these instruments are critical because they are used to identify potentially flammable or explosive atmospheres. Key parameter to understand when selecting and using an instrument include:

- How the instrument reacts when either in a flammable/explosive atmosphere or an oxygen deficient atmosphere. The meter on some devices may "peg out" on the high end and then drop to or below zero in an atmosphere that is above the upper explosive limit (UEL). Some will not function without adequate oxygen, which can be misread as a low LEL reading.

- The difference in response between the gas used to calibrate the instrument versus the flammable gas or vapor in the atmosphere. All of these devices are calibrated by the user using a known gas source. These instruments typically react based on the heat energy in the gas, and therefore they will show different readings when encountering gases with different heat values from the calibration gas. To ensure safety, a calibration gas should be used that will cause the percent LEL of other common gases to be overestimated rather than underestimated. For this reason, either pentane or hexane are often used for calibration, since with either of these two calibrating gases the level of other common gases and vapors will be overestimated. Conversely, methane would be a poor choice as a calibration gas because it can cause underestimation of the hazard from other gases (Table 1.5).

- Whether there are any compounds that will interfere with accurate readings or that can damage the instrument. Some compounds in high enough concentrations will impede the combustion of the flammable vapors at the sensor. Other compounds may

TABLE 1.5. Choice of Calibration Gas Is Critical for Safety

Gas Being Sampled	Actual Concentration in Air of "Gas Being Sampled" when Meter Reads "100% LEL"	
	with Hexane Calibration	with Methane Calibration
Acetone	70% LEL	170% LEL
Benzene	70% LEL	190% LEL
Ethylene	60% LEL	130% LEL
Methane	40% LEL	100% LEL
Methanol	50% LEL	110% LEL
Toluene	90% LEL	210% LEL

either overheat the sensor or coat it so it does not operate properly. Some devices use a "catalyst" to permit combustion to occur at a lower temperature; these may be subject to poisoning by certain compounds. An understanding of other airborne materials that may be present will aid in equipment selection and proper operation.

Instruments for Specific Gases and Vapors

ELECTROCHEMICAL OR METAL OXIDE SEMICONDUCTOR DEVICES. There are many instruments with sensors that detect a single or multiple specific compounds (Figure 1.4). Generally these have either electrochemical or metal oxide semiconductor detectors. In some devices, each sensor is specific for a single compound, so a device that measures multiple compounds has several sensors, while in other devices a single sensor can measure different gases, depending on sensor voltage and other operating parameters. A *simplified* explanation of each sensor is:

- *Electrochemical*—the sensor has two electrodes in a chemical matrix. The composition of the electrodes and matrix is chosen to sense the compound of interest, which is usually oxygen or a contaminant. When

the compound contacts the chemical matrix, a reaction occurs that changes either the current or voltage between the electrodes (depending on the instrument). The change in current or voltage is proportional to the amount of the compound. In order to improve sensitivity and specificity, the sensor may feature a membrane that excludes interfering compounds from reaching the matrix, a catalyst that causes the reaction to proceed at a lower temperature, or a reference electrode to allow more accurate measurement of the current. If the reaction is sensitive to ambient temperature, a temperature sensor may be part of the circuitry to adjust the reading for fluctuations in ambient temperature.

- *Metal oxide*—the sensor consists of a semiconductor with a coating of a metallic oxide such as zinc, nickel or tin. As oxygen in normal air is adsorbed onto the coating, a baseline current flow develops in the semiconductor. When the oxygen is displaced by molecules of the contaminant, a change in resistance of the semiconductor occurs which is proportional to contaminant concentration. Selectivity is achieved through different mixtures of oxides and different operating temperatures. Some devices can measure several contaminants by modifying the operating temperature of a single metal oxide sensor, while others have multiple sensors.

Each type of device is available from multiple manufacturers for different contaminants in a wide variety of measurement ranges, instrument sizes, and other features. Selection of the correct instrument depends on possible interferences and other operating/use conditions.

OTHER SPECIFIC GAS/VAPOR INSTRUMENTS. Instruments have been developed to take advantage of specific reactions or responses of specific gases and vapors:

- Formaldehyde will fluoresce (emit light of a specific wavelength) when excited by a bright light source. The amount of fluorescence can be measured and is proportional to the formaldehyde concentration.
- Mercury vapors can be detected using instruments based on two characteristics of mercury: (a) Airborne mercury absorbs ultraviolet light produced by a mercury lamp; (b) mercury will form an amalgam with gold (a solid solution of gold in mercury), and the amalgam exhibits a different electrical conductivity than pure gold. Separate instruments are available using either UV light absorption or a gold foil to measure the concentration of mercury vapor in air.

Nonspecific Gas/Vapor Instruments. This is a class of instruments that respond to multiple compounds in air. Some can be calibrated to measure specific compounds in air, while others give a general concentration reading. Several are portable, smaller versions of the laboratory instruments described earlier.

DETECTORS BASED ON IONIZATION. Molecules of gases or vapors can be *ionized*

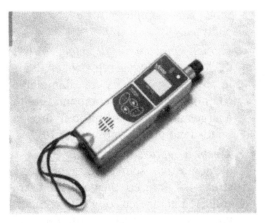

Figure 1.12. Hand-held photoionization detector instrument. (Courtesy of Industrial Scientific Corporation.)

(give an electrical charge) by a variety of means, and then the charged particles collected in an electric field where the resulting current is proportional to the concentration of molecules.

- *Photoionization detectors* (PID) (Figure 1.12) use a UV light source to ionize contaminant molecules. Chemical compounds are subject to ionization by UV light according to their ionization potential (IP), expressed in units of electron-volts (eV). Several different UV lamps are available with varying eV intensities. A lamp will ionize any compound with an IP lower than the eV rating of the lamp. If the IP for a compound is greater than the eV rating of the most intense lamp available, it cannot be analyzed using a PID device. Photoionization devices are good for spot checks in areas where the composition of the airborne contaminant has been determined by more specific procedures. For example, when removing underground gasoline storage tanks a PID is useful for periodic checks of airborne gasoline levels after the response of the PID has been "calibrated" using side-by-side comparison of PID readings and charcoal tube air samples for laboratory analysis.

- *Flame ionization detector* (FID) devices use an oxygen/hydrogen flame to ionize molecules of flammable compounds. Organic compounds with a large number of carbon-to-hydrogen bonds tend to respond well to FIDs.

- *Electron capture detectors* (ECD) use a radioactive source to generate a stream of ionized particles that are collected at an electrode. The contaminant molecules pick up some of the electrons as they flow past the radioactive source, thus reducing the current flow to the electrode. ECDs are sensitive to halogen compounds and those containing nitrogen or oxygen. A common application is measuring sulfur hexafluoride (SF_6) during tests of laboratory hood performance.

GAS CHROMATOGRAPH, GC/MASS SPECTROPHO-TOMER, AND INFRARED INSTRUMENTS. These devices operate on the same principles as the larger, fixed instruments used for laboratory analysis. Recent advances in microchip technology have made these instruments more powerful in that the signal processing and data library features have been improved, and the battery weight and size of the units have been reduced. However, although these devices have added functionality, they are more complex and difficult to operate. This can be a significant issue if the instruments are not used regularly, since the air sampling practitioner may lack the hands-on competency for trouble-free operation in the field.

MULTIPLE SENSOR DEVICES. Equipment manufacturers offer single instruments combining several of the detection principles described above to measure the concentrations of more than one gas or vapor. Typical devices have a specialty sensor such as a photoionization detector, gas chromatograph, or infrared detector along with several electrochemical or metal oxide semiconductor sensors. These instruments are valuable for measuring the overall level of chemicals in the air plus several specific compounds.

Direct Reading Devices for Particulate Matter. Measuring airborne concentrations of particulate matter with direct reading devices is not as straightforward as measuring gases and vapors. "Chemical" analysis of the particulates is rarely possible, and so the measured value is the concentration (mg/m^3) or particle count of total airborne particulates or a size-selected fraction corresponding to respirable size or another size distribution criterion. With some devices a filter can be placed in the sampling line to permit subsequent laboratory analysis for comparison with the direct reading results.

The behavior of airborne particles depends on their diameter, shape, and density. Because of these factors, it is common to consider particles as having an *equivalent (or effective) aerodynamic diameter* (EAD), which is the diameter of a spherical particle with the same density as water that has the same aerodynamic behavior as the particle of interest. Particles that are very small act more like gas or vapor molecules than true "particles." For example, particles less than about 0.01 μm in diameter are small enough to behave like large molecules. Slightly larger particles (about 0.1-μm EAD) interact with molecules in the air rather than settling out due to gravity. As a result, particles about 0.1 μm or smaller may be collected via diffusion.

Most direct reading instruments for particulate matter use a pump to draw air into the sensor part of the device, and they function on one of these two principles:

- *Scattered Light.* The air sample goes into a chamber that has a narrow beam of light passing through it. A detector

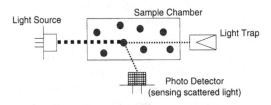

Figure 1.13. Diagram of a typical light-scattering instrument to measure airborne particulates.

Figure 1.14. Colorimetric detector tube and hand pump. (Courtesy of Nextteq, LLC.)

positioned at an angle measures the increase in light as the airborne particles scatter the light beam (Figure 1.13). The light beam may be from a laser, tungsten lamp, or other source of either visible or infrared light.

- *Oscillation Frequency of a Piezoelectric Crystal.* A piezoelectric crystal oscillates at a given frequency when in an electric field. As particulates are deposited on the crystal's surface, its oscillating frequency changes. The frequency change can be converted by the signal processing into units of dust concentration.

Direct reading particulate matter instruments are often used where other tests have been performed to characterize the chemical constituents of the airborne particulates. Where the characterization has been performed, direct-reading devices can provide valuable information on concentration and size distributions of the airborne material.

Colorimetric Direct-Reading Systems

This category describes devices that rely on the color produced by the contaminant of interest and reagents in the sampling system to indicate airborne concentration.

Colorimetric Detector Tubes. These are sealed glass tubes containing the color-producing reagent. The widest application is with a hand-operated pump for grab samples (Figure 1.14), although battery-

powered grab sampling pumps are also available. Additionally, there are long-term tubes designed to measure exposures over an extended period such as an 8-hour work shift; these operate using a battery-powered pump or as passive samplers relying on diffusion. With battery-powered pumps, several detector tubes (or detector tubes plus an adsorption tube) can be connected via a manifold to permit parallel sampling for multiple compounds.

To measure some contaminants that cannot be detected with a simple colorimetric tube, enhancements are available such as:

- A "pre-reaction" tube placed in line before the analyzer detector tube converts the contaminant of interest into a chemical compound that can be measured by the detector tube.
- A "pre-cleanse" tube before the detector tube removes interferences from the air sample so the contaminant of interest can be measured accurately.
- A "pyrolyzer" unit that uses heat to convert the contaminant to a compound that can be measured with the detector tube.
- A "trap" tube *after* the detector tube for long-term samples collected with a battery-powered pump to remove reaction products that could damage the pump.

The color-based reaction is generally "read" as the length of colored stain that develops in the tube, although some devices use an internal photometer to read and display the colorimetric change. For some tubes the concentration level (in ppm) is read directly from a graduated scale on the tube; in other cases the tube is marked with a linear scale and the user refers to a calibration chart provided by the manufacturer to convert length of stain into concentration. Many tubes can measure lower concentrations if a larger air sample is collected. With a hand pump, this is accomplished by increasing the number of pump strokes. For these tubes the calibration chart gives concentration reading based on sample volume. Long-term tubes report concentration in units of *ppm-hours*. To calculate the average airborne concentration, divide the *ppm-hours* reading by the sampling time (in *hours*).

Detector tube systems are especially good for "range-finding" measurements since they can measure more than 150 contaminants (see listing in Chapter 13). These devices are simple to use but are highly engineered by their manufacturer, and so careful conformance to operating instructions is critical to achieving the maximum accuracy and specificity. Typical operating parameters covered in the instructions include:

- Operating temperature and humidity range
- Limits of detection and quantification
- Sample volume (number of pump strokes, etc.)
- Interfering compounds
- Any accuracy or quality assurance certification standards that the tube meets
- Time duration that the stain is stable following use
- Storage recommendations
- Expiration date
- Disposal precautions or requirements for spent tubes

Because of their flexibility and ability to measure many airborne contaminants, a detector tube system is part of many air sampling practitioners' toolkits. However, before using detector tubes for repeated measurement of a single or a few contaminants, evaluate the cost of the tubes against a simple electronic direct reading instrument (if available). Many low-cost instruments are available that may be more economical in the long run than using numerous colorimetric tubes.

Other Colorimetric Devices. Colorimetric badges are dosimeter-like devices that are similar to the long-term diffusion detector tubes described in the previous section (Figure 1.15). The color change on the badge after use is read visually or by use of a color photometer.

Colorimetric tape sampler is a fixed device for area samples that functions by drawing a chemical-impregnated paper

Figure 1.15. Colorimetric passive dosimeter. (Courtesy of Assay Technology.)

tape past a sampling port where the pump-drawn air sample impinges on the tape. A photometer measures the resulting color change and reports it in units of concentration. The electronic circuitry can display the concentration, store data for later downloading, and trigger alarms if preset concentration levels are exceeded.

Liquid-based colorimetric instruments are used for a few applications where no other direct-reading device is available. With these instruments the air sample passes through a liquid scrubber/reagent where a color change develops based on the chemical's concentration in the air sample.

MONITORING RECORDS

Keeping adequate records of the monitoring is critical. This includes information on the monitoring procedure (equipment, sampling, and analytical procedures, etc.), the people being monitored, and the exposure situation. Without adequate records, it will not be possible to understand the monitoring results, and over the long-term it will be impossible to reconstruct the exposure situation and resulting exposure levels. For example, when asbestos is monitored where other airborne fibers may be present, it is important to know whether samples were analyzed using polarized light microscopy (which counts all fibers meeting size and shape criteria) or scanning electron microscopy (which can identify only the asbestos fibers). Similarly, if

"hydrocarbon vapor" measurements are made with a direct reading instrument, it is helpful to know whether a PID instrument (which responds differently to various hydrocarbon compounds) or a portable gas chromatograph calibrated for the specific hydrocarbon compounds of interest was used. As an illustration of the typical data elements to include in the record, Table 1.1 shows the monitoring records that must be maintained under the U.S. Federal OSHA Benzene Standard.

SUMMARY

This chapter presented an overview of air sampling and analytical techniques. The purpose was to give readers a common understanding of air sampling nomenclature and basic principles to enhance understanding of the remaining chapters.

REFERENCES

1. Occupational Safety and Health Administration. *Benzene*. U.S. Code of Federal Regulation (29CFR1910.1028).
2. Mulhausen, J. R., and J. Damiano. *A Strategy for Assessing and Managing Occupational Exposures*, 2nd ed. Fairfax, VA: AIHA, 1998.
3. SKC, Inc. *Comprehensive Catalog and Air Sampling Guide*. Eighty Four, PA, 2004.
4. Grunder, F. I. *PAT Program Report Background and Current Status. AIHA J.* **63**:797–798, 2002.

CHAPTER 2

HAZARDS

This chapter provides background on the types of hazards that exposure monitoring is intended to evaluate. The broadest categories are *health hazards* (which cause disease) and *physical hazards* (which cause harm from pressure, temperature, impact, or other physical effect). Chemicals are the largest group of hazardous agents evaluated by air sampling and, depending on the specific compound, can pose one or both types of hazards. Microbial agents generally are potential health hazards rather than physical hazards.

The potentially harmful effects from chemicals depend on the organ system(s) affected or the disease produced for health hazards, as well as the physical characteristic causing the harm for physical agents. As an illustration, U.S. Federal OSHA uses this definition of chemical hazard categories:

- *Health Hazards.* Chemicals that are carcinogens, toxic or highly toxic agents, reproductive toxins, irritants, corrosives, sensitizers, hepatotoxins, nephrotoxins, neurotoxins, agents which act on the hematopoietic system, and agents which damage the lungs, skin, eyes, or mucous membranes.

- *Physical Hazards.* Chemicals for which there is scientifically valid evidence that it is a combustible liquid, a compressed gas, explosive, flammable, an organic peroxide, an oxidizer, pyrophoric, unstable (reactive) or water-reactive.

Detailed descriptions and examples of these terms are in Table 2.1.

Although safety hazards related to the physical characteristics of a chemical can be objectively defined in terms of testing requirements (e.g., flammability), health hazard definitions are less precise and more subjective. Health hazards may cause measurable changes in the body, such as decreased pulmonary function. These changes are generally indicated by the

Air Monitoring for Toxic Exposures, Second Edition. By Henry J. McDermott
ISBN 0-471-45435-4   © 2004 John Wiley & Sons, Inc.

TABLE 2.1. Health and Physical Hazards of Chemicals (U.S. Federal OSHA Hazard Communication Standard)

Health Hazard

A chemical for which there is statistically significant evidence based on at least one study conducted in accordance with established scientific principles that acute or chronic health effects may occur in exposed employees. The term includes the following chemicals:
- Toxic and Highly Toxic:

Characteristic	Toxic	Highly Toxic
LD_{50} (oral)	>50–500 mg/kg	≤50 mg/kg
LD_{50} (dermal)	>200–1000 mg/kg	≤200 mg/kg
LC_{50} (inhalation)	>200–2000 ppm	≤200 ppm
	>2–20 mg/L	≤2 mg/L

- Irritant: Causes a reversible inflammatory effect on living tissue by chemical action at the site of contact, but is not corrosive.
- Sensitizer: Causes a substantial proportion of exposed people or animals to develop an allergic reaction in normal tissue after repeated exposure to the chemical.
- Corrosive: Causes visible destruction of, or irreversible alterations in, living tissue by chemical action at the site of contact.
- Carcinogen: Considered to be a carcinogen if: it has been evaluated by the International Agency for Research on Cancer (IARC) and found to be a carcinogen or potential carcinogen; or it is listed as a carcinogen or potential carcinogen in the Annual Report on Carcinogens published by the National Toxicology Program (NTP) (latest edition); or, it is regulated by OSHA as a carcinogen.
- Hepatotoxin: Produces liver damage (e.g., jaundice, liver enlargement) such as carbon tetrachloride and nitrosamines.
- Nephrotoxin: Produces kidney damage (e.g., edema, proteinuria) such as halogenated hydrocarbons and uranium.
- Neurotoxin: Produces its primary toxic effects on the nervous system (e.g., narcosis, behavioral changes, decrease in motor functions) such as mercury and carbon disulfide.
- Hematopoietic Agent: Acts on the blood or system to decrease hemoglobin function or deprive the body tissues of oxygen, etc. (e.g., cyanosis, loss of consciousness) such as carbon monoxide and cyanides.
- Agents that damage the lung: Irritate or damage pulmonary tissue (e.g., cough, tightness in chest, shortness of breath) such as silica and asbestos.
- Reproductive Toxin: Affects the reproductive capabilities including chromosomal damage (mutations) and effects on fetuses (teratogenesis) [e.g., birth defects, sterility] such as lead and dibromochloropropane.
- Cutaneous Hazard: Affects the dermal layer of the body (e.g., defatting of the skin, rashes, irritation) such as ketones and chlorinated compounds.
- Eye Hazard: Affects the eye or visual capacity (e.g., conjunctivitis, corneal damage) such as organic solvents and acids.

Physical Hazard

A chemical for which there is scientifically valid evidence that it is a combustible liquid, a compressed gas, explosive, flammable, an organic peroxide, an oxidizer, pyrophoric, unstable (reactive) or water-reactive.
- Combustible Liquid: Any liquid having a flashpoint at or above 100°F but below 200°F.
- Compressed Gas:
  ◦ A gas or mixture of gases having, in a container, an absolute pressure exceeding 40 psi at 70°F; or

TABLE 2.1. *Continued*

- ○ A gas or mixture of gases having, in a container, an absolute pressure exceeding 104 psi at 130°F regardless of the pressure at 70°F; or
  - ○ A liquid having a vapor pressure exceeding 40 psi at 100°F.
- Explosive: Causes a sudden, almost instantaneous release of pressure, gas, and heat when subjected to sudden shock, pressure, or high temperature. ᵎ
- Flammable: A chemical that falls into one of the following categories:
  - ○ Aerosol, flammable: An aerosol that, when tested by established methods, yields a flame projection exceeding 18 inches at full valve opening, or a flashback (a flame extending back to the valve) at any degree of valve opening.
  - ○ Gas, flammable: Either a gas that, at ambient temperature and pressure, forms a flammable mixture with air at a concentration of 13% by volume or less; or, a gas that, at ambient temperature and pressure, forms a range of flammable mixtures with air wider than 12% by volume, regardless of the lower limit.
  - ○ Liquid, flammable: Any liquid having a flashpoint below 100°F.
  - ○ Solid, flammable: A solid, other than a blasting agent or explosive, that is liable to cause fire through friction, absorption of moisture, spontaneous chemical change, or retained heat from manufacturing or processing, or which can be ignited readily and when ignited burns so vigorously and persistently as to create a serious hazard.
- Organic peroxide: An organic compound that contains the bivalent –O–O– structure and which may be considered to be a structural derivative of hydrogen peroxide where one or both of the hydrogen atoms has been replaced by an organic radical.
- Oxidizer: A chemical other than a blasting agent or explosive, as defined in federal regulations, that initiates or promotes combustion in other materials, thereby causing fire either of itself or through the release of oxygen or other gases.
- Pyrophoric: Will ignite spontaneously in air at 130°F or below.
- Unstable (reactive): Will vigorously polymerize, decompose, condense, or become self-reactive under conditions of shocks, pressure, or temperature.
- Water-reactive: Reacts with water to release a gas that is either flammable or presents a health hazard.

occurrence of signs and symptoms in the exposed individuals—such as shortness of breath, which is a nonmeasurable, subjective feeling.

The determination of occupational health hazards is complicated by the fact that many of the effects or signs and symptoms occur commonly in nonoccupationally exposed populations, so that effects of exposure are difficult to separate from normally occurring illnesses. Occasionally, a substance causes an effect that is rarely seen in the population at large, such as angiosarcomas caused by vinyl chloride exposure, thus making it easier to ascertain that the occupational exposure was the primary causative factor. More often, however, the effects are common, such as pulmonary disease that is indistinguishable from the effects of smoking or hay fever and other allergic reactions. The situation is further complicated by the fact that most chemicals have not been adequately tested to determine their health hazard potential, and data do not exist to substantiate these effects. It is important that employees exposed to such hazards must be apprised of both the change in body function and the signs and symptoms to signal any change that may occur.

Besides chemicals, microbial agents are probably the next largest group of hazardous agents that are evaluated through air sampling. Bacteria and fungi are unique because they are living organisms and thus can multiply and grow. Many have more than one growth stage. In the spore stage the organism is encapsulated, is somewhat indifferent to temperature, and often behaves as a particle, whereas in the growth stage it can assume many different appearances but generally must be collected on some type of growth medium. Organic dusts, derived from vegetable, soil or animal sources, can have the characteristics of both chemical and microbial agents. Examples include grain dusts (oats, wheat, barley), straw, hay, wood dusts (a wide variety of compounds with many diverse effects ranging from asthma to suspect carcinogens), cotton, sewage sludge, and animal dander. In addition to the dust particle itself, organic dusts can contain microorganisms and their toxic products (toxins), enzymes, and other agents capable of eliciting an immunological response.

The air sampling practitioner must not overlook the potential for many products to generate chemicals during use or after storage, even though they in themselves may not appear to present a chemical hazard. For example, many building construction materials, such as fabrics, wall partitions, and particleboard, can emit formaldehyde. Paper when cut can release paper dust. Carbonless paper may also release formaldehyde from its coating. Oxyacetylene cutting or welding on painted metals or plastics may produce byproducts far more toxic than the initial ingredients.

CONTAMINANTS

Physical States

Contaminants are classified in terms of their physical state, which can be a gas, vapor, or type of aerosol. *Gases* are com-pletely airborne at room temperature and can be liquefied only by the combined effect of decreasing temperature and increasing pressure such as occurs when creating liquefied nitrogen or dry ice from carbon dioxide. A substance is considered a gas if at standard conditions of temperature and pressure (25°C, 1 atm) its normal physical state is gaseous. Examples are hydrogen, helium, oxygen, formaldehyde, ethylene oxide, carbon monoxide, argon, and nitrogen oxides.

A *vapor* is the gaseous state of a substance in equilibrium with the liquid or solid state of the substance at the given environmental conditions.[1] This equilibrium results from the vapor pressure of the substance, which if high enough, causes it to volatilize (evaporate) from a liquid or sublime (evaporate directly from a solid) into the atmosphere. Vapors become airborne as the result of the evaporation of substances that are liquids at room temperature, such as styrene or acetone. The primary reason for differentiating between gases and vapors is to be able to assess the potential likelihood of a chemical being airborne in high concentrations. For sampling it is important to note that gases and vapors stay airborne for long periods.

An *aerosol* is a system consisting of airborne solid or liquid particles that is dispersed in a gas phase, usually the atmosphere. Aerosols are generated by fire, erosion, sublimation, condensation and abrasion of minerals, metallurgical materials, organic and other inorganic substances in construction, manufacturing, mining, agriculture, and transportation, among other operations. Aerosol classifications depend on the physical nature, particle size, and method of generation. Dusts, smoke, soot, particles, mist, fumes, and fog are all terms used to describe certain types of aerosols. Solid substances are defined as particles to distinguish them from liquid droplets. When there is no need to differentiate between the particle and droplet

TABLE 2.2. Types of Air Contaminants

Airborne Material	Size Range (μm)	Description
Dust (Airborne)	0.1–30.0	Generated by pulverization or crushing of solids. Typical examples are rock, metal, wood, and coal dust. Particles may be up to 300–400 μm, but the larger particles do not remain airborne.
Mists	0.01–10.0	Suspended liquid particles formed by condensation from gaseous state or by dispersion of liquids. Mists occur above open surface electroplating tanks.
Smokes	0.01–1.0	Aerosol mixture from incomplete combustion of organic matter. This size range does not include fly ash (which is larger).
Fumes	0.001–1.0	Small solid particles created by condensation from vapor state, especially volatized metals as in welding or melted plastic materials. Fumes tend to coalesce into larger particles as the small fume particles collide.
Vapors	0.005	Gaseous form of materials that are liquid or solid at room temperature. Many solvents generate vapors.
Gases	0.0005	Materials that do not usually exist as solids or liquids at room temperature, such as carbon monoxide and ammonia. Under sufficient pressure and/or low temperature they can be changed into liquids or solids.

components of an aerosol, the collective term *particulate matter* is used. The ability of aerosols to get into the body as well as the rate at which they are absorbed depends on the particle size distribution and solubility characteristics of the aerosol. In general, smaller particles tend to be deposited in the lower regions of the lungs and many occupational diseases are associated with this area and other regions of the respiratory tract. Table 2.2 lists typical sizes and characteristics of different airborne contaminants.

Types of Particulate Matter. *Dusts* are generated from solid materials by mechanical means, such as grinding, crushing, pulverizing, chipping, or other abrasive actions occurring in natural and commercial operations. Dusts can be derived from inorganic minerals, such as asbestos, silica, and limestone, and various metals or organic sources, such as wood dust, flour, and grain dust. Dust can also be derived from animal dander, insects, mites, fungal spores, and pollen. Microbials can often be found mixed with dusts, especially those that are organic-based. Airborne dusts can be quite heterogeneous in composition and often consist of larger particles, although they range from 0.01 μm to 30 μm in diameter. Most often dusts are somewhat spherical in shape.

Fibers, particles whose length exceeds their width, can be generated from minerals, such as asbestos, and humanmade sources, including fiberglass, if the composition of the material lends itself to disintegration producing such particles. For the purposes of classification, some fibers are assigned a minimum size criterion; for example, asbestos particles must be at least three times longer than they are wide to be considered a fiber for occupational sampling purposes. Fibers are thought to behave differently when in the lung than particles in the shape of a sphere.[2] Organic sources, such as hemp and animal fibers, also exist.

Mists are suspended liquid droplets. They are generated by condensation from the gaseous to the liquid state or by mechanically breaking up a liquid into a dispersed state by spraying, splashing, foaming, or atomizing. Oil mists produced from metal-working fluids during parts machining, as well as mists above electroplating tanks, are examples. Some mists can have a vapor component as well, such as paint spray mists, which contain volatile solvents. When a mist's droplets evaporate or vapor condensation occurs, the aerosol will contain higher concentrations of very small particles. *Fog* is a term used for a mist that has a particle concentration high enough to obscure visibility.

Fumes are produced by such processes as combustion, distillation, calcination, condensation, sublimation, and chemical reactions. Fumes are solids that are the result of condensation of solids from an evaporated state. When a metal or plastic evaporates, the atoms disperse singly into the air and form a uniform gaseous mixture. In the air they combine rapidly with oxygen and recondense, forming a very fine particulate ranging in size from $0.001\,\mu m$ to $1.0\,\mu m$. Many fumes have a high chemical reactivity that is thought to explain the high biological activity exhibited by some metal fumes.[3] Examples are welding fumes, ammonium chloride, hot asphalt, and volatilized polynuclear aromatic hydrocarbons from coking operations. The term *fume* is often confusing. When used to refer to exhaust fumes or paint fumes, it is inappropriate, since gases and vapors and airborne mists are not classified as fumes.

Smokes are produced by the incomplete combustion of carbonaceous materials, such as coal, oil, tobacco, and wood. Smoke particles generally range from $0.01\,\mu m$ to $1.0\,\mu m$ in diameter. Because of their small size, these particles can remain airborne for long periods of time. Situations where smoke is produced also tend to produce gases.

For the purposes of sampling, several facts are important about aerosols. Particle size is often dependent on the means in which the aerosol was generated. When airborne, particles collide and can stick together. If the purpose of sampling is simply to collect the total (mass of) dust, this factor will not influence the results to a significant degree, since it is likely that these larger particles will still be collected. If size-selective sampling is done, the results may not reflect the actual airborne dust composition, because it will appear as if there are fewer particles and they are larger in size. This consideration can be especially important when sampling near combustion sources, because thermal effects can promote coagulation of smaller particles, leading to changes in the particle number and composition of the aerosol. As a result of gravity and diffusion, particles will settle onto surfaces, thus decreasing airborne concentrations.

Despite the type and origin of the airborne particulate matter, the air sampling practitioner needs to understand that the particulates can be divided into two broad classes depending on their aerodynamic size. The *aerodynamic size* is the equivalent diameter of a sphere with the density of water that behaves like that particulate in air. It is used because different density and different shaped particles of the same size move differently in air. The two classes of particulates are:

- Particles or droplets that settle out rapidly and do not represent an airborne hazard. These are generally larger than $100\,\mu m$. Unless captured in a ventilation system, large particles settle on floors, roof beams, and machinery. Large droplets may also coat walls and make surfaces slippery or sticky. In general, these larger particulates do not present the same health hazard as smaller particles. In order to give an accurate estimate of inhalation hazard, the air sampling technique should not collect these size particles.

- Fine particles or small droplets that do remain airborne for long periods. Their aerodynamic size is less than 50 μm and usually less than 30 μm. These small particulates are often generated along with the larger particulates, especially in grinding operations. Very small particulates meet significant resistance as they move through air; as a result, they have almost no power of motion independent of the surrounding air. For example, if a 5-μm-diameter particle of sawdust is released at a height of 10 feet above the floor in perfectly still air, it will take almost a full 8-hour shift for the particle to settle to the floor. These smaller particles that present an inhalation hazard are the particles that should be collected in an air sample.

Chemical Properties

Vapor pressure is a measure of the ability of a compound to become airborne. The higher the vapor pressure (larger the number), the higher the airborne concentration possibility. Vapor pressures should be referenced to the temperature at which they were measured. Compounds that are a gas at room temperature, such as ammonia, will have vapor pressures greater than 760 mm Hg (1 atm). Table 2.3 lists the vapor pressure of common solvents and organic liquid chemicals. An example of a liquid with a high vapor pressure is pentane: 509 mm Hg at 25°C. From Table 2.3 the volatility of toluene is about three times that of xylene. This means that if both liquids are handled in the same manner, the resulting airborne concentration of toluene will be three times higher than that of xylene. For strict comparison of compounds, vapor pressures at the same temperature and units of pressure must be used.

The degree of volatility can affect the sampling method, since it often dictates the

TABLE 2.3. Vapor Pressure and TLV® for Common Hydrocarbon Compounds

Chemical	Vapor Pressure, mm Hg at 25°C	TLV®, ppm (8-hr TWA)[a]
Pentane	509	600
Hexane	152	50 (Skin)
Benzene	91	0.5 (Skin)
Toluene	28	50 (Skin)
Octane	14	300
Xylene	8	100

[a] *Skin* notation means that there is the potential for significant absorption through the skin, and so air sampling alone may not accurately measure total exposure if skin adsorption can occur.

Source: References 18 and 21.

physical state of the compound as it is found in the air. Volatile compounds have vapor pressures greater than 1 mm Hg at ambient temperature and exist entirely in the vapor phase when airborne. Semivolatile compounds have vapor pressures of 10^{-7} mm Hg to 1 mm Hg and can be present in both the vapor and particle state. Nonvolatile compounds have vapor pressures of less than 10^{-7} mm Hg and are found exclusively in the particle-bound state when in the air.[4] Some of the classes of semivolatile organic compounds of particular interest to sampling practitioners include polynuclear aromatics (PNAs), organochlorine pesticides, organophosphate pesticides, polychlorinated biphenyls, and chlorinated dibenzodioxins and furans. The challenges of sampling for these compounds are discussed further in the chapter on concurrent sampling (Chapter 8).

Occasionally a chemical will be described as being "heavier than air," suggesting the gas or vapor cloud will sink to the ground and stay low. Generally in these cases the property of vapor density, defined as the weight of a gas or vapor compared to the weight of an equal volume of air, is the point of reference. Low ambient temperature increases vapor density as does high humidity.[5] Vapor density can be important in situations where high concentrations are present due to an emergency release, buildup to combustible concentrations, or chemical release in a confined space (sludge pits, tanks, and even ditches) where there is a lack of air movement. In reality, when comparing molecular weights, most chemicals are heavier than diatomic nitrogen, the primary component of air with a molecular weight of 28 g/mole, and the most important criterion is the ratio of contaminant to available air currents and air volume for mixing. For example, the molecular weight of methyl isocyanate, the compound released in 1984 in Bhopal, India, is 57 g/mole, twice that of "air," and this compound traveled a significant distance outdoors before dispersing. Table 2.4 shows vapor densities for selected compounds.

In cases where chronic health hazard concentrations are present (generally quite low when compared to the situations discussed above), most of these compounds are going to mix with air and disperse readily throughout an area. A simple calculation can be done to demonstrate this property using 1,1,1-trichloroethane, a common degreasing solvent.[6] Assuming there is ample air for dilution, the following calculations indicate the effective vapor density of a 1000 ppm 1,1,1-trichloroethane–air mixture:

$$\text{vapor density of air} = 1$$

$$\begin{matrix}\text{vapor density of} \\ 1,1,1\text{-trichloroethane} = 4.6\end{matrix}$$

$$1000\,\text{ppm} = \frac{\begin{matrix}1000 \text{ parts of} \\ 1,1,1\text{-trichloroethane}\end{matrix}}{\begin{matrix}1,000,000 \text{ parts of} \\ \text{contaminated air}\end{matrix}} = 0.001$$

Density calculation:

$$\begin{matrix}0.001\,(4.6) = 0.0046 \\ (1,1,1-\text{trichloroethane})\end{matrix}$$

$$0.999\,(1.0) = 0.999 \quad (\text{air})$$

$$0.999 + 0.0046 = 1.0036, \text{ or}$$
approximately 1.004, which is the effective vapor density of the mixture

Therefore, the 1,1,1-trichloroethane–air mixture compared to clean air would have a ratio of 1004:1000 and not 4.6:1. Since 1000 ppm is several times the acceptable occupational exposure limit for 1,1,1-trichloroethane, a mixture normally encountered in these situations would contain much less. Thus, the effects of window ventilation, cross currents, wind, traffic, and heat are usually more important than molecular weight and vapor density.

When comparing the densities of liquids, the term *specific gravity* is used. The specific gravity is the relative weight per unit volume of a liquid compared with the weight per unit volume of pure water. Water has an arbitrarily assigned value of $1.000\,\text{g/cm}^3$ and all liquids heavier than water will have higher specific gravities.

For gases and vapors, *solubility* in water is the principal characteristic affecting penetration to the deeper areas of the respiratory tract when inhaled. The more water-soluble, the greater the likelihood of dissolving in the nasal or oral airways while

TABLE 2.4. Vapor Densities of Selected Compounds

Compound	Vapor Density (air = 1)
Ammonia	0.59
Carbon dioxide	1.52
Carbon monoxide	0.97
Chlorine	2.45
Gasoline	3.93
Hydrogen	0.07
Hydrogen sulfide	1.17
Methane	0.55
Methyl isocyanate	1.97
Propane	1.52
1,1,1-Trichloroethane	4.60
Nitrogen	0.97

less water-soluble gases can penetrate to the smaller air ways and alveoli in the lungs. Gases that are soluble in water are often upper respiratory irritants, such as hydrogen chloride. The basis for solubility of particles in the lungs of humans may be very different from that in water or other solvents or other species. The lung fluids contain lipids and proteins that modify the solubility in ways that are difficult to evaluate or predict.[7] On the other hand, when referring to the ability of a contaminant to penetrate the skin, the primary concern is solubility in fats because it often influences the ability of a material to penetrate the skin.

Compounds that dissolve in the mucous of the respiratory system and do not produce any immediate local irritation at the site of deposition can reach the bloodstream. Insoluble compounds will deposit in the respiratory system or will be carried on the mucous and cilia in the esophagus to the mouth and will be swallowed. The solubility of most compounds in the gastrointestinal tract is far less than that in the lung.

In occupational exposure standards, dusts are classified as either soluble or insoluble according to their solubility in water. Examples of insoluble dusts are minerals, metal oxides, silica, granite, and asbestos. Soluble dusts include some minerals, such as limestone and dolomite, and organic dusts, such as trinitrotoluene, flour, soap, leather, cork, wood, and plastics. One definition that has been used is 5% solubility in water; however, this definition may change in the future.[8] Table 2.5 lists soluble and insoluble chemical forms.

A characteristic related to water solubility is hygroscopicity, the ability of a com-

TABLE 2.5. **Examples of Soluble and Insoluble Chemical Forms**

Compound	Soluble Forms	Insoluble Forms
Molybdenum	Ammonium molybdate Sodium molybdate	Calcium molybdate Molybdenum trioxide Molybdenum halides Molybdenum disulfide
Nickel (inorganic salts)	Nickel chloride Nickel nitrate Nickel sulfate	
Platinum	Ammonium chloroplatinate Sodium chloroplatinate Platinic chloride Platinum chloride Sodium tetrachloroplatinate Potassium tetrachloroplatinate Ammonium tetrachloroplatinate Sodium hexachloroplatinate Potassium hexachloroplatinate Ammonium hexachloroplatinate	
Rhodium	Rhodium nitrate Rhodium potassium sulfate Rhodium sulfate Rhodium sulfite	

pound to absorb water from the air. Hygroscopic particles increase in diameter as they pass through the high humidity of the respiratory tract. A 1-μm particle may grow by water absorption to 3 μm. When aerosols of ammonium sulfate or sulfuric acid are inhaled, the individual particles grow and as a result impact more on the upper airways. Generally the effect of hygroscopicity on deposition will be less for particles composed of materials having greater densities and high molecular weights.[2]

Complex Compounds. Some solids, such as naphthalene, found in mothballs, have a vapor pressure high enough for some of this material to exist in an airborne vapor state. Therefore, when sampling these types of compounds the collection method must account for the aerosol as well as the vapor component. Compounds that have been found to exist as both an aerosol and a particulate include caprolactam, PCBs, fenamiphos, methomyl, methyl demeton, acrylamide, and diazinon.

Mixtures, such as paint and pesticide formulations, frequently have both vapor and particulate components. In these cases the air sampling practitioner must know the specific components of interest. For example, with spray painting, often the solvent is the contaminant of interest. The sampling method must collect the solvent vapors plus the solvent present in the airborne paint droplets. In some operations where spraying of resins is done, the particulates represent a significant percent of the air concentration. In these situations, particle size is important as well as the potential for vaporization and polymerization, which will alter the relative percent of particulate and vapor that are present. For example, styrene when sprayed has been shown to generate both an aerosol and a vapor with as much as 32–33% of the total styrene concentration due to the aerosol component.[9]

Sampling reactive compounds requires special techniques, because through reactions such as polymerization, hydrolysis, and oxidation the sample can be changed or lost. This situation can lead to misinterpretation about what types of compounds and concentrations are present. One group of reactive compounds is the diisocyanates, which react with water to polymerize and release carbon dioxide; thus the sampling method must prevent contact of the diisocyanates with water vapor. This is a good illustration of why the sampling protocol for each contaminant must be closely followed.

Routes of Entry

Toxic agents enter the body through the gastrointestinal tract (GI), the respiratory system, or the skin.

Contaminants entering the GI tract often produce effects only on the cells that line the tract, although absorption from the tract can also take place. For example, some compounds such as alcohols are absorbed directly through the stomach. Caustic or primary irritants can destroy the mucosal lining of the GI tract, allowing entry of other chemicals into the bloodstream.

Inhalation via the respiratory system is the most common occupational route of entry for chemicals and other agents. It is also the most susceptible site. Substances reach the bloodstream very shortly after inhalation, provided that they have the right properties to pass through the lungs. During studies to determine if there are any differences in the degree of exposure if a person favors breathing through the mouth rather than the nose, it has been noted that the oral airway is far less efficient at removing materials from inspired air than is the nasal airway.[10] Therefore, during mouth breathing, somewhat greater doses of toxic materials may reach sensitive sites in the lung than during nasal breathing.

Because larger particles when inhaled have a tendency to deposit in the upper

TABLE 2.6. Factors Influencing the Effects of Inhaled Agents

Physical Properties
Physical state
Size, shape, and density of particles, mist, or aerosol determines the site of deposition.
Solubility
   Particulates: Insoluble agents produce local effects whereas soluble compounds may produce systemic effects.
   Gases and vapors: Insoluble agents, such as nitrogen oxides, are inhaled into small air passages, whereas soluble agents, such as ammonia and sulfur dioxide, seldom pass beyond the nose and nasopharynx.
Hygroscopic particles increase in size as they travel down the respiratory tract.
Electric charge influences the site of deposition.

Chemical Properties
Acidity and alkalinity have a toxic effect on cilia, cells, and enzyme systems.
Agents such as carbon monoxide and hydrogen cyanide have systemic effects, whereas fluorine compounds may have both local and systemic effects.
Fibrogenicity.
Antigenicity stimulates antibodies.

TABLE 2.7. Comparison of Oral and Dermal LD$_{50}$'s for Selected Pesticides

Compound	Oral LD$_{50}$ (mg/kg)	Dermal LD$_{50}$ (mg/kg)
TEPP	1.1	2.4
Phorate (Thimet)	1.7	4.3
Phosdrin	4.9	4.45
Disolfoton	4.5	10.5
Demeton (Systox)	4.3	11
Thionazin (Nemaphos)	4.9	14
Ethyl parathion	8.3	14
Chlorfenvinfos	14	30
Bidrin	18	42
Methyl parathion	19	67
Dichlorovos	68	91
Azinphosmethyl	12	220
Malathion	1187	>4440

areas of the respiratory system, a certain portion of these are likely to be swallowed rather than retained in the respiratory system. The main factors that affect the rate of absorption or deposition of material from the respiratory tract into the bloodstream include solubility, particle size, and concentration.[10] Biological variability between individuals, differences in health, confounding factors such as cigarette smoking, differences that relate to age or breathing styles, as well as inherent differences in airway sizes can also impact exposures. Table 2.6 lists some chemical and physical properties of compounds that influence toxicity.[11]

Dermal exposure depends to a significant degree on the solubility in fat of the compound. In addition, the site of exposure

is a factor: Forehead, testicle, abdominal, arm, and back skin all have different permeabilities. Many solvents and pesticides are capable of penetrating the skin. Some chemicals gain entry through the skin very effectively. Table 2.7 compares the results when lethal doses of pesticides capable of killing 50% of the exposed animals (called the LD$_{50}$) were delivered both by oral and dermal routes of entry.[12] The smaller the value of the LD$_{50}$, the more toxic the compound. LD$_{50}$ is expressed in units of mg of chemical per kilogram of body weight (mg/kg).

While not considered a primary route of entry, due to the concerns about contracting hepatitis and AIDS in the health care field, exposure via injection has become an important matter. Needle sticks are one of the primary means by which health care workers are exposed to infected body fluids.

Elimination of Hazardous Materials

The ability of a compound to accumulate in the body or a target organ also depends on the overall rate at which a chemical gets into the body as well as how fast it is metabolized and eliminated. The primary way that agents are eliminated from the body is in the urine, although other modes include the feces, sweat, tears, hair, nails, and breath. Some chemicals such as ethanol are eliminated from the body very rapidly, within hours, whereas others such as lead and arsenic can persist for days. Biologic half-lives describe the length of time it takes for the concentration of a chemical in the body to decrease by 50%. Since the biological half-life is one way of estimating the rate of elimination of a compound from the body, it also represents the likelihood of a compound accumulating in the body. Body burdens can be viewed in more than one way, including peak body burden, average body burden, and residual burden once exposure has ended. The shorter the half-life, the faster the peak body burden is reached. The chapter on biological monitoring (Chapter 18) describes the collection of bodily fluids to supplement air sampling measurements.

Exposure Versus Dose

It is useful to differentiate between concentrations measured outside and inside the body. The term *exposure* refers to those levels measured outside the body and often is reserved for air sampling measurements. In most cases, these do not directly correlate with what actually gets inside the body. *Absorbed dose* refers to the portion of the exposure that actually enters the body through the skin, lungs, or other route and reaches the bloodstream to be transported to the target organs. *Deposited dose* is a term usually reserved for skin exposures and refers to the amount of material that is actually deposited on the skin's surface.[12] Actual body dose differs from the results of air sampling measurements used to approximate inhalation exposure due to many variables: degree of activity of the individual (work versus rest); changes in air

temperature (affects body metabolism); fluctuations in airborne concentrations; ability of the contaminant to be eliminated in exhaled air; accumulation in the tissue or ability to be metabolized to more or less toxic metabolites.[13] Individual biovariations, such as age, size, sex, and genetics, can also affect the dose. Exposures are expressed as milligrams per cubic meter (mg/m^3) or parts per million (ppm) and dose is expressed as the mass of contaminant per mass of body weight or mg/kg.

The primary mechanism for describing and calculating exposure limits is the 8-hour time-weighted average (TWA). Virtually all of the standards designed to be applied to workers are based on this concept. The occupational exposure TWA is calculated as follows:

$$\frac{C_1(T_1) + C_2(T_2) + \cdots + C_n(T_n)}{8 \text{ hours}}$$

where C is the concentration of a sample and T is the sampling time for that sample. Special adjustments to the acceptable exposure level may be needed for work shifts that are linger than 8 hours. As an illustration, for compounds that cause systemic toxicity increasing exposure times (>8 hours) will require decreasing the allowable exposure level in order to ensure that the total body burden will not exceed that allowed by the 8-hour standard.

Some contaminants can cause adverse effects with a short exposure; exposure to these is controlled through a *ceiling* or *short-term exposure limits*. When a ceiling is applied to an 8-hour standard, it means that the PEL can never exceed this value. Thus, the ceiling is the maximum allowable concentration over the shift. It is common to have ceiling standards for irritants.

The short-term exposure limit (STEL) is a peak concentration to which workers can be exposed continuously for a short period of time without suffering from irritation, chronic or irreversible tissue damage, or narcosis of sufficient degree to increase the likelihood of accidental injury, impair self-rescue, or materially decrease work efficiency. STELs are valid exposures provided the daily PEL–TWA is not exceeded. The STEL is not intended to be a separate independent exposure limit; rather it supplements the 8-hour TWA limit where there are recognized acute effects from a substance whose toxic effects are primarily of a chronic nature.

Exposures at the STEL should not exceed a 15-minute time period and in some exposure guidelines should not be repeated more than four times per day. In addition, there should be at least 60 minutes of unexposed time between successive exposures at the STEL. The incremental difference between the 8-hour TWA and the STEL is not always the same

for different chemicals. For example, the 1990 OSHA 8-hour TWA for benzene is 1 ppm, while the STEL is 5 ppm. The 8-hour TWA and STEL for ethyl benzene are 100 ppm and 125 ppm, respectively. Different chemicals can act over different periods of time, including less than 15 minutes, but the use of a generic period simplifies the approach of sampling for peak exposures.

Another approach for managing peak exposures that has been attempted by the National Institute for Occupational Safety and Health (NIOSH) is to select a short-term sampling period based on the biological activity of a compound. However, this method is time consuming and may not be practical as the result, depending on the acute effect identified for a given chemical, is different short-term sampling periods for different chemicals. Figure 2.1 demon-

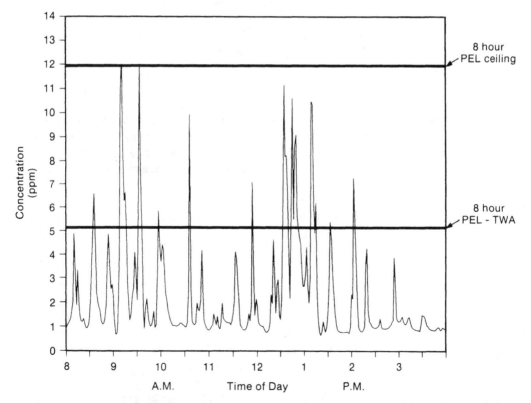

Figure 2.1. Representations of OSHA PELs. The 8-hour TWA showing typical fluctuations, and the 8-hour PEL ceiling as a maximum value not to be exceeded.

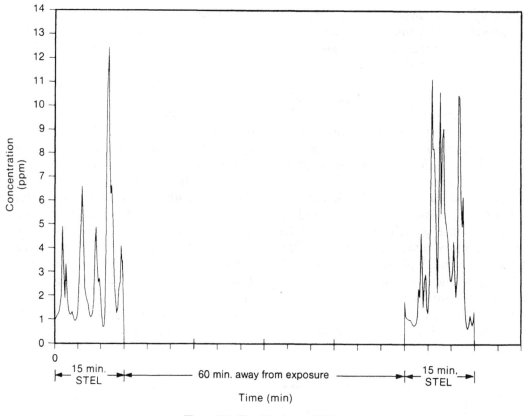

Figure 2.2. The 15-minute STEL.

strates how a TWA standard is actually the average of the varying concentration levels that usually exist over any given workday, but if the standard is a ceiling the averaging effect is ignored in favor of the maximum level detected. Figure 2.2 shows the STEL, which is also an average concentration, but over a much shorter time period.

TOXIC EFFECTS

Exposures are characterized by their time period and the concentration of the contaminant present. Acute exposures are the result of exposures to high concentrations over periods of less than 24 hours and can involve either single or repeated exposures within that period. Chronic or long-term exposures involve months or years of exposures to low levels of a contaminant. Examples of chronic health effects are silicosis and byssinosis.

The responses to a toxic exposure vary from immediate effects to delayed symptoms, such as those associated with carcinogens where there are long latency periods. When identifying chemicals that are of high hazard in an emergency release, acute toxicity is the most important criterion. On the other hand, for long-term, low-level exposures to the community or workers, chronic toxicity is the most important. For a given chemical, its acute and chronic effects can be quite different.

Not all chemicals are capable of causing acute toxicity. An example is asbestos. Even though the hazards of asbestos are well documented, it causes only chronic health effects. Even the worst exposure is unlikely to cause toxic effects for several years. Long-term, high-dose exposures tend to cause asbestosis, which could occur in a shorter period of time than the latency period associated with cancer development, but in both cases the time period is in years.

The site of action associated with toxic substances varies widely. Local effects such as a skin burn are those lesions caused at the site of first contact between the agent and the individual. Systemic effects involve the absorption and distribution of the toxic agent from the site of entry, whether via oral, dermal, or inhalation, to a distant site where the toxic response occurs. An example is mercury, which produces toxic effects on the central nervous system.

When two or more substances that act on the same organ systems are present, their combined effect is a consideration. Methylene chloride and carbon monoxide both bind with red blood cells to form carboxyhemoglobin. Most solvents affect the central nervous system; thus for operations where many solvents are used, such as paint and ink manufacturing, additive effects are a concern. Some combinations of exposures, such as smoking and asbestos, can act synergistically, meaning there is a much greater effect than would be predicted from the exposure to levels of the individual substanes.

Systemic Effects

Chemicals that have a systemic effect act on tissues other than the site of entry. As the result of exposure, a variety of adverse effects can be manifested at multiple target organ sites. Although the specificity of the systemic effect caused by exposure to the substances may vary, in general these chemicals are capable of interfering with

biological processes and impairing normal organ function.

Target Organ Effects

A compound can preferentially act on a given organ to cause a toxic effect. Some compounds such as lead can act on many organs whereas others affect primarily one organ. The ability of a chemical or its metabolite to bind at a given site often determines the target organ affected.

Substances that act on the nervous system can cause either peripheral nervous system (PNS) or central nervous system (CNS) effects or both. Motor function (muscular weakness or unsteadiness of gait), sensory function (alterations in sight or sensations of touch, pain, or temperature), and/or behavior may be affected. PNS toxicants, such as mercury, cause degeneration of the nerve and its structure. The most notable immediate effect of overexposure to CNS toxicants is narcosis (dizziness and drowsiness), although prolonged or very gross exposures may cause effects such as visual disturbances, seizures, and brain damage. Chemicals known to cause narcosis include solvents, such as heptane, toluene, and methylene chloride.

Liver effects depend on the dose, duration, and particular chemical agent involved. Acute exposures can cause lipid (fatty) accumulations in the liver cells, cell death (necrosis), and liver dysfunction. Chronic exposures may lead to cirrhotic changes and/or the development of liver cancer. The earliest and most sensitive indicator of liver toxicity is an alteration in biochemical liver functions, such as changes in specific enzyme functions. Chlorinated compounds frequently have adverse effects on the liver as do many solvents.

Depending on their site of action, chemicals acting on the kidneys can interfere with hydration, excretion of wastes, electrolytic balance, or metabolism. Because a significant degree of kidney cell damage must occur before it is reflected by measurable changes in kidney function, kidney function tests do not necessarily detect early damage. Compounds that affect the kidneys include 1,3-dichloropropene, ethyl silicate, and hexachlorobutadiene.

Effects on the cardiovascular system include cardiac sensitization, which results in disturbances in the heartbeat, and vasodilation when blood vessels expand, resulting in blood pressure and circulation losses. Chemicals that affect the cardiovascular system include carbon disulfide, some freons, and sodium azide.

The bladder can also be a site of toxic effects, most notably cancer. Compounds such as α- and β-naphthylamine cause bladder cancer.

Other Exposure Effects

Irritation. The eyes, skin, and upper respiratory system can be irritated by chemicals. The symptoms of irritation include stinging, itching, and burning of the eyes; tearing; burning sensation in the nasal passages; nasal inflammation; cough; sputum production; chest pain; wheezing and other breathing difficulties. At low levels irritation can be useful as a warning that chemicals are present. Chlorine, ammonia, and formaldehyde as well as many other compounds are irritants.

Sensitization. A sensitization reaction, also known as an allergic reaction, is defined as an adverse response to a chemical following a previous exposure to that substance or to a structurally similar one.[13] A related phenomenon is cross-sensitization. It occurs when exposure to one substance elicits a sensitization reaction, not only upon subsequent exposure to the same substance but also with exposure to a different substance with a similar structure. Common target organs for sensitization are the skin and respiratory system. Sensitivity to a chemical can persist for an individual's lifetime. Examples of sensitizers are isocyanates and poison ivy.

Respiratory Effects. The respiratory system is susceptible to many diseases including cancer. Chronic pulmonary disease can include fibrotic changes as well as other diseases such as emphysema. Fibrotic conditions in the lung result in a loss of elasticity with a decrease in the ability of the individual to utilize oxygen. Fibrosis is often considered the same as pneumoconiosis; however, pneumoconiosis is a more general term indicating the presence of a foreign substance in the lungs. Dusts, such as silica and coal, cause fibrosis.

Biochemical/Metabolic Effects. Many chemicals have the ability to interfere with the normal metabolism or biochemistry of the body. Examples are substances that act on enzymes such as acetylcholinesterase to inhibit their actions, or substances that interfere with the oxygen-carrying capacity of blood. Many organophosphate and carbonate pesticides inhibit acetylcholinesterase, resulting in signs and symptoms such as bronchoconstriction, increased bronchial secretions, salivation, and tearing; nausea; vomiting; cramps; constriction of the pupils; muscular weakness; and cardiac irregularities. If sufficiently severe, this condition can lead to death. Substances that interfere with the oxygen-carrying capacity of the blood often act by tying up the hemoglobin in the red blood cells, which are responsible for oxygen transport. Carbon monoxide and methylene chloride are examples. Chemicals with biochemical and metabolic effects are often selected for biological monitoring when their toxic effect can be measured. This type of sampling is discussed in the chapter on biological monitoring (Chapter 18).

Carcinogenic Effects. Cancer is the result of an abnormal tissue composed of

TABLE 2.8. IARC Carcinogen Categories

Group 1	The agent (mixture) is carcinogenic to human, or the exposure circumstance entails exposures that are carcinogenic to humans.
Group 2	Group 2A is for agents, mixtures and exposure circumstances that entail exposures that are *probably* carcinogenic to humans. Group 2B is for agents, mixtures and exposure circumstances that entail exposures that are *possibly* carcinogenic to humans.
Group 3	The agent, mixture or exposure circumstance is not classifiable as to its carcinogenicity to humans.
Group 4	The agent or mixture is probably not carcinogenic to humans.

cells that have been altered in such a way as to cause unrestricted cell growth invading organ systems. As a result, these cells lose the ability to function normally and the tissue that they comprise interferes with the vital functions of normal organ systems. Chemicals are categorized according to the evidence linking their ability to cause cancer in humans. The International Agency for Research on Cancer (IARC) is one of the primary agencies involved in doing carcinogenic risk assessments. For evidence of possible carcinogenicity in humans the IARC uses two sources: epidemiological studies of groups of people, often workers; and long-term (usually 2 years) animal tests. Some compounds known to cause cancer in humans are asbestos, benzene, and MOCA. Table 2.8 lists IARC categories for carcinogens.[14]

Reproductive Effects. Chemicals that cause reproductive effects can act in two different ways. They can cause sterility in either males or females or cause defects in the embryo or fetus, termed *teratogenesis*. When they affect offspring, they are called teratogens. Many chlorinated pesticides affect hormone systems and are implicated in sterility effects on females, and dibromochloropropane (DBCP) and kepone are known to cause sterility in males. Lead as well as many other compounds may cause teratogenic effects in offspring when pregnant females are exposed. Lead has also caused decreased sperm counts in males.

Nuisance Effects. High concentrations of some dusts may seriously reduce visibility; cause unpleasant deposits in the eyes, ears, and nasal passages; or cause injury to the skin or mucous membranes either due to their own properties or by the rigorous skin cleansing procedures that may be necessary to remove them. The terms *inert* and *nuisance* dusts have been applied to those dusts that are not known to produce any significant toxic effects when exposures are kept under reasonable control. However, these terms are misnomers because there is no dust that will not evoke some type of response in the lung if inhaled in sufficient amounts. Examples of compounds that have been included in this category include calcium carbonate, cellulose (paper fiber), emery, glycerin mist, graphite, gypsum, kaolin, limestone, magnesite, marble, silicon carbide, starch, and sucrose. Be aware of the fact that since toxicological information on many chemicals including dusts is lacking, the philosophy regarding a particular dust can change.

WARNING SIGNS

Many chemicals reveal their presence due to certain physical properties including odor, color, or ability to cause very mild irritation. A knowledge of these properties can provide valuable clues to levels of concentration and identification of locations to sample. For chemicals, odor is the most important warning property.

Chemicals are considered to have adequate warning properties if the odor threshold is at or below the exposure limit. Exposure limits are discussed in detail in a later section on standards in this chapter. Increases in concentration do not necessarily cause the same incremental increase in odor, and the intensity varies from one substance to another. Some odors, termed *characteristic* because they are very distinctive and tend to be associated with one particular compound, are used to describe other compounds that have a similar smell. Some chemicals, such as hydrogen sulfide, can also cause olfactory fatigue where the senses become "dulled," leading to the inability to detect the material even though it is still present.

Odor thresholds can be defined in more than one way. For example, in one case it might be detection: "I smell something." In another case it may involve recognition: "This smells like acetone." Therefore, comparing odor thresholds can be difficult as the criteria for the test can vary. Some tests determine the odor threshold of a compound mixed in water. The drawback of this data is that they are only valid when the concern is an odor associated with water and not air. There are wide individual differences in levels of odor perception, including differences in the level of odor detected when panelists come from a "clean air" background versus if they have spent time in an industrial atmosphere, as well as a decrease in the ability to detect odors with increasing age or as the result of allergies. Therefore, more reliable data are given as a range of values. Compounds with the lowest odor thresholds have the most variability in ranges of detection, whereas others with the highest thresholds have the least.

Several methods have been used to determine odor thresholds. Usually testing procedures involve panels of volunteers. A mask and tubing of chemically inert odorless material are used to expose each subject to the compound. First, a concentration known to be above the odor threshold is used. Once familiar with the odor they are attempting to detect, a zero concentration sample is presented. The concentration in the mask is then raised slowly over a period of several minutes with the subject sniffing the mask at 15-second intervals until it is detected. A concern regarding the estimation of odor thresholds is that laboratory experiments that gauge the capacity to detect and recognize small amounts of the warning agents may not accurately predict detection and recognition in the field.[15] Some test subjects repeatedly challenged by a compound grow more sensitive to the material, whereas others develop a tolerance or experience olfactory fatigue.

Odor thresholds can be useful for determining when to change respirator cartridges and are the basis for the isoamyl acetate (banana oil) respirator fit test. They can also assist in investigating potential engineering control failures. Strange odors are often the basis for indoor air concerns as well as community odor problems. One area where odor rather than testing has been strongly relied on is in locating propane and natural gas leaks. Odorants such as ethyl mercaptan and thiophane are added to the gas to enhance the likelihood of leak detection. However, it is not a good idea to use odors to attempt to estimate the quantity of a contaminant. In particular, the sampler should never sniff unknown samples of water, soil, or other materials in order to determine if chemicals are present.

Irritation responses are also measured using panels of volunteers similar to the procedures used for odor thresholds. Subjective measurements of irritation can be affected by a variety of psychological and physiological factors, including airflow over the eyes, the presence of dust particles, the

TABLE 2.9. Examples of Irritation and Odor Thresholds

Compound	Irritation Threshold (ppm)	Odor Threshold (ppm)	Odor Description
Acetic acid	10–15	0.2–24	Vinegar, characteristic
Acetone		100	Nail polish remover, characteristic
Ammonia	55–140	0.32–55	Characteristic
Benzene		4.68	Aromatic
Butyl acetate	300	0.037–20	Fruity
Carbon disulfide		0.21	Characteristic
Chlorine	1–6	0.01–5	Bleach
Chloroform	>4096	50–307	Characteristic
p-Dichlorobenzene	80–160	15–30	Mothballs
Dimethyl amine		0.047	Fishy, ammonia
Epichlorohydrin	100	10–16	Chloroform
Ethyl ether	200	0.33	Characteristic
Formaldehyde	0.25–2	0.1–1.0	Characteristic
Hydrogen sulfide	50–100	0.00001–0.8	Rotten egg, characteristic
Isoamyl acetate	100	0.002–7	Banana oil, characteristic
Isopropyl alcohol	400	7.5–200	Characteristic
Isopropyl amine	10–20	0.71–10	Ammonia
Isopropyl ether	800	0.053–300	Ether-like
Methyl alcohol	7500–69,000	53.3–5900	Characteristic
Methyl methacrylate	170–250	0.05–0.34	Characteristic
Naphtha, coal tar	200–300	4.68–100	Aromatic
Phosgene		1	Hay-like
Phosphine		0.021	Oniony, mustardish, fishy
Stoddard solvent	400	1–30	Kerosene
Styrene (uninhibited)	200–400	0.047–200	Characteristic
Sulfur dioxide	6–20	0.47–5	Characteristic
Toluene	300–400	0.17–40	Rubber cement
1,1,1-Trichloroethane	500–1000	20–400	Chloroform
Turpentine	200	200	Characteristic
Xylene	200	0.05–200	Aromatic

amount of sleep the previous night, and anticipation. The time for a response to occur is also important, and generally higher concentrations are detected sooner. Table 2.9 lists irritation and odor thresholds.[16–18]

Certain chemicals, such as metallic fumes and mercaptans, can cause an odd taste in the mouth upon exposure. Some gases are colored when present in high concentrations. For example, nitrogen oxides are red-brown and chlorine is yellow-green. A puddle on the floor, a dark stain on the ground, or piles of dusts on rafters or window sills are obvious situations where chemicals may be present. Concentrated vapors of some chemicals, such as hydrogen chloride, hydrogen fluoride, and ammonia, form visible clouds in areas where high humidity exists.

Toxicity Versus the Hazard

The toxicity of a substance, while important, should not be the sole criterion used in determining the existence of a health hazard associated with a specific situation. Many factors should be considered, such as the chemical and physical properties of this toxic substance, the ability of other toxic substances to interact with it, and the influence of environmental conditions such as temperature as well as the concentration present. Table 2.3 (right column) shows the 8-hour TWA exposure guidelines for each compound in the table. The ideal solvent would have the needed solvent properties, plus a low vapor pressure (i.e., less volatility) and a high exposure standard (i.e., lower inhalation toxicity), and not be absorbed through the skin. As an example, octane would be a good choice from Table 2.3 from a toxicological standpoint if the chemical and physical characteristics were acceptable.

The degree of hazard associated with exposure to a specific substance also depends on the conditions of use. For example, a highly toxic chemical that is processed in a closed, isolated system may be less hazardous in actual use than a low-toxicity compound handled in an open batch process. Therefore, the nature of the process in which the substance is used or generated, the possibility of reaction with other agents, the degree of engineering controls including ventilation, and the amount of personal protective equipment in use all relate to the potential hazard associated with each use of a given agent. Another factor affecting the ability of a chemical to elicit a toxic response is the susceptibility of the biological system or individual. Although individual susceptibility is useful in evaluating potential harm following an exposure, it is seldom used to determine appropriate exposure levels. Acceptable exposure standards are set to protect most people without regard to individual traits or susceptibility.

STANDARDS AND GUIDELINES FOR AIR SAMPLING

Since the results of sampling usually consist of airborne concentration values, proper interpretation requires comparing them to some type of agreed-upon standard. The reasoning and research behind standards are what make exposure results useful. For example, if studies have shown that 500 ppm of a particular substance causes dizziness in workers, 50 ppm might be selected to be a level that will provide a safe margin between exposure and this observed effect.

Many organizations recommend exposure levels; however, unless these levels are incorporated into a regulation, they are simply guidelines whereas regulations are legally binding requirements. Although most commonly the federal and state regulations are referenced, some cities also have their own air emissions regulations. Often these are implemented by local health departments or county environmental agencies.

The air sampling should be planned and conducted to yield results that can be compared to an appropriate standard. If the sample covered a full shift in an occupational setting, the result should be compared to an appropriate TWA value from OSHA, NIOSH, ACGIH or another recognized group.

While standards and guidelines are important, the ultimate criterion for a sampling professional such as an industrial hygienist is professional judgment. There may be a situation where the professional chooses to use a guideline that is lower than a standard because there is a hypersensitive worker population or it is necessary to protect individuals other than workers. It is not always possible to use the lowest value; therefore, it is essential to be practical as well as judicious. It is the ability to utilize standards and toxicity information, along with the existence of the necessary conditions of use for determining the degree of the hazard, that sets the sampling professional apart from others.

TABLE 2.10. The Reference Worker

Parameter	Value
Weight	70 kg
Height	175 cm
Age	20–30 years
Body surface area	1.8 m^2
Lung weight	1000 g
Lung surface area	80 m^2
Total lung capacity	5.6 L
Vital capacity	4.3 L
Residual volume	1.3 L
Respiratory dead space	160 mL
Breathing rate	15 breaths/min
Tidal volume	1450 ml
Inspiratory flow rate	43.5 L/min
Minute volume	21.75
Inspiratory period	2 sec
Expiratory period	2 sec

Chronic Occupational Exposures

Workplace air standards are the most numerous and cover a wide variety of chemicals. In addition, with few exceptions, occupational air monitoring methods have been developed that are chemical specific. These standards apply only to workplace exposures; assume that individuals work an 8-hour day, 40-hour work week; and are designed to be applied to the healthy working individual and do not take into account gender-related or age-related considerations. Some individuals have increased susceptibility and may not be protected by the standards. The most difficult situations are represented by individuals who have heritable (genetic) characteristics such as sickle cell anemia, preexisting diseases such as asthma, and those who become sensitized to a compound such as an amine. Habits, especially drug use, alcohol intake, and smoking, as well as age, influence the amount of exposure that can be tolerated.[19] See Table 2.10 for the physical characteristics of a reference worker used to develop standards and estimate the consequences of doses of various agents.[20] The best way to view these standards is not as fine lines between safe and unsafe conditions but as guidelines applied by professionals to protect workers.

U.S. Federal OSHA. The standards for chemicals that the Federal Occupational Safety and Health Administration (OSHA) promulgates are termed permissible exposure limits (PELs). These include regulated exposure limits for nearly 600 chemicals along with 24 more comprehensive single-substance standards. OSHA standards are considered to be the legal minimum health protection required of employers and are designed to protect against a variety of toxic effects including irritation, target organ toxicity, cancer, chronic lung disease, and biochemical/metabolic effects. Table 2.11 contains some examples of the basis upon which OSHA standards for various chemicals have been set.

An "action level" is incorporated into many of OSHA's single-substance standards that is one-half of the 8-hour TWA and triggers certain requirements such as medical monitoring, training, and repeat air sampling. The action level applies a safety factor to the 8-hour PEL by taking into account day-to-day variations, statistical error, and other aspects that influence the accuracy of sampling results, so although a respirator is not required at this level, repeat sampling increases the likelihood that an overexposure will be detected. The action level for benzene is currently 0.5 ppm calculated as an 8-hour TWA.

While the existence of PELs tends to focus attention on those chemicals with PELs, exposures to other contaminants should also be evaluated. The General Duty Clause, section 5 (a)(1) of the OSHA Act, provides that "each employer shall furnish to each of his employees a place of employment which is free from recognized hazards that are causing or likely to cause death or serious physical harm to these

TABLE 2.11. Examples of the Basis for OSHA Standards

Compound	Toxic Effect Prevented
Acrylonitrile	Liver carcinogen
Allyl chloride	Liver and kidney toxicity
Aluminium, soluble salts	By analogy to hydrogen chloride because hydrolysis to hydrogen chloride occurs (pulmonary injury)
sec-Amyl acetate	Irritation to eyes and respiratory tract
Asphalt fumes	Carcinogen
Carbon monoxide	Carboxyhemoglobinemia
Chlorodifluoromethane	No effect level for cardiac sensitization
Chlorodiphenyl, 54% (PCBs)	Systemic effects
Cobalt, metal, fume, and dust	Sensitization
Cobalt carbonyl	Systemic toxicity
Cyclohexanone	Liver toxicity
Dichloromonofluoromethane (Freon 21)	Hepatotoxicity, cardiac sensitization
Diethanolamine	No effect level for impaired vision and skin irritation
Ethyl alcohol	eye and respiratory tract irritation
Fibrous glass dust	Nuisance dust
Formamide	Testicular toxicity, teratogenicity
Hydrogen bromide	Irritation
Hydrogen chloride	Pulmonary injury
Hydrogen cyanide	Systemic toxicity including cyanide poisoning, weakness, mucosal irritation, colic, nervousness, and enlargement of the thyroid in humans
Hydrogen fluoride	Eye and nose irritation
Hydrogen sulfide	Occular effects
Isoamyl acetate	Irritation of the upper respiratory tract
Isobutyl alcohol	Irritation, narcosis
Malathion	Nuisance dust
Manganese (and compounds)	Manganism
Manganese, fume	Neuropathic
Methyl amyl ketone	Irritation
Methyl butyl ketone	Neuropathic
Methyl isocyanate	Mucous membrane irritation
Methyl parathion	Cholinesterase inhibition
Toluene	Narcosis
Xylene	Irritation

employees." Thus OSHA compliance officers can cite for an overexposure to a chemical for which no PEL exists.

Threshold Limit Values. The American Conference of Governmental Industrial Hygienists (ACGIH) was founded in 1938 and is an independent organization comprised of industrial hygienists and other health professionals from academia and government-related institutions. The threshold limit values (TLVs®) developed by ACGIH are recommended exposure limits based on a belief that there is a threshold(s) of response, derived from an assessment of the available published scientific information including studies in exposed humans and experimental animals. New chemicals are added on a regular basis and exposure levels are often adjusted, usually to new, lower levels, and for each TLV® a set of supporting documentation is published. Like the PELs, TLV–TWAs refer to 8-hour exposures and 40-hour weeks and are expressed in parts per million (ppm) at a standard temperature and pressure of 25°C and 1 atmosphere or milligrams per cubic meter (mg/M^3). TLV–TWAs refer to airborne concentrations to which nearly are workers may be exposed for 8 hours per day, 40 hours per week for a working lifetime without adverse effects. The TWA is calculated using the equation in Chapter 2.

For most substances with an assigned TLV–TWA there is not enough toxicological data available to assign an appropriate short-term exposure limit (STEL). For these substances the TLV® Committee recommends that short duration exposures follow this guideline[21]:

> Excursions in worker exposure levels may exceed 3 times the TLV–TWA for no more than a total of 30 minutes during a workday, and under no circumstances should they exceed 5 times the TLV–TWA, provided that the TLV–TWA is not exceeded.

A calculation is also available to tailor a TLV® for situations when mixtures of several chemicals are present, each of which is likely to affect the same target organ. A common example is the use of solvents. If a worker was exposed to the full TLV of each compound, it is very likely than an excessive exposure would occur; therefore, in these situations the TLV® for the individual compounds is adjusted lower by considering their effects as a group. Although the air sample is analyzed for each component, the TLV® for a mixture of exposures to an individual is calculated using the following formula[21]:

$$\frac{C_1}{TLV_1} + \frac{C_2}{TLV_2} + \cdots + \frac{C_n}{TLV_n} \le 1$$

where C_1 is the concentration of the first compound, C_2 is the concentration of the second compound, C_n is the concentration of the last compound, $TLV_1 = TLV®$ for the first compound, $TLV_2 = TLV®$ for the second compound, and $TLV_n = TLV®$ for the last compound. This formula can also be used for mixtures of biologically active mineral dusts. For example, as there are various types of silica, each with a different TLV, a calculation for additive effects should be considered if more than one type is involved in an exposure.

One area where the TLVs® are applying the latest technology is size distribution for particulate matter. Traditionally, most TLVs® and OSHA PELs have referred to total particulates as collected with a filter. In some cases the exposure standard applied to the "respirable fraction," which is an air sample collected with a small cyclone that discards particles too large to be inhaled. Now the ACGIH TLV Committee is expanding the particle size-selective TLV® categories.

The particle size-selective TLVs® take into account not only the inherent toxicity of the particles but also the particle size distribution, their patterns of deposition within the respiratory tract, and the related rate of dissolution and translocation to target tissues. They also take into consideration the diseases associated with the inhaled material, and are based on the physical characteristics of the lung; size, mass, distribution, and dynamics of particles; physical and chemical composition of particles emitted by varying processes; and other factors including dissolution rates in the lung.[22] Figure 2.3 shows the areas of the respiratory system that correspond to the size-selective TLVs.[23] OSHA uses a much simpler approach to regulate dusts based on particle size. However, this may change to reflect ACGIH's particle-size TLVs® in the future.

Inhalable particulate mass (IPM)–TLVs® are designed to represent materials

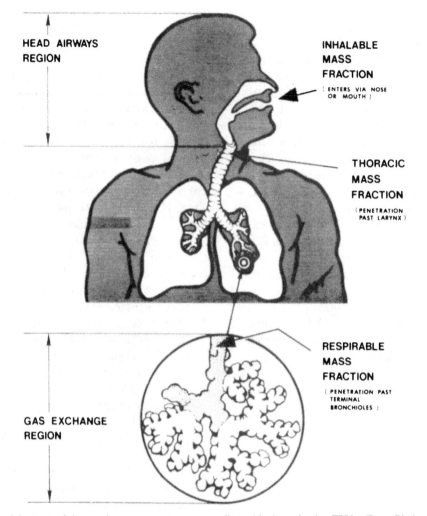

Figure 2.3. Area of the respiratory system corresponding with size-selective TLVs. (From Phalen, R. F. Rationale and recommendations for particle size-selective sampling in the workplace. *Appl. Ind. Hyg.* **1**(1):3–14, 1986; reproduced with permission of *Applied Industrial Hygiene.*)

that are hazardous when deposited anywhere in the respiratory tract. IPM–TLVs apply to particles ranging in size up to 100 μm.

The thoracic particle mass (TPM)–TLV® applies to chemicals that are hazardous when deposited anywhere within the lung airways and the gas exchange region. These particles are capable of penetrating a separator-sampler whose size collection efficiency is based on a median aerodynamic diameter of 10 μm ± 1.0 μm and a geometric standard deviation of 1.5 (±0.1).

The respirable particulate mass (RPM) limits are designed to represent contaminants that are hazardous when deposited in the gas exchange region. These particles will penetrate a separator-sampler whose size collection efficiency is based on a median aerodynamic diameter of 3.5 μm ± 0.3 μm and a geometric standard deviation of 1.5 (±0.1).

National Institute for Occupational Safety and Health (NIOSH). NIOSH functions as an agency within the Department of Health and Human Services whose principal responsibilities are to conduct research to identify hazards and to develop new techniques that relate to workplace health and safety. This duty includes conducting various types of field investigations, such as health hazard evaluations, in specific workplaces at the request of employees and employers. Health hazard evaluations are generally done when there are worker complaints that cannot be associated with any known chemical or other hazards in an operation. Most importantly, NIOSH also develops and publishes air sampling methods.

NIOSH has recommended a number of standards for chemicals, known as recommended exposure limits (RELs). These often include both 8- and 10-hour TWAs as well as ceiling limits of varying time periods. The mechanism for setting an REL is usually through publishing a criteria document. In the process of developing a criteria document an extensive set of research is done to review existing human and animal data. These documents also contain recommendations for exposure limits, warning label wording, exposure and medical monitoring, sampling and analytical methods, and recommended controls for engineering and personal protective equipment.

U.S. Mine Safety and Health Administration.
The Mine Safety and Health Administration (MSHA) oversees underground and surface mining, associated surface preparation and processing operations, mine construction, and other activities related to mining. MSHA has health standards for air contaminants, such as coal dust, silica, radon, and asbestos, that may be encountered during mining activities.

American Industrial Hygiene Association.
The American Industrial Hygiene Association (AIHA) is a professional organization that has been in existence since the 1930s. Workplace environmental exposure limits (WEELs) are developed by an AIHA committee whose purpose is to establish workplace environmental limits for chemical substances and physical stresses for which no TLV, PEL, or other limit exists. The committee utilizes all available information on toxicology, epidemiology, industrial hygiene, and workplace experience information to develop safe exposure guidelines. Like the PELs and TLVs, WEELs are in the form of 8-hour TWAs and 15-minute STELs.

Extended Workshifts

It is important to be aware of the potential limitations of applying standards based on an 8-hour exposure to situations where the shifts are longer. For example, seasonal work often involves 10- to 12-hour days or work weeks of 6 days, as does project-related work such as remediation on hazardous waste sites and jobs involving concentrated repairs over a period of time common during shutdowns of industrial facilities. An adjusted 8-hour PEL should be developed for substances that are capable of accumulating in the body to prevent the body burden from increasing over the amount the 8-hour TWA standard assumes is safe. Figure 2.4 provides an example of how body burdens resulting from 8-hour and 10-hour workdays can differ for the same chemical.[24] It should be noted that while standards may be lowered to account for increased exposure periods, they are never increased to allow for decreased exposure periods, such as a 7.5-hour day, 37.5-hour work week.

Several approaches have been recommended for adjusting standards where work schedules exceed 8 hours per day. The most widely accepted is the model by Brief and Scala[25] that calculates a reduction factor (F_R) to be applied to the 8-hour TWA acceptable exposure value:

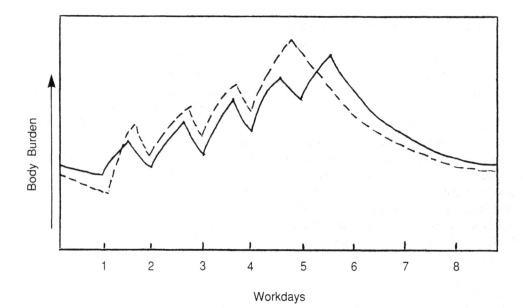

Figure 2.4. Comparisons of differences in body burdens for a chronic toxicant for 8-hour and 10-hour workdays. The solid line represents a 5-day to 8-hour workday week; the dashed line represents a 4-day to 10-hour workday week. (Reprinted with permission by the *American Industrial Hygiene Journal*, Vol. 38, p. A-614, 1977.)

$$F_R = \frac{40\,\text{hr}/\text{wk}}{T_E} \times \frac{T_N}{128\,\text{hr}/\text{wk}}$$

where F_R is the reduction factor, T_E is the time of exposure, hr/wk, and T_N is the time not exposed, hr/wk.

The RF adjusts for both the extra hours of exposure per week and the fewer hours for recovery (nonexposure), which is 128 hours/week with a normal 40-hour work week.

Before making compliance decisions for extended exposure periods, it is important to understand the approach used by the regulatory agency with jurisdiction over the workplace or other exposure situation.

Immediately Dangerous to Life or Health (IDLH) Values

The IDLH level is not a standard, but is often used for evaluating exposures. The concept of the IDLH was established by NIOSH and represents the maximum level to which a healthy individual can be exposed for 30 minutes and escape without suffering irreversible health effects or impairing symptoms.[18] The IDLH demarcates the concentration above which only respirators offering maximum protection, such as self-contained breathing apparatus, must be used. In theory if there is a potential for an IDLH atmosphere to occur, a person at risk should be provided with not only air monitoring but also respiratory protection. An existing IDLH atmosphere should never be entered unless the respirator is already donned.

When using the IDLH as a measure of a level of concern, certain factors must be taken into account:

The IDLH is based on the response of the healthy male working population and does not take into account exposures to more sensitive individuals, such as elderly, children, or people with various health problems.

The IDLH is based on a 30-minute exposure time that may not be realistic for accidental airborne releases.

IDLH values do not exist for all acutely toxic chemicals.

By using the IDLH as the level of concern, this methodology may not identify all quantities of concern that could result in serious but reversible injury.

Chronic Exposures to the General Community

The greatest concern when setting standards to protect the general populace is the wide range of ages and degrees of health that must be considered. The Environmental Protection Agency (EPA), who has the primary responsibility for community standards, has set only a few substance specific air exposure standards for chronic exposures to the general community. Their standards are set for the ambient or general community air. National *primary* ambient air quality standards define levels of air quality that provide an adequate margin of safety to protect the public health while *secondary* standards define levels of air quality that are necessary to protect against specific adverse effects of a pollutant. Allowable concentrations for secondary standards are higher than for primary standards and apply to much shorter periods of time. These air pollution standards are based on a number of different time periods, such as 1, 3, and 24 hours as well as 3-month and 1-year averages, depending on the contaminant. The EPA also publishes guidance documents on air sampling and has developed a number of air sampling methods for outdoor and indoor air.

The American Society of Heating, Refrigeration and Air Conditioning Engineers (ASHRAE) has developed guidelines for acceptable indoor air quality, published as its "Ventilation for Acceptable Indoor Air Quality" standard. These guidelines are used for indoor air situations such as office buildings in which occupants develop illnesses that appear to be of unknown origin but that can actually be due to contamination carried into the building by its HVAC system.

Acute Exposures to the General Community

The potential for a chemical release in or near a community, whether it be from an overturned railroad car or an industrial accident is very real and so a number of organizations have suggested concentrations that would trigger emergency warning measures. For example, the Emergency Response Planning Guidelines (ERPGs), developed by the American Industrial Hygiene Association, are designed to address acute exposures to the general community. Three exposure levels are set for each of a number of chemicals:

ERPG-1: The maximum airborne concentration below which it is believed that nearly all individuals could be exposed for up to 1 hour experiencing only mild, transient adverse health effects or without perceiving a clearly defined objectionable odor.

ERPG-2: The maximum airborne concentration below which it is believed that nearly all individuals could be exposed for up to 1 hour without experiencing or developing irreversible or other serious health effects or symptoms that could impair an individual's ability to take protective action.

ERPG-3: The maximum airborne concentration below which it is believed that nearly all individuals could be exposed for up to 1 hour without experiencing or developing life-threatening health effects.

Over 100 ERPGs have been issued; typical chemicals include: ammonia, benzene, carbon monoxide, chlorine, hydrochloric acid, hydrogen sulfide, and sulfuric acid.

Acute Exposure Guideline Levels (AEGLs)

Probably the most significant exposure levels for acute community exposures are the Acute Exposure Guideline Levels (AEGLs) under the auspices of the U.S. EPA. These values are intended for once-in-a-lifetime, short-term exposures to airborne concentrations of acutely toxic, high-priority chemicals. The goal is to have scientifically credible, acute (short-term) once-in-a-lifetime exposure guideline levels within the constraints of data availability, resources, and time. Eventually they hope to cover approximately 400 to 500 acutely hazardous substances.[26] The AEGL National Advisory Committee is made up of government agencies, professional associations, organized labor, academia, the private sector, and one member representing environmental justice.

AEGLs represent threshold exposure limits for the general public and are applicable to emergency exposure periods ranging from 10 minutes to 8 hours. Three levels of AEGLs distinguished by varying degrees of severity of toxic effects are developed for each of five exposure periods (10 and 30 min, 1 hr, 4 hr, and 8 hr) if data are sufficient. It is believed that the recommended exposure levels are applicable to the general population including infants and children, and other individuals who may be susceptible. The three AEGLs have been defined as follows:

- AEGL-1 is the airborne concentration [expressed as parts per million or milligrams per cubic meter (ppm or mg/m^3)] of a substance above which it is predicted that the general population, including susceptible individuals, could experience notable discomfort, irritation, or certain asymptomatic nonsensory effects. However, the effects are not disabling and are transient and reversible upon cessation of exposure.

- AEGL-2 is the airborne concentration (expressed as ppm or mg/m^3) of a substance above which it is predicted that the general population, including susceptible individuals, could experience irreversible or other serious, long-lasting adverse health effects or an impaired ability to escape.

- AEGL-3 is the airborne concentration (expressed as ppm or mg/m^3) of a substance above which it is predicted that the general population, including susceptible individuals, could experience life-threatening health effects or death.

Airborne concentrations below the AEGL-1 represent exposure levels that can produce mild and progressively increasing but transient and nondisabling odor, taste, and sensory irritation or certain asymptomatic, nonsensory effects. With increasing airborne concentrations above each AEGL, there is a progressive increase in the likelihood of occurrence and the severity of effects described for each corresponding AEGL. Although the AEGL values represent threshold levels for the general public, including susceptible subpopulations, such as infants, children, the elderly, persons with asthma, and those with other illnesses, it is recognized that individuals, subject to unique or idiosyncratic responses, could experience the effects described at concentrations below the corresponding AEGL.

Other Types of Monitoring

Biological Monitoring. Biological exposure indices (BEIs) are limits that have been established by ACGIH that apply to concentrations of chemicals inside the body. BEIs represent the levels of the determinants most likely to be observed in specimens collected from a healthy worker who has been exposed to chemicals to the same extent as a worker with inhalation exposure to the TLV®–TWA. Therefore, BEIs apply to 8-hour exposures, 5 days a week.[21] Standards have been set for urine metabolites, blood measurements, and exhaled air. Biological monitoring is considered complementary to air monitoring. Biological monitoring involves measuring specific chemical substances normally found in the body, but may increase from work experience; substances normally absent from the body; substances resulting from or affected by a biochemical response to exogenous influences.[8]

Biological monitoring can be used to substantiate air monitoring, to test the efficiency of personal protective equipment, to determine the likelihood of dermal and oral exposure, and to detect nonoccupational exposure. This last application is somewhat controversial in that this type of information may or may not be appropriate for release to persons other than the employee.

Dermal Samples. Currently there are no standards to use for comparison of wipe samples or other forms of dermal exposure samples. Therefore, each situation must be evaluated on a case-by-case basis. In some cases, the presence of any material such as a carcinogen like MOCA on work surfaces or solvents found on the inside of personal protective equipment such as gloves is a concern.

Bulk Samples. There are no OSHA or other occupational standards for bulk samples. Bulk sample results may be expressed in mg/L, μg/L, total μg, or percent. These should be converted to match whatever standard or guideline will be used. For example, a sample of paint analyzed for lead may be reported as mg/L and would be converted to percent because the Housing and Urban Development (HUD) standard for lead in paint is expressed in percent.

A TLV® can be calculated for a liquid mixture (usually solvents) when it is assumed that the atmospheric composition of the vapor above the mixture is similar to that of the mixture.[21] The percent composition (by weight) of the liquid mixture must be known:

$$\frac{1}{\dfrac{f_a}{\text{TLV}_a} + \dfrac{f_b}{\text{TLV}_b} + \cdots + \dfrac{f_n}{\text{TLV}_n}}$$

where f_a through f_n represent the percent of each material in the mixture, f_a being the percent of component a, f_b the percent of b, and so forth, and TLV_a through TLV_n are the TLV® for each material in the mixture, TLV_a being the TLV® for component a, TLV_b being the TLV® for component b, and so forth.

However, this approach will not be valid of the vapor pressures of the various components of the liquid mixture are significantly different. In these cases the exposure contribution from each component should be measured separately rather them trying to assign a TLV® for the liquid mixture.

EPA regulations contain tests used to identify materials as hazardous waste that can also be applied to samples of soil and liquid wastes.[27] Many state environmental agencies publish cleanup levels for certain situations, such as underground storage tank removal. The EPA sets standards for chemicals, microbials, and radon in drinking water via the Clean Drinking Water regulations.

EXPOSURE CONTROLS

An individual's exposure to an airborne substance is related to the amount of contaminants in the air and the time period during which the individual breathes the concentration. Any factors that interrupt the exposure pattern by reducing the amount of contaminant in the person's breathing zone or the amount of time that the person spends in the area will reduce the overall exposure. As an example, to lower occupational exposures review the contaminant source, the path it travels to the worker, and the employee's work pattern and use of protective equipment. When implementing controls, the use of engineering controls, which eliminate the exposure or remove the contaminant, is preferred, followed by work practice and administrative controls. Personal protective equipment, such as respirators, should only be used for routine exposures if a better control technique is not feasible.

Source Control

- Preventive maintenance to repair leaks or control other factors that increase emissions.
- Substitution of less toxic materials for those already in use. Carbon tetrachloride is an example of a solvent that was once in wide use but has been replaced with less toxic solvents in almost every use. The substitution of artificial abrasives for sand in abrasive blasting cleaning operations to reduce exposure to free silica is another illustration.
- A change in the process to reduce the amount of contaminant released. For example, the use of a water spray at many dust-generating processes prevents dispersion of dust. The use of large bulk sacks (Figure 2.5) for dry powders in place of individual smaller sacks can reduce the dust emissions that accompany the dumping and disposal of the small sacks.
- Local exhaust systems capture or contain contaminants at their source

Figure 2.5. Bulk sacks with a drain spout at the bottom can reduce dust emissions compared to emptying a large number of smaller bags. (Courtesy of B.A.G. corporation.)

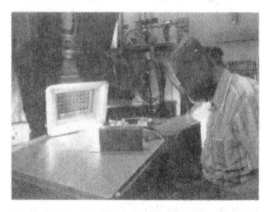

Figure 2.6. Local exhaust systems capture contaminants at their source before they enter the workroom environment.

before they escape into the workroom environment (Figure 2.6). A typical system consists of one or more hoods and ducts, an air cleaner if needed, and a fan (Figure 2.7). The big advantage

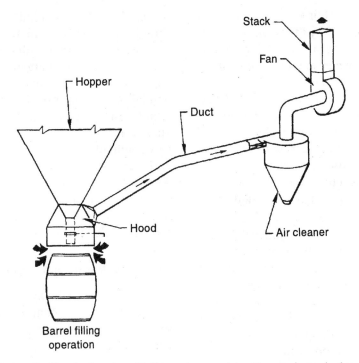

Figure 2.7. A typical local exhaust ventilation system consists of hoods, ducts, air cleaner, and fan.

of local exhaust systems is that they remove contaminants rather than just dilute them, as is the case with general wall-mounted exhaust fans. Even with local exhaust, some airborne contaminants may still be in the workroom air due to uncontrolled sources or less than 100% collection efficiency at the hoods.

Sometimes airborne levels of contaminants can be monitored continuously with a sensor that alarms when concentrations exceed an established level. When an alarm sounds, the workers or supervisors take steps to reduce airborne levels. This approach is useful to detect abnormal operating conditions that cause excessive emissions.

Other Steps to Reduce Exposures

Although steps that reduce airborne contaminant levels are usually preferred, here are some techniques to reduce exposures by changing the work patterns or protecting the employees:

- Work procedures that allow workers to do the job with minimum exposure. Procedures can cover topics such as venting equipment before opening it, flushing it with safe materials to remove chemicals, and even where to stand during the job to minimize exposure levels.

- Personal protective equipment such as respiratory protective devices. There are a variety of respirators suitable for different contaminants and levels of exposures. They are divided into two main classes: (a) air-purifying respirators that remove contaminants from ambient air and (b) atmosphere-supplying respirators that provide respirable air to the worker from an air compressor, from a large compressed air cylinder, or, in the case of a self-contained breathing unit, from an air cylinder carried by the worker. Generally, respirators are used for routine tasks only when engineering controls or work procedures do not reduce exposures to acceptable levels. Respirators may be needed during emergencies or unusual conditions. Specific legal and consensus standards cover respirator use. Follow these standards when establishing a respirator program.

- Training and education that help employees reduce exposures. If contaminants do not have a noticeable odor or other sensory warning properties, employees may not understand how exposures occur. If employees have this information, as well as knowledge of the potential health effects, they may be able to help devise methods to reduce their exposures.

Air monitoring is conducted after exposure controls are implemented in order to gauge their effectiveness and also verify that the post-control exposures are within acceptable limits. Pre- and post-implementation data are also powerful communications tools to show workers the importance of using the controls effectively. Figure 2.8 shows real-time data collected using a direct reading instrument in a welding shop during a specific welding task. The top chart shows the airborne levels of particulates without local ventilation, while the bottom chart shows the same operation with the exhaust system in operation. In each case there is a "background" level of approximately $1\,mg/m^3$ either due to releases from other operations in the shop or due to less than 100% control from the exhaust hood. The height and time duration (i.e., width) of the "peaks" in the bottom chart are significantly less than in the top chart; this is the contaminant control achieved by the local exhaust system.

These *Source Control* steps can also reduce the amount of contaminants dis-charged to the community environment. With local exhaust ventilation, air cleaners are used to remove the contaminants before the air is discharged to the community environment. Some environmental regulations set limits on the amounts of pollutants that can be discharged from specific industrial operations. For example, U.S. Environmental Protection Agency standards govern the discharge of Hazardous Air Pollutants, including asbestos and mercury, to the atmosphere. To meet these standards, a local exhaust system with the proper air cleaning device may be needed.

Exposure Pathway Modifications

The exposure pathway is the route by which the contaminant travels from the source to the worker's breathing zone. Choices include:

- Lengthening the exposure pathway by increasing the distance between source and worker. The dilution rate is influenced by drafts and other air motion, but in almost all cases the contaminant concentration decreases with increased distance from the source.
- Interrupting the exposure pathway with physical barriers such as doors, curtains, or baffles to impede the movement of contaminated air toward workers.
- Use of dilution (or general) exhaust ventilation. Either natural or mechanically induced air movement can be used to dilute contaminants. Dilution ventilation is used in situations meeting these criteria:
  - Small quantities of contaminants released into the workroom at fairly uniform rates.
  - Sufficient distance from the worker (or source of ignition for fire/explosion hazards) to the contaminant source to allow dilution to safe levels.

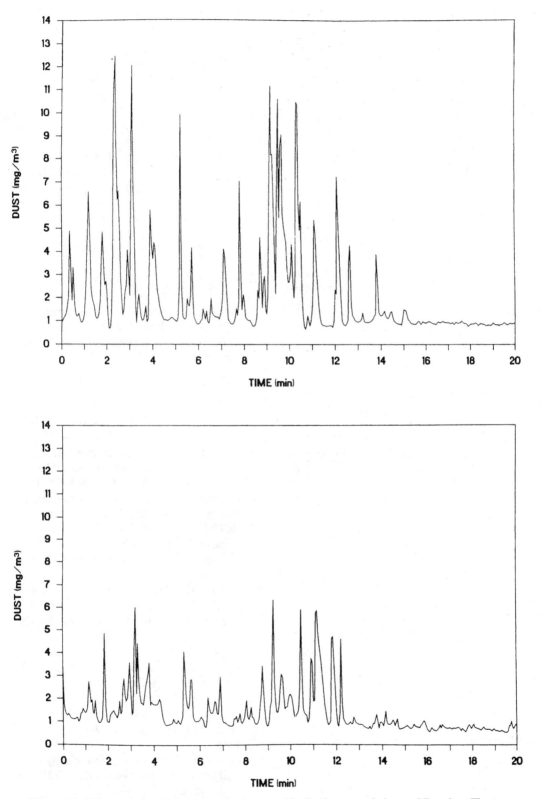

Figure 2.8. Using real-time data to view the impact of instituting controls in a welding shop. The top chart shows airborne particulate levels without ventilation; the bottom chart shows the same welding task with local exhaust ventilation. (*Source: Applied. Ind. Hyg.* **4**(9):239, 1989.)

○ Contaminants of relatively low toxicity or fire hazard so that no major problems will result from unanticipated minor employee exposure or concentration exceedances.

○ No air cleaning device needed to collect contaminants before the exhaust air is discharged into the community environment.

○ No corrosion or other problems from the diluted contaminants in the workroom air.

• Isolating the process or the worker. Usually this involves moving one or the other to a different room to minimize exposure levels and exposure time. This especially applies to workers who are not directly involved with the process releasing the contaminants. The degree of isolation required depends on the toxicity of the contaminant, the amount of contaminant released, and work patterns around the process. Often moving a unit to another room is sufficient, while in other cases a control room supplied with fresh air for the operators may be needed.

SUMMARY

This chapter provided background information on the types and properties of contaminants, toxicological principles, and source of exposure standards. An understanding of this information is vital to the air sampling practitioner in collecting valid air samples to measure exposures to toxic contaminants.

REFERENCES

1. Soule, R. Industrial hygiene sampling and analyses. In *Patty's Industrial Hygiene and Toxicology*, vol. 1, E. Bingham, B. Cohrssen, and C. H. Powell, eds. New York: John Wiley & Sons, 2000, pp. 265–316.

2. Stuart, B. O., P. J. Lioy, and R. F. Phalen. Use of size-selection in estimating TLVs. *ACGIH Ann.* **11**:85–96, 1984.

3. ACGIH Reports: TLVs for soluble and insoluble compounds. *Appl. Ind. Hyg.* **2**(6):R-4–R-6, 1987.

4. Riggins, R. M., and B. A. Petersen. Sampling and analysis methodology for semivolatile and nonvolatile organic compounds in air. In *Indoor Air and Human Health: Major Indoor Air Pollutants and Their Health Implications*, R. B. Gammage and S. V. Kaye, eds. Chelsea, MI: Lewis Publishers, 1985, pp. 351–359.

5. Wray, T. Vapor density. *Haz. Mat. World*, May 1989, pp. 68–69.

6. American Conference of Governmental Industrial Hygienists. *Industrial Ventilation, A Manual of Recommended Practice*, 24th ed. Cincinnati: ACGIH, 2001.

7. Lippmann, M. Dosimetry for chemical agents: An overview. *ACGIH Ann.* **1**:11–21, 1981.

8. Mastromatteo, E. TLVs: Changes in philosophy. *Appl. Ind. Hyg.* **3**(3):F-12–F-16, 1988.

9. Malek, R. F., et al. The effect of aerosol on estimates of inhalation exposures to airborne styrene. *AIHA J.* **47**(8):524–529, 1986.

10. Raabe, O. G. Size-selective sampling criteria for thoracic and respirable mass fractions. *ACGIH Ann.* **11**:55–56, 1984.

11. Morgan, W. K. C. The respiratory effects of particles, vapors, and fumes. *AIHA J.* **47**(11):670–673, 1986.

12. Popendorf, W. J., and J. T. Leffingwell. Regulating OP pesticide residues for farmworker protection. *Res. Rev.* **82**:125–201, 1982.

13. Klassen, C. D., ed. *Casarett and Doull's Toxicology: The Basic Science of Poisons*, 6th ed. New York: McGraw-Hill, 2001.

14. Vianio, H., and L. Tomatis. Exposure to carcinogens: An overview of scientific and regulatory aspects. *Appl. Ind. Hyg.* **1**(1):42–48, 1986.

15. Cain, W. S., and A. Turn. Smell of danger: An analysis of LP-gas odorization. *AIHA J.* **46**(3):115–126, 1985.

16. Environmental Protection Agency. *SOPs for Work on Hazardous Waste Sites*. Washington, D.C.: EPA, 1986.

17. Billings, C. E., and L. C. Jonas. Odor thresholds in air as compared to threshold limit values. *AIHA J.* **42**(6):479–480, 1981.

18. National Institute of Occupational Safety and Health. *Pocket Guide to Chemical Hazards*. Washington, D.C.: U.S. Dept. of Health and Human Services, June 1997.

19. DeSilva, P. TLVs to protect "nearly all workers." *Appl. Ind. Hyg.* **1**(1):49–53, 1986.

20. Phalen, R. F. Airway anatomy and physiology. *ACGIH Trans.* **11**:35–46, 1984.

21. American Conference of Governmental Industrial Hygienists. *Threshold Limit Values and Biological Exposure Indices for 2002*. Cincinnati: ACGIH, 2002.

22. Lioy, P. J., M. Lippman, and R. F. Phalen. Rationale for particle size-selective air sampling. *ACGIH Ann.* **11**:27–34, 1984.

23. Phalen, R. F. Introduction and recommendations. Particle size-selective sampling in the workplace. *ACGIH Ann.* **11**:85–96, 1984.

24. Hickey, J. L. S., and P. C. Reist. Application of occupational exposure limits to unusual work schedules. *AIHA J.* **38**(11):613–619, 1977.

25. Brief, R. S., and R. A. Scala. Occupational exposure limits for novel work schedules. *AIHA J.* **36**:467–471, 1975.

26. National Research Council, Subcommittee on Acute Exposure Guideline Levels. *Standing Operating Procedures for Developing Acute Exposure Guideline Levels for Hazardous Chemicals*. Washington, D.C.: National Academy Press, 2001.

27. 40 Code of Federal Regulations, Part 261. Identification and Listing of Hazardous Waste.

CHAPTER 3

EXPOSURE ASSESSMENT STRATEGY AND MONITORING PLAN

A "universal" law of exposure monitoring is that it is not possible to sample every individual for every toxic substance in their environment. In the occupational setting, the OSHA Hazard Communication Standard[1] defines "hazardous chemical" so broadly that practically all chemicals are included in the rule's requirement that Material Safety Data Sheets and container labels be provided along with worker training, a written program, and other provisions. So even deciding what chemicals pose a significant toxic threat and thus should be monitored can be a challenge. It would be even more difficult to sample all members of the general community for even one environmental toxic agent. For these reasons, air sampling should be conducted based on (a) an *Exposure Assessment Strategy*, which describes how the exposures will be measured and how conclusions on whether the exposures are acceptable will be made, and (b) a *Monitoring Plan*, which is a tactical plan for the monitoring to be performed over the next

year or so to carry out the strategy. Having a strategy in place and then following the monitoring plan is the best way to meet the goals of the monitoring program with an efficient use of resources. The opposite scenario is when a group of air samples are collected to define exposures; and when the data are compiled, they are incomplete because some key exposures were missed, there was a pattern of the same people being monitored, some results are unusually high or low and no follow-up monitoring was performed, or some acknowledged high-exposure tasks were not sampled due to logistics or other problems.

A full exposure assessment is a comprehensive process to identify similar exposure groups, define exposure profiles, judge the acceptability of exposures to environmental agents, and plan follow-up actions. Table 3.1 lists typical environmental agents. The process is part quantitative (based on numerical data) but has a heavy qualitative component and generally requires significant professional judgment where data are

Air Monitoring for Toxic Exposures, Second Edition. By Henry J. McDermott
ISBN 0-471-45435-4 © 2004 John Wiley & Sons, Inc.

TABLE 3.1. Occupational Environmental Agents

Agent	Characteristics
Raw material	Materials that are primary inputs to the manufacturing or other process such as reactant chemicals or diluents.
Intermediate	Materials that are formed during the process but that are not sold. Often these are not isolated during the process, although exposures can occur during process sampling, maintenance, or from fugitive emissions. Toxicological and physical data on these substances may be limited.
Product	The materials that are produced for sale or transfer as part of the manufacturing process.
Additive	Materials added to the process or final product to enhance the process or product. These include catalysts, inhibitors, and pigments.
Maintenance and construction materials	Materials, such as paints, insulation, and lubricants, to which maintenance and construction workers are exposed. They are not part of the basic plant processes.
Laboratory chemical	Materials used to prepare and analyze samples (process intermediates, products, waste water, etc.). May be handled in small amounts, but often are concentrated pure chemicals.
Waste material (for disposal)	Mixtures of solid and liquid materials with no further use in the plant or process. May be separated for recycling or for handling as a hazardous waste. Includes byproducts from the process as well as raw material and other packaging, off-specification products that can not be reprocessed, and construction debris.
Physical agent	Noise, radiation, heat or cold, etc.
Biological agent	Biologically active materials produced by living organisms as well as the organisms themselves.

Source: Reference 2.

not available. Although the terms may be similar, exposure assessment is not the same as exposure monitoring—in fact the assessment process prioritizes potential exposures to focus air monitoring on exposures that need to be further quantified.

EXPOSURE ASSESSMENT

A comprehensive exposure assessment attempts to accomplish eight objectives[2]:

- Characterize exposures to all potentially hazardous chemical, physical and biological agents.
- Characterize the level of exposure and time-related variability (hourly or daily variations, etc.). This is often the most difficult step since limited exposure data may show exposure levels on the days when samples were collected, but not allow a good estimate of how the exposures vary over time.
- Assess the potential harm from the exposure, which could be the risk of noncompliance with an exposure standard or the potential harm to exposed individuals.
- Clearly identify and prioritize for control any unacceptable exposures.
- Identify exposures where additional information or data are needed to judge their acceptability.
- Document the exposures and control practices, and communicate results to affected parties.

- Develop a historic record of exposures for future health studies or other purposes.
- Aid in the effective allocation of resources to exposure monitoring and control efforts.

Exposure assessment can be performed for occupational exposures, community environmental exposures through one or more exposure pathways (air, water, food), or for all environmental and behavioral risks including personal traits such as dietary choices, smoking, and so on.

The process involves the following steps:

1. Gather existing and easy-to-collect background data about the exposed population, environmental agents they are exposed to, and approximate exposure levels.
2. Define "similar exposure groups" (SEGs)—subsets of the total exposed population that can be grouped together because information indicates that the exposures to members of the group are about equal for the purposes of the assessment.
3. Define the exposure profile for each SEG—this is a description of the magnitude and variability of exposures for an SEG.
4. Judge whether the exposure to each agent and the combined exposures are within acceptable limits.
5. Plan follow-up steps to fill information gaps or control unacceptable exposures.

Sometimes the scope of the exposure assessment is narrower than the five steps described above. For example, a government sponsored heath-related exposure assessment may define the exposures and potential risks to a community population from an agent such as second-hand tobacco smoke, but stop short of judging the acceptability of these exposures and planning follow-up actions.

When planning an exposure assessment, the assessor must consider the tradeoff between a theoretical assessment and an extremely focused one that yields tangible action steps. Often the size of the population determines where on the continuum between theoretical and practical the assessment will fall. For example, if the exposure assessment is aimed at assessing the risk to all workers in the nation to a specific chemical, then gross assumptions about number of individuals exposed, exposure levels, and use and effectiveness of exposure controls must be made. Sometimes the only population data available are the number of people employed in a plant or segment of the economy. In these cases the total number of people who work in a facility where the substance is handled may be considered to be "exposed." As a result of these broad assumptions, the exposure assessment may describe general risks to the whole exposed population, not the exposure within a specific plant handling the chemical. An exposure assessment for a specific plant would be based on relevant information about the process, amount of chemical used, airborne levels, time duration of exposure, and other factors that govern exposure. Thus this specific plant exposure assessment will have more value in defining the actual exposure profile for these workers as well as the likelihood of exposures over the occupational standard, the need for controls, and other follow-up steps. Environmental exposure assessments tend to be based on very broad estimates of characteristics of the exposed population, contaminant concentrations, and exposure duration. A comprehensive environmental exposure assessment must be based on many difficult-to-quantify assumptions including how far from the source an individual lives, how much time they spend in the their community (both hours per week and years as a resident), the source and quality of drinking water as well as quantity consumed, possible uptake of water-borne contaminants from showering and brushing teeth, and how much soil a young child might eat while playing in the yard. While these studies are useful in highlighting exposure pathways needing further evaluation, they lack the specificity needed to define actual exposures and identify where control measures may be needed.

The outcome of an exposure assessment is to classify each exposure into one of these categories:

- *Acceptable*—the exposure level and variability are low enough that resulting exposures are low compared to acceptable exposure levels. Periodic reassessment may be recommended to see if the exposures have increased.

- *Uncertain*—an exposure profile cannot be judged to be either acceptable or unacceptable. This is usually due to inadequate characterization of the exposure profile, or insufficient data to set an acceptable exposure level. In many cases where the exposure profile is an issue, there is a qualitative indication of exposure but very little or no monitoring data exist. These exposures are good candidates for exposure monitoring. However, sometimes additional monitoring will not provide much help in making a decision. If the measured exposures are close to the acceptable exposure level and exhibit variability, it may not be possible to conclude that the exposure is acceptable even with a lot more data. It may be more practical to reduce the exposure level so the resulting exposures are judged to be acceptable.

- *Unacceptable*—the exposure level is high enough to present an unacceptable risk. Of course, evidence of an adverse health effect would automatically cause an exposure to be classed as unacceptable. However,

even without adverse effects an exposure that exceeded the criteria for "acceptability" would be judged to be unacceptable. The criteria for acceptability depend in part on the type of exposure assessment that is performed, and they will be discussed later in this chapter.

Exposure Assessment for "Nonroutine" Operations

"Nonroutine" exposures are those that are characterized by the factors listed in Table 3.2. Exposures that are truly "nonroutine" are difficult to assess using a structured approach since the exposures vary widely by task and over time. It is important to identify these exposures and treat them appropriately. Often the best approach is to anticipate these situations and be prepared to respond quickly with the following:

- Immediate control of suspect exposures using feasible engineering controls, work practices, and personal protective equipment. Acceptable performance in this area often requires that control devices such as portable blowers and flexible duct be readily available to meet reasonably foreseeable needs.

- Evaluation of exposures using direct reading instruments, and sample collection devices with quick turnaround laboratory analysis.

- Adjustment of the immediate controls based on the monitoring results.

TABLE 3.2. Characteristics of Nonroutine Tasks

- Short duration
- Transient work force
- Little repetition of specific tasks over time
- Work site conditions frequently change
- Variable work practices
- Multiple environmental agents
- Limited health effects data on agents and typical exposure conditions

- Adequate documentation of exposure conditions, controls, and levels.
- A process to learn from experience to permit anticipation and resolution of problem exposures.

Over time a sound process for addressing nonroutine operations should result in (a) fewer situations where unacceptable exposures may be occurring and (b) more instances where problem exposures are proactively recognized and prevented.

Also note that operations such as maintenance and repair are not considered to be "nonroutine" under this section. Although these exposures may not occur regularly and be a challenge to monitor, they can be addressed using the exposure assessment process described in this chapter.

Role of Air Monitoring

Air monitoring for toxic exposures is one part of the exposure assessment process. It is especially valuable in further defining exposure profiles that were classified as "uncertain" as part of the initial exposure assessment. Air sampling often plays a major role in occupational exposure assessments, and so the remainder of this chapter will focus on exposure assessments performed on occupational work groups. Much of the information in this chapter is based on *A Strategy for Assessing and Managing Occupational Exposures* by Mulhausen and Damiano,[2] which is a fundamental resource for occupational exposure assessments and should be consulted for further information.

PERFORMING AN EXPOSURE ASSESSMENT

Step 1: Gather Existing and Basic Information

This phase involves collecting data on the work setting, the work force and the chemical, physical and biological agents that are present. The information will be used in the subsequent steps. The following minimum information is needed:

- Processes, operations and facilities.
- Size and make-up of the work force, and how work is assigned.
- Environmental agents present, including their chemical and physical properties, and any relevant usage information.
- How and where individuals are exposed to the agents, including any existing exposure data.
- Toxicological information for each environmental agent, and the acceptable exposure level for each.

Table 3.3 lists typical sources of information for this step in the assessment.

Past Surveys as Source of Information.

Existing exposure monitoring data are especially valuable information during the initial step of the exposure assessment. If routine monitoring has not been performed, past industrial hygiene and safety surveys may be available. These survey reports often contain personal or area measurements performed during the survey. As such they offer a "snapshot" of exposures along with a description of plant conditions during the survey.

Monitoring records can be divided into four broad categories:

- Compliance monitoring—limited sampling performed to determine whether exposures exceed some type of legal (e.g., OSHA) standard, an exposure guideline adopted by an authoritative body (ACGIH, ANSI, NIOSH), or an internal standard adopted by the employing organization (company, agency, etc.). Some governmental regulations require monitoring (see Table

TABLE 3.3. Source of Basic Information for the Exposure Assessment

Source	Description of Information
Current and prior Industrial Hygiene and Safety Surveys (see description in text)	• Description of operations, processes, throughput or production rate, environmental agents, exposure controls, work assignments, employee complaints • Exposure monitoring data • Evaluation of compliance status • Recommendations for controls, worker education, etc.
Interviews of workers, managers, and technical staff (engineering and maintenance, etc.)	• Background on above items • Any differences between what was laid out in written documents and how things were actually done
Interviews of support staff professionals such as Health, Safety, Environment, Medical, Workers' Compensation, and Human Resources	• Health problems • Exposure history • Work practices affecting exposures • Approximate dates when controls were implemented
Plant records	• Chemical inventories • Production records • Material safety data sheets • Regulatory citations • Work histories • Routine exposure monitoring records • Engineering control performance tests • Biological monitoring results
Government, industry, consensus and other authoritative standards	• Past, current, and proposed exposure limits
Technical literature	• Toxicological studies • Epidemiological studies

1.1), but in many cases the requirement is simply not to exceed the exposure standard with the option of whether or not to monitor left to the employer.

• Diagnostic monitoring—samples collected to understand a "problem" situation, usually employee complaints or possible illness caused by environmental agents. The term also applies to sampling performed to evaluate the effectiveness of engineering controls.

• Baseline monitoring—samples collected as part of a systematic plan or sampling campaign to define the exposure profile to one or more environmental agents. As used in this context, the samples are collected on different days or under different operating conditions to estimate the range of exposures to that job.

• Area, source, or stack samples that measure airborne levels of contaminants but cannot be interpreted to represent exposures to people who are included in the exposure assessment.

The type and extent of monitoring data depend on the reason that the survey was performed:

Comprehensive Industrial Hygiene Survey. This is a comprehensive health hazard assessment that the employer conducts;

TABLE 3.4. Items to Consider During Workplace Health Hazard Assessments

- Health concerns and perceived hazards—based on discussions with managers, supervisors and employees.
- Systems which ensure that new and experienced employees are properly trained and motivated to follow procedures, observe precautions, and use personal protective equipment.
- Plant layout drawings, flow diagrams, work schedules, job classifications and descriptions, and known exposures which occur under normal conditions.
- Types, frequency, and duration of routine and periodic maintenance activities which may result in exposures.
- Changes, or potential changes, in processes, schedules, or working conditions.
- A current list of chemicals present at the facility, usually available in the facility's Hazard Communication Program chemical inventory.

- Chemicals that are brought on site by contractors and vendors.
- Quantities, concentrations and toxicities of chemicals present at the facility.
- Potential sources of exposure to physical agents such as noise, heat, and radiation.
- Existing engineering controls which are being used to reduce exposures.
- Environmental conditions such as wind, heat, humidity and cold, which might cause exposure levels to vary.
- Results of previous exposure monitoring at the facility.
- Results of exposure monitoring during similar activities at other facilities.
- Recommendations an actions taken from previous industrial hygiene surveys.

it is often performed at a preset interval depending on the nature and extent of hazards, changes in workplace processes and work schedules, and the results of previous assessments. Table 3.4 shows many of the items considered during a comprehensive survey. This type of survey has the most useful information for an exposure assessment since it usually contains a description of the exposures and plant processes, along with the data, and clear recommendations on where exposure controls and other follow-up steps are needed. If a routine exposure monitoring program is in place, the survey report may summarize data collected since the last comprehensive survey. It is often performed by a knowledgeable professional with the ability to recognize potential hazards and evaluate them appropriately.

Insurance Surveys. There are two types of insurance related surveys that may generate exposure monitoring data:

- A "loss control" survey that is generally designed to assist an employer to minimize the potential for occupational disease claims under the employer's workers compensation policy. An insurance carrier may request that an industrial hygiene survey be done because the employer is in a high-risk classification; otherwise the policy holder may request one. The size and nature of an operation can also determine whether or not an industrial hygiene survey will be done. Other survey issues vary with the particular employer's situation, for example, employees may be complaining of discomfort in the course of their work. Usually air sampling surveys are provided annually as part of the insurance service. These annual surveys are usually not comprehensive since only a few individuals are sampled for one or two contaminants and they are aimed at evaluating compliance status.

The reports are often brief and may be in memo form.

- Workers compensation claim investigations result in another type of insurance survey. Often the employer requests sampling of a specific job function be done to determine whether there is actually an exposure or possible overexposure. The main problem with these surveys is that they often take place some time after the individual's employment. Changes in the process, materials, and chemicals may have occurred since that time. Consequently, it may be difficult to assume that the exposure evaluated by the survey is significantly related to the claim.

Response to a Complaint or Injury. Surveys are also done in response to an employee complaint or injury. These surveys produce diagnostic data. Subjective employee complaints and injuries resulting from a nonroutine activity or unknown exposure source can be most difficult to investigate and often represent one-time conditions that do not regularly occur. For example, in some production plants the process sewers are linked throughout the entire facility. If two reactive components get into the sewer at the same time and meet downstream, the resulting reaction can form an irritating, toxic, or odorous compound. It can be difficult to diagnose the problem and prevent it from happening again.

Evaluation of Engineering and Process Controls. Air sampling conducted as part of a comprehensive insurance or other type of survey may indicate the need for engineering controls such as ventilation or process controls such as change in materials or better enclosures on process equipment. Follow-up air monitoring can determine the effectiveness of these controls and confirm that the exposures have

been reduced sufficiently. This type of sampling falls into the "diagnostic" category.

To show the impact of an engineering or process control, the best samples are before and after samples collected near the equipment. For ventilation and similar controls, the samples may be collected on the same day with the fan off and then with the system running. In contrast, controls that require changes to equipment or chemicals must be collected before the project is begun and after it is completed. Source or area samples are not very useable for an exposure assessment except as a qualitative indicator that exposures have been reduced when employees work around the piece of equipment. This may allow the priority of the exposure for follow-up personal monitoring to be reduced if the controlled source was previously the major contributor to exposure.

Source samples that show the amount of contaminant *released* per time period, rather than solely airborne concentration, can be used in mathematical modeling of occupational exposures as described later in this chapter.

Compliance Agency Surveys. These are surveys conducted by an enforcement agency to determine compliance with occupational health regulations that may or may not involve exposure monitoring. The inspection may be triggered by a programmed inspection process, an employee complaint, or a report of serious injury or illness. Any monitoring data fall into the compliance category; sampling may be for full shift or for short-term exposure over 15 minutes or the duration of the task depending on the standard being enforced. In the United States, agencies such as federal and state OSHA, the Mine Safety and Health Administration (MSHA), and the Coast Guard are typical enforcement agencies.

Often the jobs or tasks to be monitored are selected by the agency personnel to represent the most heavily exposed individuals. Because of this the resulting data are usually thought to represent the worst-case (highest) exposures rather than typical exposure patterns. One study by the Workers' Compensation Board of British Columbia (Canada)[3] on wood dust exposures in lumber mills compared data in a compliance database with monitoring results in a broader "research" database. The study showed that exposures reported in the compliance database were higher than the research data, indicating that the enforcement personnel were able to select the employees with the higher apparent exposures and the higher exposure situations on the day(s) that they collected samples. However, the differences in average exposure values from both databases were close enough that the authors concluded that the compliance database could be a good starting point for exposure assessments, epidemiological studies, and other noncompliance purposes.

Agencies such as the state OSHA consultation services and NIOSH also provide nonenforcement evaluation of potential hazards and recommendations for control. Although not an enforcement action, any monitoring data from the visit are usually aimed at comparing exposures to recognized exposure standards or guidelines.

Other Survey Situations

COMMUNITY COMPLAINTS. Nuisance complaints arising from the general public are primarily related to noxious odors, eye irritation, or in some cases, more serious illnesses. For nuisance complaints the inspection often relies more on knowledge of local sources and wind direction to identify the cause rather than air monitoring. Air monitoring can document the extent of the nuisance or health hazard if airborne concentrations are high enough to be measured. Monitoring activities are often geared to identifying "mystery" contaminants. Often the levels of contaminant mea-

surement are extremely low due to the highly sensitive nature of many individuals, especially when compared to standards for workplace exposures. These data are not generally useful in an exposure assessment since they represent brief episodes.

INDOOR AIR QUALITY ISSUES. Tight building syndrome assessments involve sampling exposures of individuals who may or may not be the client's employees. The survey may be at the request of the building management or performed by government agency personnel. The cause of irritation or illnesses may be due to outside pollution being entrained into the building's ventilation system or from a source within the building. Lately, mold and other microorganisms have been implicated in IAQ problems. As with community complaints, the diagnostic data collected often are not useful in a traditional occupational exposure assessment. However, if the complaint is due to some recognized factor such as chemicals released from a neighboring facility, enough monitoring data showing typical exposures may be available to be used in the exposure assessment.

HAZARDOUS WASTE SITES. Hazardous waste sites are potentially more dangerous than any other environment in which sampling may be needed. There are usually large amounts of concentrated and often unknown materials at these sites; such materials are often in open pools or deteriorating containers. Special personal protective equipment is often necessary. Consequently investigations, sampling, and other activities at a hazardous waste site are usually much more difficult and time-consuming than routine environmental sampling.

Under OSHA standards, several types of monitoring are performed at hazardous waste sites. Upon initial entry, general survey air monitoring is done to identify any immediately dangerous to life and health (IDLH) conditions, exposures exceeding health standards, or other dangerous conditions, such as the presence of flammable atmospheres or oxygen-deficient environments. Subsequently, periodic or continuous monitoring is to be done when any of these situations may exist or when there is indication that exposure levels may have increased since prior monitoring. Examples of situations where additional monitoring may also be required include work on a different portion of the site, handling of contaminants other than those previously identified, initiation of a new type of operation, and safeguarding employees at the greatest risk of exposure during cleanup operations.

Typical monitoring data consist of baseline personal samples to characterize exposures and determine the need for respiratory protection, followed by frequent measurements with direct reading instruments to show any changes in ambient concentrations. Generally, the direct reading instruments are not specific for the contaminants of interest, so a "working" level of total hydrocarbon vapors or another group of compounds is chosen as the exposure standard.

EMERGENCY RESPONSE. Emergency monitoring situations differ from most surveys in several ways. Since acute rather than chronic effects are the primary concern, the required detection sensitivity is less stringent. However, in many cases, the exact compounds of interest may not be immediately known and therefore a broad range of monitoring techniques must be used to make sure the toxic compounds are detected. In addition, the development of a monitoring strategy is difficult due to the need for immediate response. For more information on data collected during emergency response operations, see Chapter 4.

AREA OR STACK SAMPLING SURVEYS. In a few cases, workplace monitoring is performed for reasons other than employee health or environmental quality. One instance may be if rust is noted on machinery in a shop near several potential sources such as (a) an electroplating room that could release acid fumes and (b) a large air conditioning unit that could leak halogenated compounds that would break down in a welding arc to form toxic and corrosive compounds. If airflow studies around the plating room and leak tests of the air conditioner did not pinpoint the problem's cause, air samples might be collected to identify the corrosive compounds. These are diagnostic area samples, and they might be useful in helping to diagnose employee complaints.

Stack samples are collected in chimneys, process stack vents, and similar discharge points. The sampling practitioner uses specialized techniques to collect samples that represent the airflow within the stack. Generally, these do not represent exposures to people, although modeling programs can estimate the airborne concentration at different points away from the discharge point based on wind speed and direction, temperature, exit velocity, and other relevant parameters. The modeling results usually are more valid as a general estimate of airborne levels at some distance from the source, and so they are used for community environment assessments rather than for occupational assessments.

Table 3.5 lists the types of surveys and typical categories of data from each that are considered as part of an exposure assessment.

Existing "Baseline" and Similar Monitoring Records.

In addition to survey results, some locations may have existing monitoring data that were collected either to define the chemical exposures to selected jobs (such as operator or maintenance mechanic) or to characterize exposures to all jobs from a specific chemical (such as asbestos exposures to every individual who encounters this material in thermal insula-

TABLE 3.5. Types of Surveys and Resulting Exposure Data

Type of Survey	Type of Monitoring Data			
	Compliance	Diagnostic	Baseline	Area or Source
Comprehensive	X	X	?	
Insurance	X	X		
Compliance agency	X			
Evaluation of controls		X		X
Employee complaints	X	X		
Community complaints				X
Indoor air quality issues		X		X
Hazardous waste sites	X	X		
Emergency response	X			
Area and stack sampling				X

tion and from other sources). Some organizations undertake an extensive effort to develop an exposure monitoring database to define exposure levels to most chemicals in all jobs. Often these start with a concentrated monitoring campaign to develop initial baseline profiles (especially for job categories with higher exposures), followed by periodic monitoring to keep the information current as conditions and chemicals change. These databases are extremely valuable in performing a formal exposure assessment. See the section on *Statistical Monitoring Approaches* later in this chapter for guidelines on how to handle these data as part of the exposure assessment.

Performing a Walk-Through Survey with Range-Finding Air Sampling. Unless a very recent survey is available, a brief "walk-through" survey is helpful as part of Step 1 to put past surveys and other information in perspective. Existing information on processes, the chemical and other agents, and the way people do the jobs can be verified or updated. The status of past recommendations can be checked. Any gaps in the prior surveys can be filled. This is also a good opportunity to collect range-finding concentration readings using direct reading instruments; read Step 4 on judging exposures below for guidance on the types of readings to make.

Step 2: Define Similar Exposure Groups (SEGs)

Similar exposure groups (SEGs) are groups of workers having the same general exposure profile for the agent(s) included in the exposure assessment because of the similar nature and frequency of the tasks they perform, the materials and processes with which they work, and the similarity in the way they perform their tasks. SEGs may be determined through observation, exposure monitoring, or a combination of the two. With the sampling approach, sufficient exposure monitoring is performed to group workers into appropriate SEGs.

A more common approach to assigning SEGs is through observation of the operations, work conditions and other factors that affect exposure level and duration. A typical project to assign SEGS first divides the operation or plant into separate processes, then divides workers by job category, then by the task(s) they perform, and finally by the agents they encounter. Table 3.6 shows how this process could be applied to the various categories of workers as defined in OSHA's Hazardous Waste Operations and Emergency Response (HAZWOPER) Standard.[4]

Assigning SEGs via observation works well where variations in exposure are due

TABLE 3.6. Possible "Similar Exposure Groups" for Hazardous Waste Cleanup Site (Based on Training Categories in OSHA's HAZWOPER Standard)

Similar Exposure Group	Characteristics
General site—equipment operator	Operates equipment such as backhoe and bulldozer; drives large trucks.
General site—laborer	Moves drums, digs with shovel, and performs other manual work.
General site—supervisor	Provides supervision to general site workers.
Occasional site worker	Performs limited work such as ground water sampling or land surveying; is not expected to be exposed above the acceptable exposure limit.
Management and administrative	Little time spent in area of hazardous waste operations; little opportunity for exposure to hazardous materials.
Hazardous materials technician	Handles leaking hazardous waste drums—takes immediate action to stop leak and then repackages leaking drum.

to job-related factors such as work load, weather conditions, and the occurrence of leaks or spills. It does not work as well when variations in exposure are caused by personal factors—that is, how the person does the job. For example, if one individual is careful to use the provided local exhaust system while another person does not use it properly, they are not really in a similar exposure group even though their job titles and work assignments are identical. In these cases, one solution is to establish good work practices, train employees on the content and importance of the work procedures, and then use good supervision to provide the feedback and follow-up necessary to achieve conformance with the work practices.

The basic assumption in assigning SEGs is that monitoring performed on any member of the SEG is representative for all SEG members. Thus it is important to look for factors that will cause misclassifications. Some examples include:

- An informal or formal "seniority" system where an employee with more time on the job has different job duties than a less experienced employee in the same job classification. For example, in a chemical plant a less experienced operator may routinely collect process samples or drain equipment in preparation for maintenance while the more experienced operator works in the control room remotely monitoring process conditions. The differentiation in tasks may be due to either (a) the desirability of the task or (b) the knowledge and experience needed for more complex tasks.

- Differences in tasks due to the work shift. For example, operators on the night shift may prepare equipment for maintenance more than on day shift if most maintenance work is performed on the day shift. Tank truck loading may be performed more often by day shift operators since the truckers normally deliver to customers during the day.

In other cases what appear to be different job categories are really the same type of job. Some human resources systems may assign employees to different job classifications to account for pay differences. For example, an assembler with less than six months' experience may have a different job classification than more experienced assemblers to reflect different hourly pay rate or a provisional status even though all assemblers have similar job duties and exposures. These different classifications can be combined into a single SEG.

Assigning SEGs is an area where good professional judgment is necessary for a successful exposure assessment.

Step 3: Developing Exposure Profiles

This step involves estimating the magnitude and variability of exposures for the SEG, and it is based on existing information from Step 1 plus professional judgment. Note that air samples are not collected during this phase in an initial assessment—the sampling will be performed as follow-up to the exposure assessment to fill information gaps needed to categorize exposures as acceptable or unacceptable.

Often this step is highly qualitative: Information on the environmental agents encountered by an SEG are reviewed along with likely exposure scenarios, and a tentative decision is made as to whether any exposure could be significant. For example, if warehouse workers unload drums of chemical raw materials and there are no monitoring results for the operation, it will be helpful to know information such as how often leaking drums or drums with a chemical residue on the outside surface were received, whether these employees would handle the leaking drums (as opposed to refusing delivery or calling the plant hazardous materials specialist to take control) as well as the skin and inhalation toxicity of the chemicals. With this information it might be possible to conclude either that:

- Essentially no exposure occurs.
- The opportunity for exposure exists and insufficient data exist to define the extent of the exposure.

- Exposure appears to occur regularly; and due to the volatility of the chemical and reports of eye irritation, exposures probably exceed the acceptable exposure limit.

As can be seen, even in the absence of exposure measurements, a qualitative "exposure profile" can help to identify both negligible exposures needing no follow-up and clearly unacceptable exposures that need control.

When exposure data are available for the SEG, usually they are used to calculate average exposures and estimate how often the acceptable exposure level is exceeded. If there are more than about five exposure measurements for one chemical substance, refer to the discussion of monitoring data later in this chapter for guidelines on defining the exposure profile.

Step 4: Judging the Acceptability of Exposures

An acceptable exposure level (AEL) for each agent is required before this step can be completed unless there are adverse signs or symptoms from exposure. Generally, a regulatory standard or an exposure guideline from an authoritative body (ACGIH, ANSI, etc.) is used. If none exist or if there is evidence that the published values are not sufficiently protective, the organization may adopt their own internal exposure criteria.

The starting point for this step is the exposure profiles defined for each SEG. Building on the warehouse workers example in the previous section, the following judgments can be made:

- If essentially no exposure occurs, the risk is low and the exposure is acceptable.
- If the opportunity for exposure exists and insufficient data exist, it is not possible to judge the acceptability of the

exposure. This is the "uncertain" category. Exposure monitoring is the biggest factor for resolving "uncertain" exposures if few or no quantitative data exist.

- If the exposure appears to occur regularly and exposures probably exceed the acceptable exposure limit, the exposure is unacceptable and should be controlled.

However, even with monitoring data, selecting the criteria for acceptable and unacceptable exposures may not be straightforward. Two factors are:

- The degree of process (or exposure) "control" is important; this means how much variation occurs in exposures to the same SEG. Little variability usually indicates good control while large variability can indicate poor control. Poor control means that exposures can be very low one day and significantly higher on other days. For example, consider a case where 10 different samples for the same SEG were collected and all are below the acceptable exposure level (AEL). If some values are close to the AEL and there is a relatively wide range in the data, there is a distinct possibility that the AEL is exceeded at times even though all current results are below the AEL. With this exposure, if more samples were collected, you would expect some results to show exposures >AEL.
- Whether the chemical has chronic toxicity (effects after long-term exposure) or acute toxicity (shortly or immediately following exposure). For a chronic toxicant, the fact that a small percentage of samples are close to or exceed the AEL does not automatically mean that the exposure is unacceptable. For many chronic toxicants, an occasional exposure over the AEL is not harmful. So for toxic agents that

manifest their toxic effect over time, an exposure profile might be judged to be acceptable if the long-term average exposure was <50% of the AEL, and the AEL was exceeded on no more than 3% of the work days. Note, however, that regulatory agencies do not use a statistical approach to deciding when to issue citations—if they measure an exposure that exceeds the legal AEL, the employer is liable to be cited.

The combined effect of exposures to more than one chemical or other environmental agent must be considered during this step.

Step 5: Plan Follow-up Steps

This step involves planning actions to fill information gaps or control unacceptable exposures. The need to control unacceptable exposures is clear. The control techniques described in Chapter 2 are typical approaches to use. Remember that there is an accepted hierarchy of controls: Feasible engineering controls first, followed by work practices, with routine use of personal protective equipment as the last choice.

This leaves the "uncertain" exposures to be resolved. When an initial exposure assessment is performed, typically many of the exposure profiles fall into the "uncertain" category because of a lack of exposure data. Deciding how to fill these information gaps is an important part of the exposure assessment by prioritizing sampling needs and deciding how to approach the monitoring.

As a rule of thumb, consider monitoring those exposures in the "uncertain" category where the exposure level appears to exceed 10% of the AEL on more than infrequent occasions. This guideline is based on statistical studies that indicate a low probability (less than 5% in most cases) of an exposure exceeding the AEL when only a few monitoring results are available and all are below 10% of the AEL. If this guideline yields a lot of exposures to be monitored, then further prioritization of the effort is required:

- Place a higher priority on exposures where the long-term average exposure level appears to be between 50–100% of the AEL.
- Place less priority where exposures appear to be between 10–50% of the AEL.
- Of course, exposures above 100% of the AEL have already been identified for control.

When exposure monitoring is performed to fill information gaps, this is the most efficient way to gather data:

- Collect personal samples that cover the entire exposure period. This is the work shift for time-weighted average (TWA) samples and a 15-minute highest exposure period for short-term exposure limit (STEL) samples.
- Plan on collecting 6–10 samples for the SEG for each chemical of interest. Sampling for the material can be stopped if the first 2–3 samples are below 10% of the AEL or exposures exceed the AEL.
- Decide the monitoring strategy (worst-case, random, etc.) as described in the next section.
- Prepare and follow a written monitoring plan as outlined at the end of this chapter.

Monitoring Strategy. A strategy or systematic approach to exposure monitoring is required because it is very difficult to identify the higher exposure situations through *purely* random sampling without collecting many samples. *Purely random sampling* denotes random sampling where

TABLE 3.7. Minimum Sample Size for Purely Random Sampling to Include at Least One Sample in Highest 10% of Exposures

Size of Employee Group	Minimum Sample Size
1	1
2	2
3	3
4	4
5	5
6	6
7	7
8	7
9	8
10	9
11–12	10
13–14	11
15–17	12
18–20	13
21–24	14
25–29	15
30–37	16
38–49	17
50	18

Note: This is *not* the minimum sampling size for a statistical sampling approach.

no assumptions are made about the distribution of the data, or which employees have higher exposures. For example, with a work group of 15 employees you would need to collect at least 12 samples to be 90% confident that you were including one of the top 10% of the exposures in the sampling population. Table 3.7 shows sample size for different size work groups for *purely* random sampling. Because of the large sample size required with purely random sampling, appropriate statistical tools along with professional judgment are almost always used to reduce the sample size needed to make acceptability decisions.

Statistical Versus Worst-Case Monitoring. Years of research have shown that many exposure profiles follow a log-normal dis-

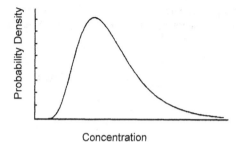

Figure 3.1. Log-normal frequency distribution.

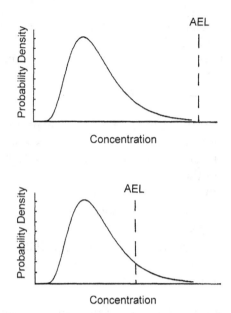

Figure 3.2. The top exposure profile is acceptable since the exposures are below the AEL, while the bottom profile is probably unacceptable since many exposures exceed the AEL.

tribution. This means that if many exposure samples were collected, the shape of the frequency distribution curve would be similar to Figure 3.1. In Figure 3.1, the vertical axis (Probability Density) represents the number of samples with the concentration value on the horizontal axis. Exposures can never be less than zero, the bulk of the exposures are in some middle range, and typically there are relatively few very high exposures on days when the work load was usually high, weather conditions were adverse, or there were releases or other rare factors that caused high exposures. Although Figure 3.1 shows the general shape of a log-normal distribution, there are many different exposure profiles depending on the level and variability of the data. For log-normal data, the acceptability of the exposure depends on where the AEL falls on the frequency plot (Figure 3.2). In the top diagram in Figure 3.2 the exposure would be judged to be acceptable since the airborne values are less than the AEL, while the exposure represented by the bottom diagram would be unacceptable since a significant number of exposures exceed the AEL.

When using a statistical approach, the number of samples collected is chosen to permit a check that the distribution is actually log-normal, and then to infer the shape of the distribution curve. Generally, 6–10 random samples are adequate for these two tasks unless the sample results are close to

the AEL and/or there is large variability (i.e., a wide range) in the measured values.

Worst-case sampling is the standard approach used by compliance agency enforcement personnel and employers where the goal to check compliance with a regulatory or authoritative AEL. Available information and professional judgment are used to select the highest exposure(s) for monitoring, and a few samples are collected during these exposures. If these samples do not exceed the AEL, then it is concluded that these as well as other, lesser exposures are also acceptable. The number of samples depends in part on the uncertainty about whether the truly worst case scenarios were identified for monitoring. There are two ways to use this approach:

- The one or two SEG(s) with the highest exposure potential in the entire plant can be selected and monitored on the days or during the tasks causing the greatest exposure. If these exposures are acceptable, it can be concluded that exposures to other SEGs are also acceptable.

• The apparent worst-case exposure for each SEG can be identified and monitored for acceptability. The advantage in this option is that some monitoring results are collected for each SEG that has a significant potential for exposure.

If the initial exposure assessment shows a need for extensive monitoring to resolve information gaps, often the first option above is given the highest priority for monitoring, with monitoring of less exposed SEGs given a lower priority.

Statistical Monitoring Approach. This section provides a simple explanation of the application of a log-normal statistical sampling approach during an exposure assessment. For more detailed information, refer to *A Strategy for Assessing and Managing Occupational Exposures* by Mulhausen and Damiano.[2] This summary is most valuable in planning future monitoring rather than in evaluating past monitoring results because there are certain conditions that must be met to ensure that the results of the statistical approach are valid, and for past monitoring it may not be clear that the conditions existed when the monitoring was performed.

In order to perform the statistical analysis of monitoring results for a specific chemical collected for a single SEG, the following conditions or assumptions must exist:

• The exposure potential for the SEG must not significantly change over the course of the sampling campaign. The process, chemicals, work assignments, and other relevant factors must not change. If they do, there are different exposure populations before and after the changes.

• A random sampling protocol must be used so that every member of the SEG has an equal change of being monitored. Although a true random approach may be difficult to follow, any systematic selection of workers, tasks, dates, and so on, must be avoided. Monitor in different months or seasons if weather, workload and other time-dependent factors affect exposures. In particular, do not attempt to focus on "typical" events to get a representative exposure profile—the random selection technique takes care of this aspect. When sampling for STEL exposures, a good approach is to use judgment to identify the tasks with the highest potential for exposure, and then use a random protocol to choose which repetitions of this task to monitor.

As an illustration of random sampling, consider a rail tank car loading operation (Figure 3.3) that professional judgment indicates will be the single biggest source of short-term exposure to asphalt fumes for terminal personnel. During a typical week at the plant, there are 100 loading operations performed and it was decided to monitor 8 of these. Table 3.8 is a simple random number table containing 100

Figure 3.3. Rail car loading with liquid asphalt exposes operator to fumes.

TABLE 3.8. Random Number Table

Column 1	Column 2	Column 3	Column 4	Column 5
6	46	89	65	33
7	43	83	21	96
59	14	71	70	31
68	51	9	41	74
73	22	54	2	35
10	3	28	38	25
81	17	90	82	49
50	57	45	29	78
23	15	84	37	62
55	18	85	98	32
69	58	80	12	77
66	91	30	27	16
48	19	99	56	26
13	63	52	86	92
94	72	39	53	97
5	60	61	8	95
1	24	76	75	44
64	93	40	11	36
20	42	87	4	100
34	47	88	79	67

numbers in random order. To decide which loadings to monitor, number each loading operation during the week starting with number 1 as the first loading on Monday, and so on. Then start at the top of column 1 in Table 3.8 and read the first 8 numbers: 6, 7, 59, 68, 73, 10, 81, and 50. These are the loading operations to sample (arranged in sequential order): 6, 7, 10, 50, 59, 68, 73, and 81. An alternate approach is to close your eyes and randomly pick a place on the table to start, and then read the first eight numbers after that point. Most electronic calculators can generate random numbers; this is often a more convenient than using a table. For selecting which full shifts to sample, number each shift during the sampling campaign and use the above process. For example, for a job filled on 3 shifts per day, 5 days per week, there are about 60 work shifts per month. If 10 samples were to be collected in one month, select the first 10 numbers that are not greater than 60 from Table 3.8.

Data Analysis. If the monitoring data truly follow a log-normal distribution, there is a simple approach to plotting the data and drawing conclusions about the exposure profile. Special graph paper called "log-probability" paper is used for the data analysis (Figure 3.4), and also computer software packages are available for performing log-normal plots; one is included with reference 2. In addition to the simple plot, mathematical statistical analysis may be performed to yield more information, but this is beyond the scope of this book.

To construct a log-probability plot follow the steps outlined below as shown in Table 3.9. Column 1 in this table lists the 10 TWA values for asphalt fumes described earlier in the section. The ACGIH TLV® for asphalt fumes is $0.5\,mg/m^3$ as a benzene-soluble aerosol based on an inhalable size-fraction sample.[5]

1. Review the individual monitoring values in column 1 including backup

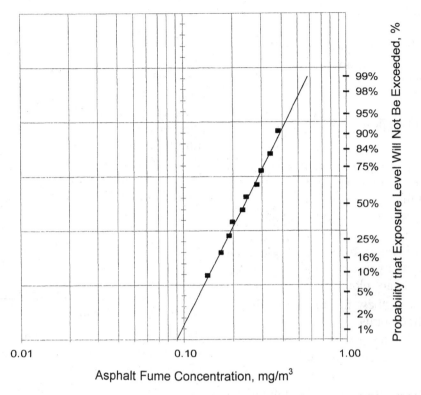

Figure 3.4. Log-normal probability plot of data in Table 3.9 prepared using commercially available computer software.[2]

TABLE 3.9. Log-Normal Probability Data Plotting

(1) Original Monitoring Data (mg/m³)	(2) Ranked Data[a] (mg/m³)	(3) Rank Order	(4) Plotting Position
0.24	0.14	1	9.1%
0.28	0.17	2	18.2%
0.23	0.19	3	27.3%
0.14	0.20	4	36.4%
0.38	0.23	5	45.5%
0.34	0.24	6	54.5%
0.17	0.28	7	63.6%
0.30	0.30	8	72.7%
0.20	0.34	9	81.8%
0.19	0.38	10	90.9%

[a] Data from column 1 arranged in ascending order.

information to ensure that they are valid and pertain to the SEG being evaluated.

2. Rank order the monitoring values (put lowest one first, then next highest, etc.) as shown in column 2. If there were any values below the limit of quantification (LOQ), they would be at the top of the list.

3. Number each value from 1 to the

total number of results (n) as shown in column 3.

4. Calculate the plotting position for each value from this equation for column 4:

$$\text{Plotting position} = \frac{\text{rank} \times 100}{(n+1)}$$

The plotting position is a statistical adjustment to account for the fact that the highest value measured is not the highest value that can occur.

5. Plot the monitoring values against the plotting position on log-normal probability paper (Figure 3.4). Ignore (i.e., do not plot) any LOQ values.

6. Draw a "best fit" line through the plotted points.

7. Check whether the data fall essentially on a straight line. If so, a log-normal distribution can be assumed. If not, either there are data from more than one "exposure group" mixed together (see below) or the data do not fit a log-normal distribution and this analysis is not valid.

8. If the log-normal assumption is valid, read the percentiles of interest from the plot. For example, in Figure 3.4 the data indicate that exposures are less than the $0.5 \, \text{mg/m}^3$ TLV® on about 98% of the shifts, while the average TWA exposure (represented by the 50% point on the plot) is about $0.25 \, \text{mg/m}^3$, which is 50% of the TLV®. This would appear to be an acceptable exposure.

If the data do not fall along a straight line when plotted on log-probability paper, it may indicate that exposures from more than one SEG are mixed together. As an illustration, consider carbon monoxide (CO) exposure measurements on cashiers at two different underground parking structures. Based on a review of job duties and likely exposures, they were considered as a single

SEG, and the eight measurements at each location were combined for analysis (Figure 3.5). These data fall on an approximate straight line, but there is some deviation. When the data are plotted separately by location (bottom diagrams), the straight line fit is noticeably better, thus indicating that the exposure is different at each structure. It can be seen that for parking structure A the geometric mean (50%) exposure level is 6 ppm and exposures exceed the ACGIH TLV-TWA® for CO of 25 ppm on about 8% of the shifts. At structure B the geometric mean is almost the same (5–6 ppm), but exposures exceed the ACGIH TLV-TWA® on only about 1% of the shifts.

Role of Modeling in Occupational Exposure Assessments

As discussed earlier, mathematical modeling of airborne levels are more typical of community environmental assessments rather than occupational exposure assessments. However, recent research indicates that modeling may provide a degree of accuracy comparable to limited air monitoring when the characteristics of the emissions source (amount of chemical released per time period), the air exchange rate between the immediate area around the source and the remainder of the work room, and the overall work room ventilation rate are all known. With this information the airborne concentration can be predicted in the room volume near the source using simulation techniques, which may permit estimations of worker exposures.[6] Although the resulting estimations can have a high degree of uncertainty, it should be understood that exposure profile estimates based on only a few air sampling results also have large uncertainties.

EXPOSURE MONITORING PLAN

Once the broad strategy has been decided, an exposure monitoring plan should be developed to execute the strategy. Expo-

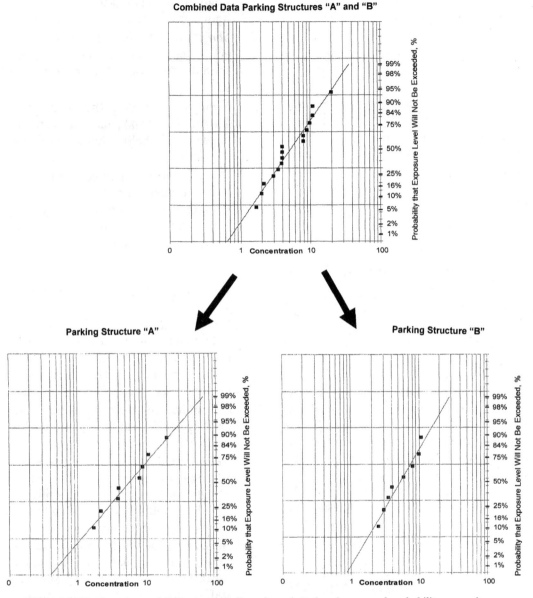

Figure 3.5. If values do not fall on a straight line when plotted on log-normal probability paper, it may indicate that data from two separate exposure profiles have been combined.

sure monitoring plans are used to focus available resources on the highest priority exposures. As monitoring results become available, plans can be revised to reflect new priorities. A well-designed and effectively implemented plan can:

• Help managers, supervisors, and employees understand potential risks created by the presence of chemical, physical, and biological agents in the workplace.

• Demonstrate that employees are protected from occupational health hazards.

• Identify areas where additional engineering controls or process improvements may be needed to reduce exposures.

- Ensure effective use of occupational health resources, including staff resources.
- Ensure compliance with applicable regulations and company policy.

The plan should address exposure monitoring during routine operations, maintenance activities, and planned shutdowns and turnarounds. Each facility should also consider what monitoring may be needed in emergency situations, and include appropriate guidelines in their facility emergency response plan.

Writing the Exposure Monitoring Plan

The exposure assessment identifies which exposures are significant, what exposure monitoring is required for compliance, and what additional exposure information will be useful. The next step is to design a plan which indicates which activities and jobs should be monitored, and how many and what types of samples should be taken. The plan can also lay out a monitoring schedule and indicate who is responsible for conducting the monitoring. An example of a typical exposure monitoring plan for a petroleum operation is shown in Table 3.10. Formats used for exposure monitoring plans will vary, but most plans will include the following basic information:

- *Location, Activity, and/or Job Title.* Information to help identify the SEG and activities involved, and to help classify the sample for future reference.
- *Source of Exposure.* The material or activity which is the source of the exposure, such as gasoline for total hydrocarbons and benzene; crude oil for H_2S; operating equipment for noise; and catalyst material for metals.
- *Exposures to be Measured.* Specific agents of concern which can be ana-

lyzed and compared to an established limit, such as benzene from gasoline.
- *Sampling Method.* Identifies the equipment and procedure which should be used to collect the sample.
- *Number and Type.* The number of TWA, STEL, Peak, and Area samples needed to help characterize the exposure.
- *Purpose of the Samples and Comments.* Explanation of why the sample is being taken, what the sampling strategy is, and notes on any special procedures or information needed.

The plan should be prepared by an individual or team with these characteristics:

- Have a good understanding of current industrial hygiene principles and practices, including regulatory requirements, exposure limits, and sampling and analytical methods.
- Be familiar with the types of exposures existing or likely to exist at the facility being reviewed.
- Be able to identify job activities where potential exposures occur and gauge which exposures may be significant, based on both health risks and compliance requirements.
- Be able to identify resources needed to execute the plan, such as equipment, consumable sampling supplies, staff time, travel expenses, and training or professional development for the sampling practitioners.

Evaluation of Monitoring Results

All monitoring results should be reviewed as they become available against the goals of the monitoring program and data gaps as described in the exposure assessment strategy. Conclusions based on monitoring results may include:

TABLE 3.10. Typical Exposure Monitoring Plan

Location, Activity and/or SEG Title	Source of Exposure	Contaminants	Sampling Method	Number and Type of Samples	Comments
Pipeline operator	Gasoline	Benzene, total hydrocarbon vapors	Organic vapor badge	2–4 TWA, 6–8 STEL	Worst-case sampling—hot days with little wind. Take STEL samples on at least two different operators when TWAs are taken.
Production operator gauging sour crude oil tanks	Sour crude oil	H_2S	H_2S meters and detector tubes	Two grab samples per tank	Evaluate need for engineering controls and PPE. Ensure that the operator is following H_2S safety guide. Take samples on hot days with minimal wind.
Welder in maintenance shop	Welding fumes	Chromium, manganese, nickel, and iron	High-flow pump, two-piece 37-mm cassette with MCE filters	2–3 TWA	Worst-case exposure during stainless steel welding on cold days when shop doors are closed. Identify welding rods and materials used.
Mechanic removing insulation	Insulation (non-asbestos)	Refractory ceramic fibers	25-mm asbestos cassette per NIOSH Method 7400	6 TWA 6 STEL 6 Area	Collect samples during equipment maintenance. Take as many samples as possible to develop baseline exposure profile.
Grease manufacturing operator	Lube oil	Oil mist	Low-flow pump with tare-weighed PVC filter	2 TWAs on each shift	Worst-case exposure on hot, calm days when several grease kettles are cooking.
Terminal operator	Blended gasoline	MTBE	Low-flow pump with charcoal tube adsorber	3 TWA 3 STEL	Worst-case sampling during MTBE blending on hot, calm days. Review initial data; if above 10ppm, advise Industrial Hygienist to plan additional monitoring.

- *Exposures Well Below Established Limits.* When representative samples are clearly low, additional monitoring is not usually needed. Additional sampling may be scheduled if there is a need for more baseline data or if operating conditions change.
- *Sampling Inconclusive.* Results may be inconclusive if they are approaching established limits. Additional samples may be needed to confirm exposures levels and/or to apply statistical analysis.

  Sampling may also be inconclusive if the results show a wide range of levels or do not appear reasonable. The sampling strategy may not have identified employees with equivalent exposures, or significant errors or other problems may have occurred during monitoring or analysis. In these cases, additional samples are needed or the sampling strategy should be redesigned.
- *Exposures Above Established Limits.* If exposure results exceed established limits, the hierarchy for reducing exposures should be instituted: (1) engineering controls, or (2) administrative controls, and then (3) personal protective equipment if necessary.

Follow-up testing should be conducted when administrative and engineering controls are instituted to ensure the controls are adequate.

Periodic Reviews

Exposure monitoring plans should be reviewed and updated periodically. The revised plan should reflect the results of previous monitoring, as well as changes in operating conditions, regulations, and exposure limits.

SUMMARY

Exposure monitoring should be performed according to a broad *exposure assessment strategy* and a tactical *monitoring plan*. This helps to ensure that the effort and expense invested in the monitoring yields information that has the maximum value for the intended purpose.

REFERENCES

1. Occupational Safety and Health Administration. *Hazard Communication Standard*. U.S. Code of Federal Regulation (29CFR1910.1200).
2. Mulhausen, J. R., and J. Damiano. *A Strategy for Assessing and Managing Occupational Exposures*, 2nd ed. Fairfax, VA: AIHA, 1998.
3. Hall, H., K. Teschke, H. Davies, P. Demers, and S. Marion. Exposure levels and determinants of softwood dust exposures in BC lumber mills, 1981–1997. *AIHA J.* **63**:709–714, 2002.
4. Occupational Safety and Health Administration. *Hazardous Waste Operations and Emergency Response*. U.S. Code of Federal Regulation (29CFR1910.120).
5. American Conference of Governmental Industrial Hygienists. *Threshold Limit Values and Biological Exposure Indices for 2004*. Cincinnati: ACGIH, 2004.
6. Nicas, M., and M. Jayjock. Uncertainty in exposure estimates made by modeling versus monitoring. *AIHA J.* **63**:275–283, 2002.

CHAPTER 4

AIR MONITORING AT EMERGENCIES INCLUDING TERRORISM EVENTS

This chapter provides guidance on performing air monitoring at emergencies including spills, releases, terrorism events, and similar situations where airborne contaminants may present a risk to responders or the public. *Terrorism events* include release or potential release of chemical (both chemical warfare and toxic industrial chemicals), biological, or radioactive agents. Terrorism events and other types of emergencies are grouped together because the preplanning and other aspects of air monitoring are similar for both.

This chapter has two purposes:

- Provide first responders and others with a defined role in terrorism or emergency response with detailed information on air monitoring that they can use in preparing response plans, developing liaisons with specialized response teams from other agencies, selecting and obtaining equipment, preparing procedures and training staff.

- Provide sampling practitioners without a formal role in incident response with enough background so that they can have an awareness of the topic and can plan for any response that might be an incidental part of their normal role in occupational or environmental monitoring.

Since many readers whose role is limited to emergency response may not have much need for the material in the book on occupational and environmental monitoring, this chapter is designed as a "stand alone" chapter. As a topic is discussed in this chapter, there are references to more detail and specific figures in later chapters. In this manner the reader can find relevant supplemental information efficiently. However, even emergency response sampling practitioners should understand the introductory and background information in Chapters 1–3.

As a working definition, *terrorism* is "violence, or the threat of violence, calculated to

Air Monitoring for Toxic Exposures, Second Edition. By Henry J. McDermott
ISBN 0-471-45435-4   © 2004 John Wiley & Sons, Inc.

create an atmosphere of fear and alarm, through acts designed to coerce others into actions they otherwise would not undertake or into refraining from actions that they desired to take. All terrorist acts are crime. This violence or threat of violence is generally directed against civilian targets. The motives of all terrorists are political, and terrorist actions are generally carried out in a way that will achieve maximum publicity."[1] Such events involve agents that are capable of producing large-scale physical destruction, widespread disruption, and/or mass casualties, which include chemical, biological, or radiological agents.

Although terrorism events and other emergencies are considered together in this chapter, air monitoring at terrorism events is typically unique for several reasons:

- The identity of the specific chemical, biological, or radioactive agent may not be known. In many cases, simply confirming or "ruling out" the presence of a terrorism agent is the first priority. With accidental releases, the identity of the material is often known either because of the source (a specific factory or tank) or from the shipping documents that accompany the bulk material in transit. For example, Material Safety Data Sheets (MSDSs) and similar information accompany rail and truck shipments so emergency responders can identify any spilled material; supplemental emergency response information is available by telephone from CHEMTREC® or similar resources.
- The need to identify chemical agents at extremely low airborne concentrations may increase the likelihood of "false positives" or misidentification of the material, both of which can be a serious problem where the public is potentially exposed to the agents.
- Since terrorism events are by definition criminal acts, sampling should

follow stringent collection, chain-of-custody, and documentation procedures if the results are to be used for criminal evidentiary purposes. Even accidental spills may result in regulatory penalty, or civil or criminal prosecution, so proper procedures and documental are needed in these situations as well.

There are certain "risk factors" that can help facility managers determine whether their facilities are at a higher risk for a terrorism event. Characteristics of higher risk facilities include:

- Uses, handles, stores, or transports hazardous materials.
- Provides essential services—for example, sewage treatment, electricity, fuels, telephone, and so on.
- Has a high volume of pedestrian traffic.
- Has limited means of egress, such as a high-rise complex or underground operations.
- Has a high volume of incoming materials (e.g., mail, imports/exports, raw materials).
- Is considered a high profile site, such as a water dam, military installation, or classified site.
- Is part of the transportation system, such as shipyard, bus line, truck transport, or airline.

All sampling practitioners and other individuals involved in emergency and terrorism response need adequate protective equipment. However, the chemical, biological, and radioactive agents present at terrorism sites may require extraordinary protective equipment and procedures because of the extremely high toxicity or danger to health from these agents (Figure 4.1).

Military response to terrorism agents often involves different circumstances and

Figure 4.1. Response to terrorism events requires stringent personal protective equipment. (Courtesy of Draeger Safety, Inc.)

Figure 4.2. Field-portable gas chromatograph/mass spectrometer unit has application in military and civilian responses. (Courtesy of Infincon, Inc.)

parameters than do "civilian" terrorism events—those with actual or potential public exposure. For example, military troops have specialized protective equipment and training, and so air monitoring is mainly aimed at detecting the presence of an agent so that protective equipment can be donned. Additionally, detection equipment must be rugged enough to withstand battlefield conditions, and, if the site is remote from typical urban/industrial areas, many of the possible interferences that may cause false positives are not expected to be present.[2] For these reasons, air monitoring for military applications is not covered in this book, although the same instruments may be used (Figure 4.2) and the military services have specially trained teams that can be called upon for assistance by civilian agencies if a terrorism event does occur. Additionally, the U.S. Army Edgewood Chemical Biological Center has tested many commercially available direct reading instruments for chemical warfare agents; the results of these tests are discussed in this chapter.

REASONS FOR AIR SAMPLING

There are several different reasons for performing monitoring, either with direct

reading instruments or with sample collection devices, at possible terrorism events or other emergency sites. These include:

- Confirming the presence of a terrorism agent or toxic chemical by identifying the specific agent(s) that are present.
- "Ruling out" the presence of a terrorism agent, either by identifying the substances present or by performing a negative scan for suspected terrorism agents.
- Confirming the identity of a released material that seasoned emergency responders have tentatively identified through sensory response (i.e., Mace and pepper spray).
- Assessing the airborne concentration of terrorism agents for:
  - Selecting proper personal protective equipment (PPE).
  - Deciding whether to evacuate members of the public.
  - Predicting extent of adverse health effects.
  - Plotting dispersion of airborne clouds.
  - Identifying which areas will require cleanup or decontamination.

TABLE 4.1. Purpose of Air Sampling at Terrorism Events

Purpose	Required Minimum Detection Level	Possible Sampling Alternatives
Confirm terrorism agent	Very low	Bulk or surface samples
"Rule out" terrorism agent	Depends on situation	Bulk or surface samples
PPE selection	Depends on situation	None
Evacuation perimeter	Very low	None
Plot dispersion of contaminant cloud	Very low	No sampling alternative, but computer modeling may be an option.
Identify areas needing cleanup	Very low	Surface samples
Post-cleanup assessment	Very low	Surface samples

— Determining that post-cleanup levels are acceptable.

Air samples are preferred for those reasons listed above that relate to inhalation dose. Air sampling is not the best choice for many of these items. For example, a sample of liquid or solid residue would be preferable for the first two reasons, while surface samples (swipe or vacuum) would be a good option for identifying areas needing decontamination and the post-cleanup contamination levels. When air sampling is performed, direct reading instruments are often preferred because of the immediate need for air sampling results. See Table 4.1 for more detail on the data needs and possible alternatives to air sampling at terrorism events.

TERRORISM AGENTS

The effects of the various terrorism agents can vary widely. Many government and medical web sites contain information of symptoms and treatment, while other references discuss means of protecting people and buildings against possible attack. A brief description of the effects of the different classes of agents is provided below. Much of the material in this section is excerpted from the NIOSH document

Guidance for Filtration and Air-Cleaning Systems to Protect Building Environments from Airborne Chemical, Biological, or Radiological Attacks.[3]

Chemical Agents

Chemical Warfare Agents. Classical chemical warfare agents include a wide variety of different compounds that can affect humans in various ways. Chemical warfare agents (CWAs) commonly exist as either a gas or a liquid aerosol. Many of the agents have low vapor pressures and are delivered as a liquid aerosol, while many other higher vapor pressure agents are gaseous. Chemical agents are classified by either their physiological action (e.g., blister agent) or their military use (e.g., incapacitating agent). CWAs also have a military designation; these are shown in parentheses:

- Blister agents, also known as vesicants, include sulfur mustard (H/HD) and nitrogen mustard (HN), arsenicals (lewisite (L)), and phosgene oxime (CX). Blister agents tend to have relatively low volatility and modest acute toxicity, compared to other CWAs. Blister agents produce pain and injury to the eyes, reddening and blistering of the skin, and, when inhaled, damage to the mucous membranes and respiratory tract. Mustard may produce major destruction of the epidermal layer of the skin.

- Nerve agents are derivatives of organophosphate esters and are among the most toxic chemicals known. This class includes materials such as O-ethyl-S-(2-diisopropyl aminoethyl) methyl phosphonothiolate (VX), ethyl N,N-dimethyl phosphoroamido cyanidate (Tabun (GA)), isopropyl methylphosphonofluoridate (Sarin (GB)), and pinacolyl methyl phosphonofluoridate (Soman (GD)). Nerve agents have a wide range of volatilities and their toxicity is approximately 100 times higher than blood and choking agents. Nerve agents inhibit the cholinesterase enzymes. The cholinesterase enzymes are responsible for the hydrolysis of acetylcholine, a chemical neurotransmitter. This inhibition creates an accumulation of acetylcholine at a cholinergic synapse that disrupts the normal transmission of nerve impulses in the central nervous system (CNS), peripheral voluntary nervous system, and autonomic nervous system.

- Blood agents are highly volatile inhalation hazards and include hydrogen cyanide (AC), cyanogens chloride (CK), and arsine (SA). These agents are transported by the blood to all body tissues where the agent blocks the oxidative processes, which prevents tissue cells from utilizing oxygen. The CNS is especially affected and excessive exposure leads to cessation of respiration followed by cardiovascular collapse.

- Choking agents (lung-damaging agents) are also highly volatile inhalation hazards. They include phosgene (CG), diphosgene (DP), chlorine, and chloropicrin (PS). These agents produce injury to the lungs and irritation of the eyes and the respiratory tract. They may also cause intractable pulmonary edema and predispose to secondary pneumonia.

- Incapacitating agents produce temporary physical or mental effects, or both. They are usually distinguished from riot-control agents by their longer period of effectiveness, which may be as long as days after exposure. Examples of incapacitating agents include 3-quinuclidinyl benzilate (BZ), cannabinols, phenothiazines, fentanyls, and central nervous system stimulants (e.g., d-lysergic acid diethyl amide (LSD)).

Toxic Industrial Chemicals. Toxic industrial chemicals (TICs) are generally considered to be chemicals that are stored, transported, or commercially or clandestinely available in the physical form and sufficient quantity that they could be used in a terrorism event. Examples are storage tanks at industrial plants or pipelines on public rights-of-way that could be breached, rail car or tank truck quantities of toxic chemicals, and drums or smaller containers that could be released into the atmosphere or another medium such as a public water supply (Figure 4.3). These chemicals are commonly categorized by their hazardous properties, such as toxicity, reactivity, stability, combustibility, corrosiveness, ability to oxidize other materials, and radioactivity. For the purposes of collecting air samples on a sorbent sampling medium, gaseous agents can be divided into the following categories: organic vapors (e.g., cyclohexane), acid gases (e.g., hydrogen sulfide), basic gases (e.g., ammonia), and specialty chemicals (e.g., formaldehyde or phosgene). Of principal concern to the responder are those TICs that have a high toxicity and are readily obtainable in adequate quantities.

Figure 4.3. The hazard from toxic industrial chemicals depends on their toxicity, physical form, and quantity. (Courtesy of RKI Instruments, Inc.)

Biological Agents

Disease-causing biological agents such as *Bacillus anthracis* (anthrax), *Variola major* (smallpox), *Yersinia pestis* (bubonic plague), *Brucella suis* (brucellosis), *Francisella tularensis* (tularemia), *Coxiella burnetti* (Q fever), *Clostridium botulinum* (botulism toxin), viral hemorrhagic fever agents, and others have the potential for use in a terrorist attack. Each of these biological agents may travel through the air as an aerosol. Generally, viruses are the smallest, while bacteria and spores are larger. In nature, biological agents and other aerosols often collide to form larger particles; however, terrorist groups may modify these agents in ways that reduce the agglomeration to form larger particles, thus increasing the number of biological agents that may potentially be inhaled. There are significant differences from one agent to another in their adverse public health impact and the mass casualties they can inflict. An agent's infectivity, toxicity, stability as an aerosol, ability to be dispersed, and concentration all influence the extent of the hazard. Other important factors include person-to-person agent communicability and treatment difficulty. Biologi-

cal agents have many entry routes and physiological effects. They generally are nonvolatile.

Another category of biological agents is *toxins*. This group includes bacterial (exotoxins and endotoxins), algae (blue-green algae and dinoflagellates), mycotoxins (tricothocenes and aflatoxins), botulinum, and plant- and animal-derived toxins. Toxins form an extremely diverse category of materials and are typically most effectively introduced into the body by inhalation of an aerosol. They are much more toxic than chemical agents. Their persistency is determined by their stability in water and exposure to heat or direct solar radiation.

The U.S. Department of Health and Human Services uses a priority approach in determining how critical each biological agent is as a terrorist weapon (Table 4.2)[4]:

- High priority (Category A) agents include organisms that pose a risk to national security because they can be easily disseminated or transmitted person-to-person; cause high mortality, with potential for major public health impact; might cause public panic and social disruption; and require special action for public health preparedness.
- Second highest priority (Category B) agents include those that are moderately easy to disseminate; cause moderate morbidity and low mortality; and require specific enhancements of CDC's diagnostic capacity and enhanced disease surveillance.
- Third highest priority (Category C) agents include emerging pathogens that could be engineered for mass dissemination in the future because of availability, ease of production and dissemination, and potential for high morbidity and mortality and major health impact.

TABLE 4.2. Typical Biological Agents in U.S. DHHS Prioritization Categories

Category A	Category B	Category C
Variola major (smallpox)	*Coxiella burnetti* (Q fever)	Nipah virus
Bacillus anthracis (anthrax)	*Brucella* species (brucellosis)	Hantaviruses
Yersinia pestis (plague)	*Burkholderia mallei* (glanders)	Tickborne hemorrhagic fever
Clostridium botulinum toxin	Alphaviruses	viruses
(botulism)	Venezuelan encephalomyelitis	Tickborne encephalitis viruses
Francisella tularensis	Eastern and western equine	Yellow fever
(tulararemia)	encephalomyelitis	Multidrug-resistant
Filoviruses	Ricin toxin from *Ricinus*	tuberculosis
Ebola hemorrhagic fever	*communis* (castor beans)	
Marburg hemorrhagic fever	Epsilon toxin of *Clostridium*	
Lassa fever	*perfringens*	
Junin (Argentine hemorrhagic	*Staphylococcus* enterotoxin B	
fever) and related viruses		

Radiological Hazards

Radiological hazards of interest for air sampling are the ionizing radiation emitted by radioisotopes that may occur as an aerosol, be carried on particulate matter, or occur in a gaseous state. These hazards can be divided into three general forms: alpha, beta, and gamma radiation:

- Alpha particles, consisting of two neutrons and two protons, are the least penetrating and the most ionizing form. Alpha particles are emitted from the nucleus of radioactive atoms and transfer their energy at very short distances. Alpha particles are readily shielded by paper or skin and are most dangerous when inhaled and deposited in the respiratory tract.

- Beta particles are negatively charged particles emitted from the nucleus of radioactive atoms. Beta particles are more penetrating than alpha particles, presenting an internal exposure hazard. They can penetrate the skin and cause burns. If they contact a high density material, they may generate X rays, also known as *Bremmstrahlung* radiation.

- Gamma rays are emitted from the nucleus of an atom during radioactive decay. Gamma radiation can cause ionization in materials and biological damage to human tissues, presenting an external radiation hazard.

There are three primary scenarios in which radioactive materials could potentially be dispersed by a terrorist: attack on a fixed nuclear facility; a nuclear weapon (device); or conventional explosives or other means to spread radioactive materials (a dirty bomb). Nuclear reactors and weapons are based on the fission reaction of radioactive material; nuclear weapons may also be of *fusion* design, but these are probably too complex for terrorist groups to obtain. A dirty bomb kills or injures because of the initial blast of the conventional explosive. In addition, it disperses airborne and surface radioactive contamination, hence the term "dirty." A study by the Federation of American Scientists found that a dirty bomb would not inflict deaths on anything like the scale of even a crude nuclear device, but it could cause widespread contamination exceeding U.S. EPA safety guidelines.[5] This could require the long-term evacuation of the contaminated area until cleanup is performed, or if cleanup is not feasible, permanent abandonment. In either case, costs for an urban area could potentially cost trillions of dollars. Such bombs could be miniature devices or as large as a truck. These bombs are not capable of sustaining a nuclear chain reaction, but they may include fissile material such as enriched uranium or plutonium.

The airborne hazard from all three radioactive material release scenarios would mainly consist of submicron-sized and larger aerosols containing the radioactive isotopes in the original material or resulting from the fission reactions within a nuclear reactor or device; there are also gaseous emissions from fission reactions such as radioactive iodine. People may be exposed to material in the dust inhaled during initial passage of the dust cloud (if they have not been evacuated); if this material is an alpha emitter, it will stay in the body and lead to long-term exposure. Additionally, anyone living in the affected area will be exposed to radioactive material from the dust that settles from the cloud. If the material contains gamma emitters, residents will be continuously exposed to radiation from this dust. If the material contains alpha emitters, any dust that is resuspended due to wind, automobile movement, or other actions may also be inhaled, adding to exposure. In rural areas, people might be exposed to radiation from contaminated crops and surface water sources.

Incident-Related Exposures to Additional Substances

In addition to the terrorism agents, responders and the public may be exposed to other potentially hazardous airborne substances following an event (Figure 4.4). These include:

- Airborne construction material (pulverized debris) from collapsed struc-

Figure 4.4. The airborne "cloud" from emergency sites may contain smoke, pulverized construction materials, organic vapors, carbon monoxide, and other hazardous compounds. (Courtesy of Draeger Safety, Inc.)

tures such as concrete dust, asbestos, and moldy debris.
- Carbon monoxide (CO) from emergency vehicle exhaust, portable generators, fires, cutting torches, and so on.
- Organic vapors and aerosols from smoldering fires.
- Other chemicals released from storage facilities or railcars, and so on, incidental to the terrorism event.
- Lead from cutting structural members that are coated with lead-based paints.

Air sampling plans should include the capability to monitor for these incident-related substances that could be present at levels of concern to responders or the public.

IDENTIFYING A TERRORISM EVENT

It may be difficult to determine whether a terrorism event has actually occurred. This is different from most of the air monitoring situations covered in the rest of this book. For example, an industrial plant has access to Material Safety Data Sheets (MSDSs) or other documents that describe the potentially toxic components of raw materials and other products that they purchase. Based on the toxicity, volatility, and amount used, a qualified industrial hygienist or other professional can decide whether exposure monitoring is indicated. In other cases a characteristic odor is an indication of exposure, or employees report headaches or other adverse effects that could be due to chemical exposure.

The same is not true for possible terrorism events. These events may be identified by:

- Limited or widespread signs or symptoms in affected people that could be attributed to a terrorism agent. Unfortunately, for chemical warfare and biological agents the occurrence of serious illness or death may be the first indication of the event.
- Sensory response (odor, irritation).
- Visual observations and reports of unusual activities (such as spraying) or abandoned empty chemical drums.
- Detonation of explosive devices (which could indicate a "dirty bomb").
- Announcement by terrorist group (pre- or post-event).
- Abnormal results of routine environmental sampling such as ambient air samples or potable water source analyses.

Once reported, local agencies would initiate their emergency response procedures to determine if an event has occurred, treat casualties, identify the agent(s), mitigate the adverse effects, apprehend the perpetrators, and prevent a future occurrence.

PLANNING FOR EMERGENCIES AND TERRORISM EVENTS

Planning is the key to success in response to an emergencies and terrorism events. Any individual with responsibility for plan-

ning air monitoring during such events should have access to a qualified industrial hygienist or other sampling professional during the planning stages. The qualified professional can help develop the monitoring strategy and plan, select equipment, write sampling protocols, identify analytical laboratories, determine equipment maintenance and calibration schedules, and identify other resources for monitoring.

Identifying and coordinating with supplemental monitoring and response resources can be especially important. In the United States, agencies such as the Federal Bureau of Investigation (FBI), U.S. Coast Guard (USCG), Department of Transportation (DOT), the Environmental Protection Agency (EPA), the Occupational Safety and Health Administration (OSHA), Center for Disease Control (CDC), and National Institute for Occupational Safety and Health (NIOSH) can assist with terrorism events. Assistance ranges from technical advice to special response teams within the military services or other government organizations that can be deployed for chemical, biological, or radiological events when their special training and equipment could play a key role. Examples include:

- U.S. EPA's Environmental Response Team can provide expert advice and also deploy specialists in decontamination, air monitoring, water supply protection, site safety, and other specialty areas. The EPA also has a Radiological Emergency Response Team with expertise in radiation monitoring, radionuclide analysis, health physics, and risk assessment. The EPA routinely operates over 250 ambient air sampling stations across the United States, which monitor radioactivity and other contaminants in various environmental media.
- U.S. Department of Energy's (DOE) National Nuclear Security Adminis-

tration can deploy various specially trained and equipped teams to perform tasks such as: aerial sampling using fixed or rotary wing aircraft; predicting the probable spread of nuclear, chemical, or hazardous material contamination in the atmosphere; emergency response to a nuclear weapon incident; search and identification of nuclear materials; technical support to render safe operations; and packaging for final disposition. Special teams are available to provide local, state and federal agencies with assistance for all types of radiological incidents.

- National Guard Weapons of Mass Destruction Civil Support Teams can be deployed to assist local first responding agencies in determining the nature of a terrorist attack, provide medical and technical advice, and coordinate identification of additional assets and assist in integrating them into the response organization.
- The Federal Bureau of Investigation (FBI) provides on-site assistance and has laboratory resources to assist in identifying and quantifying terrorism agents.

The above list is a sample of the government resources available to local responders. Each of these special teams or experts is trained to function within the Incident Command System, so they can be quickly integrated into the local response organization at an event.

In addition to government resources, local industry may be a valuable source of advice and assistance to the sampling practitioner planning for a terrorism event or other emergency. This is especially important when considering risks from Toxic Industrial Chemicals. Safety and health staff and emergency responders from local facilities are the best source of information about the identities, toxicity, amounts, and relative vulnerability of toxic chemicals used at their facilities. They usually have monitoring devices for the chemicals they handle along with trained sampling practitioners and industrial hygienists. They, along with transportation companies, may have the best information on hazardous cargoes moving through the area. Often local industries participate in mutual aid response groups so that resources from other plants are available for community monitoring if any of the plants has a chemical release. Although much of the chemical information is in "chemical inventory" reports made available to the public under community Right-to-Know regulations or facility safety and response plans, the most useful information for response planning is obtained by working directly with the local facilities.

Once the list of possible terrorism agents is compiled, those for which air sampling may be needed are identified, sampling methods and equipment are selected, and either the equipment is procured or a source for renting or borrowing the equipment is located. While local responders should have commonly used equipment on hand, it may not be feasible to procure specialized direct reading devices with low probability of being used. Sampling equipment manufacturers, equipment rental companies, and even nearby universities and national laboratories may be a source of specialized sampling equipment. Devices such as the QCM Cascade Impactor Air Particle Analyzer from California Measurements, Inc. (see Figure 14.11) described in Chapter 14 could be very useful in sampling airborne aerosols from an event, but these are not in widespread use. It will be necessary to identify nearby organizations that own devices like this and arrange to borrow it. Including the devices and skilled operators in preparedness exercises is a good way for response agencies to become aware of the capabilities and limitations of the specialized monitoring equipment.

Figure 4.5. Firefighters use various toxic gas monitors as part of their normal responses. (Courtesy of Draeger Safety, Inc.)

During the planning process, it is also necessary to identify how sampling equipment will be deployed at a site of a suspected or actual event. Most fire service first-responders have direct reading instruments such as a combination instrument (for oxygen, combustible gases, carbon monoxide, and hydrogen sulfide) or instruments to measure numerous contaminants at parts-per-million levels (Figure 4.5). These instruments are not applicable to true terrorism events because they do not measure CWAs or most TICs, and the level of detection is too high. Specialized direct reading instruments and air sampling pump systems will be needed. Detector (colorimetric) tubes are available for a variety of CWAs and TICs; these may be a good option. Some electronic direct readings may also be applicable for these response situations. In many cases the airborne concentration may be too low to be measured directly, or a direct reading instrument may not be available. Longer-term samples using sampling pumps equipped with filters to capture particulates or equipped with

adsorption tubes (e.g., charcoal) to collect organic vapors are often the best choice in these situations, and in most cases they are a good means of collecting back-up samples for laboratory analysis. It may be advisable to store the initial response sampling equipment in "quick-deploy" kits that are maintained in a charged and calibrated status. These kits are commercially available for colorimetric detector tubes and electronic instruments as described later in this chapter; for sampling pumps and collection media the kits may have to be assembled locally. Sampling practitioners can begin collecting initial samples immediately by positioning sampling pump kits at different locations upwind and downwind of the site, turning on the pumps, and recording minimum data such as date, location, and starting time. Decisions about which of the collected samples to analyze for contaminants can be made after more is known about the potential agent(s) involved and other details; but meanwhile the required air volume is being sampled to permit an accurate determination of contaminants with a low limit of detection.

Any existing routine environmental air sampling stations should also be identified in advance. These may be operated by the U.S. EPA or local air quality agencies as part of ambient air quality monitoring, or by industrial facilities and research laboratories as part of their environmental permits. In case of an actual or suspected event, the filters or charcoal adsorption canisters can be quickly analyzed for suspected terrorism agents.

In summary, during the planning stage it is important to: identify relevant terrorism agents; procure equipment or find sources of assistance; and develop processes for collecting and analyzing the samples. This will permit rapid deployment of sampling equipment or immediate calls for assistance so that air sampling is initiated in a timely manner if an event occurs.

AIR SAMPLING FOR CHEMICAL AGENTS

Specific air sampling methods depend on the type of agent, concentration, need for immediate results and whether a direct reading instrument (or other rapid method) exists and is available at the site.

Chemical Agents

This section covers air sampling for CWAs and TICs. One important characteristic of a chemical compound that is considered as a "chemical warfare agent" is that it has high acute toxicity at very low airborne concentrations or with a small amount of skin contact. This is illustrated in Table 4.3, which lists the AEGL-3 (Acute Exposure Guideline Level—possible lethal level) for a 1-hour exposure for several CWAs and also two industrial chemicals that are generally considered to be extremely toxic: phosgene and hydrogen cyanide. (See Chapter 2 for a detailed discussion of AEGLs.) As can be seen for the table, the AEGL-3 value for the CWA agents is significantly lower than for the extremely toxic industrial chemicals. Thus any monitoring technique must have a very low limit of detection to be useful in identi-

fying CWAs at concentrations that are not harmful.

Chemical agents that are released from a point source (such as a breaking or bursting device) will have a high concentration at the source with lower concentrations as the material disperses. Agents disseminated through a moving aircraft or vehicle-mounted spray device may not exist in a high concentration except immediately at the spray nozzle outlet. Blister agents tend to have low vapor pressures (i.e., low volatilities) and thus exist mainly in the liquid form, producing little airborne vapor. The ideal air sampling methods for chemical agents (CWAs and TICs) are selective to the specific compound and are able to sense or determine concentration at an appropriately low level in the atmosphere. For CWAs the level should be extremely low while for TICs the required level of detection depends of the degree of toxicity. There are four types of detection methods for chemical agents:

- Military test kits for CWAs (air sampling and paper test strips)
- Colorimetric detector tubes
- Direct reading instruments
- Sample collection devices (with laboratory analysis)

The U.S. Army Soldier and Biological Chemical Command at Aberdeen Proving Ground, Maryland has performed evaluations of numerous detection devices and direct reading instruments for CWAs such as Tabun (GA), Sarin (GB), and Mustard (HD) as representative agents.[6-21] Several of the reports are referenced in this section; before selecting instruments, check their web site (http://hld.sbccom.army.mil/downloads) for a current list of available evaluation reports.

In these tests the various detection devices currently being used by the military services and some first responder organizations were screened for their ability to

TABLE 4.3. AEGL-3 (Lethal) Value for One-Hour Exposure

Substance	AEGL-3 (Lethal) (1-hour exposure), ppm
GA	0.039
GB	0.022
GD	0.017
GF	0.018
VX	0.00091
Phosgene	0.75
Hydrogen cyanide	15

Source: National Research Council, Subcommittee on Acute Exposure Guideline Levels. *Acute Exposure Guideline Levels for Selected Airborne Chemicals*, Vol. 2, Washington, D.C.: National Academy Press, 2002.

TABLE 4.4. Military Criteria for CWA Detector Response and Personnel Exposure

	Concentration (ppm) for Specified Time Period		
Agent	JSOR[a] Instrument Response	IDLH[b]	AEL[c]
HD	0.300 in 120 sec	N/A	0.0005 up to 8 hr
GA	0.015 in 30 sec	0.03 up to 30 min	0.000015 up to 8 hr
GB	0.017 in 30 sec	0.03 up to 30 min	0.000017 up to 8 hr

[a] Joint service operational requirements for point sampling detectors.
[b] Immediately dangerous to life or health values to determine level of CW protection.
[c] Airborne exposure limit values from U.S. Army Regulation 385–61 to determine when protective masks are required. Military personnel can operate without a protective mask for up to 8 hr at concentrations below the AEL.

Source: Reference 6.

detect chemical warfare agent vapors. The approach was to seek the concentration level where a detection signal was indicated for the respective agent. The detectors were each tested with GA, GB, and HD with different concentration levels at 20°C (±2°) and low relative humidity in an attempt to determine the minimum detectable level (MDL) and to establish a response curve. Test conditions were then varied to include the assessment of temperature, humidity, and concentration effects. Each observation was confirmed with repeated trials.

The units were also qualitatively tested outdoors in the presence of common potential interferents such as gasoline, diesel fuel, jet propulsion fuel (JP8), kerosene, ethylene glycol anti-freeze, AFFF liquid (aqueous film-forming foam used for fire fighting), and household chlorine bleach vapors. Vapor from a 10% HTH slurry (chlorinating CW agent decontaminant), engine exhausts, burning fuels and other burning materials were also tested. These were field experiments involving open containers, truck engines, and fires producing smoke plumes that were sampled by the detectors placed at various distances downwind. Distances were chosen to obtain reasonable exposures for each detector: 1–2 m for fumes and 5–10 m for smokes. Finally, laboratory tests were also performed to show the CW agent

detection capability of the detection devices in the presence of high concentrations of potential interferent vapors.

The devices were evaluated against military requirements for this type of instrument, called the *Joint Services Operational Requirement (JSOR) for Point Sampling CWA Detectors*, which is based on levels requiring personal protective equipment. These levels are summarized in Table 4.4.

Military Test Kits. This category includes prepackaged test kits, such as the U.S. M256A1 *Chemical Agent Detector Kit*, and paper test strips. The M256A1 contains 12 disposable samplers/detectors with test spots for nerve, blister, and blood agents and mustard gas. Liquid reagent ampoules are crushed and allowed to wet the test spots, which turn a distinctive color if these CWAs are present in the air.

Paper Test Strips. Test strips are used to detect CWAs in liquid droplets; they are not used for air sampling but are included in this section for completeness. Most first responder units for terrorism events will have paper test kits such as the M9 and M8:

- M9 chemical detection paper is used to identify the presence of liquid chemical agent aerosols. It contains a suspension of an agent-sensitive red

Figure 4.6. M8 paper showing exposure to droplets of blister agent. (Courtesy of Nextteq, LLC.)

Figure 4.7. Response kit with paper-tape samplers for chemical warfare agents. (Courtesy of Nextteq, LLC.)

indicator dye in a paper matrix. It will detect and turn pink, red, reddish brown, and red-purple within about 30 seconds when exposed to liquid nerve agents and blister agents, but it does not identify the specific agent, nor does it detect any biological agents such as anthrax. Petroleum products also cause the paper to change color. It is commercially available in an adhesive backed 32-foot-long roll for rapidly draping over doorways, windows, or equipment. Possible interferences are organic solvents, and petroleum compounds (oils or grease) may cause a false-positive color change.

- M8 paper is a booklet of 25 test sheets with a color comparison chart for identifying G or V nerve agents or H blister agent (Figure 4.6). M8 not only detects these agents, but differentiates between them by producing a distinct, irreversible, agent-specific color for G (gold color), H (red color), or V (green color) agents. Possible interferences are ammonia and some petroleum products. A modification of M8 paper has adhesive backing for attachment to surfaces to detect the deposition of the chemical warfare agents; this is called "3-way paper."

Both M9 and M8 detectors require at least 0.02 mL of liquid and respond within 30 seconds. They respond within a temperature range of approximately 26–120°F and should not be stored in direct sunlight.

They are available as a kit (Figure 4.7) and are used together because they respond to different interferences and give different false positives. M9 is more sensitive than M8 and thus will respond first, although the initial indication may be pinpoint-size color spots rather than a distinctive color change. One manufacturer offers a proficiency certification test kit that incorporates test reagents that simulate G or V nerve agents or H blister agents for training. This is an inexpensive and safe method to ensure correct use of M9 and M8 test kits for chemical warfare agent detection (Figure 4.8).[22]

Colorimetric Detector Tubes. Detector tubes, described in detail in Chapter 13, are used for short-term (or *grab*) samples (see Figures 13.1 to 13.3). A typical detector tube is a glass tube filled with a solid granular material, such as silica gel, that has been coated with one or more detection reagents that are especially sensitive to the target substance and quickly produce a distinct layer of color change when a specified amount of air is drawn through the tube. A

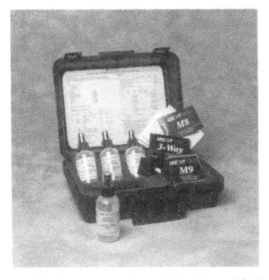

Figure 4.8. Proficiency testing kit for training/ evaluation in proper use of paper-tape samplers. (Courtesy of Nextteq, LLC.)

Figure 4.9. Battery-powered sampling pump and detector tube manifold. (Courtesy of Nextteq, LLC.)

calibration scale is printed on the side of the tube, which is used to read concentrations of the measured substances (gases and vapors) directly; in the case where a tube can be used for different concentration ranges, there may be two scales on a tube, each corresponding to a different air volume. Hand pumps (either of "piston" or "bellows" design) are commonly used with detector tubes, but battery- or line-powered sampling pumps are also used. In order to achieve the low limit of detection necessary for CWAs, for some tubes the number of pump "strokes" is very high (e.g., up to 50 strokes), so a electric pump may be the only practical choice when multiple samples are to be collected. Manifolds are also available to hold multiple tubes to reduce the time needed to perform initial sampling (Figure 4.9).

These tubes provide the ability to do direct reading measurements while in the field for a wide variety of gases and vapors. Since no laboratory analysis is required, they provide fast on-site results and are often considered along with direct reading electronic instruments when immediate

Figure 4.10. HazMat detector tube response kit. (Courtesy of Draeger Safety, Inc.)

results are needed. Hazardous material (Hazmat) response teams typically are equipped with detector tubes for common materials such as ammonia, chlorine, acid gases, and organophosphate pesticides, so responders have familiarity with these devices that can be quickly expanded to include detector tubes for CWAs[23] (Figure 4.10).

There is a military detection kit (M18) that contains detector tubes for measuring the concentration of selected airborne chemicals. The M18 comes with disposable tubes for cyanide, phosgene, lewisite, sulfur mustard, and nerve agents GA (Tabun), GB (Sarin), GD (Soman), and VX. With the pump included with this kit, tests for each take only 2 to 3 minutes but must be conducted serially.[23]

Several manufacturers offer tubes to the civilian market to detect CWAs. For example, MSA offers five separate detector tubes for CWAs:

- Blister agent: Lewisite.
- Blood and choking agents: Hydrogen cyanide (AC), cyanogen chloride (CK), phosgene (CG), and diphosgene (DP).
- Blister agent: Sulfur mustard (HD) gas only.
- Blister agents: Sulfur mustard gas (HD) and nitrogen mustard gas (HN).
- Nerve agents: Sarin (GB), Tabun (GA), Soman (GD), and VX.

A battery-powered pump system is available that can sample with four tubes at once to save time, or a single-tube manual pump can be used.

When tested by the U.S. Army Soldier Biological and Chemical Command, detector tubes generally showed good sensitivity to the representative CWA they were evaluated against, but the minimum detection level for most tubes was above the accepted exposure level.[6,7] Thus they were useful in detecting the presence of the agent, but not in determining that the airborne concentration was low enough so that respiratory protection was not needed. The tubes typically functioned well in the presence of common interfering compounds, but did not function well at low temperatures and some could not detect the CWAs at high ambient relative humid-

TABLE 4.5. Sensitivity Tests for Several Detector Tubes for CWAs

Agent	Concentration, ppm	
	MSA	Draeger
HD	3.0[a]	4.0[c]
GA	0.01[b]	0.01[d]
GB	0.01[b]	0.02[d]

[a] MSA sulfur mustard (HD) gas only tube.
[b] MSA nerve gas tube.
[c] Draeger thioether tube [sulfur mustard (HD)].
[d] Draeger phosphoric acid ester tube (nerve agents).
Sources: References 6 and 7.

ity. Table 4.5 shows the results for some detector tube sensitivity tests and Table 4.6 shows interference tests for the same tubes.

In situations where the sensitivity limitations are not a disqualifying factor, detector tubes may offer several strong advantages for the civilian sampling practitioner with a need to measure the concentration of CWAs. Detector tubes are well-known technology to most sampling practitioners, are easy to use, and have a good storage life. With detector tubes there is a modest initial purchase expense, and there is not the ongoing calibration and maintenance typical of an electronic direct reading instrument. However, a battery-powered sampling pump and manifold to collect several samples simultaneously may be a practical necessity for tubes requiring a large sample volume.

Direct-Reading Electronic Instruments

CWA-Specific Instruments. CWA-specific electronic direct-reading devices were generally developed for military applications. They operate on one of three main principles:

- *Ion Mobility Spectrometry (IMS).* As shown in Figure 4.11, with these devices a radioactive source (typically nickel-63) adds a charge to (i.e.,

TABLE 4.6. Results of Field Interference Tests for Several Detector Tubes for CWAs

Interferent	MSA Sulfur Mustard (HD) Tube	MSA Nerve Gas Tube	Draeger Thioether (HD) Tube	Draeger Phosphoric Acid Ester (Nerve) Tube
Gasoline exhaust, at idle	Negative	Negative	Not tested	Not tested
Gasoline exhaust, at revved	Negative	Negative	Negative	Negative
Diesel exhaust	Negative	Negative	Not tested	Not tested
Diesel exhaust, at revved	Negative	Negative	Negative	Negative
Kerosene vapor	Negative	Negative	Negative	Negative
Burning kerosene smoke	Negative	Negative	Negative	Negative
JP8 (jet fuel) vapor	Negative	Negative	Negative	Negative
Burning JP8 smoke	Negative	Negative	Negative	Negative
Burning gasoline smoke	Negative	Negative	Negative	Negative
Burning diesel smoke	Negative	Negative	Negative	Negative
Diluted AFFF[a] vapor	Negative	Negative	Negative	Negative
Diesel vapor	Negative	Negative	Not tested	Not tested
Gasoline vapor	Negative	Negative	Not tested	Not tested
10% HTH vapor	Negative	Negative	Negative	Negative
Bleach vapor	Negative	Negative	Negative	Negative
Burning cardboard smoke	Negative	Negative	Negative	Negative
Burning cloth smoke	Negative	Negative	Negative	Negative
Burning wood fire smoke	Negative	Negative	Negative	Negative
Doused wood fire smoke	Negative	Negative	Negative	Negative
Burning tire smoke	Negative	Negative	Negative	Negative

[a] *Aqueous film-forming foam* fire suppression agent.

Sources: References 6 and 7.

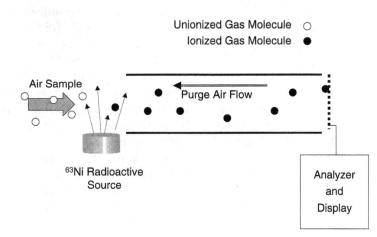

Figure 4.11. Diagram of ion mobility spectrometer (IMS) sensor for chemical warfare agents.

ionizes) chemical molecules in air that is drawn into the unit. These charged molecules move through a tube subject to an electric field until they impact an electrode with an opposite electric charge. A counter current flow of clean air carries away uncharged molecules and particles, and other contaminants in the sampled air. When the charged particles contact the electrode, an electric current is generated that is proportional to the number of molecules and the charge on each molecule. A microprocessor analyzes the electric current patterns and compares them to the profiles generated by known CWAs. When a "match" in patterns occurs, the instrument alarms. These instruments are susceptible to false positives (false alarms) because the software algorithms that compare the patterns must be set with a relatively broad "window" to ensure that no CWAs are missed; this increases the likelihood that other chemicals will trigger a false alarm. Instrument manufacturers continually upgrade the software in their instruments to improve detection capability and reduce false alarms.

• *Flame Ionization Detection (FID)*. This is described in Chapter 11 (see Figure 11.11). This type of device is also considered as a nonspecific detector and thus is susceptible to some false positives. Like the IMS instruments, manufacturers continually update the software to reduce the occurrence of false alarms.

• *Surface Acoustic Wave (SAW)*. As illustrated in Figure 4.12, the operation of a SAW detector is based on a piezoelectric crystal. This type of crystal oscillates (i.e., vibrates slightly) when it is subject to an electric current. The SAW *sensing* crystal is coated with a polymer that absorbs airborne vapors as air passes over the crystal, while a *reference* crystal is uncoated and sealed so no vapors can contact it. The basic SAW detector contains a *sensing* and a *reference* crystal along with electronic circuitry that monitors changes in oscillating frequency between the two crystals. As chemical molecules are absorbed on the coating of the *sensing* crystal, its oscillating frequency changes slightly as compared to the other (reference) crystal; the change in frequency is expressed as chemical concentration. The SAW detector *array* used in CWA instruments consists of several sensing crystals, each with a different polymer coating that absorbs only one specific CWA.

SABRE 2000 IMS Detector. One example of an IMS instrument that was tested in the U.S. Army evaluation program is the SABRE 2000 hand-held trace particle and vapor detector from Barringer Technologies, Inc.[8] It uses a sealed 15-millicurie nickel-63 radioactive source to ionize contaminant molecules. The instrument is capable of detecting HD and nerve agents by using different cartridges for either positive or negative ionization mode. Prior to operation, the appropriate cartridge (positive or negative ionization) is installed in the instrument to detect the substances of interest. The positive ionization cartridge is programmed to detect GA, GB, GD, and GF (different G nerve agents), HN3 (nitrogen mustard), and VX and V_x (two versions of V nerve agent). The negative ionization cartridge is programmed to detect HD (sulfur mustard) or L (lewisite). In addition, the instrument can be used to detect vapor in two different operating modes: "sniff" mode or "preconcentration" mode. In sniff mode, the instrument can draw a vapor sample and monitor surface contamination to produce an analysis in 10 seconds. For particle sampling in sniff mode, a textured surface sampling card is used to swipe a suspected contaminated surface. This sample card is then inserted into a slot of the instrument for thermal desorption. The desorbed vapor is drawn into the IMS cell for analysis after pressing the start button. In the "pre-concentration mode," a vapor sample

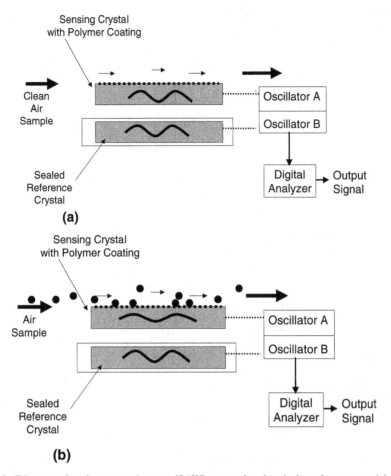

Figure 4.12. Diagram of surface acoustic wave (SAW) sensor for chemical warfare agents: (a) No agent present; (b) chemical agent present.

is drawn directly through the instrument's built-in pre-concentration cartridge for a designated period, which is then heated to drive off (i.e., desorb) the chemical for analysis. The sampling practitioner can adjust the time of pre-concentration of the vapor sample (up to 30 seconds maximum).

The SABRE 2000 weighs approximately 6 pounds (including the batteries), provides both audible and visual alarms, and can identify explosives and narcotics as well as CWAs. It can operate in the temperature range from 32°F to >115°F. A fully charged battery will last approximately 1.5 hours or process up to 80 samples. The SABRE 2000 has the ability to store the results from

analyses in a tabular form or as a graphic IMS spectrum for later retrieval using a computer. When coupled with the computer, it is possible to immediately capture the results and display the IMS spectrum and other sample information.

Data from the U.S. Army tests of the SABRE 2000 for Minimum Detectable Levels, surface contamination detection, and field interferences tests are shown in Tables 4.7 to 4.9. Several different units of the instrument were tested for comparison. The surface contamination tests (Table 4.8) show the results of liquid HD and GB using the SABRE 2000 in the *particle sniff* mode. The limited evaluation indicated that the surface swab swipe was able to produce

TABLE 4.7. Minimum Detectable Level (MDL) for the SABRE 2000 at Ambient Temperatures and 50% Relative Humidity

| | Concentration (ppm) and Response Times | | | |
| | Sniff Mode | | Pre-concentration Mode | |
Agent	Unit A	Unit B	Unit A	Unit B
GA	0.010 in 10 sec	0.010 in 10 sec	0.003 in 40 sec	0.005 in 40 sec
GB	0.010 in 10 sec	0.005 in 10 sec	0.009 in 40 sec	0.005 in 40 sec
HD	—	—	0.08 in 15 sec	0.12 in 15 sec

Source: Reference 8.

TABLE 4.8. Liquid Surface Contamination Detection Response Results for SABRE 2000

| | HD | | GB | |
| | Unit A Response (Alarms/Trials) | Unit C Response (Alarms/Trials) | Unit A Response (Alarms/Trials) | Unit C Response (Alarms/Trials) |
Surface				
Asphalt	No alarm	2/3	3/3	3/3
Concrete	1/2	1/2	2/2	2/2
Military Combat Uniform	1/2	2/2	1/2	1/2
Wood	2/2	2/2	2/2	2/2

Source: Reference 8.

the appropriate detection response when the contaminant was absorbed on the sample swab. On rough or absorbing surfaces such as asphalt and concrete, the ability to swab the contamination is hindered. The amount of sample collected from the swipe is limited when the agent has soaked into the substrate, which can result in failures to yield a consistent detection response.

HAZMATCAD Plus™ SAW Detector. One SAW instrument that was tested in the U.S. Army evaluation program is the HAZMATCAD Plus™ from Microsensor Systems Inc.[9] (Figure 4.13). This device combines an array of three SAW detectors along with electrochemical sensors to respond to a wide range of CWAs and TICs and has dual mode operation for either fast response or high sensitivity. The instrument uses a sample pump to collect and concentrate a vapor sample on the preconcentrator. After sample collection, the trap is heated and the sample is released into the SAW detector array. The preconcentration step significantly reduces the occurrence of false alarms. The HAZMATCAD™ uses four distinct mechanisms to specifically identify CWAs and differentiate them from other gases and vapors: concentrator sorbent material; thermal desorption profile; selective sensor polymer coatings; and pattern recognition software. HAZMATCAD™ will produce both an audible and visual alarm when the preset threshold levels for the CWA detection algorithm are matched. The detector simultaneously detects blister and nerve agents. The instrument has a built-in datalogger and has an RS-232 and infrared input/output port. There is no need for user

TABLE 4.9. SABRE 2000 Responses to Field Interference Testing

	Tests Using the 30-second "Pre-concentration" Mode	
Interferent	Unit A (Negative Mode) 40-Second Response Time Alarms/Trials	Unit C (Positive Mode) 40-Second Response Time Alarms/Trials
Gasoline exhaust, idle	0/3	0/3
Gasoline exhaust, revved	0/3	0/3
Diesel exhaust, revved	2/3[b]	0/3
Gasoline vapor	0/3	0/3
Diesel vapor	0/3	0/3
JP8 (jet fuel) vapor	0/3	0/3
Kerosene vapor	0/3	0/3
AFFF[a] vapor	0/3	2/3[c]
Bleach vapor	0/3	0/3
Insect repellent	0/3	0/3
HTH bleach vapor	3/3[d]	0/3
Burning gasoline smoke	0/3	0/3
Burning JP8 smoke	0/3	0/3
Burning kerosene smoke	0/3	0/3
Burning diesel smoke	0/3	0/3
Burning cardboard smoke	2/3[e]	0/3
Burning cotton smoke	0/3	0/3
Burning wood fire smoke	1/3[f]	3/3[g]
Doused wood fire smoke	0/1	1/1[c]
Burning rubber	1/2[d]	0/2

[a] Aqueous film-forming foam firefighting agent.
[b] False responses of lewisite.
[c] False responses of GA.
[d] False responses of HD.
[e] False responses of NG-N (explosive) and L2 (lewisite).
[f] False response of NGC (explosive).
[g] False responses of GA and GB.
Source: Reference 8.

calibration, but the manufacturer provides a diffusion tube containing a G-agent simulant for a semiquantitative "confidence check." The SAW array has a five-year shelf life.

Several different units of the instrument were tested for comparison. Tables 4.10 and 4.11 show the results of testing for minimum detectable levels and laboratory interferences tests. The laboratory interference tests involved exposing the device to 1% of the headspace (saturated) concentration of the materials listed in Table 4.11 (except for ammonia that was tested at 25 ppm) at ambient temperature and 50% relative humidity. If the detector false alarmed at 1% vapor concentration, it was retested at 0.1% vapor concentration.

General-Use Direct-Reading Instruments.
This category includes the direct reading devices used for industrial hygiene and environmental measurements that are covered in other chapters in this book. Since these instruments are summarized in Chapter 1 and covered in detail in Chapters 9–12, only a brief description is presented here:

• *Electrochemical* sensors have two electrodes in a chemical matrix; the composition of the electrodes and matrix are chosen to sense the chemical of interest. When the compound contacts the chemical matrix, a reaction occurs that changes either the current or voltage between the electrodes; the change is proportional to the amount of the compound. These sensors and devices are covered in Chapter 11.

• *Metal oxide semiconductor* sensors consist of a semiconductor with a coating of a metallic oxide such as zinc, nickel, or tin. When molecules of a contaminant are adsorbed onto the coating, a change in resistance of the semiconductor occurs which is proportional to contaminant concentration. Selectivity is achieved through differ-

Figure 4.13. HAZMATCAD Plus™ surface acoustic wave instrument for chemical warfare agents and other chemicals. (Courtesy of Microsensor Systems, Inc.)

TABLE 4.10. Minimum Detectable Level (MDL) for the HAZMATCAD™ SAW at Ambient Temperatures and 50% Relative Humidity

	Concentration (ppm) and Response Times with Detector in "Fast" Mode	
Agent	Unit A	Unit B
GA	0.032 in	0.032 in
	30–34 sec	42–47 sec
GB	0.14 in	0.14 in
	43–46 sec	82–89 sec
HD	0.056 in	0.056 in
	44–57 sec	69–88 sec

Source: Reference 9.

TABLE 4.11. HAZMATCAD™ Responses to Laboratory Interference Testing

Chemical Tested	Unit A (1% Concentration Level)	Unit A (0.1% Concentration Level)
1% AFFF[a]	No alarm	Not tested
Gasoline	No alarm	Not tested
JP8 (jet fuel)	No alarm	Not tested
Toluene	No alarm	Not tested
Floor wax	No alarm	Not tested
Windex (glass cleaner)	Low G	No alarm
Diesel	No alarm	Not tested
Bleach	No alarm	Not tested
Ammonia (25 ppm)	No alarm	Not tested

[a] Aqueous film-forming foam firefighting agent.

Source: Reference 9.

ent mixtures of oxides and different operating temperatures. Some devices can measure several contaminants by modifying the operating temperature of a single MOS sensor, while others have multiple sensors. These sensors and devices are also covered in Chapter 11.

- *Photoionization detectors* (PIDs) use an ultraviolet (UV) light source to ionize contaminant molecules. Chemical compounds are subject to ionization by UV light according to their ionization potential (IP), expressed in units of electron-volts (eV). Several different UV lamps are available with varying eV intensities; a lamp will ionize any compound with an IP lower than the eV rating of the lamp. These detectors are nonspecific in that the instrument will not indicate which specific compound is in the air. These sensors and devices are covered in Chapter 12.

- *Flame ionization detector* (FID) devices use an oxygen/hydrogen flame to ionize molecules of flammable compounds. These detectors are used in general direct reading instruments as well as the dedicated CWA unit described earlier. Like the PID, FIDs are nonspecific and are covered in Chapter 12.

- *Gas chromatography* (GC) instruments operate on the principle that a volatilized sample is mixed with a carrier gas and injected into a column that separates the components in the sample according to the time it takes the component to travel through the column. As the molecules emerge from the column a detector measures the amount of each material. Field portable GC instruments are covered in Chapter 13.

- *GC/mass spectrometer* (GC/MS) adds a special detector that passes the ionized molecules emerging from the column through an electric field that focuses the molecules by atomic mass. This permits more accurate identification of specific compounds using a library of data developed for the specific instrument. These devices are also covered in Chapter 13.

- *Infrared spectroscopy* is a technique that uses an optical instrument that measures the absorption of "light" in the infrared spectrum by the sample. The general principles of operation are described in Chapter 12. The *fourier transform IR* (FTIR) adds a sophisticated interferometer that adds compound identification capabilities to the basic IR detector. IR and FTIR instrument are also covered in Chapter 13.

The applicability of these general-use direct-reading electronic instruments for any given situation is governed by their ability to detect to the chemical(s) of interest and the lower detection limit for each chemical. These factors are especially critical for CWAs because of their extreme toxicity:

- Instruments that are classified as "nonspecific" above (PID and FID) will respond to many chemicals in the air; a very sensitive device will respond to many non-CWAs in the environment at low levels and give false alarms in a terrorism response. Sometimes the lack of sensitivity can help the responder reach a conclusion; for example, if there is an unknown airborne chemical contaminant present that yields a reading on a PID device but there are no observed adverse effects in unprotected people, the chemical vapor is probably not a pure CWA since people would show adverse effects below the concentrations causing the PID to respond.

- Currently available electrochemical and metal oxide semiconductor sensors cannot respond to very low levels of contaminants. These devices can be useful in TIC releases, but not for CWAs. As an example, the International Sensor Technology IQ-1000 (see Figure 10.7) with Mega-Gas Sensor™ can perform a "gas search" to scan an area for up to 100 different hazardous gases. Each of the gases is placed into one of three gas groups to permit rapid scanning. The instrument searches for each group and provides a relative reading from 0 to 100. A zero reading in all gas groups indicates that none of the gases are present above their limit of detection. The scan data will not reflect the presence or exact concentration of any specific gas; only an indication of whether one or more of the gases in that group is present. If the scan detects that a gas is present, more detailed follow-up monitoring is needed to identify the gas and measure the concentration.

- High-end field portable instruments for multiple specific gases and vapors (i.e., GC/MS and IR or FTIR) have the potential for higher compound specificity and lower limit of detection. These instruments are expensive and sophisticated, and they can be complicated to operate. They are designed to be used with a personal or laptop computer either to operate the instrument or to manage the data. The compound identification and measurement steps involve comparing the airborne sample to stored calibration or library data, and so a computer is essential for this step. They must be specially configured by the manufacturer for the specific compounds to be measured. Finally, despite their sophistication, these are not designed to identify the components of a completely unknown or "mystery" airborne mixture unless the chemicals have previously been stored in the compound library and the instrument has been calibrated with the substance. Figure 4.2 shows a field-portable GC/MS at a response site.

Instrument manufacturers continually develop their technology to add capability for emergency and terrorism response. For example, the AreaRAE™ (Figure 4.14) is a rugged, weatherproof "lunch box"-type multi-sensor equipped with a photoionization detector (PID) for parts-per-million measurement of volatile organic com-

Figure 4.14. The AreaRAE™ is a remote chemical sensor unit with PID and other sensors that can be located up to two miles from the computer controlling the system. (Courtesy of RAE Systems.)

pounds (VOCs), as well as LEL and O_2, and it has up to two electrochemical sensors for measurement of specific toxic substances such as CO and H_2S. It can be equipped with a wireless, RF (radio frequency) modem that allows real-time data transmission with a base controller located up to two miles away from the detector. Any personal computer can be used as the base station for an AreaRAE system; the software used to control AreaRAE systems is capable of monitoring the input of up to 16 remotely located sensor units. One or more units can be located throughout a very large area to sense the presence of chemicals at low concentrations.

With some direct reading instruments the "effective" lower limit of detection can be decreased (i.e., its sensitivity can be increased) by preconcentrating the air sample before analysis. With this step the ambient air is sampled with a pump and adsorption tube that traps the contaminants from a relatively large air sample in a solid sorbent. For analysis the contaminant is "blown" from the adsorber unit into the instrument by a jet of gas (often accompanied by heat) so the "sample" that is analyzed has a higher concentration of contaminant than the ambient air. This type of preconcentration step using certain equipment is referred to as *Solid-Phase Micro Extraction (SPME).*

Sample Collection Device Methods.

Sample collection device (SCD) methods involve collecting the air sample in the field using a device that is returned to a laboratory for analysis. For most occupational and environmental sampling the laboratory is a fixed "analytical laboratory" with bench-scale equipment. For emergency and terrorism response there are field and mobile laboratories where smaller or full-size versions of the analytical instruments are mounted in a vehicle (e.g., truck or trailer) or field-portable shelter. This gives sophisticated analytical capability without the

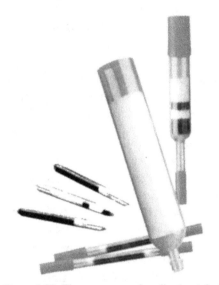

Figure 4.15. The proper sample collection tube for gas and vapor sampling depends on the chemical, the required minimum detection level, and the analytical technique to be used in the laboratory. (Courtesy of SKC, Inc.)

inherent drawback of SCD methods of the delay in getting samples back to the laboratory.

Monitoring for gases and vapors using sample collection devices are covered in Chapters 5 and 6. For *active sampling* a pump is used to draw the air through a device that captures the contaminants. Usually gases and vapors are collected in adsorption tubes containing a *sorbent* charcoal (activated carbon), silica gel, or another sorbent (Figure 4.15). A typical tube contains two sections of adsorbing material separated by an inert spacer. As contaminated air is drawn through the tube, the airborne chemical is adsorbed (deposited on the surface) of the adsorbent particles in the front section. If the front section becomes saturated with the contaminant molecules, some of the contaminant molecules "breakthrough" the front layer and are collected on the backup section. At the laboratory each section is analyzed separately; if significant breakthrough has occurred, the results will be questionable and, if possible, the sampling should be repeated. Liquid-filled *absorption* devices such as impingers or bubblers (see Figure 1.7) can be used where there are no suitable solid sorbent methods, but the need to handle liquid sampling media in the field make these techniques less desirable. For a few chemicals, specially treated filters are used for monitoring certain reactive chemicals; the airborne chemicals react with the compounds on the filter to form other chemical compounds that are stable and can be measured in the laboratory. Another choice is to collect samples in special sampling bags made of relatively inert materials such as Tedlar™ or Teflon™ (see Figure 6.6). In the laboratory the air samples can be concentrated if needed before analysis. Evacuated cylinders (see Figure 1.5) can also be used in some cases. However, when measuring very low concentrations of contaminants the bags and evacuated cylinders may not provide sufficient sample volume for low-level analysis.

Another type of sampling device for gases and vapors is a passive monitor (see Figure 1.3). It is typically a small badge-like device that contains an adsorbent similar to the sampling tubes. Sampling occurs without a pump due to diffusion of contaminant molecules into the badge where the molecules adhere to the sorbent. For emergency and terrorism response the sampling time required to collect enough chemical for analysis may be a disqualifying factor for these devices.

Monitoring for particles and droplets using sample collection devices are covered in Chapter 7. The term *aerosol* is used to describe fine solid or liquid matter suspended in a gas, usually air. Often the term *particle* is also used for convenience even though the aerosol may actually consist of a liquid droplet since its behavior resembles a discrete particle. The two biggest factors of interest with an aerosol are generally its chemical makeup and particle size distribution. The size of a particle is a very important characteristic since it governs how an aerosol behaves when airborne and where it deposits in the human body when inhaled. See Table 2.2 for a list of different types of particulate matter and their sizes.

For occupational exposure situations the size categories are usually related to where in the respiratory tract the particles will deposit: *inhalable* (anywhere in the respiratory tract), *thoracic* (lung airways and lower lung), and *respirable* (lower lung gas-exchange region). For environmental samples the size criteria are based on particles size such as *particulate matter—10 microns* (PM_{10}).

There are several choices when air sampling for aerosols:

- Total sample methods use a filter or other device to collect all of the particulate matter for analysis.
- Size-segregating sampling devices, such as a cyclone (see Figure 7.15), can be used to separate and discard particles outside the size range of interest. Only particles within the size range of interest are collected on the filter.
- Inertial impactors contain different stages that collect particles of a specific size range (see Figure 7.16). With this type of sampler the filter from each stage can be analyzed separately to determine the quantity and chemical composition of particles in each size range.

At a response site the use of a high-volume sampling pump and filter is the most convenient way to collect air samples for aerosol analysis (Figure 4.16). This allows a sufficiently large sample to be collected in a short time to identify agents and estimate initial hazard. However, the respiratory hazards over the longer term and also away from the immediate site of the incident are almost completely due to very small particles that will remain airborne and disperse through the air. Thus it may be necessary to

Figure 4.16. High-volume air sampler and filter used for aerosol sampling. (Courtesy of F&J Specialty Products, Inc.)

have the capability to collect size-selective samples using cyclones or inertial impactors. For *all* size-selective devices the sampling flowrate must be maintained within a narrow range to achieve proper size differentiation; for this reason the sampling pump and sample collection device must be selected together.

In addition to the information on sample collection device methods in Chapters 5–7, the sampling practitioner preparing for emergency or terrorism response should consider several points. The first point is that coordination with key staff at the laboratory that will be analyzing the samples is critical. They will need to specify the sampling devices to be used and the minimum sample volume needed to detect chemical agents at low levels. These parameters will partially depend on the analytical equipment that is available. They should also detail the storage and shipping methods for the samples.

The second point to consider is that because sampling at terrorism events is designed to measure very low levels of airborne chemicals, standard sampling protocols may need to be adjusted to avoid possible loss of sample while collecting sufficient sample volume. Two special situa-

tions to be considered during the planning phase are:

• Sampling rate to collect the minimum air volume quickly: Sampling pumps must be selected to draw a high enough flowrate so the minimum sample will be collected over the allowable time period. At an event there is an inherent need to obtain quick information about the possible agent(s) and the level of each. It may be necessary to select high-flowrate sampling pumps so that the required minimum air volume is collected after 30 minutes or one hour (see Figure 5.3). This will allow the initial samples to be returned to the laboratory for analysis.

• Sample loss from the collection device: As mentioned earlier, collection media can become saturated with chemicals (or water vapor in high humidity environments); at this point, breakthrough occurs. With sorbent tubes the chemical molecules that break through are captured on the backup section. In high concentration environments it is possible for the backup section to also become saturated, thus allowing some molecules to pass through the sampling tube completely. When the sampling device is an impinger, bubbler, or impregnated filter, unless a separate device is used in series as a backup, saturation will cause complete loss of part of the contaminant. Sampled material can also be lost from the collection tube through a mechanism known as *air stripping*. If air containing a chemical contaminant is sampled, and then clean air is passed through the tube for an extended period of time, some of the chemical molecules may be purged from the tube by the clean air. For normal occupational and environmental monitoring, the sampling can be repeated if sample loss or excessive breakthrough occurs. However, this is not an option for terrorism and emergency response. Preventive measures that should be considered are using large-size sampling devices and also using multiple devices connected in series so that any contaminants that break through one collector will be captured on a subsequent collection device.

AIR SAMPLING FOR BIOLOGICAL AGENTS

Chapter 16, entitled *Sampling for Bioaerosols*, covers sampling principles and devices for biological agents. The term *bioaerosols* means airborne particles derived from microbial, viral, and related agents. Some important considerations for bioaerosol sampling are as follows:

- Collect bioaerosols intact so they can be grown in a culture medium, or identified or counted using direct microscope observation.
- Some bioaerosols are collected directly onto a growth media such as an agar plate or nutrient-treated filter.
- One may need to avoid or account for the possible deposition of multiple bioaerosols directly on top of each other for analytical techniques that involve counting the resulting colony forming units (CFUs). This may require very short sampling times (possibly less than one minute for heavy concentrations).

Air sampling for biological agents has the advantage of concentrating the very low concentration of bioaerosols that often exist in the ambient environment, thus increasing the likelihood that the analysis will detect the presence of the bioaerosols. Aggressive air sampling techniques have been developed for recent anthrax responses that model EPA guidance for

clearing facilities for reoccupancy after asbestos decontamination. With this procedure, while the area is under negative pressure, all surfaces are aggressively agitated and air is continuously kept in motion using a fan while samples are collected. An air sampling method that maximizes the likelihood of detecting contamination should be used.

Air sampling and analytical guidelines for specific agents are periodically updated by the U.S. Centers for Disease Control (CDC) and the Occupational Safety and Health Administration (OSHA); consult their web sites for the latest information. As an example, according to U.S. OSHA the following techniques are available and have proven useful for air sampling for anthrax: gelatin filter (low-volume sampling); mixed cellulose ester (MCE) filter; Anderson air sampler and single-stage impactors; open agar plate; liquid impingers; and a dry filter unit (high-volume air sampler with polyester 1-µm filter).

OSHA Method 7[24] covers air sampling for anthrax spores. *Bacillus anthracis* spores in air are collected using a special dry filter unit that consists of a high-flow air sampling pump that collects airborne spores on 47-mm polyester filters with a 0.1-µm particle collection efficiency of 75% (PEF-1 filters).[25] After collection, each filter either is transferred into a 50-mL conical tube or is left in the sample filter holder and transferred into a 50-mL conical tube in the laboratory. Once out of the contaminated area, the outside of each bag containing the samples is disinfected. The sample must be kept away from UV light, including sunlight. Samples are transported at ambient temperature. When shipping the samples to the laboratory, all current federal and state regulations with regard to shipping infectious waste must be followed. In the laboratory the samples are prepared and an assay is conducted using standard biological techniques under Biosafety Level (BSL) 2 practices. The sample is prepared in a manner that will yield a suspension; aliquots of the suspension may be assayed by different microbiological techniques depending on the sensitivity needed. Traditional culture techniques provide the most sensitivity (2 CFU per sample) while a more rapid polymerase chain reaction (PCR) assay has a detection limit of 200 CFU per sample.

There are numerous research and development efforts to develop real time detectors for biological agents. These include many of the sampling devices described in Chapters 7 and 14 (filters, impactors, cyclones, etc.) combined with a rapid method, such as polymerase chain reaction (PCR), to identify the type and amount of biological agent that is present.

AIR SAMPLING FOR RADIOLOGICAL HAZARDS

There are many local, state, and federal agencies concerned with radiological hazards from a terrorist event or nuclear accident. These agencies can be an excellent source of information, and many offer on-site assistance in case of an incident such as 24-hour radiological assistance teams, aircraft and helicopters with radiation monitoring capability, and dispersion modeling capability. Localities with nuclear power plants or significant use of radioisotopes for medical, research, or educational use are expected to have comprehensive incident response plans in place as part of their licensing requirements.

The basic sampling tool for airborne radioactive materials is a high volume sampler with a filter to collect the airborne radioactive particles as shown in Figure 4.16. The filters can be counted onsite with a hand-held survey instrument such as a Geiger–Mueller counter, which with the proper probe (such as a *pancake* probe) responds to alpha, beta, and gamma radiation (Figure 4.17). There are portable gamma spectroscopy instruments for

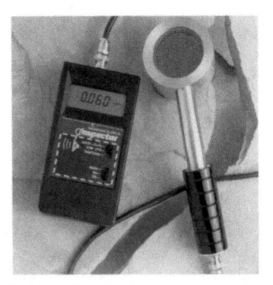

Figure 4.17. Radiation survey meter equipped with a "pancake probe" to detect alpha, beta, and gamma radiation. (Courtesy of S.E. International, Inc.)

radioactive material identification. Filters can also be sent to a laboratory for more detailed analysis including identification of the various isotopes that are present.

Gaseous radioactive iodine (^{129}I and ^{131}I) from a nuclear power plant release or detonation of a fission device is collected using special charcoal canisters in a high-volume sampling pump (Figure 4.18) for laboratory analysis. Both isotopes of iodine emit beta particles upon radioactive decay; iodine-129 has a half-life of 15.7 million years while iodine-131 has a half-life of about 8 days. Because of the relatively short half-life of ^{131}I, the date and time that the sample is analyzed relative to both the date and time of the incident and when the sample was collected may be important data elements in evaluating the hazard from the ^{131}I.

Sampling devices such as filters and charcoal canisters concentrate the airborne radioactivity. As a result the potential dose from sample collectors may be quite high. It is important to follow established procedures when handling and transporting radioactive samples so that other responders, shipping personnel, and laboratory staff are not exposed.

Figure 4.18. Canisters for use with a sampling pump to collect air samples for radioactive iodine. (Courtesy of F&J Specialty Products, Inc.)

SUMMARY

This chapter presented an overview of air sampling and analytical techniques for emergency incidents and terrorism events. Typical hazards are chemical warfare agents, toxic industrial chemicals, biological agents, and radioactive materials. Air sampling is only one facet of the response to these events, but air monitoring can provide important information about the identity of the agent(s) involved, exposure levels, and the extent that harmful agents have dispersed. For suspected terrorism events, air sampling may help to determine on whether a terrorism event has actually occurred.

There are inherent challenges in sampling for very toxic chemical warfare agents and harmful biological agents because their airborne concentration may be extremely low and the analytical technique must be capable of identifying the specific agent(s) without interference from other airborne chemicals or bioaerosols. Accurate determination of chemical warfare agents at very low levels with a direct reading electronic instrument requires specialized devices such as a dedicated ion mobility spectrometer or surface acoustic wave (SAW) monitor, or a field-portable GC/MS or similar sophisticated instrument.

Instrument manufacturers continue to improve the technology for these applications. Contact equipment suppliers, and check the U.S. Army and other evaluation test results, when considering a specific instrument.

REFERENCES

1. Second Annual Report of the Advisory Panel to Assess Domestic Response Capabilities for Terrorism Involving Weapons of Mass Destruction, *Toward a National Strategy for Combating Terrorism*, Arlington, VA: RAND, 2000.

2. Hawley, C., *Hazardous Materials Air Monitoring and Detection Devices*, Albany, NY: Delmar–Thomson Learning, 2002.

3. DHHS (NIOSH) Publication No. 2003–136, *Guidance for Filtration and Air-Cleaning Systems to Protect Building Environments from Airborne Chemical, Biological, or Radiological Attacks*, Department of Health and Human Services, Centers for Disease Control and Prevention, National Institute for Occupational Safety and Health, April 2003.

4. Statement of D. A. Henderson, M. D., Director, Office of Public Health Preparedness, Department of Health and Human Services before the Committee on Science, United States House of Representatives, December 5, 2001.

5. Testimony of Dr. Henry Kelly, President, Federation of American Scientists before the Senate Committee on Foreign Relations, March 6, 2002.

6. *Testing of Commercially Available Detectors Against Chemical Warfare Agents—Summary Report*, U.S. Army Soldier Biological and Chemical Command, Aberdeen Proving Ground, MD, February 1999.

7. *Testing of MSA Detector Tubes Against Chemical Warfare Agents—Summary Report*, U.S. Army Soldier Biological and Chemical Command, Aberdeen Proving Ground, MD, July 2000.

8. *Testing of Sabre 2000 Handheld Trace and Vapor Detector Against Chemical Warfare Agents—Summary Report*, U.S. Army

Soldier Biological and Chemical Command, Aberdeen Proving Ground, MD, January 2003.

9. *Testing of HAZMATCAD Detectors Against Chemical Warfare Agents—Summary Report*, U.S. Army Soldier Biological and Chemical Command, Aberdeen Proving Ground, MD, February 2002.

10. *Testing of APD2000 Chemical Warfare Agent Detector Against Chemical Warfare Agents—Summary Report*, U.S. Army Soldier Biological and Chemical Command, Aberdeen Proving Ground, MD, August 2000.

11. *Testing of M90-D1-C Chemical Warfare Agent Detector Against Chemical Warfare Agents—Summary Report*, U.S. Army Soldier Biological and Chemical Command, Aberdeen Proving Ground, MD, December 2000.

12. *Testing of Miran SapphIRe Portable Ambient Air Analyzers Against Chemical Warfare Agents—Summary Report*, U.S. Army Soldier Biological and Chemical Command, Aberdeen Proving Ground, MD, July 2000.

13. *Testing of UC AP2C Portable Chemical Contamination Control Monitor Collective Unit Against Chemical Warfare Agents—Summary Report*, U.S. Army Soldier Biological and Chemical Command, Aberdeen Proving Ground, MD, May 2001.

14. *Testing of CAM Chemical Agent Monitor (Type L) Against Chemical Warfare Agents—Summary Report*, U.S. Army Soldier Biological and Chemical Command, Aberdeen Proving Ground, MD, August 2001.

15. *Testing of RAE Systems ppbRAE Volatile Organic Compound (VOC) Monitor Photo-Ionization Detector (PID) Against Chemical Warfare Agents—Summary Report*, U.S. Army Soldier Biological and Chemical Command, Aberdeen Proving Ground, MD, September 2001.

16. *Testing of SAW MINICAD MKII Detector Against Chemical Warfare Agents—Summary Report*, U.S. Army Soldier Biological and Chemical Command, Aberdeen Proving Ground, MD, September 2001.

17. *Testing of the Agilent GC-FPD/MSD (Gas Chromatograph–Flame Photometric Detector/Mass Selective Detector) System Against Chemical Warfare Agents—Summary Report*, U.S. Army Soldier Biological and Chemical Command, Aberdeen Proving Ground, MD, October 2002.

18. *Testing of the VAPORTRACER Against Chemical Warfare Agents—Summary Report*, U.S. Army Soldier Biological and Chemical Command, Aberdeen Proving Ground, MD, May 2002.

19. *Testing of the IMS2000^{TM} (Bruker Daltonics GmbH Ion Mobility Spectrometer 2000) Against Chemical Warfare Agents—Summary Report*, U.S. Army Soldier Biological and Chemical Command, Aberdeen Proving Ground, MD, July 2003.

20. *Testing of Photovac MicroFID Handheld Flame Ionization Detectors Against Chemical Warfare Agents—Summary Report*, U.S. Army Soldier Biological and Chemical Command, Aberdeen Proving Ground, MD, October 1999.

21. *Testing of the Scentoscreen Gas Chromatograph Instrument Against Chemical Warfare Agents—Summary Report*, U.S. Army Soldier Biological and Chemical Command, Aberdeen Proving Ground, MD, August 2002.

22. *Product Literature*, Nextteq, LLC, Tampa, FL, 2003.

23. Committee on R&D Needs, Health Science Policy Program, Institute of Medicine and Board on Environmental Studies and Toxicology, Commission on Life Sciences, National Research Council, *Chemical and Biological Terrorism—Research and Development to Improve Civilian Medical Response*, Washington, D.C.: National Academy Press, 1999.

24. Method No. OSA7, *Bacillus Anthracis Spores (Etiologic Agent of Anthrax) In Air*, U.S. Occupational Safety & Health Administration, Washington, D.C., 2003.

25. *Product Literature*, New Windsor, NY: American Felt and Filter Company, 2003.

PART II

SAMPLE COLLECTION DEVICE METHODS FOR CHEMICALS

CHAPTER 5

INTRODUCTION TO MONITORING USING SAMPLE COLLECTION DEVICES

What distinguishes *sample collection device* (SCD) methods from other sampling techniques is the requirement for a laboratory analysis of the sampling medium. In the past, SCD samples were sometimes referred to as *integrated samples* since the airborne concentration was averaged or integrated over the sampling period. However, many direct reading instruments have data storage features that permit integrating exposures over the monitoring period, so the term *integrated* is no longer specific to SCD methods.

This chapter builds on the general description of methods and equipment in Chapter 1 and provides the information that applies to all SCD monitoring—that is, sampling for gases, vapors, and particulate matter. Subsequent chapters cover these topics in more detail.

These methods are used for both occupational and environmental monitoring; the primary difference between the two is that environmental methods generally need to collect very low levels of contaminants when sampling ambient outdoor air. For occupational exposures the usual sampling period is 8–10 hours or less, and the sampling equipment must be portable and light enough so workers can actually carry the equipment with them. For environmental exposure monitoring the equipment is generally left in a single location for the whole sample duration, and so its size and weight are less important. The sampling devices are usually larger so that sampling can cover a longer period or have a higher sample airflow rate so lower levels of contaminants can be quantified. For gases and vapors the typical sampling processes are: adsorption using a sorbent tube or diffusion "badge"; collection in a liquid medium using a bubbler or impinger; or collection in a bag or evacuated container for laboratory analysis of the sampled air. Some type of filter is the most common SCD media for particulate matter, with liquid-filled impingers used for special applications.

For particulate matter sampling, a major

Air Monitoring for Toxic Exposures, Second Edition. By Henry J. McDermott
ISBN 0-471-45435-4 © 2004 John Wiley & Sons, Inc.

factor in field sampling is often the size distribution of interest:

- Total sampling methods collect the entire mass on one filter or other device.
- Size-selective methods either (a) collect the entire mass in a manner that allows the size distribution to be determined (usually by collecting different size ranges on separate "stages" of the collector) or (b) collect only certain size fractions and discard the rest of the particles.

REVIEW OF THE METRIC SYSTEM

The *metric system* consists of the units and terms used in many scientific applications including exposure monitoring for measurements of weight (mass), volume, concentration, air flowrate, and other parameters. This system is useful because it is based upon factors of 10; that is, units can be related to other units by multiplying or dividing by factors of 10. Since the background of air sampling practitioners varies, this section provides an overview of the metric system.

Weight or Mass

In the metric system, the weight of a substance is expressed in units based upon the gram (g). For reference, there are about 454 g in a pound of any material. However, for exposure monitoring, the primary unit of interest is the milligram (mg); there are one thousand milligrams in a gram. The microgram (μg), which is one millionth of a gram, is also used for very small quantities; this unit is often seen on laboratory analytical reports. These units can be related using Table 5.1 below based upon factors of 1000.

TABLE 5.1. Converting Between Units of Weight in the Metric System

To Convert From	To	Do This
mg	g	Divide number of mg by 1000
mg	μg	Multiply number of mg by 1000
μg	mg	Divide number of μg by 1000

TABLE 5.2. Converting Between Units of Volume in the Metric System

To Convert From	To	Do This
mL	Liters	Divide number of mL by 1000.
Liters	mL	Multiply number of liters by 1000.
Liters	m^3	Divide number of m^3 by 1000.
m^3	Liters	Multiply number of m^3 by 1000.
mL	cc	Nothing, they are equivalent.

Volume

The basic unit for volume in the metric system is the liter (L). One liter is about the same volume as a quart. Another important unit is the milliliter (mL)—one thousandth of a liter, or about the volume of a fifth of a teaspoon. One mL is exactly the same volume unit as the cubic centimeter (cc). A third volume unit used in exposure monitoring is the cubic meter (m^3). A meter is about the same as a "yard" (3 feet); a cubic meter is the volume in a cube with a meter length on all sides. There are 1000 liters in a cubic meter. Table 5.2 shows how to convert from one unit to another unit using factors of 1000.

Concentration

There are two main types of units for airborne concentration of a substance: (a) parts per million and (b) milligrams per cubic meter. Parts per million (ppm) are used for vapors and gases and indicate the number of parts of a substance per million parts of contaminated air. Milligrams per cubic meter (mg/m^3) is used for particulate matter (dusts, fumes, and mists) and also gases and vapors. The unit of mg/m^3 indicates the amount of a substance (in units of weight) per cubic meter of air. Other units used for concentration are parts per billion (ppb) and micrograms per cubic meter ($\mu g/m^3$). There are 1000 ppb in 1 ppm and 1000 $\mu g/m^3$ in 1 mg/m^3.

These units can be related to each other using Table 5.3. The relationship between ppm and mg/m^3 uses the constant of *24.45* and also the molecular weight of the substance. The molecular weight (MW) is different for each substance and is an indication of the relative size and weight of the molecules for that substance. It can be found on Material Safety Data Sheets (MSDSs) or handbooks for chemicals.

Flowrate

The units of flowrate are used to express the volume of air being pulled by a pump per unit of time. The two most common units are: liters per minute (Lpm) and milliliters per minute (mL/min). An mL/min is exactly the same flowrate as a cc/min. There are 1000 mL/min in a Lpm. These units can be converted from one to the other using factors of 1000 as shown in Table 5.4.

METHOD SELECTION

SCD methods used for the air sampling of specific compounds are developed by NIOSH, OSHA, and the EPA. For occupational exposures the NIOSH *Manual of Analytical Methods*[1] is the most complete source of sampling methods. Table 1.3 shows an excerpt from an OSHA method for benzene. The selection of an SCD air sampling method depends on the physical and chemical characteristics of the air contaminant plus other factors relevant to the sampling conditions. Factors include expected concentrations, temperature, high humidity, and the presence of interfering substances. If sampling is being done for compliance purposes, the regulatory agency may specify the method to be used. The availability of laboratory analysis,

TABLE 5.3. Converting Between Units of Concentration

To Convert From	To	Do This
ppm	ppb	Multiply number of ppm by 1000.
ppb	ppm	Divide number of ppb by 1000.
mg/m^3	$\mu g/m^3$	Multiply number of mg/m^3 by 1000.
$\mu g/m^3$	mg/m^3	Divide number of $\mu g/m^3$ by 1000.
mg/m^3	ppm	Multiply number of mg/m^3 by $\dfrac{(24.45)}{MW}$
ppm	mg/m^3	Multiply number of ppm by $\dfrac{MW}{(24.45)}$

TABLE 5.4. Converting Between Units of Flow

To Convert From	To	Do This
mL/min	Lpm	Divide number of mL/min by 1000.
Lpm	mL/min	Multiply number of Lpm by 1000.
mL/min	cc/min	Nothing, they are equivalent.

suitability of the sampling method to the environment to be sampled, complexity of the method, and cost of the analysis are also important.

Prior to sampling, the minimum and maximum sample volumes allowed by the method must be determined. Collection of the minimum volume is necessary in order to collect enough contaminant to meet the detection limit of the method and measure concentration levels relevant to the acceptable exposure level (AEL) or other exposure standard. Not exceeding the maximum volume is necessary to avoid either (a) loss of contaminant from the sample through overloading or (b) collecting so much contaminant that the analytical method is no longer valid. When high sensitivity is desired, higher sampling rates or longer sampling times should be used, but the sampler must generally stay within the limits of the maximum sampling volume and flowrate of the method.

Often the minimum sample volume is stated in the sampling method. For example, from Table 1.3 the OSHA Method 12 (Charcoal Adsorption Tube) for benzene contains the following information:

- Recommended air volume and sampling rate for standard charcoal tubes containing 100 mg of charcoal in the front section and 50 mg of charcoal in the backup section of the tube: 10-L sample collected at 0.2 L/min. Air sample size is limited by the number of milligrams that the tube will hold before overloading.
- The limit of detection for this analytical procedure allows measurement of an air concentration of 0.04 ppm for a 10-L air sample.

This information allows a quick calculation of that a sample time of 50 minutes (10 L divided by 0.2 L/min) will permit detection of 0.04 ppm. For shorter sampling time the minimum LOD will be proportionately higher. For example, the LOD for a 15-minute STEL for benzene is approximately 0.15 ppm. For higher levels of benzene, the sampling rate can be reduced or a larger charcoal tube can be used.

Breakthrough information is contained in the NIOSH *Manual of Analytical Methods—Method 1501 for Aromatic Hydrocarbons*.[2] This document shows that breakthrough can occur with a standard charcoal tube when sampling more than 45 L of air containing about 47 ppm of benzene. Since the preferred sample volume is only 10 L, and the concentration should be closer to the AEL of 1 ppm, breakthrough should not be an issue.

PUMPS AND OTHER SAMPLING EQUIPMENT

SCD sampling, except for passive methods, requires a pump or other air mover to draw the ambient air through the collection device. Sampling pumps fall into two broad categories: (a) personal pump that can be "worn" by an individual and (b) high-volume pumps used for area samples. Evacuated cylinders or other containers are another method of drawing air for SCD sampling.

Personal Sampling Pumps

These are battery-powered pumps that draw between 5 mL/min and 5 L/min. Smaller pumps in this category (up to about 500 mL/min) have traditionally been referred to as "low-flow" pumps (Figure 5.1), while devices with higher air flow rates were called "high-flow" pumps. Since many pumps are now available that can operate over the full range of sampling rates, the low- and high-flow distinctions are less valid now (Figure 5.2).

Many personal pumps operate by means of a moving diaphragm that generates a

Figure 5.1. Easy-to-use low-flow battery-powered sampling pump. (Courtesy of SKC, Inc.)

Figure 5.2. Universal sampling pumps cover both low- and high-flow (5–5000 mL/min) ranges. (Courtesy of SKC, Inc.)

pulsing airflow pattern. Most pumps use some technique to smooth out the pulsations to produce near-constant airflow. The exception is pumps that operate on a stroke principle: These use a stroke counter to measure airflow. Stroke pumps are calibrated in terms of airflow/pump stroke, and the stroke counter reading is recorded in the field for later calculation of sample volume.

Key selection factors for personal pumps include:

- *Flow Range:* Typically 50–200 mL/min for low-flow pumps, 20–80 mL/min for ultra-low-flow pumps, and 5 mL/min to 3–5 L/min for multirange pumps.
- *Pressure Capability:* Pumps need a certain suction capability to overcome the resistance to airflow through filters and other sample collection devices. The resistance to airflow, also called *back pressure*, increases as the filter becomes loaded with dust. Personal pumps usually have a back pressure rating of 20–40 inches of water gauge at typical flow setting.
- *Constant Flow Control:* More sophisticated pumps have internal pressure sensors and circuitry that allows the pumps to maintain a constant airflow as the back pressure increases. Some pumps also adjust for changes in temperature and atmospheric pressure that affect sampling rate. Often the adjustment is an increase in pump motor speed as the back pressure increases. Pumps without this feature must be checked regularly during air sampling, especially when sampling particulates, to ensure that the airflow has not changed; small decrements in airflow are corrected by manual adjustments to pump speed.

- *Multi-tube Sampling Capability:* Some sampling conditions require use of several colorimetric or adsorption sampling tubes in parallel. The tubes are mounted in a manifold, usually with a suction control for each tube. Pumps with adequate back-pressure capability and constant flow control can be used for multi-tube sampling.

- *Flow Fault Feature:* This feature tracks periods of reduced flow during the sampling and warns when set criteria have been exceeded. This alerts the sampling practitioner to possible invalid samples due to reduced airflow through the pump caused by kinked tubing or a blocked SCD inlet.

- *Programmable Features:* Internal circuitry that allows a wide variety of features such as automatic start and stop times, security codes to protect data, automatic calibration using a electronic calibrator, and links to a personal computer for recordkeeping.

- *Pump Display:* More sophisticated pumps can display various parameters such as flowrate, sample volume, run time, time of day, battery status, temperature, atmospheric pressure, and flow fault on a liquid crystal display (LCD). Larger size pulseless flow pumps may have a small rotameter (flow indicator) mounted on the pump.

- *Battery Type:* Nickel cadmium (NiCad) rechargeable batteries are typical with some pumps using nickel metal hydride (NiMH) batteries for longer sampling duration. Some simple pumps use disposable alkaline batteries.

- *Battery Charging System:* Manufacturers offer charging stations that handle a single pump or several pumps at once. More complex systems discharge NiCad batteries completely to avoid the "memory effect" that shortens service life, then charges them quickly and reverts to a trickle charge to maintain a full charge until the next use.

- *Safety and Approvals:* Pumps are available that are intrinsically safe and that are listed or approved by various agencies or similar bodies.

High-Volume Pumps

These pumps are either battery-powered or operate from line current that are used for environmental sampling and indoor air quality studies (Figure 5.3). Some have rugged, weather-resistant cases for outdoor use, while others are suitable for indoor or protected environmental sampling only. Key selection parameters include:

- *Flow Range:* Typically up to about 30 L/min, although some draw 250 L/min or more.

Figure 5.3. High-volume pump for environmental sampling. (Courtesy of SKC, Inc.)

- *Pressure Capability:* These pumps usually have a back pressure rating up to about 90 inches of water gauge at typical flow setting.
- *Constant Flow Control:* More sophisticated pumps maintain a constant airflow as the back pressure increases. Some high-volume pumps can generate enough suction to maintain constant flow using a *critical orifice* (as explained below).
- *Noise Level:* Some high-volume pumps are rotary vane pumps (i.e., similar to a laboratory vacuum pump), while others operate using a moving diaphragm. Rotary vane pumps are loud, which may restrict their use indoors or in noise-sensitive areas unless enclosed in a noise-attenuating enclosure.
- *Weight:* These pumps are used for fixed sampling, so weight is not as critical as with personal sampling pumps. Most pumps weigh 15 pounds or less; however, the larger pumps can weigh up to 75 pounds. Battery-powered pumps may use a lead-acid battery similar to an automobile battery, which can add considerable weight.

Evacuated Canisters

This sampling method involves drawing a vacuum on a suitable container and then closing container's flow valve to maintain the vacuum (Figure 1.5). For sampling, the valve is opened in the sampling location to draw ambient air into the container for later laboratory analysis. The valve is closed at the end of the sampling period. As used in this context, the technique uses a special stainless steel canister that may have additional internal surface treatment depending on the compounds and concentrations of interest. For long-term samples a critical orifice or other flow control device is used to regulate the airflow into the container so that the flowrate is constant over the sam-

pling period. The flow controller could be plugged by particulate matter so the technique is only suitable for gases and vapors unless the air stream is filtered. There are several container sizes available up to about 15 L. Container size depends on sampling time; a 6-L volume usually is sufficient for a 24-hour sample.

UNDERSTANDING THE CRITICAL ORIFICE

The *critical orifice* is a basic flow control technique that can be used when adequate suction is available. Some high-volume pumps and evacuated containers can provide sufficient suction; personal sampling pumps generally cannot.

Critical Orifices

As used in air sampling, the critical orifice is a simple metal hole inserted downstream of a sampling device and upstream of the sampling pump. Its purpose is flow regulation. The term *critical* means that the air in the small nozzle throat is moving at the speed of sound. Critical orifices will pass a constant flowrate once the downstream pressure is low enough to reach critical flow even if the downstream pressure is reduced further. Devices of this type have long been used for controlling the flowrate through filter holders, cascade impactors, and impingers.

When the pressure downstream from a critical orifice is about 50% of the pressure immediately upstream, the velocity through the orifice reaches the speed of sound. Once the speed of sound has been reached, creating a lower vacuum will not increase the air speed and the volume of air moving through the nozzle will be independent of further vacuum changes.

Critical orifices are sold as separate units and also incorporated into sampling devices. Several sizes of critical orifices are available; nominal flowrates used for air sampling are 2–15 L/min. Each size represents a different flowrate and the flowrate is often stamped on the unit by the manufacturer. However, flowrates for critical orifices must be determined by calibration, because they may differ from the manufacturer's specifications. When used with other devices or filter media, the flowrate will be different and should also be checked with a rotameter or other calibration device.

Although the critical orifice is an attractive device for maintaining a constant flow rate though a sampling instrument, certain limitations exist. The suction pump must have the ability to maintain a particular negative pressure at the desired flowrate. This information can be ascertained by consulting the manufacturer. If these restrictions are strictly observed, the critical orifice is a very economical and simple technique for maintaining constant airflows.

CALIBRATION DEVICES

Since the air sample volume must be known in order to calculate the airborne contaminant concentration, air sampling pumps are calibrated before use so that the flowrate through the pump is accurately known. The flowrate is used to calculate the volume of air sampled (volume equals flowrate times sampling period). Additionally, some sampling devices such as impingers and respirable dust cyclones are designed to operate at a specified flowrate. For these devices, calibration is necessary to adjust the sampling pump to the proper flowrate.

Pumps are generally calibrated before and after sampling, with the entire sampling "train" assembled as it will be or was used in the field. The average flowrate (before and after sampling) should be used to calculate the sample volume. Differences of more than 5% in flowrate between the pre- and post-sampling calibrations indicate a problem that should be investigated and resolved.

Calibration devices that involve timing the flow of a known volume of air through the pump, such as the bubble burette, are referred to as *primary standards*. These are inherently less subject to error than calibration devices that rely on other measurement principles. Nonprimary calibration devices are calibrated against a primary standard to verify their accuracy. Additionally, critical orifices must be calibrated to verify that the flowrate specified by the manufacturer is accurate. The calibration procedures for the various devices are discussed later in this chapter.

Several devices are used to calibrate pumps. The accuracy of each of the following devices varies with manufacturer and model; consult the manufacturer's literature for exact figures. They all are sufficiently accurate for most air sampling for toxic exposures.

Soap Film Flowmeter. All pumps can be calibrated using a soap film method, also called a "bubble burette." This calibration procedure is based upon using a burette of known volume and measuring the time it takes for the pump to move a soap film in the burette over a specific volume. Because this approach relies on a volume which can be physically measured, it is a "primary standard." There are several different-size commercial devices available (Figure 5.4), or a laboratory burette turned upside down may be used.

Automated Soap Film and Piston Flowmeters. This type of calibrator resembles an automated "bubble burette" or operates on a similar principle. In a typical unit the user releases a soap film which moves past infrared sensors in a tube that measures the time duration as the film traverses a known distance. Other models feature a very low friction piston to replace the bubble film. Because the flowrate is determined electronically, this method provides better precision and is faster compared to the bubble

Figure 5.4. Bubble burette used to calibrate air sampling pumps. (Courtesy of SKC, Inc.)

Figure 5.5. Calibrating a sampling pump with an electronic calibrator. (Courtesy of SKC, Inc.)

burette, and in some cases the calibration data can be stored for electronic record-keeping. This method can be used to calibrate any pump in the flowrate range of the automatic calibrator. A typical model can measure 1–6000 mL/min for personal-type pumps, while a larger calibrate can measure up to 30 L/min. This method should not be affected by altitude. Some automated calibrators can be connected to personal computers to provide added functionality in setting pump flowrates, checking calibration, and maintaining calibration records (Figure 5.5).

Rotameter. A rotameter (Figure 5.6) is a portable calibration device in which air passes upward through a internally tapered glass or plastic tube (the top being wider than the bottom) that contains a ball or other-shaped "float." The ball stabilizes at

the position where the weight of the ball is balanced by the velocity of the upward flowing air. A printed scale on the tube corresponds to the airflow through the tube for each ball position. Some devices have two balls; each ball corresponds to a separate printed scale on the tube. This procedure can be used to calibrate any continuous-flow (nonpulsating) pump if the proper-size calibrated rotameter (flowmeter) is available; rotameters reading up to 5.0 L/min are commonly available. This is not a primary standard; airflow should be checked periodically against a primary standard. The rotameter should be recalibrated every 3 months and when it is used at altitudes above 2000 feet following the procedure described later in this chapter.

The accuracy of calibrated rotameters can depend upon the following factors:

- The rotameter must be held in a vertical position during use. A canted position may cause "drag" between the float and tube wall.
- Dirt or other contamination on the float or inside walls of the tapered tube and cracks or leaks will affect the accuracy.

Dry Gas Meters. This type of meter consists of two bags interconnected by

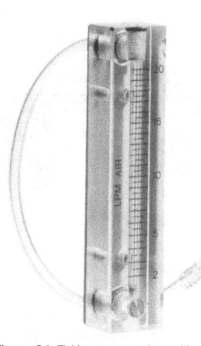

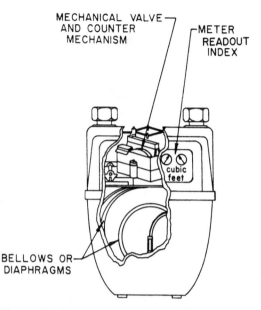

Figure 5.7. Dry gas meter. (From *The Industrial Environment—Its Evaluation and Control*, NIOSH, Washington, D.C. 1973.)

Figure 5.6. Field rotameter for calibration of constant-flow sampling pumps. (Courtesy of SKC, Inc.)

mechanical valves and a cycle counting device (Figure 5.7). The air or gas fills one bag as the other bag empties. When the cycle is complete, the valves switch and the second bag fills as the first one empties. Air or gas containing particulates should be filtered before the meter to prevent damage or premature wear. When dry gas meters are used in pressurized systems, the piping valves should be opened slowly to permit the bags to become equalized with the pressure in the piping. Abrupt fluctuations in pressure or vacuum can damage the operating mechanism. Dry gas meters are useful for preparing calibration mixtures of gases and vapors as they can be used to deliver accurate volumes to Tedlar bags. This device is a secondary standard and must be calibrated using a primary standard such as a spirometer.

Electronic Flowmeters. Electronic mass flow monitors are hot wire devices in which measurement of the flow is based on

thermal conduction. The temperature of the airstream depends on the mass rate of flow and the heat input. Some meters have heated wires and some have heated thermistors and thermocouples. Mass flowmeters are incorporated into high-volume air sampling apparatus and are also available as portable instruments for field use. The readout is in actual flow units, so no extrapolation or calculations are required. If properly calibrated against a bubble burette so that their accuracy is within ±3%, these units provide ease of use in the field, including a wide range of flowrates with a single instrument; they have a wide operating temperature range; they are battery-powered; and the miniaturized versions are very portable. Some are now available with digital readouts.

Tylan mass flowmeters are used in conjunction with high-flow sampling apparatus for environmental aerosol collection. In these sensors the temperature rise of a gas is a function of the amount of heat added, the mass flowrate, and the properties of the gas being uses. To measure the airflow rate, instruments generally use small stainless steel sensor tubes. Two external, heated resistance thermometers are wound around the sensor tube, one upstream and one downstream. When gas is flowing, heat is transferred along the line of flow to the downstream thermometer, thus producing a signal proportional to the gas flow. The greater the flow, the greater the differential between the thermometers.

If volumetric flowrate is needed, then the following calculation is necessary.

$$Q_{act} = Q_{ind}\left(\frac{d_s}{d_a}\right)$$

where Q_{act} is the actual flow, Q_{ind} is the indicated flow, d_s is the air density at standard conditions of 25°C and 760 mm Hg, and d_a is the actual air density inside the transducer.

The calibration on these instruments must be checked periodically. It can be done with a bubble burette; however, if an instrument requires substantial adjustment, it is better to send it to the manufacturer who can use sources such as a wind tunnel to calibrate. Adjustment of the calibration involves opening up the unit and adjusting potentiometers used to zero and span the various range scales on the meter. For use, most of these instruments require proper orientation so that air always flows in a specified direction. Some units incorporate "flow straighteners" on the inlet to eliminate turbulence in the airstream.

Continued use in dirty environments may necessitate periodic cleaning of the sensor. Flowmeters based on thermal conductivity are affected by dust and reactive gases, but in a different manner than rotameters and wet test meters. Corrosion will alter the electrical conductivity and lead ultimately to failure of the element, and coating with inert material will reduce heat transfer and ultimately reduce the response of the instrument. Chlorides and lead are especially destructive. These units should never be used in flammable or explosive atmospheres. Mass flow sensors should not be placed in line following bubble burettes since moisture will affect these units.

CALIBRATION PROCEDURES

Pumps

Personal pumps are small and portable, which allows their use for personal exposure monitoring. The calibration method that is used for these pumps depends primarily on whether the flow through the pump is continuous or pulsating:

- Pumps that have built-in pulse dampening devices and essentially continuous flow can be calibrated with a rotameter or other direct-reading flowmeter. See the procedure entitled "Calibration with a Rotameter" in this section. When using a rotameter, it must be periodically calibrated to ensure it is accurate. See the procedure entitled "Calibration of a Rotameter."
- Personal pumps with a piston pump without dampening have a flow with a definite pulsing pattern and cannot be calibrated with a rotameter. The rotameter ball moves up and down with each piston cycle, and the average flowrate cannot be determined. For this type of pump, an electronic automated calibrator or "bubble burette" is used for calibration. See the procedures later in this section.

High-volume pumps are calibrated either with an electronic automatic calibrator, with a mass flow electronic meter, or with an orifice calibration system in which the pressure drop through an orifice or constriction in the flow line is proportional to the airflow. U.S. EPA regulations may specify the required calibration technique.

For battery-powered pumps, the batteries should be fully charged before use or calibration. Allow a fully charged pump to operate for 10–20 minutes prior to calibration to allow the batteries to discharge their peak voltage.

Calibration with a Burette. This bubble burette procedure is required for pumps with pulsating flow pattern, but is suitable for any pump if the burette size permits accurate timing of the soap film travel. Different-size burettes are available, permitting calibration at airflows of 35 mL/min to 6.0 L/min.

Procedure

1. Assemble the pump, sampling device (charcoal or other tube), and burette as shown in Figure 5.8. Use the same type of sampling device as will be used for actual monitoring. "Bubble solution" sold as a child's toy is a convenient soap solution to use. If a charcoal tube is used, it should be changed periodically since charcoal may become saturated by carryover of moisture from the bubble solution, thereby increasing airflow resistance.

2. Start the pump and introduce a soap film into the burette by briefly touching the surface of the soap solution to the bottom of the burette. Do not keep the burette opening submerged in the soap solution since a slug of solution may be drawn into the adsorber tube or pump. Repeat until the burette walls are wet and the soap film travels the length of the burette without breaking. As an alternative, the burette can be inverted and filled with liquid to wet the walls.

3. Adjust the flowrate control on the pump to the approximate setting for the desired flowrate as specified in manufacturer's literature. The goal is

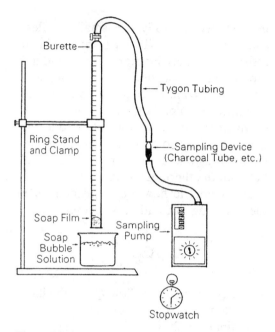

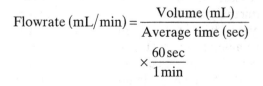

Figure 5.8. Pump calibration with a bubble burette.

to get within about 10% of the desired flowrate.

4. Perform the following steps to determine the flowrate in milliliters per minute (mL/min). Figure 5.9 shows the Pump Calibration Worksheet, which can be used to record the information from the steps below.

   • Using a stopwatch, measure the time for the soap bubble to travel a given volume (say 100 mL). Repeat this three times, recording the time to the nearest tenth of a second, and average the three readings.

   • Calculate the flowrate using the formula below:

$$\text{Flowrate (mL/min)} = \frac{\text{Volume (mL)}}{\text{Average time (sec)}} \times \frac{60\,\text{sec}}{1\,\text{min}}$$

Example: The time for the bubble to travel 100 mL on each of three repetitions was 30.3 sec, 29.8 sec, and 31.4 sec.

Pump: _____

Sampling Device: _____

Pump Setting: _____.

<u>Bubble Burette Procedure:</u>

Step 1: Record the volume the soap bubble traveled: _____ mL

Step 2: Using a stopwatch, time how long it takes for the soap bubble to travel the volume indicated in Step 1. Repeat this step three times and record below.

        A:_____ sec   B: _____ sec   C:_____ sec

Step 3: Calculate the average time by summing the three times in Step 2 and divide by three.

$$\text{Average Time} = \frac{A + B + C}{3}$$

$$\text{Average Time} = \frac{(\_\_\_\_) + (\_\_\_\_) + (\_\_\_\_)}{3} = _____ \text{ sec}$$

Step 4: Calculate flowrate.

$$\text{Flowrate} = \frac{\text{Volume Traveled}}{\text{Average Time}} \times \frac{60 \text{ sec}}{\text{min}}$$

$$\text{Flowrate} = \frac{_____ \text{ mL}}{(_____ \text{ sec})} \times \frac{60 \text{ sec}}{\text{min}} = _____ \text{ mL/min}$$

Name: _____   Date: _____

Figure 5.9. Data sheet to record pump calibration using a bubble burette.

$$\text{Average time} = \frac{30.3 + 29.8 + 31.4}{3}$$

$$\text{Average time} = 30.5 \text{ sec}$$

$$\text{Flowrate (mL/min)} = \frac{100 \text{ mL}}{30.5 \text{ sec}} \times \frac{60 \text{ sec}}{1 \text{ min}}$$

$$= 196.7 \text{ mL/min}$$

5. Calibrate the pump before and after the monitoring period. Average the two flowrates (i.e., before and after) and use this average flowrate to calculate the sample volume as follows:

$$\text{Sample volume} = \text{Average flowrate} \times \text{Monitoring time}$$

Calibration with an Automated Fixed-Volume Calibrator. Prior to use, electronic bubble meters should be cleaned. This involves removing the flow cell and gently flushing with tap water and wiping with a clean cloth. If a stubborn residue persists, the bottom plate can be removed from these units for additional cleaning. Often the flow tube is made of a resin material that can easily be scratched. These devices use a sterile soap solution to prevent residue buildup in the flow tube. Generally, these burettes should not be stored for longer than one week with soap solution in the flow cell. When transporting electronic bubble meters, remove the soap

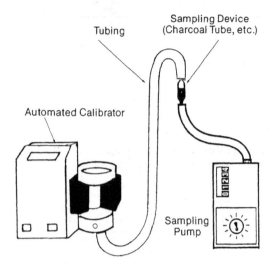

Figure 5.10. Pump calibration with an electronic calibrator.

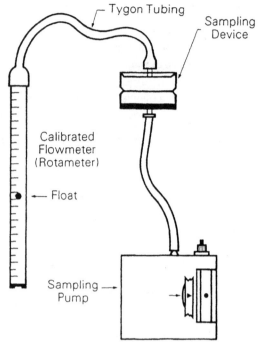

Figure 5.11. Pump calibration with a rotameter.

solution and, for air shipment, ensure that the inlet and outlet ports are not sealed to prevent a buildup of pressure.

Procedure

1. Assemble the pump, sampling device, and the automated calibrator as shown in Figure 5.10. Use the same type of sampling train as will be used for the actual monitoring.

2. Adjust the flowrate of the pump to the approximate setting for the desired flowrate.

3. Start the pump and initiate each calibration run by depressing the start button.

4. Make enough runs to ensure that the calibrated flowrates indicated on the readout are consistent; that is, the readings should not vary by more than 5%. Three calibration readings within 5% should be used to calculate the average flowrate.

5. Calculate the average flowrate as follows:

$$\text{Average flowrate} = \frac{\text{Run 1} + \text{Run 2} + \text{Run 3}}{3}$$

6. Use this final average flowrate to calculate the volume pulled by the pump as follows:

$$\text{Sample volume} = \text{Average flowrate} \times \text{Monitoring time}$$

Calibration with a Rotameter

Procedure

1. Assemble the pump, sampling device (filter, tube, impinger, etc.), and proper capacity rotameter as shown in Figure 5.11. Use the same type of sampling device that will be used for the actual sampling. Where the design of the sampling device, such as a respirable dust cyclone, does not allow in-line connection of the rotameter, use the "jar" technique illustrated in Figure 5.12.

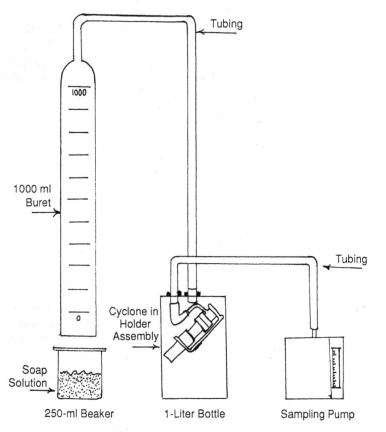

Figure 5.12. "Jar" technique for in-line calibration of a respirable dust cyclone.

2. Start the pump and adjust the pump flow setting control according to manufacturer's instructions to give the desired reading on the rotameter (read the center of the float). A rotameter must be in the vertical position to measure properly.

3. Take the rotameter to the field and use it to check the flowrate. If the "final" flowrate is different from the "initial" calibration, use the average of the two to calculate the volume sampled.

Other Flow Devices

Calibration of a Rotameter. All rotameters should be checked for proper calibration about every 3 months. Dirt and other contamination can build up on the walls, affecting the accuracy of the rotameter. Leaks and cracks in the rotameter will also affect the accuracy. The rotameter should be checked with a "primary" calibration method such as a *bubble burette* to ensure that it is still in proper calibration.

Changes of altitude or barometric pressure can affect the float position on a rotameter and the accuracy of the calibration. Rotameters used above 2000 feet must be calibrated initially prior to use and about every 3 months.

Some rotameters have the scale that reads directly in flowrate units (e.g., mL/min). Note that this type of scale does not apply for rotameters used above 2000-feet altitude; in these cases the rotameter must be used with a calibration chart that relates the scale reading to the actual airflow rate.

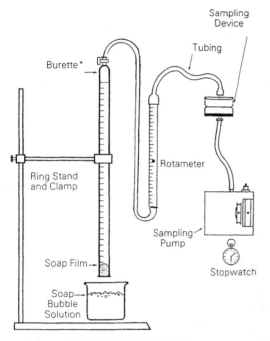

Figure 5.13. Calibration of a field rotameter.

Procedure

1. Assemble the calibration setup as shown in Figure 5.13: the pump, the sampling hose, the sampling device, the rotameter, and the burette.

2. Set the flowrate of the pump at 0.5 L/min using the rotameter.

3. Using a soap solution, "wet" or coat the inside of the burette.

4. Using a stopwatch, determine the time in seconds required for the soap bubble to travel a predetermined volume: 500 mL.

5. Repeat Step 4 two more times and determine the average time.

6. Calculate the actual flowrate (volume ÷ time).

7. Plot the rotameter reading on the vertical axis and the bubble meter flowrate on the horizontal axis of the graph on the Rotameter Calibration Worksheet shown in Figure 5.14.

8. Repeat Steps 2–7 at different airflows that cover the range of the device. Draw a line through the data points on the graph. Figure 5.15 shows an example of the resulting graph.

9. If the difference between the rotameter flowrate versus bubble meter flowrate is greater than 10%, the flowmeter should be cleaned and recalibrated.

10. To use the graph to set the pump at a particular flowrate, follow these steps:
    - Find the desired flowrate on the horizontal axis (actual flowrate from burette).
    - Follow this line up to where it intercepts the plot line and over to the vertical axis (rotameter reading).
    - Read the value on the vertical axis to indicate the setting for the rotameter to achieve the desired pump flowrate.

11. This procedure can be accomplished using an automated calibrator in place of the burette in Figure 5.13. Follow Steps 1–10 to check the rotameter with the automated calibrator.

Example: In order to collect a respirable dust sample using a cyclone (1.7 L/min) at 5000-feet altitude, the rotameter was calibrated according to instructions above. The data are plotted on the graph in Figure 5.15. To use the graph, locate 1.7 L/min (1700 mL/m) on the horizontal axis (bubble meter). Follow this line up to where it intercepts the plot line. Follow the intercept line to the left to determine the indicated flowrate on the vertical axis. This shows 1.4 L/min (1400 mL/m) on the rotameter. Thus, to set the pump at 1.7 L/min, the rotameter should read 1.4 L/min.

Rotameter Calibration Worksheet

Rotameter Ser. No. _____   Sampling Device _____

Date _____   Name _____

Rotameter Setting (ml/min.)	Volume Pulled By Pump (ml)	Time (Seconds) 1st Trial	Time (Seconds) 2nd Trial	Time (Seconds) 3rd Trial	Average Time[1]	Flowrate[2]
500						
1000						
1500						
1700						
2000						

(1) Average time = (1st time + 2nd time + 3rd time)/3 = secs.
(2) Actual Flowrate = vol. pulled by pump ÷ ave. time × (60 secs./min.) = ml/min.

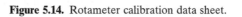

Figure 5.14. Rotameter calibration data sheet.

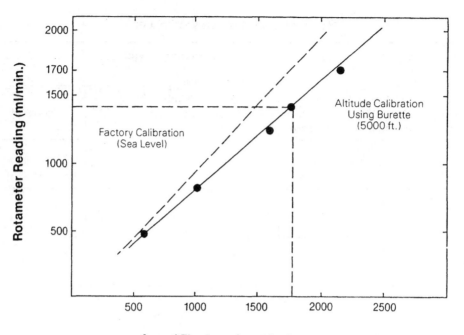

Figure 5.15. Example rotameter calibration plot.

Calibration of a Critical Orifice

Procedure

1. Select a critical orifice, note its number, and hook it via tubing to a vacuum source with a manometer in line. Attach the other end of the orifice to a bubble burette. The vacuum source and the manometer must be downstream from the critical orifice.

2. Turn on the vacuum gradually and increase its flow rate in increments, and measure the amount of time in seconds for the bubble to transverse the burette. This action should create a critical flow through the orifice.

3. Calculate the flowrate of each trial as related to increasing downstream pressure. Record pressure, temperature, and flow rate.

4. The flowrate of air should increase in a linear fashion with the increasing downstream pressure, up to the point of sonic velocity. At this point, the flowrate will stabilize and any further increase in vacuum will not affect the stable flowrate.

SAMPLE IDENTIFICATION AND CHAIN OF CUSTODY

A process is needed to ensure that the identity and integrity of the sampling devices is maintained during all handling by the sampling practitioner, shipping service, analytical laboratory, and others who handle the sample. Generally, all sampling devices (tubes, cassettes, impingers) should be labeled with a unique identification code before placing them in the

Figure 5.16. Use of duplicate numbered labels to clearly identify sampling device and data sheet.

sampling train. The method for numbering depends on the complexity of the sampling handling process; a simple method is to use the date, sample practitioner's initials, and a sequential number for the sample collected on that date (e.g., 031204-CHF-05 would be the fifth sample collected by CHF on March 12, 2004). If sampling records from different facilities in the same company are stored in a central database, it may be necessary to add a location code to the sample identification to ensure that each sample designation is unique. The sample identification code is then entered on the sampling record, sent along with the SCD to the analytical laboratory, and used in summary reports. Another option for assigning sample numbers is to purchase two-part adhesive labels with the same number on each label. One label goes on the device and the other is pasted on the sampling data record (Figure 5.16).

Maintaining sampling integrity can range from simply keeping the sample device intact and avoiding mix-ups to establishing a legally defensible chain-of-custody procedure involving tamper-proof adhesive seals, transfer of custody forms, and other records. Any sample that could be used for evidentiary purposes should meet the appropriate chain-of-custody standards. Some regulatory agencies may have similar requirements for samples collected to register a pesticide, to demonstrate compliance with an abatement order, or for other regulated purposes.

Even with a largely informal sample integrity system, adhesive seals can play an important role in preventing sample contamination. Seals can keep the end caps from coming off of sorbent tubes and can keep the plugs from falling out of the inlet and outlet of a filter cassette. Seals minimize the likelihood of leakage from vials used to ship impinger liquids. If there is any chance that the solvent vapors from the seal's adhesive compound could interfere with the sample analysis, have the laboratory evaluate possible solvent contamination of the sample device before using the seals. Note that it is generally not good practice to ship bulk samples of contaminant in the same shipping container as the air samples to avoid contaminating the SCD, even if the bulk sample container is taped or sealed closed.

DOCUMENTING EXPOSURE MONITORING

Adequate documentation of monitoring is as important as using the proper equipment and procedures to collect the sample. The results will have lasting value only if the sampling conditions and work activities are recorded along with the numerical result. Current OSHA standards require that exposure measurements and the background information needed to interpret sample results be retained for 30 years and be made available upon request to employees and their representatives.[3]

This section discussed the data elements that may be recorded to document the monitoring. Where an organization uses a comprehensive database to store these records, a detailed sampling record form with key elements (such as sampling device and analytical procedure) coded to help

Workplace Exposure Monitoring Record

	Sampling Time (24 Hour Clock)	
	Start	End

Sample Number | | | | | | | | |

Location

Location Description _____   Facility Name: _____

Sample Collected by Name _____

Sampling Medium Used (Select Only One) [ ]

F = Filter   Y = Cyclone and Filter   C = Charcoal Tube   S = Silica Gel Tube   A = Other Adsorption Tube   D = Detector Tube
P = Passive Organic Vapor Badge   B = Other Passive Badge   I = Impinger/Bubbler   R = Direct Read Instrument
N = Noise Dosimeter   L = Sound Level Meter   O = Other (Describe Below)

Collector No. _____   Describe _____

Sampling Information

Sample Date | | | | | | | |
Month   Day   Year

Sample Type [ ]
P = Personal   A = Area   R = Source
B = Blank   S = Spike   K = Bulk

Purpose of this Sample [ ]
T = TWA   P = Peak   S = Partial
(for personal samples only)

Substances Sampled/Results

Substance Name	L/G	Result
		•
		•
		•

Concentration [ ] mg/m³   [ ] µg/m³   [ ] ppm   [ ] ppb   [ ] Fibers/cc

[ ] Other | | | | | | |

Average Flowrate | | | • | | | [ ] cc/Min.   [ ] L/Min.

For Pumps with Stroke Counter:

Final Count | | | | | | | Net Count | | | | | | |

Sample Volume | | | | • | | | [ ] cc   [ ] L

Initial Count | | | | | | | cc/Stroke | | | • | | |

Job Description (Personal Samples Only)

Name (First-MI-Last) _____   ID: _____

Employee? [ ] Yes [ ] No   If No, Name of Company | | | | | | | | | | | | | | | | | | | |

Job Code _____   Job Title _____

Work Period (24 Hours: 0001-2400)   Start Time | | | | Stop Time | | | |

Protective Equipment Used (Personal Samples Only) (Use Respirator Code for Air Samples, Hearing Protection Code for Noise Samples)
| | | | Other Protection (Describe) _____

Sampling Conditions

Operating Conditions [ ] (Select One)   Source of Contaminant: _____

N = Normal   A = Abnormal   W = Worst Case Conditions   U = Unknown

Abnormal Conditions [ ] (Select no more than two)

E = Operating Equipment Malfunction   P = Process Stopped or Interrupted During Shift   T = Turnaround/Shutdown Operation
S = Startup   M = Major Spill Occurred   J = Job Rotation Occurred During Shift
U = Unusual Weather Conditions (Describe Below)   O = Other (Describe Below)

Estimated frequency that these conditions occur. (%) | | |

Exposure Measured On This Day Was [ ]   N = Normal   L = Lower Than Normal   H = Higher Than Normal

Task Description _____   Comments _____

Other Comments   Void Sample: [ ] Yes [ ] No   (Explain reason for voiding sample in comments)

Figure 5.17. Exposure monitoring data sheet suitable for electronic data storage.

achieve consistency will allow efficient data sorting and analysis (Figure 5.17). A simpler form can be used with a manual records system (Figure 5.18).

When conducting a focused monitoring campaign to evaluates exposures to a single substance or during specific operations throughout the organization, it is often

INDUSTRIAL HYGIENE
SAMPLING FORM

		PAGE 1 OF 1	SAMPLED BY KLM	SAMPLE NO. 200

EMPLOYEE (NAME & NO.) Jack Johns	PLANT/FACILITY Printed circuit fabrication	ACCOUNT NO. 123-47	DATE 9/29/98

JOB CLASSIFICATION Technician	CONTAMINANT 1,1,1-Trichloroethane	TYPE SAMPLE ▶ [X] PERSONAL [ ] AREA [ ] SOURCE

AREA SAMPLED
Fabrication Room including solvent spray cleaner and general work area

WEATHER CONDITIONS	WIND. NA	TEMPERATURE. NA
	BAROMETER. NA	HUMIDITY. NA
SOURCE DISTANCE	TO GROUND NA	TO EMPLOYEE NA

TIME STARTED	WORK TASKS	TIME AT TASK
8:00-9:55A	Sample 1 – Cleaned 4 boards	1hr,55min
9:55-12:05P	Sample 2 – Cleaned 6 boards	2hr,10min
12:05-2:00P	Sample 3 – Cleaned 3 boards	1hr,55min
2:00-4:00P	Sample 4 – Cleaned 7 boards	2hr
2:13	Peak sample – duplicate sampling to Sample 4 while cleaning one board	12min

INSTRUMENT TYPE & NO. Low Flow		FLOW RATE 40 cc/min	VOLUME SAMPLED --
CALIBRATED BY KLM	DATE 9/29/98	METER READING INITIAL: -- FINAL: --	DETECTOR TUBE READING --

REMARKS: Strong odor of solvent around cleaner unit.

SKETCH AREA: SHOW TEST LOCATIONS, PERSONNEL, SOURCES, WIND DIRECTIONS, VENTILATION, NORTH DIRECTION.

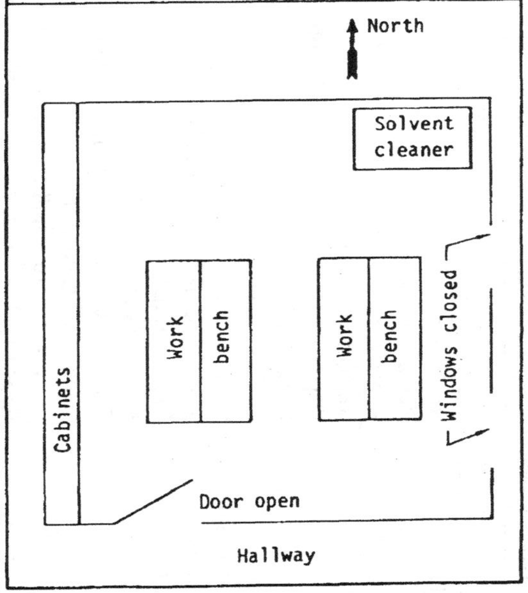

North

Cabinets

Work bench

Work bench

Windows closed

Solvent cleaner

Door open

Hallway

LAB ANALYSIS			DATE 10/2/98	BY Peters
	PPM	mg/m3		
See below				

REMARKS: Sample 1 – 127 ppm Peak = 243 ppm
Sample 2 – 153 ppm
Sample 3 – 98 ppm
Sample 4 – 175 ppm

Figure 5.18. Example of a simple exposure monitoring data sheet.

147

TABLE 5.5. Example Work Tasks to Achieve Consistent Data in Monitoring Campaign

Equipment	Task
Service station gasoline dispenser—inside of cabinet	Replace filter
	Replace strainer
	Replace diaphragm
	Replace meter seal
	Repair/replace supply tubing or coupling
	Repair/replace piping or union
	Calibrate meter
Service station gasoline dispenser—outside of cabinet	Replace hose
	Replace breakaway shut-off valve
	Replace nozzle
	Replace swivel
	Drain product from vapor hose (Stage II vapor recovery)
Gasoline storage tanks	Gauge tank/check for water
	Drain water from spill containment area
	Repair/replace drain valve for spill containment area
	Install or replace fill adapter
	Install tank identification tag
	Remove level gauge
	Repair/replace vapor recovery poppet valve
	Repair/replace vent cap
	Repair/replace check valve
	Repair/replace sump pump/utility cover gasket
	Check sumps for product/water
	Install/repair fill cap

helpful to customize the data form by giving specific definitions and codes for expected tasks and exposure conditions. For example, Table 5.5 shows typical tasks encountered when evaluating exposures to benzene in a petroleum operation. Having the sampling practitioner use these specific codes when completing the sampling data form allows the results to categorized for later sorting and decision making.

Typical Data Elements

This section describes the data elements on Figure 5.17.

Sample Identification Code: A code assigned by the air sample practitioner to uniquely identify the sample. The same code should be used on the sample collection device and the data form.

Location Information: Sufficient information so the sampling location can be identified.

Sample Practitioner's Name and/or Identification Number: This allows identification of the person who collected the sample.

Sampling Medium Used: The type of sampling device used to collect the sample.

Collector No.: If appropriate, the identifying number of the sampling device—for example, the serial number of the respirable dust cyclone, passive dosimeter, and so on.

Collector Description: A detailed description of the sampling device *when appro-*

priate (e.g., the solution used in an impinger, the type of direct reading meter, the type of "other passive badge," the type of filter).

Sample Date: The date the sampling was performed. Use the date the sampling started if the sampling period extends beyond midnight.

Sample Type: Indicate if the sample was personal, area, source sample, blank, spike, or bulk.

- *Personal:* A sample collected to evaluate an individual's exposure. This is normally collected in the individual's breathing zone (1 foot from nose/ mouth).
- *Area:* A sample collected to evaluate the concentration of a substance in a particular area. Generally, this is *not* an accurate indication of an employee's exposure when working in the area.
- *Source:* A sample collected to determine the concentration of a substance at the source—for example, near a pump responsible for releasing a vapor into the work area. This type of sample is normally only collected to evaluate the effectiveness of controls—that is, the concentrations "before" and "after" controls are installed.
- *Blank:* A blank sampling device should be submitted with each group of samples submitted for analysis. A sample data form should be completed for blank samples. A blank should be a sampling device which is from the same lot as the fields samples, but no air is drawn through the device.
- *Spike:* A sampling device which has had a known amount of a substance placed on the sampling media. These samples can be submitted with samples to evaluate the laboratory's analysis procedure.

- *Bulk:* A sample of material located in a work area being evaluated. These samples may be analyzed to determine the presence of specific substances in the bulk sample. A sample data form should be completed to document the location from which the sample was collected.

Purpose of This Sample: For personal samples only, indicate the purpose of the monitoring.

- *TWA:* The sample was collected to evaluate the individual's full shift TWA exposure.
- *Peak:* The sample was collected to evaluate the individual's 15-minute (STEL) exposure or the 30-minute asbestos excursion limit.
- *Partial Shift:* A sample was collected to evaluate a job task, or other portion of a shift when a certain operation is taking place.

Monitored Time Period: The length of time the sample was collected (in minutes).

Substances Sampled/Results: The chemical name for each substance monitored (e.g., benzene, dicloroethane, carbon tetrachloride), and the result for each substance monitored.

Concentration Units: The desired units for the results.

Average Flowrate: The average flowrate calculated from the flowrate at the beginning of the sample and the flowrate at the end of the sample. These should be based upon calibration information.

Pumps with Stroke Counters: The initial reading on the stroke counter at the start of the monitoring period, the final count at the end of the monitoring period, and the net count. Also enter the cc/stroke calibration for the pump.

Sample Volume: The sample volume determined from the average flowrate and the monitoring time period as follows:

$$\text{Sample volume} = \text{Average flowrate}$$
$$\times \text{ Monitoring time}$$

Example: The sample volume for a pump with an average flowrate of 1.7 L/m and a monitoring time of 370 minutes would be

$$\text{Sample volume} = 1.70\,\text{L}/\text{min} \times 370\,\text{min}$$
$$\text{Sample volume} = 629\text{L}$$

Volume Units: The volume units as either liters or cubic centimeters (same as mL).

Personal and Job Information (Personal Samples Only):

- *Name:* The name of the individual sampled.
- *Employee Identification Number:* Enter the individual's identification number.
- *Company Employee:* Indicate if the employee is a company employee. If the individual is not a company employee, enter the name of the contracting or other company for which the person works.
- *Position Title:* The job title performed by the individual on the day of the monitoring, even if this was not the employee's normal job title.
- *Work Period:* The work period for the employee on the day of the monitoring even if this was not the employee's normal work period. Use the 24-hour clock format.

Protective Equipment (Respirators) Used: It is critical to record the use of any respiratory protective equipment for personal samples. Often the terms *air sampling* and *exposure monitoring* are used almost interchangeably. This is only valid when the individual is not wearing a respirator. If respirator use is not recorded, any high measured values could be misinterpreted as overexpo-

sures when in fact the employee was adequately protected. The protection afforded by a respirator depends on three factors:

- The face-to-facepiece seal that prevents inward leakage of contaminated air into the respirator facepiece as the individual inhales. The assigned protection factor is an indicator of the sealing efficiency of a respirator. Table 5.6 shows typical examples of assigned protection factors.[4-7]
- For air-purifying respirators (that remove contaminants from ambient air through filters or adsorption cartridges), the efficiency of the cartridge in removing contaminants. It may be useful to record the type of cartridge, including filter designation.
- Whether the respirator use followed a comprehensive respiratory protection program, including employee training, proper fit testing, and maintenance of the equipment. A brief review or evaluation of the respiratory protection program is usually included in the summary report of the exposure monitoring.

Table 5.7 shows typical respirator codes that can be listed on the sampling data form. These codes contain enough detail to permit identification of the assigned protection factor.

Other types of protection worn can also be described (e.g., gloves, if monitoring was performed for a substance that can be absorbed through the skin).

Sampling Conditions:

- *Operating Conditions:* The type of operating conditions on the day the monitored was performed.
- *Frequency of Occurrence:* The frequency that these type of conditions

TABLE 5.6. Assigned Protection Factors for Respiratory Protective Devices

		Assigned Protection Factors		
Respirator	NIOSH	OSHA Benzene Standard	OSHA Lead Standard	OSHA Asbestos Standard
Dust mask	5			
Half facepiece air-purifying respirator with appropriate gas/vapor cartridge or particulate filter	10	10 (with organic vapor cartridges)	10 (with high-efficiency particulate air filters[a])	10 (with high-efficiency particulate air filters[a])
Full facepiece air-purifying respirator with appropriate gas/vapor cartridge or high-efficiency particulate filter (HEPA, N100, R100, P100)	50	50 (with organic vapor cartridges)	50 (with high-efficiency particulate air filters[a])	50
Powered air-purifying respirator with tight-fitting full facepiece and appropriate gas/vapor cartridges or filters	50	100 (with organic vapor cartridges)	1,000 (with high-efficiency particulate air filters[a])	100
Full facepiece supplied-air respirator in pressure demand or other positive-pressure mode	2,000	1,000	2,000	1,000
Full facepiece supplied-air respirator in pressure demand or other positive-pressure mode with auxiliary positive-pressure self-contained breathing apparatus	10,000	1,000+	2,000+	1,000+
Full facepiece self-contained breathing apparatus in pressure demand or other positive-pressure mode	10,000	1,000+	2,000+	—

[a] Equivalent to N100, R100, or P100 filters certified by NIOSH under 42 CFR 84.

Source: References 4–7.

Example: A respirator with an APF of 10 can be used in atmospheres up to 10 times the allowable exposure level so long as the levels do not exceed the immediately dangerous life or health (IDLH) level for that substance.

TABLE 5.7. Typical Respiratory Protection Codes for Sampling Records

Code	Description
Air Purifying	
AFE	High efficiency—full face
AHE	High efficiency—half face
ALF	Low efficiency—full face
ALH	Low efficiency—half face
APH	Half facepiece—other
APF	Full facepiece—other
Powered Air-Purifying	
PAT	Tight-fitting
PAY	Loose helmet, Tyvek
PAP	Loose helmet, plastic
PLE	Low-efficiency filter
Supplied Air	
SPP	Positive pressure
SNP	Negative pressure
SCF	Continuous flow—full face
SCH	Continuous flow—helmet
SPS	Positive pressure/escape bottle
Self-Contained Breathing Apparatus (SCBA)	
CBP	Positive pressure
CBN	Negative pressure
No Respirator	
OOO	No respirator

occur as a percentage. For example, if the conditions only occur once a week, this would be a frequency of 1 out of every 7 days or one-seventh of the time. One-seventh is about 14%.

- *Abnormal Conditions:* If the type of operating conditions was abnormal, indicate the type of abnormal conditions. Comments on abnormal conditions can be made in the *Comments* section.
- *Exposure Measured on This Day:* The sampling practitioner can indicate if the employee's exposure on the day of the monitoring was normal, higher than normal, or lower than normal for these operating conditions.
- *Task Description:* A brief summary of the task or tasks performed by the employee during the monitoring time period.

Comments: Any comments on the sampling conditions or tasks performed that would be helpful in understanding the results.

Source Name: The name of the source of the substances being monitored when this information would be useful. The source is usually a chemical, product, or process stream. For example, if an employee's exposure to benzene was due to handling gasoline, enter "gasoline" as the source of the exposure.

Quality Assurance: Any problems which occurred during the monitoring, shipping, or analysis of the sample. This information is used to identify sample results with quality assurance problems when presented on reports.

PERFORMING THE EXPOSURE MONITORING

After the preparatory steps have been completed, sampling using SCD techniques should be conducted using the following general procedure:

- Identify the contaminants to be monitored and the operation or job assignments to be evaluated based upon the Monitoring Plan (Chapter 3).
- Arrange the monitoring with field location personnel to assure that the operations to be monitored or other sources will be running.
- Select the type of SCD to be used. These are discussed in more detail in the next few chapters. If applicable, obtain a copy of the sampling procedure from the relevant agency or other source. Appendix A contains typical sampling procedures.
- Gather the sampling equipment specified in the Monitoring Procedure and calibrate it according to instructions in this chapter.

- Estimate the number of samples to be collected to cover the entire exposure period. If the concentration is expected to be high, it may be necessary to replace the sampling device during the shift to avoid overloading. In addition, if job tasks are different during the shift, it may be beneficial to use a different sample for each job task to measure exposures during each task separately.

- For personal samples, select the individuals to be monitored. This may be based on a random selection method or be those who are expected to receive the greatest exposures due to job task or operating conditions (i.e., worst case).

- Perform the monitoring according to the appropriate procedure. Collect *breathing zone* samples to evaluate inhalation exposure. The *breathing zone* is defined as being within 12 inches of the nose and mouth.

- Document the monitoring on a sample data form.

- Submit the appropriate number of "blanks." The purpose of blanks is to identify sampling and analytical errors. There are two types of sampling blanks: field and media blanks. Field blanks, the most common, are clean sample media taken to the sampling site and handled in every way as the samples, except that no air is drawn through them. Media blanks are new, unopened SCDs that are sometimes sent with the samples to the laboratory to evaluate recovery efficiency or identify any background interferences on the unused collectors. During analyses, field blanks are used to measure the contribution of any contamination that may have occurred during handling, shipping, and storage before analysis. The specific method being used should be consulted concerning the actual number and type of blanks required. A general rule is at least one blank for each day or shift during which sampling is conducted. A typical recommended practice for the number of field blanks is 2 field blanks for each 10 samples, with a maximum of 10 field blanks for each sample set.

- Send the samples to a laboratory accredited by the American Industrial Hygiene Association (AIHA)[8] or other authoritative group for analysis as described in the next section. If you want the laboratory to calculate the airborne concentration, provide the sample volume for each sample.

- Since exposure levels may vary from day to day due to weather conditions, leaks, spills, and operating conditions, monitor each exposure on several days to estimate the range of exposures if required by the Monitoring Plan.

- Review monitoring results periodically to identify problem exposure areas and to determine the necessity for engineering controls or additional monitoring. Update the Monitoring Plan as required.

LABORATORY ANALYSIS

Air samples collected with a sample collection device, which integrates the sample over a period of time (e.g., a charcoal tube, or filter), must be sent to a laboratory for analysis. This section will discuss the principal considerations for analysis of samples and quality assurance programs to obtain reliable results.

Sending Samples to a Laboratory

In order to obtain useful results from the laboratory analysis, the laboratory must be given the necessary information to properly analyze the sample. There may be special shipping considerations needed to preserve the samples. These items are described below:

- *Sample Identification:* This is a unique sample identification number or code for each sample sent to the laboratory, usually the sample identification from the sampling data form. The laboratory will use this number as part of their sample integrity process, although the laboratory will also assign its own identification number to the sample.

- *Sample Volume:* This value must be given to the laboratory if they are expected to report the sample results in concentration units (ppm, mg/m^3, etc.).

- *Substances to Be Determined:* These are called *analytes*, and they are the contaminants or other compounds that the laboratory is to look for in the samples. These should be chemical names—for example, inorganic arsenic, 1,2-dichloroethane, and so on.

- *Analytical Method:* This is the specific "approved" analytical method from a recognized body such as NIOSH, OSHA, or the EPA to be used for the analysis. Prior to sampling, the laboratory should be contacted to verify that they can perform the particular method for the samples collected.

- *Units for Results:* Indicate the desired units for the laboratory to use in reporting the sample results—for example, ppm, ppb, mg/m^3, μg/m^3, or fibers/cc, and so on.

- *Interferences:* If there are any suspected interferences in the sample which will affect the results, indicate these to the laboratory. In some cases these could be chemicals with a similar structure to the analyte, or a completely different compound that experience has shown can interfere with the desired analysis. For example, for GC analysis it could be a compound with a similar elution time (time it emerges from the column) as the chemical of interest.

- *Special Shipping Instructions:* Some samples will require special procedures such as shipping in dry ice, or using expedited shipping to reach the laboratory within a certain timeframe. Some samples may require special labeling or handling to meet hazardous material shipping requirements. Discuss any special arrangements with the laboratory or shipping service prior to sampling.

- *Other Instructions:* If expedited analysis or other special handling is required, these should be noted.

- *Contact Information:* Name and contact information for the sampling practitioner or other person for the laboratory to contact with questions about the samples, and the person to receive the results.

If monitoring for a substance has not been performed before, it is always a good idea to discuss the monitoring with the laboratory in advance to anticipate any special problems or special requirements (e.g., bulk samples along with the airborne samples).

Quality Assurance Programs

Samples sent to a laboratory for analysis should be included in a quality assurance program to ensure that the results obtained are accurate and reliable. Without such programs, it is difficult to be sure that the results are valid. This is a program adopted by the organization performing the sampling, and is separate from the laboratory's quality assurance efforts. A typical program consists of the following parts:

- *Selecting Accredited Laboratories:* Laboratories which are used should be accredited by the American Industrial

Hygiene Association (AIHA) or a similar recognized body for the specific analyses being requested. A laboratory achieves accreditation by meeting the standards of the certifying body and maintaining acceptable performance in periodic proficiency tests.

- *Preliminary Screening Using Spikes:* Prior to sending samples to a laboratory for analysis, the laboratory is "screened" using spikes. A spike is a sample on which a known amount of the substance to be analyzed is added. The laboratory is sent a series of spikes for the substances of interest, and the results are compared with the known amounts on the spikes. The laboratory's performance is expressed as a percentage recovery of the spike amount. For example, if the spike had 10 µg of benzene, and the laboratory reported 7 µg, the percentage recovery would be 70%. A laboratory's performance must be within an acceptable tolerance range to be accepted for actual air samples. Normally this range is about 75% to 125%, depending on the analytical method. Based upon this screening process, a laboratory is determined to be acceptable or not acceptable for further use.

- *Ongoing Quality Assurance Checks:* After an accredited laboratory has been selected and screened, checks on the laboratory's performance may be conducted. One way to check performance is by including SCDs "spiked" with the contaminant at known levels that should result in concentrations in the range of one-tenth to two times the acceptable exposure level (AEL). The spiked samples are made to resemble regular air samples—they are given identification codes and hypothetical "sample volumes" are reported to ensure they are treated the same as the actual air samples. Spiked charcoal

tubes for common contaminants can be purchased commercially; it may be necessary to prepare other spikes using micropipettes. These ongoing checks provide an indication of the laboratory's performance for each individual set of samples. Although a laboratory's performance overall may be very good, it is still possible to have problems with a particular set of samples. In addition, these spikes provide a means of tracking the laboratory's performance over time. When used, spiked samples may be included either with selected batches or with every batch sent to a laboratory.

As a minimum, all analytical reports must be reviewed for completeness and to see if the results "make sense" based on sampling conditions and the sampling practitioner's knowledge of the exposure scenario. For example, the reported results of a "worst-case" 15-minute STEL sample should always be higher than the 8-hour time-weighted average exposure sample collected on the same individual. If not, possible reasons include switching SCDs either in the field or in the laboratory, incorrect transcription of sample identification codes, or a gross misidentification of the purported "worst-case" STEL exposure. Often any inconsistencies or outright errors can be identified and resolved if they are recognized and investigated immediately after the results are received.

VOIDING SAMPLES

Occasionally a sample is damaged, is lost, or cannot be analyzed because of shipping or other problems. In other cases the sample is analyzed but, for a legitimate reason, the sample result is determined to be invalid. For these cases a method of voiding the sample is needed if a "permanent" record of the sample has

TABLE 5.8. Typical Reasons for "Voiding" Samples

Sample lost in shipping
Sample lost in analysis
Sample lost during monitoring
Sampling equipment malfunction
Incorrect sampling device
Sample tampered with
Flowrate changed >10%
Nonstandard sampling method
Sample breakthrough
Sample not analyzed
Nonstandard analytical method
Interference by other compounds
Laboratory recovery was outside range: 75%
   to 125%
No spike submitted to laboratory
Fibers too dense to count

been created through assigning it a sample identification code or entering the record into a database. The voiding procedure should require a review of the sample and reasons by a knowledgeable person, along with a signed statement or other record clearly indicating the reason why the sample was voided. Typical reasons for voiding a sample are listed in Table 5.8. If records are stored in a database, voided samples are not generally included in routine summaries of the data, although they can be accessed if needed. They are not deleted from the database.

EXAMPLES: CALCULATING AIR MONITORING RESULTS

These examples illustrate the method of calculating full-shift time-weighted average results (TWA) based upon full-shift monitoring once the sampling has been performed and the analytical results returned by the laboratory.

There are three steps in calculating these final exposure results:

1. Calculating the volume of air sampled.

2. Calculating the concentration of the contaminant in the collected sample.
3. Calculating the final exposure result.

If the laboratory has reported results in concentration units (ppm, mg/m³, etc.), only Step 3 is needed.

Step 1: Volume of Air Sampled

The volume of air drawn by the pump during the monitoring time period is calculated based upon the number of minutes the pump ran and the flowrate of the pump as follows:

$$\text{Sample volume} = \text{Average flowrate} \times \text{Monitoring time}$$

Example 1. Assume that a pump used to monitor an employee's exposure was determined to be running at a flowrate of 200 mL/min and the pump ran for 300 minutes. What was the volume of air drawn by the pump?

$$\text{Sample volume} = \text{Average flowrate} \times \text{Monitoring time}$$

$$\text{Sample volume} = 200 \text{ mL/min} \times 300 \text{ min}$$

$$\text{Sample volume} = 60,000 \text{ mL}$$

In order to calculate concentration (Step 2), the volume of air should be expressed in units of either liters (L) or cubic meters (m³). The correct volume units will depend upon the units of weight used in the laboratory results. The units to use are as follows:

- If the laboratory report uses mg to express weight, then use m³ as the unit of volume.
- If the laboratory report uses μg to express weight, then use L (liters) as the unit of volume.

Table 5.2 in the *Review of the Metric System*, earlier in this chapter, shows how to convert from mL to L and from L to m³. For this example, we will convert 60,000 mL to m³ as follows:

$$60,000\,\text{mL} \div 1000 = 60 \text{ liters (L)}$$
$$60\,\text{L} \div 1000 = 0.06 \text{ cubic meters (m}^3)$$

Thus the volume of air collected with the pump can be expressed in three equivalent units: 60,000 mL, 60 L, or 0.06 m³.

Step 2: Calculating the Concentration

The concentration of the contaminant is calculated based upon the volume of air pulled by the pump (Step 1) and the amount of contaminant from the analytical laboratory results. The formula for the calculation is

$$\text{Concentration} = \frac{\text{Weight}}{\text{Volume}}$$

Example 2. Assume the laboratory reports the weight of the substance collected with the pump in Step 1 as 0.11 mg. What is the concentration?

$$\text{Concentration} = \frac{\text{Weight}}{\text{Volume}}$$
$$\text{Concentration} = \frac{0.11\,\text{mg}}{0.06\,\text{m}^3}$$
$$\text{Concentration} = 1.9\,\text{mg/m}^3$$

If the substance sampled was a gas or vapor, the concentration would normally be expressed in units of ppm. To convert from mg/m³ to ppm, use the following formula from Table 5.3 in the *Review of the Metric System*:

$$\text{ppm} = \text{mg/m}^3 \times \frac{24.45}{\text{MW}}$$

The MW (molecular weight) is a number specific to the substance sampled. For this example, assume that benzene is the analyte. The molecular weight (MW) of benzene is 78.

$$\text{ppm} = 1.9\,\text{mg/m}^3 \times \frac{24.45}{78}$$
$$\text{ppm} = 0.6\,\text{ppm}$$

Thus, the concentration for the employee's exposure based upon the volume of the sample and the laboratory result is 0.6 ppm after converting from mg/m³ to ppm.

Step 3: Calculating the Final Exposure Result

The final time-weighted average (TWA) results for full-shift samples, or samples collected to represent the full shift, are calculated as follows:

$$\text{TWA} = \frac{C_1 T_1 + C_2 T_2 + C_3 T_3 + \cdots + C_n T_n}{480\,\text{min}}$$

where C is the concentration result for the monitoring time period and T is the exposure time associated with the concentration result.

Example 3. Assume in the example that the concentration of 0.6 ppm calculated in Step 2 for 300 minutes occurred in the morning. Assume that in the afternoon a second sample was collected to monitor the same employee for 180 minutes and the concentration was determined to be 1 ppm. Based upon these two results, what is the TWA result for the shift for the employee monitored?

Using the equation above, C_1 is 0.6 ppm and C_2 is 1 ppm. There is no C_3 since C_1 and C_2 accounted for the whole shift. T_1 is 300 minutes and T_2 is 180 minutes.

$$\text{TWA} = \frac{\begin{array}{c}(0.6\,\text{ppm})(300\,\text{min})\\ +(1\,\text{ppm})(180\,\text{min})\end{array}}{480\,\text{min}}$$
$$\text{TWA} = 0.75\,\text{ppm} = 0.8\,\text{ppm}$$

Thus, the TWA exposure for the employee monitored was 0.8 ppm based upon the two samples collected in the morning and the afternoon.

If only one sample is collected to evaluate an employee's full-shift exposure, C_1 will be the concentration result for that sample. T_1 will be the complete exposure time for the employee; that is, the complete length of time during the shift that the employee was exposed at the concentration C_1. This time period should include any exposure time outside the monitoring period. C_2 will be zero since all of the employee's exposure has been accounted for in C_1 and no other exposure occurred. T_2 will be the remainder of the shift—that is, the portion not accounted for by T_1.

Example 4. Assume that an employee was monitored for 400 minutes and the concentration result was 1.0 ppm. The employee, however, was actually exposed for a total of 420 minutes. What is the employee's TWA exposure?

C_1 is 1.0 ppm from the sample concentration result. T_1 is 420 minutes, the length of the exposure time. C_2 is zero since there was not exposure outside the exposure time. T_2 is 60 minutes (the remainder of the shift).

$$TWA = \frac{(1.0\,\text{ppm})(420\,\text{min}) + (0\,\text{ppm})(60\,\text{min})}{480\,\text{min}}$$

$$TWA = 0.88\,\text{ppm} = 0.9\,\text{ppm}$$

Thus the TWA exposure is 0.9 ppm.

Example 5. Assume that an employee was monitored for the full shift length of 480 minutes and the concentration result was 1.0 ppm. What is the employee's TWA exposure?

C_1 is 1.0 ppm for the sample concentration result. T_1 is 480 minutes. C_2 and T_2 are both zero since the full shift has been covered.

$$TWA = \frac{C_1 T_1 + C_2 T_2}{480\,\text{min}}$$

$$TWA = \frac{(1.0\,\text{ppm})(480\,\text{min}) + (0\,\text{ppm})(0\,\text{min})}{480\,\text{min}}$$

$$TWA = 1\,\text{ppm}$$

Thus the final TWA exposure result for this employee is 1.0 ppm.

The final TWA exposure result in Examples 4 and 5 represents the average exposure level over the entire shift. Note that each result is based upon exposure time. In most cases, the monitoring should last the full shift in order to accurately represent the complete exposure the employee receives. If the monitoring does not last the full shift, the actual exposure time must be estimated for the calculation as shown in Example 4 above. It is important that this estimation be accurate and not underestimate the length of exposure.

COMPARING RESULTS TO EXPOSURE LIMITS

Exposure monitoring results are usually compared with exposure limits to determine if the exposures are unacceptably high. Exposure limits have been established by regulatory agencies such as OSHA and the EPA and by authoritative bodies such as the American Conference of Industrial Hygienists and AIHA. These are described in Chapter 2 (Hazards).

Chapter 3 (Exposure Assessment Strategy and Monitoring Plan) contains guidelines on comparing exposure measurements to acceptable exposure levels (AELs).

SUMMARY

Sample collection device (SCD) methods require laboratory analysis of the sampling medium. Successful monitoring using SCD methods requires careful planning since the sampling results are not available from the laboratory until some time after the sampling has been performed. Key elements to be considered include method selection, equipment selection, equipment calibration, sampling according to proper procedures, analysis using an accredited laboratory, and adequate quality assurance for the entire process.

The next several chapters discuss specific SCD approaches in more detail.

REFERENCES

1. National Institute for Occupational Safety and Health. *Manual of Analytical Methods*, 4th ed., 1994.

2. National Institute for Occupational Safety and Health. *Manual of Analytical Methods*, No. 1501. *Hydrocarbons, Aromatic*, 4th ed., August 15, 1994.

3. Occupational Safety and Health Administration. *Access to Employee Exposure and Medical Records*. U.S. Code of Federal Regulation (29CFR1910.1020).

4. Plog, B. A., and P. J. Quinlan, eds. *Fundamentals of Industrial Hygiene*, 5th ed. Washington, D.C.: National Safety Council, 2002.

5. Occupational Safety and Health Administration. *Benzene*. U.S. Code of Federal Regulation (29CFR1910.1028).

6. Occupational Safety and Health Administration. *Lead*. U.S. Code of Federal Regulation (29CFR1910.1025).

7. Occupational Safety and Health Administration. *Asbestos*. U.S. Code of Federal Regulation (29CFR1910.1001).

8. Grunder, F.I. PAT Program Report Background and Current Status. *AIHA J.* **63**: 797–798, 2002.

CHAPTER 6

SAMPLE COLLECTION DEVICE METHODS FOR GASES AND VAPORS

Contrary to high dust situations where visible deposits can make it easy to estimate where concentrations are likely to be high, the presence of gases and vapors is often much more complex to detect. Since most gases and vapors cannot be seen, and levels of concern are often below or just at odor thresholds, they are generally detected using sampling methods. Surveys for gases and vapors are rarely ever simple situations involving a single compound; instead most often there is a mixture of contaminants in varying quantities and proportions.

When airborne, gases and vapors are in the form of individual molecules and thus tend to diffuse easily into the air and spread rapidly throughout an area. For air sampling purposes the properties of solubility, vapor pressure, and reactivity can be applied to selecting the proper collection procedure. There are two basic methods for collecting integrated samples of gases and vapors: The first involves a battery-powered pump actively pulling air through an appropriate medium. In the second method the medium is simply exposed to contaminated air, allowing the compounds of interest to passively diffuse or permeate it. The collection medium can be an adsorbent, such as a sorbent in a tube, or an absorbent, such as a solution, in which the gas or vapor dissolves and/or converts to form another compound. Gases and vapors can also be collected in bags. Additionally, some higher-boiling-point vapors can be collected on specially treated filters; these are discussed in Chapter 7.

In general, sorbent tube methods are chosen over liquid-based sampling procedures to avoid the inconvenience of handling liquids in the field. Passive methods are often preferred over active methods since no pump is needed.

ACTIVE SAMPLE COLLECTION DEVICE MONITORING

Active Sample Collection Device (SCD) gas and vapor methods for both occupa-

Air Monitoring for Toxic Exposures, Second Edition. By Henry J. McDermott
ISBN 0-471-45435-4 © 2004 John Wiley & Sons, Inc.

tional and environmental sampling use sorbent tubes, impingers, and bags. The primary difference is that environmental sampling trains are often more complex; for example, impingers may be in an ice bath and environmental sampling periods are much longer, often 24 hours rather than the 8-hour period for occupational samples. In addition, other methods are used for environmental sampling of gases and vapors: stainless steel canisters and cryogenic trap collection. Canisters are used for very low levels (trace) of volatile organic compounds (VOCs) while cryogenic methods are used to collect compounds that are very volatile or otherwise hard to stabilize on sorbents. As cryogenic sampling requires special techniques that are not widely applicable to most types of survey work, they will not be discussed further.

For a summary of the types of media used for collecting occupational and environmental samples of many gases and vapors, see Table 6.1. This chapter describes the media and methods commonly used to collect integrated samples of gases and vapors (Figure 6.1). Other chapters of interest on gas and vapor sampling describe real time instrument methods, color change methods, and collection of bulk (or grab) air samples.

Solid Sorbent Sampling

Sampling with sorbent tubes is done for both occupational and environmental exposures. The sampling media are often different as well as the methods and the calculations. Generally charcoal, silica gel, and chromosorbs are used for occupational air sampling and Tenax and carbon molecular sieves are preferred for environmental samples. Since the levels expected to be present are much lower in environmental samples, the collection period is much longer. Typically environmental samples are collected as a 24-hour average. Sorbents for occupational sampling are often

contained inside glass tubes, while those for environmental sorbents are often packed inside of glass or metal cartridges.

Solid sorbents are specific for groups of compounds and one sorbent will not work with all compounds. Most solid sorbents do not differentiate between compounds during collection so unwanted compounds as well as the target compounds are collected. There is no standard nomenclature for sorbent tubes. Different manufacturers have different names for their sorbent materials and thus XAD and ORBO refer to manufacturers' brands and not a specific material. In the case of the ORBO brand, the tubes are numbered, with the number referring to the type of sorbent used in the tube. For example, ORBO-22 is used for formaldehyde and contains a styrene divinyl benzene polymer coated with *N*-benzylethanolamine; ORBO-32 contains charcoal granules.

When more than one compound is present that can be collected on a sorbent tube, the amount of any individual component that can be collected is reduced because each tube has a certain number of active sites. A reduction in sampling time or volume may be required because of the higher overall concentration being presented to the tube. Other factors that affect the ability of sorbent tubes to be efficient gas and vapor collectors include the size and mass of granules.

The biggest concern in collecting material on a sorbent tube is whether breakthrough can occur. Breakthrough occurs when the front section of a tube is saturated and enough compound accumulates in the backup section that it begins to exit the tube with the airstream. Breakthrough is usually defined as the presence of 25% or more of a contaminant in the rear portion of a sorbent tube. Figure 6.2 shows a typical breakthrough curve for an organic vapor. Almost no breakthrough occurs until "Saturation" is achieved, then breakthrough increases quickly.[1] In this situation,

TABLE 6.1. Media Used in NIOSH and EPA Methods for Gases and Vapors

Compound	Method Number[a]	Media
Acetaldehyde[b]	2538	XAD-2 tube, 225/450
	3507	Girard T Reagent in a bubbler
	TO-5	Dinitrophenyl hydrazine solution in an impinger
Acetic acid[b]	1603	Charcoal tube, 100/50
Acetic anhydride[b]	3506	Alkaline hydroxylamine solution in a bubbler
Acetone[b]	1300	Charcoal tube, 100/50
Acetone cyanohydrin	2506	Porapak QS tube, 100/50
Acetonitrile[b]	1606	Charcoal tube, 400/200
Acrolein	2501	2-(Hydroxymethyl) piperdine on XAD-2 tube, 120/60
	TO-5	Dinitrophenylhydrazine solution in an impinger
Acrylonitrile[b]	1604	Charcoal tube, 100/50
	TO-2	Carbon molecular sieve
Allyl alcohol[b]	1402	Charcoal tube, 100/50
Allyl chloride[b]	1000	Charcoal tube, 100/50
	TO-2	Carbon molecular sieve
Ammonia[b]	6701	Sulfuric acid solution in a passive sorbent badge
Amyl acetate, *n*-, *sec*-[b]	1450	Charcoal tube, 100/50
Aniline	2002	Silica gel tube, 150/75
Anisidine	2514	XAD-2 tube, 150/75
Arsine	6001	Charcoal tube, 100/50
Benzaldehyde	TO-5	Dinitrophenylhydrazine solution in an impinger
Benzene[b]	1500, 1501	Charcoal tube, 100/50
	TO-1	Tenax GC
	TO-2	Carbon molecular sieve
Benzyl chloride[b]	1003	Charcoal tube, 100/50
	TO-1	Tenax GC
Biphenyl	2530	Tenax GC tube, 20/10
Bromoform[b]	1501	Charcoal tube, 100/50
Butyl acetate, *n*-, *sec*-, *tert*-[b]	1450	Charcoal tube, 100/50
Bromotrifluoromethane	1017	Two charcoal tubes in series, 400/200 and 100/50
1,3-Butadiene[b]	1024	Charcoal tube, 400/200
Butyl alcohol, *n*-, *sec*-, *iso*-[b]	1401	Charcoal tube, 100/50
tert-Butyl alcohol[b]	1400	Charcoal tube, 100/50
Butyl cellosolve[b]	1403	Charcoal tube, 100/50
p-tert-Butyl toluene[b]	1501	Charcoal tube, 100/50
Camphor[b]	1301	Charcoal tube, 100/50
Carbon disulfide	1600	Charcoal tube, 100/50 and sodium sulfate tube, 270 (moisture removal)
Carbon tetrachloride[b]	1501	Charcoal tube, 100/50
	TO-2	Carbon molecular sieve
Chlorobenzene[b]	1501	Charcoal tube, 100/50
Chlorobromomethane[b]	1501	Charcoal tube, 100/50

TABLE 6.1. *Continued*

Compound	Method Number[a]	Media
Chloroform[b]	1501	Charcoal tube, 100/50
	TO-2	Carbon molecular sieve
Chloroprene	1002	Charcoal tube, 100/50
	TO-1	Tenax GC
Chloroacetic acid	2008	Silica gel tube, 100/50
Cresol, o-, m-, p-	2001	Silica gel tube, 150/75
	TO-8	Sodium hydroxide solution in an impinger
Cumene[b]	1501	Charcoal tube, 100/50
Cyclohexanone[b]	1300	Charcoal tube, 100/50
Cyclohexane[b]	1500	Charcoal tube, 100/50
Cyclohexene[b]	1500	Charcoal tube, 100/50
Cyclohexanol[b]	1402	Charcoal tube, 100/50
1,3-Cyclopentadiene	2535	Maleic-anhydride-coated chromosorb 104, 100/50
Diacetone alcohol[b]	1402	Charcoal tube, 100/50
Diazomethane	2515	Octanoic acid-coated XAD-2 tube, 100/50
Dibromodifluoromethane	1012	Two charcoal tubes in series, 150 each
2-Dibutylaminoethanol	2007	Silica gel tube, 300/150
Dichlorobenzene, o⁻, p⁻[b]	1501	Charcoal tube, 100/50
	TO-1	Tenax GC
1,1-Dichloroethane[b]	1501	Charcoal tube, 100/50
1,2-Dichloroethylene[b]	1501	Charcoal tube, 100/50
Dichlorodifluoromethane	1018	Two charcoal tubes in series, 400/200 and 100/50
sym-Dichloroethyl ether[b]	1004	Charcoal tube, 100/50
Dichlorofluoromethane	2516	Two charcoal tubes in series, 400 and 200
1,1-Dichloro-1-nitroethane	1601	Petroleum charcoal tube, 100/50
1,2-Dichloropropane[b]	1013	Petroleum charcoal tube, 100/50
Dichlorotetrafluoroethane	1018	Two charcoal tubes in series, 400/200 and 100/50
2-Diethylaminoethanol	2007	Silica gel tube, 300/150
Diisobutyl ketone[b]	1300	Charcoal tube, 100/50
Dimethylacetamide	2004	Silica gel tube, 150/75
n,n-Dimethylaniline[b]	2002	Silica gel tube, 150/75
Dimethylformamide[b]	2004	Silica gel tube, 150/75
n,n-Dimethyl-p-toluidine	2002	Silica gel tube, 150/75
Dimethylsulfate	2524	Porapak P tube, 100/50
Dioxane[b]	1602	Charcoal tube, 100/50
Epichorohydrin[b]	1010	Charcoal tube, 100/50
Ethanol[b]	1400	Charcoal tube, 100/50
Ethanolamine	2007	Silica gel tube, 300/150
2-Ethoxyethyl acetate[b]	1450	Charcoal tube, 100/50
Ethyl acrylate[b]	1450	Charcoal tube, 100/50
Ethyl benzene[b]	1501	Charcoal tube, 100/50
Ethyl bromide[b]	1011	Charcoal tube, 100/50
Ethyl butyl ketone[b]	1301	Charcoal tube, 100/50
Ethyl chloride	2519	Two charcoal tubes in series, 400 and 200
Ethyl cellosolve[b]	1403	Charcoal tube, 100/50

TABLE 6.1. *Continued*

Compound	Method Number[a]	Media
Ethylene chlorohydrin	2513	Petroleum charcoal tube, 100/50
Ethylene dibromide[b]	1008	Charcoal tube, 100/50
Ethylene dichloride[b]	1501	Charcoal tube, 100/50
	TO-2	Carbon molecular sieve
Ethylene glycol dinitrate	2507	Tenax GC tube, 100/50
Ethylene oxide	1607	Two charcoal tubes in series, 400 and 200
Ethylene oxide[b]	1614	Hydrogen-bromide-coated petroleum charcoal tube, 100/50
Ethyl ether	1610	Charcoal tube, 100/50
Formaldehyde[b]	2541	2-(Hydroxymethyl) piperidine-coated XAD-2 tube, 120/60
	3500	1 μm PTFE membrane filter, 37 mm, followed by two impingers containing sodium bisulfite in series
	TO-5	Dinitrophenylhydrazine solution in an impinger
Furfural[b]	2529	2-(Hydroxymethyl) piperidine-coated XAD-2 tube, 120/60
Glycidol	1608	Charcoal tube, 100/50
n-Heptane[b]	1500	Charcoal tube, 100/50
Hexachloro-1,3-cyclopentadiene	2518	Two Porapak T tubes in series, 75 and 25
Hexachloroethane[b]	1501	Charcoal tube, 100/50
n-Hexane[b]	1500	Charcoal tube, 100/50
2-Hexanone[b]	1300	Charcoal tube, 100/50
Hydrazine[b]	3503	Hydrochloric acid solution in a bubbler
Hydrogen bromide	7903	Washed silica gel tube, 400/200
Hydrogen chloride[b]	7903	Washed silica gel tube, 400/200
Hydrogen fluoride[b]	7903	Washed silica gel tube, 400/200
Iodine	6005	Alkali-treated charcoal tube, 100/50
Isoamyl acetate[b]	1450	Charcoal tube, 100/50
Isoamyl alcohol[b]	1402	Charcoal tube, 100/50
Isbutyl acetate[b]	1450	Charcoal tube, 100/50
Isocyanate	5521	Impinger with solution of 1-(2-methoxyphenyl)-piperazine in toluene
Isophorone[b]	2508	Petroleum charcoal tube, 100/50
Isopropyl alcohol[b]	1400	Charcoal tube, 100/50
Mesityl oxide[b]	1301	Charcoal tube, 100/50
Methanol	2000	Silica gel tube, 100/50
Methyl n-amyl ketone[b]	1301	Charcoal tube, 100/50
Methyl bromide	2520	Two petroleum charcoal tubes in series, 400 and 200
Methyl cellosolve[b]	1403	Charcoal tube, 100/50
Methyl chloride	1001	Two charcoal tubes in series, 400/200 and 100/50
Methyl chloroform[b]	1501	Charcoal tube, 100/50
	TO-2	Carbon molecular sieve
Methyl cyclohexane[b]	1500	Charcoal tube, 100/50
Methylcyclohexanone	2521	Porapak Q tube, 150/75

TABLE 6.1. *Continued*

Compound	Method Number[a]	Media
Methylene chloride[b]	1005	Two charcoal tubes in series, 100/50
	TO-2	Carbon molecular sieve
Methyl ethyl ketone peroxide	3508	Dimethyl phthalate solution in an impinger
5-Methyl-3-heptanone	1301	Charcoal tube, 100/50
Methyl iodide[b]	1014	Charcoal tube, 100/50
Methyl isoamyl acetate	1450	Charcoal tube, 100/50
Methyl isobutyl carbinol[b]	1402	Charcoal tube, 100/50
Methyl isobutyl ketone[b]	1300	Charcoal tube, 100/50
α-Methyl styrene[b]	1501	Charcoal tube, 100/50
Mevinphos	2503	Chromosorb 102 tube, 100/50
Naphthalene[b]	1501	Charcoal tube, 100/50
Naphthas	1550	Charcoal tube, 100/50
Nickel carbonyl	6007	Low-nickel charcoal tube, 120/60
Nitric acid[b]	7903	Washed silica gel tube, 400/200
Nitrobenzene	2005	Silica gel tube, 150/75
	TO-1	Tenax GC
Nitroethane	2526	Two XAD-2 tubes in series, 600 and 300
Nitrogen dioxide[b]	6700	Palmes tube with three triethanolamine-treated screens (passive sampler)
Nitroglycerin	2507	Texan-GC tube, 100/50
Nitromethane	2527	Chromosorb 106 tube, 600/300
2-Nitropropane	2528	Chromosorb 106 tube, 100/50
N-Nitrosodimethylamine	TO-7	Thermosorb/N
n-Octane[b]	1500	Charcoal tube, 100/50
1-Octanethiol	2510	Tenax GC tube, 100/50
Pentachloroethane	2517	Poropak R tube, 70/35
n-Pentane[b]	1500	Charcoal tube, 100/50
2-Pentanone[b]	1300	Charcoal tube, 100/50
Phenol	OSHA 32	Polyurethane foam (PUF), 76 mm
	TO-8	Sodium hydroxide solution in an impinger
Phosphorus trichloride	6402	Distilled water in a bubbler
n-Propyl acetate[b]	1450	Charcoal tube, 100/50
n-Propyl alcohol	1401	Charcoal tube, 100/50
Propylene oxide[b]	1612	Charcoal tube, 100/50
Pyridine[b]	1613	Charcoal tube, 100/50
Stibine	6008	Mercuric-chloride-coated silica gel tube, 1000/500
Styrene	1501	Charcoal tube, 100/50
1,1,2,2-Tetrabromoethane	2003	Silica gel tube, 150/75
1,1,1,2-Tetrachloro-1,2-difluoroethane	1016	Charcoal tube, 100/50
1,1,2,2-Tetrachloro-2,2-difluoroethane	1016	Charcoal tube, 100/50
1,1,2,2-Tetrachloroethane[b]	1019	Petroleum charcoal tube, 100/50
Tetrachloroethylene[b]	1501	Charcoal tube, 100/50
	TO-1	Tenaz GC
Tetraethyl lead	2533	XAD-2 tube, 100/50
Tetraethyl pyrophosphate	2504	Two Chromosorb 102 tubes in series, 100/50
Tetrahydrofuran[b]	1609	Charcoal tube, 100/50
Tetramethyl lead	2534	XAD-2 resin tube, 400/200

TABLE 6.1. *Continued*

Compound	Method Number[a]	Media
Tetramethyl thiourea	3505	Distilled water in an impinger
Toluene[b]	1500	Charcoal tube, 100/50
	TO-1	Tenax GC
	1501	Charcoal tube, 100/50
Toluene-2,4-diisocyanate	2535	Tube with N-[(4-nitrophenyl)methyl]- propylamine coated on glass wool
o-Toluidine	2002	Silica gel tube, 150/75
Trichlorofluoromethane[b]	1006	Charcoal tube, 400/200
1,1,2-Trichloroethane[b]	1501	Charcoal tube, 100/50
Trichloroethylene[b]	1022	Charcoal tube, 100/50
	TO-1	Tenax GC
	TO-2	Carbon molecular sieve
1,2,3-Trichloropropane[b]	1501	Charcoal tube, 100/50
1,1,2-Trichloro-1,2,2-trifluoroethane	1020	Charcoal tube, 100/50
Turpentine	1551	Charcoal tube, 100/50
Vinyl bromide[b]	1009	Charcoal tube, 400/200
Vinyl chloride[b]	1007	Two charcoal tubes in series, 150
	TO-2	Carbon molecular sieve
Vinylidene chloride[b]	1015	Charcoal tube, 100/50
	TO-2	Carbon molecular sieve
Vinyl toluene[b]	1501	Charcoal tube, 100/50
Xylene[b]	1501	Charcoal tube, 100/50
	TO-1	Tenax GC
2,4-Xylidine	2002	Silica gel tube, 150/75

Note: All charcoal tubes are coconut charcoal unless otherwise specified.
[a] TO methods refer to EPA methods. Other methods are NIOSH unless otherwise indicated.
[b] Passive methods are available; consider evaluating for suitability.

it is very likely that sample has been lost through the tube. Some chemicals are prone to breakthrough, such as methylene chloride, which must be taken into consideration when planning sampling strategies. When results indicate breakthrough, the best interpretation is that actual concentrations were higher. Migration is a phenomenon similar to breakthrough, wherein an amount of the contaminant migrates from the front to the rear section of the tube while stored prior to analysis. This situation cannot be differentiated from actual breakthrough.

When selecting sorbents the potential for water absorption must be considered if high humidity conditions are present at the site to be sampled. Competition for the adsorbent's surface will occur between the compound being sampled and the water, with the result that contaminant breakthrough may occur.

The flow rate will also impact the ability to retain gases and vapors on a sorbent above the optimum flowrate usually specified by a method. Increasing the flowrate above this level may result in a proportional decease in the ability of the sorbent to retain contaminants. The variability of the pressure drop on a tube is dependent on the particle size of the sorbent.[2] Sorbents composed of fine adsorbent particles

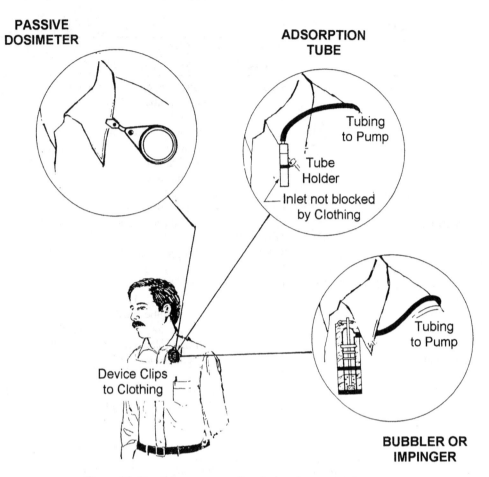

Figure 6.1. Breathing zone sampling devices for gases and vapors.

sample most efficiently, but their resistance to airflow is inversely proportional to particle size. Another problem with solid sorbents is that changes in air resistance due to swelling, shrinking, or channeling of the adsorbent are sometimes encountered. Table 6.2 lists factors affecting the collection behavior of solid sorbents.[3]

Gas chromatography/mass spectroscopy (GC/MS) is recognized as a powerful analytical method for sorbent samples. However, even this technique cannot be used for all compounds. For example, although GC/MS will detect many organics, it does not detect inorganics. Depending on the sorbent material, some organic compounds may have a poor affinity and others

too great an affinity. They will either not be collected or once they are collected they cannot be released from the media for detection. Also, libraries of compounds used by these instruments, while large, do not contain all compounds. Although suitable for many applications, GC/MS or any other single technique will not provide all the answers.

Charcoal Sorbents. Charcoal is one of the most commonly used sorbents and is useful for sampling a wide variety of organic gases and vapors, including several different compounds at a time. Both the National Institute of Occupational Safety and Health (NIOSH) and Occupational

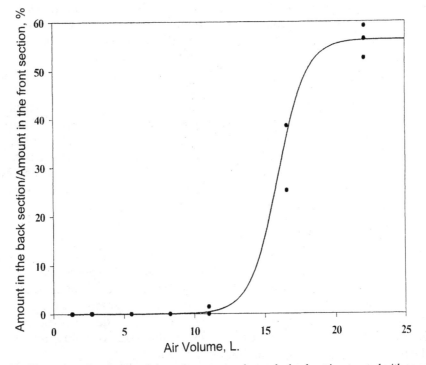

Figure 6.2. Illustration of typical breakthrough curve as a charcoal adsorbent is saturated with an organic vapor. Since breakthrough depends on the mass of chemical adsorbed as well as the effects of water vapor, test results usually specify the sample volume, vapor concentration, and relative humidity.

TABLE 6.2. Factors Affecting the Collection Characteristics of Solid Sorbents

Factor	Effects
Temperature	Adsorption is reduced at higher temperatures. Increased temperature is proportional to increased breakthrough. Reaction rates are increased at higher temperatures.
Humidity	Water vapor is adsorbed by polar sorbents, increasing the likelihood of breakthrough for chemicals. Increased humidity is proportional to increased breakthrough.
Flow rate	Higher sampling flow rates can lead to breakthrough at lower volumes. This varies with the type of sorbent.
Concentration	Breakthrough likelihood increases (and breakthrough volume decreases) with increasing concentrations of contaminant.
Mixtures	When two or more compounds are present, the compound most strongly held will displace the other compounds, in order, down the length of the tube. Polar sorbents hold compounds with the largest dielectric constant and dipole moment most strongly. Nonpolar sorbents prefer compounds of a higher boiling point or larger molecular volume.
Particle size	Decreases in sorbent particle size are proportional to increases in sampling efficiency and drop in pressure.
Size of tube	A doubling of the sorbent volume doubles the concentration required for breakthrough.

Safety and Health Administration (OSHA) methods specify the use of charcoal tubes for a wide variety of organic compounds.

Charcoal can be derived from a variety of carbon-containing materials, but most of the charcoal used for air sampling is coconut- or petroleum-based. NIOSH recommends coconut charcoal, which has a higher capacity than petroleum charcoal for most compounds, although in some cases such as ethylene oxide collection a different type of charcoal is used. Ordinary charcoal is "activated" by steam at 800–900°C, causing it to form a porous structure. Activated charcoal is not useful for sampling certain types of reactive compounds, such as mercaptans and aldehydes, due to its high surface reactivity with these compounds. Inorganic compounds, such as ozone, nitrogen dioxide, chlorine, hydrogen sulfide, and sulfur dioxide, react chemically with activated charcoal. Glass sampling tubes vary in size ranging from 150 mg to greater than 1000 mg of charcoal. Another type of tube that is also carbon-based is a carbon molecular sieve. This material has a different structure than the conventional charcoal discussed here and is discussed in greater detail in the section on molecular sieves.

The most common charcoal tube in use for occupational sampling was designed according to NIOSH recommended specifications to include fabrication of glass tubing 7 cm long with a 6-mm outside diameter and a 4-mm internal diameter. The charcoal is 20/40 mesh-sized. A 100-mg front section is separated from a 50-mg backup section by a piece of urethane foam. A second piece of foam sits at the outlet to prevent granules from being sucked out of the tube during sampling. A plug of glass wool is in the very front of the tube. An unused tube will have both ends flame sealed. This structure and dimension is common to many tubes, although specialized tubes as well as scaled-up (proportionately larger) versions are also available.

Analysis of charcoal tubes is generally done using the solvent carbon disulfide to desorb (remove) the adsorbed contaminants.

The adsorbing capability of activated charcoal varies from batch to batch in commercial production. Charcoal tube collection efficiency for various hydrocarbons may be affected by such variables as sampling rate, vapor concentrations, and total adsorbed hydrocarbon mass. Nonpolar compounds, preferentially sampled on charcoal, will displace polar compounds in charcoal media. Polar compounds are those that contain oxygen and have high dielectric constants (e.g., alcohols and ketones). Competitive adsorption also occurs among polar compounds. Factors that may contribute to the affinity of the molecules for charcoal include hydrogen bonding, molecular size, volatility, and the dipole moment.[4]

Breakthrough volumes are variable and a function of the carbon, temperature, humidity, storage times, and pollutant. It has been estimated that for every 10°C increase in temperature, breakthrough volume is decreased by 1% to 10%.[5]

Charcoal tubes have been shown to be affected by high humidity. Nonpolar compounds, such as toluene and 1,1,1-trichloroethane, are the least affected. However, toluene has been shown to break through at lower sampling volumes if high humidity is present rather than in dry air. Reliable samples can be collected by limiting sample volumes where high humidity exists.[3] Polar compounds, such as ethyl cellosolve and dioxane, are the most susceptible.[6] Water vapor may also carry ethanol through the charcoal, for example. Charcoal has a higher affinity for water vapor than carbon molecular sieves.[7]

The type of compound may affect the flow rate selected for collection. A study of benzene sampling in charcoal determined that benzene was collected more efficiently at flow rates of 2 Lpm and higher. It is the opposite of what would be expected when using traditional charcoal tube sampling methods that specify flow rates of 50–200 mL/min.[8] The desorption efficiency for a particular compound can vary from one batch of charcoal sorbent to another. It may also be affected by the rate of loading, the total loading applied to the sorbent, and the distribution of material within the sorbent.[9]

Even though many compounds can be collected on a single charcoal tube, laboratories often prefer that a separate tube be used for each similar group of compounds sampled. For example, aromatics such as benzene, toluene, and xylene would be collected separately from chlorinated compounds. When sampling air containing mixtures of organic vapors of unknown concentrations, it may be useful to use two tubes in series, so that the backup tube is present in the case of breakthrough. When the mixture contains both polar (especially ethanol and methanol) and nonpolar compounds, the backup tube should contain silica gel as the collection medium.[10]

Silica Gel Sorbents.

Silica gel is considered a more selective sorbent than activated charcoal, and gases and vapors are more easily desorbed from it.[11] Silica gel is an amorphous form of silica derived from the interaction of sodium silicate and sulfuric acid. It is the adsorbent recommended for collecting organic amines, both alkyl and aromatic, such as aniline and o-toluidine.

Factors that affect the dynamic adsorption of materials onto silica gel include the size range of the gel particles, tube diameter and length, temperature during sampling, concentration of contaminant being sampled, air humidity, and duration of sampling. Since the polarity of the adsorbed compound determines the binding strength on silica gel, compounds of high polarity will displace compounds of lesser polarity. Therefore, when attempting to collect relatively nonpolar compounds, the presence of coexisting polar compounds may interfere with collection on silica gel. The following compounds provide an example of the order of preferential adsorption of polar materials onto silica gel: water, alcohols, aldehydes, ketones, esters, aromatics, olefins, paraffins.[12]

Silica gel will show an increase in breakthrough capacity with increasing humidity. Therefore, under high humidity conditions the sample will be lost due to saturation with water vapor. The tendency of silica gel to absorb water vapor and displace collected components is its chief disadvantage.

Some methods specify that silica gel be washed to rid it of any impurities. Washing may be done using distilled water or in some cases inorganic acids. An example is NIOSH Method S138 that specifies sulfuric-acid-washed silica gel for the collection of butylamine.

Molecular Sieves.

A carbon molecular sieve adsorbent is the carbon skeletal framework remaining after pyrolysis of the synthetic polymeric or petroleum pitch precursors. The result is a spherical, macroporous structure. The choice of starting polymer or pitch dictates the physical characteristics of the sieve, such as particle size and shape and sieve pore structure. The diameters of the micropores and their number are responsible for the differences in tube retention volume, adsorption coefficient, and equilibrium sorption capacity for a given compound.[7] The limiting factor with molecular sieves is humidity.

Spherocarb, Carboxen, Carbosieve, and Purasieve are examples of molecular sieves. Carbon-based molecular sieves are often called graphitized sieves. Nitrogen dioxide is collected on a triethanolamine (TEA)-impregnated molecular sieve. Carbon molecular sieves are most commonly used to collect environmental samples of highly volatile nonpolar organic compounds.

Porous Polymeric Sorbents. Porous polymers are another class of sorbent used for air sampling. These include Tenax GC, Porapak, Chromosorb, and XAD tubes. Their wide variety offers a high degree of selectivity for specific applications. Most porous polymers are copolymers in which one entity is styrene, ethylvinylbenzene, or divinylbenzene and the other monomer is a polar vinyl compound. Limitations include displacement of less volatile compounds, especially by carbon dioxide; irreversible adsorption of some compounds, such as amines and glycols; oxidation, hydrolysis, and polymerization reactions of the sample; chemical changes of the contaminant in the presence of reactive gases and vapors, such as nitrogen oxides, sulfur dioxide, and inorganic acids; artifacts arising from reaction and thermal desorption; limited retention capacity; thermal instability; and limitations of sampling volume, flow, and time.[12]

The Porapaks are a group of porous polymers that exhibit a wide range of polarity. The least polar member, Porapak P, is used in gas chromatography columns, and the most polar, Porapak T, can separate water and formaldehyde. Porapak QS is used to collect acetone cyanohydrin, Porapak P is used for dimethyl sulfate, and Porapak T is recommended for sampling hexachloro-1,3-cyclopentadiene.

The Chromosorbs are similar to the Porapaks. Chromosorb 101 is the least polar and Chromosorb 104 is the most polar. Chromosorb 106 is a cross-linked polystyrene porous polymer that has been used to sample nitromethane and 2-nitropropane. Chromosorb 102 has been recommended for collecting mevinphos, an organophosphate, and tetraethyl pyrophosphate.

XAD resins are a brand name referring to a number of different porous polymer types. XAD-2, the most commonly used, is equivalent to Chromosorb 102 and is used to collect anisidine and tetraethyl lead. XAD-4 has been used to collect organophosphorus pesticides. Amberlite XAD-2 is a styrene-divinylbenzene polymer.

Tenax. Tenax is one of the most widely used porous polymers, especially for environmental sampling. It has been used for studies of volatile organic compound (VOC) levels in indoor air for comparison with outdoor air levels. Compared to industrial situations, these levels can be expected to be very low. Tenax is a polymer of 2,6-diphenyl-p-phenylene oxide and can be used to collect organic bases, neutral compounds, and high-boiling compounds. Tenax is used mostly for sampling low concentrations of volatile compounds.

Tenax GC has a high thermal stability, being able to withstand temperatures of up to 350°C, which permits it to be used for thermal desorption. Thermal desorption involves heating the sorbent tube and "blowing" the contaminant into the analytical instrument with a gas. Thermal desorption is desirable because the entire sample is introduced into the analytical system, whereas the alternative, solvent extraction, dilutes the sample and allows only a portion of it to be injected. A limitation with thermal desorption is that the sample can be injected only once.

Tenax has other properties besides its temperature stability that make it useful for concentrating compounds of medium volatility. It is relatively inert and has low, but not zero, affinity for water vapor, and due to extensive use, its advantages and limitations have been well characterized. Multicomponent qualitative analysis by GC/MS and quantitative analysis with standards can be done.[13]

However, retention or breakthrough volumes for a variety of highly volatile compounds on Tenax are low.[14] Other limitations of Tenax include a laborious cleanup procedure to control blank problems, a short useful half-life after sample

TABLE 6.3. Types of Solid Sorbents

Solid Sorbent	Characteristics
Activated charcoal	Very large surface area : weight ratio.
	Reactive surface.
	High absorptive capacity.
	Breakthrough capacity is a function of the source of the charcoal, its particle size, and packing configuration in the sorbent bed.
Silica gel	Less reactive than charcoal.
	Polar in nature.
	Very hygroscopic (water adsorbing).
Porous polymers	Less surface area and less reactive surface than charcoal.
	Absorptive capacity is also lower.
	Reactivity is lower as well.
Molecular sieves	Zeolites and carbon molecular sieves that retain adsorbed species according to their molecular size.
	Water may displace organics.
Coated sorbents	A sorbent upon which a layer of reagent has been deposited.
	Adsorptive capacity determined by the capacity of the reagent to react with the particular analyte.

collection (approximately one month), and a tendency to decompose during sampling to produce acetophenone and benzaldehyde. Other problems that have been noted are as follows: In the case where Tenax is contaminated, blank corrections may not apply equally to all cartridges in a set; very high humidity in the air being sampled may affect sample retention; high concentrations or high numbers of compounds may exceed the retention capacity of the Tenax bed; and artifact formation may occur due to chemical reactions during sampling and/or thermal desorption. Tenax is not effective for low-molecular-weight hydrocarbons (C_4 and below) and mid-range (C_{5-12}) highly polar compounds.[15]

Tenax also reacts with strong oxidizing agents, such as chlorine, ozone, nitrogen oxides, and sulfur oxides, to form benzaldehyde, acetophenone, and phenol. Other documented chemical transformations on Tenax include oxidizers reacting with organics, such as styrene, to produce compounds, such as benzaldehyde and chlorostyrene. One way of removing oxidants before they reach Tenax is to use sodium-thiosulfate-impregnated glass fiber or Teflon filters in front of Tenax in the sampling train.[15]

Other Sorbents. Other sorbents occasionally used for integrated gas and vapor sampling include alumina gel and florisil. Alumina gel, a form of aluminum oxide, is rarely used as an air sampling sorbent except for special applications. Alumina gel permits several thousandfold concentration of pollutants. It selectively absorbs polar and higher molecular weight compounds, with the degree of polarity determining binding strength of a compound on this gel.[13] It has an affinity for water, like all polar sorbents. Florisil, based on silicic acid, is used for polychlorinated biphenyl (PCB) collection as well as for some pesticides. It is sometimes acid washed to remove residual magnesium. Table 6.3 lists types and uses of solid sorbents.[12]

Use of Sorbent Tubes

Occupational Sampling. Collection of gases and vapors on sorbent tubes is one of the most common technique used in occupational sampling. The most frequently used sorbent is the charcoal tube and analysis is almost always done by gas chromatography. Most often quantitative sampling is done to identify concentrations to which employees are exposed. However, an added advantage of sorbent sampling is that samples can be collected for qualitative as well as quantitative analyses. Qualitative samples are discussed further in the chapter on Bulk Sampling Methods in the section on bulk air samples.

If a single contaminant such as benzene is present at concentrations up to the PEL, a single 50/100-mg charcoal tube is sufficient. On the other hand, if several solvents in the same family are present, such as benzene, toluene, xylene, and ethylbenzene, it may be better to use a larger tube. The NIOSH or OSHA sampling method will dictate what media, flow rate, and time to sample; however, at a typical flow rate of 100 mL/min sampling would be done for 1.5 hours to collect a minimum volume of 10 L. Passive monitors can be used instead of active techniques. These are described in a later section of this chapter.

Procedure. This procedure assumes a pump with a rotameter is being used. Stroke volume pumps can also be used for sampling. Their use is described in Chapter 5.

1. A low-flow pump (10–200 mL/min) or a combination flow pump set at the low flow range is used. Hook up tubing to the pump inlet. If the pump is designed to be used with a specific tube holder (e.g., it contains a flow restricter), that holder must be attached to the end of the tubing.

2. Break the ends off a sorbent tube of the type to be used for sampling and insert it into the holder. If the adsorbent tube is inserted directly inside of tygon tubing without a special

holder, duct tape should be wrapped around both ends to minimize leaks and protect employees from the jagged glass tube opening.

3. If the sampling protocol requires the use of two tubes in series, or a prefilter in front of the tube, then these must be in the calibration train. The airflow to the pump should be in the direction of the arrow on the tube. There is usually a glass wool plug at the inlet end of tubes which can help achieve proper orientation.

4. Calibrate the pump using an electronic calibrator or rotameter as described in Chapter 5. The flowrate recommended in the procedure for the vapor(s) to be sampled should be used, usually about 100 mL/min with a solid sorbent tube in line, but falling within a range of 50–200 mL/min. Some methods give a range of flow rates. Higher rates are usually used when collecting short-term samples.

5. Determine whether high humidity is present. If so, a drying tube might be added to the front of the sampling train to absorb water. This is more important for silica gel and molecular sieves than for charcoal tubes.

6. Immediately prior to sampling, break off both ends of a new tube to provide an opening at least one half of the internal diameter at each end. There are tube breakers available or small needle-nosed pliers will work. Do not use the charging port or the exhaust outlet of the pump to break the ends of sorbent tubes to avoid damaging the pump. Label the tube with a unique number.

7. Connect the tube to the calibrated sampling pump. The smaller section of the tube is used as the backup section and is positioned closest to the pump.

8. Select the employee(s) or area(s) to be sampled. Discuss the purpose of sampling and advise the employee not to remove or tamper with the sampling equipment. Inform the employee when and where the equipment will be removed.

9. Attach the pump to the worker's belt, waist line, or the inside of a shirt pocket. Place the collection device on the shirt collar or otherwise near the employee's breathing zone. (See Figure 6.1.) Run the tubing either up the worker's back or under the arm. The sorbent tube should be placed in a vertical position, with the inlet either up or down during sampling to avoid channeling and premature breakthrough. Use duct tape to keep the tubing out of the employee's way so the tubing is not pulled off the pump.

10. Turn on the pump and record the start time on the sampling data form. Get the employee's full name, employee number, position or title, or record a description of the area in which the pump is placed. Include the pump number and the number on the sample, beginning sample time, air temperature, relative humidity, and atmospheric pressure if the elevation is above sea level.

11. Observe the rotameter on the pump for a short time to make sure it is pulling air. Mark down the setting of the rotameter ball (calibrated mark). As a minimum, check the pump after approximately the first half hour, hour, and every hour thereafter to make sure it is running. Ensure that the tubing is still attached to both the pump and the collection device and that it is not pinched. Make sure the sorbent tube is still in the same position.

12. Collect the sample for a sufficient time to meet the minimum sample volume requirements for the method. For example, organic vapors on a 150-mg charcoal tube require a 10-L sample volume; a 600-mg tube requires 40 L.

13. Periodically monitor the employee throughout the sampling period to ensure that sample integrity is maintained and job activities and work practices are identified.

14. If several tubes will be collected in sequence on the same employee, before removing the pump at the end of the sample period, check to see that the rotameter ball is still at the calibrated mark. If the ball is not at the calibrated mark, the flowrate of the pump must be adjusted prior to inserting another sorbent tube.

15. Record the ending time. Immediately following sampling the tube should be capped off on both ends. Solvent vapors may penetrate (diffuse) through plastic caps on charcoal tubes; therefore, it is important to store collected samples away from all sources of chemicals.[16] Following sampling at the end of the day the pumps must be recalibrated. Compare the flow rate at the beginning of the sampling period with this postcalibration.

16. Prepare the field blanks about the same time as sampling is begun. These field blanks should consist of unused solid sorbent tubes from the same lot used for sample collection. Handle and ship the field blanks exactly as the samples (break the ends and seal with plastic caps), but do not draw air through the field blanks. NIOSH recommends 2 field blanks be used for each 10 samples, with a maximum of 10 field blanks per sample set. In addition to the

sample tubes and field blanks, it is also recommended that 2 unopened tubes be shipped to be used as media blanks, so that desorption efficiency studies can be performed on the same lot of sorbent used for sampling.

Occupational Sampling with Multiple Tubes. Using the same techniques described for collecting gases and vapors on a single tube, it is possible to sample using several tubes at once. It is often done in indoor air situations where a battery of screening samples are collected to find if any contaminants are present at detectable levels. This technique is also useful for parallel sampling using tubes for different contaminants, including tubes of different size and composition. Long-term colorimetric tubes may be used in one or more ports.

Multiple tube holders are available that allow for sampling with up to four tubes at a time, but only one pump is required (Figure 6.3). Variable flow controllers attached to each tube allow for regulation of different flow rates for each tube. Tube flows are additive and the total cannot exceed the pump's capacity, which is frequently 500 mL/min for low flow pumps. The sampling rate of the pump is controlled by the characteristics of the limiting orifice assembly built into the tube manifold. The maximum flow rate that can be distributed among the tubes depends on the pump and should be determined ahead of time. This procedure assumes a combination (high–low) constant flow pump is being used.

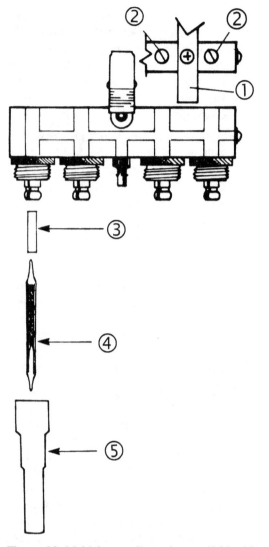

Figure 6.3. Multiple sampling tube manifold with adjustable flows. 1, Protective cover for flow adjustment; 2, flow adjustment screw (each tube has its own adjustment; 3, rubber tubing to hold tube to manifold; 4, sorbent tube; 5, protective cover. (Courtesy of SKC, Inc.)

Procedure

1. Select the correct tube manifold for the maximum number of tubes to be collected; do not plan to sample with empty slots. Flow rate selection for each tube depends on the minimum sampling volume for each method. It is generally best to set up tubes so

that all can be collected for the same time period. This often requires some preparatory calculations in order to match up all the tubes. Since setup can take some time, if repeat sampling is to be done it may be best to dedicate a pump/tube manifold combination for this purpose rather than configuring the system on each sampling day.

2. Break the ends of each tube and install the tube into the tube holder housing, with the arrow of the tube facing toward the manifold. Select the proper size cover for each tube.

3. If appropriate, follow the procedures provided by the pump manufacturer for converting combination pumps to low flow. Once this is done, hook up the tube manifold to the pump using Tygon tubing.

4. The end of a tube is hooked up with tubing to a calibration device such as a rotameter or bubble buret. Some tube covers require a fitting at the front end of the housing in order to attach tubing for calibration whereas for others the cover is removed. Tube covers used to hold tubes in place should be screwed down tightly on each end to make sure tubes are correctly positioned. Put a unique sample identification code on each tube as it is installed inside the holder.

5. At the base of the manifold there is a screw that is a fine flow adjustment for each tube. This screw is generally visible or underneath a protective cap. Ideally this is the only adjustment that will be necessary, but when as many as four tubes are in use, sometimes the gross flow adjustment on the pump has to be turned up.

6. Ideally, once the individual tube holders are set up, changes in the other holders should not affect the system, but as a practical matter they can. Thus, when adjusting the flowrate in one tube, it is best to double-check the flows of other tubes. Make sure the flow is stable for each tube. Shake the tube while the pump is running and watch the rotameter installed on the pump. If the flow through the tube is stable, the pump's flow rate will not change.

7. Once flows are set, if sampling is not to begin right away put caps on the ends of the tubes/tube holders and set the pump manifold system aside while the other pumps are being calibrated.

8. For sampling the manifold is installed with the tubes positioned vertically. The same procedures are followed as with the use of a single sorbent tube (steps 8–16).

Environmental Sampling. As noted, the Environmental Protection Agency's (EPA) air sampling methods for gases and vapors differ significantly from NIOSH/OSHA ones. When monitoring for low levels of contaminant in ambient air, such as a boundary line at a hazardous waste site or outdoor samples to compare with those collected inside a building suspected of having tight building syndrome, EPA techniques that provide enhanced sensitivity over those used for occupational sampling are needed. Most analyses are done using thermal rather than solvent desorption, allowing for an entire sample to be analyzed rather than a small portion.

Two examples of specific EPA methods for gases and vapors are for nonpolar organics.[17] Depending on the volatility of the compounds, different sorbent media are used. For highly volatile compounds having boiling points in the −15°C to 120°C range, carbon molecular sieve (CMS) tubes are used and the method is referred to as TO2. For less volatile compounds in the 80°C to 200°C range, Tenax tubes are used and the method is referred to as TO1. A significant precaution is the need to rigorously avoid contamination of exterior surfaces when cartridges are used to contain the sorbent as the entire surface is subjected to the purge gas stream during the desorption process. Modified versions of these methods allow glass tubes similar to occupational sorbent tubes to be used instead of cartridges.

Sorbent cartridges should be stored in Teflon-capped culture tubes wrapped in foil to limit exposure of sampling cartridges to ultraviolet (UV) light prior to sampling and during shipment. When sampling outdoors, meteorological information is very important. Following is a list of volatile organic compounds for which the CMS adsorption method has been evaluated.[17]

Acrylonitrile

Allyl chloride

Benzene

Carbon tetrachloride

Chloroform

1,2-Dichloroethane

Methylene chloride

Toluene

1,1,1-Trichloroethane

Vinyl chloride

Vinylidene chloride

Ambient air is drawn through the sample cartridges or tube(s) for up to 24 hours. For the 80°C to 200°C range a single Tenax tube is used, and a pair of CMS-filled cartridges are used for the −15°C to 120°C range. Most inorganic atmospheric compounds will pass through the CMS cartridges.

Prior to sampling, perform the following calculations. The maximum total volume (V_{max}) that can be sampled is calculated as

$$V_{max} = \frac{(\text{Retention vol.})(\text{wt. of Tenax, gm})}{1.5}$$

The maximum useable flowrate (Q_{max}) is calculated as

$$Q_{max} = \frac{V_{max}(1,000)}{\text{Sampling time}}$$

TABLE 6.4. Retention Volume Estimates for Compounds on Tenax

Compound	Estimated Retention Volume at 100°F L/g
Benzene	19
Bromobenzene	300
Bromoform	150
Carbon tetrachloride	8
Chlorobenzene	150
Chloroform	8
Cumene	440
1,2-Dichloroethane	10
1,2-Dichloropropane	30
1,3-Dichloropropane	90
Ethyl benzene	200
Ethylene dibromide	60
n-Heptane	20
l-Heptene	40
Tetrachloroethylene	80
Toluene	97
1,1,1-Trichloroethane	6
Trichloroethylene	20
Xylene	200

Table 6.4 lists retention volume estimates for compounds on Tenax[17] in units of liters per gram of Tenax.

Procedure

1. The sampling system consists of a pump, with a rotameter, flow regulator, and sampling cartridge. The EPA recommends the use of mass flow controllers instead of needle valves as flow regulators.

2. Calibrate the system for the flow rate to be used before sampling. If the sampling interval exceeds 4 hours, the flow rate should be checked at an intermediate point during sampling as well. The rotameter on the pump allows the flowrate to be checked without disturbing the sampling process. For more information on performing flow rate calibrations, see Chapter 5.

3. Remove fresh cartridges from their sealed container just prior to sampling and hook up using Tygon tubing. Use two cartridges in series; the second is a backup in case of breakthrough. If glass tubes are used, handle with polyester gloves and make sure they do not come into contact with other surfaces. Tape the sample number to the pump and to the cartridge container.

4. If high dust levels are present place a filter in a cassette on the inlet to the cartridges. Glass cartridges are connected using Teflon ferrules and threaded compression fittings.

5. Start the pump and record the following parameters: date, sampling location, time, ambient temperature, barometric pressure, relative humidity, flowrate, rotameter reading, and sample number.

6. At the end of the sampling period remove the tubes (cartridges) one at a time and replace in the original container (use the gloves for the glass cartridges). Seal the cartridges or culture tubes in a friction-top can containing a layer of charcoal and store at reduced temperature (20°C) before and during shipment.

7. If the flowrates at the beginning and end of the sampling period differed by more than 10%, mark the cartridges as suspect.

8. Calculate the average flowrate for each set of cartridges:

$$Q_a = \frac{Q_1 + Q_2 + \cdots + Q_n}{n}$$

where $Q_1, Q_2, \ldots, Q_n$ = flow rates determined by beginning, intermediate, and end points during sampling.

9. Calculate the total volume (V_m) for each cartridge:

$$V_m = \frac{(\text{Sampling time})(\text{average flowrate})}{1000}$$

10. Adjust the volume (V_s) for differences in temperature and pressure from STP at the sampling site.

$$V_s = \frac{V_m (P_a)(298)}{(760)(273 + t_a)}$$

where P_a = pressure at sampling site (mm Hg) and t_a = temperature at sampling site, °C.

11. During each sampling event at least one set of parallel samples (two or more samples collected simultaneously) should be collected, preferably at different flowrates. If agreement between parallel samples is not generally within ±25%, the user should collect parallel samples on a much more frequent basis.

Impingers

The basic design of the all-glass midget impinger was developed in 1944.[18] Originally the impinger was used primarily for collecting aerosols, especially for particle sizing, but its current application is for gas and vapor sampling. The midget impinger (Figure 6.4) is designed to contain 10–20 mL of liquid, while earlier impingers such as the Greenburg–Smith impinger held much larger volumes. The function of these absorbers is to provide sufficient contact between the sampled air and the liquid surface to provide complete absorption of the gas or vapor.

These devices are usually made of glass with an inlet tube connected to a stopper fitted into a graduated vial such that the inlet tube rests slightly above the vial bottom. A measured volume of absorber liquid is placed into the vial, the stopper inlet is put in place, and the unit is then connected to the pump by flexible tubing.

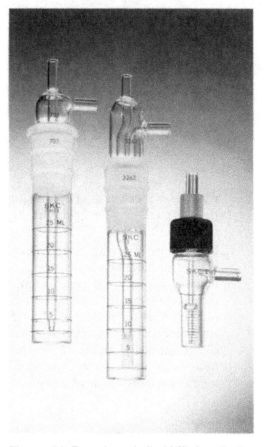

Figure 6.4. Examples of liquid-filled collection devices. *Left* to *right*: a midget impinger, a midget bubbler, and a spill-proof impinger. (Courtesy of SKC, Inc.)

When the pump is turned on, the contaminated air is channeled down through the liquid at a right angle to the bottom of the vial. The airstream then impinges against the vial bottom, mixing the air with the absorber liquid; the necessary air-to-liquid contact is achieved by agitation. Some impingers are made of Teflon while some glass units are coated with a plastic film to hold the glass pieces together in the event of breakage.

Fritted bubblers are similar in use and appearance to an impinger, but generally are used when a higher degree of air-liquid mixing is desired. The fritted glass tip is at the end of the impinger tube where the air goes into the solution. With these devices, the contaminated air is forced through porous glass, breaking the airstream into numerous small bubbles. The frits are categorized as fine, coarse, or extra coarse, depending on the number of openings per unit area. The selection of what size frit to use depends on the ease of collection, with coarse frits used for gases and vapors that are soluble and frits of fine porosity used for gases that are difficult to collect. However, the finer the frit, the higher the pressure drop. Airflow through frits must be carefully controlled to avoid the formation of large bubbles. The size of the bubbles depends on the size of these openings as well as the type of absorber solution.

Impingers are suitable for collecting nonreactive gases and vapors that are highly soluble in the absorbing liquid, such as hydrogen chloride, as well as those that react rapidly with a reagent in the solution, such as occurs in the neutralization of strong acids and bases. The reagent is added to enhance solubility or to react with the contaminant once it is in the solution to stabilize it, thus reducing volatility and minimizing losses. Although most absorbing solutions are mixed in water, sometimes solubility in another compound such as a solvent is used. For example, toluene diisocyanate (TDI) is soluble in toluene, but due to the ease with which toluene vaporizes, this solvent would be used only in situations where other TDI sampling methods could not be used. An example is a heated, humid process, such as sampling a vent on a wet scrubber, to determine its efficiency. Fritted bubblers are used for minimally soluble gases and vapors such as chlorine and nitrogen dioxide.

Variations on the basic midget impinger shape have also been developed, including one with a membrane and another with a rounded shape designed to minimize the potential for spills. The membrane device has a hydrophobic Teflon membrane at either end of the tube held in place by a polypropylene cap to which tubing and a pump can be hooked. The membrane retains the sampling solution yet allows air to pass through it. The units are designed to operate at 1 Lpm, the same rate used for traditional occupational impinger sampling. The main concern in the use of these units is to measure the pressure drop on a unit to assure there are no significant variations in the filter membrane pore structure.

The most important factors to consider when sampling with impingers/bubblers are sampling flow rate, solubility of the contaminant in the absorbing solution, rate of diffusion of the contaminant into the solution, contaminant vapor pressure (volatility), volatility of the absorbing solution, and reactivity of the contaminant being collected with the absorbing solution.[19] The size of the bubble determines how much contact there is between the gas being sampled and the absorbing solution. The number of the air bubbles being released into the solution is also important. In general, the lower the sampling rate, the more complete the absorption.

Problems with the use of impingers/bubblers include condensation of material in the sampling lines and losses by adsorption or volatilization from the equipment. Impingers are bulky, and if used for personal sampling, they must be watched carefully as actions such as bending may spill liquid into the sampling lines and the precautions will limit the workers' mobility. If the liquid reaches the pump, the pump will usually break down and require repair. The use of traps in between impingers and pumps will usually minimize this type of damage to pumps, but some sample may still be lost in the lines or the trap. Glass impingers are susceptible to breakage. In some cases, the method requires immediate processing in the field in order to stabilize the collected material and often impinger methods suffer from interferences.

Frits can also create problems, such as retention of the contaminant, excessive frothing and foaming, variable pressure drop, and nonphysical uniformity from unit to unit in the same batch,[19] which precludes interchangeability without recalibration, fragility, and particulate retention not encountered with the use of impinger-type absorbers. Frits can accumulate particles that are not removed by conventional cleaning procedures after repeated use.

Occupational Sampling. Occupational sampling with impingers is being phased out due to the need to carry reagents and the potential for the impingers to spill when worn by workers. Most commonly it is replaced by the use of sorbent methods; however, there may be a situation where the sorbent tube is not adequate such as high humidity or only an impinger method exists for the contaminant of interest. As analysis of liquid solutions is often done using colorimetric techniques, it is especially important to identify any interferences that might be present. The laboratory should report any off-colors that occur during the analysis as this is suggestive of interferences.

The sampling methods for midget impingers and fritted bubblers are the same. Most standard methods rely on one or two impingers to absorb contaminants; however, some may require several connected in series. Two impingers or bubblers connected in series will increase the overall collection efficiency. If the amount collected in the second impinger is greater than 10%, sample breakthrough is a concern. Table 6.5 displays the effect of two bubblers in series on the collection efficiency of various compounds.[19]

There may be cases where there is a significant amount of particules present. Particules may interfere with analysis, clog orifices, and interfere with collection. In this case, a prefilter in a cassette should precede the impinger. The filter and cas-

TABLE 6.5. Effect of Two Bubblers in Series on Collection Efficiency

| Vapor | Solvent | Maximum Air Volume in Liters for 95% Recovery | |
		1 Bubbler (10 mL)	2 Bubblers (10 mL each)
Acetone	Water	0.8	5.4
Methanol	Water	9.0	62.0
n-Butanol	Water	10.0	68.0
Chloroform	Isopropanol	0.8	5.4
Carbon tetrachloride	Isopropanol	0.5	3.4
Methyl chloroform	Isopropanol	0.5	3.4
Trichloroethylene	Isopropanol	0.9	6.2

sette material should be nonreactive and nonabsorbing to the gases being sampled. For more information on filters, see the next chapter.

With impingers the sampling liquid is transferred to a vial or small bottle after sampling. If a light-sensitive reagent is being used, brown bottles will be needed. Even when using traps between impingers and pumps it is a good idea to pack extra pumps. For example, if the wrong part of the impinger is hooked directly to the pump, liquid will be sucked into the hose and can saturate the trap. Also pack extra impingers as the bottles and the glass tubes inside break easily. Take care when handling tubes with frits on the end as the fritted portion may snap off if pressure is applied to it. If bubbles do not appear in the impingers/bubblers when the pump is turned on, there is a leak. Often this is due to the tubing not fitting tightly on the impinger. In this case, wrap 1-inch strips of duct tape tightly around these joints. Generally only 10 mL of solution is added to an impinger flask, and absolutely no more than 20 mL should be added.

Procedure

1. Fill the impinger/bubbler with 10 mL of the sampling solution. If two impingers in series are to be used,

then these must be in place. Care should be taken to see that frits or tips are not damaged and that parts are secured tightly. The impingers should be labeled with the sample identification code.

2. Place a tygon tube over the outlet stem on the impinger. Ensure you have identified the outlet stem. If the wrong stem is selected, liquid could be sucked into the pump. In virtually every case this problem will result in having to have the pump repaired, usually at the factory. One way of identifying the correct stem is to see where the glass tubing goes inside the impinger. The correct one does *not* go into the solution.

3. A second tube is run from the first impinger to the bubble buret or rotameter. This one should be hooked to the stem of the glass tube that goes into the solution of the impinger.

4. A trap containing charcoal granules and glass wool at each end should be placed in between the impinger(s) and the pump using tubing.

5. If no bubbles appear once the pump is turned on, it is likely that there are leaks in the system and the fittings should all be checked. Because the

tygon tubing must fit tightly over the glass tubing stems of the impingers, or leaks will result, it is useful to tape over these connections with duct tape.

6. Once the system is working correctly, the same method for calibrating with the bubble buret or rotameter, as described in Chapter 5, should be followed. A flowrate of 1 Lpm is generally used with impingers for gas and vapor sampling.

7. If high levels of dust are present, there is a danger that the orifice in the impinger or the interstices in the bubbler frit might get clogged, so a glass fiber filter in a cassette should be put in front of the impinger. This must also be in the sampling train during calibration.

8. The impinger(s) is placed in an impinger holster and affixed to the employee's clothing. Tape can be used or a large "safety pin." If tape is used, it is very important that the employee does not bend over or otherwise tilt the impinger, which can result in reagent flowing into the tube. The purpose of the trap is to catch any liquid that might get into the tubing. Follow steps 8–16 as outlined in occupational sorbent tube sampling for basic procedures during the sampling period.

9. It is important to note that spillproof impingers are available that minimize the likelihood of spillage.

10. In some cases, it will be necessary to add additional solution during the sampling period to make up for losses. The amount of solution should not drop below one-half of the original amount before more is added.

11. Following sampling, have a labeled (with the sample number) glass bottle ready. Rinse the absorbing solution adhering to the outside and inside of the stem directly into the flask with a small amount (1 mL or 2 mL) of the sampling reagent. Pour the contents of the flask into the sample bottle.

12. Some sampling solutions are light sensitive. In this case, the impingers should be covered with duct tape and the sample bottles should be dark brown glass. The solutions should also be in dark brown glass bottles and should be kept inside a bag, box, or cooler when not in use.

13. If the impingers must be reused, carefully rinse them with the sampling solution using chemist's technique: Pour in about 5 mL of solution and carefully rotate the impinger on its side so that the entire interior is rinsed. If a water-based sampling solution is being used, the impinger can be rinsed using a distilled water and drained thoroughly to preserve the amount of sampling solution. If possible, the glass tubing should be rinsed and drained as well.

Environmental Sampling. Environmental samples are collected using the same types of impingers used for occupational samples; however, the units are kept in an ice bath during sampling. While occupational sampling methods using impingers often rely on colorimetric methods for analysis that suffer from interferences, environmental methods use high-pressure liquid chromatography (HPLC), a more sophisticated technique that is capable of detecting many specific compounds with few interferences. As with other environmental methods, impinger techniques are designed to detect lower concentrations than occupational methodologies.[20]

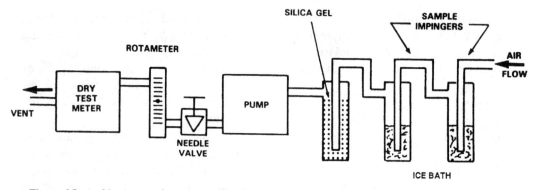

Figure 6.5. Ambient gas and vapor sampling with impingers. (From *Compendium of Methods for the Determination of Toxic Organic Compounds in Ambient Air*, EPA/600/4-87/066 [supplement to EPA/600/4-84-041].)

An example of the sampling technique is contained in EPA Method TO5 for Aldehydes and Ketones (see the following list of those compounds for which the TO5 Method has been evaluated).[17]

Acetaldehyde

Acetone

Acrolein

Benzaldehyde

Crotonaldehyde

Formaldehyde

Hexanal

Isobutyraldehyde

Methyl ethyl ketone

Pentanal

Propanal

o-Tolualdehyde

m-Tolualdehyde

p-Tolualdehyde

Ambient air is drawn through a midget impinger containing 2,4-dinitrophenylhydrazine (DNPH reagent) and isooctane over a 2-hour sampling period. To estimate a 24-hour exposure, several samples are collected during this period. Aldehydes and ketones readily form stable 2,4-DNPH derivatives. After sampling, the impinger solution is returned to the laboratory for analysis. Reversed phase HPLC–UV (high-pressure liquid chromatography with an ultraviolet detector) analysis at 370 nm determines the DNPH derivatives.

Procedure

1. The sampling train is assembled (Figure 6.5). It consists of a pump, needle valve, rotameter, and ice bath with two impingers in series containing DNPH reagent followed by a silica gel tube to act as an adsorber to prevent moisture from reaching the pump. All glassware must be rinsed with methanol and oven-dried prior to use. If the pump has a built-in rotameter, a second one is unnecessary.

2. Prior to sample collection the flow rate is calibrated using techniques described in Chapter 5. In general, flowrates of 100–1000 mL/min are useful. Flowrates greater than 1000 mL/min should not be used because impinger collection efficiency may decrease.

3. Record the flowrate at the beginning of the sampling period. If the sampling period exceeds 2 hours, the flow

rate should be measured at intermediate points during the sampling period. If a rotameter is incorporated into the sampling train, the flowrate can be observed periodically.

4. To collect an air sample, two clean midget impingers are each filled with 10 mL of purified DNPH reagent and 10 mL of isooctane. The impingers are connected in series to the sampling system and immersed in an ice bath as shown in Figure 6.5. Sample flow is started. The following parameters are recorded: date, sampling location, time, ambient temperature, barometric pressure, relative humidity, flow rate, rotameter setting, DNPH reagent batch number, and pump identification numbers.

5. The sampler is allowed to operate for the desired period with periodic recording of the variables listed above. The total volume should not exceed 80 L. Both the ice bath and, if necessary, the solution in the impingers should be replenished. The purpose of the ice bath is to slow the evaporation of the sampling solution. At least 2–3 mL of isooctane must remain in the first impinger at the end of the sampling interval.

6. At the end of the sampling period the flow rate is again checked. If the flowrate at the beginning and end of the sampling period differ by more than 15%, the sample should be marked as suspect.

7. Immediately after sampling the impingers are removed from the sampling system. The contents of the first impinger are emptied into a clean 50-mL glass vial having a Teflon-lined screw cap. The first impinger is then rinsed with the contents of the second (backup) impinger and the rinse solution is added to the vial. The vial is then capped, sealed with Teflon tape,

and placed in a friction top can containing 1–2 inches of granular charcoal and stored in a cooler.

8. Calculate the average flowrate (Q_a) in the same manner as in environmental sampling with sorbents. The total volume is then calculated using the following equation:

$$V_m = \frac{(\text{Time}_2 - \text{Time}_1)Q_a}{1000}$$

Collecting Integrated Bag Samples

In addition to occupational sampling and calibration applications, bags are used to collect gases and vapors for high-resolution analysis of environmental samples. This method is commonly used for collecting air samples in open fields or from vapor wells. For more information on this technique see the chapter on bulk sampling methods (Chapter 20).

When used for integrated sampling, bags can collect (1) certain types of gases and vapors for which other methods are unsuitable due to very low breakthrough volumes on sorbent tubes; (2) gases and vapors for which no alternative method exists; and (3) mixtures where method incompatibilities exist. They can also be used in conjunction with real-time instruments for collecting time-weighted average (TWA) samples and for the transport of calibration standards.[21] Although most often bags are used to collect grab or area samples, it is also possible to use a bag to collect a personal integrated sample (Figure 6.6). This is done when no other suitable method exists, and the concentration in the ambient air is above the Limit of Detection for the analytical instrument, as for waste anesthetic gases.

Bags come in different sizes, shapes, and materials. Most bags have a valve to allow for filling using a pump and a septum for injections using a syringe. Septums can be

Figure 6.6. A special sampling bag, when connected to the outlet of a pump, can be used to collect personal air samples. An employee can carry the bag in a backpack with a sampling hose extending to their breathing zone. (Courtesy of SKC, Inc.)

made of silicone or neoprene rubber and valves commonly are nickel-plated or stainless steel. Stainless steel valves are less likely to corrode than nickel plated ones. Fittings on bags are often multipurpose. For example, the fitting may contain both a syringe port with Teflon-coated septum and a hose connection. The fitting also acts as a shutoff valve for the hose connection. A Teflon-coated septum is often used and should be replaced after each use. When replacing the septum, be sure the Teflon side is toward the inside of the bag.

The material of construction is an important consideration to protect the integrity of the sample and to ensure that no part of the sample is lost through permeation through the walls of the bag. These bags can be of plastics but more often are of materials, such as Tedlar, Mylar,

Scotchpak, and Teflon. Often materials are laminated to aluminum to seal the pores in the plastic and reduce the likelihood of loss of a gas sample by permeation through the walls of the bag. For example, one five-layer material has aluminum foil at its core, but because aluminum reacts with many compounds, this foil is protected on both sides. The exterior is covered by a layer of polyvinylidene chloride followed by a film of polyester. The interior surface is coated with a film of polyamide followed by a layer of polyethylene. Plastic bags are transparent, allowing detection of condensation inside the bag. However, they can stretch, altering their volume size, and they are prone to residual contamination from prior samples.

The sampling period is determined by the purpose for sampling. A personal sample will be collected over 4–8 hours. For a grab sample, the maximum flow rate available on the pump can be used to fill a bag. Precise calibration of the pump is not necessary; however, if a specific sampling period is desired, the sized of the bag and the flow rate are important. NIOSH methods recommend Tedlar bags and using new bags for each sample. Since a single use may prove too expensive to be practical, the sampler can take steps to guard against bags becoming contaminated, thus preventing their reuse.

Factors that can affect the suitability of bags include the material from which they are constructed, presence or absence of other compounds, humidity, concentration of the sample in the bag, size of the bag, preequilibration, and bag-to-bag variation. Sources of sample losses from bags include leaks either through valves used in filling and sampling from the bag or through poor seam seals during bag manufacture; chemical reactions of the sample with the bag walls or other compounds collected along with it; adsorption of the sample into the bag material; and permeation of the sample through the walls of the bag.

As bags may develop a memory (retain traces of compounds previously collected), new bags should be used for taking samples of new materials unless there is proof that the bag has been adequately cleaned. This criterion is especially important in the case where air is being collected to analyze for unknowns.[22]

Bags should be leak tested, cleaned, and preconditioned by flushing with the gas to be sampled (if known) or zero air prior to use. If vapor condenses inside of a sampling bag, the bag can be purged by placing it in hot water with the valve open.

Procedure

1. For occupational sampling, the bag should be attached to the exhaust outlet of a low flow pump, with a flowrate of 20–200 mL/min. The sampling period determines the size of the sample bag and the flowrate of the pump. For example, using a 5-L bag and a sampling rate of 200 mL/min will fill a bag in 25 minutes. Sampling at 20 mL/min will fill the bag in 4 hours.

2. Attach the pump to the employee's belt and place the bag over the shoulder or in a backpack. Open the valve on the bag and turn on the pump.

3. Record start and stop times. After the sample is collected, the bag should be disconnected from the pump.

4. The sample is now ready to be sent to a laboratory for analysis. In some cases a direct reading instrument or colorimetric tube can be used for field analysis.

Environmental Sampling for VOCs Using Stainless Steel Canisters

Collection of ambient air samples in stainless-steel canisters provides convenient integration of ambient samples over longer time periods, such as 24 hours; remote sampling and central analysis; ease of storing and shipping samples; unattended sample collection; analysis of samples from multiple sites with one analytical system; and collection of sufficient sample volume to allow assessment of measurement precision and/or analysis of samples by several analytical systems.

Volatile organic compounds (VOCs) enter the atmosphere from a variety of sources, including petroleum refineries, synthetic organic chemical plants, natural gas processing plants, and automobile exhaust. VOCs are generally classified as those organics having saturated vapor pressures greater than 10^{-1} mm Hg at 25°C. Conventional methods for VOC determination use solid sorbents such as Tenax, described previously in the section on environmental sampling with sorbents. Following is a list of VOCs that can be detected with EPA canister methods TO-14.

Benzene
Benzyl chloride
Carbon tetrachloride
Chlorobenzene
Chloroform
1,2-Dibromoethane
Dichloromethane
m-Dichlorobenzene
o-Dichlorobenzene
p-Dichlorobenzene
1,1-Dichloroethane
1,2-Dichloroethane
1,2-Dichloropropane
cis-1,3-Dichloropropane
trans-1,3-Dichloropropene
Ethyl benzene
Ethyl chloride
Freon 11
Freon 12
Freon 113
Freon 114

Hexachlorobutadiene
Methyl bromide
Methyl chloride
Methyl chloroform
Styrene
1,1,2,2-Tetrachloroethane
Tetrachloroethylene
Toluene
1,2,4-Trichlorobenzene
1,1,2-Trichloroethane
Trichloroethylene
1,2,4-Trimethylbenzene
1,3,5-Trimethylbenzene
Vinyl chloride
Vinylidene chloride
m-Xylene
p-Xylene
o-Xylene

Figure 6.7. Evacuated sampling canisters can be used to collect environmental samples. A critical orifice is often used to control flowrate. (Courtesy of SKC, Inc.)

Canisters can be used both indoors and outdoors. An air sample is drawn from the ambient air into an initially evacuated stainless steel canister that has been treated internally with an inert coating. One type of coating is the SUMMA® process, in which chrome-nickel oxide is formed on the canister's interior. Another interior surface treatment prevents sulfur compounds from reacting with the interior surface of the canister, making it suitable for collecting low-level sulfur VOCs. Two sampling modes are used: passive and pressurized. In passive sampling an evacuated canister is opened to the atmosphere and the differential pressure causes the canister to fill. For pressurized sampling, an initially evacuated canister is filled by the action of the flow-controlled pump from near atmospheric vacuum to a positive pressure not to exceed 25 psig (pounds per square inch gauge). Commercial canister units are usually sold with a filter that traps particulate material entering the sampling inlet. A mass flowmeter or valve is used to control airflow into the canisters.

Care must be exercised in cleaning and handling sample canisters and associated sampling apparatus to avoid losses or contamination of the samples. Contamination is a critical issue with canister-based sampling. Cleaning of canisters can be done by a cycle of first evacuating them to a pressure of 5 mm Hg followed by pressurizing them to 40 psig with clean dried air. Cleaning should be verified by injecting a sample of canister air into a GC. Figure 6.7 shows ambient gas and vapor sampling with a stainless steel canister. A field sampling data sheet for use with canisters is shown in Figure 6.8.

Passive Sampling[22]

Procedure
1. This technique may be used to collect grab samples (duration of 10 or 30 seconds) or time-integrated samples taken through a flow-restrictive inlet, such as a mass flow controller or a critical orifice. For passive collection the canister is first evacuated at a laboratory. When opened to the

A. General Information

Site Location: _____   Shipping Date: _____
Site Address: _____   Canister Serial No.: _____
_____   Sampler ID: _____
_____   Operator: _____
Sampling Date: _____   Canister Leak
                                   Check Date: _____

B. Sampling Information

Temperature

	Interior	Ambient	Maximum	Minimum	Canister Pressure
Start			✕	✕	
Stop					✕

Pressure — Canister Pressure

Sampling Times / Flow Rates

	Local Time	Elapsed Time Meter Reading	Manifold Flow Rate	Canister Flow Rate	Flow Controller Readout
Start					
Stop					

Sampling System Certification Date: _____
Quarterly Recertification Date: _____

C. Laboratory Information

Date Received: _____
Received by: _____
Initial Pressure: _____
Final Pressure: _____
Dilution Factor: _____
Analysis Results*: _____
_____
_____
_____
_____
_____
_____
_____
_____

_____
Signature/Title

*Attach Data Sheets

Figure 6.8. Canister sampling field data sheet.

atmosphere containing the VOCs to be sampled, the differential pressure causes the sample to flow into the canister without the need for a pump.

2. In its simplest version, a manual sampling apparatus consisting of the evacuated canister and a valve preset for the desired flowrate is used. The valve is screwed on the canister top for sampling. The canister valve is opened and a timer, if in use, is programmed for the scheduled sample period and set to the correct time of day and date. The elapsed time meter should be set to zero. Several canisters attached to a manifold with separate solenoid sampling valves permit sequential sampling.

3. In a more complex version, a pump draws sample air through the sample inlet and particulate filter to purge and equilibrate these components. The inlet line begins to heat at this time as well to 65–70°C. At the start of the sampling period solenoid valves are activated, stopping the purge flow and allowing sample air to be collected.

4. At the end of the sample period, the second timer stops sample flow and seals off the canister or, alternately, the valve is manually shut.

Pressurized Sampling. Pressurized sampling is used when longer-term integrated samples or higher-volume samples are required. In pressurized canister sampling, a metal bellows-type pump draws in ambient air from the sampling manifold to fill and pressurize the sample canister to a 15 to 30-psig final pressure (Figure 6.9). For example, a 6-L evacuated canister can be filled at 10 mL/min for 24 hours to achieve a final pressure of about 21 psig.

A flow control device is chosen to maintain a constant flow into the canister over the desired sample period. This flowrate is determined so the canister is filled over the desired sample period. The flowrate can be calculated by

$$F = \frac{(P)(V)}{(T)(60)}$$

where F is the flowrate in mL/min, P is the final canister pressure, atmospheres, V is the volume of the canister in mL, and T is the sample period in hours.

For automatic operation, the timer is programmed to start and stop the pump at appropriate times for the desired sample period. The timer must also control the solenoid valve, to open the valve when starting the pump and to close the valve when stopping the pump.

The connecting lines between the sample inlet and the canister should be as short as possible to minimize their volume. The flowrate into the canister should remain relatively constant over the entire sampling period.

Procedure

1. Insert a "practice" canister into the sampling system. A certified mass flowmeter is attached to the inlet line of the manifold, just in front of the filter. The canister is opened, the sampler is turned on, and the reading of the mass flowmeter is compared to the sampler mass flow controller. The values should agree within ±10%. If not, the sampler mass flowmeter must be recalibrated or there is a leak in the system.

2. Adjust the flowrate to the proper value (e.g., 3.5 mL/min for 24 hours, 7.0 mL/min for 12 hours). Record this final flow rate on the sampling data sheet.

3. Turn off the sampler and reset the elapsed time meter to 000.0 minutes. Any time the sampler is turned off, at least 30 seconds should pass before it

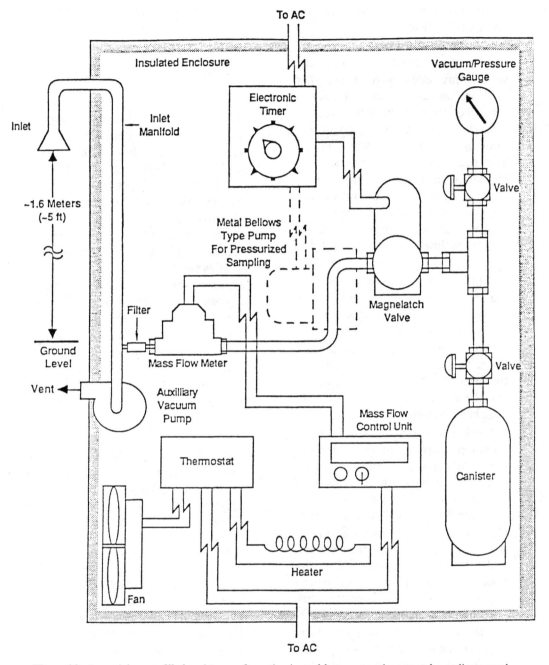

Figure 6.9. A special pump-filled canister configuration is used for some environmental sampling according to U.S. EPA methods. (From *The Determination of Volatile Organic Compounds (VOCs) in Ambient Air Using Summa® Passivated Canister Sampling and Gas Chromatographic Analysis,* U.S. EPA, May 1988.)

is turned on again. Disconnect the "practice" canister and replace it with a certified clean canister.

4. Open the canister valve and vacuum/pressure gauge valve. The pressure/vacuum in the canister is recorded on the canister sampling field data sheet as indicated by the sampler vacuum/pressure gauge. The gauge is then closed and the maximum–minimum thermometer is reset to the current temperature. The time of day and elapsed time meter readings are also recorded. Set the electronic timer for the start and stop time.

5. At the end of the sampling period, the vacuum/pressure gauge valve on the sampler is briefly opened and closed and the pressure/vacuum is recorded on the sampling field data sheet. The pressure should be close to the desired pressure. The time of day and the elapsed time meter readings are also recorded.

6. Following sampling, the maximum, minimum, current interior temperature and current ambient temperature are recorded on the sampling field data sheet. The current reading from the flow controller is recorded.

7. Close the canister valve, disconnect the sampling line from the canister, and remove the canister from the system. The final flowrate is measured and recorded on the canister sampling field data sheet and the sampler is turned off. Attach an identification tag to the canister containing the canister serial number, sample number, location, and date.

PASSIVE COLLECTORS FOR GASES AND VAPORS

One of the most important changes in air sampling for toxic compounds over the past decade has been the improved technology and also increased acceptance of passive monitors for gases and vapors. The increased use of passive monitors can be traced to several factors:

- Literature reports and validation studies that scientifically compared passive monitor performance with results obtained by other sampling techniques. These studies typically used laboratory-generated atmospheres of known concentration or side-by-side chamber exposures in the work environment of two or more different sampling devices. These sound studies replaced earlier informal evaluations where the employer placed a passive monitor alongside the "accepted" active sample collection system, and then judged the passive monitor's accuracy by how close it came to the conventional method.

- User experience with the technology that confirmed the results of experimental studies showing the validity of this approach.

- Publication of the ANSI/ISEA 104-1998, "American National Standard for Air Sampling Devices—Diffusive Type for Toxic Gases and Vapors in Working Environments,"[23] which provided a consistent method of testing the devices for performance-related factors such as effects of reverse diffusion and the impact of wind direction and sampler orientation on the device's accuracy in measuring airborne levels. This standard provides a framework for the manufacturer to test their monitors for each contaminant so the user has a measure of confidence that relevant factors have been considered and documented.

- Employers' recognized need to collect more air samples than previously was their practice. One factor was the increased understanding of sampling strategy principles (as discussed in Chapter 3) that a certain minimum number of samples is needed to adequately judge the acceptability of exposures for any given job category.[24] A second factor was legal standards such as OSHA's benzene rule that required initial monitoring of every job category on every work shift exposed to benzene from liquid mixtures containing 1% or more benzene.[25] The net impact of these factors was to increase the number of air samples collected, which led employers to seek valid approaches that required less sampling equipment and hands-on attention by skilled staff.

These devices have the potential for replacing some active sampling methods using battery powered pumps along with sample collection devices. The obvious advantages to their use are simplicity and convenience; generally they must only be slipped out of their sealed packaging and clipped onto an individual's collar or suspended in the environment to be measured. When used for personal monitoring, they do not interfere with the person's activity or change the individual's behavior pattern to the same degree as wearing a pump, sampling device and connecting tubing. They are also safe for use in potentially flammable locations since they have no electrical components or moving parts. They all require a minimum air velocity past the monitor and thus are not suitable for stagnant fixed locations such as closets or confined spaces. Most of these units are dosimeters; that is, they accumulate the airborne vapors over the sampling period and give an integrated measurement. All of the devices discussed in this section require that the passive monitor be sent to a laboratory for analysis using the methods listed in Table 6.1. There are some passive monitors that are designed to change color in proportion to the airborne concentration.

These are covered in Chapter 13 on colorimetric systems.

Table 6.6 lists contaminants that can be sampled using passive monitors. Most passive monitors are "clip-on" badges in a variety of configurations (Figure 6.10). The most commonly used passive monitors contain charcoal while others contain a treated charcoal or a solid adsorbent such as tenax or molecular sieve. However, some badge-like passive devices use special collection media, such as:

- Gold film for collection of airborne mercury.
- Glass fiber filter impregnated with 2,4-dintrophenylhydrazine (DNPH) for collection of formaldehyde.
- Silica impregnated with 2-(hydroxymethyl)piperidine (2-HMP) for collection of acrolein.

The special collection media function by reacting with the contaminant of interest, thus "locking up" the contaminant as a stable compound and reducing the potential for subsequent loss due to back diffusion or other storage losses.[24]

A few commercially available passive monitors use a design other than the badge-type configuration. These are discussed later in this chapter.

Diffusion and Permeation Principles

Basically these devices rely on one of two basic collection principles: diffusion or permeation. Practically all commercially available passive monitors are diffusion devices. These typically consist of a charcoal or other collection pad or wafer within a badge-like housing. A grid with precision-sized sampling ports controls the diffusion area and, hence, the sampling rate, and it also acts as a wind screen. A cover protects the device for shipping (Figure 6.11). With diffusion-controlled monitors, the mass uptake of the monitor is controlled by the

TABLE 6.6. Typical Compounds Sampled on Passive Dosimeters

Acetaldehyde	Dimethylaniline
Acetic acid	Dioxane
Acetone	Dipropylene glycol methyl ether
Acetonitrile	Enflurane (Ethrane)®
Acrolein	Epichlorohydrin
Acrylonitrile	Ethanolamine
Allyl alcohol	Ethoxyethanol (Cellosolve)
Allyl chloride	Ethoxyethyl acetate (ethyl Cellosolve)
Ammonia	Ethyl acetate
Amyl acetate	Ethyl acrylate
Benzaldehyde	Ethyl alcohol (ethanol)
Benzene	Ethyl-2-cyanoacrylate
Benzyl chloride	Ethyl ether
Butadiene	Ethyl lactate
Butanol	Ethylamine
Butoxyethanol (butyl Cellosolve)	Ethylbenzene
Butyl acetate	Ethylene chlorohydrin
Butyl acrylate	Ethylene dibromide
n-Butylamine	Ethylene glycol dimethyl ether
Butyl(n)glycidyl ether	Ethylene oxide
Butyraldehyde	Ethylenediamine
Camphor	Fluorotrichloromethane (CFC11)
Carbon tetrachloride	Formaldehyde
Chloro(o)styrene	Glutaraldehyde
Chloro(o)toluene	Halothane
Chlorobenzene	Heptane
Chlorobromomethane	Heptanone (methyl amyl ketone)
Chloroform	Hexanaldehyde
Chloroprene	Hexane
Cumene	Hexanone (MBK)
Cyclohexane	Hydrogen peroxide
Cyclohexanol	Isoamyl acetate
Cyclohexanone	Isoamyl alcohol
Cyclohexylamine	Isobutyl acetate
Desflurane (Suprane)®	Isobutyl alcohol
Diacetone alcohol	Isocyanates
Dichloro-1-fluoroethane (HCFC141b)	Isoflurane (Forane)®
Dichlorobenzene	Isophorone
Dichlorodifluoromethane (CFC12)	Isopropyl acetate
Dichloroethane	Isopropyl alcohol
Dichloroethyl ether	Isopropyl ether
Dichloroethylene	Isopropyl glycidyl ether (IGE)
Dichlorofluoromethane (CFC21)	Isopropylamine
Dichlorotetrafluoroethane (CFC114)	Limonene (as dipentene)
Dicyclopentadiene	Mercury
Diethanolamine	Mesityl oxide
Diethylamine	Methoxyethanol (methyl Cellosolve)
Diethylenetriamine	Methoxyethyl acetate
Diisobutylketone	Methoxyflurane
Dimethyl formamide (DMF)	Methyl acrylate
Dimethylamine	Methyl alcohol (methanol)

TABLE 6.6. *Continued*

Methyl bromide	Perchloroethylene
Methyl chloroform	Phenyl ether
Methyl cyclohexane	Propylene glycol methyl ether acetate
Methyl ethyl ketone (2-butanone)	Propionaldehyde
Methyl formate	Propyl acetate
Methyl isoamyl ketone	Propyl alcohol
Methyl isobutyl carbinol	Propyl bromide
Methyl isobutyl ketone (hexone)	Propylene dichloride
Methyl methacrylate	Propylene glycol methyl ether
Methyl propyl ketone (2-pentanone)	Propylene oxide
1-Methyl-2-pyrrolidinone	Pyridine
Methyl styrene	Sevoflurane (Ultane)®
Methyl tertiary butyl ether (MTBE)	Styrene
Methylal (dimethoxymethane)	Tetrachloro-2,2-difluoroethane
Methylamine	Tetrahydrofuran (THF)
Methylcyclohexane	Tolualdehyde
Methylene chloride	Toluene
Mineral spirits (Stoddard solvent)	Trichloroethane (methylchloroform)
Morpholine	Trichloroethylene (TCE)
Naphtha VM&P	Trichlorotrifluoroethane (CFC113)
Naphthalene	Trimethylbenzene (mesitylene)
Nitrogen dioxide	1,2,4-Trimethylbenzene
Nitrous oxides	Vinyl acetate
Nonane	Vinyl bromide
Octane	Vinyl chloride
Ozone	Vinyl toluene (methyl styrene)
Pentane	Vinylidene chloride
2-pentanone	Xylenes

geometry of the badge opening and internal cavity, as well as by the physiochemical properties of the contaminant. The badge exhibits an equivalent "sampling rate" which is a function of the diffusion coefficient of the chemical and the geometry of the device.[26] As molecules pass through the draft shield, they are collected on an adsorbent material such as charcoal. A concentration gradient is created within the cavity of stagnant air and the amount of gas or vapor transferred is proportional to the ambient vapor concentration. The sorbent surface area, path length from the badge's surface to the sorbent material, and badge sampling rate along with the desorption efficiency are used to calculate the final concentration of material collected.[26]

Each compound has a unique diffusion coefficient for each type of badge. Diffusion coefficients can be determined experimentally or calculated; however, since diffusion rate is related to molecular size for the contaminant, rates can often be predicted with verification using limited experimental studies. As the diffusion coefficient is necessary to calculate the final concentration after analysis in the laboratory, this limits sampled materials to those for which this value is established.

When permeation is used for sampling, the mass uptake of the monitor is controlled by the physicochemical characteristics of the membrane and the contaminant. The mass uptake is a direct function of the badge permeation sampling rate, the

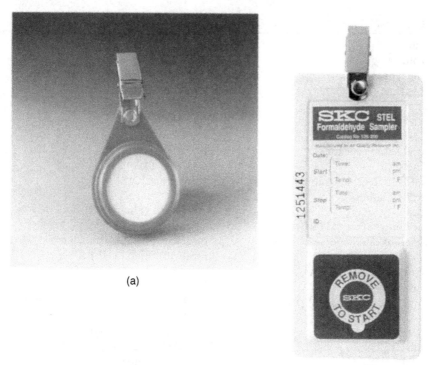

Figure 6.10. Passive dosimeters for (a) organic vapors (photo copyrighted 3M 2003) and (b) a specific chemical (formaldehyde). (Courtesy of SKC, Inc.)

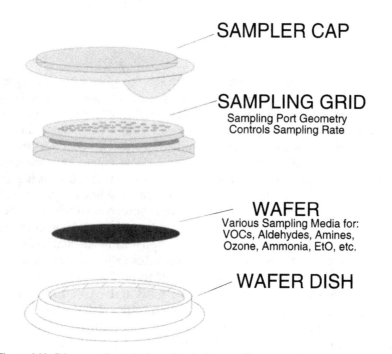

Figure 6.11. Diagram of a typical passive dosimeter. (Courtesy of Assay Technology.)

ambient concentration, and the sampling time.[26]

On permeation dosimeters, the gaseous contaminants dissolve in a polymeric membrane and are then transferred to a collection medium, such as a solution. Permeation across the membrane is controlled by the solubility of the gas or vapor in the membrane material and by the rate of its diffusion across the membrane under a concentration gradient. The permeation constant (ppm-hours/ηg), mass of compound collected, and exposure time determine the concentration collected.[26] These constants vary depending on the design of the monitor and the contaminants involved. Factors influencing permeation include thickness and uniformity of the membrane, affinity of the membrane for the contaminant, swelling or shrinking of the membrane, and possible etching by corrosive chemicals. The efficiency of these devices depends on finding a membrane that is easily permeated by the contaminant of interest and not by all others. Therefore, permeation dosimeters where they exist would be useful to selectively sample a single contaminant from a mixture of possible interfering contaminants due to the selectivity of a properly chosen membrane.

As with diffusion monitors, sampling rates must be determined for each analyte and type of permeation monitor. Difficulties in making such determinations include the need to use thin and fragile membranes to obtain practical sampling rates; the long response times to concentration fluctuations potentially resulting in TWA measurements; and the fact that sampling rate and degree of permeation may be affected by changes in temperature or ambient humidity.[27]

Factors Affecting Passive Monitor Performance

The accuracy and precision of passive monitors depends on sampling time, air currents, and temperature and humidity effects. Liquid sorbent badges are also affected by low relative humidity as moisture may evaporate from the badge, reducing the effectiveness of collection.[28]

In the presence of mixtures passive monitors may be affected by competition between compounds for adsorptive sites, resulting in displacement of one compound by another or preferential adsorption of one compound over another.

All passive monitors require a certain minimum air movement to be present during sampling in order to prevent "starvation" at the sampler surface. In a perfectly stagnant environment, molecules are sampled from the region adjacent to the sampler face, resulting in a reduction of the concentration. If a slight airflow is present, the concentration at the sampler face is continuously renewed and is equal to the ambient concentration. Badge samplers have a high area-to-length ratio, and therefore high uptake rates, requiring a minimum face velocity of 0.05–0.1 m/sec, depending on the type. In most cases, the body's natural movements will provide sufficient air motion for sampling. The orientation of the badge's face to the direction of air movement may affect collection.

In general, because the effective sampling rate is much lower for passive samplers than for an active method, a smaller quantity of contaminant is collected; therefore, special care must be taken to evaluate the lower limit of detection for the sampler and analytical method over the work shift or other monitoring period.

Passive monitors must be handled with care in the field as a penetration in the membrane would cause variations in the amount of contaminant collected. Membranes should not be touched with the fingers, due to oil that will clog the pores and decrease the results. A splash of oil or reagent on the surface of the membranes may increase results if it has volatiles in them. Obvious dirt or discoloration on the

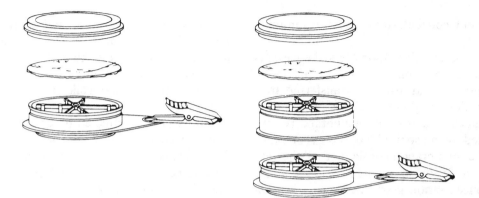

Figure 6.12. Organic vapor monitors with and without a backup section. (Courtesy of 3M Corporation.)

surface of the membranes will make the sample suspect.

Saturating the collection media by sampling for too long a period or too high a concentration will result in erroneous data since some contaminant will not be collected. One manufacturer offers a two-part badge with a main section and a separate backup section (Figure 6.12). During sampling the contaminant will be collected on the main section; if the main section becomes saturated, some contaminant will migrate to the backup section. Immediately after sampling is completed, the two sections are sealed in separate containers and returned to the laboratory for analysis. If the backup section is found to contain more than a nominal amount of contaminant, the possibility of overloading must be considered. Since it is normal to find some contaminant on the backup section, refer to the device's instructions for detailed guidance. Another manufacturer provides flexibility in choosing an equivalent sampling rate by offering badges with a different number of holes on the device's sampling grid (Figure 6.13). Using the unit with fewer holes reduces the sampling rate by 75%. This is important for high concentrations or for extended sampling

Figure 6.13. Effective "sampling rate" in a passive diffusion monitor is controlled by the open area of the badge. (Courtesy of Assay Technology.)

periods for indoor air quality or environmental sampling. However, the most common technique for avoiding saturation is to adjust the sampling time per badge based on (a) compound-specific retention data from the monitor's manufacturer and (b) estimates or direct-reading instrument screening readings of the airborne concentration.

After the sampling is completed the passive sampler is sent to a laboratory for analysis. Factors affecting accuracy during this phase include the effects of storage time on recovery of the analyte (i.e., contaminant) from the collection media and the availability of a suitable analytical technique.

Overall, the factors to be considered in evaluating the suitability of diffusion monitoring include[29]:

- Suitability of the collection medium for the contaminant(s).
- Any need to collect multiple contaminants on a single device.
- Concentration range of the contaminant(s).
- Sampling time.
- Required or desired accuracy and precision.
- Ambient conditions such as temperature, humidity and atmospheric pressure.
- Reactivity/stability of contaminant following collection.
- Presence of interfering compounds in the air.
- Collection efficiency of the device.
- Storage stability.

American National Standard for Diffusive Air Sampling Devices (ANSI/ISEA 104)[23].

With all of the performance-related and quality control issues discussed above, it is important to have a consistent method for evaluating diffusion dosimeters for any potential application. A consistent method permits both the manufacturer and end-user to evaluate a specific passive monitor for performance against relevant criteria. ANSI/ISEA National Standard 104 (first issued in 1998), which is targeted at passive dosimeters used for assessment of personal exposure to contaminants found in the workplace, provides that framework.

TABLE 6.7. Typical Evaluation Categories in ANSI/ISEA 104-1998 for Diffusion Passive Samplers

- Desorption efficiency
- Effect of concentration/time on sampler accuracy
- Bias due to reverse diffusion
- Background (blank) determination
- Effects of air velocity and orientation
- Effect of temperature and humidity
- Effect of storage
- Sampler integrity
- Interferences
- Shelf life
- Lot-to-lot variation

The evaluation parameters chosen for test methods in the standard (Table 6.7) will help define the operating limits of the sampler, and provide information on its conditions for use. Devices tested and marketed under this standard have uniform package labeling which informs the user of appropriate applications, controls and limitations for the device. Additionally, instructions are provided for correct use to obtain results that meet the NIOSH requirements for test accuracy.

A common application of ANSI/ISEA 104 is to standardize passive monitor evaluation studies by manufacturers. Using the criteria in the standard, they test their devices and summarize the findings in a report for potential users. Included in the report is a description of the test apparatus and method. This section describes how the standard gas mixture was generated and how the exposure chamber or other test equipment was configured. A typical test method is to generate test concentrations by static dilution from 100% of the analyte, mixed volumetrically with input air pumped at a pre-set flow rate through a chamber containing the samplers being tested. Flow is often verified by an in-line rotameter, and analyte concentrations are verified by charcoal tube samples continuously drawn from locations in the chamber bracketing the devices under test. Typical evaluation criteria and results discussed in the summary report include[30]:

- *Desorption Efficiency:* Data on tests of analyte recovery and desorption efficiency determined by analysis of charcoal wafers "spiked" from standard analyte solutions in carbon disulfide. Samplers are typically tested at several spike levels corresponding to levels expected for 8-hour sampler exposures at 0.5–2.0 times the OSHA PEL or other exposure standard.

- *Effect of Concentration/Time on Sampler Accuracy:* Data from tests to determine the "efficiency" of devices in collecting the analyte at different concentrations and various exposure times. Typical sampling conditions are in the range 1–8 hours and 0.1–2.0 times the OSHA PEL. Data are reported as how well the results reflect the "ppm-hours" exposure within the test chamber.

- *Bias Due to Reverse Diffusion:* Data from tests to determine the effect of "clean air" in drawing the analyte off the adsorbent pad. Samplers are typically subject to a "pulse" of analyte greater than the OSHA PEL over time period less than 50% of the recommended sampling time, followed by a zero exposure period for the duration of the recommended sampling time. The recovery of analyte from samplers analyzed immediately following the "pulse" is compared with analyte recovery from identically exposed samplers analyzed following the zero exposure period. The difference between these two recoveries is taken as the extent of reverse diffusion (i.e., evaporative loss as *% of sample*) from the sampler under the experimental conditions chosen.

- *Background (Blank) Determination:* Results from unexposed samplers that are analyzed to determine background analyte levels (if any) on the sampler prior to sampling.

- *Effects of Air Velocity and Orientation:* Data from tests to determine whether the ambient air velocity or orientation of the sampler's face as compared to wind direction effect collection efficiency. Typically, samplers are exposed to atmospheres of the analyte for 2–4 hours at 1–2 times the OSHA PEL in a chamber with three zones of different cross-sectional areas such that linear velocities of 15, 50, and 150 cm/sec, respectively, are generated. Samplers are placed in each zone with 50% of samplers placed normal to and 50% of samplers perpendicular to the flow direction. When data are compared from the six locations (representing normal air velocity and orientation variation in workplaces), any significant differences among the six groups will indicate an effect of air velocity and orientation on sampling rate for velocities in the range 15–150 cm/sec. Generally a test conducted with one analyte will apply to other vapors when the same sampler is used.

- *Effect of Temperature and Humidity:* Samplers are typically exposed to atmospheres of the analyte for 2–4 hours at 1–2 times the OSHA PEL in several chamber runs in which nearly identical exposures were applied with variations in temperature and humidity. Typical temperature/relative humidity conditions are: 22°C/50%RH, 10°C/50%RH, 30°C/30%RH, and 30°C/70% RH. Any variations in tests at the four conditions (representing normal temperature and humidity variation) will show any significant effects of temperature and humidity on sampling rate in the range 10–30°C and 30–70% RH. Generally a test conducted with one analyte will apply to other vapors when the same sampling device is used.

- *Effect of Storage:* Results of tests to determine the effect of storage time on analyte recovery. Typically the test involves exposing identical sets of samplers previously exposed to > 80% RH overnight to analyte concentrations at the OSHA PEL for at least 50% of the recommended sampling time at 20–25°C. One set of samplers is analyzed immediately, and the others are analyzed at prescribed intervals up to 14 days. Any significant variations in recovery over storage time will show that there is an effect of storage time on recovery.

- *Sampler Integrity:* Data from tests to determine any contamination of the sampler from exposure to atmospheres containing the analyte before the sampler is removed from its protective packaging. The test involves exposing the sampler within its sealed package to the analyte for several hours. If results from the analysis are significantly different from results for unexposed samplers (blank values), the integrity of sampler packaging may not be adequate. Often a test is performed with a compound (such as ethylene oxide) which has a high permeability through plastics and pinholes as compared to other analytes, and these results are considered to be applicable to all samplers packaged in the same manner by the manufacturer.

- *Interferences:* These data demonstrate that various compounds (e.g., organic vapors) collected on the sampler can be adequately differentiated by the analytical technique recommended for the analyte. For example, for organic vapors an analytical technique such as capillary gas chromatography (similar to OSHA Method 7) can separate and analyze several hundred different organic vapors. This method involves co-injection of analytical sample onto dual, high-resolution capillary columns providing identification of each analyte from its characteristic emergence time on two analytical columns and quantitation of analytes.

- *Shelf Life:* This test evaluates whether storage of new, sealed samplers affects their ability to accurately measure airborne analyte levels. Typically, two groups of samplers (one group freshly manufactured and one group manufactured 16 months previously) are subject to three exposure tests for 2–4 hours at 1–2 times the PEL. Any significant differences found between the two groups would indicate an effect on sampler stability when stored at room temperature for up to 16 months. Generally, a test conducted with one analyte will apply to other vapors when the same sampler is used.

- *Lot-to-Lot Variation:* Data from tests to evaluate the ability of samplers from different manufacturing lots to perform comparably. Generally, three groups of samplers from separate manufacturing lots are subject to exposure tests for 2–4 hours at 1–2 times the PEL. When data from the three lots are compared in each of the different exposure conditions, any significant differences among the groups indicate that there are differences among different lots of samplers. Generally, a test conducted with one analyte will apply to other vapors when the same sampler is used.

Passive Monitors Other than "Badges"

While most passive monitors are of the badge-type design, there are other configurations that have been developed to meet specific needs.

Figure 6.14. SKC vial formaldehyde sampler. (Courtesy of SKC, Inc.)

Indoor Air Formaldehyde Passive Sampler. The indoor air formaldehyde passive sampler from SKC (Figure 6.14) is a device specifically designed for accurate indoor measurements of low formaldehyde levels in the home, office, or industrial environment over a 5- to 7-day period. The sampler was field-validated by the Indoor Air Quality Program, Lawrence Berkeley Laboratories at the University of California, Berkeley.

Inside the vial is a disk of paper impregnated with sodium hydrogen sulfite. Formaldehyde combines with the reactive media on the disk and forms the stable compound sodium formaldehyde bisulfate. Detection limit is 0.01 ppm ± 30% or 0.02 ± 15%. There are no known interferences for the method; formaldehyde may be accurately measured in the presence of other substances such as phenol, aldehydes, and aromatic hydrocarbons.

For use, the unit is uncapped and hung from a mounting surface (door jamb or ceiling) by pushing the mounting pin through the ribbon and into the mounting surface. The sampler must be at least 24 inches away from any wall and away from outside doors and windows. After allowing the sampler to hang undisturbed for 5–7 days, recap the sampler and record finish time and date on the identification label. The sampler is then sent to an accredited laboratory for chromotropic acid assay (CTA) analysis. There are some values that are specific to the device [uptake rate (µg/ppm-hr) and Standard deviation (µg)] that appear on the product label. This information must be sent to the analytical laboratory.

Palmes Tube for Nitrogen Dioxide. A well-known diffusion nitrogen dioxide sampler, the Palmes, uses a tube with three stainless steel grids coated with triethanolamine (TEA) (Figure 6.15). For use the sampler is attached to a support with flanged cap down. Sampling is started by removing the flanged cap. The sampling time should be estimated such that the amount of NO_2 collected is in the range 1.2 to 80 ppm-hr (0.13 to 8.5 µg NO_2). Sampling is terminated by replacing flanged cap. In the laboratory a reagent made up of sulfanilamide, water, and *N*-1-naphthylethylenediamine dihydrochloride (NEDA) is added to the vial. After the color develops, the concentration is determined by colorimetric analysis and comparison to a calibration curve developed using laboratory standards.

A study on the Palmes sampler found it to be accurate for use in atmospheres with fluctuating concentrations over 8-hour sampling periods, but recommended caution when using it for short-term exposure limit (STEL) samples.[31] A laboratory chamber study performed by NIOSH in 1983 indicated the Palmes tube successfully sampled nitrogen dioxide at levels of 0.5, 5,

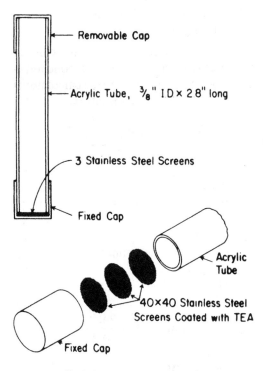

Removable Cap

Acrylic Tube, $\frac{3}{8}$" I D × 2 8" long

3 Stainless Steel Screens

Fixed Cap

Acrylic Tube

40×40 Stainless Steel Screens Coated with TEA

Fixed Cap

Exploded View of Sampler Bottom

Figure 6.15. Passive monitor for nitrogen dioxide. (Drawing from E. D. Palmes NO$_2$ and NO$_x$ diffusion techniques, *Am. Conf. Govt. Ind. Hyg. Ann.* **1**:263–266, 1981, reproduced with permission of the American Conference on Governmental Industrial Hygienists.)

and 10 ppm.[32] Another study found that reduced humidity led to reduced sampling rates and it was concluded that the reaction products of TEA and nitrogen dioxide appeared to be different under wet and dry conditions.[33]

Use of Passive Monitors for Occupational Sampling

Passive monitors play an important role in occupational exposure monitoring. Like all techniques, they have certain advantages and limitations.

The use of the devices in the field is straightforward. People with much less training than typical for a qualified

sampling practitioner can be taught how to use these devices. This approach has merit for screening studies and periodic monitoring of low exposures. However, it is important that the work be conducted under the direction, and the data reviewed by, a competent sampling practitioner or other professional. It is also important that regular workplace observations be performed during the monitor to ensure that the monitor is still in place and not covered by other clothing, and to observe workplace conditions that could affect exposure levels.

The following procedure is generic in nature since the specific instructions for the various passive monitors differ as to specifics.

Procedure

1. Before monitoring record the start time, sampling date, employee name, and other information, such as temperature and relative humidity, on the container. Open the sample container when sampling is to start.

2. Some badges have covers that must be removed and replaced with a membrane. The covers are used for shipment later. Other badges come in more than one piece and must be assembled for sampling.

3. Hook the unit to the collar or shirt neck if a worker is to be sampled (Figure 6.10). Ensure that the open face of the sampler is facing toward the environment and exposed for the entire sampling period. For area samples, attach it at least 1 meter above the floor in the area to be sampled. Ensure that the ambient air velocity at the sampler position is above the minimum velocity recommended by the manufacturer. Avoid sampling stagnant areas, such as against walls or in corners of rooms.

4. The minimum sampling time is governed by the sampling rate and the sensitivity of the analytical method. The maximum sampling time is determined by the sampling rate and by the adsorptive capacity of the charcoal adsorbent. When the calculated maximum sampling time is less than the desired sampling period, two or more samplers should be used in sequence to accommodate the desired exposure period.

Minimum sampling time (min) =

$$\frac{\text{Minimum detection limit, } \eta g}{(0.2)(\text{PEL}, \text{mg}/\text{m}^3) \times Q}$$

Maximum sampling time (min) =

$$\frac{\text{Sampler capacity, } \eta g}{(2)(\text{PEL}, \text{mg}/\text{m}^3) \times Q}$$

where Q is the sampling rate, cm^3/min.

5. At the end of the sampling period, remove the sampler, separate sampling and backup layers if necessary, and record the end time on the container. If a backup section is incorporated into the sampler, separate it immediately after sampling and put into its own container. If a disposable section is present, remove and discard it and cover both sides of the sampling section. Seal the sampler with the cover and refrigerate. It is critical that covers on badges be replaced immediately and securely following sampling, or the collected material may begin to diffuse out of the badge.

6. Send to the laboratory for analysis as soon as possible. Prepare a blank for each set of monitors at the monitoring site by removing a monitor from its sealed container, taking off the cap or seals, and replacing it with the shipping cover. Label the monitor as a blank and submit the blank along with the exposed monitors to the laboratory.

7. The following calculations can be used to determine badge concentrations. In most cases, these calculations are done by the laboratory.

$$\frac{\text{mg}}{\text{M}^3} = \frac{(W)(10^6 \text{ cm}^3/\text{m}^3)}{(r)(K_o)(t)}$$

where W is the corrected weight in milligrams, r is the recovery (desorption) efficiency, K_o is the monitor sampling rate, cm^3/min, and t is the sampling time, minutes. Converting to ppm at STP of 298 °K and 760 mm Hg:

$$\text{ppm} = \frac{\text{mg}}{\text{M}^3} \times \frac{24.45}{\text{MW}}$$

where MW is the molecular weight. Temperature correction if sampling temperature was significantly different than 298 °K:

$$C_0 = \frac{\text{mg}}{\text{M}^3} \times \frac{(298 \text{ °K})^{1/2}}{(T_s)}$$

$$C_0 = \text{ppm} \times \frac{(298 \text{ °K})^{1/2}}{(T_s)}$$

where T_s = temperature at sample site in degrees kelvin. This correction eliminates an error of approximately 1% for every 3°C increment above or below 24°C.

Use of Passive Monitors for Residential and Ambient Monitoring

Badges are potentially useful for ambient air monitoring where much lower levels are expected. This usefulness should be evaluated for each compound because some compounds such as 1,2-dichlorethane and chlorobenzene are not effectively sampled at very low levels. A criteria for determining whether a badge is suitable is to assume that the lowest useful airborne concentration for sampling is 10 times the median blank or 10 times the badge's detection limit, whichever is greater. Sometimes extending the length of the sampling period will increase the detection limit. For example, organic vapor badges have been used with good results for residential monitoring using sampling periods of 2 weeks or longer.[34]

Procedure

1. Select at least two rooms in the home for monitoring, preferably high-activity areas such as the living room and bedroom. Each floor of the home should have at least one monitor. Because of significant humidity fluctuations, do not select the bathroom as a monitoring site.

2. Place a monitor in each room so that it is well positioned at least 0.3 meter from any wall. The monitor should not be placed in any nonrepresentative ventilation area, such as directly in front of an air conditioner or heating vent.

3. Due to room condition variations, measurements should be taken for at least 24 hours for each monitoring point. In some cases, monitors are designed to be left for up to 7 days. During the monitoring period, the room temperature should be maintained above 21°C and 50% relative humidity. In addition, no smoking, cooking, or other combustion sources should be allowed in the room during the monitoring period.

SUMMARY

This chapter covered sample collection device methods for gases and vapors. These methods are the mainstay for exposure monitoring for the large number of gases and vapors encountered in the occupational setting and general ambient environment.

The defining characteristic of these methods is that the sampling device must be returned to a laboratory for analysis after the sampling is completed. This step can be an inconvenience in that it causes a time delay between the sampling period and when the results are available. However, laboratories have access to instruments and procedures that often provide more sensitivity and accuracy as compared to direct reading instruments used in the field.

There are two major categories of SCD methods: active sampling that involve a pump, and passive methods that collect the contaminant via diffusion or (in fewer cases) permeation.

REFERENCES

1. Shin, Y. C., G. Y. Yi, Y. Kim, and N. W. Paik. Development of a sampling and analytical method for 2,2-dichloro-1,1,1-trichloroethane in workplace air. *AIHA J.* **63:**715–20, 2002.

2. Burnett, R. D. Evaluation of charcoal sampling tubes. *Fundamentals of Analytical Procedures in IH*. Akron: AIHA, 1987, pp. 64–72.

3. Saalwaechter, A. T., et al. Performance testing of the NIOSH charcoal tube technique for the determination of air concentrations of organic vapors. *Fundamentals of Analytical Procedures in IH*. Akron: AIHA, 1987, pp. 363–373.

4. Fraust, C. L., and W. R. Hermann. Charcoal sampling tubes for vapor analysis by gas chromatography. *AIHA J.* **27:**68, 1966.

5. Nelson, G. O., A. N. Correia, and C. A. Harder. Respirator cartridge efficiency studies. VII. Effect of relative humidity and temperature. *AIHA J.* **37:**280, 1976.

6. Rudling, J., and E. Bjorkholm. Effect of adsorbed water on solvent desorption of

organic vapors collected on activated carbon. *AIHA J.* **47**(10):615–620, 1986.

7. Betz, W. R., et al. Characterization of carbon molecular sieves and activated charcoal for use in airborne contaminant sampling. *AIHA J.* **50**(4):181–187, 1989.

8. Levine, M. S., and M. Schneider. Flowrate associated variation in air sampling of low concentrations of benzene in charcoal tubes. *AIHA J.* **43**(6):423–426, 1982.

9. Feigley, C. E., and J. B. Chastain. An experimental comparison of three diffusion samplers exposed to concentration profiles of organic vapors. *AIHA J.* **43**:(4):227–234, 1982.

10. Goller, J. W. Displacement of polar by nonpolar organic vapors in sampling systems. *AIHA J.* **46**:170–173, 1985.

11. Standard Recommended Practices for Sampling Atmospheres for the Analysis of Gases and Vapors. ASTM D-1605, 1990.

12. Crisp, S. Solid sorbent gas samplers. *Ann. Occup. Hyg.* **23**:47–76, 1980.

13. National Institute of Occupational Safety and Health. *NIOSH Manual of Analytical Methods*, P. M. Eller, ed. Cincinnati: NIOSH, July 1994.

14. Brown, R. H., and C. J. Purnell. Collection and analysis of trace organic vapor pollutants in ambient atmospheres. *J. Chromatogr.* **178**:79, 1979.

15. Gordon, S. M. *Tenax Sampling of Volatile Organic Compounds in Ambient Air. Advances in Air Sampling.* ACGIH, Chelsea, MA: Lewis Publications, 1988, pp. 133–142.

16. Dharmarajan, V., and R. N. Smith. Permeation of toluene through plastic caps on charcoal tubes. *AIHA J.* **42**(9):691–693, 1981.

17. Environmental Protection Agency. Compendium of Methods for the Determination of Toxic Organic Compounds in Ambient Air. EPA-600/4-84-041, 1984.

18. Linch, A. *Evaluation of Ambient Air Quality by Personnel Monitoring.* Cleveland: CRC Press, 1974.

19. Dinardi, S. R., ed. *The Occupational Environment—Its Evaluation and Control,* Fairfax, VA: AIHA Press, 1997.

20. Riggin, R. M. Technical Assistance Document for Sampling and Analysis of Toxic Organic Compounds in Ambient Air. EPA-600/4-83-027, 1983.

21. Possner, J. C., and W. J. Woodfin. Sampling with gas bags. I: Losses of analyte with time. *Appl. Ind. Hyg.* **1**(4):163–168, 1986.

22. Environmental Protection Agency. *Compendium Method TO-14: The Determination of Volatile Organic Compounds (VOCs) in Ambient Air Using Summa Passivate Cannister Sampling and Gas Chromatographic Analysis.* Quality Assurance Division. Environmental Monitoring Systems Laboratory, U.S. EPA, Research Triangle Park, NC, January 1999.

23. American National Standards Institute/International Safety Equipment Association Standard 104-1998. *American National Standard for Air Sampling Devices—Diffusive Type for Toxic Gases and Vapors in Working Environments.* ANSI: New York, 1998.

24. Manning, C. R. Experience in the USA with novel personal monitoring systems. In *Clean Air at Work*, R. H. Brown et al., eds. Cambridge, United Kingdom: Royal Society of Chemistry, 1992.

25. Occupational Safety and Health Administration. *Benzene.* U.S. Code of Federal Regulation (29CFR1910.1028).

26. Lautenberger, W. J. Theory of passive monitors. *ACGIH Ann.* **1**:91–99, 1981.

27. Miksch, R. R. A "Passive Bubbler" Personal Monitor Employing Knudsen Diffusion: Development and Application to the Measurement of Formaldehyde in Indoor Air. Air Technology Labs, Inc., 548 E. Mallard Circle, Fresno, CA 93710.

28. Standard Practice for Sampling Workplace Atmospheres to Collect Gases or Vapors with Liquid Sorbent Diffusional Samplers. ASTM D-4598-87.

29. Coulson, D. M., et al. Diffusional sampling for toxic substances. In *Sampling Techniques.* New York: Wiley, 1977.

30. Assay Technology. *Air Sampler Evaluation (ANSI 104-1998): Summary of Test—Benzene, Toluene and Xylene.* May 2000.

31. Bartley, D. L. Passive monitoring of fluctu-ating concentrations using weak sorbents. *AIHA J.* **44**(2):879–885, 1983.

32. Douglas, K. E., and H. J. Beaulieu. Field validation study of NO₂ personal passive samplers in a "diesel" haulage under-ground mine. *AIHA J.* **44**(10):774–778, 1983.

33. Palmes, E. D., and E. R. Johnson. Explana-tion of pressure effects on a nitrogen dioxide (NO₂) sampler. *AIHA J.* **48**(1):73–76, 1987.

34. Shields, H. C., and C. J. Weschler. Analysis of ambient concentrations of organic vapors with a passive sampler. *JAPCA* **37:**1039–1045, 1987.

CHAPTER 7

SAMPLE COLLECTION DEVICE METHODS FOR AEROSOLS

The term *aerosol* is used to describe particulate matter suspended in a gas, usually air. The particulate matter is some discrete unit of fine solid or liquid matter, such as dust, fog, fume, mist, smoke, or spray. Table 2.2 defines the size ranges and other characteristics of particulate matter. For convenience, the term *particle* is also used in the literature to describe an aerosol even though it may actually consist of a liquid droplet since its behavior resembles a discrete particle. This chapter covers the basic concepts and devices for sample collection device (SCD) methods for aerosols. For more detailed information on bioaerosol sample collection, see Chapter 16.

Occupational sampling for aerosols is in a state of flux since the criteria for particle size as it relates to sampling and thus respiratory hazards has been redefined, but the toxicological data and sampling methods do not always reflect this redefinition. The problem is that new instruments and methods are becoming available that reliably measure new criteria, but current

dust standards are based on the old criteria. This matter is further discussed in the sections on inhalable, thoracic, and respirable sampling in this chapter. For the time being, the sampling practitioner must be willing to remain flexible during this period of change. As this change occurs, comparisons should be made between methods so that old results can be compared with new ones and interpreted in terms of new standards.

An advantage of SCD methods for occupational sampling of aerosols is that there are methods that can be used to identify, quantify, and differentiate between specific contaminants whereas real-time methods are nonspecific and tend to measure particle counts or mass. The primary difference between occupational and environmental samples for aerosol is the flow rate. For the most part, environmental methods use high (flow rate) volume pumps to collect large volumes of samples because the levels of concern in ambient air are much lower. Sampling periods for environmental

Air Monitoring for Toxic Exposures, Second Edition. By Henry J. McDermott
ISBN 0-471-45435-4 © 2004 John Wiley & Sons, Inc.

samples are also correspondingly longer than occupational sampling periods.

Results of air sampling measurements for aerosols are expressed in mg/M³. For fibers the standard concentration unit is fibers/cc of air. The historic dust counting unit of millions of particles per cubic foot (mppcf) is largely obsolete.

CHARACTERIZING AEROSOLS

The two biggest factors of interest with an aerosol are generally its chemical make-up and particle size distribution. The chemical composition is determined through laboratory analysis, and so the collection method must be appropriate for the analytical technique.

The size of a particle is a very important characteristic since it governs how an aerosol behaves when airborne and where it deposits in the human body when inhaled. Table 2.2 lists different types of particulate matter and their sizes. Particles suspended in air can range from 0.001 to 100 μm in diameter. Airborne particles have two major opposing forces acting on them. They are pulled downward by gravity; if they are given enough time in still air, they will eventually settle out. The downward motion is opposed by frictional drag caused by movement through the air. As a result of these forces, particles larger than about 100 μm usually settle out rapidly unless there is a lot of turbulence or the particle density is very low. Particles smaller than about 5 μm exhibit less drag than would be expected based on their size because they are small enough to "slip" through the air molecules. Another force that acts on small particles is *diffusion*, caused by the constant collision between the particle and air molecules. These collisions result in an irregular wiggling motion called *Brownian motion*. Diffusion is most significant for particles <0.01 μm; the magnitude is inversely proportional to the actual diameter squared. Very fine particles (less than about 0.001 μm) act as gas molecules rather than as a particle—they are essentially independent molecules.

As described in Chapter 2, the particle's shape and density are also critical to how it behaves when airborne. A higher-density material will cause the particle to settle faster and also give it more momentum when moving. A "jagged" particle shape will encounter more resistance when moving through air and so will exhibit a lower settling velocity than a spherical particle of the same size. Figure 7.1 shows how

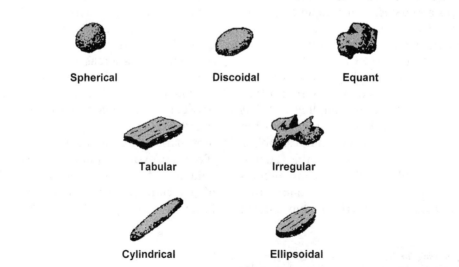

Figure 7.1. Particle shapes.

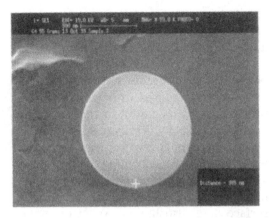

Figure 7.2. Typical spherical particles (diameter = 1 μm). (Courtesy of Dr. E. C. Kimmel, U.S. Naval Health Research Center/Environmental Health Effects Laboratory, Wright-Patterson AFB, OH.)

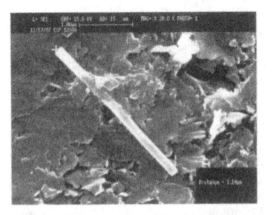

Figure 7.4. Typical straight, cylindrical fiber (length = 3 μm). (Courtesy of Dr. E. C. Kimmel, U.S. Naval Health Research Center/Environmental Health Effects Laboratory, Wright-Patterson AFB, OH.)

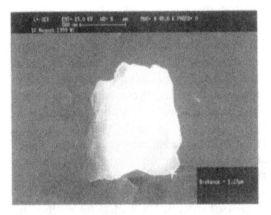

Figure 7.3. Typical small, equant-shaped particles (diameter ≈ 1 μm). (Courtesy of Dr. E. C. Kimmel, U.S. Naval Health Research Center/Environmental Health Effects Laboratory, Wright-Patterson AFB, OH.)

particle shapes are characterized, while Figures 7.2 to 7.4 illustrate microphotographs of actual particles. To adjust for shape and density, particle size is often described using the term *aerodynamic size*; this is the equivalent diameter of a sphere with the density of water that behaves like the particle of interest in air. Unless otherwise specified, future references to particle size or diameter in this book will mean the *aerodynamic size*. True fibers (those at least 10 times longer than their width) do

not behave like more spherical particles. Airborne fibers tend to align themselves with the direction of flow, thus reducing drag, and so their velocity in air is a function of their width (diameter) rather than length.

An aerosol with very little variation in the size of individual particles is called a *monodisperse* distribution, while an aerosol made up different sized particles is called *polydisperse*. A typical monodisperse aerosol is a special dioctyl phthalate (DOP) aerosol generated to test the efficiency of high efficiency particulate air (HEPA) filters. A HEPA filter will remove at least 97.97% of 0.3-μm-diameter monodisperse particles.

Most aerosols encountered by the sampling practitioner in the workplace or community environment are polydisperse distributions, and so the particle size distribution is an important consideration when selecting a sampling and analytical method.

Particle sizes can be determined either through physical or dynamic measurements with each measurement method measuring somewhat different quantities in the case of irregular-shaped particles. The resulting diameters are often called statistical diameters since large numbers of particles must be measured and the results must be averaged if the values are to have much significance.

Physical size measurements are associated with particle geometry, such as the diameter of a sphere or the length and width of a fiber, and are generally done under the microscope. Optical sizing is limited to particles >0.2 µm. Scanning electron microscopy can be used for particles >0.02 µm; Transmission electron microscopy can be used for all particulate matter. As measurements that identify a single dimension are difficult and time consuming to do and subject to much difference in definition, *projected area* diameters are most commonly used for optical size measurements. With this method the observer tries to visualize the size of an irregular particle in terms of the diameter of an equivalent area circle on a graticule mounted in the eyepiece. Using this comparison, an estimate is made of the particle's size.

Dynamic size measurements refer to an observed aerosol property that can be related to particle size and are usually done with sampling equipment such as cascade impactors and cyclones. The type of instrument will determine what property is to be measured. In impactor collection the range of sizes collected on each stage depends on the density and shape of the particles as well as the diameter, and therefore represents the aerodynamic behavior of the particles. Cyclones separate particle sizes based on the velocity at which the air is being collected. For example, the 10-mm nylon cyclone when operated at 1.7 Lpm matches the ACGIH curve used to describe the inhalation of particles less than 10 µm in size. Another method that can size particles is light scattering, which uses real time instruments for measurements. These instruments are discussed in Chapter 14.

It is important for the sampling professional to select the definition of particle size that is appropriate for the exposure standard being evaluated and use a method

sensitive enough to measure the size range of interest. Results from different techniques may not correspond.

After aerosol size data have been obtained for a specific system, statistical methods are required to impart quantitative meaning and utility to the measurements. Usually the first step is to obtain a frequency (number-size) distribution or at least classification according to size range. Three types of distribution useful for examining aerosols are number, mass, and surface area distribution. As an example, the *mass distribution* tells how the total airborne mass is distributed among the various particle sizes. When plotted on log-probability paper, the mass-median diameter (MMD) or the size below or above which half of the mass of the particles would occur can be obtained. The MMD identifies the middle of the distribution of mass; that is, the size for which half the total mass is contributed by smaller particles and half by larger particles. When using aerodynamic diameter methods, sometimes the mass median aerodynamic diameter (MMAD) is identified for an aerosol. The size mass distribution is used for predicting the actual dose to the lung resulting from inhalation of a given amount of dust or the weight of material collected by a filter or other collection device that is efficient only for particles larger than a given size. Figure 7.6 shows size distribution data for the smoke shown is Figures 7.2 to 7.5. The MMAD for this smoke sample is approximately 2 µm.[1]

For a practical sampling system to simulate the collection by the respiratory tract, the sampler must make approximations and assumptions in order to be able to deal with the very complex interaction of variables governing actual respiratory tract deposition. In measurement, particles are collected by an instrument that has the same efficiency (collects the same aerosol size fraction) as a certain portion of the respiratory tract, and thus the justification exists for expressing aerosol exposure standards in terms of concentrations of the size particles.

When considering the size of particulates, it is important to remember that almost all of the contaminants capable of causing a respirable hazard are so small that they have little independent action of their own when in the ambient air. They move with the air around them. The particulates that are too large to remain airborne are the ones that can move by themselves when suspended in ambient air and thus pose little potential respiratory tract hazard since they do not remain airborne for extended periods.

Inhaled particulates are deposited in different regions of the respiratory tract depending on their aerodynamic size. However, most OSHA standards and Threshold Limit Values® refer to the total airborne level of contaminant, not just the portion that remains airborne and can be inhaled. The reason is that there was no easy way to differentiate between the respirable and nonrespirable particulates when much of the exposure data supporting current standards were collected. Today

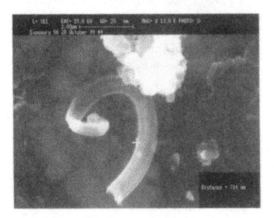

Figure 7.5. "Corkscrew" shaped fiber (length ≈ 10 μm). (Courtesy of Dr. E. C. Kimmel, U.S. Naval Health Research Center/Environmental Health Effects Laboratory, Wright-Patterson AFB, OH.)

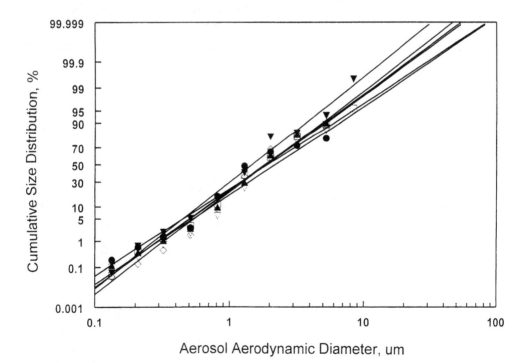

Figure 7.6. Aerodynamic diameter size distribution for particles in Figures 7.2 to 7.5. Each line represents a separate repetition of the size distribution analysis. (Courtesy of Dr. E. C. Kimmel, U.S. Naval Health Research Center/Environmental Health Effects Laboratory, Wright-Patterson AFB, OH.)

more information is available, and the ACGIH TLVs® recognize three categories of size-selective sampling criteria to recognize the effect of particle size on deposition site within the repiratory tract, and the tendency for many occupational diseases to be associated with material deposited in particular regions. The categories are:

- Inhalable particulate matter *TLVs* for those materials that are hazardous when deposited anywhere in the respiratory tract.
- Thoracic particulate mass *TLVs* for those materials that are hazardous when deposited anywhere within the lung airways and the gas-exchange region.

- Respirable particulate mass *TLVs* for those materials that are hazardous when deposited in the gas-exchange region.

As an illustration of the importance of particle size, only particles 10 μm or smaller are deposited in the lungs. Particulates larger than 10 μm are deposited in the nasopharyngeal region, removed to the throat by hair cells in the passage linings, and swallowed. Thus the toxicity of the material in the gastrointestinal tract must be considered where appropriate. Table 7.1 lists the size characteristics of each category in more detail. Sampling for each size category is covered under *Size-Selective Sampling* later in this chapter.

TABLE 7.1. Air Sampler Sizing Characteristics for Size-Selective TLVs®

Particulate Aerodynamic Diameter[a] (μm)	Percent of Mass of that Particle Size in the Air Retained in the Sampling Device		
	Inhalable Mass	Thoracic Mass	Respirable Mass
0	100	100	100
1	97		97
2	94	94	91
3			74
4		89	50
5	87		30
6		80.5	17
7			9
8		67	5
10	77	50	1
12		35	
14		23	
16		15	
18		9.5	
20	65	6	
25		2	
30	58		
40	54.5		
50	52.5		
100	50		

[a] Assumes unit density sphere.

Source: Reference 2.

AEROSOL COLLECTION MECHANISMS

Collecting airborne particles in an SCD is based on the physical parameters of the particle (size, shape, and density) and the interaction of these physical properties with the forces acting on the particles (settling and diffusion).[3] Particles are collected using more or more of these five mechanisms:

- *Interception*—occurs when a particle following an air stream contacts a surface, such as a filter fiber, and adheres to the surface.
- *Impaction*—occurs when a particle, due its inertia, cannot adjust quickly enough to the changing direction of the air stream, and contacts a surface and adheres to it. The particle's inertia is a function of its size, density, and velocity, and so impaction is a very important collection mechanism for larger particles.
- *Gravitational Settling*—results when the settling velocity of a particle is sufficient to overcome motion due to air velocity. This mechanism, also called sedimentation, is more important than impaction only with relatively high particle velocities (generally 10 cm/sec or greater).
- *Diffusion*—occurs when Brownian motion is great enough to bring particles into contact with adjacent surfaces. It is more important for very small particles.
- *Electrostatic Attraction*—results when particles are attracted to a surface because of opposing electrical charges on the particles and surface. This can be due to natural charges or charges induced on the particle or collecting surface. In general, naturally occurring charges are not an important collection mechanism for SCDs.

In most situations, interception and impaction predominate for particles larger than about 0.5 µm where aerodynamic diameter is the most important parameter. Diffusion is the predominant collection mechanism for particles <0.5 µm where actual diameter is the most important physical parameter. Particles that are about 0.2 µm in diameter are most difficult to collect since they fall between the sizes collected by interception/impaction and the sizes collected by diffusion.

These collection mechanisms find application in the various SCDs depending on the size particles to be collected:

- Filters rely on all five collection mechanisms.
- Impactors, cyclones, and liquid impingers use impaction.
- Elutriators (to separate out very large particles before a sampling filter) operate by gravitational settling.
- Electrostatic precipitators (sometimes used for environmental monitoring) are based on electrostatic attraction.

Collecting Representative Aerosol Samples

Collecting a sample of a polydisperse aerosol that accurately reflects the airborne material is a serious challenge, and it depends in part on:

- Efficiency in aspirating (or transferring) the particles from the ambient air into the collection device. For any given particle size, this depends on sampling flowrate, ambient wind speed, sampler orientation, and sampler size and shape.
- Particle "loss" within the device due to deposition on the internal walls, internal air leakage, or for other reasons. In this case the "collection mechanisms" described above function to interfere

with an accurate sample. For example, particles may impact the internal walls of the device and adhere rather than continuing on to the filter or other collection surface. If airflow generates an electrostatic charge in materials of construction of the sampling device, particles may "collect" on the internal walls due to electrostatic attraction and so be lost before the collection surface.

One important characteristic is whether the sample is collected under *isokinetic* conditions. *Isokinetic* means that the air entering the sampling device is traveling in the same direction and at the same velocity as the ambient air flowing past the device's inlet. Under these conditions, all of the particles in the flowing airstream enter the collector inlet. If the air entering the device is at a lower velocity than the surrounding ambient airflow, the excess air will curve away from the inlet port while the larger particles will continue in a straight line because of their inertial and will enter the sampler. This will result in the sample overstating the mass that is in the aerosol and also skew the particle size distribution by overstating the number of larger particles. If the air velocity entering the sampling device is greater than the ambient velocity, some "extra" air adjacent to the inlet will be drawn into the device, while the larger particles in that air will continue in a straight line and miss being collected. This will understate the mass and size distribution. Note that perfect isokinetic conditions can probably not be achieved in collecting ambient samples, but device developers spend considerable effort maximizing factors that lead to isokinetic conditions. Every sampling device should be evaluated for collection efficiency by the manufacturer. Any parameters such as sampling flowrate and orientation that affect its performance should be clearly identified.

Aerosol Sampling Using Filters

Although there are several different sample collection methods, the use of some type of filter is the most common method. For this reason an overview of filters is presented before more detailed information on SCD methods for aerosol collection.

Aerosol samplers for occupational sampling typically consist of a cassette containing a membrane or fiber filter on which the aerosol is collected. Cassettes and filters come in many sizes, and some cassettes have an extension cowl such as is used in asbestos sampling to decrease electrostatic effects. In some cases, the cassette is contained within another piece of sampling equipment, such as a cyclone, to provide size-selection characteristics.

Environmental samplers also use filters, but the apparatus is very large and typically designed to be used outdoors. Some size selective sampling is also done and all techniques utilize high-volume pumps usually collecting for a 24-hour period.

Filters come in a variety of materials and are prepared through different techniques, although there are two basic types used for aerosol sampling: (a) compacted or felted fibers and (b) membranes. Collection characteristics are different for each type of filter medium. For example, because fiber filters offer comparatively low resistance to airflow, their use is recommended when a large volume of aerosol must be sampled. The choice of filter medium depends on the physical and chemical properties of the aerosol to be sampled, the sampler, and the analysis to be performed.

For a given aerosol and a given filter, the collection efficiency varies with the face velocity and the particle size. With an appropriate filter, the air will pass through, leaving the particles behind. If the filter is too porous, some particles penetrate the filter and are lost while others become embedded within the filter, making recovery and later analysis difficult. If the filter is too impervious, airflow resistance (pressure drop) may be too great. Ideally, particles should be trapped at the front surface of the filter if size properties are to be maintained.

The type of filter medium is specified by the method, so in most cases the sampling practitioner does not have to understand how to select a given filter medium. However, it is important to understand the limitations of each type of filter and why particular types are preferred for various uses.

Important filter characteristics are collection efficiency, pressure drop, mechanical strength, hygroscopicity, chemical purity, and possible artifact production (gas to particle conversion).[4] Additionally, all filters may contain at least trace contaminants that should be considered during filter selection.

Membrane filters are made from cellulose, PVC, or polytetrafluoroethylene (PTFE). Since most particle collection takes place at or near the surface, membrane filters are useful for applications where the collected particles will be examined under a microscope. Since the mixed cellulose ester filters (MCE) dissolve easily with acid, they are also used for collection of metals for atomic absorption analysis. Pore sizes of 0.45 μm and 0.8 μm are the most commonly used. With respect to cellulose, the main concern is their susceptibility to water. PVC membrane filters are used for sampling silica, nuisance dusts, and zinc oxide. PTFE filters come in a variety of pore sizes, 1 μm, 2 μm, and 5 μm, and are used for pesticides, alkaline dusts, and many other compounds.

The most common fiber filter is made of fiberglass. Fiber filters are available with and without binders. For occupational sampling, glass fiber filters are used to collect pesticides; 2,4-D is sampled with binderless glass fiber filters. They are also used in high-volume environmental sampling for atmospheric aerosols and lead. Environmental

filter media are much larger, often 8-by-10-inch sheets. Quartz fiber filters are used for high-volume environmental particulate sampling. An advantage of this type of filter material is minimal artifact generation. The biggest limitation is their fragility.

Another type of filter medium, polycarbonate filters, also known as Nucleopore, have superior strength and chemical and thermal stability. They are essentially transparent with a slight green tinge, and they average 5 cm to 10 cm in thickness. Polycarbonate filters are recommended for asbestos sampling for transmission electron microscopy (TEM) analysis.

When impregnated with special reagents, filters can be used to collect certain vapors, such as isocyanates, as well as particles. Table 7.2 provides a description of media used in NIOSH methods for aerosols.

As discussed, filters are most often contained is plastic holders called *cassettes* (Figure 7.7). There are two common cassette designs:

- A two stage cassette consists of a bottom piece, which contains a pad that the filter rests on, that has a small opening for connection to the sampling pump and a top piece. The top piece is a cover with a small hole to allow air to enter the cassette.
- A three-stage cassette has the same top and bottom pieces as the two-stage unit plus an "O-shaped" insert between the top and bottom. The O-shaped insert holds the filter in place so the top piece can be removed from the cassette for "open-face sampling." Open-face sampling allows the contaminant to be evenly collected on the filter for microscopic analysis.

The most straightforward analysis is called *gravimetric* analysis. It involves just weighing filters, assuming they only contain the desired contaminant, and dividing the

TABLE 7.2. Media Used in NIOSH Methods for Aerosols

Compound	Method Number	Media
Alkaline dusts	7401	1-μm PTFE membrane filter, 37 mm
Aluminum and compounds	7013	0.8-μm MCE filter, 37 mm
p-Aminophenylarsonic acid	5022	1-μm PTFE filter, 37 mm
Asbestos	7400	0.8–1.2-μm MCE filter, 25 mm with cowl
Azelaic acid	5019	5-μm PVC membrane filter, 37 mm
Barium	7056	0.8-μm MCE filter, 37 mm
Benzoyl peroxide	5009	0.8-μm MCE filter, 37 mm
Beryllium and compounds	7102	0.8-μm MCE filter, 37 mm
Boron carbide (respirable and total)	7506	5-μm PVC membrane filter, 37 mm
Bromoxynil	5010	2-μm PTFE membrane filter, 37 mm
Bromoxynil octanoate	5010	2-μm PTFE membrane filter, 37 mm
Cadmium and compounds	7048	0.8-μm MCE filter, 37 mm
Calcium and compounds	7020	0.8-μm MCE filter, 37 mm
Carbaryl	5006	Glass fiber filter (type A), 37 mm
Carbon black	5000	5-μm PVC membrane filter, 37 mm
Chlorinated diphenyl ether	5025	0.8-μm MCE filter, 37 mm
Chlorinated terphenyl	5014	Glass fiber filter (Gelman no. 64877), 37 mm
Chromium and compounds	7024	0.8-μm MCE filter, 37 mm
Chromium (VI)	7600	5.0-μm PVC membrane filter, 37 mm
Coal tar pitch volatiles	5023	2-μm PTFE membrane filter (Zeflour), 37 mm
Cobalt and compounds	7027	0.8-μm MCE filter, 37 mm
Copper (dust/fume)	7029	0.8-μm MCE filter, 37 mm
2,4-D	5001	Binderless glass fiber filter (type AE), 37 mm
o-Dianisidine	5013	5-μm PTFE membrane filter, 37 mm
Dibutylphosphate	5017	1-μm PTFE filter, 37 mm
Dibutyl phthalate	5020	0.8-μm MCE filter, 37 mm
Di(2-ethylhexyl) phthalate	5020	0.8-μm MCE filter, 37 mm
Dimethylarsenic acid	5022	1-μm PTFE filter, 37 mm
EPN	5012	Glass fiber filter (type AE), 37 mm
Ethylene thiourea	5011	5-μm PVC or 0.8-μm MCE filter
Hydroquinone	5004	0.8-μm MCE filter, 37 mm
Lead	7082	0.8-μm MCE filter, 37 mm
Lead sulfide (respirable)	7505	10-mm nylon cyclone and 5-μm PVC membrane filter, 37 mm
Malathion	5012	Glass fiber filter (type AE), 37 mm
Mineral oil mist	5026	Membrane filter, 37 mm, 0.8-μm MCE or 5-μm PVC, or glass fiber filter
Methylarsonic acid	5022	1-μm PTFE filter, 37 mm
Total dust	0500	5-μm PVC filter, 37 mm
Respirable dust	0600	10-mm nylon cyclone and 5-μm PVC
Paraquat	5003	1-μm PTFE membrane filter, 37 mm
Pyrethrum	5008	Glass fiber filter, 37 mm
Rotenone	5007	1-μm PTFE membrane filter, 37 mm
Silica, amorphous (respirable)	7501	10-mm nylon cyclone and 5-μm PVC filter, 37 mm
Silica, crystalline (respirable)	7500, 7601	10-mm nylon cyclone and 5-μm PVC filter, 37 mm
Strychnine	5016	Glass fiber filter, 37 mm
Parathion	5012	Glass fiber filter (type AE), 37 mm

TABLE 7.2. *Continued*

Compound	Method Number	Media
2,4,5-T	5001	Binderless glass fiber filter (type AE), 37 mm
o-Terphenyl	5021	2-μm PTFE filter, 37 mm
Thiram	5005	1-μm membrane filter, 37 mm
o-Toluidine	5013	5-μm PTFE membrane filter, 37 mm
Tungsten and compounds	7074	0.8-μm MCE filter, 37 mm
2,4,7-Trinitrofluoren-9-one	5018	0.5-μm PTFE membrane filter, 37 mm
Vanadium oxides (respirable)	7504	10-mm nylon cyclone and 5-μm PVC membrane filter, 37 mm
Warfarin	5002	1-μm PTFE membrane filter, 37 mm
Welding and brazing fume	7200	0.8-μm MCE filter, 37 mm
Zinc and compounds	7030	0.8-μm MCE filter, 37 mm
Zinc oxide	7502	0.8-μm PVC membrane filter, 25 mm

Figure 7.7. For sampling with two-piece cassettes (*left*), the small plug is removed. With a three-piece cassette (*right*), the top section is removed, which exposes the entire filter surface to obtain an even distribution of contaminants on the filter surface.

weight by the sampling volume to obtain mg/M³. Examples are total (nuisance) dust, respirable dust, welding fume, and oil mist. Filters must be weighed and inserted in the cassettes in advance of sampling. Matched weight filters in cassettes are sometimes used instead. In this case, there are two filters placed inside each cassette whose difference in weight does not exceed 0.1 mg. The analysis involves weighing both filters after sampling and comparing the weights. The weight difference between the two matched filters is due to the collected contaminants. The advantage is that preassembled match weight filter cassettes can be purchased and stored. If 0.1 mg is sufficient sensitivity for the contaminant to be sampled then the use of matched weight filters in cassettes is usually more convenient than preweighed single filters.

There are two major sources of filter cassette leakage: (1) external, when air enters the cassette through the tapered joints between the stages; (2) internal, when air enters the cassette through the inlet, but travel around the edge of the filter, bypassing the surface, and causing uneven distribution of aerosols.[5] The mating pieces of the cassette must seal securely to prevent air leakage around the filter or to the outside. Use of a "shrink wrap" seal around the cassette or a similar seal can help to maintain tight joints and prevent leakage.

POTENTIAL PROBLEMS

Successful aerosol sampling requires that the SCD draws a sample that is representative of the ambient environment and then maintains the collected material so the sample can be analyzed properly.

Artifacts and changes can occur to aerosols during collection on filters due to the potential for agglomeration (smaller particles sticking together) and shattering (breaking larger particles into smaller particles).[6] Potential contributors to bias during air sampling are inlet effects, filter efficiency, self-dilution, electrostatic effects, resuspended dust, and variability of concentrations within the breathing zone.[7]

Some particles may fail to enter the inlet of the sampling device because the inertia of these particles prevents them from accelerating, decelerating, or changing their direction fast enough to move with the air. Very large particles, which do not move with the surrounding air, will not be collected using conventional flowrates when the particles are far away from the sampling inlet. The three primary variables affecting the sampling efficiency of an inlet are the inlet's diameter, the suction velocity created by the flow rate of the pump, and the diameters of the particles being sampled.[8] In one experiment, larger inlets were found to be capable of sampling larger particles with less sampling bias. For inlets with long sampling tubes before the filter cassette, some particles may be lost within the sampling tube and not reach the collector.

Exceeding the sampling capacity of a filter can cause problems. For example, when sampling for asbestos a heavy dust loading will mask fibers, thus making an accurate count impossible.

While most dusts, mists, and fumes are relatively inert and remain stable after collection, they can also contain volatile or chemically reactive particles that are very susceptible to undergoing changes during sampling. Sulfuric acid droplets can increase in size due to water absorption and can react with ammonia, so accurate measurement of these droplets involves monitoring humidity during sampling and taking precautions to preserve the chemical integrity once it is on the sampling media.[7]

For gravimetric analysis, a strict protocol of drying the filters prior to each weighing may be needed to avoid inaccurate due to retained water vapor or chemical vapor.

TOTAL AEROSOL SAMPLERS

Total aerosol samplers are generally used for occupational air samples if the air contaminant is capable of passing through the lungs to the bloodstream. For example, when a lead pigment is inhaled, some particles might be carried out of the lungs by pulmonary clearance mechanisms, and subsequently swallowed. Another application of total dust sampling is for nuisance dusts, which are certain types of biologically inert dusts that in high concentrations may seriously reduce visibility; cause unpleasant deposits in the eyes, ears, and nasal passages; or cause injury to the skin or mucous membranes. OSHA also terms these dusts physical hazards. Nuisance dust standards apply to both organic and inorganic dusts, but they cannot be applied to chemicals that have specific toxic effects.

The total sample mass collected in a two- or three-stage filter cassette has in the past been widely used to characterize occupational exposures to many mineral and metallic compounds. However, this type of sample does not reflect exposures when compounds are deposited at various sites in the upper respiratory tract or tracheobronchial tree.[9] *Size-selective sampling* is the term used to describe sampling that collects only specified particle sizes. In the future, size-selective particulate mass sampling may replace total dust methods because it will eliminate the current concerns about the variability of sampling characteristics of these total dust samples and the accuracy of predicting aerosol health hazards.[10] Until then, however, "total dust" sampling methods are important for correctly comparing sample results to OSHA and other exposure standards based on those methods.

Environmental methods for sampling total atmospheric and lead aerosols involve using high-volume pumps inside enclosures. Wind and moisture have a significant impact on this type of sampling. Samples are generally collected for 24 hours or more. As opposed to occupational dust sampling where a specific process and source of dust are usually involved, atmospheric particles have a much broader composition, and may include carbon, metal, and mineral dusts.

Monitoring Occupational Exposures to Total Aerosols

For occupational exposures to total aerosols the sampling is very simple: a portable pump with a built-in rotameter connected via tubing to a filter in a cassette. Collection of a representative sample of the aerosol through the inlet of a filter cassette depends on physical factors, such as particle size, inlet size, sampling velocity, sampler shape and orientation, and ambient air velocity.[11] Generally closed-face cassette sampling techniques where only a small plug is removed from the top and bottom of the cassette prior to sampling have been used for the majority of aerosols in total dust sampling. The closed-face technique lends protection to the filter and its accumulated contaminant.[12]

When sampling with filter cassettes is done closed-face, the collected material is concentrated in the center of the filter near the inlet port. However, it has been found that collection accuracy decreases as particle size increases and as the angle of the cassette increases away from the direction of the wind.[13] One of the primary concerns raised regarding the use of the closed-face cassette is that the basis for its design is to facilitate ease of handling and analysis of the collected dust samples rather than to represent any particular set of collection characteristics. When facing into the wind, oversampling has been reported if the flow rate is different than the wind velocity. Undersampling occurs when the inlet is at an angle to the wind.[12]

The alternative to closed-face cassettes for total dust sampling is open-face cassettes. The open-face mode uses three-stage cassettes so the entire top can be removed, exposing the entire filter, thus allowing even distribution of the contaminant during sampling (Figure 7.7). This situation is vitally important for analysis such as fiber counts for asbestos that assume an analysis of a portion of the filter is representative of the entire filter. Three-stage cassettes must be used for open-face sampling because the second stage holds the filter in the cassette when the pump is turned off, cutting off the suction.

A comparison of open-faced and closed-faced sampling methods indicates a significant difference in their ability to collect certain types of particles. Differences in particle sizes, particle densities, sampling flow rates, inlet radii, and ambient airstream velocities may be responsible for the differences between open- and closed-face cassette sampling techniques. It has been determined that closed-face filter cassette sampling techniques are significantly less efficient than open-face filter cassette sampling techniques for paint spray mist, chromic acid mist, portland cement dust, grain dust, and wood dust.[12]

There are four documented sources of sampling error in total dust sampling using filter cassettes: sedimentation, geometric orientation of cassettes, airflow rate variation associated with personal pumps, and ambient wind speed and direction.[12]

Traditionally samplers for personal exposures are placed on the collar or lapel. Some studies have suggested that samplers placed on the lapel may result in higher results than if they were placed on the forehead or nose;[14] however, for practical purposes the collar or lapel is the best position to simulate the breathing zone. Dust deposited on a worker's clothing may be released and collected by a lapel-mounted sampler, although this resuspended dust is not necessarily carried into the breathing zone of the worker. The job being performed and individual work practices will also influence the results.

The worker's breathing zone consists of a hemisphere of 300-mm radius extending in front of the face, and measured from a line bisecting the ears. The sampling device should be placed within this zone for personal samples.[15]

The orientation of the cassette will affect results. The best position for the cassette during sampling is with the inlet pointing horizontally and slightly downward to prevent very large particles from settling into the inlet.

Procedure

1. Calibrate a sampling pump with a representative filter in line at a flow rate between 1 and 3 Lpm (usually 2 Lpm) according to the procedures in Chapter 5.

2. Take the plugs out and connect a preweighed, labeled, two-stage filter cassette containing the appropriate filter to the pump by attaching the tubing to the outlet end. If the cassette outlet end is sufficiently long, the tubing can be put directly over it. If not a "Leur taper adapter" can be used to connect the cassette outlet to the tubing. If leur adapters are used, the calibration should be done with them in the sampling train to account for any leakage. The inlet end of the filter cassette is the end farthest away from the filter. Make sure the plug is out of the inlet prior to turning on the pump. Preweighed filters should have a sample number attached to the cassette that has been assigned by the laboratory.

3. Select the employee or area to be sampled. Discuss the purpose of sampling and advise the employee not to remove or tamper with the sampling equipment. Inform the employee when and where the equipment will be removed. Clip the filter holder onto the worker's collar, T-shirt neck, or area where sampling is to be done and hook the pump onto the worker's belt or waistband. The inlet orifice should be in a downward vertical position to avoid contamination. Use duct tape to keep the tubing out of the employee's way. Plan the monitoring to collect at least the minimum sample volume. Do not put any tubing in front of the filter. Also start to prepare field blanks as described in step 11.

4. Turn on the pump and record the start time. Record pertinent field data including area, the employee's full name, job, position or title, or put a description of the area in which the pump is placed. Record start and end times, initial and final air temperatures, relative humidity and atmospheric pressure, or elevation above sea level and pump rotameter setting. Include the pump number and the number on the collection device. In some cases, the employee's company identification number must also be recorded.

5. Check the pump flow rate by visually observing the rotameter for any changes at least hourly. If there is a change, check the flow rate using a field rotameter.

6. As a minimum, check the pump rotameter after the first half hour, hour, and thereafter every 2 hours. Ensure that the tubing is still attached to both the pump and the collection device and that it is not pinched. Make sure the cassette is still in the same position.

7. Periodically check the pumps and work operations throughout the workday to ensure that sample integrity is maintained, and cyclical activities and work practices are identified. Change the filter if excessive loading is noted or if a significant change in the flow rate has occurred.

8. Before removing the pump at the end of the sample period, check the flow rate to ensure that the rotameter ball is still at the calibrated mark. If the ball is not at the calibrated mark, the flow rate of the pump must be checked prior to inserting another cassette.

9. Disconnect the filter after sampling and immediately cap the inlet and outlet using plugs. Turn off the pump, and record the ending time and other information in the sample data form.

10. At the end of the day the pumps must be recalibrated.

11. Prepare field blanks at about the same time sampling is begun. These field blanks should consist of unused filters and filter holders from the same lot used for sample collection. Handle and ship these field blanks exactly as the samples, but do not draw air through them. Two field blanks are recommended for each ten samples with a maximum of ten field blanks per sample set. In addition to these, two unopened filter cassettes from the same lot are often included as media blanks.

Environmental Collection of Total Aerosols. There are two different types of high-volume sampling: one for general ambient air samples, such as might be compared to the Environmental Protection Agency (EPA) primary and secondary particulate standards, and the other is the type associated with asbestos clearance sampling. The sampling set up for ambient air is much more complicated than that used for clearance. For more information on clearance sampling, see the section on asbestos in Chapter 17. For the balance of this section, high-volume filter sampling will refer to those techniques used for ambient air collection rather than clearance sampling.

In high-volume sampling (Figure 7.8) air is drawn through an 8-in. by 10-in. or round filter installed within a large sampling enclosure at 40 cfm (cubic feet per minute). The filter is weighed prior to and after sampling. The primary purpose of the enclosure is to protect the filter from dust fallout. The sampling effectiveness varies depending on the orientation of the inlet with respect to wind direction and wind speed. A typical "peaked roof" enclosure inlet

Figure 7.8. Typical high-volume sample pump for area sampling. (Courtesy of F&J Specialty Products, Inc.)

does not correspond to any specific particle size distribution, but in general over 50% of all particles less than 30 μm to 50 μm pass through such an inlet. The addition of special in let fittings allows for size-selective sampling, such as PM_{10}, an environmental measurement that corresponds to the thoracic sampling criteria. For survey work the most likely use of this technique will be indoor air concerns in buildings where an exterior source of contaminants is suspected or for boundary line monitoring of hazardous waste sites.

Generally high-volume samples are collected over a 24-hour period. Filter media can be cellulose fiber, glass fiber, quartz fiber, Teflon-coated glass fiber, and Teflon membrane. The selection of filter type depends on the purpose of the sampling. Quartz fiber filters are often used, although they are very fragile. Filters must be conditioned before and after the sampling period for 24 hours; therefore, there is a minimum built-in delay before results as available. For ambient air monitoring, inlet openings are usually 1–5 meters above ground level. Following is the basic procedure for a high volume sampler installed in an enclosure.

High-Volume Sampler Procedure

1. Tilt back the inlet of the enclosure and secure it. Calibrate using the orifice calibrator available for this unit. Record the flow rate.

2. Place the sampler and filter holder in the servicing position by raising up both the sampler motor/blower unit and filter holder until the filter holder is above the top level of the shelter. Then rotate the unit one quarter turn so that the filter holder hangs in the rectangular hole in the sampler support pan.

3. Remove the faceplate by loosening the 4 wing nuts. Allow the swing bolts to swing down out of the way.

4. Carefully center a new filter, rougher side up, on the supporting screen. Properly align the filter on the screen so that when the faceplate is in position the gasket will form an airtight seal on the outer edges of the filter.

5. Secure the filter with the faceplate and four swing bolts with sufficient pressure to avoid air leakage at the edges.

6. Rotate and lower the filter holder and blower/mover assembly to its normal operating position.

7. Wipe any dirt accumulation from around the filter holder with a clean cloth.

8. Close the lid carefully and secure with the aluminum strip, and plug all cords into their appropriate receptacles to start sampling of this type. Record the start time and other relevant data.

9. When sampling is complete, the motor is turned off and the time is recorded.

10. Reversing the procedure, the filter is removed and carefully placed in a container to return to the laboratory for weighing.

11. The final volume is calculated by multiplying the flow rate by the sampling time.

12. The weight in mg is divided by the volume in M^3 to get the results.

PARTICLE SIZE-SELECTIVE SAMPLING

Since the location in the human respiratory tract where an aerosol deposits depends on its size, the problems with characterizing human exposure to aerosols using total aerosol measurements has always been recognized as problematic. Thus there has been long interest in size-selective particle sampling.

The traditional criteria for occupational exposures to respirable dust in use by OSHA since 1970 as earlier defined by the ACGIH were based on a size selection curve developed by the U.S. Atomic Energy Commission (AEC) and applied to particles less than 10 μm in diameter. A device that sampled according to the original size-selection curve rejected all particles 10 μm and larger, and collected 50% of 3.5-μm particles and all particles ≤2 μm. The U.S. Mine Safety and Health Administration and international groups used slightly different size selection curves to represent the particle sizes felt to be deposited in the respiratory tract.

Over succeeding years, refinements were made in the various size selection curves and agreement was reached between international organizations such as CEN (Committee European de Normalization), ISO (International Organization for Standardization), and ACGIH. These new criteria are expressed as curves which relate the probability that particles of a certain size will either be inhaled or, if inhaled, will deposit in different parts of the respiratory tract.

As described earlier, the current definitions for particle size-selective sampling are three, progressively smaller, categories:

- Inhalable—the mass fraction which enters the nose and/or mouth during inhalation.
- Thoracic—the subfraction of the inhalable mass that that can penetrate the respiratory tract past the larynx.
- Respirable—the subfraction of the inhalable mass that penetrates down to the alveolar region (gas transfer area) of the lung.

Figure 2.3 illustrates the different regions; Table 7.1 shows the size distributions for each size category.

Inhalable Fraction. The inhalable particulate mass (IPM) fraction of an aerosol is the fraction of the ambient airborne particles that can enter the uppermost respiratory system compartments, the nose and mouth. Airborne material that deposits here may be absorbed and/or swallowed, although some may be expelled directly from the body by bulk cleaning mechanisms, such as sneezing, spitting, or nose blowing.[10] Inhalability depends on wind speed and direction, breathing rate, and whether breathing is by nose or mouth.[14]

It has been recommended that IPM sampling replace the present methods of total dust sampling using closed-face filter holders. The primary concern is that there are several methods of collecting total dust samples whose collection characteristics vary. Therefore, comparison as well as making an interpretation of the biological consequence of a sample are difficult. For example, the traditional open-face filter cassette does not actually measure total dust due to several collection problems, such as blunt face design, wind speed sensitivity, changes related to orientation, and particle size collection characteristics.[17] However, several samplers have been developed that are capable of meeting the criteria for IPM collection for both personal and area samples and so can overcome some of the limitations of total aerosol samples.

There are at least three general classes of chemical compounds for which sampling should be done for all inhalable particles: (1) highly soluble materials that can quickly enter the blood becoming available to pass through membranes in many regions and exhibit their toxicity, such as nicotine and soluble metal salts; (2) materials that can be toxic after oral ingestion, such as many metals including lead; (3) compounds that can exert toxic effects at their deposition site, such as hardwood dusts and acids.[10] Important categories of aerosols that are capable of exerting toxic effects following inhalation and deposition anywhere in the body include pesticides, many metals, and acids. An example of an exposure that would be sampled using an inhalable particulate sampler is a determination of the potential for nasal carcinoma from wood dust. In the case of large-sized particles or resin-impregnated sawdust, the materials would be expected to deposit and remain in the nasal passage.[18]

In many work environments airborne particle sizes are within the thoracic particulate mass (TPM) fraction rather than the IPM because larger particles are often removed via filtration or deposition on surfaces; however, in situations where there is a source of aerosol, such as sprayers, cutting or abrasive machinery, or easily resuspendable material, there may be a significant IPM concentration.[10]

Thoracic Fraction. As noted, the thoracic particulate mass (TPM) fraction represents those airborne particles that are capable of entering the upper respiratory area and trachea during mouth breathing. It has been described as representing "the worst-case potential exposure of the whole lung to particles." The TPM size-selective sampling criterion has been established as a tolerance band consisting of those particles that can penetrate a separator whose size collection has 50% of its particles (50% cutoff) 10 μm in size and a geometric standard deviation of 1.5 ± 0.01.[19]

A substance for which it would be useful to measure the TPM fraction is asbestos. Fiber levels are related to their potential to cause bronchogenic cancer since the types of fibers causing this effect deposit in the tracheobronchial and gas exchange regions.[18]

Another compound for which TPM sampling has been recommended is sulfuric acid aerosol.[20] The human health effects of major concern for this compound are bronchospasm in asthmatics and bronchitis, both of which affect the TPM region. The

bronchospasm is the result of acute exposure, and the bronchitis is the result of chronic exposure. Due to the potential for hygroscopic growth of sulfuric acid droplets in the airways, the particle size favors deposition taking place within the upper respiratory tract, trachea, and larger bronchi. However, it should be noted that most compounds that absorb water may not be good candidates for collection using TPM or respirable particulate mass (RPM) methods because of the unpredictability of where these compounds will actually penetrate in the airway as they grow in size after encountering the humidity in the respiratory tract.[19]

Due to a concern regarding a higher than normal incidence of bronchitic symptoms in miners, a study was conducted to determine whether compliance based on measurements using the criteria for the respirable standard for dusts <10 μm in size equally protected against thoracic dust exposures that would affect the bronchial region. It was found that thoracic dust levels were five to seven times higher than respirable levels depending on the area of the mine being sampled. As a result, it was determined that the separate TPM samples were useful and additional dust control techniques were required.[21]

Respirable Fraction. The respirable particulate mass (RPM) fraction is considered the portion of an aerosol available to the gas exchange region during nose breathing. It is based on ACGIH's previous recommendations with a tolerance band added. The 50% cutoff is the same as it was previously, 3.5 μm, and the geometric standard deviation is the same as for the other fractions, 1.5 ± 0.1.[19]

Iron oxide is an example of an insoluble compound that causes disease in the gas exchange regions of the lungs including pulmonary fibrosis. Therefore, the appropriate sampler for this exposure is an RPM sampler because iron oxide particles that might deposit in the nose or tracheo-bronchial region are most likely to be swallowed.[18]

SIZE-SELECTIVE SAMPLING DEVICES

The key criterion for these sampling devices is their accuracy in collecting particles according to the selection curve for three categories. The accuracy of a device depends on the factors discussed earlier plus any changes in collected particles within the collector due to breakdown of particles or agglomeration due to impaction or other forces.

Devices are available that collect just one size category. Other devices collect more than one category or the entire aerosol on different stages within the device for later calculation of mass within each of the size-selective categories. The devices described below represent a selection of available samplers; because of rapidly improving technology, manufacturers' catalogs should be checked for newer innovations.

IOM (Institute of Occupational Medicine) Sampler

The IOM sampler (Figure 7.9) was developed at the Institute of Occupational Medicine (IOM) in Edinburgh, Scotland and collects aerosol in a manner that matches how particles are inhaled into the human body. Thus the IOM device meets the sampling criteria for inhalable particulate mass. The device was designed using wind tunnel tests to closely simulate the particle collection behavior of the nose and mouth. A key feature is the filter cover cap (the ring with extension tube in Figure 7.10) that has a 15-mm circular orifice that functions as the sampling inlet for the device. The geometry of the sampling inlet, which protrudes slightly out in front of the main sampler body, prevents interference from the body of the sampling device in collecting an

Figure 7.9. Worker wearing the IOM® Sampler for particulates. (Courtesy of SKC, Inc.)

accurate sample. The inhalable sample is collected on a 25-mm-diameter filter held within a filter holder at a sampling rate of 2 L/min. The chemical hazard and method determines the type and pore size of the 25-mm filter.

Some IOM-type samplers have a modification in which polyurethane foam (PUF) plug filters are positioned within the device's inlet. Particle penetration through porous foams depends on the particle aerodynamic diameter, foam porosity, plug dimensions, and flowrate. The porosity of these materials is specified in units of pores per inch (ppi) with 30, 45, 60, and 90 ppi foams being readily available. Proper selection of porosity, plug diameter, and depth will allow collection of specific-size particles. Using this technology, manufacturers have developed a PUF disk that collects larger-size particles within the inhalable

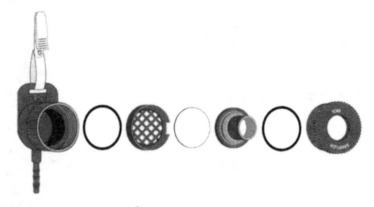

Figure 7.10. Exploded view of the IOM® Sampler. The device's configuration collects particles according to the inhalable mass criteria. (Courtesy of SKC, Inc.)

range while allowing the respirable size fraction to reach the 25-mm filter. Thus both inhalable- and respirable-size fractions can be measured with a single air sample; the inhalable fraction is the sum of the material on the PUF plug and filter. In some cases, two different PUF plugs can be used in series to separate the larger particles on the first plug and smaller particles on the second plug, with the very small particles penetrating both plugs to reach the filter. This permits measurement of all three size-specific fractions (inhalable, thoracic, and respirable) on a single sample.

When taking a sample with the IOM sampler, the filter/filter holder is weighed as a single unit before and after sampling. Thus, all of the dust both on the filter and on the internal walls of the filter holder unit is reflected in the measurement; there is no loss of particles from internal deposition within the unit. After weighing, chemical analysis may be performed on the filter and any adhered particles may be recovered from the interior of the filter holder.

Guidelines for Using the IOM Sampler

1. Follow normal gravimetric sampling protocol, including pre- and post-sampling weighing procedures:
   - Wear gloves when handling filter holders and use forceps when handling filters to prevent transferring moisture or dust particles to the filter holder or filter.
   - Weigh the filter/holder assembly as a single unit. Before pre- and post-weighing, wipe the external surface with a clean lint-free paper or cloth or with a soft lint-free brush.
   - For plastic filter holders, the filter/filter holder assemblies should be equilibrated overnight in a balance room under controlled humidity conditions before weighing. Maintain a stable humidity level in the balance room. Stainless steel filter holder assemblies may be desiccated.
   - For most accurate results when using plastic filter holder at low filter loadings, field blanks may be used to correct weights. A field blank is a pre-weighed filter/filter holder assembly that is taken out into the field, but not used. It should be removed from any packaging and exposed to the ambient conditions. Any change in pre- and post-weights are used to adjust actual field sample results.
   - The same balance should be used for both pre- and post-weighing

procedures. Allow the filter/filter holder assemble to stabilize on the balance before taking a reading.

2. To load the filter in the filter holder, unscrew the cover cap and remove the filter holder from the unit. Separate the two halves of the filter holder and place a filter into the holder's bottom portion (on the support grid) and snap the top onto the filter holder.

3. Wipe the external surface of the assembly with a clean lint-free paper or cloth or with a soft lint-free brush. Then, pre-weigh the filter/filter holder assembly as a single unit.

4. Place the filter assembly into the sampler body and screw on the cover cap. If the PUF disk is used, position that just under the cover cap before it is screwed on. Connect the assembled IOM unit to the inlet of the sampling pump using flexible tubing.

5. Calibrate the sampler flowrate with the loaded IOM in line. Either use the "jar method" described in Chapter 5 or, for some commercial IOM samplers, a snap-on calibration adapter is available. Adjust the pump flow rate to 2 L/min. Use the provided transport cap to prevent dust from entering the device while it is transported to the sampling site.

6. Connect the IOM to the sampling pump and sample at 2 L/min for the recommended sampling period as specified in the sampling method.

7. Remove the IOM filter/filter holder assembly. Wipe the external surface of the assembly with a clean lint-free paper or cloth or with a soft lint-free brush. Use the provided transport clip to prevent dust from entering the assembly while transporting it to the laboratory.

8. Post-weigh the IOM filter/filter holder assembly as a single unit to determine the amount of inhalable dust. To correct sample weights using a field blank, if the post-weight is greater than the pre-weight, subtract the difference from all results. If the post-weight is less than the pre-weight, add the difference to all results. If chemical analysis is to be performed, recover any adhered particles from the filter holder and send with the filter for analysis.

9. The filter holder can be cleaned, reloaded with a fresh filter, and reused.

10. If using the PUF filter, weigh it separately from the filter/filter holder assembly. The inhalable fraction is the sum of the weights, while the respirable fraction is represented by the weight of material on the 25-mm filter. Unless the manufacturer's instruction indicate otherwise, discard the PUF disk after use; it is *not* designed to be cleaned and reused.

Button Aerosol™ Inhalable Sampler

The Button Aerosol™ Sampler (Figure 7.11) is a patented, reusable filter sampler that closely follows the ACGIH/ISO size-selective sampling criteria for inhalable particulate mass. It features a porous curved-surface inlet that improves the collection characteristics of inhalable dust. The conductive stainless steel inlet reduces electrostatic effects and contains evenly spaced holes that act as sampling orifices for multidirectional sampling capability with low sensitivity to wind direction and velocity. The sample is collected on a 25-mm filter, which is located just behind the inlet to minimize internal particle losses. The sample may be analyzed gravimetrically or using a chemical analysis technique. The sampler operates at 4 L/min;

Figure 7.11. The Button Aerosol™ Sampler closely follows the ACGIH/ISO size-selective sampling criteria for inhalable particulate mass. (Courtesy of SKC, Inc.)

sampling accuracy is within ±30% at flows ranging from 2 to 5 L/min. It shows the same sampling efficiency when standing freely as it does when attached to a worker. A snap-on calibration adaptor is available, as is a special inlet shield to protect the filter from damage or overloading when the device is used to sample abrasive blasting operations.

Guidelines for Using the Button Aerosol™ Sampler. Wear gloves and use forceps when handling filters.

1. Place the appropriate 25-mm filter in the conductive plastic transport case and weigh. Record the weight.
2. Remove the sampler inlet by unscrewing it counterclockwise. Remove the weighed filter from the transport case and place it on top of the support screen inside the sampler.
3. Place the Teflon O-ring on top of the filter, replace the inlet, and very gently turn clockwise until moderately tight.
4. Calibrate the device by placing the calibration adaptor over the inlet and connecting it to a primary airflow standard. Connect the device's outlet to a suitable pump and calibrate the pump to 4 L/min using the procedures in Chapter 5.
5. Disconnect the calibration adapter and replace the filter used during the calibration step with a fresh weighed filter for sample collection.
6. To conduct the sampling, connect the outlet of the sampler to the inlet of a sampling pump using flexible tubing. For personal sampling, clip the sampler onto a worker's collar or pocket near the breathing zone. Clip the pump onto the worker's belt or place it in a protective pouch. Start the pump and record the start time, worker information, and flowrate on the sampling data form. For area sampling, position the sampler to avoid cross-drafts and direct projection of the particles into the inlet.
7. At the end of the sampling period, stop the pump and record the stop time and other relevant data.
8. When sampling is complete, very gently unscrew the sampler inlet and use forceps to remove the O-ring seal. Then, using forceps, carefully remove the filter from the sampler.
9. For gravimetric analysis, place the filter in the conductive plastic transport case. If a chemical analytical technique is to be used, place the filter in a sealed glass vial to eliminate losses during transport.
10. Ship the plastic transport case or glass vial with pertinent sampling information to an accredited laboratory for analysis.

Cyclone Samplers for Respirable Particulate Matter

The cyclone is a centrifugal separator that is typically conical or cylindrical in shape,

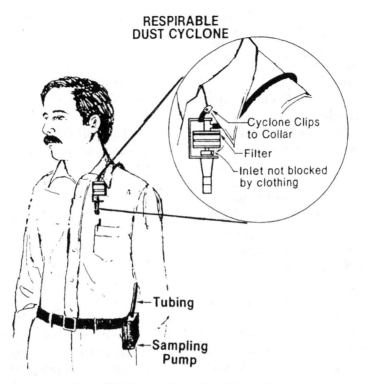

Figure 7.12. Personal sampling with a cyclone.

with one or more opening through which particle-laden air is drawn along a concentrically curved channel (Figure 7.12). Larger particles impact against the interior walls of the unit due to inertial forcess and drop into a grit chamber in the base. The lighter particles do not impact the walls and thus continue through and are drawn up through the center of the cyclone, where they are collected on a filter cassette. Cyclone samplers are based on fluid mechanics (centrifugal force) and thus separate by aerodynamic diameter, not actual diameters of the particles.

Historically the most commonly used personal respirable dust sampling device in the United States was the 10-mm "Dorr–Oliver" nylon cyclone. Much of the published exposure data and documentation for respirable exposure standards are based on monitoring performed with the 10-mm nylon cyclone. This device sepa-

TABLE 7.3. Old NIOSH/OSHA Criteria for Respirable Dust Collectors

Particle Size (μm)	% Passing Selector
<2	90
2.5	75
3.5	50
5.0	25
>10	0

rated particles according to the old respirable dust distribution curve used by the ACGIH and OSHA (Table 7.3). This size distribution is slightly different from the respirable particulate matter curve used today (Table 7.1). With a sampling rate of 1.7 Lpm, the 10-mm cyclone passes 50% of 3.5-μm aerosol particles. Its dynamics of collection were well tested with silica (a mineral) dust, but may not be accurate for other dusts without additional testing and verification. For example, with organic dusts such as wood dust, the particle size and density may cause the size separation characteristics to differ from the expected distribution at an airflow of 1.7 Lpm; in some of these cases a different sampling rate may match the desired separation curve more closely.[22] One drawback of the nylon cyclone is that is can accumulate a static charge due to the air passing over the nylon, and when highly charged aerosols are being sample this can lead to variability in results.[23] The standard Dorr–Oliver nylon cyclone has one inlet for sampled air, which can lead to sampling variations due to inlet orientation and also particle behavior within the cyclone, especially impaction on the interior wall opposite the single inlet.

A series of respirable particulate matter (RPM) cyclones (Figure 7.13) have been developed to match the current ACGIH/ISO/CEN curve and to address limitations of earlier cyclones. Typically these cyclones feature:

- Body made of metal or conductive plastic to avoid buildup of a static charge.
- Multiple air inlets into the cyclone to smooth out air movement within the cyclone.
- Provision for sampling using an open-face filter cassette (rather than closed-face) to distribute the collected material more evenly across the filter surface.

For any cyclone device the manufacturer provides the preferred air sampling rate, which is chosen to provide the optimum size separation when compared to the entire size distribution curve.

Accurate monitoring with a cyclone requires careful attention to the cleaning and assembly protocol relevant to the specific device. Cyclones typically use a snap-on or screw-on "grit pot" to collect the oversize particles. The pot must be removed and emptied before each use, and the entire device must be washed and dried so the interior is clean. Careful reassembly using any provided O-ring gaskets is necessary to prevent leaks that will interfere with the size separation performance of the cyclone. Cyclones are designed to be used with the grit pot in place even though only the discarded particles are collected in the pot.

Procedure

1. Remove the cyclone's grit pot and top pieces prior to use and inspect the cyclone interior. If the inside is visibly scored, discard this cyclone since the dust separation characteristics of this cyclone might be altered. Clean the interior of the cyclone to prevent reentrainment of large particles.

Figure 7.13. Cyclone for respirable dust sampling made of conductive plastic. (Courtesy of SKC, Inc.)

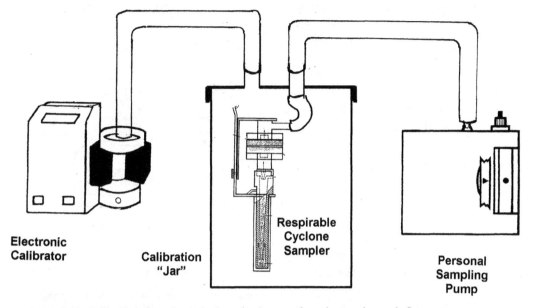

Figure 7.14. Calibration setup for a air size-selective sampler using an electronic flowmeter.

2. Insert the proper 25-mm or 37-mm filler cassette as specified in the sampling protocol. For gravimetric analysis, use a pre-weighed filter or matched pair filters. Check and adjust the alignment of the cassette and cyclone inside the cyclone body to prevent leakage. It should be tight. Connect the outlet of the cyclone to the pump via tubing.

3. Calibrate the pump using the "jar" method (Figure 7.14) designed for cyclones just prior to use. Note the point on the pump's rotameter that corresponds to the specified flowrate. A piece of cellophane tape can be wrapped at this point and marked.

4. Clip the cyclone assembly to the worker's collar or T-shirt neckline and hang the sampling pump on the worker's belt or waistband. Ensure that the cyclone hangs vertically. Make sure the sampling inlet is not blocked. Duct tape may be useful in securing it in place.

5. Turn on the pump, recording the flowrate and time. At the end of the sampling period, record the final flowrate and the time. Replace the filter cassette caps. Calibrate the pump again following sampling.

For respirable silica samples, laboratory results are usually reported under one of four categories:

1. *Percent quartz (or cristobalite):* Applicable for a respirable sample in which the amount of quartz in the sample was confirmed.

2. *"Less than or equal to"* (≤) *in units of percent:* Less or equal to values are used when the adjusted 8-hour exposure is found to be less than the permissible exposure limit (PEL), based on the sample's primary diffraction peak. The value reported represents the maximum amount of quartz that could be present. However, the presence of quartz was not confirmed using secondary and/or tertiary peaks in the sample since the sample could not be in violation of the PEL.

3. *Approximate values in units of percent:* The particle size distribution in a total dust sample is unknown and error in the X-ray diffraction (XRD)

analysis may be greater than for respirable samples. Therefore, for total dust samples an approximate result is given.

4. *Nondetected:* A sample reported as nondetected indicates that the quantity of quartz present in the sample is not greater than the detection limit of the instrument. The detection limit is usually 10 μg for quartz and 50 μg for cristobalite.

Cyclone Devices for Environmental Sampling

Cyclone air samplers are also used for environmental sampling. One application is as a low-maintenance size separator for $PM_{2.5}$ sampling. The BGI Very Sharp Cut Cyclone (VSCC)™ is designed to be used in place of an impactor (described later in this chapter) and has received U.S. EPA "Equivalency" approval for this application. The flow path through the device is shown in Figure 7.15. The aerosol enters the inlet tangentially after being preseparated by in a PM_{10} separation device before the cyclone. Particles greater than $PM_{2.5}$ (mass median diameter–aerodynamic diameter) are removed to the grit pot on the side of the unit and the $PM_{2.5}$ sample is ducted through the cyclone separator to the filter holder. Other models are designed to be used as inlet devices for direct-reading particle counters; in these devices there is no collection filter and the sample enters the photometer or other direct-reading instrument. Field and laboratory tests indicate that the device can operate up to a period of 90 days between cleaning frequencies, even in high urban loading sites. The typical approved sampling rate for this device is 16.7 Lpm.

Impactors

Impactors can separate an aerosol into a particle size distribution that allows the

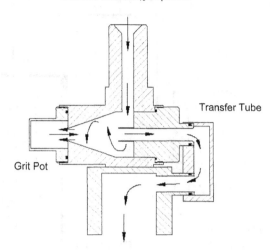

Aerosol from PM_{10} Separator

Transfer Tube

Grit Pot

$PM_{2.5}$ Sample to Filter

Figure 7.15. Cross section of the BGI Very Sharp Cut Cyclone. (Courtesy of BGI, Inc.)

sampler to characterize the primary size ranges according to their aerodynamic diameters. Impactors have been widely used for occupational measurements, including personal and area sampling, environmental air pollution studies, and microbial sampling (with modified instruments). A major use of impactor data is determining the complete particle size distribution of a sampled aerosol, which allows the particle mass concentration in any size range to be calculated, including the inhalable, thoracic, and respirable fractions. Chemical characteristics of the aerosol must be considered when selecting aerosols to sample with impactors. For example, corrosive and combustible aerosols require special techniques and should not be sample with impactors without special precautions. For more information on microbial sampling techniques, see the chapter on sampling for bioaerosols (Chapter 16).

Impactors separate particulates in an airstream by directing them toward a coated flat surface. In one design, called a *cascade* impactor, particles enter the inlet jet and pass through a series of progressively smaller jets with which there is an associated collection surface (plate) usually at right angles to it. The aerosol stream passes through the first jet, flows around the impaction surface that is obstructing its flow, and then through the next jet and its associated impaction surface, and so on. As the aerosol moves through the plates, larger particles are deposited on the top stages and smaller ones are deposited near the bottom. The open area of the radial slots are smaller for each succeeding stage; thus, the jet velocity is higher for each succeeding stage. After the last impactor stage, remaining fine particles are collected on a filter or plate.

In another design, smaller particles are diverted from the air entering the device and thus are collected on the top stage. The sampled air is directed downwind though a central vertical tube and successive larger particle surges are diverted at each stage until the largest particles are collected on the final stage.

When an impactor has been properly calibrated to define the aerodynamic media (cut) size characteristic of each stage, a size analysis may be made by calculating the percent by weight on each stage using weighings, radioactivity, or chemical analysis to determine the amount of deposited material. The cutpoint refers to the particle size for 50% collection. Cutpoints can be calculated or found in manufacturer's literature.

Cascade Impactors.
These devices collect the larger particles in the top stages and smaller particles on the lower stages. They come in different sizes. The larger ones are used for area samples and the smaller impactors can be used for personal samples. Individual impactor stages may be of the single jet or multijet variety. The latter are often preferred because resuspension of deposited particles is minimized and collection of larger samples is possible. Particles deposited on each stage may be examined microscopically. Impactors can

be designed over a wide range of flow rates and can be operated in any orientation.

The range of sizes collected by each stage of the cascade impactor depends on the density and shape of the particles as well as their diameter, and therefore represents their aerodynamic behavior of the particles. The flow rate determines the relative distribution collected on each stage.

Generally cascade impactors are very good at collecting large particles as these particles tend to impact on the plates because it is harder for them to make the turn necessary to go around the edge of the plate. Under standard conditions, this includes particles with aerodynamic diameters greater than $0.2\,\mu m$. The range can be extended to $0.05\,\mu m$ with micro-orifice impactors.

Typical problems that result in differences between impactors are particles lost to bounce, reentrainment due to overloading of the sample on a particular stage, inaccurate calibration, and lack of a sharp cutoff for each stage. Solid particles are capable of bouncing from the collection surfaces of impactors and being carried to subsequent stages or the backup filter, the result being that the size distribution is distorted toward the smaller sizes. Another source of error when comparing impactors with total dust samplers is interstage losses. Adhesive may lose its effectiveness during sampling, leading to the nonuniformity of the deposit. Sample collection can also be biased by inlet configuration and sampling flowrate.[30]

The best prevention for particle bounce is the use of silicone grease, oil, or a similar adhesive on the surfaces of the collection stages. Grease-coated surfaces have the limitation of being good for particle retention only when there is less than one monolayer of impacted particles. As particle loading increases on grease-coated impaction surfaces, the incoming particles no longer strike the grease layer but impact on the already collected particulate matter, and may bounce.

Cascade Impactor (Area Samples)

Cascade impactors (Figure 7.16) are typical of this type of collector, and they come in a variety of configurations. For example, the Thermo Andersen Series 20–800 Mark II has eight jet stages to classify aerosols from 9 µm and above to 0.4 µm and allows airborne particulates to impact upon stainless steel impaction surfaces or a variety of media substrates. At each stage, an aerosol stream passes through the jets and passes around the impaction disks. Entrained particles with enough inertia collect on the impaction surfaces while smaller particles remain entrained to be deposited upon subsequent stages. Higher jet velocities enable smaller particles to be characterized

Figure 7.16. Inertial impactor for area samples. (Courtesy of Thermo Electron Corporation.)

efficiently. The aerodynamic diameter of the collected aerosol depends upon the jet orifice velocity within each stage, the distance between the jets and impaction surface, and the collection characteristics of the preceding jet stage. A final filter collects all particles smaller than 0.4 µm.

Depending on the sampling conditions, the collection substrates may be made of stainless steel, mylar, glass fiber, or other materials. Typically the backup filter to collect the very small particles is a 34-mm-diameter PVC filter. The impactor is designed to operate at several flowrates; the higher rates permit classification of smaller particles. Table 7.4 illustrates the cut points for different sampling rates. In addition, there is an accessory that allows 3 Lpm flowrates, enabling classification of particles to 35 µm.

Personal Cascade Impactor

These devices operate on the same principles as the larger eight-stage impactors for area samples. The Thermo Andersen 290 Personal Impactor (Figure 7.17) can be configured with two to eight stages. With eight stages the cut points (µm) are 21 and above, 15, 10, 6.5, 3.5, 1, 0.7, 0.4, and the final filter. The options for collection substrates are the same as for the larger eight-stage impactor. The overall size is practical for a personal sampler: approximately 2.2 inches wide, 1.6 inches deep, and from 2.2 to 3.4 inches high depending on the number of stages. The weight is less than 8 ounces.

Prior to sampling, collection substrates

TABLE 7.4. Cut Points (µm) for Eight-Stage Andersen Cascade Impactor

| Flowrate (Lpm) | Cascade Impactor Stage | | | | | | | |
	1	2	3	4	5	6	7	8
28.3	9.0	5.8	4.7	3.3	2.1	1.1	0.7	0.4
60	8.6	6.5	4.4	3.3	2.0	1.1	0.54	0.25
90	8.0	6.5	5.2	3.5	2.6	1.7	1.0	0.43

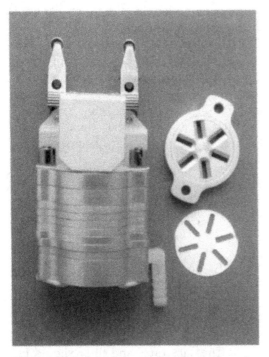

Figure 7.17. Inertial impactor for personal samples. (Courtesy of Thermo Electron Corporation.)

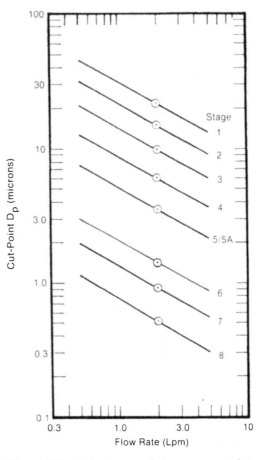

Figure 7.18. Typical personal impactor cutpoints versus flowrate for air at 25°C and 1 atm. (Courtesy of Andersen Instruments, Inc.)

and backup filters are weighed and placed in the impactor. The sampling flowrate of the personal sampling pump is set at nominally 1.7 Lpm. The impactor's personal mounting bracket is attached to the lapel or pocket. After sampling, the substrates and filter are weighed. Weight increase on each substrate is the mass of particles in the size range of that impactor stage. The total weight of particles on all stages and filter is added and the percent particle mass in each size range is calculated. Respirable particle mass fraction is determined from the particle size distribution.

An example of an application for the personal impactor is a study that was done to determine air-lead particle sizes during battery manufacturing. As a result, it was determined that where lead particles are predominantly <5 mm, a greater hazard would exist than is normally assumed using total lead collection techniques, and more stringent exposure controls are needed.

Also, to minimize contamination at the site, the impactors were prepared off site and transported inside sealed plastic bags. New bags were used for each day of sampling. A daily field control impactor was prepared, transported, and unloaded in the same manner as the samplers. Following sample collection, each substrate and backup filter was placed in a screw-capped vial, previously cleaned with nitric acid, for transport to the laboratory.[24]

Occupational Sampling with the Personal Cascade Impactor. The flowrate is critical with these devices (Figure 7.18). If it is not constant and calibrated correctly, the collecting efficiency will be hampered. New impactors should be examined for imperfections, for example, small deviation in the jet diameters such as burrs, and the jet diameter should be verified before use.[25]

As the flowrate is usually fixed for a given instrument in order to maintain the desired size distribution, the orientation of the inlet probe is the primary variable. The best approach is to align the probe parallel to the air currents in the area. Critical orifices can be used to maintain a constant flowrate.

Calculate the estimated mass to be collected ahead of time, since collection of an insufficient or excessive mass can cause erroneous results. Too much deposit will result in overloading the stages and can cause particle bounce, particle reentrainment, and changes in particle collection characteristics. Insufficient mass will result in poor precision in the analysis of deposited material on each stage and will lead to large uncertainties in interpreting the particle size distribution. A general rule is that material depositing on each stage should be >0.5% of the total mass collected on all stages.[25]

Impaction grease is applied as a suspension or solution of 10–20% grease in a solvent such as toluene. The mixture is applied to the substrate with a brush, eyedropper, or sprayer. Place the substrate on the bottom plate of the template with the two locating pins through opposite perforations. Place the top plate on top and apply the solution within the slots in the template. Note that the outer edge and center of the substrates must be devoid of grease or else the substrate will stick to the upstream impaction stage. Allow any solvent to evaporate. After drying, a cloudy white film will be visible. The final greased substrate should be tacky but not slippery, with a film thickness about equal to the diameter of the particles to be captured, 1–10 μm thick. Figure 7.19 is an example of an impactor sampling data sheet used during field monitoring.

After using the cascade impactor, it is critical that the head be opened properly and not inverted. If it does become inverted, the sample will be lost. The plates that do not use filters are more difficult to weight than samplers that do use them, because of the potential for loss of sample.

Procedure

1. Select the sampling flowrate based on the particle cutpoints desired and the vacuum capability of the sampling pump.

2. If the collection substrates are to be greased, use the procedure described above. Pre-weight all the substrates and 34-mm, 5-μm PVC backup filters. Record the weights. The substrates and filter should be equilibrated with the laboratory environment for 24 hours at a relative humidity of 50% or less before being weighed and passed over a static eliminator if they have a static electricity charge.

3. Assemble the cascade impactor. Each impactor stage is numbered along its edge. In assembly, all numbers should line up along one of the two threaded studs in ascending order from inlet to exit. No number should be upside down. The two threaded studs orient the impactor stages so that the slots in each stage are staggered from the slots in adjacent stages. The substrates are placed using a tweezer on the top surface of each impactor stage so that the perforations in the substrates match the slots in the impactor stages.

4. The unit is attached to a personal sampling pump and calibrated to 2 Lpm using the "jar" method described in Chapter 5.

Figure 7.19. Impactor data worksheet. (Courtesy of Andersen Instruments, Inc.)

The worksheet contains the following fields and table:

Test No. _____ Date _____
Name of Subject _____
Plant _____
Location _____

Sampling Flow Rate Q (Lpm) _____ Impactor Model No. _____
Starting Time T_1 (Hrs.) _____ Impactor Serial No. _____
Ending Time T_2 (Hrs.) _____
Total Sampling Time $T_2 - T_1$ (Hrs.) _____
Sampling Volume V $(M^3)^{(1)}$ = _____
Total Mass Concentration C_{tot} $(mg/M^3)^{(2)}$ _____

Stage Number	Stage Cut-Point D_p [3] (microns)	Initial Weight W_1 (mg)	Final Weight W_2 (mg)	Particulate Weight $W^{[4]}$ (mg)	Concentration $\Delta C^{[5]}$ (mg/m^3)	$\log_{10}D_p$	$\Delta\log_{10}D_p$ [6]	$\Delta C/\Delta\log_{10}D_p$ $(mg/m^3/log\ micron)$	$GMD^{[8]}$ (microns)	$\dfrac{W}{W_{tot}}$ (%)
1										
2										
3										
4										
5										
6										
7										
8										
Back-Up Filter										
$W_{tot}^{[7]}$	—	—	—			—	—		—	100

NOTES: (1) $V = \dfrac{1}{16.7}\ Q\,(T_2 - T_1)$

(2) $C_{tot} = W_{tot}/V$

(3) Cut points are defined for each stage of an impactor.

(4) $W = W_2 - W_1$

(5) $\Delta C = W/V$

(6) $\Delta_i\log_{10}D_p = \log_{10}D_{p_{i-1}} - \log_{10}D_{p_i}$

(7) W_{tot} = sum of particle masses for all stages including back-up filter.

(8) $GMD_i = \sqrt{D_{p_i}\,(D_{p_{i-1}})}$

239

5. Connect tubing to the sampling pump and set the flowrate on the pump.

6. Attach the impactor to the lapel or pocket of the worker, the pump to the belt, and the interconnecting tubing to the impactor.

7. Turn on the pump. Record the start time and other data on the sampling data sheet.

8. Turn off the pump at the end of the sampling period. Record the time.

9. During sampling the unit is attached to the worker's collar. After sampling the impactor is disassembled and the substrates and filter are removed and weighed. The weight change on each substrate represents the mass of particles in the size range of that impactor stage. The total weight of particles on all stages and the filter are added together and the percent particle mass in each size range is calculated. A graph is developed for the cumulative size distribution showing the percent of particle mass smaller than the aerodynamic article diameter. The respirable particle mass fraction is determined from the particle size distribution.

10. For best results, return the impactor to the laboratory fully assembled in the upright position and with the inlet sealed to prevent sample contamination.

Impactor Data Analysis. A major use of impactor data is determining the complete particle size distribution of sampled aerosol. Given the complete particle size distribution, the particle mass concentration in any size range can be calculated, including the inhalable, thoracic, and respirable fractions.

The most common data analysis involves developing a cumulative mass distribution, graphing this distribution, and using the graph to identify the mass median diameter and other parameters of interest. However, the sampling practitioner should be aware that correct interpretation of cascade impactor data requires experience and an understanding of what the calculations mean. The following is an overview of some typical uses for impactor data; however, there are other references that can provide more comprehensive information and the sampling practitioner is encouraged to consult them.[25]

The differential particle size distribution is a plot on log–log graph paper of the particle mass concentration in each particle size band (stage) versus the geometric mean diameter. The differential particle size distribution gives the details, or "fine" structure, of the particle size distribution.

The cumulative particle size distribution is a plot on log-normal (or log-probability) graph paper of the total particle mass smaller than a given particle size. This cumulative distribution gives an overall picture of the size of the particles.

The mass mean diameter of the log-normal distribution is the particle size where 50% of the particle mass is contained in particles larger than the already-defined MMD, and 50% of the particle mass is contained in particles smaller than the MMD.

The geometric standard deviation is the ratio of the MMD to the particle size for which 16% of the mass is borne by particles smaller than this particle size. It is a measure of the "spread" in the particle size distribution. If the geometric standard deviation equals one, all of the particles are of the same size, otherwise known as monodispersed. Table 7.5 is an analysis of impactor data.[26]

TABLE 7.5. Analyzing Impactor Data

Distribution	Formula
Differential particle size (plot on log–log graph paper)	$\dfrac{\Delta C_i}{\Delta_i \log_{10} D_{p_i}}$ (log) vs. GMD_i (log)
	$GMD_i = \sqrt{D_{p_i}(D_{p_{i-1}})}$
Cumulative particle size (plot on log normal paper)	$\dfrac{\sum_{i=j-1}^{N} W_j}{W_{tot}}\%$ (normal) vs. D_{p_i} (log)
Mass mean diameter	$\dfrac{\sum_{i=j-1}^{N} W_j}{W_{tot}} = 50\%$
Geometric standard deviation	$\dfrac{MMD}{D_{p16\%}} = \dfrac{D_{p84\%}}{MMD}$

ΔC_i = Particle mass concentration in each particle size range

D_{p_i} = Particle size or cutpoint of each individual stage

GMD_i = Geometric mean diameter of each size interval

W_j = Particle mass on impactor stage i, where i = size interval for that stage

W_{tot} = Sum of particle masses on all stages plus that on the backup filter

MMD = Mass mean diameter

TSI RespiCon™ Virtual Impactor

This device is described as a *virtual* impactor. It functions by separating the airflow into two streams, with different particle-size fractions carried by each of the two streams. It is similar to a conventional impactor, but the impaction surface is replaced with a *virtual space* of stagnant or slow-moving air where smaller particles in the airstream remain while larger particles are captured in a collection probe rather than impacted onto a surface.

Figure 7.20 is a schematic diagram of a virtual impactor. The aerosol passes through an accelerating nozzle and is directed toward a collection probe. At this point a major portion of the flow is diverted

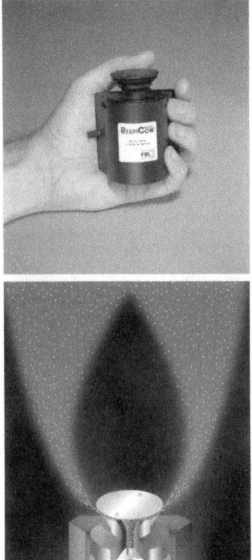

(a)

(b)

Figure 7.20. TSI RespiCon™ Virtual Impactor: (a) Unit ready for use; (b) airflow pattern through the device. (Courtesy of TSI Inc.)

90° away from the collection probe. This is where the particle-size separation takes place. Small particles with low inertia follow the flow streamlines and are carried away radially with the major flow. Large particles with greater inertia deviate from the flow stream and continue moving down the collection probe with the lesser airflow. The separation efficiency curve is determined by the ratio of the major and lesser flows and the physical dimensions of the nozzle and collection probe.

The TSI RespiCon™ Particle Sampler is a two-stage virtual impactor featuring three collection filters and two virtual impactors. The total flow is controlled with a personal air sampling pump, and the split at each stage is controlled by a flow orifice. The inlet of the sampler is designed to sample inhalable particles according to the ACGIH/ISO/CEN definition for inhalability. In the first virtual impactor stage, the total flow of 3.11 Lpm is passed through a center accelerating nozzle and directed toward the collection probe. The major flow of 2.66 Lpm is diverted away, and the particles smaller than the cut size are collected on the stage-one filter. Particles larger than the cut size are carried in the lesser flow of 0.44 Lpm to the next stage. The second virtual impactor stage separates the flow into 0.33 Lpm for the major flow and 0.11 Lpm for the lesser flow. Particles in the major flow are collected on the stage-two filter, and particles suspended in the lesser flow are deposited on the final stage-three filter. The filters at stages one and two have a central hole to allow for the collection probes. The final filter has no center hole.

The various size fractions are determined as follows:

- The 50% particle cut size for the first stage of the RespiCon™ is 4-μm aerodynamic diameter. The separation efficiency curve matches the respirable curve as defined by ACGIH/ISO/CEN. Particles collected on the first filter stage represent the *respirable* fraction of inhalable particles.
- The 50% cut size for the second stage is 10-μm aerodynamic diameter, and the efficiency curve matches the thoracic curve. Particles collected on both the first and second filter stages represent the *thoracic* fraction of inhalable particles.
- The final filter stage collects all remaining particles. Particles collected on all three filters represent the *inhalable* fraction of total suspended particles.

One characteristic of a virtual impactor is that particles smaller than the cut size of the impactor remain in both the major and lesser flows. Therefore, if the lesser flow is 10% of the total flow, then 10% of the small particles will remain with the lesser flow. Another characteristic is that particles larger than the cut size become concentrated in the lesser flow. The concentration factor is the ratio of the total flow to the lesser flow. If the lesser flow is 25% of the total flow, then the concentration factor is 4. Even though a small percentage of particles smaller than the cut size will be transported with the lesser flow and collected on later filter stages, it does not compromise the accuracy of the calculated mass concentration in each size fraction. The particle mass concentration in each size fraction can be accurately calculated using simple "correction" equations.

The RespiCon™ Particle Sampler can be used for both personal and area monitoring. For personal breathing-zone measurements, the sampler easily attaches to a chest harness. For area monitoring, the sampler is placed on a flat surface or thread-mounted onto a tripod.

Impactors for Environmental Sampling

There are impactors designed for long-term ambient air monitoring for environmental quality purposes. In the United States, many of the impactor applications are to evaluate the Particulate Matter— 10 μm (PM_{10}) and Particulate Matter— 2.5 μm ($PM_{2.5}$) airborne levels according to U.S. EPA criteria. These devices are generally stationary rather than portable and can be classified as low volume (<20 Lpm), medium volume (20–150 Lpm), or high volume (>150 Lpm).

PM_{10} sampling involves use of a sampling apparatus similar to the type used for total suspended particulate matter, except that a size-selective inlet and impactor are fitted into the large, stationary unit. For $PM_{2.5}$ sampling, an additional impactor or a cyclone is fitted to further size-categorize the airflow from the PM_{10} impactor.

In a typical application, suspended particles in the air are sampled at 40 cfm through this inlet accelerating the particles through multiple impactor nozzles. By virtue of their larger momentum, particles greater than the 10-μm impactor cutpoint impact onto the greased impaction surface. PM_{10} particles smaller than 10 μm are carried vertically upward by the air flow and down multiple vent tubes to an 8-in. by 10-in. quartz fiber filter, where they are collected. The large particles settle in the impaction chamber on the collection shim and are removed/cleaned during prescribed maintenance periods. The filter is weighed before and after sampling, and the increase represents the mass of particles smaller than 10 μm. Sampling is done for 24 hours. Impactors are described in more detail in later section. Actual use of these monitors is described in detail in the manufacturers' operating manuals.

SAMPLING FOR SPECIFIC AEROSOLS

Coal Mine Dust

Coal dust exposures are primarily related to mining, either underground, strip mining, or augering, with the highest concentrations related to underground mining. There are four types of coal mined: lignite, subbituminous, bituminous, and anthracite.

Coal dust causes coal workers pneumoconiosis (CWP), which can range in severity but is characterized by fibrosis and is found predominantly in the upper lobes of the lung. Symptoms of CWP are indistinguishable from those typical in other chronic obstructive lung diseases, especially chronic bronchitis. Most often there is an exposure to other mineral compounds, such as silica and talc, during coal mining.[27]

Currently, MSHA regulations require that respirable dust samples collected in coal mines should be sampled using a 10-mm nylon cyclone containing a 5-μm PVC filter operated at a flowrate of 2 Lpm rather than 1.7 Lpm, with the results multiplied by a factor of 1.38 to convert the results to the equivalent concentration that would be measured by an instrument meeting the criteria of the Mining Research Establishment of the National Coal Board of England.[28] The ability of this method to be a good estimator of the respirable dust concentration has been questioned and in the future may change, but for the present for compliance sampling in coal mines this method must be used.[29]

Prior to use, the rotameter on the pump should be calibrated at 1.6, 1.8, and 2.0 Lpm. The point at which the pump with the cyclone connected is pulling 2 Lpm should be marked on this rotameter so that the flowrate of the pump can be monitored during the survey. Otherwise, sampling methods are similar to those previously described for use of the cyclone.

The MSHA standard calls for designated occupational and area samples to be collected in mines and submitted to MSHA for analysis on a schedule that requires a minimum of six sample collection periods over the year (Table 7.6).[28] The standard specifies the specific work location where

TABLE 7.6. Locations for Respirable Coal Dust Sampling Under MSHA Standards

Mining Section	Sampler Location
Conventional section using cutting machine	Cutting machine operator
	On cutting machine within 36 inches of operator
Conventional section—shooting off the solid	Loading machine operator
	On loading machine within 36 inches of operator
Continuous mining (other than thin auger)	Continuous mining machine operator
	On machine within 36 inches of operator
Continuous mining—auger type	Jacksetter nearest the work face on the return air side of the continuous mining machine
	A location representing the maximum concentration to which the miner is exposed
Scoop section—using cutting machines	Cutting machine operator
	On cutting machine within 36 inches of operator
Scoop section—shooting off the solid	Coal drill operator
	On coal drill within 36 inches of operator
Longwall section	Miner nearest the return air side of the longwall working face
	On the working face on the return air side within 48 inches of the corner
Handloading section using a cutting machine	Cutting machine operator
	On cutting machine within 36 inches of operator
Handloading section—shooting off the solid	Hand loader exposed to highest concentration
	A location representing the maximum concentration of dust to which miner is exposed
Anthracite mining section	Hand loader exposed to the greatest dust concentration
	A location representing the maximum concentration to which miner is exposed

samples need to be collected for each of 10 different mining sections. A minimum of 5 samples must be collected in each location. Occupational samples can be collected on the worker (personal) or by placing the sampling apparatus near the normal working position of the miner. The production must be at normal levels and the sample must be collected for either the full shift or a minimum of 8 hours. The purpose of the area samples is to identify sources of respirable dust generation. The above description is a somewhat simplified view of the sampling strategy, and if results exceed the standard, sampling must be repeated with more frequency and may involve more positions and more attempts to randomize the samples.

Variation in coal dust concentrations can be due to changes in the ventilation used to control dust, the cutting speed, and the amount of rock being cut above or below the coal seam.[30] In mines large concentration gradients have been known to occur in many prospective sampling locations. At such locations, moving a sampler a foot one way or the other may affect the dust-level readings more than any other factor. These concentration gradients are impacted greatly by airflow patterns. An example of an area with a steep concentration gradient is a long wall, depending on the distance between the source and sampling point. When sampling is conducted downwind of mining machines, the measured concentration is not always a reliable indicator of the amount of dust produced by that machine. Ideally, sampling should be conducted only where gradients are low; however, since steep dust gradients are often unavoidable, the only recourse is to use multiple sampling points.[31]

The presence of rock bands—an irregularly occurring band of harder, noncarbonaceous material such as shale—layered within the coal seam will cause a wide variation in dust levels. When equal amounts of material are cut, the dust from rock is an order of magnitude greater than coal, so even a relatively minor rock band will cause dust levels to double. Variations in production will also cause average dust-level changes due to such events as equipment breakdowns and low production due to hard cutting caused by rock intermingled with coal, or individual work habits on the part of machine operators. One recommendation for such situations is to use both integrated and short-term sampling methods to provide a comprehensive set of data along with close monitoring of variables such as airflow, water flow, and differences in work practices.[31] In order to collect samples of respirable coal dust to satisfy MSHA standards, a sampling practitioner must be certified by MSHA.

Cotton Dust

OSHA's cotton dust standards were developed from health studies that frequently used a sampler that has become known as a vertical elutriator. This sampler also became part of the cotton dust standard. Although alternate samplers are also allowed, equivalency to the vertical elutriator must be established and currently none have been approved by OSHA. This sampling method is an example of a situation where no personal sampling method exists, and area samples are collected using a sampling strategy designed to approximate the employee's exposure. Another sampling approach used to evaluate exposures to cotton dust involves the collection of gram-negative bacteria and associated endotoxin since bract and leaf from cotton plants contain large numbers of these. These microbial materials have been implicated as possible causative agents of byssinosis, the lung disease attributed to cotton dust exposure.[32] Fungal and actinomycete spores as well as dust have been identified during cotton dust sampling.[33] For more information on techniques for sampling bioaerosols, see the chapter on Sampling for Bioaerosols.

Prior to the survey, the plant is surveyed in order to determine where to set up the instruments. Questions are asked about the employees' work activities. For example, in cotton ginning, the ginner and helper spend most of their time in the gin stand area and only a little time in the seed cotton-cleaning and lint-cleaning areas. The head pressman and crew spend most of their time operating the bale press. Some press crews may trade off duties with the suction-pipe operator and the yardman.[34] Approximately four to five elutriator measurements are used to determine an employee's exposure in most mill operations. In some areas, if an employee's work area is limited, two elutriators will be sufficient. Generally this technique is limited to sampling employees who work in specific areas rather than those who roam throughout the plant, such as maintenance or supervision people. In general, placement of vertical elutriators depends on the purpose of the study, the number of elutriators available, areas where employees spend most of their time, access to electric power, and best locations for instrument survival during emergencies such as during cotton fires.[34]

The instruments should be placed as close to the employees' work area as possible. However, the elutriators must not be located in areas with strong air currents. Cotton ginning requires a significant amount of processing air. Much of this processing air is drawn from within the building and must be replaced by air from outside through vents or doorways into the building. As a result, it is not uncommon for many areas to have air drafts whose velocity exceeds 33 m/min, which can interfere with the elutriator's performance.

Vertical Elutriator. Elutriators separate out particulates of varying sizes by gravitational effects under low velocities, and are commonly classified as horizontal or vertical based on their design. The vertical elu-

triator was selected as the appropriate device to sample cotton dust by OSHA.

The elutriator by definition is a separator or purifier. This sampler uses a cylinder standing in a vertical position as a separator; hence, the term *vertical elutriator*. The vertical elutriator was designed to collect dust particles that are less than 15 μm in aerodynamic diameter. The particle size cutpoint ranges from 9.7 μm to 10.5 μm AED.[35] The unit (Figure 7.21) has an inverted cone as a bottom section. A pump

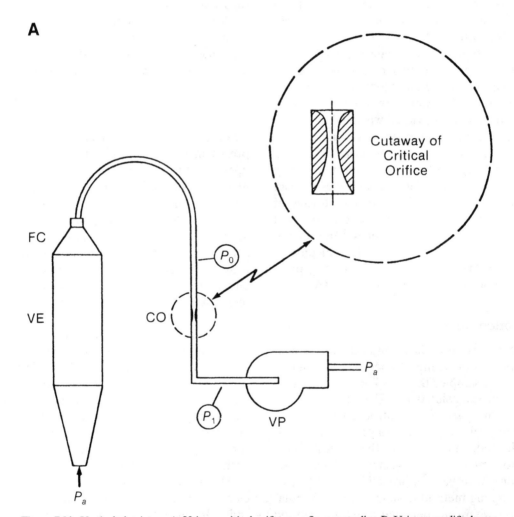

Figure 7.21. Vertical elutriators. A: Using a critical orifice as a flow controller. B: Using a modified pressure regulator as a flow controller. VE, vertical elutriator; FC, filter cassette; CO, critical orifice; RG, regulator; VP, vacuum pump; RM, rotameter; MV, manual valve; P_a, atmospheric pressure; P_o, pressure at input of regulator; P_1, pressure at output of regulator; P_2, pressure at input of manual valve. (From OSHA Field Service Memo No. 8: *Vertical Elutriator Cotton Dust Samples.*)

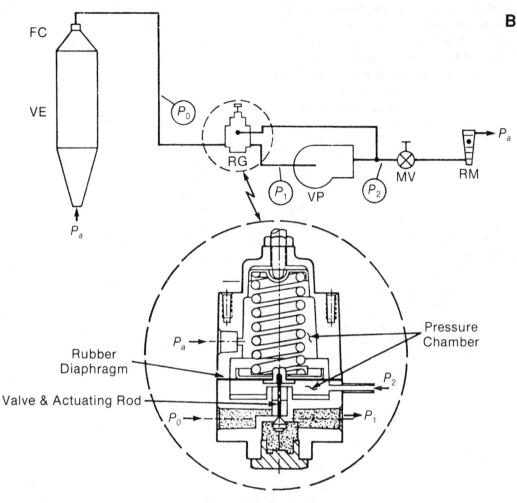

Figure 7.21. *Continued*

and flow control device is used to draw air into the small end of the cone at 7.4 Lpm. Basic samplers use a critical orifice as a flow controller.

The velocity of the air is highest at the small cone entrance, and gets progressively slower as it moves onto the wide cone area and into the cylinder. Heavier pieces of airborne material will slow down and fall back while the smaller-size particles will be carried to the collection filter. In this process, the elutriator separates cotton dust by size. As long as filter loading does not exceed 0.5 to 1 inch Hg, this device will operate as intended; however, higher loading is a problem because the orifice is located downstream of the sampler filter, which makes it susceptible to upstream pressure changes.

Another type of flow controller available consists of two valves: a manual one and a pneumatically operated one. This unit (Figure 7.21B) is capable of handling filter loading up to 8 inches of Hg, so it is better in situations where heavy loading is expected.

One study reported that the coefficient of variation when vertical elutriators sampled side-by-side in a cotton mill was 34–43%.[36] Factors contributing to precision problems in vertical elutriator sampling include low sampling rate, a penetration curve with a shallow slope, and inverted filter, and variability of performance with environmental factors. For example, there can be a loss of sample during removal of the inverted filters. The low flowrate either causes problems with reproducibility in weighing the filters or it forces the operator to sample for inordinately long periods of time.[35]

Vertical elutriators have been found to be durable under field conditions if proper precautions are taken. Vertical elutriators tend to be top heavy and prone to falling due to their construction and operating height. Therefore, they should be provided with bases that are as compact and stable as possible. The critical orifice calibrations were stable unless the orifices suffered mechanical damage or became partially plugged with lint or dirt. A vertical elutriator should not be operated under very dirty conditions without a filter cassette attached to avoid plugging the critical orifice. Routine checking and cleaning of orifices before each sampling period is recommended.

The pressure-relief valves on the vertical elutriators usually require no readjustment to their initial setting during a normal 6- to 8-hour running period. Filter loadings under normal ginning conditions are not heavy enough to cause airflow restrictions that require relief valve adjustment.[34] Ginning variables that affect vertical elutriator results arise from three primary sources: input material, layout of the gin, and management practices. Although gin layouts are standardized to some extent, dust produced by each machine varies from gin to gin. The location of machinery within the gin and the structure of the building affects the velocities, quantities, and direction of cotton flow and of ventilating air into the building. The quantity and type of foreign matter, such as bract, stems, leaf particles, and dirt in the input material

being processed, have an effect on the overall dust level within the gin. However, the inherent dust content of the raw cotton may have less effect on actual dust concentrations than dust leakage from individual machines.[34]

Procedure
1. Obtain three-piece filter cassettes 37 mm in diameter with 5-μm PVC filters inside. Plan to have enough so that 10% can be field blanks.

2. Calibrate the elutriator to a flowrate of 7.4 ± 0.2 Lpm. Secure the rubber stopper made for calibration checks into the elutriator body inlet. Wet the inside of a large bubble buret and set up. Allow the elutriator to run 1 minute prior to taking readings.

3. Properly grounded electrical power is required for operation of this instrument. Ensure that the current load on the circuits that will power the samplers will not overload them. These samplers typically draw 4 amps of current.

4. Because of space requirements, elutriators are generally transported disassembled. While carrying the assembled elutriator to the sampling location, the filter should not be in it.

5. At the sampling location, run a clean, lint-free cloth through the elutriator body to remove any dust that may have accumulated inside.

6. Insert a cassette after removing its plugs into the opening at the top of the elutriator as is done for collecting a sample. This cassette is used while the sampler is run 2 to 3 minutes, to ensure that any extraneous dust is flushed out that may have been left in the elutriator after the cleaning.

7. Stop the sampler. Remove the first cassette and insert another cassette for collecting the sample. Since the cassettes can be loosened by vibration, the cassette must be inserted securely and taped to the vertical elutriator body using electrical tape to prevent leakage during sampling.

8. Secure tubing to the pipe with tape, especially at the top where the weight of the tubing may cause a pinch at the cassette connection.

9. Changing filters during the sampling shift depends on the dust concentration (see Table 7.7).

10. When changing the filter, the vertical elutriator pump must be turned off because the filter may be ruptured while working under negative pressure.

11. The filter can be changed by tilting the elutriator body. For convenience, one end of the tape for sealing the cassette can be stuck to the body of the vertical elutriator so that it will be handy to reach while holding the instrument in the tilted position.

12. Make checks on the sampling operation every hour. Look to see if the cassette is taped down tightly, if all tubing is connected, and if the rotameter indicates the proper flowrate.

13. Following sampling clean the inside of the elutriator body with a damp cloth. This should be done after each day of sampling. Use a cloth or a brush to clean the dust from the pump motor and body.

The 6-hour sampling period is representative of the full-shift exposure. Therefore, there is no adjustment for unsampled time unless the worker was not present in the locations covered by the elutriator measurements (such as in the lunchroom). The individual elutriator measurements collected in each employee work area are averaged arithmetically. This average value is then converted to a time-weighted average (TWA) to determine the employee's exposure. For example, if five samples are collected in one area throughout a day (6 hours), the results are added up and divided by 5. The average result is then multiplied by the *actual* workday, which is the time the workers were in the area, and divided by 8.

Observations during sampling are also important for interpreting results. For example, differences in fungal spore concentrations were found to be related to the quality of cotton being processed and to the cleanliness of the mill in one study. In the same study, it was also determined that the distribution of cotton fibers and dust particles varied with the stage of processing, with the carding and speed frame stages having the highest levels of cotton fibers. Dust particles were most numerous during carding and opening.[33]

Wood Dust

In general, except for some sanding operations, studies indicate that more than 80% of wood dust is represented by particles greater than $10\,\mu m$.[37] Recent studies have determined that the ACGIH IPM criteria may be more appropriate than classical methods to use for assessing general wood dust concentrations in furniture, cabinet making, and other industries where wood dust is present. To do otherwise would

TABLE 7.7. Filter Changes for the Vertical Elutriator During Cotton Dust Sampling

Dust Concentration	Changes
<300	0
300–600	1
>600	2

TABLE 7.8. Toxic Effects of Different Types of Wood

Type of Wood	Toxic Effect
Redwood	Allergic alveolitis in the lung
Western red cedar	Allergic alveolitis in the lung
Oak	Nasal cancer, skin and eye irritation, allergic skin response, asthma
Beech	Nasal cancer, skin and eye irritation, allergic skin response, asthma
Mahogany	Nasal cancer, asthma
Maple	Nasal cancer
Walnut	Nasal cancer
Teak	Nasal cancer
Birch	Nasal cancer
White cedar	Skin and eye irritation, allergic skin response
Douglas fir	Allergic skin response
Tropical woods	Skin and eye irritation, allergic skin response
Exotic woods	Asthma
Hardwood (general)	Rhinitis, nasal dryness, pulmonary function changes
Softwood (general)	Pulmonary function changes
Fungal spores associated with wood	Skin and eye irritation, allergic skin response, asthma

exclude from the sample the major source of particles depositing in the nose. IPM sampling collects all particles that would enter the nose or mouth in a typical work situation. Consequently, IPM sampling will be the environmental measurement that is most closely related to the risk of developing nasal cancer from hardwood dust exposure. The furniture woods associated with nasal cancer are oak, beech, mahogany, maple, walnut, teak, and birch.

Different types of wood dust cannot be treated as a uniform entity. For example, data suggest that the use of hardwoods may lead to more dust of respirable size ($<15\,\mu m$) than the use of softwoods, and that very little dust from any given sample of wood dust is in the inhalable ranges. Therefore, it can be postulated that the primary sites for toxic effects are the upper, larger airways.[38] Different types of woods have been known to cause different types of toxic effects.[14] Some woods, particularly those considered tropical woods, have insecticidal and antibacterial compounds in them. It has been recommended that wood dusts be categorized as toxic, biologically active types and those with low biological action.[35] Table 7.8 gives toxic effects of different wood types.[39,40]

Since wood is an organic medium, it can provide a conducive environment for the growth of microbials. Certain types of diseases associated with wood are actually due to microorganisms that tend to grow on the wood. Therefore, if sampling is being done to evaluate exposures to this wood, microbial sampling may also be necessary in certain situations. For more information on these methods, see the chapter on Sampling for Bioaerosols.

For furniture operations, sanding departments have been shown to have the highest maximum and average dust levels. Assembly areas and finish mills where detailed wood machining is done have relatively high dust levels, but in general they are likely to be substantially lower than sanding areas. In some areas, such as assembly, the only source of exposure is reentrainment of often heavy dust deposits on furniture parts. In plywood and plywood products operations, less dust is generated than in many furniture manufacturing operations.[37]

It should be noted that all of the aforementioned operations use dry wood and generate dry wood dust. In saw mills the wood dust is wet because the wood has not been dried. Wet wood affects the density and the likely size of the particles to be generated.

The personal cascade impactor, as discussed previously, has been evaluated for personal sampling for wood dust and found to provide a good description of the size ranges associated with wood dust sampling.

SUMMARY

Collecting accurate and representative aerosol samples using *Sample Collection Devices* can be a challenge for the sampling practitioner. For total aerosol samples it is important to select the proper filter or other collection media, and to collect the samples in a manner that reflects actual airborne particulate levels.

Collecting size-selective samples adds another degree of complexity. The practitioner must choose between sampling devices that collect only the desired size fraction, or to use a cascade impactor or other method that separates the dust into numerous size fractions that can be mathematically combined to yield the desired information.

REFERENCES

1. Kimmel, E. C., and D. L. Courson. Characterization of particulate matter in carbon-graphite/epoxy advanced composite material smoke. *AIHA J.* **63**:413–429, 2002.

2. American Conference of Governmental Industrial Hygienists. *Threshold Limit Values and Biological Exposure Indices for 2002.* Cincinnati: ACGIH, 2002.

3. SKC, Inc. *Comprehensive Catalog and Air Sampling Guide.* Eighty Four, PA, 2004.

4. John, W. Sampler efficiencies: Thoracic mass fraction. *ACGIH Ann.* **11**:75–79, 1984.

5. Frazie, P. R., and G. Tivoni. A filter cassette assembly method for preventing bypass leakage. *AIHAJ.* **48**(2):176–180, 1987.

6. Furst, M. W. Air sampling. *Appl. Ind. Hyg.* **12**(3):F-12, 1988.

7. Knutson, E. O., and P. J. Lioy. Measurement and presentation of aerosol size distributions. In *Air Sampling Instruments*, 6th ed. Cincinnati, OH: ACGIH, 1983.

8. Agarwal, J. K., and B. H. Liu. A criterion for accurate aerosol sampling in calm air. *AIHA J.* **41**:191–197, 1980.

9. Lioy, P. J., et al. Rationale for particle size-selective air sampling. *ACGIH Ann.* **12**:27–34, 1984.

10. Soderholm, S. Size-selective sampling criteria for inspirable mass fraction. *ACGIH Ann.* **11**:47–52, 1984.

11. Hosketh, H. E. *Fine Particles in Gaseous Media*, 2nd ed. Chelsea, MI: Lewis Publications, 1986.

12. Beaulieu, H. J., et al. A comparison of aerosol sampling techniques: "Open" versus "closed-face" filter cassettes. *AIHA J.* **41**:758–765, 1980.

13. Buchan, R. M., et al. Aerosol sampling efficiency of 37 mm filter cassettes. *AIHA J.* **47**:825–831, 1986.

14. Martinelli, C. A., et al. Monitoring real time aerosol distribution in the breathing zone. *AIHA J.* **44**:280–285, 1983.

15. ASTM. Standard test method for respirable dust in workplace atmospheres. ASTM D4532-85, 1985.

16. Ad Hoc Working Group to Technical Committee 146-Air Quality, International Standards Organization. Recommendations on size definitions for particle sampling. *AIHA J.* **42**:A-64–A-68, 1981.

17. Hinds, W. M. Sampler efficiencies: Inspirable mass fraction. *ACGIH Ann.* **11**:67–74, 1984.

18. Stuart, B. O., et al. Use of size-selection in establishing TLVs. *ACGIH Ann.* **1**:85–96, 1984.

19. Raabe, O. G. Size selective sampling criteria for thoracic and respirable mass fractions. *ACGIH Ann.* **11**:53–65, 1984.

20. Lippmann, M., et al. Basis for a particle size-selective TLV for sulfuric acid aerosols. *Appl. Ind. Hyg.* **2**(5):188–199, 1987.

21. Potts, J. D., M. A. McCawley, and R. A. Jankowski. Thoracic dust exposures on longwall and continuous mining sections. *Appl. Occup. Env. Hyg.* **5**(7):440–447, 1990.

22. Aerosol Technology Committee, AIHA. Guide for respirable sampling. *AIHA J.* **3–4**:133–137, 1970.

23. Lippman, M. Size-selective health hazard sampling. In *Air Sampling Instruments*, 6th ed. Cincinnati, OH: ACGIH, 1983.

24. Hodgkins, D. G., et al. Air–lead particle sizes in battery manufacturing: Potential effects on the OSHA compliance mode. *Appl. Occup. Env. Hyg.* **5**(8)518–526, 1990.

25. Lodge, J. P., and T. L. Chan. *Cascade Impactor: Sampling and Data Analysis.* AIHA Monograph Series, Akron, OH: AIHA, 1986.

26. *Series 290: Instruction Manual, Marple Personal Cascade Impactors.* Bull. No. 290I.M.-3-82, Andersen Samplers, Atlanta, GA.

27. Levy, S. A. Occupational pulmonary diseases. In *Occupational Medicine, Principles and Practical Applications*, C. Zenz, ed. Chicago, IL: Year Book Publishers, 1975.

28. 30 CFR Mine Safety and Health Administration, Subchapter O.

29. Corn, M., et al. A critique of MSHA procedures for determination of permissible respirable coal mine dust containing free silica. *AIHA J.* **46**(1):4–8, 1985.

30. Treaftis, H. N., et al. Comparison of particle size distribution data obtained with cascade impaction samplers and from Coulter counter analysis of total dust samples. *AIHA J.* **47**(2):87–93, 1986.

31. Kissell, F. N., et al. How to improve the accuracy of coal mine dust sampling. *AIHA J.* **47**(10):602–606, 1986.

32. Fischer, J. J., et al. Environment influences levels of gram negative bacteria and endotoxin on cotton bracts. *AIHA J.* **43**:290–292, 1982.

33. Lacey, J., and M. E. Lacey. Micro-organisms in the air of cotton mills. *Ann. Occup. Hyg.* **31**(1):1–19, 1987.

34. Hughs, S. E., et al. Methodology for cotton gin dust sampling. *AIHA J.* **42**(6):407–410, 1981.

35. McFarland, A. R., et al. A new cotton dust sampler for PM-10 aerosol. *AIHA J.* **48**(3):293–297, 1987.

36. Suh, M. W., et al. Statistical analysis of CAM/LVE data in the major cotton manufacturing process areas. *4th Special Session on Cotton Dust Research Proc.* Technical Research Service/National Cotton Council, Memphis, TN, pp. 103–111, 1980.

37. Whitehead, L. W., et al. Suspended dust concentrations and size distributions, and qualitative analysis of inorganic particles, from woodworking operations. *AIHA J.* **42**(6):461–467, 1981.

38. Whitehead, L. W. Health effects of wood dust—relevance for an occupational standard. *AIHA J.* **43**(9):674–678, 1982.

39. Hinds, W. C. Basis for particle size-selective sampling for wood dust. *Appl. Ind. Hyg.* **3**(3):67–72, 1988.

40. Hausen, B. *Woods Injurious to Human Health: A Manual.* Walter deGruyter, New York, 1981.

CHAPTER 8

CONCURRENT SAMPLING FOR VAPORS AND AEROSOLS

Vapors and aerosols may be present concurrently when (1) semivolatile organic compounds have vapor pressures that allow them to exist in the atmosphere as a vapor and/or aerosol under certain conditions; (2) different compounds generated from the same element as a result of different processing methods occur within the same operation; (3) different compounds are generated as a result of combustion of a single material; (4) mixtures contain both volatile and nonvolatile compounds that are sprayed; (5) certain gases and vapors are capable of existing in the gaseous phase and adsorbing to particles.

Some solids, termed *semivolatile*, have a high enough vapor pressure that evaporation losses occur during the sampling period when traditional filtration methods are used. During sampling, air is continuously passing through the collected aerosol; thus, organic aerosols with significant vapor pressures could be partially or totally lost during the sampling process unless special precautions are taken. Semivolatile organic

air pollutants include polynuclear aromatic hydrocarbons (PAHs) with four or fewer fused rings and their nitro derivatives, chlorobenzenes, chlorotoluenes, polychlorobiphenyls, organochlorine, and organophosphate pesticides, and the various polychlorodibenzo-*p*-dioxins. These are the most common groups found in both the vapor and particulate phases. Some semivolatile organic compounds have vapor pressures that allow them to exist in the atmosphere as vapors, in the condensed aerosol phase, or in both phases depending on the ambient conditions at the time of sampling;[1] for example, certain solids such as arsenic trioxide and benzidine sublime, meaning they go directly from the solid phase to the vapor phase without going through a liquid phase.

It has been suggested that different states of a semivolatile compound, particles or vapors, will have different effects on the body.[2] For example, if a particle is surrounded by a thin shell of essentially saturated vapor, and it gains entry to the

Air Monitoring for Toxic Exposures, Second Edition. By Henry J. McDermott
ISBN 0-471-45435-4   © 2004 John Wiley & Sons, Inc.

respiratory passages, it may have a more potent effect on the mucosal surface than the same substance in a pure vapor state, which is more likely to be diluted by room air. Therefore, a particle cloud will have a greater effect than the same weight of a vapor. The effect of these particles can be distinguished from the vapor only if they can be collected and analyzed separately.

Multiple species of the same element may exist in the same industrial environment. Inorganic antimony compounds are all possible workplace contaminants during the lead-acid battery manufacturing process and may coexist. Antimony fume can be generated during torching or welding on aluminum and other scrap; stibine (antimony hydride) gas can be generated during the battery boosting and forming process as a result of the antimony-contaminated lead plates coming into contact with sulfuric acid and nascent hydrogen gas. A concern in this case is the varying degrees of toxicity each compound might possess. Dusts of antimony compounds cause irritation of the mucous membranes and a variety of systemic effects, and stibine gas causes hemolysis of red blood cells.

Another situation of interest in which sampling involves both vapors and mists is when a solvent-based mixture is sprayed. The toxic effects of mixtures, such as paints and pesticide formulations, will depend on the components of the mixture. When polyurethane paint containing lead pigment is sprayed on automobiles there are several toxic concerns: the isocyanate, sometimes toluene diisocyanate; the pigment, often lead chromate; and the solvent, usually a mixture of several.

Toluene diisocyanate is a lung sensitizer, and based on animal data, is suspected of being a carcinogen. Lead acts on many different organs including the kidney and liver. Solvents affect the central nervous system. Depending on the conditions of exposure (type of respiratory protection, effectiveness of engineering controls such as a paint spray booth, concentrations, length of exposure), a variety of acute or chronic effects could result.

Some gases and vapors are capable of adsorbing to particles. Many dusts are highly adsorbent and often can contain a significant amount of a gaseous component, such as sulfur dioxide or formaldehyde. The adsorption of gases and vapors onto particulate matter can also create a potential for reactions as a result of catalytic effects on the surface of particulates.[3]

COLLECTION METHODS FOR SEMIVOLATILE COMPOUNDS

Occupational Sampling

For occupational sampling of semivolatile compounds, most often a filter cassette is used in front of a sorbent tube. The filter collects the aerosol; the tube collects the vapor.[4] When filters precede sorbent tubes to collect aerosol, they are often very small in diameter—13 mm—compared to those used strictly for aerosol collection. Once the sampling train is assembled, the procedures are the same as described in Chapters 5–7 on *sample collection devices*. The flow rate is determined by the method and should be acknowledged. Calibration must be done with the entire train assembled. Table 8.1 describes media used in NIOSH methods for semivolatile compounds.

Examples of compounds that would be collected using occupational methods are PAHs. These compounds may consist of both an aerosol and gaseous portion. They can be a concern in an active industrial process, for example, the manufacturing of coke, as well as on hazardous waste sites, such as coal gassification plants. Sampling would most likely be done outdoors in these instances. Vapor pressure data for a number of PAHs (Table 8.2) indicate that

TABLE 8.1. Media Used in NIOSH Methods for Combination Aerosol and Vapor

Compound	Method Number	Medium
Acenaphthene	5506, 5515	2-μm PTFE filter, 37 mm, and washed XAD-2 tube, 100/50
Acenaphthene	5506, 5515	2-μm PTFE filter, 37 mm, and washed XAD-2 tube, 100/50
Aldrin	5502	Glass fiber filter (Gelman type AE) in combination with isooctane in a bubbler
Anthracene	5506, 5515	2-μm PTFE filter, 37 mm, and washed XAD-2 tube, 100/50
Arsenic trioxide	7901	Sodium carbonate-impregnated 0.8-μm MCEF, 37 mm
Benz(a)anthracene	5506	2-μm PTFE filter, 37 mm, and washed XAD-2 tube, 100/50
Benzidine	5509	Glass fiber filter (type AE), 13 mm, and silica gel, 50
Benzo(b)fluoranthene	5506, 5515	2-μm PTFE filter, 37 mm, and washed XAD-2 tube, 100/50
Benzo(k)fluoranthene	5506, 5515	2-μm PTFE filter, 37 mm, and washed XAD-2 tube, 100/50
Benzo(ghi)perylene	5506, 5515	2-μm PTFE filter, 37 mm, and washed XAD-2 tube, 100/50
Benzo(a)pyrene	5506, 5515	2-μm PTFE filter, 37 mm, and washed XAD-2 tube, 100/50
Benzo(e)pyrene	5506, 5515	2-μm PTFE filter, 37 mm, and washed XAD-2 tube, 100/50
Chrysene	5506, 5515	2-μm PTFE filter, 37 mm, and washed XAD-2 tube, 100/50
Cyanides	7904	0.8-μm MCE filter, 37 mm, and potassium hydroxide in a bubbler (Filter collects cyanide salts; bubbler collects HCN.)
Demeton (Demeton is a liquid with a low vapor pressure—0.001 mm Hg.)	5514	2-μm MCE filter and XAD-2 tube, 150/75 (Filter collects aerosol; tube collects vapor.)
Dibenz(a, h)anthracene	5506, 5515	2-μm PTFE filter, 37 mm, and washed XAD-2 tube, 100/50
Diborane	6006	PTF filter and oxidizer-impregnated charcoal tube, 100/50 (Filter removes boron-containing particulates.)
Dibutyl tin bis(isooctyl mercaptoacetate)	5504	Glass fiber filter, 37 mm, and XAD-2 tube, 80/40
Ethylene glycol (Ethylene glycol is a liquid with a low vapor pressure—0.1 mm Hg.)	5500	Binderless glass fiber filter, 13 mm, and silica gel tube, 520/260
Fluoranthene	5506, 5515	2-μm PTFE filter, 37 mm, and washed XAD-2 tube, 100/50
Fluorene	5506, 5515	2-μm PTFE filter, 37 mm, and washed XAD-2 tube, 100/50

TABLE 8.1. *Continued*

Compound	Method Number	Medium
Fluorides	7902	0.8-μm MCE filter, 37 mm, and sodium carbonate-treated cellulose pad, 37 m in series (Fluoride salts liberate HF in the presence of acids; the method collects both aerosol and gaseous fluorides.)
Indeno(1,2,3-*cd*)pyrene	5506, 5515	2-μm PTFE filter, 37 mm, and washed XAD-2 tube, 100/50
Kepone	5508	0.8-μm MCE filter, 37 mm, and impinger with sodium hydroxide (May have something to do with the conversion of mirex to kepone, since mirex is converted to kepone in the presence of sodium hydroxide.)
Lindane	5502	Glass fiber filter (Gelman type AE) in combination with isooctane in a bubbler
Mercury	6000	Glass fiber prefilter, 13 mm, and silvered Chromosorb P, 30 (Filter collects particulate mercury; tube collects mercury vapor.)
Naphthalene	5506, 5515	2-μm PTFE filter, 37 mm, and washed XAD-2 tube, 100/50
Naphthylamines	5518	Glass fiber filter, 13 mm, and silica gel tube, 100/50
PCBs	5503	Glass fiber, 13 mm, and florisil tube, 100/50
Pentachlorobenzene	5517	PTFE fiber mat filter, 13 mm, and amberlite XAD-2 tube, 100/50
Phenanathrene	5506, 5515	2-μm PTFE filter, 37 mm, and washed XAD-2 tube, 100/50
Phosphorus	7905	0.8-μm MCE filter, 37 mm, and Tenax GC tube, 100/50
Pyrene	5506, 5515	2-μm PTFE filter, 37 mm, and washed XAD-2 tube, 100/50
Sulfur dioxide	6004	0.8-μm MCE filter, 37 mm, and cellulose filter impregnated with potassium hydroxide, 37 mm, in series (Sulfur dioxide is collected on the treated filter and sulfuric acid, sulfate salts, and sulfite salts are collected on the front filter as particulates.)
1,2,4,5-Tetrachlorobenzene	5517	PTFE fiber mat filter, 13 mm, and amberlite XAD-2 tube, 100/50
Tributyl tin	5504	Glass fiber filter, 37 mm, and XAD-2 tube, 80/40
Tributyl tin chloride	5504	Glass fiber filter, 37 mm, and XAD-2 tube, 80/40
1,2,4-Trichlorobenzene	5517	PTFE fiber mat filter, 13 mm, and amberlite XAD-2 tube, 100/50
Tricyclohexyltin hydroxide	5504	Glass fiber filter, 37 mm, and XAD-2 tube, 80/40

TABLE 8.2. Vapor Pressures of Selected PAHs

Compound	Vapor Pressure (at 25°C), torr
Phenanthrene	1.7×10^{-4}
Anthracene	2.5×10^{-5}
Pyrene	6.8×10^{-6}
Fluoranthrene	4.9×10^{-6}
Benzo(a)anthracene	1.1×10^{-7}
Chrysene	9.0×10^{-9}
Benzo(a)pyrene	5.5×10^{-9}
Benzo(e)pyrene	5.5×10^{-9}
Perylene	4.3×10^{-9}
Benzo(g, h, i)perylene	1.0×10^{-10}
Coronene	1.5×10^{-12}

compounds less volatile than chrysene would exist primarily in the particulate state.[5]

The following conclusions reached during a study that evaluated factors affecting the sampling of airborne PAHs show how important it is to identify the characteristics and physical behavior of a semivolatile contaminant prior to sampling: (1) A fraction of the more volatile PAHs (three- and four-ring) may exist in the vapor phase at ambient temperatures, with the percentage increasing as temperatures elevate; (2) even though it is assumed that higher molecular weight PAHs will be collected and retained on a filter at ambient temperatures, a vapor trap is needed as a backup to the filter; (3) care must be taken when conducting particle-size analyses on PAH-containing aerosols using cascade impactors, since PAHs may be stripped from the material collected on one stage and be readsorbed by particles on subsequent stages, and compounds in the vapor phase may condense on certain stages of the sampler as a result of cooling during sampling; (4) care should be taken during sampling, since PAHs may react with ozone or may be degraded by sunlight and ultraviolet light; (5) losses of PAHs can occur from filters during storage.[6]

When sampling PAHs, NIOSH recom-mends a combination of 2-μm-pore-size polytetrafluoroethylene (PTFE) filters in 37-mm cassettes in series with a 100/50 XAD-2 resin sorbent tube at a flowrate of 1.7 Lpm.[4]

Environmental Measurement Methods

For environmental collection of semivolatile compounds, the most common type of medium is polyurethane foam (PUF). It has low resistance to airflow; thus, large volumes of air can be collected on it. It is easy to purify and handle; however, it has been shown to have a poorer retention for the more volatile pesticides and polychlorinated biphenyls (PCBs) compared to XAD-2 or Tenax resins, and it does not differentiate between the vapor and particulate phases. The addition of a Tenax cartridge or tube to the sampling train (following the PUF) may be useful in collecting the vapor phase. There is also a concern that mutagenic artifacts may form during sampling.[5]

The primary semivolatiles of concern in environmental (outdoor) measurements are PCBs, PAHs, and pesticides. In EPA Methods TO4, TO9, and TO13, an ambient air sample is drawn over a 24-hour period through a glass fiber filter with a PUF backup absorbent. The PUF adsorbent specified by the EPA method is a polyether-type polyurethane foam (density no. 3014, or 0.0225 g/cm^3). This type of foam is used for furniture upholstery. It is white and yellows upon exposure to light. The PUF inserts are 6-cm cylindrical plugs cut from 3-inch sheet stock, and should fit with slight compression in the glass cartridge, supported by the wire screen.[7] The filter and backup foam are returned to the laboratory for analysis. PCBs and pesticides are recovered by Soxhlet extraction and then subjected to GC/ECD (gas chromatography/electron capture detector) or other analysis. The method is capable of detecting >1 μg/m^3 using a 24-hour sampling

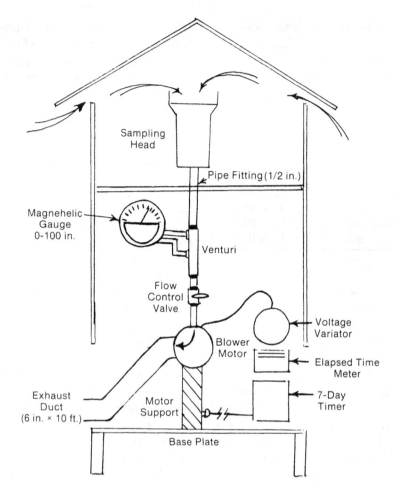

Figure 8.1. High-volume ambient air samplers set up for semivolatile sampling. (Courtesy of General MetalWorks, Inc.)

period. Figures 8.1 and 8.2 show semivolatile sampling equipment.

Procedure

1. The airflow through the sampling system is monitored by a venturi/Magnehelic® assembly. A multipoint calibration of this assembly must be conducted every 6 months using an audit calibration orifice.

2. Prior to calibration, a blank PUF cartridge and filter are placed in the sampling head and the high-volume

sampling pump is turned on (Table 8.3). The flow control valve is fully opened and the voltage is adjusted so that a sample flow rate corresponding to approximately 110% of the desired flow rate is indicated on the magnehelic (based on the 6-month multipoint calibration curve). The motor is allowed to warm up for approximately 10 minutes, and then the flow control valve is adjusted to achieve the desired flow rate. The ambient temperature and barometric pressure should be recorded.

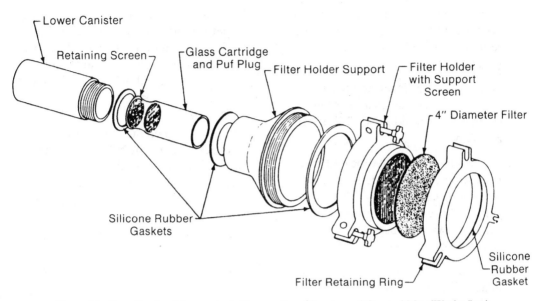

Figure 8.2. Sampling head for semivolatile sampling. (Courtesy of General MetalWorks, Inc.)

TABLE 8.3. **Selected Components Determined Using High-Volume PUF Sampling Procedure**

Compound	Air Concentration ($\eta g/m^3$)	% Recovery
Aldrin	0.3–3.0	28
4,4′-DDE	0.6–6.0	89
4,4′-DDT	1.8–18	83
Chlordane	15–150	73
Chlorobiphenyls		
4,4′-Di-	2.0–20	62
2,4,5-Tri-	0.2–2.0	36
2,4′,5-Tri-	0.2–2.0	86
2,2′,5,5′-Tetra-	0.2–2.0	94
2,2′,4,5,5′-Penta-	0.2–2.0	92
2,2′,4,4′,5,5′-Hexa-	0.2–2.0	86

3. The calibration orifice is then placed on the sampling head and a manometer is attached to the tap on the calibration orifice. The sampler is momentarily turned off to set the zero level of the manometer. The sampler is then switched on and the manometer reading is recorded. Once a stable reading is achieved, the sampler is then switched off.

4. The calibration curve for the orifice is used to calculate sample flow from the data obtained in the previous step, and the calibration curve for the venturi/Magnehelic® assembly is used to calculate sample flow from the data obtained during the 6-month calibration. If the two values do not agree within 10%, the sampler should be inspected for

damage and flow blockages. If no obvious problems are found, the sampler should be recalibrated (multipoint).

5. A multipoint calibration of the calibration orifice, against a primary standard, should be obtained annually.

6. The sampler should be located in an unobstructed area, at least 2 meters from any obstacle to airflow. The exhaust hose should be stretched in the downwind direction to prevent recycling of air.

7. A clean sampling cartridge and quartz fiber filter are removed from sealed transport containers and placed in the sampling head using forceps and gloved hands. The head is tightly sealed into the sampling system. The aluminum foil wrapping is placed back in the sealed container for later use.

8. The zero reading of the Magnehelic® is checked. Ambient temperature, barometric pressure, elapsed time meter setting, sampler serial number, filter number, and PUF cartridge number are recorded.

9. The voltage and flow control valve are placed at the settings used with the calibration orifice and the power switch is turned on. The elapsed time meter is activated and the start time is recorded. The flow (Magnehelic® setting) is adjusted, if necessary using the flow control valve.

10. The Magnehelic® reading is recorded every 6 hours during the sampling period. The calibration curve is used to calculate the flow rate. Ambient temperature and barometric pressure are recorded at the beginning and end of the sampling period.

11. At the end of the desired sampling period, the power is turned off and the filter and PUF cartridges are wrapped with the original aluminum foil and placed in sealed, labeled containers for transport to the laboratory.

12. The Magnehelic® calibration is checked using the calibration orifice. If the calibration deviates by more than 10% from the initial reading, the flow data for that sample must be marked as suspect and the sampler should be inspected and/or removed from service.

13. At least one field blank is to be returned to the laboratory with each group of samples.

14. The total sample volume is calculated from the average of the flowrates determined at the beginning, intermediate, and end points during sampling (Lpm).

An alternate, more portable, sampling setup for semivolatiles involves the use of a PUF cartridge and a personal air sampling pump (Figure 8.3).

COLLECTION OF MULTIPLE SPECIES: ARSENIC

Industrial operations where multiple species can be generated either during the same process or in processes within the same operation require an understanding of the chemistry of each phase of the process, as well as the properties of each compound that might be formed, or sampling will not fully evaluate the situation.

Multiple species of inorganic arsenic can coexist in the lead-acid battery manufacturing process.[8] Arsenic fume can be generated during torching or welding on aluminum and other scrap; arsenic trioxide vapor may also form around weld-

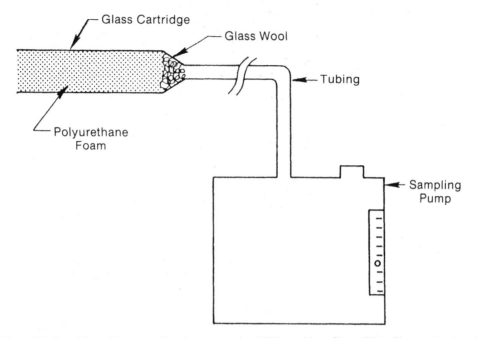

Figure 8.3. Portable ambient sampling apparatus using PUF cartridges. (From EPA, *Characterization of Hazardous Waste Sites—A Methods Manual*, Vol. 2: *Available Sampling Methods*. EPA-600/4-83-040, September 1983.)

ing torches, and arsine gas might be released during the battery forming and boosting processes as a result of arsenic-contaminated lead plates coming into contact with sulfuric acid and nascent hydrogen. Arsenic trioxide can exist in both particulate and vapor phases at ordinary room temperatures. The equilibrium vapor pressure of this compound is influenced by the temperature. Concerns in this situation are the varying degrees of toxicity associated with each compound and the need to used different collection methods to identify different compounds or the degree of the hazard may be underestimated.

As an illustration of the importance of selecting the correct sampling method, the following problems were identified in the course of selecting sampling methods to use in a lead-acid battery manufacturing operation. Arsenic trioxide vapor generated from torching or welding operations could condense, forming a fume that coexists with the vapor phase. When collected

on a filter, the arsenic trioxide fume would ordinarily be interpreted as particulate arsenic due to the need to use atomic absorption for analysis of this medium. If the arsenic trioxide migrated through the filter to the backup pad, it would not be identified at all, since backup pads are not routinely analyzed. When collected on charcoal tubes, arsenic trioxide would be identified as arsine due to the limitations of this analytical method. In this case, a separate method must be used to collect the arsenic trioxide. The arsenic solids can be trapped on a 13-mm filter cassette and the arsine vapor can be adsorbed on a charcoal tube. Following analysis, these results are adjusted to deduct the amount of arsenic trioxide present.[8]

Other inorganic arsenic-containing compounds, such as arsenic tribromide, arsenic triiodide, and arsenic monophosphide, also have equilibrium vapor concentrations (as does arsenic) in excess of $1\,\mu g/m^3$ at the existing air temperature. Therefore, in

TABLE 8.4. Percent Distribution of Cigarette Smoke

Material (% by Vol.)	Weight (mg/cigarette)	Weight of Total Effluent (%)
Particulate matter	40.6	8.2
Nitrogen (67.2)	295.4	59.0
Oxygen (13.3)	66.8	13.4
Carbon dioxide (9.8)	68.1	13.6
Carbon monoxide (3.7)	16.2	3.2
Hydrogen (2.2)	0.7	0.1
Argon (0.8)	5.0	1.0
Methane (0.5)	1.3	0.3
Water vapor	5.8	1.2
C_2-C_4 hydrocarbons	2.5	0.5
Carbonyls	1.9	0.4
Hydrogen cyanide	0.3	0.1
Other known gaseous materials	1.0	0.2
Total	500	100

these cases a sampling approach that collects both aerosol and vapor must also be used.

COMBUSTION PROCESSES: CIGARETTE SMOKE COLLECTION

Combustion processes can occur both outdoors and indoors. Due to their complexity, sometimes sampling is done only for the compounds expected to exist in the highest or most toxic concentrations. Cigarette smoke would be of concern in an indoor air situation, and the methods would be a combination of environmental and occupational sampling techniques, depending on what was to be sampled.

Cigarette smoke is a complex mixture of liquid droplets containing a large variety of organic and inorganic chemicals that are dispersed in a gaseous medium. The smoke aerosol is a continuously changing entity, and aging results in changes in its physical and chemical properties. The composition of the smoke from a given product will depend on the chemical composition and physical properties of the tobacco leaf. These in turn depend on the genetic makeup and environmental factors, such as mineral nutrition, soil properties, moisture supply, temperature, and light intensity during the growth cycle of the leaf. Finally, during the manufacturing process many modifications of cigarettes are possible, including the use of different parts of the tobacco plant (leaves, stems, vines), all of which vary in composition, as well as additives and papers. Table 8.4 shows distribution of cigarette smoke[9] and Table 8.5 lists toxic agents in the gas phase of cigarette smoke.[9]

Reasons for monitoring cigarette smoke are listed below.[10]

- Certain toxic agents in tobacco products and/or their smoke may also occur in the workplace, thus increasing exposure to the agent. These include carbon monoxide, acetone, acrolein, aldehydes, arsenic, cadmium, hydrogen cyanide, hydrogen sulfide, ketones, lead, methyl nitrite, nicotine, nitrogen dioxide, phenol, and polycyclic aromatic compounds.

- Workplace chemicals may be transformed into more harmful agents by smoking due to the heat generated by

TABLE 8.5. Major Toxic Agents in the Gas Phase of Cigarette Smoke (Unaged)

Agent	Concentration in One U.S. Cigarette
Dimethylnitrosamine	13.0 ηg
Ethylmethylnitrosamine	1.8 ηg
Diethylnitrosamine	1.5 ηg
Nitrosopyrrolidine	11.0 ηg
Hydrazine	32.0 ηg
Vinyl chloride	12.0 ηg
Urethane	30.0 ηg
Formaldehyde	30.0 μg
Hydrogen cyanide	110.0 μg
Acrolein	70.0 μg
Acetaldehyde	800.0 μg
Nitrogen oxides	350.0 μg
Ammonia	60.0 μg
Pyridine	10.0 μg
Carbon monoxide	17.0 μg
Acrylonitrile	10.0 μg
2-Nitroproane	0.92 μg

burning tobacco. Polymer fume fever from degradation of heated Teflon® has been reported as a result of cigarette smoking.

- Tobacco products can become contaminated by chemicals used in the workplace, thus increasing the amount of toxic chemicals entering the body.
- Smoking may contribute to an effect comparable to what can result from exposure to toxic agents found in the workplace, thus causing an additive biological effect. For example, combined exposure to cigarette smoke and chlorine can enhance the damage done by chlorine alone. Other agents that can act additively with tobacco smoke include cotton dust, coal dust, and beta-radiation.
- Smoking can act synergistically with toxic agents found in the workplace to cause a much more profound effect than that anticipated simply from the separate influences of the occupational exposure and smoking. The synergistic effect of smoking and asbestos exposure that enhances the likelihood of an individual developing cancer is well known. Other chemicals and physical agents that appear to act synergistically with tobacco smoke include radon progeny, gold mine exposures, and exposures in the rubber industry.

Due to the complexity of cigarette smoke, usually monitoring involves sampling for a specific constituent. Often carbon monoxide is selected, due to the ease in monitoring for it and the low levels expected to be present without smoking. However, for comparison, samples should be taken outside at the same time to account for background levels from exhausts that have been entrained into the building's heating, ventilating, and air conditioning (HVAC) system. Carbon monoxide can be sampled using real-time instrumentation and long-term detector tubes. For more information, see Chapters 10 and 13. Another approach is to measure nicotine as it is considered a carcinogen.

COLLECTION OF MIXTURES

In the case in which a mixture is being sprayed, the type of collection methods will vary significantly and depend on the compounds present. When liquid paint is sprayed from an atomizing nozzle, the generated aerosol conditions are dependent on the liquid properties, spray-nozzle flow conditions, and droplet evaporation. Initial droplet size directly affects droplet evaporation rate. Once the nozzle hardware and paint have been chosen, the aerosol concentration and particle size distribution are primarily dependent on the flow rates of the liquid paint, nozzle air, and diluent air. As the volatiles evaporate, the aerosol is changing from "wet" to "dry," and a steady particle-size distribution can be attained only after the evaporation is complete. Paint spray nozzles generate droplets typically in the range of 20 µm to 100 µm in diameter. The droplets form as a result of the shearing action of the air jet in the nozzle on the liquid film. The diameter of the droplets decreases as evaporation progresses, causing an unstable particle-size distribution that changes over time.[11]

An example of the complexity of paint spray sampling is illustrated by a study to measure the epoxy content of paint spray aerosol.[12] The purpose for sampling the epoxy was to determine the concentration of the reactive epoxide during spraying. Compounds containing the epoxide group have demonstrated a wide number of toxic effects. Aerosol epoxide content was collected using midget impingers containing dimethyl formamide, which stops the epoxide reaction with the amine from the curing agent to preserve the epoxide for analysis. Nonvolatile aerosol mass samples were collected on 1-µm PTFE filters and a cascade impactor was used to characterize the size distribution of the generated aerosol. The sampling location was selected to keep the instruments out of the direct spray from the spray guns and to remain clear of the paint hoses as the painters moved across the work area. In this case, a wide range of aerosol epoxy concentrations was observed and the aerosol generated during spray finishing was found to be unstable, changing in physical and chemical properties while it was airborne.

The primary sources of variation that can affect the results of paint spray sampling are the type and effectiveness of the ventilation systems, and where the spraying takes place relative to the positioning of the ventilation system; the size of the item being painted; and the amount of pigment in the paint. Paint pigment formulation could have an effect as well, because the percentage of pigment in paints varies. One study found a variation of 0.2% to 8.9% lead by weight alone.[13]

REFERENCES

1. Clements, J. B., and R. G. Lewis. Sampling for organic compounds. In *Principles of Environmental Sampling.* Washington, D.C.: American Chemical Society, 1987, pp. 287–296.
2. Chemical Substances TLV Committee Study Paper. *ACGIH Ann.* **4**:153–157, 1983.
3. Soderholm, S. C. Aerosol instabilities. *Appl. Ind. Hyg.* **3**(2):35–40, 1988.
4. National Institute for Occupational Safety and Health. *NIOSH Manual of Analytical Methods,* 3rd ed. Cincinnati: NIOSH, 1984.
5. Riggin, R. M., and B. A. Petersen. Sampling and analysis methodology for semivolatile and nonvolatile organic compounds in air. In *Indoor Air Quality,* H. Kasuga, ed. New York: Springer-Verlag, 1989, pp. 351–359.
6. Leinster, P., and M. J. Evans. Factors affecting the sampling of airborne polycyclic aromatic hydrocarbons—a review. *Ann. Occup. Hyg.* **30**(4):481–495, 1986.
7. Environmental Protection Agency. Compendium of Methods for the Determination of Toxic Organic Compounds in Ambient Air. EPA-600/4-84-041, 1984.
8. Costello, R. J., P. M. Eller, and R. D. Hull. Measurement of multiple inorganic arsenic species. *AIHA J.* **44**(1):21–28, 1983.
9. Surgeon General's Report on Cigarette Smoking. 1975.
10. National Institute for Occupational Safety and Health. *Current Intelligence Bulletin 31.* Adverse Health Effects of Smoking and the Occupational Environment. NIOSH, Cincinnati, February 5, 1979.
11. Beaulieu, H. J., et al. A comparison of aerosol sampling techniques: "Open" versus "closed-face" filter cassettes. *AIHA J.* **41**(10):758–765, 1980.
12. Herrick, R. F., M. J. Ellenbecker, and T. J. Smith. Measurement of the epoxy content of paint spray aerosol: Three case studies. *Appl. Ind. Hyg.* **3**(4):123–128, 1988.
13. Ackley, M. W. Paint spray tests for respirators: Aerosol characteristics. *AIHA J.* **41**(5):309–316, 1980.

PART III

REAL-TIME MEASUREMENT INSTRUMENTS

CHAPTER 9

INTRODUCTION TO MONITORING USING REAL-TIME METHODS

Real-Time methods and *sample collection device* (SCD) methods are the two major categories for exposure monitoring. SCD methods (covered in Chapters 5–8) require laboratory analysis of the sampling medium. In contrast, *real-time* methods give the results right in the field as the monitoring is performed. While the opportunity to receive immediate results is an immense advantage, there are many challenges and selection criteria to consider for successful use of real-time methods. Under the heading of real-time methods, there are two types of devices:

- Direct-Reading Electronic Instruments. Electronic devices that measure concentration or exposure level. Advances in sensor and microprocessor technology have vastly increased the options available to sampling practitioners.
- Colorimetric Systems. These indicate concentration or exposure level by a color change in the sampling medium—often the "length of stain" in a tube where the length is proportional to the airborne concentration of contaminant.

This chapter presents an overview of real-time methods; subsequent chapters cover these topics in more detail. With both classes of devices, there are ones that can measure the instantaneous concentration or reflect an *integrated sample* that averages or integrates the airborne concentration over the sampling period, or both. Some real-time devices can be configured to also collect a sample on a filter or adsorption tube for later SCD analysis; this can be used to verify or obtain further analytical detail on the real-time results.

A major plus of electronic direct reading instruments is the ability to use a *data logger* (either built into the device or external) to record concentration values and other parameters over the sampling period. This permits later computer analysis and plotting of the data, including calculation

Air Monitoring for Toxic Exposures, Second Edition. By Henry J. McDermott
ISBN 0-471-45435-4   © 2004 John Wiley & Sons, Inc.

of TWA, short-term, and ceiling exposures. The data, along with field notes, may also help to identify the circumstances that caused any high exposures and the effectiveness of exposure controls.

Real-time methods are used for both occupational and environmental monitoring. For occupational exposures, there are personal devices that are small and light enough to be worn by the worker, larger portable instruments that can be carried by the sampling practitioner, and fixed systems for area monitoring. For environmental exposure monitoring the equipment is generally left in a single location for the whole sample duration, and so its size and weight is less important. Even when the sampling equipment is compact, the electrical power requirements for an extended sampling period usually require line connection or a heavy battery pack.

DIRECT-READING INSTRUMENTS

Direct-reading instruments can be used for personal monitoring, for conducting surveys in work sites to identify sources such as leaks and potential levels of contaminants, and for continuous fixed station monitoring. Documentation of the effects of work practices on exposures can be done with these instruments, and these results can in turn be used as an educational tool with workers to get them to modify their practices in order to reduce exposures. The documentation can also be an evaluation of the effectiveness of control systems. One leak-testing application is fugitive emissions. Another common application is testing specific equipment such as ethylene oxide sterilizers for leaks.

In some cases, direct-reading instrumentation is the best available method for monitoring a chemical. While there are many specific methods available, some chemicals are very difficult to collect on media. In addition, the trend is to identify immediate

chemical levels over short time periods rather than using an integrated measurement period of 2, 4, or 8 hours to provide an average value. A problem has been that many field instruments lack the specificity and detection limits of laboratory methods. Although commercial technology is not perfected, significant progress is being made. The biggest consideration for the sampling practitioner in the future will not be the availability of equipment but the cost: As instruments become more tuneable or specific, their reliability increases, and their detection capability becomes more sensitive to lower levels, their cost increases. Some high-grade field instruments exceed the price range of a luxury car.

Direct-reading instruments fall into several categories: those that are specific for a single gas or vapor; general survey instruments, which respond to many different gases and vapors at once, but cannot differentiate between them; more sophisticated analyzers such as gas chromatographs (GC) that can sample for several specific gases and vapors at once, or infrared (IR) instruments that can be tuned in the field to provide a specific measurement for a number of different gases and vapors, one at a time; and instruments that monitor for aerosols. In each case, one or more different principles of detection are involved. Most monitors dedicated to a single specific chemical are electrochemical or IR-based.

General survey instruments can be classified as catalytic combustion, photoionization (PID), flame ionization (FID), and metallic oxide semiconductor (MOS) and are used to measure both combustible and health hazard levels. GCs and IRs are examples of laboratory analytical instruments that have been modified for field use. Real-time aerosol instruments do not differentiate between types of aerosols, but are useful for developing sampling strategies and represent an area of monitoring that is expanding.

There is a wide variety of technology used in direct-reading instruments. Since the catalytic combustion-type sensor was developed in 1923, there are instruments based on this principle that were developed more than 30 years ago and are still sold. Other instruments have been around for some time, but only recently have come into popular use. An example is the PID general survey instrument. Prior to the concerns of health and safety in hazardous waste work, this instrument was rarely used due to its extreme nonspecificity. With the advent of hazardous waste work, nonspecificity became a virtue, making these units highly desirable for detecting a wide range of contaminants. In other cases, changes in sampling needs have also dictated changes in equipment needs. For years, IR was the method of choice for detection of ethylene oxide, but when the Occupational Safety and Health Administration (OSHA) standard was lowered to 1 ppm, the portable GC with a PID gained popularity because of its increased sensitivity.

Microprocessor capability has allowed miniaturization of real time monitors so that they can be worn for personal monitoring. Many are dosimeters, meaning an 8-hour time-weighted average (TWA) can be integrated. Many have their own data-logging capability and can provide a comprehensive printout of the day's exposure. Others can be attached to a data logger while being worn by the individual. Some instruments for gases and vapors have alarms for high levels to provide warning capabilities that are essential for work in confined spaces, repair of tanks, process vessels, and pipelines. In some cases, these units do not have a readout, only an alarm.

As more instruments become microprocessor-based and data-logger compatible, sampling practitioners will need to become familiar with the basics of using printers and computers if they want to be able to utilize all the features and advantages of these units.

There are three different levels of real-time monitoring: (1) stationary fixed sensing systems that can involve several sensors and cover a number of different areas of a plant; (2) small personal monitors, many of which store data for later output to a computer; and (3) portable survey instruments that are used for taking a variety of different types of samples.

Advantages and Disadvantages of Direct-Reading Equipment

Direct-reading instruments are an important method for measuring airborne levels and exposures to many compounds. Like all sampling methodologies, they have certain advantages and disadvantages. Some very important advantages of real-time instrumentation include:

- The ability to measure concentrations of contaminants on-site with almost instantaneous results.
- Continuous fixed station monitoring to alert personnel to emergency situations via audible and visual alarms.
- The ability to collect very short samples and therefore break down a task or process and analyze emissions at each step.
- Identification sources of problems, such as leaks, and the opportunity to evaluate the effectiveness of solutions as soon as they are in effect.
- Ongoing records of contaminant levels (or nonlevels) to protect companies from false allegations.
- Less work on the part of the sampling professional once the instrument is set up when compared to conventional integrated methods.
- Reduced laboratory costs.
- Immediate samples without concerns over loss of samples during storage and shipment.

- Sample results without having to wait out the standard 10-day laboratory turnaround time or pay a surcharge for quicker turnarounds.

Some potential disadvantages of direct reading instruments include:

- While these devices can be very good for quantifying exposures to known contaminants, they are very limited for identifying of unknown substances. For this application, SCD methods followed by laboratory analysis are better.
- Use of an instrument for the wrong application, and then the data are incorrectly interpreted. For example, a direct-reading photo-ionization detector (PID) device will not respond well to benzene vapors, and so its use for this compound can yield erroneous information.
- The tendency to rely on an instrument as long as it appears to be functioning properly. These devices are susceptible to malfunctions like any sensitive instrument, and they often receive rough treatment in the field. Periodic performance checks are required to confirm that the instrument is still working correctly.
- Inability to calibrate the instrument adequately. All direct-reading instruments require periodic calibration checks using the gas or vapor of interest. A common obstacle is the restrictions on air transport of toxic substances, which prevent proper calibration with gas mixtures during field use of the instrument.
- Interfering compounds that give an erroneous reading. These are described in further detail below.

Interferences. Most sensors have at least one known compound that will interfere with their ability to accurately measure the compounds they are designed to monitor, and there may be many others for which sensors have not been tested. Therefore, interferences must be considered when selecting an instrument, including interferences by other chemicals on measurements of specific chemicals and ones such as humidity or the lack of it. Often these data are available from manufacturers. Sometimes these data are presented as ratios, for example, 300:1. The larger the ratio, the less significant the interference. Calculation of these ratios is done by dividing the amount of interferent used in the test by the change in the meter reading, for example, a challenge of 150 ppm of carbon monoxide to an instrument designed to measure hydrogen sulfide caused an increase of 2 ppm in the response of the meter to 10 ppm hydrogen sulfide:

$$\frac{150 \, \text{ppm CO}}{2 \, \text{ppm change in reading}} = 75:1$$

Interference charts are most often constructed by exposing instruments to single gases and then determining their response. Manufacturer's data on interferences may be misleading as they may not contain information on all potential interferents. Only those compounds for which the manufacturer has tested are listed, usually 6 to 10 compounds. The manufacturer's literature should note whether measurements were made with or without a filter in place.

Radio-frequency interference (RFI) from sources such as high-voltage lines and radio transmitters can be a problem for many instruments. Some instruments have cases specifically designed to protect against RFI. In order not to lose this protection, often the case must not be opened by the user.

If there are interferences present, it must be determined whether the levels are high enough to give unacceptable measurements or require increased maintenance.

There may be trade-offs in this area: Non-specific instrumentation may be inexpensive and have fast response time, but if instruments have an unacceptably high number of false alarms, production costs could be unacceptably high. Therefore, before purchasing instrumentation the sampler should know what other chemicals can be released into the plant atmosphere in addition to the compound of interest.

Stationary Monitoring Systems

Although the focus of this book is personal and portable monitoring survey equipment, it is useful for the sampling practitioner to have an understanding of the purpose of stationary monitoring systems. Often due to the system's perceived complexity, sampling practitioner will be overwhelmed at the prospect of selecting, evaluating, or using data from a stationary monitoring system, especially one that is multipoint. The biggest problem with stationary monitoring systems is a lack of regular scheduled maintenance and calibration on the part of many facilities. Once installed, the units tend to be forgotten and taken for granted.

Stationary monitoring systems in actual use vary in complexity from simple single-point systems, used for warning only, to complex multipoint systems with data acquisition, data averaging, and exposure estimation. Some systems have the capability to switch on fans or other electrical devices at preset levels (Figure 9.1). The complexity of the system depends on the use to which the data will be put. Data recording allows the data to be used to identify problem areas and assist in the implementation of controls. If the system is to be used for warning only, then the outputs from the system can be very simple and no data storage is necessary. In choosing a system, it is important to know how the data will be used so that the proper system components can be chosen.

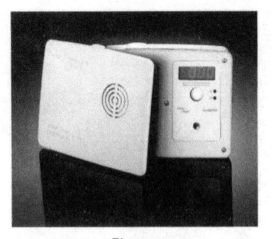

Figure 9.1. AirAware™ stand-alone monitor warns of hazardous gas concentrations and can also actuate external devices such as remote alarm, ventilation, or shutdown systems. (Courtesy of Industrial Scientific Corporation.)

Continuous multipoint monitoring of workplace air (Figure 9.2) has demonstrated a potential for solving a wide variety of workplace exposure problems, and is valuable in both identifying problems and by keeping workers and management aware of pollutant concentrations in the workplace. These types of monitoring systems are an important part of workplace controls in terms of measuring exposure concentrations over time in many areas of a plant, thus developing a historical information base about concentrations and the sources contributing to them in these areas. Since personal monitoring and portable direct reading surveys are performed much less often, they often provide much less information than these types of systems.

Examples of industries where stationary monitoring units have been installed include plastics and resins, textiles, parking garages, steel production, ethylene oxide sterilizers, wastewater treatment in industrial facilities, processes using isocyanates, and semiconductor manufacturing.[1]

Design and Selection. There are three main components of a stationary monitoring system: the sensor, the sampling system,

(a)   (b)

Figure 9.2. Beacon™ fixed system gas monitoring system is (a) a microprocessor controller for up to 8 sensors and (b) a diffusion sensor for a specific toxic gas, oxygen, or combustible vapor. (Courtesy of RKI Instruments, Inc.)

and the data acquisition and display box. The main configurations for multipoint monitoring systems are: a single instrument with multiple sensors, a single analyzer and sensor that can sequentially analyze from different points via sampling lines, and multiple instruments. In a single-instrument multisensor system, there are several sensors, each continuously monitoring a certain point. With a single analyzer /sensor system, a sample switching system is used to time-share the analyzer among the various sampling points. In the multiinstrument system an analyzer is used for each sampling point.[1]

The number of sensors required, their detection principle and specificity in the environment to be monitored, as well as the components of the sampling system must be considered when selecting a stationary monitoring system. The sample collection system must be able to take data on samples from sensors at each sampling point or pull contaminant through sampling lines to a centralized analyzer. In the latter case, the multipoint system must be able to draw in the samples reliably with minimum degradation, and to switch from one sample to the other without mixing samples. The most important aspect of collection, regardless of the type of system, is that it be done in a timely manner. If the process being monitored is a constant operation with little variation, readings can be taken with the gas monitor every half hour, and these readings should be recorded if the sampler is taking compliance measurements. If the process is variable, then it may be necessary to take the readings as often as every 5 minutes or less.[1]

Where the system provides an alarm function, it is important that the alarm be heard and seen in both the areas where the samples are collected and in some central location, such as the control room, where supervisory personnel can take appropriate actions.

The location of the sampling points and the alarm level must be chosen together to ensure that the warning alarm is given before overexposures occur. For example, the sensors should ideally be located between the employees and the expected contaminant source to permit the sensors to pick up the released chemical before workers are exposed to high levels. Where the equipment configuration does not permit identification of likely contaminant sources and locating the sensors near these sources, another solution is to set the alarm point low enough to detect any increase in typical workplace concentrations. In these cases the alarm point does not reflect an occupational exposure level, but is merely selected to give very early warning of an increased level so follow-up measures can be taken. As an illustration, consider a system to warn of high carbon monoxide (CO) levels where the normal concentration was measured at 5 ppm. The ACGIH TLV® for CO is 25 ppm time-weighted average over the work shift. Under ACGIH guidelines, brief excursions up to three times the TLV® may be permissible—in this case 75 ppm. If the sensors for the alarm system can be located near the expected sources, it may be acceptable to set the alarm point as high as 50 ppm if no employees will be working between the source and sensor. However, if the source cannot be identified or the workers may be close to the source when the release occurs, it may be prudent to set the alarm point at 10 or 15 ppm. This is well below even the 8-hour TLV® of 25 ppm, but may be needed to give adequate warning of any unusual releases that result in concentrations above the usual 5 ppm level. A 10- or 15-ppm alarm set point may result in numerous "false alarms" that do not represent true abnormal concentration situations, but is necessary to avoid overexposures if a significant release did occur.

To achieve the performance needed from a stationary system, careful attention to selection factors is required. Generally the likelihood of success is highest when people knowledgeable in the system (representing the supplier) and the workplace conditions (representing the employer) participate in system selection. Factors to consider include[2]:

- Sensitivity, specificity, and accuracy along with being rugged enough for use in the workplace.
- Placement of the sampling probes.
- Selection of a detection method.
- For single-instrument systems, a reliable purging and sample switching mechanism. If the sample lines are not purged quickly enough, a long time delay between sampling and analysis will result.
- Compatibility of the material from which the sample lines and valves of the system are made with the substance to be measured. Adsorption, absorption, or condensation of the components of the sample onto the surfaces of tubing will cause loss of sample.

System expandability is a desirable feature. Some systems may allow the addition of more sample points at modest expense, and may allow several substances to be monitored without major system modifications. Other systems may be very specific, but cannot be expanded to monitor more than one substance without major modification. A system with a great deal of computer data acquisition and

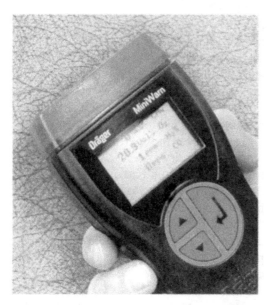

Figure 9.3. Draeger MiniWarn is equipped with four different toxic gas sensors. (Courtesy of Draeger Safety, Inc.)

control may allow more flexible formats for data acquisition and also allow data to be analyzed in a number of different ways. Vendor support, supplies, and spare parts to keep the system running must also be considered.

Personal Monitoring

Advances in personal direct reading instruments have provided the sampling practitioner with many more options to sample collection device (SCD) methods. Direct-reading devices are available for many gases and vapors. The earliest devices generally measured common gases such as carbon monoxide, chlorine, hydrogen, hydrogen cyanide, hydrogen sulfide, hydrazine, oxygen, phosgene, sulfur dioxide, and nitrogen dioxide, all using electrochemical sensors for detection as well as for oxygen deficiency monitoring. Modern devices can measure many more compounds. For example, the Draeger MiniWarn (Figure 9.3) can be equipped with four sensors selected from over 40 different compounds (Table 9.1).

TABLE 9.1. Interchangeable Sensors Available for the Draeger MiniWarn

Compound	Concentration Range
Acetaldehyde	0–200 ppm
Acetylene	0–100 ppm
Acrylonitrile	0–100 ppm
Ammonia	0–300 ppm
Arsine	0–10.0 ppm
Bromine	0–20.0 ppm
Butadiene	0–100 ppm
Carbon dioxide	0–5.00% vol.
Carbon monoxide	0–2000 ppm
Carbon monoxide	0–10,000 ppm
Chlorine	0–20.0 ppm
Chlorine dioxide	0–20.0 ppm
Combustible gases	0–100% LEL
Diborane	0–1.00 ppm
Diethylamine	0–100 ppm
Diethyl ether	0–200 ppm
Dimethylamine	0–100 ppm
Dimethyl sulfide	0–40 ppm
Ethanol	0–300 ppm
Ethylene	0–100 ppm
Ethylene oxide	0–200 ppm
Formaldehyde	0–200 ppm
Fluorine	0–20.0 ppm
Germane	0–20.0 ppm
Hydrogen	0–2000 ppm
Hydrogen cyanide	0–50.0 ppm
Hydrogen selenide	0–1.00 ppm
Hydrogen sulfide	0–100 ppm
Hydrogen sulfide	0–1000 ppm
iso-Butene	0–300 ppm
iso-Propyl alcohol	0–300 ppm
Methanol	0–200 ppm
Methyl amine	0–100 ppm
Nitric oxide	0–100 ppm
Nitrogen dioxide	0–50 ppm
Oxygen	0–25.0% vol.
Phosphine	0–10.0 ppm
Phosphine	0–1000 ppm
Propylene	0–100 ppm
Propylene oxide	0–200 ppm
Silane	0–10.0 ppm
Styrene	0–100 ppm
Sulfur dioxide	0–50.0 ppm
Tetrahydrothiophene	0–40 ppm
tert-Butyl mercaptan	0–40 ppm
Triethylamine	0–100 ppm
Trimethylamine	0–100 ppm
Vinyl acetate	0–100 ppm
Vinyl chloride	0–100 ppm

Because of the low power requirements and small size, the electrochemical sensor is often used for alarm and dosimeter systems designed to monitor for specific chemicals and for measuring oxygen levels. Some units claim as much as +5% or +3 ppm accuracy, whichever is greater. One ppm is often the minimum detection level. The principle of detection for these systems is discussed in the chapter on Monitoring Instruments Dedicated to a Single Chemical.

Small personal monitors are also being used to detect combustible gases. These units can be worn in a confined space and monitor continuously, drastically improving the protection to the worker over the past practice, which was to monitor briefly prior to entering the space. Because hydrogen sulfide is such a hazard in confined spaces, several continuous monitoring personal-sized units are available for this contaminant. The need arises from the fact that hydrogen sulfide causes olfactory fatigue, that is, the odor-detecting cells in the nasal passages become tired after a period of exposure. As a result, the worker may think that the gas has gone away or may not be aware of an increase in concentration.

Personal Dosimeters for Gases and Vapors. When concentration readings can be stored in the personal monitoring device it can be used as a *dosimeter*. The trend is to incorporate the ability to store data into personal-sized monitors. As a result, real time dosimeters are capable of collecting and averaging a worker's dose, making the measurement comparable to the OSHA permissible exposure limit (PEL). Dosimeters frequently read out in ppm-hours; therefore, to obtain the reading in ppm the dose must be divided by the number of hours in the measurement period.

Monitors used as dosimeters may have electronic circuitry that provides a TWA, but not necessarily a continuous readout of the concentration. Personal units are available that read out directly, allowing them to provide instantaneous results as well as cumulative average. For some monitors with storage capabilities to download data, a separate unit is required to collect and print out the data that is specific for that manufacturer and application. Some units provide a display of the current TWA that can be updated periodically or at the user's command. The Draeger MiniWarn described earlier can function as a dosumeter. It can log 50 hours of one-minute concentration average values, and it can download the data to a personal computer using either an RS-232 or infrared interface. Graphics software is available that allows these data to be stored and manipulated in a computer. While monitoring, it will provide a digital ambient concentration display, a total average display, and a minute peak exposure.

This area of monitoring is likely to continue to grow. The data-logging capabilities of many of these units are opening up new monitoring possibilities, such as monitoring inside vehicles and in other situations where the ability to make a continuous record of measurements was once limited by the need for alternating current (AC) voltage or by the bulk of the equipment needed.

Personal Alarm Monitors for Gases and Vapors. The term *go/no go* is often applied to units that have only alarms and no concentration display. The alarm circuit is designed to activate whenever the pre-selected value is reached or exceeded for a predetermined time. Some units can be set for two alarm points. Alarms are often preset at the factory and cannot be changed by the user. Some units are more flexible and can be adjusted. Alarms can be audible, visual, or vibratory. As an illustration, the MSA Cricket® (Figure 9.4) is a maintenance-free single gas alarm available for oxygen, carbon monoxide, or hydrogen sulfide. It contains a battery and

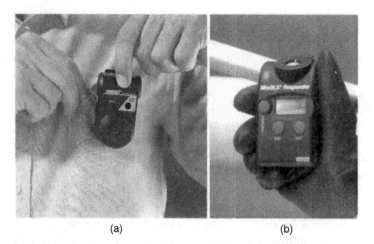

(a)   (b)

Figure 9.4. The MSA Cricket and Responder personal-size direct-reading instruments. (Courtesy of The Mine Safety Appliances Company.)

sensor with at least a one-year service life; once the device is activated (turned on), it runs continuously and only requires periodic sensor response tests. It has two warning levels and both audible and visual alarms. The MSA Responder® (Figure 9.4) is a more sophisticated single gas monitor for chlorine, chlorine dioxide, or ammonia in addition to the three gases listed above. It has a digital readout, adjustable alarm points, and audible, visual, and vibrating alarms.

Calibration of these monitors with the gas or vapor being measured is necessary. Also important is the determination of field conditions, such as temperature, humidity, and air velocity, and potential interferences encountered during sampling as well as interpretation of the effect of these variables on the measurements.

Personal Dosimeters for Airborne Particulates. The choice of personal particulate dust monitors is much more limited than for devices that measure gases and vapors. However, there are several choices available. For example, the SKC HAZ-DUST™ (Figure 9.5) is a real-time area dust monitor designed for industrial hygiene, indoor air quality, or hazardous

Figure 9.5. Haz-Dust™ direct-reading instrument for airborne particulate matter. (Courtesy of SKC, Inc.)

materials investigations. It has a dual range that allows measurements of dust particulate concentrations between 0–20 mg/m³ and 0–200 mg/m³, with an ultimate sensitivity down to 0.01 mg/m³. It can be factory-calibrated with a thoracic dust particulate mass with a cut point of 10 μm or a respirable dust particulate mass (Arizona Road Dust) with a cut point of 3.5 μm (+10%). It weighs three pounds, which is light enough to be carried by a worker. The instrument calculates airborne dust concentrations and has a built-in alarm that provides an audible warning signal at a user-set level. It can be coupled with a rechargeable data logger.

These devices typically use the principle of near-forward light scattering to measure the concentration of airborne dust particles. An infrared light source is positioned at a 90° angle from a photodetector, and airborne dust particles entering the infrared beam scatter the light. The amount of light received by the photodetector is directly proportional to the aerosol concentration.

Advantages, and Other Considerations of Personal Monitors.

The use of personal real-time monitors offers a number of advantages if done correctly. Below are advantages of these devices.[2,3]

- They are portable, can be conveniently worn by workers, and can be used in confined areas.
- They can collect data under a wide variety of changing environmental conditions at high and low levels with typically better than +15% accuracy.
- In most cases, the measurement is specific for the chemical being measured.
- They can collect rapidly fluctuating contaminant concentrations with precision and accuracy.
- In operation they are independent of vibration and orientation, and thus follow the workers about their normal duties.
- With data logging capability they offer the ability to obtain a detailed contaminant exposure profile through downloading of the monitoring data to a personal computer.
- Some instruments offer a communication package to link two devices together. This is useful in confined space entries since the entrants have one instrument and the outside attendant can monitor conditions within the confined space with a separate instrument linked with the communication package.

Disadvantages include the limited number of chemicals for which personal monitors are available and increased maintenance.

When using personal real-time monitoring instruments, it is essential that not only the sampling practitioner understand the use of the instrument but the worker as well. The worker should be informed of the concentration at which the alarm goes off as well as what actions to take as a result. The purpose for the monitor as well as the toxic symptoms of the substance being monitored should also be reviewed.

The location of the readout on the device is important. If it is on the front or bottom, a worker may find it difficult to view while wearing the monitor. Some monitors have extension cords that allow the sensor to be placed near the breathing zone while the electronics remain in the shirt pocket or on a belt. Since the sensor is often placed near the collar, the user must take care not to block the sensor inlet. A disadvantage of the miniaturization of sensors is the increase in drift and lower accuracy that can result from lower signal–background ratios when compared to conventional survey-sized units, which is due to the fact that sensor output is a direct function of the sensing electrode area.

In some cases, several single-point stationary full-time area monitors are more useful than personal alarm units. For example, if there are many workers in an area, the area monitors would warn many individuals rather than one at a time. Also, there would be fewer monitors to maintain as it is not practical to fit large numbers of workers with these units. On the other hand, in the case of a confined space entry, the personal units are the best choice.

Instrument Characteristics

Direct reading equipment can be specific for chemicals or very nonspecific, and the selection of an instrument will depend on the sampler's needs. In some cases, such as sampling on hazardous waste sites with unknown contaminants and concentrations, a nonspecific instrument is needed. In other cases, such as using an instrument to measure employee exposure to a known contaminant, a very specific instrument is needed.

Regardless of the type or brand, each instrument has certain characteristics which make that device unique. Understanding these characteristics allows the prospective user to compare instruments and to select the one most appropriate for the intended use. If the user already owns a device, understanding its operating characteristics can be very helpful in determining the types of applications for which the device is best suited. For example, some instruments have incorporated speakers that feature clicking noises similar to a Geiger counter. As the concentration increases, so does the pace of the clicking, which may be useful in situations where the meter is difficult to read—not unusual for liquid crystal screens in dimly lit areas. Most speakers and other optional features, including alarms, can be deactivated.

The design of an instrument can make it easier to use in the field. For example, a probe with the readout integrated into it allows the sampler to make measurements and observe the readout without being distracted from where the probe is being pointed, which is especially important when conducting leak testing. The advantage of probes attached via tubing to an instrument is that they can be raised to collect a sample or extended to get closer to a source while the readout is still at eye level.

The minimum detectable limit must also be taken into consideration. If the instrument is to be used to provide a warning for occupational health purposes, then its minimum detectable limit should be well below OSHA's PEL. If the instrument cannot respond to the highest observed concentrations, then a total exposure estimation will not be possible and the instrument can only be used for warning purposes.[4]

Important instrument characteristics that should be considered include:

Ability to Store or Transfer Data: The ability to store or transfer data is extremely important since microprocessors are being incorporated into more and more instruments.

Alarm: The alarm is usually audible, but it can also be a visible and/or vibrating signal that is activated at a predetermined level of contaminant or reading of the meter. Some devices have preset alarms, but the ability to set the alarm points based on local conditions of use can be very useful.

Ease of Use: Features such as portability, ergonomic design, nonbulky shape, location of the readout, and ability to carry while leaving the hands free are important.

Electronics: This section consists of the amplifier and the associated electronic circuitry necessary for the instrument to function. It receives the signal from the sensor, amplifies it, and displays it visually. This signal can be stored for subsequent readout using data loggers.

Filters: Filters are often incorporated into an instrument to remove particulate and interfering gases before the sample reaches the sensor. Filters can affect the response time of an instrument. Sometimes the choice of a filter requires some trade-offs between specificity and response time.

Fuel Source: Some instruments also require a source of fuel, for example, those with flame ionization detectors use hydrogen gas.

Function "Switches": These are switches with associated dials or "windows" on a scroll-down menu used to adjust the electronics to properly operate the sensor and to process the signal. Common adjustments are span adjust, scale select, and zero adjust.

Lag Time: It is the period of time between when the meter is exposed to the contaminated atmosphere and a step change is seen in the meter response.

Latching: It refers to alarm instruments. If a high concentration triggers the alarm, it will not turn off until the concentration goes below the alarm level. The instrument must often be reset.

Linearity: It is the deviation between actual meter readings and the readings predicated by a specified straight line. Some instruments are not linear over their entire scale, meaning that actual concentrations are not represented by the readout on certain portions of the meter scale. If a meter has multiple scales, it should be linear of each scale or at least calibrated on each scale. Instruments are often linear over only the portion of the concentration range they are designed to monitor. The actual range of linearity will vary from chemical to chemical.

Locking Out: Locking out is the same as latching.

Peak Hold: A peak hold function is useful for situations when an instrument is lowered into a confined space or when it is difficult to see the readout. In this case, the maximum value is stored and can be read when returning the unit to better light.

Power Source: It can be batteries or alternating current. The source provides the electric power for the electronics, pump, sensor, detector, and other accessories. The power going to

the instrument may have to be conditioned if the instrument manufacturer has not made provisions for line voltage fluctuations of the magnitude of those that can be encountered in the workplace.

Probe: Many instruments have probes that are extensions to enable sampling of air at a distance away from the instrument. The distance varies from a few feet to so or more feet depending on the instrument.

Pump: A pump allows the instrument to pull air into it rather than having to passively wait for a sample, which is particularly important for those instruments that have sampling lines. Devices that utilize the diffusion principle for sample collection do not require a pump.

Range of Concentration: Some instruments have several ranges, which is an advantage when sampling for a variety of concentrations. On devices with a meter readout, the meter may have several readout scales—one corresponding to each range. With a meter, the user must be aware of what scale is in use. Otherwise, erroneous results will be recorded and dangerous exposures may occur. The monitoring range of an instrument will determine what situations it can be used to sample. For example an instrument set up for ambient air monitoring will be far too sensitive and have too low a range to use for industrial monitoring. The range on a scale can determine the accuracy of a reading. For example a scale of 0 to 500 marked at 5 ppm intervals will make individual readings more difficult than a scale of 0 to 10 or 0 to 50 with marks to indicate one-half or one-tenth of a ppm. Devices with LCDs (liquid crystal displays) avoid many of the problems with reading meters.

Range of Detection: The lower limit of detection or sensitivity is the lowest concentration that an instrument is capable of accurately reading. The detection range is important when selecting an instrument, because if the lowest level the instrument is capable of measuring is greater than the concentration of interest, it will not be useful. An example is using a combustible gas indicator (CGI) to measure health hazard levels of hexane. Since the GCI measures in percent, it will not detect low ppm levels (1% = 10,000 ppm).

Readout: The readout is the display of the chemical's concentration in units appropriate to the measurement. For example, the combustible gas indicator (CGI) displays percent of lower explosive limit (LEL). Most gas and vapor instruments measure in ppm, and dust monitors display fibers/cc *or* mg/M^3.

Recovery Time: It is used to describe the time it takes the needle to return to zero after the instrument has been removed from the contaminated atmosphere. Recovery time is normally longer than response time, and depends on the concentration the instrument has been exposed to, especially if saturation has occurred.

Response Factor: General survey instruments can usually only be calibrated with one chemical at a time; therefore, the response of the instrument to any other chemical that is detected will be relative to the calibration chemical.

Response Time: Direct reading instruments making continuous measurements do not perfectly identify every concentration fluctuation since every instrument requires a certain amount of time in order to collect enough sample to respond. This period is known as the response time. The result is a certain amount of averaging, even in real time instruments; however, since response times are on the average of seconds, averaging is minimal compared to the length of time required for integrated measurements. Response time is defined as the time it takes for the instrument to give a 90% (of the actual concentration) reading after the sample collection has begun. Some samplers break down this time into lag time, the time interval between introduction of the contaminant to the instrument and the first observable response, and rise time, the time period between the initial response and 90% of the final instrument reading.

Rise Time: It is the period after a lag time until full response is reached.

Sample Outlet: This feature allows samples to be collected from an exhaust port in the instrument for further analysis. This is only possible with nondestructive detection methods.

Saturation: It occurs if the instrument is exposed to a higher concentration than it is designed to measure and purge under normal circumstances. The result can be sustained deflection of the needle or readout. Usually the instrument must be removed from the contaminated atmosphere and purged with clean air in order to get it to return to zero, unless this high exposure has damaged the sensor or contaminated the instrument's internals.

Scrolling Mode: It has become common for the displays on instruments with multiple sensors to "scroll" or sequence the display every few seconds to display the measurement of a different sensor.

Sensor: The sensor is the component that responds to a chemical or physical property of the contaminant. The terms *detector* and *sensor* are used interchangeably. In a sensor, an electrical signal proportional to the chemical concentration is generated and transmitted to the electronics section, which in turn displays it to the observer. Sensors can be either destructive or nondestructive, meaning the sample may or may not be "consumed" or destroyed over the course of the measurement. With nondestructive instruments, often the sample can be collected following measurement for additional laboratory analyses.

Sensor Life: It is highly variable and differs from instrument to instrument. It is dependent on the type, frequency of instrument use, presence of contaminants (to the sensor), concentrations of chemicals measured, and storage conditions.

Stability: An instrument is considered to be stable if it maintains its electronic balance over the time required for sampling. *Zero drift* or *span drift* are terms also used to refer to stability. Drift refers to a gradual change in the response that is unrelated to the contaminant being measured, the result being that the reading becomes inaccurate. An unstable instrument may require frequent zeroing and calibration, which is an important consideration for extended periods of monitoring.

Time Constant: The shorter the time constant of a direct reading instrument the more readings it makes in any given period giving a closer approximation of real time concentrations. An example is for an instrument to continuously indicate a reading for a short time after it has been removed from the area where the contaminant(s) are present. This

is a function of the principle of detection of the instrument as well as its electronic circuits. Some instruments have the capability of varying the time constant for their measurements.

Active and Passive Collection. Active and passive collection are the two basic methods for introducing a test atmosphere to a sensor. The active method uses a battery-or-line-powered vacuum pump that draws the air through the external sampling system to the instrument sensor. Diffusion head sensors are characterized by relatively low cost, ease of installation, and simple maintenance. Sensor heads can be removed from the electronics housing and replaced by the user in most cases. In general, these sensors are the best choice for sampling in remote locations because they can be linked electronically to a remote monitoring station. Passive sampling depends on air currents and convection to push the gas, vapor, or aerosol through the detection region.

The advantages of actively drawing the air to the sensor include[5]:

- Hard-to-reach sample locations are more easily sampled. For example, a probe can be inserted into a duct or other restricted space, and the air can be drawn to the sensor.
- The temperature and humidity of the test atmosphere can be controlled more readily by drawing the air over a water reservoir or through a drying tube. This feature is important when the air being sampled is outside the operating temperature or humidity range of the sensor.

Instruments using the passive sampling method rely on the principle of gaseous diffusion to provide contact of the test atmosphere with the sensor. The advantages of the diffusion method are[5]:

- Instruments are lighter and more compact.
- The pump, its maintenance, and the increased power needs of associated airflow system are eliminated.
- It is possible to sample remotely in the case of confined space entry. This sampling is accomplished by placing the sensor in the test atmosphere and connecting it to the remotely located readout. When the proper electronics and cable are used, a rapid readout of the contaminant levels is possible without the delay of pumping the gases to the sensor, thus reducing the response time.

Temperature changes can result in exaggerated error in diffusion instruments. A lower signal output can occur with this mode, because in the pump-drawn mode, where a directed stream of gas flows across the electrode, the sensor output is much higher, since the gas molecules are replaced quicker than they can be consumed. For a diffusion mode sensor, the number of gas molecules available per unit time is much smaller. The lower current occurs because the base is consumed faster than it can be replenished. The depletion effect gives an even lower signal–background ratio.[5]

Hazardous Atmospheres. Never take a direct reading instrument or any other instrument into a potentially explosive atmosphere without checking with the manufacturer to ensure that a specific instrument meets the appropriate instrinsic safety requirements of Underwriters Laboratory (UL), Factory Mutual (FM), or the Mine Safety and Health Administration (MSHA). UL is an independent testing and approval agency that examines electrically operated equipment and accessories, primarily from the standpoints of safety and freedom from hazard. Approved equipment carries the UL label. FM is a combination of insurance companies who have formed a testing and approval agency for fire protec-

tion and other industrial safety equipment. Products submitted to them are tested for safety in performance and, if acceptable, receive the FM stamp. Hazardous atmospheres may contain volatile flammable liquids and gases, combustible dusts and easily ignitable fibers, and other materials as defined by various classes within the National Electrical Code. Some instrument manufacturers, rather than having a UL or FM rating, have chosen to get approval by the Canadian standard for hazardous atmospheres. The term *intrinsically safe* is often applied to instruments that have been certified for use in hazardous atmospheres.

Calibration and Maintenance of Real-Time Instruments

The primary concerns for quality control when using direct reading instruments are understanding and minimizing sources of error through proper calibration, maintenance, and use of the instrument. Initial checks and adjustments consist of battery check and zero and calibration procedures, all of which should be performed at least once a day. Of these, the most critical first step is proper calibration. Two different types of calibrations are often required for real-time instruments: flowrate and chemical concentration. There are different levels of chemical calibration, ranging from factory calibration with several concentration ranges of different gases, to a specified concentration of a single gas done by the user, to a field check with an unknown gas concentration.

Units are sometimes calibrated electronically rather than with a gas or vapor. This is often the case with very small instruments, such as those used for personal monitoring and for instruments for which the calibration gas is difficult to generate by the typical user, such as mercury. For the electronic calibration, generally a voltmeter is used or a button is pushed to generate a readout adjusted to match a number on the cell via turning a screw. A variation of an electrical calibration is used in tape-based detection instruments that often utilize cards with colored spots to serve as calibrations. This is because the instrument interprets the degree of discoloration as concentration.

Most instruments using vacuum pumps are designed to sample at a specified flowrate. Know that rate and what rate would represent a significant shift higher or lower than that flow. The instrument's manual may indicate what effect these changes could have on the instrument's response. Flowrates of portable instruments should be checked every 40 to 50 hours of operation with at least a secondary standard (rotameter), which in turn is compared frequently with a primary standard. For a detailed description of techniques for using the bubble burette or rotameter, see Chapter 5. For instruments with pumps, a leak test will confirm that the sample is being drawn into the sensor.

The preferable way to calibrate concentrations is with the specific contaminant that is going to be present during the survey. The precision of a direct reading measurement is a function of the concentration of the contaminant present and the range of its linear measurements. The concentration(s) of calibration gas should be similar to the level(s) of contaminant expected to be found in the field. Calibration before and after each use should be performed for personal monitoring instruments designed to measure specific chemicals using a gas standard that provides a concentration at or near the OSHA PEL of the compounds to be measured. If calibrating a general survey meter, then a gas or vapor should be selected that best fits the user's need. This selection depends on the application. It is often recommended that the calibration of most survey instruments be carried out using two or more concentrations of the reference gas in order to verify the response over the full range of concentrations anticipated in use.

If a single concentration is used, calibration is commonly performed using a span gas having a concentration approximately equal to the midscale value of the meter. Never connect a gas cylinder directly to the intake port of most instruments, as doing so will pressurize the instrument and cause serious damage to the sensor. Instead, fill a sample bag with the calibrating gas and attach it to the intake port of the unit. For more information see the discussions included with each type of detection method on calibrating a particular type of instrument.

Although calibration with known concentrations of the chemicals of interest is the best method for calibrating real time instruments, preparations of these materials are not always available or they are problematic. For example, formaldehyde is not stable as a gas preparation and instead the gas must be derived from crystals of paraformaldehyde. A similar problem exists with reactive gases such as isocyanates. In some situations, the best calibration gas is toxic, so another gas that causes a similar response is used. For example, most PIDs are calibrated with isobutylene and adjusted to be equivalent in response to benzene, a carcinogen.

Aerosols of appropriate concentrations for calibrating instruments are even more difficult to generate than gas and vapor atmospheres; therefore, unless the sampler has extremely good skills, knowledge, and equipment, calibrations of this type are best achieved by sending the instrument to the factory. Aerosol monitoring instruments based on light scattering must be calibrated against a dust sample of the specific dust compound that has been collected using integrated methods. Changes in aerosol size distribution or optical properties of the dust being measured will necessitate a new calibration.[6] The user can change the calibration constant of an aerosol monitoring device for a specific type of aerosol by performing integrated sampling on a filter concurrently with

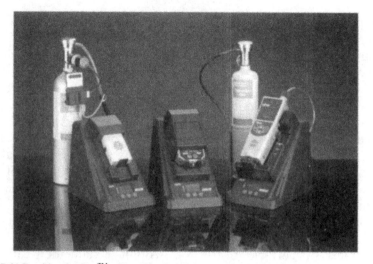

Figure 9.6. DS2 Docking Station™ technology for direct-reading instruments provides two way wireless and/or ethernet connectivity among up to 100 stand-alone stations to manage calibration, record keeping, battery recharging, and instrument diagnostics. (Courtesy of Industrial Scientific Corporation.)

taking measurements with the instrument. The ratio of the two concentration values can then be used to adjust the instrument's response.

A zero calibration gas may or may not be required. Zero gas is a high-purity air. Most manufacturers do not require calibration with zero gas to make the zero setting adjustment. Instead an adjustment is used to electronically balance the circuitry to zero an instrument. Some instruments have an electronic zero as well as a zero potentiometer that can be adjusted with a screwdriver. The use of an electronic zero provides the capability to accurately set an instrument to zero in the presence of atmospheres containing background contaminant levels.

Another technique used in the field is to check the response of the meter with a known substance without necessarily knowing its exact concentration. It is known as a field check, and although not useful for calibration, it will show if there are any serious malfunctions in an instrument.

Instrument manufacturers often offer their own calibration kits. Many of these kits are specially set up for a given instrument, and using different equipment may cause problems. For example, a regulator on a gas cylinder may be set up to deliver a very low pressure, which is important if the cylinder is attached directly to an instrument. Use of the wrong cylinder–regulator combination could result in pressurizing an instrument. Always heed the manufacturer's instructions when selecting calibration methods for any given instrument. The user should determine prior to purchasing from another vendor if the combination will be incompatible or result in less accuracy. Generally calibration kits contain a gas cylinder with a regulator, tubing, and a bag. Some manufacturers offer an integrated charging and calibration "station" (Figure 9.6) that greatly simplifies keeping the instruments ready for use.

Prior to calibrating any instrument it is a good practice to review the manufacturer's manual. Most manuals give step-by-step instructions for use with their specific instrument. In addition, any idiosyncrasies or problems often encountered during calibration are often noted. While the instrument's manual will provide guidance in calibration frequency, the user should make a determination based on experience. The following guidelines are representative of instruments in general.[7]

- Calibrate before and after each use to provide assurance of measurement accuracy.
- Always calibrate an instrument when it is first purchased regardless of what was done in the factory.
- Calibrate whenever the instrument has been "bounced around," such as can occur during shipping, heavy usage, climbing ladders, and so forth, and after an instrument has been exposed to potential sensor poison and severe weather conditions.
- Calibrate following repairs or replacement of any of the instrument's parts.
- The frequency of calibration should be increased when a sensor nears the end of its service life or when abnormally high variations are observed. The frequency of tests should also be increased when a new instrument or sensor is installed. Sensors often exhibit higher shift during the first month of operation.

Generally, most instruments are a complete system and parts should not be interchanged with other instruments or substituted for those provided by the manufacturer. Very few malfunctions can be easily corrected in the field; thus, users should check out an instrument prior to a survey. A preventive maintenance program will also prevent or minimize problems in the field. The manual's troubleshooting section may be helpful; if not, call the manufacturer. When filters on inlets are discolored or heavily clogged with dust they should be changed. The exact schedule will depend on the instrument.

Consult the manufacturer's instructions for when to use interference filter, organic filters, and humidity adapters. For continuous monitoring applications in excess of 3 hours, electrochemical and solid-state devices often require an external source of humidity. This source usually consists of a bottle with distilled water installed in-line with the intake port. If a filter is incorporated into the inlet of the instrument, it should be removed because it will trap the water vapor intended for the sensor. The air filter should be reconnected to the front of the humidity control bottle.

Most chemical sensors depend on a specific chemical reaction to generate a signal. This reaction may occur on a solid or liquid surface, light, or flame. The reactions involved in these sensing mechanisms may or may not be reversible, and exposure to other contaminants that can degrade the sensor surface also occur, with the result that the sensor material is consumed during use. The result can be sensor drift or a high readout even in a clean atmosphere. Therefore, sensors or certain parts of them will have to be replaced periodically. Sensor replacement is simple in most personal units; the sensor just plugs into the top of the unit. Sensors generally must be replaced at the end of their service life or at the interval recommended by the manufacturer. The sensor should be removed from the monitor whenever it will be out of service any length of time. For proper readings batteries must have a sufficient level of charge. Usually it is not a good idea to operate an instrument while it is being charged.

Data Displays and Output

Most instruments generate a voltage proportional to the concentration of compounds detected by the sensor. Initially almost all instruments used an analog scale where readings were made by viewing the movement of a needle on a scale reading ppm or compared to a calibration graph to get the concentration. Analog output is defined as a continuously variable voltage or current that is proportional to the instrument's measurement, and it is suitable for use with a recorder and alarm system. Modern instruments use digital readouts and storage capabilities that make it convenient to link to personal computers and similar digital devices.

Most readouts are linear, that is, there is an equivalent increment in the display reading for equal increments in concentration. For example, the needle will move the same distance on a 0 to 200 scale if the concentration of a compound increases from 10 ppm to 40 ppm as it will during an increase of 130 ppm to 160 ppm. Linear scales usually have two to four concentration ranges, because a scale that would provide sufficient space to adequately distinguish between measurements by 1 ppm would require a large space on the front of an instrument. Also, usually at the lower ranges it is more important to differentiate between fractions of a ppm whereas at higher ranges it is not. On a few monitors the readout is presented as a range rather than as a specific value. Some instruments have a logarithmic scale that, like log paper, compresses the range at higher concentrations, the result being that the sampler does not need to switch between ranges. These are most common in instruments designed to measure leaks where high concentrations are expected.

A comparison of logarithmic readouts to linear readouts follow[8]:

1. A single meter setting is required for a logarithmic readout whereas different scales (ranges) are required for linear readouts.

2. A logarithmic meter may be more compatible with measurement of air contaminants in typical workroom situations, because their concentrations are considered to fit a lognormal statistical distribution.[3]

3. The zero can be set on the linear scale at the background level, allowing the instrument to automatically subtract this value during measurements.

4. If a device produces a logarithmic output current, logarithmic scales do not suffer from a lack of linearity at low (approximately 10%) scale readings because there is no conversion necessary, whereas linear scales have this problem. If the readout on an instrument is indirect, that is, in units such as percent absorbance or percent transmission, it must be converted to concentration and a calibration curve must be consulted in order to determine the actual concentration.

Often when making measurements needles or digital values will jump around because of fluctuations in airborne concentrations and air currents. In these situations, deciding where to make a reading can be difficult. Even in the case of digital instruments, it is not uncommon for the readings to change rapidly unless the instrument is set to average over a certain period. Some instruments allow the user to control the frequency of changes in the readout, for example, displaying a 10-second average instead of a 1-minute average measurement, but most have their own individual response times.

In some instruments more than one averaging period is available and in this situation a slower period is desirable. A conservative rule of thumb would be to select the maximum value over a given time period. Another option would be to record the range for the values, for example, 56 to 72, or 64 to 74, and so on. In general, the upper and lower 10% of any instrument's range should be avoided, and if possible the sampling practitioner should operate it only in the middle 50% of the overall range.

For instruments with meter, parallex can cause readings on scales to appear higher or lower unless the user is looking directly down on the needle. The use of a mirror in the background of the meter can eliminate this problem, so that the user is assured of looking straight down at the needle by lining up the needle with its mirror image.

In strong sunlight a digital display on a liquid crystal panel can turn black, making it unreadable. Although after a period of time the panel returns to normal, it can be a nuisance when trying to make measurements. Therefore, precautions should be taken when doing measurements outdoors, for example, on roofs and balconies or in areas brightly illuminated by skylights or windowed walls.

Data Loggers. A data logger is a simple computer-like device with limited functions that can collect, store, transmit or analyze information from the direct reading instrument. The advent of data loggers has made it much more practical to use direct-reading instruments for personal monitoring and continuous monitoring. Previously, direct-reading instruments were equipped with a strip chart recorder or similar device to record the instrument's readings over time. Small-scale battery-powered recorders were developed that could be worn on a worker's belt along with the monitoring instrument for personal exposure monitoring. Larger instruments could be connected to a full-size strip chart recorder to yield a stationary monitoring system.

The next step was development of stand-alone data loggers that connected to the recorder output jack on the direct-reading instrument. In some cases these were made to connect to a specific monitoring instrument, but generic loggers were also made to work with many direct-reading instruments. When using the generic loggers, extreme care was needed to ensure compatibility between the devices in communicating data, proper calibration of the logger and, finally, that the logger could print out summary tables and graphs that matched the actual data. During this time period, both the technology and user's knowledge of personal computers were more limited than today, and so generic stand-alone data loggers often did not meet their potential.

Today, data logging capability is built into many direct reading instruments, and the higher-end instruments have impressive logging capability. For example, one four-gas monitor from BW Technologies features a logger based on the Flash MultimediaCard storage device. More than two months of data (collected at five second intervals) can be stored on a standard card. Up to one year of data can be stored on a large-capacity card with a longer interval between samples. All events and occurrences are recorded including sensor readings, alarm set points, alarm conditions, TWA and peak exposures, calibration, pump operation, event flags, instrument status, and instrument serial number with time and date stamp. Two versions of the instrument are offered:

- *User downloadable data logger* using on an off-the-shelf card reader to transfer the information to a personal computer for analysis using standard database or spreadsheet applications.
- *"Black box" data logger*, where the data can only be downloaded by an authorized factory service center. This option allows a continuous, tamper-proof method of collecting data that can be downloaded from the instrument in the event of an occurrence or incident.

Other instrument manufacturers offer a dedicated data logging unit (one is shown in Figure 10.9) when the instrument does not have a built-in logger. The interface programs with personal computers have also been improved to the point where prior compatibility and communications problems have all but been eliminated. There are some "generic" data loggers available (Figure 9.7) that can be purchased by the sampling practitioner to be connected to a direct reading instrument,

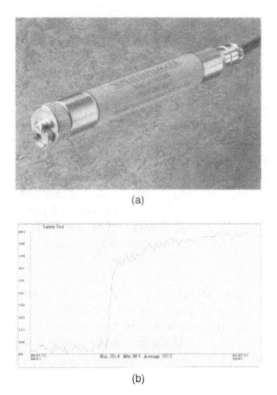

(a)

(b)

Figure 9.7. Datalogging devices are available for instruments without internal logging capability: (a) Nomad Datalogger™; (b) typical output plot showing concentration versus time of day. (Courtesy of Interscan Corporation.)

but these are much less common than instruments with built-in data loggers.

A modern data logging option is the GrayWolf DirectSense™ multi-gas probe that connects to a mobile computer such as a pocket or notebook computer (Figure 9.8). Up to five toxic gas monitors plus a temperature sensor can be configured into single sensing probe. Gas sensors include sulfur dioxide, nitrogen dioxide, nitric oxide, carbon monoxide, ammonia, hydrogen sulfide, hydrogen cyanide, ethylene oxide, chlorine, and oxygen. Up to five gas sensors (in addition to a temperature sensor) can be configured into a single sampling probe. Optional accessories include a digital camera and computer-assisted drafting (CAD) unit, which allows complete documentation and analysis of airborne concentration measurements and related data and information using both standard and customized software packages.

Data loggers function by storing the voltage response signal from the direct-reading instrument. Since the data are stored digitally, analysis can be performed by the logging unit and sent to a printer, or

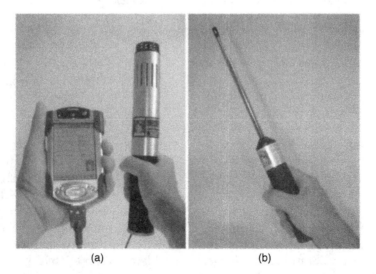

(a)   (b)

Figure 9.8. (a) GrayWolf IAQ monitoring probe with pocket computer links an advanced direct-reading instrument directly to computer technology. (b) Optional air velocity probe. (Courtesy of GrayWolf Sensing Solutions.)

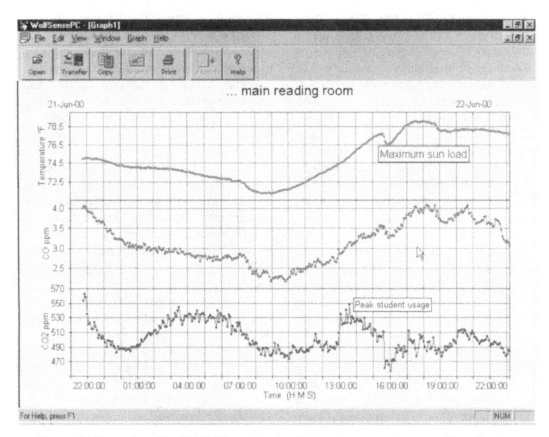

Figure 9.9. Output from GrayWolf IAQ monitor showing temperature, carbon monoxide, and carbon dioxide levels as a function of time. (Courtesy of GrayWolf Sensing Solutions.)

the data can be transferred to a personal computer for analysis and long-term storage. In the case where data-logging capability has been incorporated into the instrument, the capability, storage, and other aspects can vary considerably. Some instruments can record one value—typically the average or peak reading for each measurement—whereas others can store the maximum, minimum, and average concentrations every second or so. For personal monitors with built-in loggers, the premium on small size and weight causes a trade-off between data storage and analysis capability and the added size and weight. Depending on the logger's design, choosing a short sampling interval and storing the maximum number of data elements for each interval can reduce the overall time period of logging before the logger's memory becomes full.

Transferring Data from the Logger. Most data loggers can send data in three different modes: formatted or compressed reports or a raw data dump. Formatted reports are designed for direct printing on a printer in an understandable, readable format. Compressed reports are designed to be sent to a computer for further processing. A compressed output report contains no descriptive headers or graphs and is significantly shorter than a formatted report. It is necessary that the computer receiving a compressed report have knowledge of the data format so that it can be properly interpreted, which usually involves sending the data to a software program. Figure 9.9 shows typical data output from a direct-reading instrument and data logger after computer analysis.

Data signals are transmitted using either parallel or serial connections. Serial transfer is much slower than parallel, but more practical for sending data over long cables and long distances due to inherent problems in parallel data transfer. Parallel transfer is used for hooking up devices such as printers and computers that are likely to be close together. Serial transfer is used when long cables are required and for transferring over phone lines via modems.

In addition to the analog ports for the instrument's connection, some loggers have serial ports and others have both serial and parallel ports. In the case of a logger that is designed to dump to a printer only, in order to utilize a computer a program may have to be written by the user. Generally, when downloading to a printer or computer, a logger will use an RS-232 serial port. Most data loggers are capable of sending a formatted output to a printer. If the output were not formatted, it would be just a string of numbers with no indication of what each number represented or when it was collected.

Each instrument has certain expectations as to how it needs to send and receive data, including the following: parity, start bits, data bits, stop bits, and baud rate. Baud rate is the speed at which characters are sent and received by the data logger and other devices that use serial communications. If the baud rates of each device do not match, the transmission will not work. The baud rate can vary from 150 to 9,600 depending on the instrument. Some instruments are flexible and any of several rates can be selected. Others will operate only at one specified baud rate.

Often a separate software package is required in order to transfer data and store it in a usable and manipulatable format in a computer. Sometimes these are software programs available from the data-logger manufacturer that allow the data to

be saved, viewed, plotted onscreen, or exported in an ASCII format for use with other programs. The type of graphics capability of these programs often varies. Sometimes loggers can also be dumped into a commercial data base program. Since most loggers can send information in ASCII code, most any data base program can be used.

In some cases, when information is downloaded to a computer it is erased from an instrument's memory; in other cases, the instrument will continue to store data until the user purges them, or the number of allowable data storage events is exceeded and the instrument starts to record over the earlier data.

Many loggers have the capability of being operated off-site via a modem connection to a computer. Long-term monitoring is possible, and via the modem the data logging unit can be told to stop, dump data, and clear. There will be certain specific commands to use to access the unit. Usually the commands involve telling the unit to start/stop logging and send data in a given report format. These remote request commands can be different for each data logger.

In general, when in the printer mode instruments respond to commands from their own keypad to send information (characters) out of the RS-232 port to a printer (peripheral device). When in a computer mode the instrument responds to commands from the computer to send and receive data and perform other tasks.

Important Facts for Stand-Alone Data Loggers. As mentioned earlier, stand-alone data loggers are not commonly used because so many instruments feature integral loggers. However, here are some important considerations about stand-alone data loggers.

Battery Time: Nine-volt batteries are common in data loggers and often provide up to 100 hours of operation. Most data loggers will store data once a battery is removed or has lost its charge for a given period; however, this period may range from a few hours to a few days. For safety, change battery packs on data loggers that are used regularly and always keep a spare battery pack available.

Clock Capability to Record Time Intervals: Some manufacturers have incorporated the ability to pause the unit for breaks and lunch periods, which can save memory space. The timer function should continue so that future measurements are attributed to the correct time periods.

Communication Protocols: In order for data loggers to send data to computers, the specific requirements of each instrument must be identified and set on the computer. These include baud rate, number of data bits, and parity.

Ease of Programing: In most cases, as part of the initial setup the user must input the date and time and then choose the desired input voltage range, TWA storage period, scale, range, calibration, and units of the stored data. The user also programs which data to store. They can range from all measurements, to the maximum value detected in each measurement period (or average value), to the minimum value, or all three. Some data loggers are tutorial and have self-prompting and error messages whereas others require extensive study of the user manual. An example of a self-prompting message is: Connect Span Gas Then Press Enter. Another technique is to have flashing letters symbolizing the opportunity to enter various data.

Due to the need to keep instruments small, many keypads rely on the use of +/− for a variety of functions, including changing numbers in dates and times and entering other numerical parameters. Units that use arrow ($\uparrow$, $\downarrow$) keys to toggle among digits use a cursor to indicate which digit is ready to be changed. These designs can take a lot longer to program than keypads incorporating the alphabet and numbers. However, keypads with the alphabet and numbers take up more room on the instrument.

Format of Printout: The format of datalogger printouts varies widely. The simplest is a numerical listing of values. Some plot data as minimums, averages, and maxima for each sampling interval. Other formats include graph and histogram construction. In some cases, there is only a single format; in others there is a choice of different formats.

Amplitude Distribution: An amplitude distribution report contains a graph showing what percent of the data is distributed within certain ranges. Often this graph indicates concentration, the number of samples collected at that concentration, and the percent of the total number of samples represented. Therefore, the most frequently occurring concentration will have the highest plot and percentage.

Cumulative Average: Some units can also provide a cumulative average; thus, an 8-hour cumulative average would represent an 8-hour TWA.

Time History Report: These reports often show a minimum, maximum, and average value stored for each sampling period. The average will depend on the number of samples made over a given period.

Modes: Most data loggers have three modes: display, program, and output. The display mode allows the user to observe the input reading and may include date/time, alarm settings, and other data. The program mode is for programing parameters. The output mode is used to output the input readings, any statistics the unit is capable of computing, and other data such as time history, amplitude distributions, and graphs.

Number of Channels: Most current stand-alone data loggers designed to be attached to portable survey instruments have just one channel, that is, they can log from only one instrument or one sensor at a time. However, multichannel data loggers are available. Some log for instruments that have more than one sensor; in such cases, they are generally dedicated to these instruments. Another has the capability of being hooked up to four different instruments at a time, which of course limits the portability of the system. For stationary monitoring, more sophisticated data-logging units have been available for some time.

RS-232 Connection: This is a standard interface through which devices can communicate.

Sampling Periods: Sampling periods represent the periods over which the data logger averages or integrates the sample points it receives. They also determine how fast storage memory will be filled. As an example, if the user selects a sample period of 1 minute, and the unit is sampling at 1 sample/second every minute, the unit will average the sixty 1-second readings it has received and store them as one sampling period. The shorter the sampling period, the closer the unit comes to storing real-time data, or instaneous measurements.

Separate Readout Device: Most current data loggers and data-logging instruments have enough processing capability that they can be independently hooked up to a printer/computer without having to go through a secondary processor. However, some units still require the use of an intermediary device in order to transmit data to a printer or computer, because such data loggers only compute and store averages. Therefore, in order to do further calculations and analyses of the data, the data must be transferred to a computer with a data base program. In some cases, companies have developed specific software programs for inputting computer data.

Software to Interface with Computers: When converting information to a computer where it can be stored on a disk and analyzed further, two types of software are needed. The first is a communications program, allowing the instrument to "talk" to the computer. The other is a database or other program to allow the user to put the data into some sort of usable report format, develop graphs, and so forth. Otherwise, the information must be stored in ASCII (American Standard Code for Information Interchange), which is a universal text that most computers can understand.

Storage Memory: When comparing data loggers for storage capability, the amount of room required (in bytes) for each measurement as well as other bytes necessary to store other information, such as parameters, should be considered. Data loggers will either record continuously, and after the maximum number of readings are collected begin to overwrite the existing data, or stop recording once memory is full. Therefore, the user must plan to dump data at regular intervals and be aware of the maximum sampling period available.

Stored Points: The total number of data readings that can be stored along with the length of required sampling time will often determine the sampling intervals. Typical sampling rates for data loggers are one sample per second or four samples per second. The rate represents the number of sampling points received from an instrument. For example, a unit with a maximum of 700 storage points collected at 1-second intervals will use up its storage in 11.7 minutes; if the collection interval is increased to 1 minute, the unit will store data for 700 minutes, or approximately 11.7 hours; and if the storage interval is further increased to 5 minutes, the length of data logging can be extended to 58.3 hours. When selecting data to log, some instruments can log either points (instantaneous values) or average values at period intervals. Often preselection is necessary.

Tag Numbers: Tag numbers can be used to label data with a unique code for a given sensor. They are also required to separate various events. In this way, a data logger can be moved from instrument to instrument in a single survey if tag numbers for sensors are used, which allow distinguishing data during the test.

COLORIMETRIC SYSTEMS

These devices rely on the color produced by the contaminant of interest and reagents in the sampling system to indicate airborne concentration. They are widely used because of the simplicity of operation and general low cost, but like any real-time method they have limitations on accuracy, specificity, and possible interferences. These systems are covered in more detail in a separate chapter.

Colorimetric Detector Tubes

These are sealed glass tubes containing the color-producing reagent (Figure 1.14). These are probably the most commonly

used real-time monitoring device because tubes for various contaminants can be kept on hand and used with a hand pump for range-finding samples and to identify the presence of specific compounds. Battery-powered grab sampling pumps are also available, and there are long-term tubes designed to measure exposures over an extended period such as an 8-hour work shift—these operate using a battery-powered pump or as passive samplers relying on diffusion. With battery-powered pumps, several detector tubes (or detector tubes plus an adsorption tube) can be connected via a manifold to permit parallel sampling for multiple compounds.

The color-based reaction is generally "read" as the length of colored stain that develops in the tube, although some devices use an internal photometer to read and display the colorimetric change. Many tubes can measure lower concentrations if a larger air sample is collected. With a hand pump, this is accomplished by increasing the number of pump strokes. For these tubes the calibration chart gives concentration reading based on sample volume. Long-term tubes report concentration in units of *ppm-hours*. To calculate the average airborne concentration, divide the *ppm-hours* reading by the sampling time (in *hours*).

Other Colorimetric Devices

In addition to colorimetric tubes, other systems based on color change include the following:

- *Colorimetric badges* are dosimeter-like passive devices. The color change on the badge after use is read visually or by use of a color photometer.
- *Colorimetric tape sampler* is a fixed device for area samples that functions by drawing a chemical-impregnated paper tape past a sampling port where the pump-drawn air sample impinges on the tape. A photometer measures

the resulting color change and reports it in units of concentration. Some devices can store data for later downloading, trigger alarms if preset concentration levels are exceeded, and display the concentration value.

SUMMARY

Real-time methods have a distinct advantage over *sample collection device* methods (which require laboratory analysis) because of the obvious benefit of having results immediately. Advances in sensor and data storage/processing technologies have made direct-reading instruments a viable option for many monitoring applications. Colorimetric systems have a long history of successful use, and continuing improvements increase their value.

However, all real-time methods have certain limitations. The sampling practitioner must ensure that the method is suitable for the contaminants of interest, considering factors such as sensor response, accuracy, and the possible presence of interfering compounds. Selection criteria for real-time device include:

- Need for portability
- Amount of time and expertise available for training
- Need for high sensitivity (low detection levels) and accuracy
- Conditions of use—heat, humidity, use of a cart versus a shoulder strap
- Appropriateness for the application
- Purpose for sampling
- Instrument availability and complexity of use
- Specificity
- Personal choice based on past experience
- Measurement period required

- Preparation time available prior to use of an instrument
- Minimum detectable limit
- Required dynamic range
- Time constant
- Presence and classification of hazardous (explosive) locations
- Type of data acquisition and display required
- Interferences
- Personnel for operation and maintenance

The next several chapters discuss specific *real-time* methods in more detail.

REFERENCES

1. Smith, J. Uses and selection of equipment for engineering control monitoring. *AIHA J.* **44**(6):466–472, 1983.
2. Stetter, J. R., and D. R. Rutt. Instrumental carbon monoxide dosimetry. *AIHA J.* **41**(10): 704–712, 1980.
3. Langhorst, M. L. Comparative laboratory evaluation of six chlorine monitoring devices. *AIHA J.* **48**(5):347–361, 1982.
4. Smith, J. Uses and selection of equipment for engineering control monitoring. *AIHA J.* **44**(6):466–472, 1983.
5. Shaw, M. More Straight Talk About Toxic Gas Monitors. Interscan Corp., P.O. Box 2496, Chatsworth, CA 91313-2496.
6. Willeke, K., and S. J. Degarmo. Passive versus active aerosol monitoring. *AIHA J.* **3**(9): 263–266, 1988.
7. Long, S. E., and R. W. Lawrence. Understanding Combustible Gas Sensors—Types, Installation Considerations, Costs and Maintenance Costs. Mine Safety Appliance, Inc., Pittsburgh, PA.
8. *Organic Vapor Analyzer (OVA) 108 Operating Manual.* Foxboro, MA: Foxboro Instrument Co., Inc.

CHAPTER 10

INSTRUMENTS WITH SENSORS FOR SPECIFIC CHEMICALS

Direct-reading electronic instruments are available that function in several different ways. For example, here are three major classifications:

- Instruments designed with one or more sensors that respond to *specific* chemicals.
- Units that respond to broader categories of airborne chemicals such as combustible gases or hydrocarbon vapors.
- Devices that contain a sensing system that can be "tuned" to respond to many different chemicals. Infrared (IR) and gas chromatography (GC) units typify this group.

However, there is significant overlap among these three categories. For example, some IR devices are set to measure just one specific gas or vapor, and some sophisticated "single sensor" solid-state units can be adjusted in the field so the unit can measure the concentration of many differ-

ent gases and vapors. This chapter builds on the general information in Chapter 9 and discusses the first category of instruments listed above—those instruments that are designed with sensors to measure specific chemicals. Any exceptions and overlaps to the specific chemical approach are noted in the text. The following chapters cover the other types of direct-reading electronic instruments and colorimetric systems.

Generally, instruments with sensors that respond to a single chemical are less expensive and easier to use than instruments with detection systems that can be used to detect many different chemicals, such as GC and IR devices. Instruments with broader measurement capabilities can also be dedicated to the measurement of a single chemical; these can be cost effective when the unit can be constructed more simply than those with broader monitoring capabilities.

The instruments with sensors for specific chemicals are most usable in situations where:

Air Monitoring for Toxic Exposures, Second Edition. By Henry J. McDermott
ISBN 0-471-45435-4 © 2004 John Wiley & Sons, Inc.

- A "pure" gas or vapor, or known mixture, is used under circumstances where no chemical reactions occur to produce unknown contaminants. An example of this use is operations within the semiconductor industry involving gases such as arsine and phosphine.
- A known gas or vapor of relatively high toxicity is released or generated, and a real-time device is needed to monitor levels of this contaminant. Examples are carbon monoxide (CO) from internal combustion engines or hydrogen sulfide (H_2S) released from certain crude oils.

They are not as commonly used where source of the hazard is a gas or vapor mixture with many different compounds *and* the composition can change over time since the sensor, which responds to a specific chemical, may not adequately "measure" the hazardous components under these changing conditions.

Advances in sensor technology and electronics have increased the number and capability of the chemical-specific sensor instruments available to the sampling practitioner. Manufacturers now offer a wide range of options. Probably the simplest are "disposable" single gas monitor alarms that are designed to operate over their guaranteed life span (often one year or more) with no maintenance beyond a periodic calibration check. Typical instruments (Figure 9.3) have multiple sensors—one for each contaminant—to measure several gases plus the concentration of combustible gases and vapors in the atmosphere that can be changed by the user to monitor other gases (Figure 10.1). The more complex instruments have features such as: sophisticated datalogging and data management capability; multiple warning alarm levels; a pump to draw samples through tubing from some distance from the unit; and the option to include a broad response

Figure 10.1. Sensors can be easily changed or replaced by the user on many direct-reading instruments. (Courtesy of Industrial Scientific Corporation.)

sensor such as a photoionization detector (PID) [as described in Chapter 12] along with the chemical-specific sensors.

Beyond the specific-chemical sensor definition, but employing the same technology, are units with a metal oxide sensor that can be "tuned" in the field primarily by modifying the sensor temperature to respond to over 100 different specific chemicals for quick evaluation of potentially hazardous atmospheres. For completeness, one of the devices is described in this chapter.

Sensor technology has evolved to a point where even the nomenclature traditionally used by manufacturers and sampling practitioners is no longer a precise description of the various sensors and measurement principles. For example, a broad category of measuring devices use *electrochemical* sensors—a type of sensor where a reaction between the contaminant of interest and a chemical electrolyte within the sensor generates an electrical current that is proportional to the contaminant concentration. There are other categories of sensors, such as solid-state sensors, that measure concentration based on changes in the sensor's *electrical* properties as the *chemical* contaminant contacts the sensor. So from a technical standpoint these could also be described as electrochemical sensors even though the principle of operation, selection factors, advantages, and limitations are completely different from the sensor group commonly referred to as *electrochemical*. To avoid confusion, it is important to understand the working definition of the various sensor and instrument categories as described in this chapter[1]:

- *Electrochemical Sensors.* These contain an electrolyte that reacts with the vapor or gas of interest to produce an electrical signal that is proportional to the concentration.
- *Metal Oxide Sensors.* Gas or vapor of interest changes the electrical properties of a heated metal oxide surface; the change is proportional to concentration.
- *Infrared (IR).* Gas or vapor of interest absorbs IR radiation; the resulting increase in temperature within the sample cell or reduction in source strength due to absorption is proportional to concentration.
- *Ultraviolet (UV).* As used in these instruments, the gas or vapor of interest absorbs UV radiation of a certain wavelength; the resulting reduction in source strength is proportional to concentration.
- *Chemiluminescence.* Reaction between gas or vapor of interest and another chemical causes a reaction product that is "excited," and emits a light of characteristic wavelength that can be measured. The amount of light is proportional to the concentration.
- *Gold Film.* Mercury vapor is absorbed on a gold film; the change in electrical resistance over the sampling interval is proportional to airborne mercury concentration.

The majority of available instruments are based on either electrochemical sensors or metal oxide sensors, so these will be covered in more detail than the other categories. Since the main use of the information in this chapter is to assist with instrument selection, Table 10.1 lists some key selection parameters. Measurement techniques for several specific chemicals are described at the end of this chapter.

It is important to note that innovations by manufacturers continually lead to (a) the introduction of new devices and (b) improvements in existing sensors that either make them usable for more compounds or overcome some of the prior limitations to the technology. For this reason,

TABLE 10.1. Selection Parameters for Specific-Chemical Direct Reading Instruments

Chemical(s) to be measured
Concentration range to be measured
Specificity of sensor for chemical of interest
Interfering chemicals (false reading or sensor damage)
Acceptable environmental range (oxygen, relative humidity, and temperature)
Whether sensor exposure to contaminant is constant or intermittent
Whether ambient atmosphere is corrosive to sensor
Whether ambient moisture might condense on sensor
Possibility of high concentration peaks above expected levels
Whether intrinsically safe for explosive atmospheres
Interference from radio-frequency and/or magnetic fields
Need for remote sampling (hose or remote sensor)
Reasonable operating instructions
Maintenance and calibration requirements
Ruggedness
Size and weight[a]
Energy consumption (battery life)[a]
Initial and ongoing cost

[a] Mainly applicable for portable instruments.

any description of characteristics or limitations for a type of sensor should be considered to be a general statement. Check with specific manufacturers to determine if their products overcome any of the typical limitations described in this chapter.

CALIBRATION

All of these instruments require calibration using a known concentration of the gas or vapor of interest. When available, a low-pressure cylinder containing an appropriate concentration of the gas or vapor in *air* is usually the most convenient calibration technique; these can be supplied as part of a calibration kit with a regulator and other accessories (Figure 10.2). The chemicals indicated by a superscript *a* in Table 10.2 indicate substances for which calibration mixtures can be obtained from commercial suppliers; many other gas mixtures can be custom ordered. Two or more gases may be provided in the same cylinder to calibrate multi-sensor instruments. Most sensors require oxygen to function properly so it is imperative not to use a gas mixture using

Figure 10.2. Calibration kit consisting of gas cylinder and pressure regulator facilitates sensor calibration. (Courtesy of Industrial Scientific Corporation.)

an inert gas such as nitrogen as the diluent. Many sensors cannot withstand pressure so connecting the gas bottle directly to the sensor may cause damage. Additionally, many sensors require some moisture in the air for proper operation; compressed gas mixtures are dry and so moisture will have to be added.

Consult the instrument instructions for specific information. General guidelines are covered in this chapter; Appendix B contain instructions for preparing calibration gas mixtures when a gas-in-air calibration mixture is not available.

ELECTROCHEMICAL SENSORS

Electrochemical sensors are commonly used for the detection of about 40 gases and vapors. Typical compounds are the specific chemical substances listed in Table 9.1. Instruments are available for survey and personal measurements (Figure 10.3). In many instruments the sensors are readily changed by the sampling practitioner. Some sensors have an internal memory that stores data such as measuring ranges, calibration, and alarm set points so the information is not lost when the sensor is removed from the instrument. Electrochemical detectors use a number of different types of operating principles. However, these methods have not arisen logically over the years from a single branch of electrochemistry, and therefore there has been no systematic approach for naming them. Many different commercial products use similar detection principles, but because no single nomenclature is used consistently for the description of these devices, the commercial product literature can be confusing.

Electrochemical sensors can be differentiated according to which parameter of Ohm's Law (voltage = current × resistance) is the dependent variable of a system's operating principle, and thus categorized as

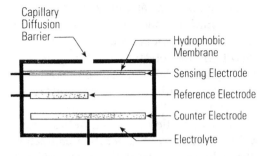
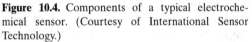

Figure 10.4. Components of a typical electrochemical sensor. (Courtesy of International Sensor Technology.)

Figure 10.3. Microprocessor technology permits design of direct-reading instruments that are extremely small in size. (Courtesy of RKI Instruments, Inc.)

either potentiometric, conductimetric, or amperometric. Amperometric devices are further divided into the subclasses of coulometric and voltammetric, these two having totally different operating principles.

In most electrochemical gas sensors the gas diffuses through a membrane and a layer of electrolyte separating the membrane and the working electrode (Figure 10.4). In some cases, the electrolyte is a liquid; in other cases, it is a gel or it is immobilized in a solid matrix. In order to make instruments specific, the electrolytes and electrodes in each electrochemical cell must be unique.

There are three types of electrodes: working (sensing), reference, and counter, although sensors can be constructed of either two or three electrodes. In a three-

electrode sensor all three electrodes are constructed of the same material. It is only the way they are connected to the circuit that determines which are sensing, counter, and reference electrodes. The sensing electrode interacts with the gas to be monitored. The purpose of the counter electrode is to act as an electron sink for anodic sensing reactions, and as an electron source for cathodic sensing reactions, and simultaneously to complete the circuit for the sensor. The reference electrode is a floating electrode that maintains the proper voltage at the sensing electrode despite changes to it over time due to chemical reactions. The three-electrode sensor is less expensive to make than the two-electrode sensor.

Various modifications of sensors are possible among manufacturers. For example, one manufacturer may choose to slightly acidify the hydrogen peroxide electrolyte in a sulfur dioxide conductimetric sensor to decrease the effects of certain interferents. The gas exposure path can be modified so that the differing amounts of gas reach the electrode, thus affecting the magnitude of the signal. In units with active sampling, high flowrates are used to pull gases to the sensing electrode in order to obtain the highest sensitivity. Control of the electrode exposure area is important and in practical designs as much catalyst as possible is packed into this area (by using materials with high surface areas) to ensure adequate reacting capability and high sensitivity. Each chemical species has its own unique interaction at an electrode. But for almost all the electrochemical cells, the primary variables are current, working electrode potential, analyte concentration, and time.

The construction materials of the sensor will also influence its operating characteristics. Choosing construction materials and sensor geometry is critical and has a profound influence on the accuracy, precision, response time, sensitivity, background, noise, stability, lifetime, and selectivity of the resulting sensor.[2] For example, selection of a gold rather than a platinum electrocatalyst for the sensing electrode allows for selective determination of hydrogen sulfide in the presence of carbon monoxide.

The main purpose of a membrane is to keep the electrolyte in proper contact with the electrodes, thus eliminating the need for a flowing electrolyte solution across the electrode. The membrane may also control the amount of gas or vapor molecules reaching the electrodes, and protect the sensing electrode by filtering out particulate matter. However, if a membrane has a low permeability to air, the sensor will have a slower response time. Materials used for membrane construction are typically Teflon and high-density plastics like polypropylene, because such materials must be compatible with reactive gases and corrosive electrolytes.

It is extremely important to select the proper electrolyte for each sensor as the electrolyte composition can affect the solubility and the rate of diffusion of the reactant gas to the electrode (catalyst) surface. Electrolyte composition can also alter the chemical being monitored before it reaches the electrode surface. For example, the use of an acidic electrolyte to detect ammonia causes the formation of ammonium ion, which may not be as electrochemically active as ammonia under the conditions of the cell. The electrolyte profoundly influences the response characteristics observed for sensors with strongly acidic and basic electrolytes.

Another means of chemically controlling the sensor properties is by altering the composition of the electrocatalyst. Each catalyst formulation will have unique properties. The activity of platinum for carbon monoxide oxidation has been found to be 10^3 to 10^6 times better than that of gold. Although the reactions occur on both metals, one is orders of magnitude faster than the other.

The selectivity of the sensor can also be improved by controlling the electrochemical potential of the working electrode. For example, proper selection of a gold electrode potential will allow the determination of nitrogen dioxide in the presence of nitric oxide.[1] Response time, linearity, zero drift, repeatability, sensor stability, and even sensor life are dependent on sensor design and methodology.

As noted, gas sensors are often covered with a membrane that is selectively permeable to a given contaminant. This membrane minimizes the likelihood of poisoning of the electrodes by an electroactive or surface-active species. The resolution of these systems is also enhanced if other contaminants (interferents) that would also undergo electron transfer at the electrode can be excluded by tuning the circuitry. The appropriate choice of electrocatalyst can help in achieving selectivity in some applications. However, while the proper choice of membrane or electrolyte to achieve selectivity is helpful, it is not going to eliminate all possible interferents, since virtually any compound for which a similar type of detection method is used is a potential interferent.

Conductimetric Sensors

Conductimetric sensors are those in which the gas being measured reacts chemically with an electrolyte, changing its conductivity. Many older-design gas sensors of this type require frequent maintenance to replace reagents and parts, and in general this principle is rarely used.

Potentiometric Sensors

If a cell undergoes a change in the electrical potential of the sensor, it is classified as potentiometric. Potentiometric devices have been well studied and can serve as sensors for a variety of gases, such as ammonia, hydrogen cyanide, carbon dioxide, hydrogen sulfide, and sulfur dioxide. Potentiometric devices include ion-selective electrodes (of which the pH electrode is a special example), and at least one commercial toxic gas monitor. These operate on the basis of the Nernst equation, in which electrode potential is a function of concentration:

$$E = E^* - \frac{RT}{z}\log C$$

where E is the potential difference (in volts), E^* is the electromotive force for the cell (in volts), R is a constant, T is the temperature, C is the concentration (a product of the activities of each species that can be considered as effective concentrations), and z is the charge number for the cell reaction.

Potentiometric sensors for gases utilize the direct chemical reaction of the gas with the electrolyte, thereby changing the potential of the sensing electrode. A change in potential produces a change in current and thus a change in voltage. Thus, the potentiometric sensor observes the potential difference between the sensing electrode and the counter electrode that occurs when the chemical species of choice is detected. The magnitude of the change in potential is related to concentration. The output is logarithmically dependent on the concentration of the species being detected. When such a sensor is exposed to a relatively high level, it takes time for the electrolyte to recuperate. With a very high level the electrolyte may never fully recuperate.

Amperometric Sensors

As noted earlier, amperometric sensors can be subdivided into two classifications: *coulometric* and *voltammetric*. However, often the term *amperometric* also is used to describe a voltammetric instrument, while it is never used to describe a coulometric monitor. The indicating electrode is polarized against a reference electrode to a constant potential. The sample flows through the cell and the electric current proportional to the concentration of the species to be measured is recorded. A voltammetric sensor uses a constant voltage circuit, that is, the sensing electrode is held at a fixed voltage. Commercial coulometric monitors are constant current devices. The process that occurs in these sensors is electrooxidation or electroreduction of the species to be analyzed, with the result generating a current. Some require oxygen to function and others do not.

Amperometric sensors are important in portable instrument design because they are relatively small, inexpensive, and lightweight, and use very little power to generate significant signals. The parameters most frequently observed to influence the characteristics of amperometric sensors include sample flow rate, working electrode composition, electrolyte, membrane type, and electrochemical potential of the sensing electrode.[2] Accuracy is higher than other methods such as potentiometry. These sensors exhibit fast responses, can detect ppm levels of electrochemically active gases and vapors, can be engineered to have significant selectivity, and can be operated over a wide range of temperatures.

Coulometric Sensors. Coulometric detectors measure the quantity of electricity (in coulombs) that passes through a solution during the occurrence of an electrochemical reaction. During coulometric sensing, the gas being monitored is consumed by electrolysis during passage through the sensor. The current is controlled by the feed rate of the sample and corresponds to the charge passed in a given time unit:

$$Current = \frac{Charge}{Time}$$

It is generally a wet chemical titration method where one of the reactants is generated in the test cell by electrolysis of a solution. These detectors can be made specific by adjusting concentration, pH, and composition of the electrolyte. An important feature of the coulometric sensor is that the current is, within certain limits, independent of the change in the working electrode's activity; thus within this range the electrode characteristics are unimportant. Electrodes may be constructed from platinum foil, wire, or mesh. The accuracy of the sample flow rate controls the accuracy of measurement and must be kept as constant as possible; the signal is, however, independent of the temperature. Coulometric cells are reusable when periodically cleaned and recharged with a fresh electrolyte solution. This technique is used almost exclusively for ozone.

Voltammetric Sensors. A typical voltammetric sensor consists of six major parts: filter, membrane, working or sensing electrode, electrolyte, counter electrode, and reference electrode. The gaseous species of interest is transported (by pump or diffusion) across the membrane to the sensing electrode of the cell; it then migrates to the electrolyte boundary, dissolves in the electrolyte, diffuses to the electrode surface, and reacts electrochemically. The products of the reaction diffuse away from the electrocatalyst surface.[3]

The signal from the sensor is sent to a main current amplifier, and is read out as a voltage across the feedback resistor of the amplifier. Different monitor ranges use different feedbacks. Only three-electrode voltammetric sensors in which the counter electrode acts as an oxygen electrode require oxygen. In voltammetric sensors designed for gas monitoring, electrode characteristics are very important, including the size of a sensing electrode. Voltammetric sensors are constructed in such a way that very little depletion of the gas being monitored occurs.

As a way of understanding this method of detection, a typical voltammetric curve is shown in Figure 10.5. Oxidation and reduction current is represented on the $+y$ and $-y$ axis, respectively, while voltage is plotted along the x axis. Below the reaction potential of a given electrooxidizable gas there is no reaction, and therefore the current is zero (except for a small charging effect). As the reaction potential is approached and exceeded, the current rises sharply until it reaches a maximum limited only by the diffusion of the gas to the reaction site. The magnitude of this current is directly proportional to the number of electrons per mole, the Faraday constant, the diffusion coefficient, the surface area of the reaction site, and the concentration of the gas, and is inversely proportional to the diffusion gradient resulting from the varying concentration of reacting gas that ranges from the highest concentration that is found in the electrolyte to zero concentration at the electrode surface. There is a linear relationship between the current generated and the ppm of the gas.

If the voltage is increased beyond the limited diffusion current, which is represented by the a to b section of the curve in Figure 10.5, the current will increase again due to electrooxidation of the electrocatalyst comprising the electrode, and a further increase of voltage results in a very high current due to electrooxidation of water

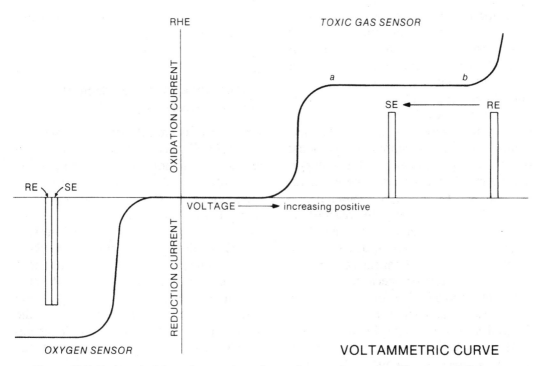

Figure 10.5. Basic principles of operation of a voltammetric sensor. (Courtesy of Interscan Corporation.)

to oxygen gas, and no detection takes place.

To have a working device, the sensing (working) electrode must be held at a potential within the region of the limited diffusion current. This is represented by SE in Figure 10.5. A reference electrode (RE) that has a lower or higher oxidation potential than that of SE is used to maintain the sensing electrode in the region of the limited diffusion current. The potential of the reference electrode must not change with the passage of current. Depending on whether the potential of RE is higher or lower than SE, it will have a voltage that is biased in either the negative or positive direction, respectively. The height of the diffusion plateau from the voltage axis is linear with the ppm of the gas. This is an example of an externally biased sensor.

It is also possible to have an internally biased sensor, an example being the oxygen sensor containing potassium hydroxide and

a lead anode. The lead–lead hydroxide reference electrode potential has a value coinciding with the limited diffusion current for oxygen reduction. It is therefore unnecessary to apply an external bias. The sensing electrode in this case is immediately polarized from its rest potential to assume the potential of the reference electrode. This is shown to the left of the y axis in Figure 10.5.

Typical features of voltammetric sensors are fast response, linear response in a broad concentration range, nonlinear dependence on the flow rate, and the temperature dependence of the signal.

Limitations of Electrochemical Sensors

Most manufacturers of electrochemical sensors specify the lower temperature limits for operation, usually 0°C to 10°C, and upper limits, typically 50°C to 60°C. These do not present a problem for most applications, but the user should be aware that lower temperatures tend to result in longer response times. When used outdoors in cold climates, ambient temperatures should be monitored to assure that the lower limits are not exceeded.

Electrochemical sensors designed to measure toxic gases may have cross sensitivities to other compounds. Response specificity is determined by the semipermeable membrane selected, the electrode material, and the retarding potential (the potential used to retard the reaction of gases other than that being sampled). Major interferences are those gases for which other electrochemical sensors are designed to monitor, because they are easily electrooxidizable or electroreducible. Gases that cannot be oxidized or reduced, such as methane, do not interfere.

Methods of eliminating interferences are to use selective chemical filters or to prescrub the sampled atmosphere. Both of these methods have been used by manufacturers for some applications. For example, activated carbon is a common filter material. However, not all interferences can be eliminated and the sampling practitioner needs to be familiar with those that may pose a problem.

Poor stability of a voltammetric sensor is primarily due to the effect of temperature on the inherent background current present in every sensor. This current is due to internal electrooxidation or electroreduction. This accounts for the zero drift found in such sensors. Most passive electrochemical sensors should not be exposed to the air when not in use as the water loss by evaporation through the porous membrane will shorten sensor life. Most commercial diffusion mode sensors have a limited volume to hold water.

Sensor longevity is a consideration. Since most chemical sensors depend on a specific chemical reaction to generate a signal, the reactions involved in these sensing mechanisms are often not suffi-ciently reversible, so that the sensitive material (electrolyte) is consumed during the life of the sensor.

Calibration and Maintenance of Electrochemical Units

For electrochemical sensors used to monitor chemicals, such as hydrogen sulfide, carbon monoxide, nitrogen dioxide, and sulfur dioxide, stable cylinders of calibration gases in the concentration range of interest are best to use for calibration. These sensors may also be zeroed in fresh air or zero air from a compressed gas cylinder or ambient air treated to remove contaminants. The frequency of calibration cannot be prescribed exactly, but a good rule is to calibrate at least once a day at the start of a shift. One manufacturer recommends the following calibration intervals:

- At least every six months
- Whenever the instrument has been exposed to a high concentration of gas
- When an instrument has been stored for more than one month without use
- When changing sensor
- If instrument display directs the user to perform a calibration

A special cup designed to fit over a sensor is available from at least one manufacturer. It allows a direct hookup of a gas bag to instruments that have a protruding sensor.

Calibration is best performed using a span gas with a concentration about equal to the mid-scale value of the meter. Never connect a gas cylinder directly to the intake port of an electrochemical instrument, as doing so will pressurize the instrument and cause serious damage to the sensor. Instead, fill a sample bag with the calibrating gas and attach it to the intake port of the unit.

For continuous monitoring applications in excess of 3 hours, some electrochemical devices require an external source of humidity. This setup often consists of a bottle with distilled water installed in-line with the intake port. If a filter is incorporated into the inlet of the instrument, it should be removed because it will trap the water vapor intended for the sensor. The air filter should be reconnected to the front of the humidity control bottle.

The accuracy of most of these sensors is dependent on the age of the sensor. Electrochemical sensor life typically ranges from about one to three years. Sensors should be replaced when they can no longer be calibrated or zeroed easily. Sensor life is highly variable and differs among instruments. Most sensors are given a rated "use life" in hours by the manufacturer. The actual life will depend on the type of sensor, frequency and duration of use, concentration levels measured, presence of compounds that also react with or contaminate the sensor, temperature, humidity, maintenance effort, and storage conditions for the instrument. For example, a sulfur dioxide sensor exposed to 5 ppm continuously may last 10,000 hours, or more than a year. If the concentration is lower or higher than 5 ppm, a longer or shorter life will result. It is important to remember that estimates of sensor life are for normal usage, and misuse can greatly shorten the life of the cell. Adequate warmup time is necessary following sensor replacement; from 5 to 24 hours may be necessary to allow the sensor to stabilize.[4] Some spare sensors can be stored in a special holder with the proper electrical bias maintained with a battery. Sensors stored in this manner need less time to stabilize when installed in the instrument.

Additional signs indicating a need to replace a cell include readings that become erratic or are strongly affected by moving the instrument. If a cell starts to leak, it should be immediately removed from the instrument and properly disposed of while wearing protective gloves.

For some units, proper readings can only be obtained when the battery has a sufficient level of charge. Removal and separate packaging of the sensor prior to shipment is often necessary, especially if it contains a wet electrolyte. This packaging is also needed prior to long-term storage. When shipping electrochemical instruments, remove the cell from the instrument, wrap it in a plastic bag, and pack it with the instrument. The instrument should be in an upright position when shipped. Handle electrochemical sensors carefully; they are sealed units that contain electrolytes that are often acidic or caustic. If electrolyte is leaking from a sensor, exercise caution so that the electrolyte does not contact skin, eyes, or clothing, thus avoiding burns. If contact occurs, rinse the area immediately with a large quantity of water. In case of contact with the eyes, immediately flush eyes with plenty of clean water for at least 15 minutes and contact a physician.

METAL OXIDE SENSORS

Metal oxide (MO) sensors are the most common solid-state device. MO sensors (Figure 10.6) are very small and consist of a metal oxide film made up of mixed metal oxides (iron, zinc, tin) coated on a heated ceramic substrate fused or wrapped around a platinum wire coil. A heater and collector are imbedded along with two electrodes into a metal oxide material in the sensor cell and a voltage is applied between the electrodes. Inside the cell two matched elements, one active and one inactive, are installed in a Wheatstone bridge circuit similar to the configuration discussed in the next chapter for hot wire combustion sensors. Like other semiconductor sensors, the signal or sensing mechanism of MO sensors is based on changes in the concentration of electrons that in turn alters the conductivity of the semiconductor.

Absorbed oxygen from the ambient air

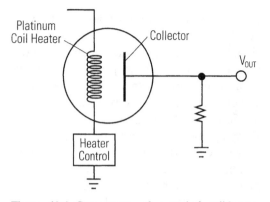

Figure 10.6. Components of a typical solid-state (metal oxide) sensor. (Courtesy of International Sensor Technology.)

establishes an equilibrium reaction with conduction layers of the metallic oxide, giving a base level of electron conduction through the sensor. The partial pressure of oxygen in air is nearly constant; heating the bead to a fixed temperature results in an amount of oxygen being absorbed in a stable manner. When a combustible or otherwise reducing gas such as carbon monoxide or hydrogen sulfide adsorbs on the sensor surface of the heated semiconductor bead, it reacts with the adsorbed oxygen layer and an electron charge transfer occurs between the surface of the semiconductor and the adsorbed molecule because of the difference in electron energy between them.

As gas molecules displace oxygen molecules, the surface of these sensors is altered, causing changes in its resistance. When the gas disappears, the sensor returns to its original resistance level. In order to promote the reactions of contaminants with the sensor surface, the temperature on the sensor surface (usually between 150°C and 300°C) must be carefully controlled by applying a specific voltage across the heating coil. The most widely used semiconductors in commercial gas sensors include SnO_2, ZnO_2, WO_3, and NiO.[5]

Selectivity is made possible through different mixtures of oxides and selected tem-

peratures of operations. Thus, the semiconductor coatings both enhance and accelerate oxidation processes as high as 20 times that of its conductivity in air.

By being very selective with the metal oxide coatings and the Wheatstone bridge temperature on MO sensors, the number of interferents can be minimized by reducing the response of the sensor to other gases and vapors. For example, in one instrument the same sensor can be used for detecting either carbon monoxide or methylene chloride. For methylene chloride detection, the temperature is increased to the point where carbon monoxide molecules burn off faster than the sensor can detect them.

Even when conditioned electronically to respond to a particular gas, the MO sensor will still respond to a variety of other gases and vapors, mainly hydrocarbons. For example, a sensor designed to measure carbon monoxide may also be sensitive to nitrogen oxide, sulfur dioxide, hydrogen sulfide, and unburned hydrocarbons. Activated carbon or another material is sometimes used as "filter" to remove interfering compounds. Of course, MO sensor response to gases other than those for which it is calibrated will not result in accurate readouts of concentrations of such gases. For example, a sensor calibrated to activate an alarm in a concentration of 20% methane may also alarm in the presence of a different concentration of gasoline.

After repeated exposures to gases, MO sensors can saturate. These sensors can also be attacked by corrosive gases, such as chlorine, bromine, and fluorine, and other compounds similar to those found in cigarette smoke.[6] Most sensors will not recognize the presence of inert gases, unless the concentrations are sufficient to displace the level of oxygen to below 10% in air. This potential for saturation and susceptibility of certain deleterious compounds is the reason many MO detectors have a purge switch to "burn" off contamination on the sensor. Eventually the sensor may saturate and the purge will not be effective and the sensor will need to be replaced. In general, if an instrument begins to read high in clean air, or if the unit is unused for several weeks or months, it must be purged by powering it up in a clean, well ventilated environment for 1 to 3 hours. During this time it may alarm as contaminants burn off the sensor. Solid-state sensors may last for many years in clean environments. However, with high levels of contaminants regular replacement may be needed, regardless of the concern about saturation, because the reactive material on the sensor's surface is consumed over time.

MO sensors cannot be operated in an absence of oxygen, such as can occur when monitoring confined spaces. A minimum of 10% relative humidity is generally required for accurate calibrations and field measurements; high humidity may adversely affect the sensors. Solid-state sensors also are affected by temperature in that their resistance increases in elevated temperatures, so these instruments are not desirable for making measurements on hot processes.

The major advantages of solid-state sensors are their simplicity in function, small size, and low cost. The major disadvantages are lack of selectivity as well as insufficient sensitivity for certain purposes. Due to the elevated sensor temperature, irreversible reactions with some gaseous impurities in the atmosphere may affect stability and cause baseline drift.[7]

Solid-state devices with external (passive) sensors should not be stored in a contaminated environment, because the sensor may be "fouled" due to its susceptible location outside the instrument's case. Examples are closets with oily rags, operations where airborne contaminants are generated, and chemical storage areas near sources of combustion such as gas stoves and engines. Oil mist and dust can coat the interior of these sensors.

As an example of these devices, Inter-

Figure 10.7. The IQ-1000 can detect over 100 different toxic and combustible gases using a single solid-state sensor, and it provides datalogging and data management capability to store months of data. (Courtesy of International Sensor Technology.)

national Sensor Technology offers two configurations of a four-sensor unit with sophisticated data-logging and data management capability allowing the storage of months of data. An RS-232 port allows the data to be sent to a computer or printer. There are three alarm set-points for each gas or vapor. One unit, the IQ-1000 (Figure 10.7), can be equipped with their Mega-Gas Sensor™ that can detect over 100 different toxic and combustible gases using a single solid-state sensor. The second unit, the TLV Panther, is equipped with a photoionization detector (PID) for rapid screening of volatile organic compounds. Both instruments accept electrochemical, catalytic bead, infrared (IR), or solid-state sensors in the second port and accept two electrochemical sensors in the remaining sensor ports. Table 10.2 lists the gases and vapors that can be measured with the available solid-state sensors. These devices represent sophisticated instruments that combine solid-state and other sensor technology to give the sampling practitioner multiple options. Features of these instruments that illustrate typical capabilities of high-end direct-reading instruments include[8]:

- *Alarms:* There are three separate levels for each sensor set by the user that can trigger audible or visual alarms. There is also the option to set

TABLE 10.2. Gases Measured Using Metal Oxide Sensors

Gas	Full-Scale Ranges
Acetic acid	100, 200 ppm
Acetone[a]	100, 200, 500, 1000, 5000 ppm; % LEL
Acetonitrile[a]	100 ppm
Acetylene[a]	50 ppm; % LEL; 3% by volume
Acrolein	50 ppm
Acrylic acid	100 ppm
Acrylonitrile[a]	50, 60, 80, 100, 200, 500 ppm; % LEL
Allyl alcohol	% LEL
Allyl chloride	200 ppm
Ammonia[a]	50, 70, 75, 100, 150, 200, 300, 400, 500, 1000, 2000, 2500, 4000, 5000 ppm; 1%, 2%, 10% by volume, 10%, 25%, 100% LEL
Anisole	100 ppm
Arsenic pentafluoride	5 ppm
Arsine	1, 10 ppm
Benzene[a]	50, 75, 100, 1000 ppm; % LEL
Biphenyl	50%, 100% LEL
Boron trichloride	500 ppm
Boron trifluoride	500 ppm
Bromine	20 ppm
Butadiene[a]	50, 100, 3000 ppm; % LEL
Butane[a]	400, 1000 ppm; 100%, 200% LEL
Butanol	1000 ppm, 100% LEL
Butene[a]	100% LEL
Butyl acetate	100 ppm; % LEL
Carbon disulfide	50, 60, 100 ppm; 5% by volume
Carbon monoxide[a]	50, 100, 150, 200, 250, 300, 500, 1000, 3000, 5000 ppm; 3%, 5% by volume, % LEL
Carbon tetrachloride	50, 100, 10,000 ppm
Cellosolve acetate	100 ppm
Chlorine[a]	10, 20, 50, 100, 200 ppm
Chlorine dioxide	10, 20 ppm
Chlorobutadiene	100% LEL
Chloroethanol	200 ppm
Chloroform	50, 100, 200 ppm
Chlorotrifluoroethylene	100% LEL
Cumene	100% LEL
Cyanogen chloride	20 ppm
Cyclohexane[a]	100 ppm, 100% LEL
Cyclopentane[a]	50 ppm
Deuterium	50%, 100% LEL
Diborane	10, 50 ppm
Dibromoethane	50 ppm
Dibutylamine	100% LEL
Dichlorobutene	1% by volume
Dichloroethane (EDC)[a]	50, 100 ppm, % LEL
Dichlorofluoroethane	100, 1000 ppm
Dichloropentadiene	50 ppm
Dichlorosilane	50, 100 ppm
Diesel fuel	50 ppm; 100% LEL
Diethyl benzene	100% LEL
Diethyl sulfide	10 ppm

TABLE 10.2. *Continued*

Gas	Full-Scale Ranges
Difluorochloroethane	100% LEL
Difluoroethane (152 A)	100% LEL
Dimethyl ether	100% LEL
Dimethylamine (DMA)	30, 50 ppm
Epichlorohydrin	50, 100, 500, 1000 ppm
Ethane[a]	1000 ppm
Ethanol[a]	200, 1000, 2000 ppm; % LEL
Ethyl acetate[a]	200, 1000 ppm; % LEL
Ethyl benzene[a]	200 ppm; % LEL
Ethyl chloride[a]	100 ppm; % LEL
Ethyl chlorocarbonate	1% by Volume
Ethyl ether	100, 800, 1000 ppm; % LEL
Ethylene[a]	100, 1000, 1200 ppm; % LEL
Ethylene oxide	5, 10, 20, 30, 50, 75, 100, 150, 200, 300, 1000, 1500, 2000, 3000 ppm; % LEL
Fluorine	20, 100 ppm
Formaldehyde	15, 50, 100, 500, 1000 ppm
Freon-11[a]	1000, 2000, 5000 ppm
Freon-12[a]	1000, 2000, 3000 ppm
Freon-22	100, 200, 500, 1000, 2000 ppm
Freon-113[a]	100, 200, 500, 1000, 2000 ppm; 1% by volume
Freon-114[a]	1000, 2000, 20,000 ppm
Freon-123[a]	1000 ppm
Fuel oil or kerosene	100% LEL
Gasoline	100, 1000, 2000, 20,000 ppm; % LEL
Germane	10, 50 ppm
Heptane[a]	1000 ppm, % LEL
Hexane[a]	50, 100, 200, 2000, 2500, 3000 ppm, % LEL
Hexene[a]	% LEL
Hydrazine	5, 10, 20, 100, 1000 ppm, 1% by volume
Hydrogen[a]	50, 100, 200, 500, 1000, 2000, 5000 ppm; 3%, 5% by volume, 2% to 100% LEL
Hydrogen bromide	50 ppm
Hydrogen chloride	50, 100, 200, 400, 500, 1000 ppm
Hydrogen cyanide[a]	20, 30, 50, 100, 200, 1000, 10,000 ppm
Hydrogen fluoride	20, 50, 100, 200 ppm
Hydrogen sulfide[a]	5, 10, 20, 30, 50, 100, 300, 1000 ppm; % LEL
Isobutane[a]	1000, 3000 ppm, % LEL
Isobutylene[a]	% LEL
Isopentane[a]	1000 ppm
Isoprene	% LEL
Isopropanol[a]	200, 400, 500, 1000 ppm; % LEL
JP-4 (jet fuel)	1000 ppm; % LEL
JP-5 (jet fuel)	1000, 5000 ppm, % LEL
Methane[a]	100, 200, 1000, 1500, 2000, 5000 ppm; 1%, 2% by volume, 100%, 200% LEL
Methanol	200, 300, 400, 500, 1000, 2000, 5000 ppm; 15%, 30%, 100% LEL
Methyl acetate	30 ppm
Methyl acrylate	60 ppm
Methyl bromide	20, 50, 60, 100, 500, 1000, 10,000; 40,000 ppm
Methyl butanol	% LEL

TABLE 10.2. *Continued*

Gas	Full-Scale Ranges
Methyl cellosolve	% LEL
Methyl chloride[a]	100, 200, 300, 2000, 10,000 ppm; % LEL
Methyl ethyl ketone	100, 500, 1000, 4000 ppm; 100% LEL
Methyl hydrazine	5 ppm
Methyl isobutyl ketone	200, 500, 2000 ppm; 50%, 100% LEL
Methyl mercaptan	30 ppm
Methyl methacrylate	100 ppm; % LEL
Methyl *tert*-butyl ether	100% LEL
Methylene chloride[a]	20, 100, 200, 300, 400, 500, 600, 1000, 2000, 3000, 5000 ppm; % LEL
Mineral spirits	200, 3000 ppm; % LEL
Monochlorobenzene	100% LEL
Monoethylamine	30, 100, 1000 ppm
Morpholine	500 ppm
Naptha	1000 ppm, 100% LEL
Natural gas	1000, 2000 ppm; 2%, 4% by volume, % LEL
Nitric oxide[a]	20, 50 ppm
Nitrogen dioxide[a]	20, 50, 100 ppm
Nitrogen trifluoride	50, 500, 1000 ppm
Nonane	2000 ppm
Pentane[a]	200, 1000 ppm, % LEL
Perchloroethylene	200, 1000, 2000, 20,000 ppm
Phenol	100 ppm
Phosgene	50 ppm
Phosphine[a]	3, 5, 10, 20, 30, 50 ppm
Phosphorus oxychloride	200 ppm
Picoline	% LEL
Propane[a]	100, 1000 ppm; 100% LEL
Propylene[a]	100, 200, 1000, 5000 ppm; % LEL
Propylene oxide	100 ppm; % LEL
Silane	10, 20, 50 ppm
Silicon tetrachloride	1000 ppm
Silicon tetrafluoride	1000 ppm
Styrene[a]	200, 300 ppm; % LEL
Sulfur dioxide[a]	50, 100 ppm
Tetrahydrofuran	200, 300, 1000 ppm; % LEL
Tetraline	100 ppm
Toluene[a]	50, 100, 200, 500, 2000, 5000 ppm; % LEL
Toluene diisocyanate	15 ppm
Trichloroethane[a]	50, 100, 500, 1000 ppm; 1% by volume
Trichloroethylene[a]	50, 100, 200, 300, 500, 1000, 2000 ppm; % LEL
Triethylamine (TEA)	100 ppm
Trifluoroethanol	25, 100 ppm
Trimethylamine (TMA)	50 ppm
Tungsten hexafluoride	50 ppm
Turpentine	% LEL
Vinyl acetate	1000 ppm; % LEL
Vinyl chloride[a]	20, 50, 100, 200, 400, 500, 1000, 4000, 10,000 ppm; 10%, 100% LEL
Vinylidene chloride[a]	50 ppm
Xylene[a]	100, 200, 300, 1000 ppm, 1% by volume

[a] Indicates calibration gas-in-air mixture commercially available.

Sources: Sensor information is taken from International Sensor Technology,[8] and calibration gas information is taken from Scott Specialty Gases.

a level that the concentration must drop below before an alarm will shut off even after the original alarm condition has disappeared. There is also an optional "confidence beep," which is a periodic beep that verifies to the user that the instrument is functioning properly.

- *Calibration:* All gases have preprogrammed default calibration values for zero (clean air) and span (known concentration gas) calibrations. For improved accuracy the instrument should be zeroed in clean air and then calibrated with a known concentration of each gas of interest. For best results the instrument should be powered on for at least 30 minutes before being zeroed, longer if the instrument has been off for an extended period. If the instrument is to be used indoors, it should be zeroed in a nonsmoking room which is considered to contain clean air, or else in clean air outdoors in the shade. For calibration a sample bag must be used; the instrument can be damaged if connected directly to a high pressure gas cylinder. The instrument has a built-in temperature compensation circuit that offsets the effects of temperature; before calibration the instrument should be operated for at least 10 minutes at the ambient temperature in which it will be calibrated. The unit stores the date and type of calibration for each gas, and it displays this on the LCD readout:

○ Precisely calibrated (P)—the sensor has been calibrated individually using the specific gas of interest.

○ Needs calibration (N)—appears in the display when the number of days between calibrations that was programmed for that gas and range has expired. The calibration interval is user settable.

○ Uncalibrated (U)—The sensor has not been precisely calibrated for the gas and concentration range, and instead it is operating from the default calibration values.

○ Low sensitivity (L)—indicates that the sensor had a low sensitivity for the gas when last calibrated. This could be due to using either (a) a calibration gas concentration lower than the calibration value that was input into the instrument or (b) a sensor approaching the end of its service life.

○ High sensitivity (H)—indicates that the sensor had a high sensitivity for the gas when last calibrated. This could be due to using a calibration gas concentration higher than the calibration value that was input into the instrument.

- *Warm-up Time:* Both solid-state and electrochemical sensors require warm-up time in order to reach a stable operating temperature and function properly. When first powering the unit on, solid-state sensors should be allowed to stabilize for 20 minutes or more before use. Some electrochemical sensors such as hydrogen chloride (HCl) will require two hours of warm-up time if they have been powered off for any significant length of time.

- *Recovery Time:* During calibration or exposure to gas it may take the sensor five minutes or more to completely return to zero. Typically sensors will recover fairly rapidly to about 10 or 15 percent of full scale, and have a lingering effect thereafter as the remaining gas molecules slowly dissipate.

- *Sampling System:* The instrument is equipped with a sampling pump, which must be on whenever the unit is calibrated or taking readings. It is important to avoid drawing in a sample from a hot, steaming source. Prolonged exposure to these sources can damage the sensors. Additionally, care must be taken to ensure that no liquid of any type is drawn into the sampling system. Also avoid exposing electrochemical sensors to high concentrations of gases for any longer than necessary in order to prolong the service life of the sensor.

- *Multi-gas Sensors:* When used with the Mega-Gas solid-state sensor, the instrument stores the setup and calibration data for each gas in its memory. These data include the gas response curve, the sensor operating temperature, and the zero and span values. When a gas is selected from the device's menu, the IQ-1000 automatically uses these data to configure itself properly for the gas selected, allowing the unit to provide an accurate reading of the gas concentration. To change gases at any time, the new gas is selected from the menu. However, it is important to note that a multi-gas MO sensor is nonspecific, meaning that it will respond to other gases besides the one that the instrument is set to measure.

In addition to looking for a single gas, a "gas search" feature allows the use of the Mega-Gas Sensor to scan an area for hazardous gases. Each of the gases is placed into one of three gas groups. Grouping the gases permits a full scan to be performed in a short time interval. The instrument searches each group and provides a relative reading from 0 to 100. A zero reading in all gas groups indicates that none of the gases are present. The scan data will not reflect the presence or exact concentration of any specific gas; only an indication of whether one or more of the gases in that group is present. If the scan detects that a gas is present, more detailed follow-up monitoring is needed to identify the gas and measure the concentration.

There is also a multi-gas electrochemical sensor available that is capable of detecting arsine, carbon monoxide, carbonyl sulfide, diborane, ethylene, hydrogen, hydrogen bromide, hydrogen chloride, hydrogen cyanide, hydrogen sulfide, methyl mercaptan, nitric oxide, phosphine, silane, sulfur dioxide, and vinyl chloride.

OTHER DETECTION PRINCIPLES

There are several detection principles used in chemical-specific instruments in addition to electrochemical and solid-state (metal oxide) sensors:

- *Infrared (IR) Detectors.* These are used for compounds that absorb IR radiation. Carbon dioxide (CO_2), described later in this chapter, is the most common application of IR in a single-chemical device. See Chapter 12 for a more extensive discussion of IR detection principles.
- *Ultraviolet (UV) Detectors.* These are used for chemicals that absorb a specific wavelength of UV light. Instruments for ozone and mercury illustrate typical applications. Monitors that measure ozone convert it to oxygen, which in turn absorbs UV. Since mercury absorbs strongly in the UV spectrum, no chemical conversion is necessary. Any compound that absorbs UV light of the same wavelength as the vapor or gas of interest is a potential interference. Filters are used to avoid interference from these compounds. UV instruments are discussed further in the section on mercury. Also see Chapter 12 for a more extensive discussion of UV detection principles.
- *Chemiluminescence.* A reaction between gas or vapor of interest and another chemical causes a reaction product that is "excited" and that emits a light of characteristic wave-

length that can be measured. The amount of light is proportional to the concentration. This technique is used to measure ozone and nitrogen dioxides.

In a typical instrument a known volume of ambient air is introduced into a sample chamber where it reacts with a specific reagent that will then cause radiant energy to be released. The light that is emitted may be detected by an instrument calibrated for the specific contaminant. The use of narrow-band optical filters enhances the selectivity of these instruments because the impact of interferences is decreased. One chemiluminescence analyzer for nitrogen dioxide (NO_2) uses ozone as the reagent. In this device the NO_2 must be converted to nitric oxide (NO), which subsequently reacts with the ozone to produce infrared radiation at a wavelength in the 0.5- to 3-μm range. This type of instrument is more complex, and therefore usually more expensive, than other chemical-specific instruments for the same contaminant.

- *Gold Film.* Mercury vapor is absorbed on a gold film; the change in electrical resistance over the sampling interval is proportional to airborne mercury concentration. The most common application of this principle is mercury monitoring, although it has also been applied to hydrogen sulfide.

- *Surface Acoustic Wave Detection.* This is a relatively new type of sensor receiving attention because of application to detecting chemical warfare agents at very low levels. This device uses chemically coated piezoelectric crystals that absorb target gases. The absorption causes a change in the resonant frequency of the crystal that is proportional to the concentration and can be measured by a microcomputer. By creating an "array" of separate crystals (each coated with a different absorbing chemical) within a single sensor, this type of instrument is able to identify and measure many chemical warfare agents simultaneously. The surface acoustic wave detector can also be used as the detector in a gas chromatograph, thereby using the GC column for added separation of airborne chemicals to aid identification and concentration measurement. These devices are described in more detail in Chapter 4.

- *Colorimetric Electronic Instruments.* These rely on a color change between the gas or vapor of interest and a chemical-impregnated paper tape or liquid solution. The color change is proportional to the concentration. Typically, the degree of color change is measured at a specific wavelength in the visible light region using an optical spectrophotometer that is balanced using a "blank" of the colorimetric agent. These are covered in Chapter 13 along with other colorimetric detections systems.

SPECIFIC CHEMICALS

Oxygen

Oxygen measurements are commonly made in conjunction with combustible gas measurements usually for confined space entry. For a discussion of combustible gas measurement techniques, see the next chapter.

Oxygen monitors generally use galvanic electrochemical cells. This cell could be described as a small battery with two electrodes, a noble metal and a base metal, in an electrolyte, covered by a permeable Teflon membrane. Typically, the noble electrode is gold or platinum, and the base metal is lead or zinc. The reactions within the cell lead to oxidation of the base metal, and generate a corresponding current.

The galvanic cell generates a current at the two-electrode terminals of the cell directly proportional to the rate of oxygen diffusion into the cell. At the end of the useful life of the galvanic cell, its output current drops dramatically, which is one way of determining the need for cell replacement. No refurbishing of the sealed fuel cell is possible. Therefore, oxygen cells are inherently self-consuming and must be replaced periodically. The life of the cell is shortened by exposure to heat or dry gases. Operating hours do not influence cell life, since the cell is constantly exposed to ambient air and constantly generating its output current, regardless of whether the instrument switch is on or off.

The role of water vapor as an air diluent often goes unrecognized as a cause for error when testing warm spaces for oxygen content using a diffusion-type instrument. An oxygen indicator is usually set to read 21% on atmospheric air, which contains a certain amount of water depending on the temperature and relative humidity (RH). A closed vessel or highly confined space, however, can have close to 100% RH. If the temperature in this space is substantially above the ambient, the oxygen will be noticeably diluted; therefore, at 43°C, for example, the highest reading to expect on a tank at 100% RH is 19%.

Oxygen cells are temperature-dependent as well. This dependence is compensated for by design to some degree, but when first exposed to a change in temperature, the thin membrane may respond to the change before the thermistor compensator has time to do so, and the reading will temporarily overshoot. The sensors are generally temperature-compensated. The manufacturer's literature will specify the temperate compensation range for a specific instrument.

Electrochemical sensors for oxygen provide either a continuous or on-demand display of the present percent oxygen in the atmosphere, and the alarm circuitry is designed to activate when the concentration drops to 19.5% oxygen. Some models, designed to be used in hospitals or as area monitors, have both upper and lower alarm levels so that oxygen-enriched atmospheres may also be monitored. Many of these models have output suitable for a printer or computer, so that a permanent record of the oxygen in the atmosphere may be maintained. The output of these devices is in percent oxygen.

Strong oxidants such as fluorine, chlorine, bromide, and oxone may lead to erroneously high oxygen readings when these substances are present in significant concentrations. Sulfur dioxide and nitrogen oxides can also interfere. Acid mists or other corrosive atmospheres can poison oxygen probes, decreasing their sensitivity. Examples are mercaptans and hydrogen sulfide in high concentrations (>1%), which can occur in confined spaces. Acid gases, such as carbon dioxide, will shorten the service life of an oxygen sensor.

Calibration and Maintenance of Oxygen Monitors.

Due to the temperature dependency described earlier, it is important to avoid large temperature changes between the calibration and the usage conditions. Try to calibrate the instrument immediately before testing, and at nearly the same temperature as the testing conditions.

The most simple example of field calibration of an electrochemical sensor is the oxygen monitor. Calibration can be done by placing it in fresh outdoor air and adjusting the calibration potentiometer to make the readout at the specific oxygen percent indicated in the manufacturer's literature, usually 20.8% or 20.9%. To determine if it responds to oxygen deficiency, a person should hold his or her breath for a few seconds, then slowly exhale, directing the exhaled breath to the sensor. Air that has been in contact with the lungs for 5 seconds or so will be depleted of oxygen down to about 16%. If it is functioning properly, the meter will deflect downscale and the alarm circuit will be activated.[9] A need for more frequent and larger calibration adjustments indicates the need for sensor replacement. This calibration check is not precise, but it is a quick and reliable test.

Because oxygen monitors are dependent on the partial pressure of oxygen, readings may not be valid if the total pressure differs from one atmosphere. For example, the MSA Model 245 oxygen indicator, when calibrated at sea level, will indicate 20% in fresh air at 1000 feet, 19.3% at 2000 feet, 18.6% at 3000 feet, and so forth.

Routine instrument maintenance is important. Check the sensor cell diffusion barrier for dirt and moisture. Clean with a soft cloth or tissue if needed. Replace the sensor cell as needed. Store the cell in nitrogen when not in use to prolong its life. Some models have nondisposable, user-rechargeable sensors that require charging every 2 to 4 weeks. Check batteries and replace when needed.

Use of Oxygen Monitors. Always check the atmosphere for explosives prior to activating the oxygen indicator. In sample areas where the temperature is not constant (it changes by more than 17°C), or in sampling atmospheres that differ in temperature from that of calibration air (by more than 17°C), the fresh-air reading (calibration) should be rechecked every hour to obtain the greatest accuracy possible. The operating range for RH is 10% to 90%. Avoid touching the sensor with the hand or sharp objects, since the membrane is easily damaged.

Sampling conditions that lead to condensation of moisture on the sensor face will cause erroneously low oxygen readings. These conditions, such as taking a cool sensor into a warm, moist atmosphere, produce a film of water on the sensor face

that decreases the transport of oxygen from the atmosphere to the inside of the sensor. To minimize this problem, the sensor should be kept as warm or warmer than the sample area—before and in the intervals between sampling.[9]

Procedure
1. Start the monitor as the manufacturer recommends. Allow sufficient warm-up/equilibration time if the monitor has been in a different environment. If a remote sensor is located in a different area than the monitor's electronics, allow extra time to reach thermal equilibrium.

2. For ambient air monitoring, place the monitor in the atmosphere to be analyzed. Allow it to equilibrate. Record the initial percent oxygen readout. Allow the instrument to continue monitoring. Observe any downward trends in percent oxygen levels. If a concentration of 19.5% oxygen is approached, ventilate with fresh air if possible. If the level falls below this, evacuate the area.

3. For measurements in a hose or piping system, place the sensor in an air line adapter. Allow it to equilibrate. If the line is pressurized, bleed off a stream to the sensor at atmospheric pressure in order to not pressurize the sensor. Record the percent oxygen.

4. Periodically throughout the shift perform an on-site calibration at 20.9% oxygen by exposing the sensor to fresh air or to a compressed air cylinder. Replace the sensor cell when the instrument can no longer be adjusted to read 20.9% oxygen in fresh air. Take care not to pressurize the sensor by a direct hookup to a cylinder, as doing so will cause high readings and could do irreparable damage to the sensor cell.

5. Perform a downscale calibration by exposing the sensor to 16% to 19% oxygen (in nitrogen or a similar gas) in accordance with the manufacturer's instruments.

6. Check for alarm function by exposing the unit to 16% to 19% oxygen in nitrogen from a cylinder or by blowing self-exhaled breath directly onto the sensor. Following the manufacturer's instructions for this procedure, adjust the readout on the meter to agree with the calibration standard.

Interpretation of Oxygen Measurements. If the oxygen content is below 19.5%, according to the Occupational Safety and Health Administration (OSHA) definition, the air is oxygen-deficient and appropriate atmosphere-supplying respirators are required to work in the area. If the oxygen content is greater than 25%, an oxygen-rich situation exists and there is a potential for explosions. If toxic contaminants are present, a pressure demand self-contained breathing apparatus (SCBA) is required for less than 19.5%, and a cartridge respirator may be worn for greater than 19.5% as long as the concentrations do not exceed the cartridge respirator's capability to protect the user. A low oxygen reading in an enclosed atmosphere indicates that some other gas has displaced much of the air or some process has consumed much of the available oxygen.

Changes in barometric pressure due to altitude will have an effect on the meter reading. If the instrument is calibrated at sea level, it will indicate a lower percentage of oxygen by volume at higher altitudes. However, adequate oxygen to sustain life is dependent on partial pressure rather than percentage by volume, and a lower reading at a higher altitude is acceptable. It is best to compensate for this situation ahead of time.

Carbon Monoxide

Carbon monoxide (CO) most commonly occurs as a byproduct of combustion. The most common detection methods for CO are electrochemical cells, solid-state sensors, and IR methods. With the proper mixture of metal oxides and the appropriate operating temperature, solid-state sensors can be made relatively specific for CO and can have a significantly reduced response to other gases and vapors. IR-based instruments when dedicated to the measurement of CO are frequently stationary devices. For a discussion of these devices see Chapter 12.

The most common specific operating principle for detecting CO is electrochemical oxidation at a potential-controlled electrode. The current generated by this electrochemical reaction is then directly proportional to the CO concentration. The basic components of most instruments are a three-electrode electrochemical sensor, a sampling system (pump), a sample preconditioning unit (water bottle, filter), an electronic control circuit (potentiostat), and a measuring circuit. The potential of the sensing electrode is controlled by the potentiostat. The sensor consists of a sensing electrode, a counter electrode, a reference electrode, a housing containing sulfuric acid solution, and two face plates (Figure 10.8). The overall cell reaction is

$$2CO + O_2 \rightarrow 2CO_2$$

Common measurement ranges are 0–100 ppm and 0–500 ppm. Fast response and recovery are a property of these electrochemical sensors as well as minimal drift. Interference filters must be used in the intake port to remove oxides of nitrogen and other compounds. Other compounds known to significantly interfere with CO measurements by electrochemical detection include hydrogen, hydrogen sulfide, sulfur dioxide, ethane, methane, ammonia, and propane. The degree of

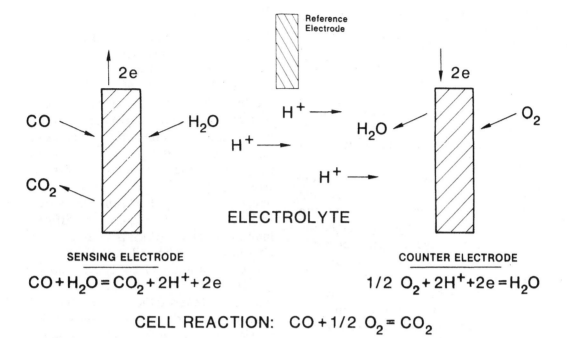

Reference Electrode

2e

CO   H$_2$O

CO$_2$

H$^+$ →

H$^+$ →   H$_2$O   O$_2$

H$^+$ →

2e

ELECTROLYTE

SENSING ELECTRODE

$CO + H_2O = CO_2 + 2H^+ + 2e$

COUNTER ELECTRODE

$1/2\ O_2 + 2H^+ + 2e = H_2O$

CELL REACTION: $CO + 1/2\ O_2 = CO_2$

Figure 10.8. Schematic of an electrochemical sensor for carbon monoxide. (Courtesy of ACGIH.)

interference depends on each manufacturer's instrument.

Hydrogen Sulfide

Hydrogen sulfide is commonly monitored using real time instruments rather than integrated measurement techniques due to its prevalence, especially in confined spaces, and its high acute toxicity. Hydrogen sulfide causes olfactory fatigue; that is, the odor-detecting cells in the nasal passages become tired after a period of exposure to hydrogen sulfide. As a result, the worker may think that the gas has gone away or may not be aware of an increase in concentration. Measurements are often made in conjunction with measurements for combustible gases and oxygen deficiency.

Instruments that measure hydrogen sulfide are based on several different principles, the most common being solid-state and wet electrochemical sensors. There is

also an instrument available based on gold film.

Most instruments designed to measure health hazard levels are capable of detecting hydrogen sulfide over a range of 0.1 ppm to 50 ppm. All hydrogen sulfide meters should have an audible alarm that can be preset at a desired level due to the high acute toxicity of this compound.

The primary types of monitors available for hydrogen sulfide are small personal units capable of continuous monitoring usually set to alarm at 10 ppm and combination units where a hydrogen sulfide sensor is included along with a combustible gas and oxygen sensor. Personal monitors are generally based on wet electrochemical sensors.

A similar electrochemical cell to that used to detect carbon monoxide but of a different material and different voltage is used for hydrogen sulfide. A common hydrogen sulfide electrochemical cell utilizes a reaction in which hydrogen sulfide is electrooxidized to sulfuric acid in an aqueous electrolyte at a catalytically active electrode according to the following equation:

$$H_2S + 4H_2O \rightarrow H_2SO_4 + 8H^+ + 8e^-$$

The potential at which the electrode is maintained is such that neither the electrooxidation of water nor the electroreduction of oxygen occurs at a measurable rate. Therefore, the current measured in the detector is a result of the electrooxidation of hydrogen sulfide, and is proportional to the partial pressure of hydrogen sulfide in the gas sample. The current generated in the sensor is amplified and displayed in units of ppm.

Hydrogen sulfide is a very reactive gas and tends to absorb in or react with many materials; therefore, the best material for sampling hoses is Teflon. The most common interferents for hydrogen sulfide electrochemical sensors are mercaptans, although some cells are sensitive to low levels of sulfur dioxide, propane, nitric oxide, ethylene, acetylene, or ethanol.

There are three calibration methods available for hydrogen sulfide: permeation tubes, cylinders of span gas, and ampoules. For more information on the use of permeation tubes, see Appendix B. The ampoule test kit allows the sampler to perform an operational test check prior to use. It consists of a plastic bottle and H_2S test ampoule. The ampoule is broken by vigorous shaking of the bottle and the sensing element is inserted. For this kit, an instrument is considered operational if a reading of greater than 10 ppm is obtained.

It has been recommended that a daily (or prior to each use) calibration check be performed on instruments that measure hydrogen sulfide due to the highly toxic nature of this gas. Instruments should not be zeroed in the field wherever there is the possibility that hydrogen sulfide could be present, due to the need for accurate measurements.[10]

Sulfur Dioxide

Sulfur dioxide can be detected using electrochemical (conductimetric, coulometric, voltammetric), solid-state, and infrared methods. An example of a voltammetric cell for sulfur dixode functions as follows: A current is generated by charge transfer of sulfur dioxide at the surface of an externally biased electrode. This current is proportional to the concentration of sulfur dioxide in the region of the electrode. The linear relationship between cell current and sulfur dioxide requires a diffusion limiting condition, which is a charge transfer whose rate is limited by the concentration gradient in the electrolyte enveloping the electrode (the arrival rate of sulfur dioxide molecules at the electrode surface).

Conductimetric sensors detect sulfur dioxide by trapping the sulfur dioxide in a dilute solution of hydrogen peroxide. The air sample is bubbled through the solution in the electrolytic cell, and sulfur dioxide in the air reacts with the peroxide solution to form sulfuric acid. The conductivity change of the cell is proportional to the sulfuric acid concentration. Thus, the sulfur dioxide in the sampled air is determined by measuring the conductivity before and after sampling and the total sample volume. Conductivity instruments often require the meter response (in conductivity units) to be converted to concentration using a calculation involving initial and final conductivity, sampling time, and a calibration constant, or on some units, a given percent of the scale is proportional to ppm. Sulfur dioxide calibration can be done using permeation tubes or a mixture in gas cylinders. Ozone is a common interferent in the electrochemical and conductimetric techniques.

Nitrogen Oxides

Real time measurement of nitrogen oxides (NO_x) can be done using chemiluminescence, electrochemical, solid-state or tape-based detection methods. Chemiluminescent analyzers can monitor for either nitric oxide or nitrogen dioxide. To measure nitric oxide concentrations, the gas sample being analyzed is blended with ozone in a flow reactor. The resulting light emissions are monitored by a photomultiplier tube. To measure total oxides of nitrogen, the gas sample is diverted through an NO_2 to NO converter before being admitted to the flow reactor. To measure nitrogen dioxide, the gas sample is intermittently diverted through the converter, and the NO signal is subtracted from the NO_x signal. Some instruments utilize a dual-stream principle with two reaction chambers.

$$NO + O_3 \rightarrow NO_2^* + O_2$$

$$NO_2^* \rightarrow NO_2 + h\nu$$

The chemiluminescent detection of NO_x with ozone is not subject to interference from any of the common air pollutants, such as ozone, nitrogen dioxide, carbon monoxide, ammonia, and sulfur dioxide. Possible interference from hydrocarbons is eliminated by means of a red sharp-cut optical filter.

When these instruments are operated in the nitrogen dioxide or NO_x modes, any compounds converted to nitrogen oxide in the thermal converter are potential interferences. The principal compound of concern is ammonia; however, it is not an interferent for converters that are operated at less than 300°C. Other nitrogen compounds, such as peroxyacetyl nitrate (PAN) and organic nitrates, decompose thermally in the converter to NO and may represent interferences in some polluted atmospheres or in smog chambers. Chemiluminescent analyzers often have very low detection limits of 0.002 ppm to 0.01 ppm, with maximum ranges of 0–5 ppm to 0–10,000 ppm and accuracies of ±1–2%.

Most electrochemical monitors are usually designed to detect nitrogen dioxide only and work similarly to those discussed under carbon monoxide. Sulfur dioxide and hydrogen sulfide have been known to interfere with measurements of nitrogen dioxide by these instruments.

Ozone

Due to the difficulties in the use of integrated sampling methods to measure ozone, the sampling practitioner usually relies on real time instruments to measure this compound. Ozone can be monitored using chemiluminescence, UV, and coulometric methods. Ozone can be generated by electric sources such as plate makers, X-ray machines or UV generators, arc welding equipment, electric arcs, mercury vapor lamps, linear accelerators, electrical discharges, and photocopy machines. It is also used in the organic chemical industry, in cold storage rooms as a disinfectant for food, in water purification, in textile and paper pulp bleaching, in aging liquor and wood, in the perfumery industry, in treating industrial wastes, in the rapid drying of varnishes and printing inks, and in the deodorizing of feathers. It is a common outdoor pollutant and occurs naturally in the higher atmosphere where it absorbs solar UV radiation.

In this regard it is often lumped into a group called total oxidants. In a situation where oxidation is suspected, such as when metals are rusting, measurement of total oxidants may be preferable to the use of an ozone-specific monitor. Other known oxidants commonly included in this term are chlorine, bromine, fluorine, iodine, nitrogen dioxide, nitric oxide, hydrogen peroxide, peracetic acid, peroxyacetyl nitrate (PAN), and chlorine dioxide.

In a typical chemiluminescent monitor for ozone the system provides a flow of ethylene concentric with a continuous air sample into a reaction chamber. The two streams merge and react, and the light emitted by the reaction is sensed by a photomultiplier tube. The output of the tube is amplified and displayed on a meter. Some of these monitors can detect ozone down to 0.001 ppm.

It is generally advisable to expose a new chemiluminescence analyzer or one that has been out of service to a relatively high concentration of ozone (1 ppm or greater) for a short period of time before use or calibration. This procedure will passivate the surfaces to ozone. The period takes from 5 to 30 minutes, depending on how long it has been since the monitor was used. Chemiluminescence monitors have cylinders of ethylene that must be kept fully charged. Portable instruments have their own internal supply of ethylene.

A typical coulometric oxidant meter is one that detects ozone through the oxidation reduction of potassium iodide (KI). The sampled air is brought into contact with a chemical sensing solution containing the proper amount of KI as it is metered into the sensor at the cathode of a unique cathode–anode electrode support structure by way of a solution supply tube. The solution flows in a fine film down the electrode support, upon which are wound many turns of a fine wire cathode and a single turn of a wire anode. An air sample is pumped through the sensor, where it comes into contact with the solution contained on the electrode support, and exits through a precision vacuum pump. This reaction takes place on the cathode portion of the electrode support and is as follows:

$$O_3 + KI + H_2O \rightarrow O_2 + I_2 + 2KOH$$

In this region, any ozone in the air sample reacts with the sensing solution. At the cathode, a thin layer of hydrogen gas is produced by a polarization current. When a voltage is applied to the electrodes, the hydrogen layer builds up to its maximum and the polarization current ceases to flow. As free iodine is produced by the reaction with ozone, it immediately reacts with the

hydrogen, reducing it to produce hydrogen iodide. The removal of the hydrogen from the cathode causes a repolarization current to flow in the external circuit, reestablishing equilibrium. Thus, for each ozone molecule reacting in the sensor, two electrons flow through the external circuit. The rate of electron flow, or current, is directly proportional to the mass per unit time of ozone entering the sensor. This reaction is based on ozone chemistry, but the halogens react similarly. Therefore, the detection of other oxidants such as chlorine, bromine, fluorine, and iodine is also possible.

In a coulometric sensor for oxidants, reducing compounds such as hydrogen sulfide and sulfur dioxide are negative interferents while oxidizing compounds are positive interferents. Most coulometric instruments require a source of current other than a battery and can also require long warm-up periods if the meter is turned off for several hours. Replenishment of the consumable sensing solution must be done in a laboratory.

Calibration of ozone monitors is done using an ozone generator, which is usually a mercury arc emitting 190-μm radiation.

Mercury

The primary use of mercury vapor monitors is to detect mercury spills. Many devices still contain mercury, such as thermometers, sphygmomanometers, barometers, and manometers. Many older switches also contain mercury. The problem with any mercury spill is that mercury easily vaporizes at room temperature, where it can be breathed or absorbed through the skin. Some of the most dangerous mercury leaks occur in ovens or when mercury is dropped on a floor and gets into a heating vent, because when the material is heated the concentration of airborne mercury increases as the spilled material volatilizes. Dental offices, laboratories, and hospitals may be prone to mercury spills.

During a spill, fugitive mercury droplets can roll everywhere. It is not uncommon in these situations for mercury to be tracked all over the room on shoes and the wheels of carts, or for mercury to absorb into porous surfaces such as concrete. In the haste that can follow a mercury spill, inappropriate cleanup procedures are often used, such as sweeping or using a vacuum cleaner that does not have a charcoal filter at its exhaust port. It takes very little mercury to create an unsafe environment. Quantities as low as 1 mL can evaporate over a period of time and contaminate millions of cubic feet of air to levels in excess of allowable limits.[11]

In addition to being used for spill detection, mercury monitors are also used for determining if contamination has occurred to shoes, cart wheels, and other mobile items. It is also possible to monitor smaller articles for contamination as described in Chapter 20.

Industrial operations for which mercury-specific instruments are useful include:

- Manufacture and repair of electrical meters, mercury arc rectifiers, and dry-cell batteries
- Mercury cells in chlor-alkali plant production of chlorine
- Manufacture of neon signs, mercury arc lamps, and electronic tubes
- Mining and refining of cinnabar and gold and silver ores

There are two main principles of detection used for mercury: UV light and gold film, although an instrument based on atomic absorption is also available.

Mercury absorbs heavily in the UV region of the spectrum, thus allowing this principle to be used to detect it. In a typical UV-based instrument, air to be sampled is drawn into an absorption chamber where a selective 253.7-μm UV light is absorbed by the sample. At the other end of the chamber, a photo-resistive element measures the intensity of radiation passing through the intervening space. The presence of mercury vapor will reduce the radiation energy reaching the photoresistive element in proportion to the vapor concentration. The change affects a photoresistive element, which is connected as one arm of a Wheatstone bridge, creating an unbalanced condition that is detected and displayed on the meter as the mercury vapor concentration in mg/M^3. With the low exposure levels currently of concern for mercury, UV-based instruments are generally used for spill and leak detection and not for verification that PELs are being met.

It should be emphasized that any organic compound that absorbs at the same UV wavelength that can be pulled along with the mercury vapor into the gas cell will be a positive interferent, including particulates such as cigarette smoke. Various hydrocarbons such as acetone, benzene, acetylene, and gasoline, and hexane, trichloroethylene, water vapor, and sulfur compounds also interfere. Exposure to chlorine and mercury together or sequentially may produce a negative bias in a UV-based instrument's readings.[12] Due to the potential for water vapor interference, it is best to zero the instrument at the same humidity as the atmosphere to be tested.

In the gold film technique for direct mercury detection a thin gold film that is part of a piezoelectric sensor undergoes an increase in electrical resistance proportional to the mass of mercury in the sample.

The Jerome 431-X (Figure 10.9) mercury vapor analyzer uses a patented gold film sensor in a portable hand-held unit. It measures mercury levels from 0.003 to $0.999 \, mg/m^3$ in seconds. The gold film sensor is inherently stable and selective to mercury with few, if any, interferences. When the sample cycle is activated, the internal pump draws a precise volume of air over the sensor. Mercury in the sample

Figure 10.9. Direct-reading mercury vapor instrument with dedicated data logger. (Courtesy of Arizona Instrument.)

is adsorbed and integrated by the sensor, registering it as a proportional change in electrical resistance. The internal microprocessor ensures a linear response throughout the entire range of the sensor. The instrument computes the concentration of mercury in milligrams per cubic meter or nanograms, and it displays the final result in the LCD readout.

Generation of specific mercury vapor concentrations requires a mercury vapor generator setup and considerable experience. Usually it is done by laboratories with specific expertise in this area, such as those operated by equipment manufacturers. Therefore, for most users, it will be best to send the unit back to the factory for calibration.

Prior to each use, the sampling professional should check an instrument's general response to mercury and the condition of the charcoal in the zero filter. The procedures below are followed or a kit can be purchased from the manufacturer for the response check.

1. Place one drop of mercury in a small glass container or thermos and cover while working inside a laboratory hood.

2. Turn on the instrument and allow it to warm up for 15 to 25 minutes.

3. Zero the instrument with the zero adjustment, using the filtered air setting if one exists. In this setting the air is pulled through a mercury absorbent often made of iodine-treated charcoal.

4. Set the knob on the sampling position, and lifting the cover insert the probe into the glass container so that mercury vapor is drawn into the intake port. Observe that the meter pointer deflects upscale and then drops toward zero as the mercury vapor concentration is reduced in the container over this sampling period.

5. Check the condition of the absorbent in the filter by using the "filtered air" position and exposing the probe to the mercury in the glass container. Note whether the meter rests at zero; if not, the absorbent is saturated with mercury and must be replaced.

Hydrogen Cyanide

Hydrogen cyanide can be found in electroplating plants and in reclamation areas of coke oven batteries. Since this compound is acutely toxic, the best type of monitoring is to use a small personal real time monitor mounted on a worker and running at all times during work. These instruments are generally electrochemical or solid state and are similar to those described for hydrogen sulfide.

Carbon Dioxide

Carbon dioxide is a byproduct of human respiration. It is often measured in indoor air to determine the effectiveness of ventilation systems. The most common instrument used for measuring this compound is the IR. Usually monitoring is done in several areas of a building, covering the full course of a day, with a constant recording of the results. If levels are seen to gradually increase, a lack of air change may be the problem.

SUMMARY

Direct-reading electronic instruments with one or more sensors that respond to *specific* chemicals are a valuable asset to the sampling practitioner. Generally, instruments with sensors that respond to a single chemical are less expensive and easier to use than instruments with detection systems that can be used to detect many different chemicals, such as gas chromatograph and infrared devices.

The instruments with sensors for specific chemicals are most usable in situations where either (a) a "pure" gas or vapor, or known mixture, is used under circumstances where no chemical reactions occur to produce unknown contaminants or (b) a known gas or vapor of relatively high toxicity is released or generated, and a real-time device is needed to monitor levels of this contaminant. They are not as commonly used where source of the hazard is a gas or vapor mixture with many different compounds *and* the composition can change over time since the sensor, which responds to a specific chemical, may not adequately "measure" the hazardous components under these changing conditions.

Typical sensor and instrument categories include electrochemical, metal oxide, infrared, ultraviolet, chemiluminescence, and gold film. The majority of available instruments are based on either electrochemical sensors or metal oxide sensors.

Innovations by manufacturers continually lead to (a) the introduction of new devices and (b) improvements in existing sensors that either make them usable for more compounds or overcome some of the prior limitations to the technology. Check with specific manufacturers to determine if their products overcome any of the typical limitations described in this chapter.

REFERENCES

1. Chou, J. *Hazardous Gas Monitors, A Practical Guide to Selection, Operation and Applications*. New York: McGraw-Hill, 2000.

2. Stetter, J. R. Electrochemical sensors, sensor arrays, and computer algorithms for the detection and identification of airborne chemicals. In *Detection of Airborne Chemicals*. Washington, D.C.: American Chemical Society, 1986.

3. Stetter, J. R. Instrumentation to monitor chemical exposure in the synfuel industry. *Am. Conf. Ind. Hyg. Ann.* **11**:225–248, 1984.

4. Shaw, M. More Straight Talk About Toxic Gas Monitors. Interscan Corp., P.O. Box 2496, Chatsworth, CA 91313-2496.

5. Stetter, J. *Instrumentation in Synfuel Industry*. Annals of the American Conference of Governmental Industrial Hygienists, Cincinnati, 1984.

6. American Industrial Hygiene Association. *Manual of Recommended Practice for Portable Direct-Reading Carbon Monoxide Indicators*. Akron, OH: AIHA, 1985.

7. Gentry, S. J. *Catalytic Devices, Chemical Sensors*, New York: Routledge Chapman and Hall, 1987, pp. 259–274.

8. IST IQ-1000 Operating Instructions. International Sensor Technology, Riverside, CA, 2003.

9. *Miniguard II Manual*. Mine Safety Appliances, Inc., Pittsburgh, PA.

10. Thompkins, F. C., and J. H. Becker. *An Evaluation of Portable, Direct-Reading Hydrogen Sulfide Meters*. Cincinnati, OH: NIOSH, July 1976.

11. Easton, D. N. *Management and Control of Hg Exposure*. American Laboratory, July 1988.

12. McCammon, C. S., and J. W. Woodfin. An evaluation of a passive monitor for mercury vapor. *AIHA J.* **38**:378–386, 1977.

CHAPTER 11

GENERAL SURVEY INSTRUMENTS FOR GASES AND VAPORS

This chapter describes general survey instruments, which are in the category of direct-reading instruments that respond to broad groups of airborne chemicals such as combustible gases or volatile organic compounds (VOCs). These instruments are different from the devices described in the previous chapter (that have one or more sensors that respond to *specific* chemicals) and the next chapter (that contain a sensing system that can be *tuned* to measure the concentration of many different specific chemicals). As noted in the previous chapter, there is some overlap among these three categories of instruments. For example, the solid-state sensor that can be set to scan for the presence of over 100 chemicals described in Chapter 10 can be a considered a general survey instrument when used in that mode. However, since its main use is as a sensor for specific chemicals, it was described in Chapter 10.

The general survey instruments find use where the hazard comes from a broad cat-egory of gases or vapors where analyzing for specific combustible gases may not adequately assess the hazard, or where the components of the airborne contaminant mixture are not known. They are also useful where there is no specific sensor for the gases and vapors that are known to be present since the general instrument can be used as a screening tool to determine if more selective sampling using sample collection device methods and subsequent laboratory analysis are needed. General survey instruments are capable of detecting a large number of contaminants but generally cannot distinguish between them. In addition, the instrument's response will be different for each chemical measured and so the results must be interpreted carefully. Typically, these instruments are used for area measurements, although occasionally a personal exposure measurement may be taken by putting the sampling probe in the breathing zone of an individual.

Air Monitoring for Toxic Exposures, Second Edition. By Henry J. McDermott
ISBN 0-471-45435-4   © 2004 John Wiley & Sons, Inc.

The two main types of general survey instruments covered in this chapter are:

- Combustible gas indicator (CGI), which indicates the presence of combustible gases or vapors as compared to the compound(s) used for calibration. The typical readout is in units of *percent of lower explosive limit* (LEL)—the LEL is the minimum concentration in air of the compound that will ignite when tested under standard conditions. The LEL is almost always much higher than an acceptable exposure standard (OSHA PEL, ACGIH TLV®, etc.) based on toxicity or health effects. Originally developed to detect methane in underground mines, CGIs are extensively used prior to confined space entry or before doing "hot work" such as welding. They are also used to detect flammable gas leaks or accumulations in basements, excavations, trenches, and other enclosed spaces.
- Flame ionization or photoionization instruments that can measure many common organic compounds and other gases and vapors at parts per million or parts per billion levels. Because of their sensitivity to low concentrations, they are used for screening measurement in the occupational setting for potential health hazards, especially at hazardous waste sites, and also for environmental monitoring. A common environmental application is to "sniff" around process equipment and piping systems for fugitive leaks.

Although the general survey devices are very useful to the sampling practitioner, an inherent limitation that *must* always be considered is their varying response to different chemical molecules. Each of these instruments is calibrated using a specific gas or mixture, and all measurements of airborne concentrations are relative to the instrument's response to the calibration gas molecule(s). For example, a CGI calibrated using methane would *underreport* the actual LEL by half if the actual airborne vapor was toluene. This means that toluene actually at 140% LEL (in the explosive range) would only indicate 70% LEL on the methane-calibrated CGI. Similarly, a photoionization device calibrated using benzene will indicate 30% or less of the actual concentrations of vapors such as hexane, and it may not indicate the presence of methane at all. General survey instruments have other limitations or operating considerations, which will be described in the section of the specific instruments. Considerations include the following:

- They may not detect the presence of some hazardous or extremely toxic compounds such as hydrogen sulfide.
- Most require an adequate concentration of oxygen.
- With some devices a "zero" reading may occur even though a high concentration of the gas is present. This is a characteristic of many CGI designs.
- General survey instruments may not be reliable indicators of exposure to chemicals with extremely low acceptable exposure standards since other, less toxic compounds may mask the response to the higher toxicity material.
- While CGIs are tested and rated as *intrinsically safe* for use in potentially flammable atmospheres, other types of general survey instruments may not be so rated.

In all cases the operating instructions for the specific instrument provide the best source of information. Often more than one type of direct-reading instrument is needed to adequately assess potential hazards in any environment.

MEASUREMENT OF EXPLOSIVE ATMOSPHERES: COMBUSTIBLE GAS INDICATORS

Combustible gas indicators (CGIs) are used for measuring explosive levels of gases, often in confined spaces, mines, and excavations. Other terms for the CGI are *explosimeter* and *heat of combustion* analyzer. CGIs were the first electronic direct-reading instruments to be developed for their application in underground mines. The first meter design, which is still available, operates by drawing air with a squeeze bulb into the sensing chamber which contains a resistive wire element. The resistance of the wire element changes as flammable vapors are oxidized; the change in resistance is shown on the meter as *percent of LEL* (Figure 11.1).

Basis for Combustible Gas Detection

There are many variations available in CGIs. Some instruments measure both LEL and explosive percentages while others are designed to measure LEL percentages and also lower ppm levels. Some instruments utilize selector switches that change the temperature of the sensor. In one position an instrument operates with the detector hot enough to burn methane along with any other combustibles present, and when the switch is changed to the petroleum vapor setting, the sensor temperature is reduced below the point at which methane is burned, allowing for detection of only those combustibles having lower combustion temperatures. Some instruments can detect hydrogen whereas others cannot, yet hydrogen is a

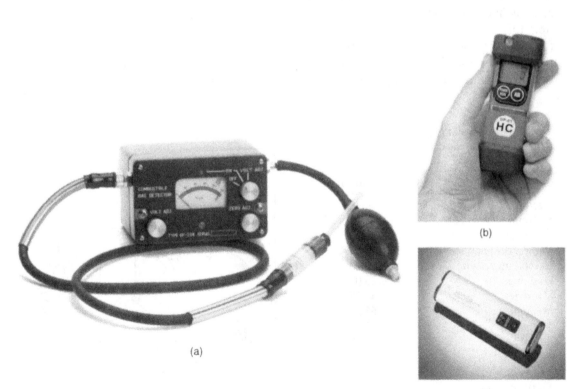

(a)

(b)

(c)

Figure 11.1. Combustible gas indicators [(a) and (b) Courtesy of RKI Instruments, Inc.; (c) courtesy of Industrial Scientific Corporation.]

very explosive gas. Therefore, it is important to be familiar with the principle of detection for any instrument along with the type of atmospheres it can measure as well at its limitations.

The detector can be inside of an instrument if it has a vacuum pump to pull air inside or outside if the instrument operates on the diffusion principle. The diffusion sensing elements are often capable of being taken off the instrument and attached to the end of a cable to be lowered into an enclosed space such as a utility vault or tank. This is an advantage over pump methods that must pull a sample to the sensor as the air to be sampled surrounds the sensor.

Heat of combustion refers to the heat released by the complete combustion of a unit mass of combustible material. It is a measure of the maximum amount of heat that can be released by a certain mass of a combustible chemical. Heats of combustion for various chemicals are available (Table 11.1).[1] Some sources give two different values: one for the heat of combustion for a material in its liquid state and

another for the gross heat of combustion. The heat of combustion for the liquid will always be less than the gross heat; however, in some cases the values are the same.

Catalytic Sensors. Virtually all heat of combustion instruments in use today are based on catalytic combustion. The use of a catalyst in conjunction with a basic Wheatstone bridge allows combustible gases to combine with oxygen (oxidize) at much lower temperatures than would be required for normal combustion. The sensing circuit consists of a pair of opposing platinum sensors, placed to form two legs of a Wheatstone bridge, with one element exposed to the gas being sampled in a combustion chamber and the reference element either sealed or else treated to prevent contact with the atmosphere (Figure 11.2).

In this design one filament is coated with a catalyst (usually platinum or palladium) that initiates combustion (oxidizes the gas mixture) on its surface, thereby increasing its temperature and consequently increasing its resistance. The catalytic filament is connected in series with a second, uncoated

TABLE 11.1. Heats of Combustion for Some Representative Compounds

Compound	Gross Heat of Combustion (mJ/kg)
Aceton	30.76
Benzene	41.83
n-Butane	49.5
Carbon disulfide	10.32
Carbon monoxide	10.10
Gasoline	43.0[a]
Hydrogen	141.79
Methane	55.50
Methanol	22.70
n-Octane	47.89
n-Pentane	48.64
Phosgene	1.72
Propane	50.35
Toluene	42.43

[a] Heat of combustion is for the liquid.

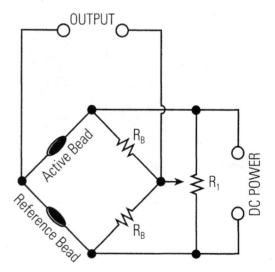

Figure 11.2. Diagram of combustible gas indicator *Wheatstone Bridge* circuit. (Courtesy of International Sensor Technology.)

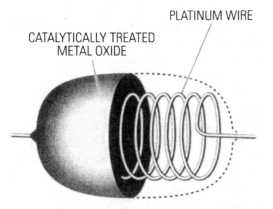

Figure 11.3. Platinum filament bead used in many combustible gas indicator instruments. (Courtesy of International Sensor Technology.)

Figure 11.4. Platinum bead sensor. (Courtesy of International Sensor Technology.)

reference filament that operates at the same voltage but does not cause oxidation and therefore no temperature increase occurs on it. This inactive compensator filament acts to offset any electrical changes caused by fluctuations in flow conditions, sample temperature, pressure, and/or humidity. The voltage applied to the two filaments in series is divided so that the greater voltage drop across the exposed filament unbalances the Wheatstone bridge. Coiling the platinum filament to form a "bead," a design many instruments have, protects it from contact with materials that could damage it (Figure 11.3). Older designs, utilize hot wire filaments in a Wheatstone bridge without the catalytic coating.

The catalytic bead requires oxygen levels usually above 15% to sustain combustion; therefore, in an oxygen-deficient space erroneous readings may result. Heating causes a slow deterioration of the catalyst with the result that the detector must be periodically replaced. Since catalytic bead sensors operate at lower temperatures, they need less power than other sensors, they have a longer, more stable sensor life, and they have reduced zero drift than simple hot wire sensors.

Some heat of combustion instruments are compensated and others are uncom-pensated. In an uncompensated instrument, the active and reference elements are similar or identical, but gas is exposed only to the active element while the reference rests in an isolated cavity. In compensated designs (Figure 11.4), both the active and the reference elements are exposed to the gas, but the reference element is noncatalytic. This design gives better stability under conditions of pressure, temperature, and background inert gas variations. For example, the uncompensated element, when exposed to 100% gas, responds to the cooling effect of that gas, which produces a substantial downscale reading. The compensated detector, on the other hand, experiences the same cooling effect on both elements, and they cancel out. Thus, the catalytic activity continues to the point where all of the oxygen is gone, and the reading stays at or above zero.

Thermal Conductivity Detectors.

Thermal conductivity is another detection method used for explosive concentrations that uses the specific heat of combustion of a gas or vapor as a measure of its concentration in air. It is used in instruments designed to measure very high concentrations, namely percent of gas as opposed to percent of the LEL of a gas, which most CGIs indicate on their readouts.

When an instrument has dual scales for both percent LEL and percent gas, both a catalytic combustion and thermal conductivity sensor are incorporated into the instrument. The thermal conductivity filament is substituted via a selector switch into the Wheatstone bridge. Combustibles in the sample then cool this filament, decreasing its resistance and unbalancing the bridge, which is just the opposite of the catalytic combustion techniques. This filament is not as susceptible to "poisoning" or oxygen levels as the catalytic filament. This method is nonspecific and not sensitive to low levels; consequently, it is used in CGIs that measure in total percent of combustibles. Because this instrument is capable of measuring extremely high concentrations, it must be used with extreme caution to ensure that the user is not exposed to an explosive environment.

Semiconductor Sensors. Semiconductor (metal oxide) sensors are also used for detecting combustible gas levels. These types of instruments are discussed in greater detail in Chapter 10.

Instrument Safety and Trends

In order to be used in a potentially combustible atmosphere, instruments must be qualified as intrinsically safe. They must pass the testing requirements of the Underwriters Laboratory (UL), Factory Mutual (FM) or other recognized testing/certification bodies. Following this testing, the codes for the type of atmospheres for which the instrument can be safely used according to the class division and group classification system of hazardous atmospheres of the National Fire Protection Association (NFPA) are incorporated into the label on the instrument. The more stringent the classification given to an instrument, such as Class 1 (usage for gases and vapors) and Division 1 (usage in areas of ignitable concentrations), the broader its use. Therefore, prior to selecting an instrument, the types of explosive atmospheres most likely to be encountered must be considered.

Many instruments combine catalytic combustion sensors to detect combustible gases with individual electrochemical sensors for oxygen, carbon monoxide, and hydrogen sulfide. Because OSHA requires that oxygen deficiency be measured at the same time as combustible gases when entering a confined space, the trend is for monitors to provide both of these measurements. For more information on oxygen, carbon monoxide, and hydrogen sulfide monitors, see Chapter 10.

Limitations

All CGIs, regardless of their detection principle, have limitations associated with their conditions of use. Before making measurements with any CGI, sampling practitioners should be aware of the limitations associated with the specific instrument they are planning to use. Some of these limitations are discussed here. Others may be described in the manual provided by the manufacturer.

CGIs should not be used for health hazard determinations or to measure the head space of soil or water samples, since there is a very large difference between percent as measured by CGIs and ppm levels generally of concern for health hazards and ppb (parts per billion) levels of concern in soil and water contamination. This difference becomes more apparent when it is realized that 1% is equivalent to 10,000 ppm.

$$1\% = \frac{1}{100}$$
$$= \frac{1,000}{100,000}$$
$$= \frac{10,000}{1,000,000}$$
$$= 10,000\,\text{ppm}$$

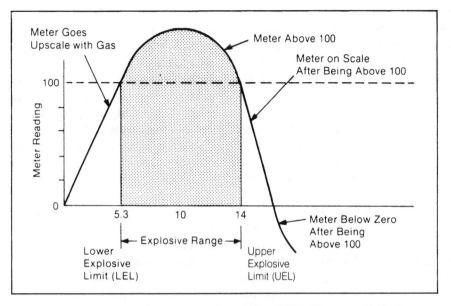

Figure 11.5. Sensor response range from LEL to UEL. (Courtesy of MSA.)

Using a similar calculation it can be shown that 1% is equivalent to 10,000,000 ppb.

Many instruments require oxygen to function properly. For example, catalytic bead systems require oxygen levels usually above 15% to sustain combustion. Oxygen-enriched atmospheres (22% or greater) will also change the response of an instrument that was calibrated in the presence of air with a normal concentration of oxygen. Some sources have indicated that flashback arrestors may not function in this type of environment. Since the level of oxygen can effect readings, only instruments that have both oxygen and combustible gas detection should be used whenever measurements are being done in confined spaces, and it is important that the oxygen measurement be done first. Similar precautions are needed in potentially oxygen-enriched environments.

Since oxidation of the combustible gas is used to generate the signal within the instrument, if the concentration of oxygen in the air being sampled is too low for complete oxidation of the combustible gas to occur, then the signal obtained will be low.

One situation that can cause oxygen deficiency is high concentrations of a combustible gas being present, generally above the UEL, displacing the available oxygen otherwise known as being too rich to burn. As an example, units based on the hot wire and catalytic combustion principle have been known to "peg" at maximum reading and then return to zero when used in an atmosphere that is greater than the upper explosive limit (UEL). Therefore, an instrument's display should be observed whenever entering an area where combustible vapors may be present starting at a location where it is known that there are no airborne vapors. Once the UEL is reached, the curve drops off almost immediately, the result being a return to zero on the meter in spite of the hazardous levels of gas that are present (Figure 11.5).

Thermal conductivity detectors are designed for high concentrations and do not have this limitation. However, thermal conductivity detectors are less sensitive than catalytic combustion detectors. The most caution must be taken when making measurements with dual scale units, since a failure to note that measurements were being made on a less sensitive scale (higher range) could lead to interpreting results as being lower than are correct. Care is most critical with units that can measure both percent LEL and percent gas.

Catalytic combustion instruments may be affected by high humidity. When testing heated spaces or atmospheres that contain high levels of alcohols, high humidity may be present. Therefore, the use of a moisture trap to keep water from getting inside the instrument may be necessary.[2]

Although CGIs can measure a wide variety of flammable gases and vapors, they cannot measure all of them. Since often they are used to measure unknown atmospheres this is a concern. Not all instruments can measure methane or hydrogen, which are both explosive gases. Catalytic combustion units must have sensor temperatures sufficient to exceed the ignition temperature of methane, which is relatively high. High levels of some inorganic compounds can also be explosive, for example, hydrogen sulfide has an LEL of 4.3%. Therefore, instruments used to measure in atmospheres that may contain inorganics should be evaluated for their ability to detect the specific inorganic compounds. CGIs designed to measure explosive gases should not be used to indicate the presence of explosive or combustible mists or sprays, such as lubricating (mineral) oils, or for measuring the percent of flammable vapors in steam or explosive (grain) dust levels.

Many compounds, such as silanes, silicones, and silicates, have been known to "poison" catalytic combustion instruments by coating the catalyst so it does not function properly, the result being a decreased response of the instrument. In atmospheres with high concentrations of halogenated hydrocarbons, the thermal decomposition products generated by these compounds may corrode the sensor, causing readings to be low. Corrosive gases will deteriorate sensing elements, causing circuitry resistance changes and instability. Hydrogen sulfide is a common corrosive gas. Others include phosphate esters, reduced sulfur compounds, high levels of organic acids such as acetic acid, acid gases, and nitro compounds, such as nitromethane, nitroethane, and nitropropane.[2] If leaded gasoline is suspected of being present, tetraethyl lead will reduce the ability of the filament to detect other compounds, especially those like methane, that have high ignition temperatures. Oxidation of tetraethyl lead can cause deposition of a solid lead combustion product on the filament.

Arsenic and other heavy metals, as well as dusty atmospheres, can interfere with sensors. This poisoning is usually only partial, with the first loss occurring with methane because of its high heat of combustion. Therefore, if the element still has a normal methane response, it will have a normal response to other gases. An alternative approach when testing in the presence of leaded gasoline vapors is to use an inhibitor filter. This filter promotes a chemical reaction with tetraethyl lead vapors that yields a more volatile combustion product, thereby preventing contamination of the filament. One way that instruments prevent "poisoning" is by having a high enough filament temperature to prevent condensation of products.

If the material being sampled is hot, then its flash point is a consideration. If propane, with a low flash point, is being sampled it will pass through a cool hose without changing from its gaseous state. A vapor having a high flash point may condense readily as it passes through the hose, and the instrument's reading will not be representative of what is in the sampled air. As an example, a tank containing isophorone (flash point of 184°F [84°C]) that is heated to 250°F (121°C) could have an explosive atmosphere inside; however, if the sample is withdrawn and cooled to room temperature, the maximum airborne reading would be around 5% LEL, with most of the sample having condensed in the hose. Heated materials with high flash points are best handled with a diffusion-type detector on a cable.

Atmospheres with pressures that differ significantly from normal atmospheric pressure may also affect the accuracy of the reading, for example, in tunneling work. Stray magnetic fields may be encountered in manholes and electric utility vaults. These magnetic fields can affect meter movement without adequate compensation. These situations are specialized ones for which outside assistance should be obtained before proceeding with work.

High air velocities tend to cool sensing elements and adversely affect their performance. The use of protective baffles or deflectors permits sensors to be positioned in fast airstreams as well as in locations where water streams may be encountered during cleanup operations. Additionally, wind currents may affect the readings of diffusion detectors.

Since CGIs are used to detect situations where an explosive atmosphere could be present, these measurements have life-and-death potential. Individuals should never be assigned to sampling duties without training in the use and limitations of the instrument as well as the correct way to interpret the results. Some companies and agencies have instituted certification programs that require individuals to attend a course and pass a test prior to being able to perform CGI measurements.

Calibration and Maintenance

There is no single preferred gas for calibration of CGIs; instead selection depends on the intended use of the instrument. A single CGI calibrated to one gas cannot necessarily be used for all situations where flammable and explosive compounds are present.

Since the heat of combustion varies from one compound to another, the CGI should be calibrated in terms of the compound for which survey information is required if known. Alternatively, if any of several gases and vapor may be present in the space being monitored, the sensor should be calibrated on the least sensitive of the gases to provide a wide margin of safety. In other words, the gas requiring the highest concentration to get a 20% LEL response should be selected. Thus, other gases will cause the same or a greater response at lower concentrations. See Table 1.5 for an illustration of this concept.

Manufacturers calibrate CGIs with a variety of gases, although the most common are methane, pentane, propane, and hexane. Each manufacturer selects the gas with which an instrument is calibrated according to preference. However, this will affect how the instrument responds to other gases. Some gases, such as pentane and hexane, make an instrument read "high" when measuring methane and certain other gases with higher heats of combustion, whereas the reverse is true when calibrating to methane or propane. The fact that a compound has a high heat of combustion means higher concentrations are required in air for a potentially combustible concentration to exist, whereas others such as toluene have both low heats of combustion and low LELs. Pentane is often selected as a calibration gas where petroleum vapors are of concern, since it has a relatively low LEL and its heat of combustion puts it right in between methane and most aromatic compounds.

A problem when calibrating at higher concentration levels is that the sampler is potentially dealing with a flammable atmosphere. This type of calibration can only be done in a laboratory that is properly equipped. A similar situation exists when trying to verify the functioning of a flashback arrestor. The purpose of the flashback arrestor is to act as a barrier to ensure that the flame in the sensor chamber during measurements of explosive range concentrations does not travel upstream through the sample tube to the source of explosive gas.[2] The instrument must be exposed to an atmosphere between the LEL and the UEL. Do not introduce 100% methane gas to an instrument located near a source of ignition; otherwise an explosion may occur. There are private laboratories available that perform calibrations, as well as flashback arrestor and other function tests.

For field use, instruments should be calibrated weekly, if used regularly, or prior to use if used infrequently. Calibration should be done under the same circumstances as field use. For example, if a long sampling line will be used, then the same length of line should be used during the calibration check. The response time should be noted. Check the flowrate of the instrument if it has a vacuum pump, and compare it to the manufacturer's specifications using a precision rotameter or bubble burette. Variations in the flowrate will result in an inaccurate calibration. For instructions in flow rate calibrations, see Chapter 5.

Some instruments have replaceable batteries and others require charging. Generally the only maintenance permitted is replacement of batteries, flashback arrestors, particulate filters, and the sensor unit or filaments in some meters. The two filaments in a Wheatstone bridge are usually matched, one filament being a reference and the other being the operational detector, and thus should be replaced as a pair to maintain proper operation. Note that if the sensor unit or filaments are replaced a recalibration is necessary.

Never interchange parts of different makes or models of CGIs. In general, it is better for repairs to be done at the factory so that the instrument can be calibrated and inspected. Inspect the instrument for any possible signs of damage after each use. If the instrument is not used regularly, removal of the batteries may be necessary, and some types of sensor cells and other parts may require special handling prior to storage.

Field Use

A quick field check to be sure an instrument is responding should also be performed prior to making measurements, but away from the atmosphere to be tested. One way is to expose the instrument to an unlit butane lighter. Press the top lever down, but do not spin the sparking device.

Response times can vary from 15 seconds to 2 minutes, depending on the instrument. It is critical that the user know the response time prior to taking an instrument into the field. The use of a remote monitoring line also increases this time. Never directly insert a probe into a liquid. If liquid is sucked into the inside of an instrument, it will require repair and maybe even replacement of the sensors. If sampling under dusty conditions, use a filter in front of the probe. Instruments with aspirator bulbs should be purged by squeezing the bulb 8 to 10 times in clean air. The bulb should inflate completely between squeezes. These instruments usually have a "ready" indicator. The "ready" indicator must be on prior to making measurements. To use an instrument with an aspirator bulb, squeeze the bulb 7 to 8 times to draw in the sample. When the needle stabilizes, the meter indicates the concentration of gas in air.

Probes permit samples to be taken in areas that cannot be reached with a sampling line. Utility vaults, sewers, and spaces behind obstructions or areas accessible only through narrow openings can be examined by connecting the probe to a sampling line. Probes can be made of steel, brass, or plastic (dielectric nonconducting). Do not use a metal probe where shock hazards exist. Instead the dielectric plastic probe should be used in these situations.

Some manufacturers provide dilution tubes (1:1, 10:1, 20:1) with their instruments to allow for measurements above the LEL. When using these, it is necessary to calculate the actual gas concentration. Generally, since scales on the meter are not provided for use with dilution tubes, it is necessary to calculate the actual gas concentration. An instrument should never be stored or left sitting around with the dilution tube installed if there is any possibility that an inexperienced user might pick it up.

When monitoring in an open area, such as a tank farm, or when monitoring for leaks, use slow sweeping motions to assure "hot pockets" are not bypassed. Some instruments are designed to be lowered into a tank or other confined space. In this case, if the instrument has a "peak hold" function it will store the highest concentration it encounters. Other methods for collecting a sample from inside a confined space include using an extension probe to reach farther into it or dropping a piece of extension tubing. In both cases, an instrument must have a pump so that it will pull the contaminated air up to the detector. Some passive instruments have detachable detectors that attach to the instrument via a long extension cable, allowing the sampling practitioner to observe the instrument's output while standing on the outside of the confined space. The most difficult situation to monitor is that in which the entry way is at the side of a large tank since adequate coverage of the space is difficult to achieve with the probe or remote sensor.

Most instruments are capable of measurements over a limited temperature range, so if a measurement is made outside of this temperature range, it may not be accurate. As a result, some instruments are temperature-compensated. Condensation of vapors can occur in sampling tubes if used at cold temperatures, resulting in a lower readout than the actual concentration present. As described earlier, sampling heated atmospheres can also be a problem, because if the probe, the sampling line, or the instrument is at a lower temperature than the gas being sampled, a portion of the sample may condense within the sampling line, thus reducing the concentration of combustibles reaching the sensor.

Some instruments have a safety "latch" that is activated by exposure to high concentrations of combustible gases. When the readout indicates greater than 100% of the LEL, the readout "latches." When latched, the LEL readout blanks. This feature is a warning that the gas concentration has exceeded the LEL and that all personnel must be evacuated from the area.

Procedure

1. Before starting and after finishing the measurements, the instrument settings should be as follows: All on/off switches should be in the off position. The scale select switch should be in the least sensitive position, meaning on the highest range (e.g., 1 to 10,000). All accessory tubing and cables, including the battery charger, should be properly connected or disconnected as appropriate.

2. Inspect the scale on the meter and determine whether it reads % LEL, % gas concentration, or both. Select the scale of interest.

3. Turn on the instrument and check that the battery power is adequate. Next pull air into it from a clean area. This procedure is called zeroing the instrument. If the instrument has standby and battery check settings, make sure the dial is not on one of these. In some instruments the circuit zero is automatically adjusted during the warm-up, but the span setting (meter reading during calibration) is adjusted on a potentiometer (variable resister) that may be located inside the instrument. If no clean air

TABLE 11.2. Understanding Combustible Gas Indicator Measurements

Meter Readout	% Methane	% LEL
20	20% Methane, 200,000 ppm; for most other gases concentration will read lower; this instrument is always calibrated to methane	20% of the LEL for the compound the instrument was calibrated with. If methane (LEL = 5%), 20% of 5% = 1% or 10,000 ppm. The concentration of other gases measured with this instrument will depend on the responses of the specific gas to the methane calibration.

is available, a plastic bag should be filled beforehand with uncontaminated air and it can be used to zero the instrument. A charcoal tube can be fitted to a piece of tubing at the instrument's inlet, provide that it utilizes a vacuum pump. Charcoal will remove larger hydrocarbons (pentane and up) molecules but not methane and its relatives or carbon monoxide. Compressed air from a breathing air source can also be used. It should be similar in humidity content to the air at the site.

4. Do a field check using an "unlighted" butane lighter to be sure the instrument is responding. Allow the instrument to return to zero in order to make sure all connections are tight if using a diffusion sensor on a remote sampling line.

5. Put the probe into the atmosphere to be tested. Attention should be kept on the meter readout at all times during the measurement. If for any reason the needle pegs and returns to zero, do not allow anyone to go into the space until the reason is understood and the cause is resolved.

INTERPRETATION OF MEASUREMENTS OF EXPLOSIVE ATMOSPHERES

CGIs cannot differentiate between individual compounds; thus, the readout for a mixture is unlikely to be specific for any of the individual components including the gas for which it was calibrated if more than one combustible gas is present. Most units are set to alarm at 20% to 25% of the LEL. Depending on the calibration and the compounds to which the instrument is exposed, this may or may not provide an adequate safety margin. It should be noted that any time an instrument detects a percent of the LEL, the potential exists for an immediately dangerous to life and health (IDLH) atmosphere to be present.

Some instruments read in percent, others in decimals. Digital instruments as well as those with a needle and scale for the readout have minimum and maximum detectable quantities. However, unlike a meter whose scale is readily apparent, it is necessary to read the manual to know the digital monitor's maximum detectable level. In some instruments, when the concentration reaches a certain point, for example, 100% of the LEL, the meter simply displays the term "over."

Since these instruments read in percentages, they cannot be used to determine health hazards expressed in ppm levels. (Table 11.2 explains CGI measurements.) For example, the LEL for methane is 5.3%, and therefore 25% LEL is 1.3%, or 13,000 ppm. This level is much greater than any threshold limit valve (TLV®) or permissible exposure limit (PEL). For OSHA compliance, a reading in excess of 25% of the LEL of the combustible gas meter indicates an explosive hazard.

When measuring contaminants in known atmospheres, the manufacturer's response curves can assist in determining the actual concentration. As shown in Figure 11.6, a typical catalytic combustion device calibrated with a mixture of 50% of the LEL for pentane (curve 3) will read 85% when exposed to a 50% mixture of

No.	Compound	Formula	LEL
1	Methane	CH_4	5.0
2	Acetylene	CH_1CH	2.5
3	Pentane	C_5H_{12}	1.5
4	Ethyl Chloride	C_2H_5C1	3.8
5	1,4-Dioxane	$OCH_2CH_2OCH_2CH_2$	2.0
6	Xylene	$C_4H_4(CH_2)_2$	1.1

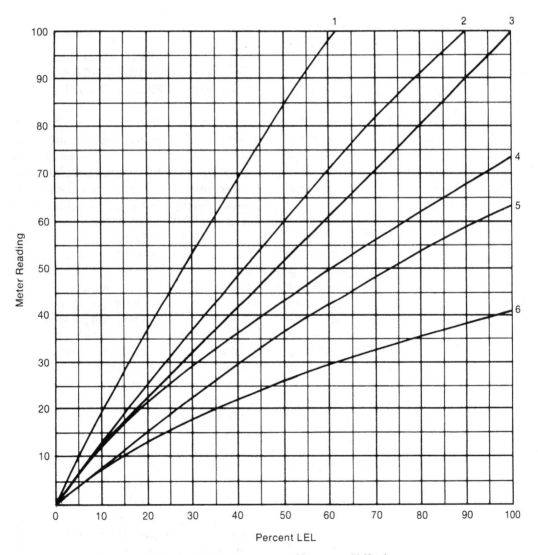

Figure 11.6. Response curves. (Courtesy of MSA.)

methane (curve 1). If the same meter were exposed to a 50% LEL of xylene (curve 6), it will read approximately 26%. Adding the fact that the typical error for these instruments is 10% due to temperature and humidity constraints, the reading for a 50% xylene LEL concentration on an instrument calibrated with pentane would be 23–29%, while the reading for a 50% LEL concentration of methane on the same instrument would be 76–94%. Correspondingly, if the instrument were calibrated to methane, the readout for the other compounds would appear less than the actual concentrations. For this reason, unless methane is actually likely to be present, it is preferable to calibrate with a compound that has a lower heat of combustion, such as pentane.

Some manufacturers provide conversion factors rather than response curves. In this case, a calculation must be performed. For example, if a meter calibrated on pentane is used to test for propane, the conversion factor for a particular unit might be 0.9. If the readout indicates 50% LEL, the true propane concentration is

$$(50\%)(0.9) = 45\% \text{ of the LEL}$$

When response curves or conversion factors are used, it is a good practice to assume an accuracy tolerance of ±25% in the interpretation of any meter response.

An oxygen deficiency generally indicates there is a high concentration of one or more other compounds that are displacing the oxygen normally present in air. These may or may not be combustible. One practice used in unloading tank trucks containing flammable and explosive compounds is to "inert" the tank by deliberately displacing the oxygen-containing air with an inert gas, such as nitrogen. Thus, the risk of an explosive atmosphere while moving the truck to a cleaning station is decreased. A similar practice is used when removing underground storage tanks; in this case, dry ice is used to create an atmosphere of carbon dioxide, which also displaces oxygen, thus allowing the tank to be removed while minimizing the risk of explosion.

If the meter response initially reads off scale and then goes downscale rapidly, it is a sign that the concentration of gas is very high, possibly exceeding the UEL, because the gas absorbs heat from the filament, increasing its resistance. If the gas concentration appears to be above the UEL, a potential explosion hazard still exists as the addition of air to the gas/air mixture will create a concentration in the explosive range.

MONITORING FOR HEALTH HAZARD LEVELS OF VOLATILE ORGANIC COMPOUNDS: FIDS AND PIDS

General survey instruments are used to monitor health hazard levels in situations where immediate results are needed even though the actual compounds present may be unknown or a mixture is present. The primary detectors used are photoionization detectors (PIDs) and flame ionization detectors (FIDs). These instruments can be used for general surveying on hazardous waste sites to determine compliance with EPA action levels and to establish site work zones. They are used for screening of air, soil, water, and drum bulk samples to establish priorities for laboratory analysis, usually for VOCs, due to the high cost of this type of analysis. They can be used to determine if decontamination procedures are effective. They are often used in a qualitative manner just to see if any volatiles are present, for example, when an odor is present in a basement. Leak detection is another common use. Another application involves boundary line sampling for total hydrocarbons to detect airborne releases from an industrial operation or hazardous waste site to the surrounding areas, especially in the case where residences, schools, or other sensitive populations are nearby. Perimeter monitoring is often done by placing fixed analyzers in north, south, east, and west locations along with a roving unit. Continuous monitoring utilizing a data logger or periodic checks at the stationary points by a sampling specialist are possible.

Response Factor Concept

Any user of a PID and FID general survey instrument must remember that the response of these instruments varies for different airborne chemical compounds. As a simple illustration, a PID calibrated with 100 ppm of isobutylene-in-air (the usual calibrating gas) will indicate (a) 120 ppm if acetone is the only airborne vapor when 100 ppm of acetone is actually present and (b) 55 ppm of benzene when 100 ppm of benzene is actually present. In situations where the sampling practitioner can be relatively certain that only a single contaminant is present, a PID or FID may either be (a) calibrated for that compound or (b) calibrated using another gas and a single response factor applied to readings in order to yield the concentration of the contaminant of interest. For other situations where very little is known about what contaminants are likely to be present, these instruments can be used for area measurements and leak detection—the readings are considered to be strictly qualitative information and a warning that something is present.

However, in many cases the airborne contaminant is a mixture where the approximate composition is known or can be determined. For example, the composition of airborne gasoline vapor has been summarized in various studies.[3] In other situations a liquid sample of the vapor source can be sent to a laboratory for "headspace" analysis—a composition analysis of the vapor space above the liquid. A composition analysis of the liquid

material *cannot* be used without further adjustment since the composition of the vapor is a function of the volatility (vapor pressure) of each component as well as its concentration in the liquid. Often it is also possible to collect an airborne sample using a charcoal tube or other sample collection device (SCD) method and to analyze for composition information in the laboratory. With knowledge of the composition, an approximate overall "calibration factor" of the PID or FID device can be calculated either from the individual response factors provided by the instrument's manufacturer or by actually developing response factors for the major components in the laboratory. It is also possible to develop a very rough calibration factor through "side-by-side" sampling using the SCD method and survey instrument if the airborne concentration remains about constant over the SCD sampling period. However, before relying on the PID or FID for field measurements, it is prudent to understand the relative response of the instrument for the different components in the airborne gas or vapor mixture. In this manner the sampling practitioner will have some degree of confidence that the instrument is adequately sensitive to the more toxic components in the contaminant mixture and that changes in the contaminant mixture will not cause an unrecognized hazardous exposure or environmental compliance issue.

A "response factor" is defined as the instrument observed reading for a specific chemical as a percent of the actual concentration of that chemical. For example, if the instrument measures the actual airborne concentration, the factor is 1.0; if it measures only half of the actual concentration, the response factor is 0.5. A response factor for a PID depends on the instruments sensitivity to a specific chemical, which is primarily a function of (a) the type of ionization lamp in the instrument and the *ionization potential* or *ionization energy* of the specific chemical of interest and (b) the chemical used to calibrate the instrument. For an FID instrument the relative response to specific chemicals primarily depends on the response of the instrument, which is a function of the number of carbon atoms in the chemical of interest compared to the number in the chemical used to calibrate the instrument.

A case study on determining the "response factor" is contained in the next section on Photoionization Detector instruments; the same concepts can be applied to FID instruments.

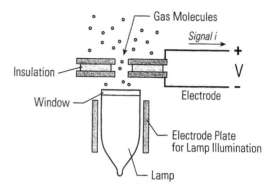

Figure 11.7. Diagram of a photoionization detector. (Courtesy of International Sensor Technology.)

Photoionization Detectors (PIDs)

Most organic chemicals and some inorganic compounds can be ionized when they are subjected to ultraviolet (UV) light. UV has shorter wavelengths and higher frequencies than visible light or infrared radiation, and thus has higher energies. UV radiation is expressed in units of electron volts (eV). The PID consists of UV lamp that ionizes the compounds in the air sample. The lamp is filled with a low-pressure inert gas that is excited or energized by a high-voltage electric current passed through the gas using electrodes inside the lamp, or external electrodes or an electromagnetic field (the latter two are "electrodeless" lamps). The wavelength of the UV light emitted by the lamp depends on the inert gas and the window material in the lamp. Generally, three different lamps are used: 9.5 eV, 10.6 eV, and 11.7 eV. Each lamp will ionize chemicals that have an *ionization energy (IE)* [also called *ionization potential (IP)*] lower than the lamp's eV output. The ionization energy or ionization potential is the energy necessary to remove an electron from the neutral atom. The PID instrument converts the concentration of ionizable chemicals in a sample to an electric signal; the ion current produced is proportional to mass and thus concentration (Figure 11.7).

Because of their flexibility and conve-nience of use, PIDs are the most popular instrument for measuring the presence of volatile organic compounds (VOCs) in industrial hygiene and environmental applications. Some PID devices also feature electrochemical and solid-state sensors as described in Chapter 10 for added flexibility. Typical applications for PID instruments include:

- Confined space entry
- Emergency response (chemical spills)
- Fugitive emissions (leak detection)
- Headspace VOC measurements in soil or water
- Arson investigations (trace accelerants)
- Hazardous waste (site entry, plus water and soil screening)
- Indoor air quality
- Quality control (residual solvents in products, etc.)

PID devices are typically calibrated with isobutylene gas-in-air, and the response to other gases and vapors is obtained by multiplying the reading by a correction factor provided by the instrument manufacturer. For very accurate measurement the user should calibrate the instrument for each contaminant of interest since the instrument response varies by lamp and even the condition of a specific lamp. The PID features a relatively linear response under about 200 ppm of contaminant, and a less linear between 200 and 2000 ppm. Above about 2000 ppm the sensor saturates, and response drops off about above this level.[4] To make measurements at levels above 2000 ppm, many instruments can be equipped with a dilution fitting that uses ambient air to dilute the sample. These are useful when sampling high concentration areas such as headspaces in containers or process gas streams.

The PID is a good general detector for organic and some inorganic compounds. These instruments can detect aliphatic and aromatic hydrocarbons, halo hydrocarbons, alcohols, ketones, aldehydes, ethylene oxide, arsine, phosphine, total reduced sulfur compounds including hydrogen sulfide, and glycol ether solvents. In general the response is lower for short-chain hydrocarbons, methyl cyanide, ethylene, carbon tetrachloride and methyl alcohol. The lamps used in these devices do not ionize the major components in air such as nitrogen and oxygen, nor carbon monoxide, carbon dioxide, and water vapor, thus eliminating interference from these compounds and allowing measurements to be made in ambient air. However, some of these compounds do interact with UV light, making it imperative that the instrument be "zeroed" in clean air under the same conditions as actual measurements will be made.

As a general rule, PIDs are more sensitive to complex molecules and less sensitive to simpler ones. Table 11.3 lists the *ionization energy (ionization potential)* and the approximated correction factors for some typical chemical compounds. The correction factors shown are based on an isobutylene-in-air calibration mixture and use of a 10.6-eV lamp in the PID. To use the correction factors, the instrument reading is *multiplied* by the correction factor to yield the actual airborne concentration of the

chemical of interest. Some key instrument parameters affecting PID sensitivity include lamp intensity, lamp seal, and flow rate through the detector. In terms of chemical structure, PID sensitivity depends on the following parameters: carbon number, functional group, and type of bond. Sensitivity increases as carbon number increases. The following sensitivity relationships have been observed.[5]

> alkanes < alkenes < aromatics
>
> alkanes < alcohols < esters < aldehydes < ketones
>
> cyclic compounds > noncyclic compounds
>
> branched compounds > nonbranched compounds
>
> fluorine-substituted < chlorine-substituted < bromine-substituted < iodine-substituted

As noted, the ionization source in the PID is a gaseous discharge lamp. These lamps contain a low-pressure gas in which a high-potential electrical current is passed. By varying the composition of the gas in the detector lamp, manufacturers are able to produce lamps with different energy (ionization potential) levels. Lamps are manufactured in several energy levels: 8.5, 9.5, 10.0, 10.2, 10.6, 11.4, and 11.7 electron volts (eV). However, since lamps must be purchased from the manufacturer of the instrument, not all lamps are available for all instruments.

Currently 11.7 eV is the highest-energy lamp commercially available that has the widest sampling range, and therefore it is the most logical one to use; however, its lamp window is manufactured from lithium fluoride (LiF) due to the need to transmit such high-photon energy, and this material degrades rapidly in the presence of humidity. LiF windows often cannot be cleared with organic solvents since they may contain trace amounts of water. This

TABLE 11.3. Ionization Energy (IE) [Ionization Potential (IP)] and PID Response Factor for Selected Chemicals

Compound	IE (eV)	Response Factor (10.6-eV lamp)	Compound	IE (eV)	Response Factor (10.6-eV lamp)
Acetaldehyde	10.23	6	Isoamyl acetate	9.90	2.1
Acetic acid	10.66	22	Isobutane	10.57	100
Acetic anhydride	10.14	6.1	Isopropyl acetate	9.99	2.6
Acetone	9.71	1.1	Isopropyl alcohol	10.12	6.0
Acrolein	10.10	3.9	Isopropyl ether	9.20	0.8
Acrylonitrile	10.91	NR	Methane	12.61	NR
Allyl alcohol	9.67	2.4	Methanol	10.85	NR
Allyl chloride	9.9	4.3	Methyl acetate	10.27	6.6
Ammonia	10.16	9.7	Methyl acrylate	(9.9)	3.7
Aniline	7.72	0.48	Methyl amine	8.97	1.2
Benzene	9.25	0.53	Methyl bromide	10.54	1.7
Benzyl chloride	9.14	0.6	Methyl t-butyl ether	9.24	0.9
1,3-Butadiene	9.07	0.85	Methyl chloride	11.22	NR
n-Butyl amine	8.71	1.1	Methylene chloride	11.32	NR
Carbon disulfide	10.07	1.2	Methyl ethyl ketone	9.51	0.9
Carbon monoxide	13.98	NR	Methyl isocyanate	9.25	0.45
Chlorine	11.48	NR	Methyl mercaptan	9.44	0.54
Chlorine dioxide	10.57	NR	Mustard	—	0.6
Chlorobenzene	9.06	0.40	Naphthalene	8.13	0.42
Crotonaldehyde	9.73	1.1	Nitric oxide	9.26	5.2
Cyclohexane	9.86	1.4	Nitrobenzene	9.81	1.9
Cyclohexanol	9.75	—	Nitroethane	10.88	NR
Cyclohexanone	9.14	0.9	Nitrogen dioxide	9.75	16
Cyclohexene	8.95	0.8	Nitromethane	11.02	NR
1,1-Dichloroethane	11.04	NR	2-Nitropropane	10.71	NR
1,2-Dichloroethylene	9.66	0.8	Octane	9.82	1.8
Dimethyl amine	8.23	1.5	Pentane	10.35	8.4
Ethyl acetate	10.01	4.6	Phenol	8.51	1.0
Ethyl alcohol	10.47	10	Phosphine	9.87	3.9
Ethyl amine	8.86	0.8	Propane	10.95	NR
Ethyl benzene	8.77	0.52	n-Propyl alcohol	10.22	5
Ethylene oxide	10.57	13	Propylene glycol	<10.2	5.5
Ethyl ether	9.51	1.1	Propylene oxide	10.22	6.6
Ethyl chloride	10.98	NR	Sarin	—	–3
Ethylene chlorohydrin	10.90	NR	Styrene	8.43	0.4
Ethyl formate	10.61	NR	Sulfur dioxide	12.32	NR
Ethyl mercaptan	9.29	0.56	Sulfur hexafluoride	15.3	NR
Formaldehyde	10.87	NR	Tabun	—	0.8
Formic acid	11.33	NR	Toluene	8.82	0.50
Heptane	9.92	2.8	1,1,1-Trichloroethane	11	NR
Hexane	10.13	4.3	Trichloroethylene	9.47	0.54
Hydrogen	15.43	NR	Triethylamine	7.82	0.9
Hydrogen cyanide	13.6	NR	Vinyl chloride	9.99	2.0
Hydrogen peroxide	10.54	NR	Water	12.61	NR
Hydrogen sulfide	10.45	3.3	m-Xylene	8.56	0.43

Notes: NR, no response; (••), Approximate Value.
Source: RAE Systems, Inc. (http://www.raesystems.com).

TABLE 11.4. Characteristics of Typical PID Lamps

Energy (eV)	Wavelength (nm)	Window	Gas	Typical Service Life
9.5	130.0	MgF$_2$	Xenon	1 year
10.6	116.5	MgF$_2$	Krypton	2–3 years
11.7	106.0	LiF	Argon	1–2 months

high-photon energy also produces radiation damage to the lithium fluoride. As a result, the lamp is delicate and often has a short service life. The 10.6-eV lamp has a magnesium fluoride window that is much more stable than the lithium fluoride; therefore, lamps in this range tend to be used more often. The 11.7-eV lamp is useful for compounds with ionization energies (IEs) greater than 10.7 eV, which includes many chlorinated compounds.

The spectral output of a lamp depends on the gas fill and the transmission characteristics of the window material. Table 11.4 shows typical gas fill and window materials for PID lamps. The actual spectral output for any lamp is not at a single energy since the fill gases have more than one emission line. For example, krypton has two distinct emissions: the major one at 123.9 nm (corresponding to 10.0 eV) and a minor one at 116.9 nm (corresponding to 10.6 eV) (Figure 11.8). For a 10.6-eV lamp utilizing krypton, the relative response will be much higher for airborne compounds with an IE of 10 eV and less, although the instrument will respond to compounds with IE's up to 10.6 eV such as ethylene oxide (IE = 10.57 eV).

For this same reason, some lamps exhibit emissions at wavelengths above their nominal eV ratings, and thus they will respond to compounds with an IE above the lamp's eV rating.[6] For example, there is a hydrogen-filled lamp available with a nominal rating of 10.2 eV that will respond to compounds up to 10.9 eV because of the broad emission spectrum of this lamp.

Lamps of different energies also have different sensitivities to a given compound.

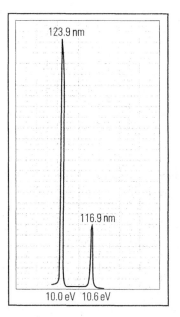

Figure 11.8. Wavelength of UV emissions from krypton, the most popular type of 10.6-eV PID lamp. (Courtesy of International Sensor Technology.)

For example, an 11.7-eV lamp has only one tenth the sensitivity to benzene that a 10.2- or 10.6-eV lamp has, even though it has a much broader detection capability than the other lamps.

Selectivity can be introduced into PID measurements to some degree by the use of different energy lamps. For example, the 9.5-eV lamp allows for the selective detection of aromatics in the presence of alkanes or oxygenated hydrocarbons, and the determination of mercaptans in the presence of hydrogen sulfide. Since aromatics, amines, and organic sulfur compounds have low IEs, these compounds can be determined with greater selectivity using a

TABLE 11.5. Comparison of the Relative Responses of the 9.5- and 10.6-eV Lamps

Compound	10.6 (eV)	9.5 (eV)	9.5/10.6 Response × 100(%)	IE (eV)
Xylene	112	112	100	8.4
Benzene	100	100	100	9.2
Styrene	100	100	100	8.5
Toluene	100	100	100	8.8
Phenol	75	77	102	8.5
Aniline	35	39	111	7.7
Pyridine	30	22	73	9.3
Methyl ethyl ketone	57	29	51	9.5
Acetone	63	6.5	10	9.7
Methyl methacrylate	30	<6	—	—
Heptane	17	<2	<10	10.1
Hexane	22	0	0	10.2
Ammonia	2	0	0	10.2

9.5-eV lamp instead of a 10.2-eV lamp. A 9.5-eV lamp has a high sensitivity for the amines but will not detect ammonia, which has an IE of 10.15 eV.[9] The 10.9-eV lamp has a wider range of response than the 10.2-eV lamp and can detect formaldehyde, formic acid, and other compounds that are difficult to measure with most instruments. Some PID instruments feature "snap on" lamp modules so changing lamps is easy, although the lamp selection is usually restricted to three choices: 9.2–9.8, 10.6 and 11.7 eV.

For example, Table 11.5 shows relative responses to some common compounds for 10.6-eV and 9.5-eV lamps. The third column of this table shows the difference in response between the two lamps. When sampling an unknown atmosphere using a PID instrument with interchangeable lamps, similar readings with both lamps might suggest that the contaminants were aromatic compounds (top rows in table) rather than the ketones or other compounds listed in the bottom half of Table 11.5. Of course, the sampling practitioner needs to have some understanding of the possible composition of the airborne contaminant before applying general guidelines as described in this example.

PIDs are sensitive to humidity extremes, possibly due to the lamp fogging and to the fact that the IE for water vapor is 12.61 eV. It has also been theorized that water molecules may collide with a photoionized contaminant is molecules inside the detector and deactivate them. In one test, the presence of 90% relative humidity (RH) caused a PID instrument to take twice as long to stabilize than at 0% RH. When compared to measurements at 0% RH, the response of this instrument to the same concentration was much lower at 90% RH. In another test, it was found that while stable readings were attained after 10 seconds in dry atmospheres, 5 minutes were required to stabilize when organics at 90% RH were sampled. It was also concluded that 90% RH decreases the response of the 10.2-eV lamp by a factor of 2 for most compounds as compared with the response under dry conditions.[7] For the 11.7-eV probe, humidity is especially a problem due to the sensitivity of the lamp material to water vapor.

Typical PID Instruments

There are a wide variety of PID-based instruments available. Recent technology innovations include a "plug-in" PID sensor,

Figure 11.9. Hand-held HNU DL-102 PID instrument featuring "snap-on" modules with different intensity lamps and sensors. (Courtesy of PID Analyzers, LLC.)

similar in appearance to the electrochemical sensors described in Chapter 10 that can give direct-reading instrument manufacturers another option in offering a photoionization detector along with other specialty sensors. As an illustration, one model of the International Sensor Technology (IST) instrument (Figure 10.7) is equipped with a photoionization detector (PID) for rapid screening of volatile organic compounds. Two instruments based on PIDs are described in this section.

The HNU Model DL-102 PID is a hand-held unit (Figure 11.9) from the same company that introduced the original field-portable PID device in the mid-1970s. The Model DL-102 features changeable "snap-on PID" heads so the user can easily switch from the standard 10.2-eV lamp to a 9.5-eV or 11.7-eV lamp. Once the PID module is

changed, the instrument can be quickly reconfigured from the keyboard so it is ready for calibration. In addition to the PID module, the instrument can be equipped with three additional sensors selected from 12 electrochemical sensors, infrared, relative humidity/temperature, air velocity/temperature, and a thermal conductivity detector.

The linear range of the instrument is 0.1 ppm to 3000 ppm; an optional dilution probe extends the linear range to >30,000 ppm for leak detection. Response time to reach 90% of the final reading is approximately 1 second. It has an internal library of response factors for more than 200 compounds and has data logging capability for more than 7000 points along with an RS-232 interface so that data can be sent to a printer or personal computer. The environmental operating conditions for the instrument are 5–40°C and 0–95% relative humidity (noncondensing). The typical zero drift is <1% and the span drift <2% over 24 hours. The sample flowrate is >0.2 Lpm to permit concurrent sampling with an adsorption tube or other sampling collection device. It is designed for a Class 1, Division 1 electrical classification (where ignitable concentrations of flammable gases or vapors can exist all of the time or some of the time under normal operating conditions).

A "personal-size" PID is available from RAE Systems. The *ToxiRAE plus PID*™ is small enough to fit in a worker's pocket and weighs only six ounces (Figure 11.10). It can operate for up to 10 hours on rechargeable batteries and has factory preset and user programmable alarms. The standard lamp is 10.2 eV, with 9.8-eV or 11.7-eV lamps optional. It has data-logging capability for more than 4000 points with a programmable sampling interval as short as 1 second, and it stores time-weighted average (TWA), short-term exposure limit (STEL), and peak information for specific compounds. When set on the lower

Figure 11.10. The ToxiRAE Plus Personal Gas Monitor is a pocket-size PID device. (Courtesy of RAE Systems, Inc.)

concentration range (0–99.9 ppm), the resolution is 0.1 ppm and on the high range (100–2000 ppm) the resolution is 1 ppm. Response time to reach 90% of the final reading is approximately 1 second on both scales. The environmental operating conditions for the PID unit are −20–45°C and 0–95% relative humidity (noncondensing). It is classified as intrinsically safe in Class 1, Division 1, Groups A, B, C, and D hazardous locations.

Calibration. All instruments are calibrated in the factory by adjusting a span setting that controls instrument sensitivity by varying the gain on the amplifier, so that the instrument will read correctly for a defined concentration of a specific vapor. However, as the lamp fatigues or becomes contaminated, the factory calibration becomes inaccurate, so each instrument should be recalibrated regularly.

If the compounds most likely to be present are known, calibrating the instrument to one of these will make the readings in the field more relevant.

Another strategy that is sometimes used is to adjust the response of an instrument calibrated with one gas to the sensitivity or response needed for another gas or vapor. For example, if benzene is the contaminant of interest, an instrument being calibrated with 100 ppm isobutylene-in-air would be adjusted to read some higher value during calibration to account for the response

factor. From Table 11.3, the response factor for benzene is 0.53, and so the adjusted reading should be

Adjusted reading
$$= \frac{\text{Concentration of calibration gas}}{\text{Correction factor the contaminant}}$$
$$= \frac{100\,\text{ppm}}{0.53} = 187\,\text{ppm}$$

So if the instrument is adjusted to read 187 ppm during calibration with 100 ppm isobutylene-in-air, it will read the actual concentration of benzene-in-air.

A third option is to use the response factor chart in the field and adjust the instrument readings for the contaminant of interest. This can be done "manually," or more sophisticated instruments store the response factors and can call them up and apply them automatically.

A response factor for an airborne mixture (RF_{mixture}) can be estimated from the response factors for individual components and the "mole percent" (X_i) of each component in the mixture using this equation:

RF_{mixture}

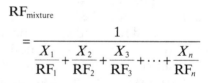

$$= \frac{1}{\dfrac{X_1}{RF_1} + \dfrac{X_2}{RF_2} + \dfrac{X_3}{RF_3} + \cdots + \dfrac{X_n}{RF_n}}$$

Table 11.6 illustrates this approach.[8] Column 2 in the table contains composition data for gasoline vapor from tank truck loading operations (X_i values in the above equation).[3] Column 2 shows the response factor (where readily available from the manufacturer), while column 4 shows the value in column 2 divided by the value in column 3. Column 4 represents the [$X_i \div RF_i$] factor in the above equation. Adding the values in column 4 and dividing that value into 1 yields the response factor for the mixture:

TABLE 11.6. Calculation of "Response Factor" for Airborne Gasoline Vapor

Gasoline Vapor Component	Vapor Volume, Mole Percent	PID Response Factor (10 eV)	Mole Percent / Response Factor[c]
Alkane Compounds			
Propane	0.01	NR[a]	—
Isobutane	0.05	100	0.001
n-Butane	0.38	67	0.006
Isopentane	0.23	8.2	0.028
n-Pentane	0.07	8.4	0.008
Cyclopentane	0.01	NR	—
2,3-Dimethylbutane	0.01	—[b]	—
2-Methylpentane	0.02	—[b]	—
3-Methylpentane	0.02	—[b]	—
n-Hexane	0.02	4.3	0.005
2,3-Dimethylpentane	0.01	—[b]	—
Methylcyclopentane	0.01	—[b]	—
2,2,4-Trimethylpentane	0.01	—[b]	—
Alkene Compounds			
Isobutylene	0.01	—[b]	—
2-Methyl-1-butene	0.02	—[b]	—
cis-2-Pentene	0.01	—[b]	—
2-Methyl-2-butene	0.02	—[b]	—
Aromatic Compounds			
Benzene	0.01	0.5	0.02
Toluene	0.02	0.5	0.04
Xylene (p, m, o)	0.01	0.5	0.02
Totals	0.95		0.128

[a] NR indicates the PID (with 10.6-eV lamp) does not respond to that vapor.
[b] The response factor was not readily available.
[c] Entry in column 4 is value in column 2 divided by value in column 3.

$$RF_{mixture} = \frac{1}{0.128} = 7.8$$

Based on this calculation, measurements of gasoline vapor with a PID that was calibrated with 100 ppm isobutylene-in-air should be multiplied by 7.8 to estimate the concentration of gasoline vapor.

A closer review of Table 11.6 shows some limitations to this approach that can occur. First, response factors were not readily available for many of the components. When this occurs the PID manufacturer should be consulted to determine if they can supply any missing values. With the gasoline vapor there were response factors available for the more prevalent components. Second, the PID's low sensitivity to n-butane (which constitutes 38% of the total vapor) indicates that a relatively large part of the vapor is not being measured by the PID. Conversely, the three aromatic compounds that are present in very low concentration have a large impact of the overall response factor. Situations like this require the sampling practitioner to evaluate the toxicity and other hazards of each component to ensure that the PID readings will adequately measure the potential hazard and will give an adequate warning should an immediately hazardous condition develop.

Generally, a direct calibration of the PID using the specific compound(s) or mixture of interest is preferred over calculated response values.

Under some working conditions it may be impossible to obtain clean background air to "zero" the instrument, or especially clean "zero" air may be needed for very accurate low-level results. In such cases, one approach is to use a "zero air zero" with a specific gas calibration. In these cases, a Tedlar bag of zero-grade air is used to obtain an accurate zero followed up with exposure to a standard of the compound of interest to adjust the span setting. This procedure assures that the instrument is calibrated at both ends of the scale. The zero air is essential, because if the instrument is zeroed with contaminated air and subsequently encounters a cleaner atmosphere, negative readings could result.

Procedure. The procedure described in this section outlines the typical steps to calibrate a PID instrument. The microprocessor in many instruments automates the calibration process. The display prompts the user on each step, and the microprocessor performs the calibration adjustments.

1. Identify the lamp intensity by its label and ensure that the instrument is set up for that lamp.
2. Set the span adjustment to the proper value for the lamp being calibrated.
3. Check the IE of the calibration gas to be used. The IE of the calibration gas must be below the rating of the lamp.
4. Set the unit on standby. In this position the lamp is off and no signal is generated. Set the zero point with the zero set control.
5. For calibrating for general survey work on lower ranges, such as 0ppm

to 20ppm, only one calibration point is required. Turn the switch to the proper concentration range and note the readout. Adjust the output reading settings as required to read the ppm concentration of the standard. For each lamp, compound, and concentration combination a different span setting will be required. Recheck the zero on the readout and adjust if necessary. This gives a two-point calibration: zero and the gas standard concentration.

6. For calibrating on the higher ranges, such as 0–200ppm and 0–2000ppm, consider the use of two or more concentrations. At a minimum, use one that will give a response of 70% to 85% of full scale and other for 25% to 35% of full scale. First calibrate with the highest standard. Then calibrate with the lower standard, using the zero adjust. Repeat these several times to ensure that a good calibration is obtained.

Maintenance. During periods of operation, dust or other foreign matter can be drawn into the instrument, forming deposits on the surface of the UV lamp or ion chamber. This condition is indicated by meter readings that are low, erratic, unstable, nonrepeatable, or drifting, or show apparent moisture sensitivity. When exposed to levels of gases and vapors higher than an instrument's detection capability, it is possible to saturate an instrument. Some contaminants are "sticky" and will remain in the instrument for long periods of time. The instrument should be inspected monthly for this condition, if in regular use. A typical cleaning procedure is described below.

Procedure
1. Very carefully disassemble the probe or instrument and remove the lamp and ion chamber.

2. During the course of normal operation a film will build up on the window of the UV lamp. First check the lamp window for fouling by looking at the surface at an incident angle because deposits, films, or discoloration may interfere with the ionization process. The rate at which the film develops depends on the type and concentration of the gas and vapor being sampled. For magnesium-fluoride lamps, the windows are cleaned by rubbing gently with lens tissue that has been dipped in a detergent solution. If further cleaning is needed, use the cleaning compound supplied by the manufacturer and spread it evenly over the surface with a lens tissue. Wipe off the compound and rinse the surface with warm water (27°C), or use a damp tissue to remove all traces of grit or oils and any static charge that may have built up on the lens. Dry carefully with clean tissue. For lithium-fluoride lamps, cleaning is done using water-free chlorinated solvents or a fine powder compound according to the manufacturers instructions. Never clean this lamp with water or any water miscible solvents, such as methanol or acetone, as they will damage the lamp.

3. Inspect the ion chamber for particulate deposits. If present, the chamber should be cleaned. A tissue or cotton swab, dry or wetted with methanol, can be used to clean off any stubborn deposits. Ensure that the chamber is absolutely dry prior to reassembly.

4. Reassemble the probe or instrument and check the analyzer's operation by zeroing and calibrating the instrument.

5. If these steps do not restore performance, the lamp may require replacement.

Field Measurements. While there are differences in the operation of all instruments, the following section contains some general information on survey PIDs. Be alert for strong sources of electromagnetic fields, such as power lines, transformers, and radio wave transmissions, since these may affect readings. Never look directly at the light source from closer than 6 inches or less without wearing proper eye protection; continued exposure to UV light will damage the eyes. Prior to field use, the instrument should be calibrated for the range of interest. If the range is unknown, calibrate the instrument at three points over the instrument's range.

As a field check, a solvent-based marking pen can be used to see if the instrument is responding. Never use automobile exhaust to check the status of an instrument as it contains condensable organic compounds and particulates that can deposit in the probe and foul the lamp.

A typical application for a PID is at a hazardous waste site. For an initial site investigation, walk the perimeter of the site first and record measurements. The PID has a relatively long response time, and therefore the sampling practitioner must walk slowly when using this type of analyzer, or areas of high gas and vapor concentrations may be entered inadvertently before the instrument has a chance to respond. Some PIDs are also highly directional and must be held close to a source in order to get an accurate reading. High humidity and wind can affect the accuracy of the meter.

If a reading is unstable, a lower span setting may be necessary. Sampling in a windy location can also cause the reading to jump, so the inlet should be sheltered in these situations. If the chemical concentration in the air is fluctuating, so will the readout.

Procedure

1. Make sure prior to field use that the instrument is fully charged. The

charger should be disengaged from the wall before disconnecting from the instrument. With a full charge, most instruments should last 8 hours, although colder temperatures can decrease this time.

2. Depending on the energy (eV) of the lamp, the span may have to be changed. Set the span to the value specified for the lamp on the calibration gas cylinder if available.

3. Adjust the zero control to "zero" on the instrument. If the span adjustment setting is changed after the zero is set, the zero should be rechecked and adjusted if necessary. Wait 15 to 20 seconds to assure that the zero reading is stable.

4. Set the instrument to the appropriate operating range. Start with the highest position and then switch to the more sensitive (lower) ranges. In this position, the UV light source should be on.

5. If using the PID to screen drum contents or other containers in which there may be high concentrations of chemicals, set the instrument on the highest-range setting first and move gradually closer to the top of the open bung hole. If wanting to approximate exposure concentrations, take a measurement at shoulder level. Do not stick the instrument's probe directly into bore holes, drum bungs, or other situations unrealistic of exposures as doing so could saturate the instrument and in any case could result in a reading above the instruments upper limit. Saturating an instrument can result in having to wait long periods of time to flush it out, or may cloud the lamp during a time when sampling is critical.

6. If high levels of dust are present attach a filter to the front of the inlet to prevent dust from entering the detection chamber and causing incorrect readings.

7. Follow the manufacture's recommendations on spare bulbs. It is handy to have at least one spare lamp along on surveys in the event that the lamp is scratched during cleaning, damaged by deposits of nonvolatile compounds, or by physical shock such as dropping the instrument or it simply wears out. However, the short shelf life of some bulbs may make this impractical.

Collection of Follow-Up Samples. Some PIDs have an exhaust port that allows simultaneous collection of a sample on a sorbent tube. A sample can be collected for laboratory analysis to confirm identities of the suspected airborne contaminants because the PID is a nondestructive detector. To collect such a sample, one or two charcoal tubes or another sorbent material are hooked to the exit port of the sampler. If the flowrate is high, two tubes may be necessary to split the airflow of the unit so that half goes through each tube, although only one has to be analyzed. The PID's flowrate should be calibrated with the tubes "in line" prior to use. For more information on flowrate calibrations, see Chapter 5.

Flame Ionization Detectors

Flame ionization detectors (FIDs) use a hydrogen flame to ionize organic molecules in the sample as it flows through the detector. In addition to its use in general survey instruments, the FID is also a widely used detector in gas chromatographs (GCs) to measure the concentration of compounds as they emerge from the separation column used in the GC. For a general survey FID instrument the entire sample is ionized as it burns in the hydrogen flame. The FID is constructed so the burner tip serves as the cathode. When organic compounds are

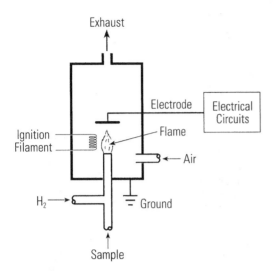

Figure 11.11. Diagram of a flame ionization detector. (Courtesy of International Sensor Technology.)

introduced into the hydrogen flame, positively charged ionized fragments of the original molecules form that are driven by an electric field to be collected by a negatively charged electrode (Figure 11.11). The positive ions generate an electric current that is proportional to number of ionized carbon atoms, which in turn is a function of the number of ionizable carbon atoms in the contaminant molecules and the concentration of the molecules in the sample.

General survey FIDs respond to almost all organic compounds, but the response is greatest to aliphatic hydrocarbons including short-chain molecules such as methane. While FIDs do not respond to atoms other than carbon, other atoms can alter an instrument's sensitivity to carbon by altering the chemical environment of the carbon atom. For example, compounds containing oxygen, such as alcohols, ethers, aldehydes, and esters, and nitrogen-containing compounds, such as amines, amides, and nitriles, have a lower response than that observed for hydrocarbons. It can also detect halogenated hydrocarbons, such as trichloroethylene, chloroform, and 1,1,1-trichloroethane, although its sensitivity to

these is low compared to most combustibles. The FID is insensitive to water, inert gases, and inorganic compounds and has a negligible response to carbon monoxide and carbon dioxide, which due to their structure do not produce appreciable ions in the detector flame.

The response of an FID relative to the calibration gas varies for each type of detector; there is no "universal" response factor for each compound equivalent to the *ionization energy* (also called *ionization potential*) value used with PIDs. For this reason, each manufacturer calibrates their flame ionization detectors for specific contaminants and supplies these factors with the instrument. As an illustration, Table 11.7 shows some relative response data for a typical portable FID. The FID reading is multiplied by the factor in Table 11.7 to give the actual airborne concentration of the contaminant.

The FID's ability to respond to methane can be a distinct advantage for certain environmental and other measurements. Some environmental regulations, especially for landfills, require evaluation of non-methane volatile organic compounds. This category of VOC vapors can be measured by taking a background reading (total VOCs), then adding a charcoal filter to the FID and repeating the reading. Since the charcoal removes most organic compounds but not methane, the difference between the two readings represents the concentration of non-methane VOCs. Typical field FID units perform this calculation automatically using their data-logging function.

FIDs are usually calibrated with methane, and many instruments can measure concentrations up to 50,000 ppm. The response curve is not linear at high concentrations, and so the manufacturer often provides several calibration points over the concentration range of the detector (Figure 11.12). Sophisticated instruments can store the calibration data and then use the calibration point closest to the

TABLE 11.7. Relative Response of an FID to Different Chemicals (calibrated to methane)

Compound	Relative Response (%)	Compound	Relative Response (%)
Acetaldehyde	6.9	Heptane	1.3
Acetone	2.7	Hexane	1.6
Acetonitrile	1.0	Isoprene	2.2
Acrolein	6.9	Isopropyl alcohol	2.4
Acrylonitrile	1.3	Methane	1.0
Ally chloride	2.7	Methanol	23.8
Aniline	3.0	Methyl ethyl ketone	1.9
Benzene	0.7	Methyl isobutyl ketone	1.9
Benzyl chloride	1.2	Methyl methacrylate	2.8
n-Butane	1.9	Methyl tert-butyl ether	2.0
1,3-Butadiene	2.7	Methylene chloride	1.4
n-Butanol	2.6	Nonane	1.1
Carbon tetrachloride	25.9	n-Pentane	1.6
Chlorobenzene	0.8	Propane	1.8
Chloroform	3.5	Styrene	1.2
Cumene	1.0	1,1,1,2-Tetrachloroethane	1.1
Cyclohexane	1.4	1,1,2,2,-Tetrachloroethane	1.8
o-Dichlorobenzene	0.7	Tetrachloroethylene	2.9
trans-1,2-Dichloroethylene	2.7	Toluene	0.9
Dimethyl formamide	2.3	1,1,1-Trichloroethane	1.4
p-Dioxane	4.6	1,1,2-Trichloroethane	1.7
Ethanol	5.2	Trichloroethylene	2.8
Ethyl acrylate	2.7	Triethylamine	1.1
Ethyl benzene	1.0	Vinyl acetate	4.4
Ethyl cellosolve	4.3	Vinyl chloride	2.1
Ethylene	2.2	Vinylidene chloride	2.6
Ethylene dibromide (1,2-dibromoethene)	2.0	o-Xylene	1.1
		m-Xylene	1.2
Ethylene dichloride (1,2-dichloroethane)	1.7	p-Xylene	1.2

Source: Photovac, Inc.

airborne level to improve the accuracy of the reading. For concentrations above 50,000 ppm, a dilution fitting can be used to dilute higher concentrations samples with ambient air. The dilution fitting can also be used to introduce oxygen from ambient air into the flame when measuring samples with less than about 15% oxygen.

Typical applications for FIDs include:

- Fugitive emissions monitoring—especially useful where the instrument must be calibrated to methane and measurement levels of 10,000 ppm or higher are required.
- Landfill monitoring—where differentiating between methane and non-methane organic vapors is needed.
- Arson investigations for the presence of accelerants since most are commercially available volatile liquids such as acetone, alcohol, gasoline, paint thinner, mineral spirits, or other petroleum-based chemicals.

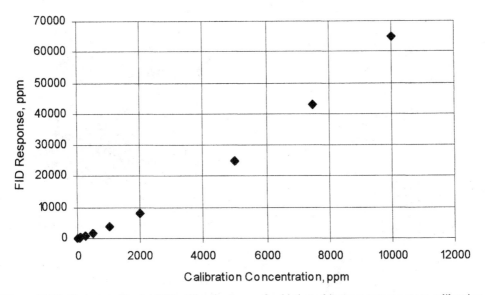

Figure 11.12. Typical multipoint FID calibration curve. Sophisticated instruments can store calibration data for the individual points, and they can use the calibration point closest to the field reading in order to maximize accuracy.

- Natural gas pipeline leak detection—where the FID's ability to measure methane is important.
- Confined space entry involving petroleum products.
- Headspace screening for soil/water samples, because of the wide measurement range (0.5–50,000 ppm).

Typical FID Instruments

Photovac's MicroFID is a hand-held instrument weighing about 8 pounds (Figure 11.13). The internal refillable hydrogen cylinder holds 9.2 liters, which allows up to 12 hours of field use. The operating range of the instrument is 0.5–2000 ppm on the low range and 10,000–50,000 on the high range. An optional dilution probe extends the range even higher for leak detection. It has data-logging capability to store date, time, and minimum, average, and maximum readings during user-selectable intervals.

The data logger can also be set to the U.S. EPA Method 21 mode to record background, sample, and difference readings for nonmethane organic vapor results. An RS-232 interface allows data to be sent to a printer or personal computer. The environmental operating conditions for the instrument are 5–45°C and 0–100% relative humidity (noncondensing). It is designed for Class 1, Division 1, Groups A, B, C, and D.

A combination FID/PID unit is available from ThermoEnvironmental Instruments. The TVA-1000B (Figure 11.14) weighs 12 pounds and operates for 8 hours on the rechargeable batteries. Minimum detection level for benzene is 100 ppb using the PID and 2000 ppb for the FID. The dynamic range is 0–2000 ppm with the PID and 0–50,000 ppm for the FID. The PID allows measurement of compounds such as ammonia, carbon disulfide, formaldehyde, and hydrogen sulfide, which cannot be measured with a FID.

Figure 11.13. Hand-held MicroFID instrument. (Courtesy of Photovac, Inc.)

Calibration

The calibration procedure for a FID instrument is similar to that for a PID device. Methane is generally used as the calibration gas. A gas-in-air mixture of about 100 ppm is often used. Because of the wide operating range of the FID, calibration at higher or lower levels may be appropriate for some applications but extreme care is needed not to use mixtures approaching flammable concentrations. Some manufacturers supply multipoint calibration data for some instruments; the instrument stores

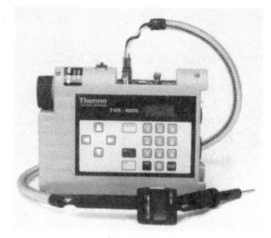

Figure 11.14. The TVA-1000 is a combination FID/PID instrument. (Courtesy of Thermo Electron Corporation.)

the different response factor values and selects the calibration value closest to the actual field measurement level. General steps for calibrating a FID are described below.

1. Ensure that the hydrogen cylinder is filled prior to operation. Before beginning calibration, ensure that a reliable source of zero air and calibration gas is available. If the ambient air is clean, it may be used as the source of zero air.

2. Select the CALIBRATION mode from the keypad. If the instrument can store calibration data for different compounds using different memory locations, choose the appropriate contaminant. Typically, the data stored are Response Factor, zero, sensitivity, and alarm level.

3. Key in desired Response Factor and hit "Enter."

4. For multi-range instruments, select the proper measurement range. For example, this might be Low Range for concentrations between 0.5 and 2000 ppm (methane equivalents), and High Range for concentrations

between 10 and 50,000 ppm (methane equivalents).

5. If using a independent supply of zero air, connect the gas bag adapter to the inlet. Open the gas bag and press "Enter." If using room air for zero, press "Enter." Wait while the instrument "zeros"—often 60–90 seconds.

6. When prompted, enter the known span gas concentration and then connect the span gas. Press "Enter" and device will set the span value (sensitivity).

7. Press the "Alarm" key and enter the alarm level for the selected contaminant.

Hydrogen Safety

Since the FID uses a small internal cylinder of hydrogen, extreme care is needed to avoid hazardous leaks or concentration buildup. For example, storing an instrument in a closed carrying case or other closed box with its internal hydrogen reservoir fully or partially charged may result in a potentially hazardous situation. In such a case, there exists the possibility that hydrogen gas may leak into the confined space and, when the case or box is opened, a spark or other source of ignition could cause the trapped hydrogen gas to explode. For this reason an FID should not be stored in its carrying case unless it is ventilated, or in its original shipping container or any container which is unventilated if the instrument is fully or partially charged with hydrogen gas. All portable flame ionization instruments should always be stored in a well-ventilated area when charged with hydrogen.

Under no circumstances can the FID instrument be safely shipped by air or ground when it is partially or fully charged with hydrogen. Hydrogen should be transported separately in an appropriate cylinder accompanied by a properly completed Shipper's Declaration form. Consult a qualified shipping company for complete instructions.

Although many FID instruments are rated as suitable for use in Class I, Division 1 hazardous locations, precautions must always be taken when using hydrogen. Always ensure that the hydrogen shut-off valve is in the closed position for storage. Consult the FID instrument's User's Manual for additional information.

Field Operation. Ensure that the hydrogen tank is full and that the instrument is properly calibrated according to the manufacturer's instructions.

Turn on the instrument, let it warm up, and then light the hydrogen flame. Perform a field "zero" check in clean air, using a known "zero" air or with a charcoal filter on the sample inlet.

If using the audible alarm it must be set for the range on which measurements are going to be made. As an example, if making survey measurements on a site where 5 ppm has been selected as the level to upgrade from an air purifying mask (level C) to a supplied air respirator (level B), then the alarm should be set for this level.

As a field check a butane lighter (button pushed but not lit) can be used to see if the instrument is responding. Never use automobile exhaust to check the status of the instrument as it contains condensable organic compounds and particulates that can deposit in the probe, partially plug the filters, and possibly damage the detector.

Survey the areas of interest while listening for the audible alarm. Walk slowly or a high concentration could be encountered before the instrument responds, since it does not respond fast enough to detect vapors at a rapid walking pace.

In general, the probe should be positioned at chest level while taking readings. If trying to approximate an individual's exposure, hold the probe near the person's breathing zone. If taking a headspace vapor measurement above a bulk sample, start on the highest range and be careful not to draw any liquid into the probe.

COMPARISON OF FID AND PID FOR GENERAL SURVEY USE

As has been discussed, the differences between the detection principles of the PID and the FID will impact measurements. Since the FID uses a hydrogen flame as the means to ionize organic vapors, it responds to virtually all compounds that contain carbon–hydrogen or carbon–carbon bonds. The PID can respond to some inorganic compounds and most but not all organic compounds.

The FID response (sensitivity) to most organic vapors is based on breaking chemical bonds, which requires a set amount of energy. For the PID, the response is based on the ionization energy (IE) of a compound and the ease in which an electron can be ionized (displaced) from the molecule. This mechanism is variable and highly dependent on the individual characteristics of a particular substance, resulting in a variable response factor for most organics that are ionizable. Therefore, in general, large sensitivity shifts are not seen between different substances when using an FID as compared to those seen with a PID.[9]

The relative responses of the FID and PID to a given compound also vary due to the fact that different compounds are commonly used to calibrate these instruments. PIDs are commonly calibrated with isobutylene and the FIDs are calibrated to methane. This difference sometimes causes problems if trying to compare results from simultaneous measurements of an FID with those of a PID.

An advantage of PIDs is that they are generally easy to use compared with many others, including FID units. There is no gas reservoir to keep filled and no flame to light. In addition, in the general survey mode these instruments can be made to respond to a narrower range of compounds than FID monitors in that the lamps can be changed to allow for some selectivity.

FIDs have a much wider range of operation than PIDs and also respond to more compounds including methane.

INTERPRETATION OF GENERAL SURVEY MEASUREMENTS FOR HEALTH HAZARDS

The limitations of each instrument are important. General survey instruments should not be used where a compound with an extremely low *immediately dangerous to life or health* (IDLH) value might be present, for example, acrolein, which has a TLV of 0.1 ppm and an IDLH of 5 ppm. Similarly, monitoring for compounds with acceptable TWA exposure limits of 1 ppm and less is better done with instruments capable of greater specificity and sensitivity. Although a PID can measure concentrations from 1 ppm to 2000 ppm, its response is not linear over this entire range. Generally measurements greater than about 600 ppm are considered to represent a higher airborne concentration than the readout value due to the nonlinearity of the instrument.

General survey instruments do not differentiate between specific chemicals, as has been discussed previously. Since their response varies, depending on what compounds are being sampled, readings are not specific for any one compound, unless the situation is such that the compound present is the only one and the instrument has been calibrated with that chemical. Since the types of compounds that the instrument can potentially detect are only a fraction of the chemicals potentially present in a spill incident or on a hazardous waste site, a zero reading does not necessarily signify the absence of air contaminants.

All readings in a contaminated area

should be compared to upwind, background readings. If used for OSHA compliance measurements, the PID and FID are only allowed for use as screening instruments. A negative reading on a PID may indicate that a sample has fewer total ionizables than the zero reference air. It is an advantage to use both instruments when methane may be present, since the response of the PID to toluene and gasoline has been shown in one study to be decreased by as much as 30% in the presence of 0.5% methane (10% of the LEL).[10]

Using an FID along with a PID in situations involving clay soil, cement, or asphalt, which are diffusion barriers allowing biogenic levels of methane to build up, will allow a better estimation of what is present. In this situation, if a measurement of a particular contaminant is desired, it may be necessary to use a gas chromatograph instead. One example is soil gas measurements. Another alternative would be the use of integrated measurement techniques, such as charcoal tubes, also sometimes used for soil gas measurements.

On uncontrolled hazardous waste sites (sites where the hazards and potential levels are unknown), the FID and PID are often used to determine what level of personal protection is required by providing data for setting arbitrary concentrations at which workers would either upgrade or downgrade their respiratory protection. This concept is called the "setting of action levels."

On typical approach on a site where the hazards are unknown is to require level C equipment (half mask respirator minimum) whenever levels exceed background and up to 5 ppm. Level B self-contained breathing apparatus (SCBA) is required for levels exceeding 5 ppm and up to 500 ppm above background. Level A protection (impermeable bodysuit with SCBA) is required whenever levels exceed 500 ppm over background. While this practice is not univerisally accepted, it illus-trates one approach. Qualified occupational safety and health specialists must review site hazards and determine appropriate action levels.

SUMMARY

This chapter describes general survey instruments, which are in the category of direct-reading instruments that respond to broad groups of airborne chemicals such as combustible gases or volatile organic compounds (VOCs). The two main types of general survey instruments are:

- Combustible gas indicators (CGIs), which indicate the presence of combustible gases or vapors as compared to the compound(s) used for calibration.
- Flame ionization or photoionization instruments that can measure many common organic compounds and other gases and vapors at parts per million or parts per billion levels.

Although the general survey devices are very useful to the sampling practitioner, an inherent limitation that *must* always be considered is their varying response to different chemical molecules. Each of these instruments is calibrated using a specific gas or mixture, and all measurements of airborne concentrations are relative to the instrument's response to the calibration gas molecule(s).

General survey instruments have other important limitations or operating considerations, which include the following:

- They may not detect some hazardous or extremely toxic compounds.
- Most require an adequate concentration of oxygen.
- A "zero" reading may occur with a CGI even though a high concentration of the gas is present.

- General survey instruments may not be reliable indicators of exposure to chemicals with extremely low acceptable exposure standards since other, less toxic compounds may mask the response to the higher toxicity material.
- Not all devices in this category may tested and rated as *intrinsically safe* for use in potentially flammable atmospheres.

REFERENCES

1. Cote, A. E., ed. *Fire Protection Handbook*, 19th ed. Quincy, MA: National Fire Protection Association, 2003.

2. American Industrial Hygiene Association. *Manual of Recommended Practice for Combustible Gas Indicators and Portable, Direct Reading Hydrocarbon Detectors*. Akron, OH, 1980.

3. McDermott, H. J., and S. E. Killiany, Jr. Quest for a gasoline TLV. *AIHA J.* **39**:110–117, 1978.

4. Chou, J. *Hazardous Gas Monitors, A Practical Guide to Selection, Operation and Applications*, New York: McGraw-Hill, 2000.

5. Langhorst, M. L. Photoionization detector sensitivity of organic compounds. *J. Chromatogr. Sci.* **19**. February 1981.

6. Burroughs, G. E., and J. L. Woebkenburg. *Effectiveness of Real-Time Monitoring*. ACGIH: Advances in Air Sampling, pp. 243–250, Cincinnati, OH, 1988.

7. Barsky, J. B., et al. An evaluation of the response of some portable direct-reading 10.2 eV and 11.8 eV photoionization detectors, and a flame ionization gas chromatograph for organic vapors in high humidity atmospheres. *AIHA J.* **46**(1):9–14, 1985.

8. Diaz, K. J., and H. J. McDermott. A Cost Effective Method for Evaluating the Accuracy of Direct Reading Instruments for Specific Hydrocarbon Vapors. American Industrial Hygiene Conference and Exposition, 1994.

9. Driscoll, J. N., and J. H. Becker. *Industrial Hygiene Monitoring with a Variable Selectivity Photoionization Analyzer*. American Laboratory, November 1979, pp. 69–76.

10. Nyquist, J. E. et al. Decreased sensitivity of photoionization detector total organic vapor detectors in the presence of methane. *AIHA J.* **51**(6):326–330. 1990.

CHAPTER 12

INSTRUMENTS FOR MULTIPLE SPECIFIC GASES AND VAPORS: GC, GC/MS, AND IR

This chapter describes direct-reading instruments that can identify and measure the concentration of specific compounds in an airborne mixture. Basically this includes two types of instruments:

- Gas chromatograph (GC) or GC/mass spectrometer (GC/MS) devices that separate the airborne compounds so that each can be identified and its concentration measured as the separated compounds pass through the detector.
- Infrared (IR) instruments where the wavelength of the emitted IR radiation can be adjusted to match the wavelength that specific chemical compounds absorb the IR radiation. Since many chemicals have a characteristic absorption wavelength, this method can provide an accurate method of identification. Additionally, since the amount of IR radiation absorbed is proportional to the concentration of the chemical, these instruments provide concentration data.

As noted in previous chapters, there is some overlap between these instruments and the devices covered in earlier chapters. Specifically, many IR devices are available that are set to emit just one wavelength of IR radiation and thus are specific to one airborne compound. As an illustration, direct-reading instruments for carbon dioxide (CO_2) often are based on the IR principle, and they use the absorption of IR radiation at a wavelength of $4.72\,\mu m$ to measure the concentration of CO_2 in the air. However, these instruments are covered in detail in this chapter (instead of Chapter 10, "Instruments with Sensors for Specific Chemicals") as part of the overall discussion of IR instruments.

The GC, GC/MS, and IR instruments that can determine the concentrations of multiple specific compounds in an airborne mixture are high-end devices that are expensive and sophisticated and can be complicated to operate. Generally, they are only used when less expensive and easier to use devices are not adequate for the monitoring job.

Air Monitoring for Toxic Exposures, Second Edition. By Henry J. McDermott
ISBN 0-471-45435-4 © 2004 John Wiley & Sons, Inc.

While these instruments have important differences, they have several characteristics in common:

- All are designed to be used with a personal or laptop computer either to operate the instrument or to manage the data. The compound identification and measurement steps involve comparing the airborne sample to stored calibration or library data, and so a computer is essential for this step. They also generate large amounts of data for each measurement, and these must be processed in a computer.

- While many configurations and options are available, the sampling practitioner generally describes their needs (compounds to be sampled, possible interferences, operating conditions, etc.) to the manufacturer, who first determines if their technology is a suitable fit for the application. If so, the manufacturer's technical staff then selects and configures the instrument to meet the user's specific needs. For this reason, there is less need for the sampling practitioner to understand the various instrument options with the goal of selecting these options on their own, as is typical for less sophisticated direct-reading instruments.

- Even though these instruments are the most sophisticated direct-reading instruments available, they still are not designed to identify the components of a completely unknown or "mystery" airborne mixture. The instruments achieve their accuracy in identifying and measuring the concentration of individual chemical compounds by being calibrated with the specific compounds they are measuring. For all of these devices, daily or more frequent calibration using a known calibration gas mixture containing each of the components of interest is recommended for optimal performance.

- Because of their size and complexity, they are not routinely used for direct occupational exposure measurements of breathing zone concentration. More typically they are used for area or source sampling; for environmental reasons; at hazardous waste sites; for field determination of the composition of an airborne mixture for real-time respirator selection decisions; or as an aid to understanding the total VOC reading of a less sophisticated instrument such as a photoionization detector (PID). There are some NIOSH occupational sampling methods for field GCs; one is described later in this chapter. For integrated samples, a personal pump can be used to fill a sampling bag for analysis using the GC or IR instrument.

The instruments described in this chapter meet these criteria:

- They are "portable" in that they are battery-powered and can be carried around in the field or set up in a temporary location.

- They can measure airborne contaminants, although most can measure "headspace" levels (vapor collected above liquid or solid samples), and some can analyze liquid samples directly.

- They are commercially available, field-ready technology.

PORTABLE GAS CHROMATOGRAPHS (GCs)

As discussed in Chapter 1, *gas chromatography* (GC) is a common laboratory analytical method for gases and organic vapors that are collected using charcoal tubes or another sample collection device. The GC operates on the principle that a volatilized sample is mixed with a carrier gas and

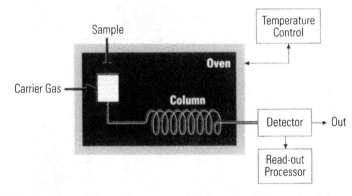

Figure 12.1. Diagram of a typical gas chromatograph. (Courtesy of International Sensor Technology.)

injected into a column that separates the components in the sample according to the time it takes the component to travel through the column (Figure 12.1). As the molecules emerge from the column, a detector measures the amount of each material. On a chromatogram, each emerging compound is represented by a "peak" based on its elution time. Component identification is made using a data "library" developed by injecting samples of known chemicals into the column and measuring the travel time for each. The concentration of each material is represented by the area under its peak on the chromatogram (Figure 12.2). For laboratory GC units, a wide choice of columns, detectors, and operating parameters such as temperature programming give the GC the ability to identify many different chemicals, especially those in mixtures.

Portable GCs as described in this chapter focus on modern, high-end portable gas chromatographs. Some earlier field GCs were relatively simple instruments that essentially added a short separation column maintained at ambient temperature to a PID instrument. These simple units have been supplanted because they do not provide much advantage beyond the direct-reading instruments with sensors for specific chemicals discussed in

Chapter 10. The GCs covered in this chapter more closely resemble laboratory instruments, with heated columns, temperature programming, and a choice of columns and detectors based on the compounds of interest and possible interferences. These instruments require a computer to control the device's operation and provide the data management system. Some require a laptop computer to be used along with the instrument in the field, while for others an office computer is sufficient to transfer information between the device and computer. The instrument manufacturer supplies a proprietary software package that operates the GC, collects, stores, and processes data, and then downloads it to the hard disk drive. These GCs measure concentrations ranging from parts-per-trillion (ppt) to percent (%) levels. Some allow continuous, unattended operation or can be controlled from a remote location using a modem and communications lines. These sophisticated, computer-controlled instruments provide the functions that are described below.

Instrument Calibration. The instrument should be calibrated with each chemical that will be measured in the field. During this mode, the system introduces a sample of a known calibration mixture into the

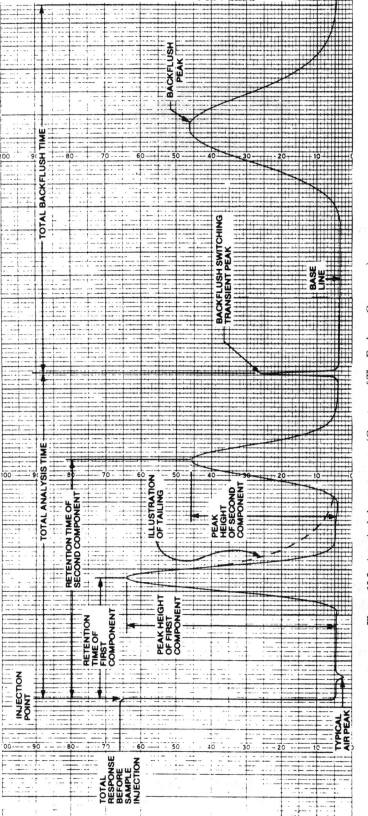

Figure 12.2. A typical chromatogram. (Courtesy of The Foxboro Company.)

362

system and performs chromatographic analysis either when the user initiates the calibration cycle or at preset frequencies. The instrument then displays this calibration chromatogram, including the name, concentration level, and retention time of each compound in the calibration mixture. The area under each peak is integrated, and the concentration level of the standard is assigned to this peak area. There are two calibration methods commonly used[1]:

- Multipoint calibration is most accurate. It involves introducing several different concentration levels of a chemical to the GC and plotting a curve of peak areas (horizontal axis) versus concentration level (vertical axis). When an unknown concentration of the same chemical is detected in the system, the area obtained is compared to the calibration curve to determine its concentration level. This method is very accurate and can cover a large concentration range.
- Single-point calibration method is often used for portable gas chromatography. Its accuracy within a reasonable concentration level range is satisfactory, and it is easy to perform. A calibration curve, similar to the multipoint curve, is drawn using only two points. One is zero with an assumed area peak count of zero. The other point is the point area obtained when a known concentration of the standard is injected at its concentration level. This curve is used to calculate analysis results. This method is relatively simple and requires only one concentration for calibration.

Single calibration is available for all instruments; multipoint calibration is an option for others. Some devices have an internal cylinder containing the calibration mixture, while others use an external cylinder or calibration bag sample.

Calibration gas mixtures can be purchased from commercial suppliers or prepared by the sampling practitioner as described in Appendix B. A "ready-to-use" calibration mixture is most convenient to use, or a higher concentration mixture that is diluted by the sampling practitioner with pure air into a calibration bag will give more calibration runs from the calibration cylinder. The concentration of each gas in the final calibration mixture should be near the concentration of interest. Check with the calibration gas supplier to determine if the compounds will be stable in the cylinder when blended at the desired concentrations.

Sample Collection and Injection. The instruments may offer up to three different ways for sample introduction:

- Direct on-column injection using a syringe. There are different syringe types for air samples and for liquid samples. Volumes of air samples injected to the column are typically 2 mL or less, while volumes of liquid samples injected to the column can range up to 2 μL. Aqueous and soil samples can be extracted with an organic solvent prior to direct injection.
- Using a *concentrator*, which is a small tube packed with an adsorbent material such as Tenax, attached to the internal sampling pump. The airborne vapors are collected on the concentrator, and then they are desorbed into the column by reversing the carrier gas flow while heating the concentrator. This process is also called *thermal desorption*. This technique allows quantification of lower airborne concentrations than can be measured using a syringe or direct (loop) injection. Sampling time using the concentrator varies from 10 seconds to 10 minutes and can measure concentra-

TABLE 12.1. Typical Sampling Times for Portable GC Instruments[1]

	Typical Sampling Time, seconds				
	Concentrator[a]				
Medium	Low (0.5–5 ppb)	Medium (5–50 ppb)	High (50 ppb–5 ppm)	Sampling Loop	Direct Injection
Air	300	120	10	10	NA
Water	200	120	NA	10	NA
Soil	200	120	NA	10	NA

[a] The longer the sampling time, the more sensitive the analysis will become due to the higher level of contaminants collected on the concentrator.

tions ranging as low as parts per trillion (ppt).

- Sample loop injection, where the internal pump draws an air sample into a fix volume loop, which is subsequently injected onto the column. The volume of the loop in a typical device can vary from 0.05 cc to 5.0 cc.

Table 12.1 shows typical sampling times for different injection methods.

An instrument with a heated injection port can handle less volatile material than can an instrument with an injection port that is maintained at ambient temperature; a heated injection port is required to analyze liquid samples.

Chromatographic Separation. In this part of the process a carrier gas (also called the *mobile phase*) is used to move the components through the column to the detector. The high-sensitivity detectors typical of portable GCs require high-purity carrier gas such as helium or hydrogen. Generally, a small internal cylinder holds enough carrier gas for 4–8 hours of operation; the internal cylinder can be refilled in the field from a larger compressed gas cylinder.

There are two types of columns: packed; and capillary. The packed column, typically 1/8 inch in diameter and up to 20 feet long, contains an inert solid support that is coated with a liquid material. The capillary

column is a very narrow tube (<1 mm in diameter) that may be over 100 meters long and contains a coating on the interior wall of the column. The function of the column geometry and coating is to control the movement of sample molecules so the optimum separation is achieved for the materials of interest as they emerge from the column (Figure 12.3). The stronger the interaction between the compound in the vapor phase and the stationary phase, the more strongly the movement of the compound will be slowed by the column, and therefore the longer its retention time. Figure 12.4 shows different column types, and Table 12.2 lists some commonly used stationary phases for GC columns. The packed column can accept a larger injection volume, but the capillary column has better separating power. As an example, Figure 12.5 shows the difference between an analysis of gasoline vapor using a packed and capillary column. In some instruments, two or more columns are installed at the same time. If a nondestructive sensor is used, the columns may be connected in series; otherwise they can be mounted parallel and a sample splitter used. Use of more than one column for a single analytical sample run permits better separation of some peaks. It should also be noted that columns for portable GCs may require special shapes and sizes, materials of construction, and other features to fit a

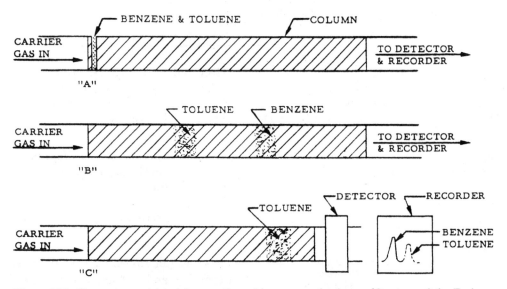

Figure 12.3. Chromatogram/pictorial separation of benzene and toluene. (Courtesy of the Foxboro Company.)

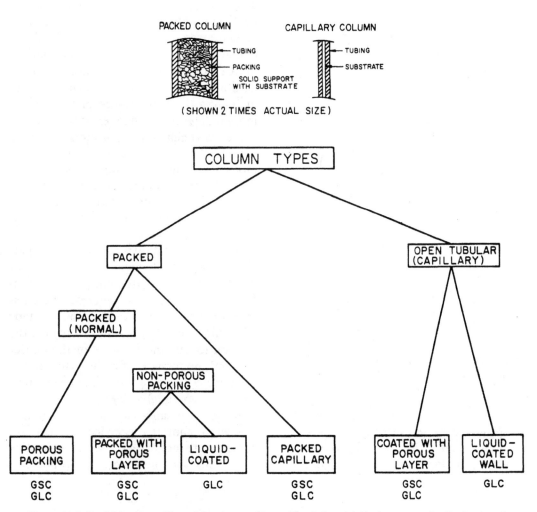

Figure 12.4. Packed and capillary GC columns. (From *The Industrial Environment—Its Evaluation & Control*, NIOSH, Cincinnati, 1973, p. 262.)

TABLE 12.2. Commonly Used Stationary Phases for GC

Phases	Applications
Liquid	
SE-30, OV-1 (methyl silicones)	Hydrocarbons, chlorinated hydrocarbons
OV-1, SE-54 (methyl/phenyl silicones)	PAHs, chlorinated pesticides, hydrocarbons
Carbonax 20 M (polyethylene glycol)	Polar compounds such as esters, alcohols
FFAP, SP-1000 (polyethylene glycol terephthalate)	Phenols, volatile acids
Solid (Packed Columns Only)	
Chromosorb, Porapak series (styrene/ divinylbenzene polymers)	Volatile alcohols, ketones, hydrocarbons, halocarbons (boiling points 30–100°C)
Carbon molecular sieves (Carbosphere, Spherocarb, Carbosieve)	C_1–C_5 hydrocarbons
Porous silica (Unibeads, Porasil A)	C_1–C_5 hydrocarbons

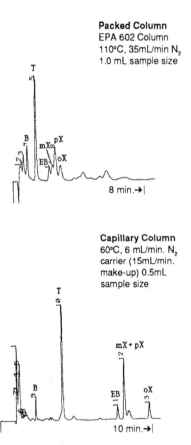

Packed Column
EPA 602 Column
110°C, 35mL/min N$_2$
1.0 mL sample size

8 min.→|

Capillary Column
60°C, 6 mL/min. N$_2$
carrier (15mL/min.
make-up) 0.5mL
sample size

mX + pX

10 min.→|

Figure 12.5. Gasoline vapor analysis on packed and capillary columns. (Courtesy of HNU Systems, Inc.)

dimensions (length and inside diameter). Custom sizes and packings can be made for special sampling situations.

In most high-end portable GCs, either a constant-temperature heated column or temperature programming is used to enhance separation. Temperatures above ambient temperature increase the volatility of the contaminant molecules; and temperature programming, which involves increasing the temperature of the column in a predetermined manner as the analysis proceeds, enhances separation of many complex mixtures because the lower-molecular-weight compounds move through the column at the lower temperature, while heavier molecules begin to move more quickly as the column temperature is increased.

The ideal situation is to choose the operating parameters to yield sharp and narrow peaks that are easy to identify and quantitate for the materials of interest. The separation is primarily dependent on the type of column and its packing or coating, the column length, the flowrate of the carrier gas, and the temperature characteristics of the system. These variables can be changed to achieve optimal separation of the components in an airborne mixture, but it is critical to calibrate the device with a known gas mixture under each set of operating conditions and then store the calibration

specific application. In most cases, columns are provided by the instrument manufacturer and are described in terms of column type, the packing or coating, and their data in a data library so the proper calibration data are used for each sample run.

One feature available on some field GC units is the ability to backflush the column to shorten the analysis run time and also prevent heavier compounds from reaching the analytical column. Backflushing requires two columns: a shorter pre-column and a longer analytical column. They are usually connected in series with a "tee" fitting. During an analytical run, the sample is carried through the pre-column toward the analytical column. When the compounds of interest have cleared the pre-column, a valve arrangement switches the direction of the carrier gas flow so it flows through the "tee" connector to both columns. The components that have already passed the junction continue through the analytical column to the detector, while compounds before the "tee" are flushed back out of the pre-column and vented. It is important to select the correct time to reverse the flow through the pre-column, or compounds of interest may be lost or undesirable compounds may be driven through the analytical column.

Identification and Integration of Peaks.
During this step a detector senses the presence and amount of compounds as they elute from the column. Table 12.3 lists characteristics of several common types of detectors; some portable GCs feature additional detectors for specific applications. The instrument's software compares each peak as it elutes to the calibration data and shows the name, concentration level, and retention time of the compounds identified during calibration. Compounds that are detected, but which do not match compounds identified during calibration, are listed as "unknowns" and often may be identified by the software by comparing the sample analysis results with other calibration results stored in the instrument's memory or by scanning various compound libraries that may contain several hundred compounds. The compound's concentration is calculated from the area under its peak, a step called *peak integration*.

Some instruments can be programmed so there is a delay between when the first peaks emerge from the column and when the compound identification and peak integration begins. This feature, called the *inhibit time*, is helpful in reducing the amount of data generated when the initial, lighter molecules do not represent compounds of interest.

Data Display and Management.
The laptop computer and proprietary software stores and displays relevant data such as the sample chromatogram, calibration chromatogram, retention times, concentration, and operating parameters. Data management is critical with portable GCs because compound identification is based on the principle that when a specific column is maintained at a specific constant temperature and carrier gas flow conditions, the retention time of each compound is constant. Thus all the operating parameters (column type, temperature programming cycle, carrier gas flowrate, etc.) must be stored along with the calibration data so that proper identification of peaks in actual samples is achieved. Additionally, the instrument must be recalibrated frequently, and so adequate data storage capability is needed to store the calibration and sample data.

Portable GC Instruments

This section describes several commercial portable GC instruments that illustrate features typically available with this category of instrument.

Photovac Voyager Portable GC.
The Voyager Portable GC (Figure 12.6) is designed for environmental site characterization and exposure monitoring. Weighing

TABLE 12.3. Types of GC Detectors

Type	Principle	Compounds Detected
Flame ionization detector (FID)	A stainless steel burner in which hydrogen is mixed with the sample stream; combustion air feeds in and diffuses around the jet (burner) where ignition occurs. The current carried across the gap of a platinum-loop collector electrode is proportional to the number of ions generated by burning the sample. Not as sensitive as many other detectors. Large linear range. Response is relatively uniform from compound to compound.	Almost all organics; response is greatest to hydrocarbons; decreases with increasing substitutions of other elements: O, S, Cl. Not sensitive to water, the permanent gases, and most inorganic compounds.
Photoionization detector (PID)	A sealed UV light source emits photons with an energy level high enough to ionize the compounds in the sample. In a chamber exposed to the light source containing a pair of electrodes, ions formed by the absorption of photons are driven to the collector electrode where a current is produced that is proportional to the concentration. Large linear range; high sensitivity; selectivity can be introduced by using different energy lamps. Response varies from compound to compound.	Organics, some inorganics, lower response to low-molecular-weight hydrocarbons. Can detect aliphatic, aromatic, halogenated hydrocarbons. Arsine, phosphine, hydrogen sulfide. Sensitive to water. No response to methane. Especially good for aromatic hydrocarbons.
Electron capture detector (ECD)	Utilizes a radioactive source such as ^{63}Ni to supply energy to the detector in the form of β radiation. Intensity of the electron beam arriving at a collection electrode is monitored. When an electron-capturing species passes through the cell the intensity of the electron beam decreases, sending out an electronic signal. Response varies widely from compound to compound. Highly sensitive and selective. Potential problems include excessive heat, which could vaporize some of the source; "aging" of the coated foil, which must be replaced; gradual loss of the source's activity.	Compounds containing halogens, cyano or nitro groups; ECD has a minimal response to hydrocarbons, alcohols, and ketones.

15 pounds, it is equipped with an internal refillable carrier gas cylinder and rechargeable battery for independent operation for up to 8 hours. Sampling practitioners can view the chromatograms or tabular data using the built-in LCD or by downloading to a computer in the field or office.[2]

The Photovac Voyager GC can be used to analyze air samples, soil gas, or the headspace above water or soil samples. Gaseous

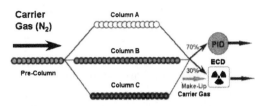

Figure 12.7. Photovac Voyager™ multi-column configuration allows separation of up to 40 compounds in a single sample. (Courtesy of Photovac, Inc.)

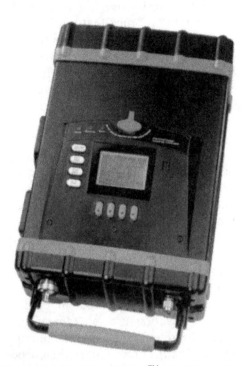

Figure 12.6. Photovac Voyager™ portable gas chromatograph. (Courtesy of Photovac, Inc.)

samples can be injected manually by syringe, or sampled using a built-in pump and loop injection. Loop injections are ideal for fence-line, ambient, confined space, emergency response, leak detection, and similar applications. Syringe injections are used for applications such as headspace analysis of waste streams, ground water, and soil extracts. The sampling practitioner can choose from a manual or continuous loop injection sampling mode. The instrument can be set to sample at user-defined intervals, thereby providing a time-history profile of concentration levels for site-specific compounds.

The Voyager uses two detectors: a photoionization detector (PID) operating at 10.6 eV for volatile organic compounds (VOCs) and an electron capture detector (ECD). The ECD is the most sensitive detector available for the analysis of electrophilic compounds such as chlorinated

hydrocarbons and other halogen-containing organic compounds. A Voyager with a PID and an ECD requires ultrazero- or zero-grade nitrogen carrier gas. The nitrogen must be 99.999% pure and contain less than 0.1 ppm of hydrocarbon contamination. When only a PID is used, the instrument can be run with nitrogen or zero-grade air as the carrier gas. Like nitrogen, the air must be 99.999% pure and contain less than 0.1 ppm of hydrocarbon contamination.

The instrument employs an innovative arrangement of analytical columns, preprogrammed temperatures, and flowrates to optimize the separation of complex VOC mixtures found in specific industries. One pre-column and three parallel analytical columns (Figure 12.7) make up this configuration. Using three separate columns allows separation of up to 40 compounds in a single sample. Excellent compound resolution on at least one of the three column phases generally occurs. A backflushing feature can be used to reduce total analysis time and column contamination by preventing the heavier "nontarget" analytes from entering the analytical columns.

The Voyager GC has a built-in total VOC (TVOC) operational mode which allows a quick screen of samples for the presence of VOCs. In a TVOC run, the sample passes directly into the PID without going through the analytical columns. If a preset alarm level is exceeded, a full GC analysis may be performed to determine which compounds are present. A TVOC analysis is sent through the PID only—a TVOC sample cannot be run through the ECD.

Proprietary software (SiteChart™) is provided with this instrument. The software sets up the preprogrammed parameters for different calibration and analytical runs (called *assays*) and is also helpful for data review and printing of the data including chromatograms and tables of peak values. Each assay automates the setup of the instrument operating conditions and data reduction parameters pertinent to identifying and quantifying a predefined set of compounds. The preset operation parameters ensure the best separation and detection possible for the compounds in a sample. The Voyagers are equipped with preprogrammed assays to detect a set of target analytes for the following specific industries: environmental (i.e., the 40 compounds on the U.S. EPA's TO-14 priority pollutants list); petrochemical/ refining; rubber/plastic; pulp & paper; surfactants/sterilants; and latex polymers. In addition, each instrument is furnished with a unique assay disk developed from the setup and calibration performed at the factory to meet the user's specified needs by running target compounds on that instrument and saving the respective retention times and peak areas in the library. The sampling practitioner can also use the SiteChart™ software to set up and store additional assays.

The Voyager portable GC can be run in the field using a personal computer with SiteChart™, or using the keypad on the instrument as long as the desired assay has been downloaded to the instrument from the computer using the SiteChart™ software. The Voyager data logs every completed analysis. Up to 40 GC log entries can be stored in the instrument's memory, which is sufficient for an 8-hour day based an analysis every 10 minutes. TVOC runs occupy less memory than GC runs, so the device will typically be able to store 10 times more TVOC runs than GC entries. Once the memory is full, the data may be downloaded or else deleted from memory so that additional runs may be performed. Each GC record includes a complete peak report and chromatogram. Once data have been downloaded to the computer, they can be reviewed by scrolling through the list of completed runs, then selecting, analyzing, and printing a specific chromatogram or data summary.

Like all GC instruments, proper calibration is critical. When a calibration run is performed, the data are stored in a library in the Voyager's memory. The library contains the retention time, peak area, and concentration of each target analyte. When a standard, preprogrammed method is changed (e.g., pressure or temperature change), the retention times for the compounds will shift. A new library then needs to be created to identify the target analytes by running the compounds of interest with the current method conditions and saving the retention time in the respective library.

While the manufacturer strongly recommends that the Voyager be calibrated using a calibration mix that contains every compound in the column's library, it is possible to calibrate it using a calibration mixture containing only a few of the compounds. Called a *ratiometric* calibration, the instrument performs a calibration with each compound in the calibration gas mixture and then calculates the approximate retention times and concentrations settings for the compounds that were not present in the calibration mixture. After a ratiometric calibration, the instrument's compound library is updated to reflect the actual and calculated information. Since the approximation may affect compound identification and the accuracy of the compound concentration reported by the instrument, a ratiometric calibration is not considered as good as an actual calibration with a known standard of each compound of interest.

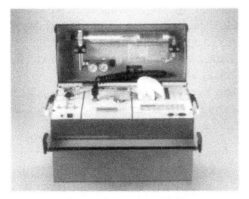

Figure 12.8. HNU Model GC-311 Portable GC can analyze gases, liquids, and soil extractions using a variety of columns and detectors. (Courtesy of PID Analyzers, LLC.)

HNU Model GC-311 Portable GC. This microprocessor controlled portable instrument (Figure 12.8) can analyze gases (ambient air, headspace, stack samples, process gases, etc.), liquids (water, solvents), and either headspace or liquid extractions of solid samples such as soils. It has a heated injector for manual injection of gases or liquids and has dual detector capability chosen from six interchangeable detectors: Photoionization detector (PID), flame ionization detector (FID), electron capture detector (ECD), flame photometric detector (FPD), thermal conductivity detector (TCD), or far ultraviolet absorbance detector (FUVD). The instrument is provided with HNU's proprietary software called PeakWorks™ for Windows® Chromatography Software for data management and control of the GC.[3]

Analyses can be run in series with nondestructive detectors such as PID or FUVD or in parallel with the other detectors to improve or verify compound identification. A built-in concentrator/thermal desorber is available as an option for indoor air quality and fenceline measurements where very low levels (ppb to ppt) are to be measured.

The instrument features a large oven so any manufacturers' columns can be used. The oven can be operated at a constant temperature (isothermal) or with temperature programming to increase temperature as the sample run progresses (up to 200°C) for faster analysis, or analysis of volatile organic compounds (VOCs) and semivolatile organic compounds such as polynuclear aromatic hydrocarbons (PAHs), some pesticides, and plasticizers in the same sample. Typical columns that are used include:

- *Packed Columns.* 1/4″, 1/8″, or 1/16″ (micropack) 2–3 meters in length with 300–500 plates per meter. The typical packing material is porous polymer, liquid phase (1–3%) on diatomite.
- *Capillary columns.* 0.53-, 0.32-, 0.20-, 0.15-mm column with liquid phase bonded to the fused silica; available in fused silica-lined stainless steel with the liquid phase bonded to the silica; efficiency >1000–3000 plates/m with typical length 15–30 meters.

Sentex Scentograph PLUS II. The Scentograph PLUS II, from Sentex Systems, Inc., is designed to operate as a portable unit or in a fixed location (Figure 12.9). It is capable of analyzing air, soil, and water samples with concentrations ranging from parts-per-trillion (ppt) to percent (%) levels. This is accomplished by combining different ways of sample introduction with different types of detectors. Samples can be analyzed by direct on-column injection, using a concentrator or using a sample loop.[1]

The detectors offered are micro argon ionization detector (MAID), argon ionization detector (AID), electron capture detector (ECD), thermal conductivity detector (TCD), and the photoionization detector (PID). The AID is sensitive to organic compounds having ionization potential of 11.7 eV or lower. These compounds include the halomethanes and

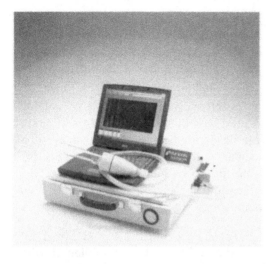

Figure 12.9. The Scentograph PLUS II field-portable GC unit. (Courtesy of Inficon, Inc.)

haloethanes, carbon tetrachloride, and 1,1,1-trichloroethane, which are poorly detected by other common field detectors. This detector is capable of detecting these compounds as well as other hydrocarbons down to ppb levels. The MAID is a smaller, more sensitive version of the AID. It is ideal for use with capillary columns and detects organic compounds to below 1 ppb level. The instrument can contain two detectors at any one time; however, only one can operate at a time. The operating temperature can be set between 30 to 179°C; a temperature above 50°C is recommended for optimum performance.

The internal carrier gas cylinder is filled with argon (99.999% pure) for the AID or helium (99.999% pure) for the ECD, PID, or TCD. The carrier gas cylinder will allow a minimum of 8 hours of operation and is easily refilled. It is also equipped with an internal calibration cylinder, which supplies gas directly to the instrument's internal calibration system. Calibration gas from the internal cylinder flows through a regulator directly to the sample loop or concentrator during calibration. The calibration cylinder is easily refillable and provides a minimum

of 8 hours' supply of calibration gas. There is also a calibration port that is used to calibrate from a sampling bag, headspace of an external container, or other external source at ambient pressure.

The Scentograph PLUS II can use the following methods to identify and quantify components in samples:

- Using current calibration data, which is based on direct calibration [either single point or multipoint (up to 5 points)] with certified standards for up to 48 chemicals. During analysis, the instrument compares the retention times obtained with those of the calibration run. When there is a match, the name of the compound and the calculated concentration level is displayed on the monitor. Peaks with no match are identified as "unknown." This method is the most accurate and is identical to that used in laboratory gas chromatographic analysis.

- Using previous calibration information stored in data files, which is similar to the previous method with the exception that the calibration data is recalled from the system's memory for comparison to the analysis data. For example, suppose a calibration run was performed under certain parameters and the values were stored in the instrument's memory. If an analysis is conducted at a later time under the same parameters, it is possible to recall the previous calibration and compare it with that analysis. If there is a match, the peaks will be named and their concentrations will be determined.

- Using a computer library search, which is based on the assumption that those relative retention times of compounds separated on a column remain unchanged, provided that the operating conditions of the analysis are kept the same. During an analysis run, the system will first "look" for compounds that were defined in the calibration run. "Unmatched" compounds are then screened using data stored in different internal libraries. The library search method is a convenient procedure for analysis. However, library values must be checked frequently in order to maintain their accuracy. The Scentograph PLUS II can contain several libraries (i.e., databases), each containing up to 48 compounds.

The instrument uses menu-driven software to control its operation, process its data, and store its chromatograms. The systems can be operated manually or set to analyze and calibrate at chosen intervals, storing the results on disk or hard drive for future review. A detachable laptop computer controls the operation and provides the data system. The proprietary Sentex software operates the GC and collects, stores, and processes data.

Typical Operation. The operation of any portable GC depends on the features of the instrument and the manufacturer's instructions. A typical operating sequence for portable use of the Scentograph PLUS II is as follows[1]:

1. Verify that the batteries are fully charged by charging overnight particularly after a prolonged period of nonuse.

2. Turn on the carrier gas valve and the calibration gas valve (if the internal calibration cylinder is to be used at the calibration source). Verify that both gas cylinders are full by inspecting the pressure gauges when valves are opened. Refill cylinders if needed before field use.

3. If calibration from an external source is desired, the internal calibration cylinder gas valve should be turned off. The instrument will automatically draw a sample through the external calibration port.

4. Inspect the column pressure using the column pressure gauge. Adjust column pressure if necessary. Typical operating pressure should be 20–30 psi for packed columns and 8–15 psi for capillary columns.

5. Connect the RS-232 connector from the instrument to the computer, turn on the computer and start the software program. The computer will automatically power the GC unit, initiate the start-up cycle, and display relevant parameters and prompts.

6. Select the calibration or analysis mode as appropriate. Change parameters, select features as required, and manipulate data as prompted by the computer.

7. When using direct syringe injection of samples, use these guidelines:

   • *Gas Samples.* Sample volumes up to 2 cc can be injected. A gas-tight syringe should be used for handling the samples. The syringe needle should be 22–26 gauge with either a 22° bevel point or a side port tip. The volume to be injected is dependent on the sample concentration. The higher the sample concentration, the smaller the sample injection volume must be to avoid overloading the detector. For sample concentrations between 500 and 5000 ppm, a sample size of 0.5 cc or less should be used. If sample concentration is unknown, start with a smaller size injection. Increase the injection volume to increase the analysis sensitivity. For the lower ppm range (1 to 10 ppm), an injection size of 1 cc can be used. Parts per billion (ppb) analyses of gas samples would require a sample injection of up to 2 cc.

   • *Liquid Samples.* The sample size is much less than that used for a gas sample injection, typically 4 μL or less, depending on the sample concentration. The capacity of the syringe should range between 1 μL and 10 μL, and it should have a 22- to 26-gauge needle with either a 22° bevel point or a side point tip.

8. After use, follow the computer prompts to shut down the instrument. When the instrument is not in use, it should be connected to the charger at all times.

Portable GC Maintenance and Use

In addition to the material already covered in this chapter, there are several maintenance and use considerations for portable GCs. Routine maintenance involves steps such as:

- Replacing injection port septum every 2–5 injections.
- Replacing inlet filter approximately once/week, depending on sample dust content.
- Cleaning UV lamp window as needed.
- Replacing UV lamp when needed.
- Cleaning injection port as required.
- Conditioning column when changing column or if contamination is suspected.
- Purging internal carrier gas cylinder before shipping instrument or if cylinder becomes completely empty.

Any column can become contaminated with compounds having long retention times. If contamination of a column is suspected, installing a new column is one way to check. Contamination may be prevented by using the backflushing feature if available, or reduced by flushing the column with carrier gas for a period of time after every analysis. However, a disadvantage of purging the column after each analysis is the time involved since it will decrease the number of samples that can be analyzed in a day. For a contaminated column, simple flushing may be used or baking the column in a drying oven at a temperature recommended by the manufacturer while passing nitrogen or another gas through the column. The baking time and temperature depends on the type of column being cleaned.

When changing columns, a conditioning period is required. If the ends of the column have been kept capped during storage, the length of conditioning time is reduced. When installing the column, avoid touching the unprotected column ends as contamination may result.

Although portable GCs have extensive data-logging and data management capability, it is usually helpful to keep a separate record of key information. Figure 12.10 shows typical information elements to be recorded for use of a portable GC instrument.

It is also important to consider the appropriate shipping regulations concerning compressed gases if a survey is planned at a distant location. Generally, it is best to have carrier gas and calibration gas cylinders shipped to the survey location ahead of time or purchased locally, since there are many restrictions on shipping compressed gas cylinders.

Storage Requirements

Proper storage of portable GCs is important and the requirements vary, depending on how soon the instrument should be ready for use. For example, one manufacturer gives these storage guidelines[2]:

- Maintain a flow of carrier gas through the column to prevent contamination. For longer storage periods, connect the instrument to an external supply of carrier gas.
- Do not allow the internal carrier gas cylinder to empty completely to

Plant _____ Date _____

Location _____

General information

Source temperature (°C) _____	Columnar temperature:
Probe temperature (°C) _____	Initial (°C)/time (min) _____
Ambient temperature (°C) _____	Program rate (°C/min) _____
Atmospheric pressure (mm) _____	Final (°C)/time (min) _____
Source pressure (^{11}Hg) _____	Carrier gas flow rate (ml/min) _____
Absolute source pressure (mm) _____	Detector temperature (°C) _____
Sampling rate (liter/min) _____	Injection time (24-hour basis) _____
Sample loop volume (ml) _____	Chart speed (mm/min) _____
Sample loop temperature (°C) _____	Dilution gas flow rate (ml/min) _____
	Dilution gas used (symbol) _____
	Dilution ratio _____

Components to be analyzed	Expected concentration
_____	_____
_____	_____
_____	_____
_____	_____
_____	_____
_____	_____

Suggested chromatographic column _____

Column flow rate _____ml/min   Head pressure _____mm Hg

Column temperature:

Isothermal _____ °C

Programmed from _____ °C to _____ °C at _____ °C/min

Injection port/sample loop temperature _____ °C

Detector temperature _____ °C

Detector flow rates: Hydrogen   _____ ml/min.

head pressure _____ mm Hg

Air/Oxygen   _____ ml/min.

head pressure _____ mm Hg

Chart speed _____ inches/minute

Compound data:

Compound	Retention time	Attenuation
_____	_____	_____
_____	_____	_____
_____	_____	_____
_____	_____	_____
_____	_____	_____

Sampling considerations

Location to set up GC _____

Special hazards to be considered _____

Power available at duct _____

Power available for GC _____

Plant safety requirements _____

Vehicle traffic rules _____

Plant entry requirements _____

Security agreements _____

Potential problems _____

Site diagrams. (Attach additional sheets if required).

Figure 12.10. GC field sampling data sheet.

prevent contamination of the internal cylinder and columns. If the internal cylinder empties completely, it must be purged before refilling. The column must then be flushed, and the instrument must be allowed to warm up for 30 minutes. The baseline of the instrument must be allowed to stabilize before it is ready for use.

- Store the instrument in a location that is free of volatile organic vapors and gases, with ambient temperature between 32°F and 105°F.
- If the battery has been discharged, connect the instrument to the AC adapter and recharge the battery.
- For immediate availability, the following steps are recommended:
  ○ Leave the instrument turned on.
  ○ Calibrate once every 24 hours.
  ○ Connect the instrument to a carrier gas cylinder, or refill the internal cylinder every 6–8 hours.
  ○ Connect the instrument to the AC adapter when not in use.
- For availability within 30 minutes:
  ○ Turn instrument off.
  ○ Connect it to an external cylinder of carrier gas.
  ○ Connect it to the AC adapter.
  ○ To use the GC, turn it on and allow a 30-minute warm-up period. Then calibrate it before use.
- For availability over a longer period:
  ○ Follow the storage recommendations for 30-minute availability, but let the instrument stabilize for at least 12 hours before calibration. This will result in optimum performance.

NIOSH Method for Ethylene Oxide By Portable GC

One established method using a portable GC is NIOSH *Manual of Analytical Methods* (NMAM) Number 3702[4]:

Procedure

1. Set up a portable GC with a PID using the following parameters: column = 1.2 m × 3 mm OD PTFE, packed with Carbopak BHT 40/100 mesh. Carrier gas flow rate = 15 ml/min. Also assemble portable computer or integrator, battery charger, regulator, and any other peripherals necessary for individual instruments. Allow the instrument to warm up and equilibrate for 30 minutes.

2. Check the sampling equipment to prevent contamination. Use different syringes for sampling and for standard preparation. Identify each syringe with a unique number. Segregate bags used for sample collection from those used for calibration standards.

3. Calibrate the GC daily in the field. Prepare bag standards by adding a known volume of ethylene oxide to a known volume of clean air in a bag. By adding a known volume in μL to a specified number of liters of air, a known concentration in ppm will be created. *Note:* Because ethylene oxide is a flammable gas, the shipment of the compressed gas must comply with 49CFR 171–177 regulations regarding shipment of hazardous materials.

4. Evacuate a 5-L to 10-L bag completely by drawing the air out with a large 1-L to 2-L syringe.

5. Draw clean air (or oxygen or nitrogen) from a supply cylinder into the syringe for measured transfer into the bag. Alternately, if a clean air supply is not available, draw room air through charcoal sorbent into the syringe. Repeat until the bag contains 5 L of air.

6. Add a known amount of pure ethylene oxide or standard ethylene oxide mixture to the bag by means

of a gas-tight syringe. *Example:* Using a gas-tight syringe, take 50 μL from a cylinder of pure ethylene oxide and inject it into 5 L of air to create a 10-ppm standard. Alternately, 200 μL of a 27% v/v ethylene oxide mixture (e.g., 88/12 w/w Freon 12 and ethylene oxide) can be added to the 5 L of air to obtain a 10.8-ppm standard.

7. Allow the bag to equilibrate, with occasional kneading, for at least 5 minutes.

8. Analyze aliquots of various sizes to establish a calibration graph. For each point three replicate samples should be done. For example, using a high instrument attenuation (low sensitivity), injections of 0.2 mL, 0.4 mL, 0.6 mL, 0.8 mL, and 10 mL, might be possible. These amounts would correspond to injections of 2 ηL, 4 ηL, 6 ηL, 8 ηL, and 10 ηL. On a more sensitive attenuation, injections of 0.02 mL, 0.04 mL, 0.06 mL, 0.08 mL, and 0.10 mL would be typical. Results will vary from instrument to instrument, and from time to time on the same instrument.

9. Plot ηL of ethylene oxide versus peak height or area if the GC cannot do this automatically. This plot should be a straight line.

10. Periodically throughout the day, check the calibration by repeating some of these injections. Ideally, each sample would be bracketed, before and after, with injections of standards, although this situation is seldom practical. Some GCs are capable of doing automatic periodic calibrations with programming.

11. Collect samples by drawing air from the contaminated area directly into a syringe or filling bags for an integrated TWA sample.

12. For syringe sampling, draw air directly into a gas syringe. Collect syringe samples by first purging a gas-tight syringe several times with clean air to remove any residual ethylene oxide from previous samples; then draw air into the syringe at the time and location of interest. If larger syringe samples are desired, such as 20 mL of air, it will be necessary to transfer some of the sample to the smaller syringe being used in the chromatograph. It is essential that the sizes of all grab sample injections be the same if concentrations are to be compared. A rubber cap placed over the end port of the larger syringe can serve as a septum for the smaller (500 μL) syringe.

13. For integrated air samples for TWA determinations, a clean bag of plastic or other material must first be evacuated. A personal air sampling pump is used at the highest airflow available.

14. Attach the plastic bag via tubing to the outlet of a personal sampling pump and pump air from the contaminated area into the bag at a rate calculated to fill the bag over the sampling period. This rate will be between 20 mL/min and 500 mL/min. Terminate sampling before the bag is 80% full. The pump's flow rate must be within ±5% of the initial setting throughout the sampling period.

15. Analyze the bag sample within 2 hours after completion of the sampling to minimize loss of analyte by adsorption and permeation as follows: Fill a gas-tight syringe, purged several times with sample, from the sample bag. Then empty it to the desired volume, and inject that volume into the chromatograph with a quick firm motion. Record the number of the syringe. Use replicate analyses to determine the repeatability of the analysis. If no

estimate of concentration is available, use an injection volume 10 μL to 25 μL at a high attenuation to reduce the possibility of column and detector overload. Depending on the results of this injection, larger volumes and/or more-sensitive attenuations may be selected.

16. Calculate the concentration of ethylene oxide from the calibration graph (ηL) and the injection volume (mL):

$$ppm = \frac{\mu L(gas)}{L(gas)} = \frac{\eta L(gas)}{mL(gas)}$$

Interpreting Measurements with the Portable GC

As noted previously, it is possible for more than one compound to have the same retention time, and in certain circumstances where unknowns may be present, it will mean that a peak appearing at or near a given retention time is not 100% confirmation of a given compound's presence. However, if the instrument is operating properly, a lack of peaks will usually mean there are no compounds present at the detection level of the instrument. The best way to confirm the presence of a given compound is to collect additional samples for analysis in the laboratory on a GC with two different columns, each of which is capable of resolving the mixture of interest and doing sequential analyses. As this is difficult to do in most portable GCs where the results are questionable, bag samples should be collected and taken to a laboratory for analysis. The interpretation of a chromatogram requires the use of calibration reference data that have been generated through testing.

Sources of Error. When using a GC for analysis, there are several sources of error that must be understood and avoided in

order to yield accurate results. These include[1,2]:

- Peak integration errors that are due to problems in starting or stopping the peak integration by the instrument or computer software. The potential problems include baseline noise, interfering (co-eluting) peaks, or excessive tailing of the peaks of interest. These errors can occur somewhat randomly and cause different integrations for consecutive runs for the same concentration of a specific contaminant under the same operating conditions. Because of this potential for error, it is important for the sampling practitioner to review the chromatograms for obvious errors and to reintegrate where necessary using different parameters to achieve accurate results.

- Syringe injection techniques may cause a large "trailing peak" on the chromatogram that is not due to a target compound. If an injection peak is observed, it may be eliminated by setting the inhibit time one or two seconds longer than the retention time indicated for this peak.

- Sampling errors occur when the analytical sample entering the instrument is different from the airborne composition and concentration that exists at the sampling site. This can be due to reasons such as nonstandard syringe sampling and injection technique, contaminants not being completed desorbed when a concentrator is used, or contaminants condensing or otherwise absorbing on the walls of Tedlar sampling bags or in sampling lines. Careful attention to sampling techniques, including the use of heated sampling lines and bags where needed, along with rigorous laboratory evaluation of planned sampling techniques can help to identify and reduce sampling errors.

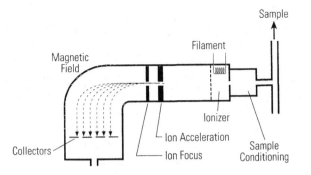

Figure 12.11. Diagram of a typical mass spectrograph. (Courtesy of International Sensor Technology.)

- Adsorption of contaminants of interest within the internals of the GC system can sometimes occur, especially when the concentration of contaminants in the sample is extremely low. If this does occur, it often affects only the first few samples since there are generally only a limited number of "absorption sites" within the instrument, and the error resolves once these sites are occupied. This type of loss can be seen by gradually increasing results when the same concentration samples are run repeatedly until the available sites are occupied. Use of a heated GC system helps to minimize this type of error.

- Detector response drift occurs when the effective response of the detector changes over time. For example, for a PID the effective UV lamp intensity may change. This drift is a time-dependent phenomenon, and it is not as compound-specific as several of the other types of error. Frequent calibration, and especially recalibration when the detector's baseline response has changed by a certain amount (often 10%), is a way to prevent this error. The calibration should be performed under similar temperature and humidity conditions as the actual sampling.

Monitoring Records. Chromatograms and information on the operating parameters and calibration should be retained as part of the record of the sampling. The amount of data that are generated by a GC instrument is very large, and so some systematic way to identify and store the relevant chromatograms and supporting information is critical. Figure 12.10 shows a typical sampling to form for recording supplemental information.

GC/Mass Spectrophotometer Instruments. A GC/mass spectrometer (GC/MS) is a GC instrument that is equipped with a special detector that passes the ionized molecules emerging from the separation column through an electric field that focuses the molecules by atomic mass (Figure 12.11). Each gas molecule has a specific mass number or weight. When the molecules are ionized in an electric field and then focused into an ion beam, the beam can be directed into a magnetic field that is perpendicular to the ion beam. The magnetic field causes the ions to scatter and deposit onto different collectors according to their mass number. The heavier molecules have more kinetic energy and thus will be influenced less by the magnetic field than lighter molecules. As a result, the heavier molecules impact the collectors that are toward the outside of the curved detector. The ions reaching each collector plate are "counted" as electrical signals.

MS sometimes provides a better means of identifying specific compounds using a library of data developed for the specific instrument than do other GC detectors. A GC/MS is especially useful when compounds emerge at the same time from the GC column (co-elute) despite efforts to adjust operating parameters to obtain good separation, or for identifying unknown compounds since the spectrographic pattern of the unknown peaks can be compared to information in the built-in data library.

GC/MS instruments have long been used in the laboratory, but their use as portable field devices is relatively new. They give the sampling practitioner new options, but are costly and complex to operate. They probably add the most value when cost savings or other benefits result from having real-time information in the field and a GC or other, less complex instrument will not suffice.

The HAPSITE by Inficon (Figure 12.12) is a portable GC/MS. It weighs about 35 pounds and contains many of the features discussed earlier for portable GC instruments such as temperature programming and an optional concentrator for very low level air samples. It uses nitrogen as a carrier gas and has an internal calibration gas cylinder. It has several built-in programs to make field use more convenient. Once the sampling practitioner selects a method, the instrument automatically acquires and analyzes the sample. The instrument monitors critical parameters, thereby verifying the tuning of the MS and mixing internal standards with the sample at the required ratios. Results can be shown on the instrument's display panel and stored for later downloading to a computer. The instrument can be used in two modes[5]:

- Survey mode where just the MS is used to give quick qualitative and semiquantitative results. This mode is good for screening samples, equipment leak checking, or fugitive emissions testing. When concentrations exceed preset threshold levels, an alarm sounds.
- Analytical mode which combines the GC and MS techniques. This permits detailed analysis of samples, and it is good for follow-up when the instrument in the Survey mode detects a compound of interest.

Typical applications of the HAPSITE include source testing for environmental permits, hazardous waste site testing, and emergency response testing. It is finding increasing use for response to terrorism events; a third-party manufacturer makes a HAPSITE "simulator" that is designed for training with the instrument without using the consumable supplies such as carrier and calibration gas.[6]

INFRARED (IR) SPECTROPHOTOMETERS

When infrared (IR) radiation is passed through a sample of gaseous molecules, the

Figure 12.12. The Inficon HAPSITE is a portable GC/MS weighing 35 pounds that brings new capability to field instruments. (Courtesy of Inficon, Inc.)

gas absorbs some specific wavelengths of the infrared radiation. This increases the energy of the gas molecules, and so their vibration or rotation increases. IR generally is that portion of the electromagnetic spectrum from 770 nm to 1000 μm. In IR spectroscopy, the term *wave number* is also used This is the number of waves in one centimeter and is the reciprocal of the wavelength, so IR ranges from about $12,900 \text{ cm}^{-1}$ to 10 cm^{-1}. Unlike PIDs and similar ionization detectors, the energy of the IR source is not high enough to cause ionization of the gas molecules. Instead, this energy transfer from the infrared radiation to the gas molecules can be seen as either (a) increased energy (temperature) in the gas molecules compared to molecules that have not been exposed to the IR radiation or (b) decreased intensity of some wavelengths of the IR radiation that is transmitted through the gas sample.

Two characteristics make IR a good method of identifying and measuring the concentration of certain compounds:

- IR absorption spectrum is unique to each gas or vapor molecules so IR can be a fairly specific detection method. Table 12.4 shows the IR absorption bands by chemical group and Table 12.5 lists the absorption wavelengths for chemicals typically measured using IR technology along with the lower detection limit.
- The amount of IR absorbed is directly related to the number of the gas molecules in the sample which gives a good measurement of concentration. Increasing the length of the IR path through the sample allows measurement of low concentrations. Internal mirrors are generally used to increase the path length to 10 meters or more a sample cell of reasonable dimensions.

Although the absorption spectrum is unique for each chemical, care must be

TABLE 12.4. Specific Infrared Absorption Bands

Grouping	Absorption Band (μm)
Alkanes	
CH_3—C—, —CH_2=	3.35–3.65
Alkenes	
—CH=CH_2	3.25–3.45
Alkyne	
—C=C—	3.05–3.25
Aromatic hydrocarbons	3.25–3.35
Substituted aromatics	6.15–6.35
Alcohols	
—OH	2.80–3.10
Acids	
—COOH	5.75–6.00
Aldehydes	
—COH	5.60–5.90
Ketones	
—C=O	5.60–5.90
Esters	
—COOR	5.75–6.00
Chlorinated compounds	
—C—Cl	12.80–15.50

Source: NIOSH: *The Industrial Environment, Its Evaluation & Control.*

taken to select the optimum wavelength for detection. For example, Trichloroethylene exhibits three absorption maxima (Figure 12.13) which occur at 10.58, 11.77, and 12.78 μm. Theoretically, any one of these could be used to detect this compound, but the 10.58-μm peak overlaps with Freon-113® and so is not suitable if Freon could be present, and the 12.78-μm peak overlaps with water vapor peaks. Therefore the 11.77-μm peak is usually the best for trichloroethylene measurement.

IR instruments vary widely in complexity and functionality. Simpler devices are dedicated to a single contaminant, while more complex instruments can measure many compounds by scanning the IR spectrum and applying sophisticated mathematical analyses to the raw data. As an illustration, an IR at 4.3 μm is commonly

TABLE 12.5. Typical Analytical Wavelength and Minimum Detectable Concentration for IR Instruments

Compound	Wavelength (μm)	Minimum Detectable Concentration (ppm)
Acetaldehyde	9.26	1.1
Acetic acid	8.72	0.3
Acetone	8.48	0.6
Acetonitrile	9.68	7.6
Acetophenone	10.70	0.3
Acetylene	3.05	1.6
Acetylene tetrabromide	8.99	1.2
Acrylonitrile	10.67	0.6
Ammonia	10.95	0.5
Aniline	9.53	0.2
Benzaldehyde	8.58	0.3
Benzene[a]	9.93	2.2
Benzyl chloride	9.54	2.5
Bromoform	8.96	0.4
1,3-Butadiene	11.10	3.9
Butane	10.40	5.1
Butyl acetate	8.33	0.6
n-Butyl alcohol	9.70	0.3
Carbon dioxide	4.72	10.2
Carbon disulfide	4.70	4.8
Carbon monoxide	4.76	2.1
Carbon tetrachloride	12.76	0.1
Cellosolve acetate	8.89	0.3
Chlorobenzene	9.40	0.4
Chlorobromomethane	8.39	1.5
Chlorodifluoromethane	9.20	1.1
Chloroform	13.12	1.1
Cresol	8.88	0.2
Cumene	9.90	2.3
Cyclohexane	3.41	0.7
Cyclopentane	11.40	9.0
Diborane	3.83	0.6
m-Dichlorobenzene	9.47	0.3
o-Dichlorobenzene	13.55	0.6
p-Dichlorobenzene	9.30	2.5
1,1-Dichloroethane	9.50	0.4
1,2-Dichloroethylene	12.30	5.3
Dichloroethyl ether	9.05	0.09
Diethylamine	8.99	1.1
Dimethylacetamide	10.10	0.4
Dimethylamine	8.79	1.2
Dimethylformamide	9.36	0.2
Dioxane	9.06	0.3
Enflurane	8.96	0.03
Ethane	12.20	10.6
Ethanolamine	12.93	2.9
Ethyl acetate	8.32	2.1
Ethyl alcohol	9.67	2.3
Ethyl benzene	9.90	3.0
Ethyl chloride	10.50	3.4

TABLE 12.5. *Continued*

Compound	Wavelength (μm)	Minimum Detectable Concentration (ppm)
Ethylene	10.70	0.5
Ethylene dibromide	8.68	0.6
Ethylene dichloride	8.37	1.3
Ethylene oxide	3.30	0.7
Ethyl ether	9.03	2.8
Formaldehyde	3.56	0.5
Formic acid	9.36	0.2
Freon 11	10.96	11.2
Freon 12	9.30	0.1
Freon 13B1	8.54	0.4
Freon 21	9.50	2.3
Freon 112	9.90	12.8
Freon 113	8.70	3.4
Freon 114	8.67	1.7
Halothane	12.46	0.3
Heptane	3.40	3.5
Hexane	3.39	3.9
Hydrazine	10.67	0.6
Hydrogen cyanide	3.03	1.7
Isoflurane	8.84	0.04
Isopropyl alcohol	8.94	1.5
Isopropyl ether	9.12	4.1
Methane	7.70	1.0
Methoxyflurane	12.10	0.2
Methyl acetate	9.7	2.3
Methyl acetylene	3.0	2.0
Methyl acrylate	8.57	0.1
Methyl alcohol	9.70	0.7
Methylamine	3.36	1.9
Methyl bromide	7.60	2.3
Methyl cellosolve	9.62	0.3
Methyl chloride	13.59	3.0
Methylene chloride	13.47	10.0
Methyl iodide	3.36	1.8
Methyl mercaptan	3.38	1.5
Methyl methacrylate	8.80	1.0
Morpholine	9.20	0.7
Nitrobenzene	11.94	0.9
Nitromethane	3.37	4.5
Nitrous oxide	4.68	0.4
Octane	3.40	0.3
Pentane	3.39	4.6
Perchloroethylene	11.10	0.2
Phosgene	11.98	0.1
Propane	3.37	6.5
Propyl alcohol	9.60	0.8
Propylene oxide	12.16	1.1
Pyridine	9.90	8.6
Styrene	11.10	0.5

TABLE 12.5. *Continued*

Compound	Wavelength (μm)	Minimum Detectable Concentration (ppm)
Sulfur dioxide	9.00	1.9
Sulfur hexafluoride	10.80	0.02
1,1,2,2-Tetrachloroethane	8.60	1.4
Tetrahydrofuran	9.40	0.6
Toluene	13.89	9.2
Total hydrocarbons	3.39	3.9
1,1,1-Trichloroethane	9.39	1.2
1,1,2-Trichloroethane	10.90	0.9
Trichloroethylene	10.84	0.4
Vinyl acetate	8.42	0.1
Vinyl chloride	11.30	0.8
Vinylidine chloride	9.40	0.3
Xylene	13.20	2.0

[a] Minimum detection level exceeds current permissible exposure standard; therefore, recommended for gross leak detection only.

Analytical wavelength: The analytical wavelength is usually the strongest band in the spectrum that is free from interference due to atmospheric water and carbon dioxide. The listed wavelengths are approximate. If more than one IR-absorbing material is present in the air in significant concentration, the use of another analytical wavelength may be necessary.

Minimum detectable limit: The concentration that would produce an absorbance equal to twice the peak-to-peak noise in a typical portable IR.

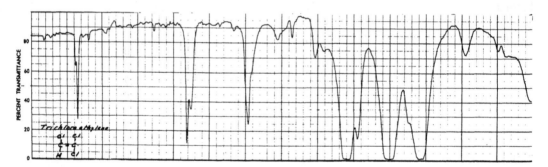

Figure 12.13. Infrared spectra of trichloroethylene. (From *The Industrial Environment—Its Evaluation & Control*, NIOSH, Cincinnati, 1973, p. 234.)

used to measure carbon dioxide (CO_2) levels during indoor air quality (IAQ) studies and in the brewing, food processing and mining industries. In IAQ investigations an increase of CO_2 above the 400–500 ppm in normal ambient air indicates that there is too little outside air being introduced in the occupied space for the number of people present. In breweries, CO_2 is produced as part of the fermentation process, and levels may exceed accept-able exposure levels in enclosed areas around vats. Figure 12.14 shows a typical infrared instrument that provides continuous auto-ranging detection of CO_2 from levels 10 ppm to 6% by volume (60,000 ppm) and features peak/hold memory, an internal sampling pump, and optional datalogging. A dedicated IR instrument is also very good for measuring carbon monoxide (CO) since there is little inference by other gases at 4.6 μm.

Figures 12.15 through 12.17 show different IR instrument configurations.[7] Figure 12.15 is a basic layout with a single cell. In some designs the sample is allowed to diffuse into and out of the cell rather than have discrete inlet and outlets for a pump-driven system. This configuration has no way to adjust for possible drift in the IR source or detector. Figure 12.16 is a single cell configuration with two detectors. The active filter/detector is chosen to measure the compound of interest, while the reference filter/detector is selected to ignore the target compound. In actual operation the reference side provides the zero point while the active side provides the signal for measuring concentration. This arrangement adjusts for changes in the IR source, and it also has the advantage of doubling the effective path length since the IR beam travels to the mirror and then back to the detectors.

Figure 12.17 shows a popular design with two cells. For instruments dedicated to a single chemical, the reference cell can either (a) contain pure target gas and serve as the baseline for total absorption or (b) contain pure reference gas (such as nitrogen) and serve as the zero baseline. For instruments that measure more than one gas, the reference cell is filled with a pure reference gas. A chopper, which is a disk with slots in it, alternately allows the light beam to pass through the sample and reference cells to the single detector.

For all IR instruments the type of detector is important to measuring low levels of

Figure 12.14. Infrared-based carbon dioxide monitor provides auto-ranging detection of levels from 10 ppm to 60,000 ppm. (Courtesy of Industrial Scientific Corporation.)

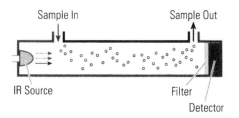

Figure 12.15. Diagram of a simple infrared detector. (Courtesy of International Sensor Technology.)

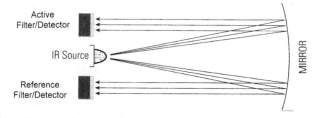

Figure 12.16. Diagram of a two-detector infrared detector configuration. (Courtesy of International Sensor Technology.)

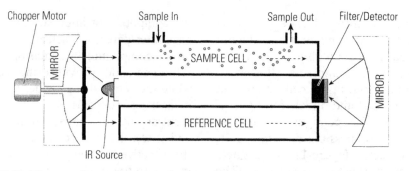

Figure 12.17. Diagram of a two-beam, single detector with chopper infrared detector configuration. (Courtesy of International Sensor Technology.)

contaminant. A variety of designs exist that are based on measuring temperature rise as the IR beam hits the detector(s) or as the temperature of the gas sample in the sample cell increases due to IR absorption. Different approaches include: (a) thermoelectric detectors or thermisters that directly convert heat into an electrical signal (b) microflow sensors or a diaphragm between the detectors in the reference and sample cells to measure the small decrease in pressure in the detector chamber due to absorption of the IR radiation transmitted through the sample cell. One novel approach is called a photoacoustic detector, which uses a sensitive microphone to measure the increase in pressure within a fixed sample cell due to the absorbed IR radiation.

Typical applications for IR instruments in addition to environmental measurements and occupational exposure monitoring include:

- *Indoor Air Quality Studies.* Measurements for compounds such as CO_2, CO, formaldehyde, or organic vapors.
- *Fume Hood/Tracer Gas Analysis.* One of the standard tests for laboratory hood performance[8] is to release a tracer gas within the hood under different operating conditions and monitor the hood perimeter for leakage back into the laboratory. A portable IR can measure commonly used test gases such as sulfur hexafluoride (SF_6).
- *Emergency Response Analysis.* Determining airborne levels of hazardous spills and releases and making real-time decisions regards personal protective equipment for responders, boundaries of "safe" zones and the need for community evacuation or shelter-in-place.
- *Leak detection.* Around equipment handling hazardous chemicals, such as medical anesthetic gases or process units, as part of routine operations or preventive maintenance.

IR Instruments for Measuring Many Gases and Vapors

Infrared instruments with the capability of measuring the concentration of many gases and vapors differ from the configurations shown in Figures 12.15 through 12.17 in that they need a way to vary the wavelength of the IR radiation directed through the sample cell to match the absorption spectra of target compounds. This can be achieved by a variable wavelength filter or by having several fixed-wavelength band-pass filters in the device. In addition, these instruments need a means to vary the path length using adjustable mirrors or another means in order to achieve adequate sensitivity for different compounds and also increase the measurement range for certain compounds.

Figure 12.18. The MIRAN SapphIRe portable infrared instrument. (Courtesy of Thermo Electron Corporation.)

MIRAN SapphIRe Analyzer. One of the most common IR devices for field measurements is the MIRAN SapphIRe series of instruments (Figure 12.18) from Thermo Electron Corporation. The SapphIRe Analyzer is available in three models with gas calibrations from one to over 100 gases. It is a microprocessor-controlled single-beam instrument that utilizes interactive programming to prompt the sampling practitioner through available options and functions.[9]

The IR wavelength is varied using a variable filter from 7.7 to 14.1 μm, along with seven fixed band-pass filters: 1.8, 3.3, 3.6, 4.0, 4.2, 4.5, and 4.7 μm. The sample is drawn into the 2.23-L sample cell by an internal pump at a rate of 15 L/min. The path length can be varied between 0.5 and 12.5 m. After the instrument is powered up and stable, sample analysis time is 20 seconds minimum and 3 minutes maximum. The response time is approximately 18 seconds to reach 90% of final reading. The SapphIRe has a digital readout of concentration in various units (ppm, percent, and mg/m^3) or absorbance units (AU). There is a RS-232 connection to use a personal computer for operating the instrument and downloading data, although the device can also be operated from the alphanumeric keypad. The device weighs 24 pounds, and battery service life is approximately four hours. Because IR devices are sensitive to ambient temperature, relative humidity (RH), and atmospheric pressure, three different operating conditions are specified:

- Reference operating conditions, within which the influence of temperature, RH, and pressure is negligible, is 23 ± 2°C, 50 ± 10% RH, and 14.7–15.3 psi.

- Normal operating conditions, within which the device is designed to operate at the specified accuracy, is 5–40°C, 5–95% RH (noncondensing), and 12.5–15.3 psi.

- Normal operative conditions, within which the device can be subjected without permanent impairment of operating characteristics, is 1–50°C, 0–100% RH (noncondensing), and 11.6–15.9 psi. Maximum temperature for intrinsically safe models is 40°C.

Fourier Transform Infrared (FT–IR) Analyzer. The Fourier transform infrared (FT–IR) device is a more complex and powerful type of analyzer when compared to a simple IR analyzer discussed above. The FT–IR uses a unit called a Michelson interferometer to measure the absorption of IR radiation at different wavelengths. The interferometer (Figure 12.19) functions as follows[7]: Radiation from the IR source is split into two equal beams in the beam splitter. One beam continues on a straight path to a moving mirror, while the other beam is deflected at an angle to a fixed mirror. When the two beams recombine after being reflected from the mirrors, they undergo a process called interference. If both reflecting mirrors are exactly the same distance from the beam splitter, both beams will be in phase with each other and the resulting recombined beam out of the unit will have the maximum amplitude. The

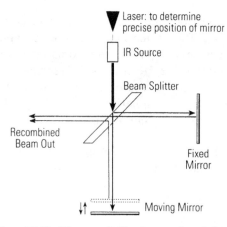

Figure 12.19. Diagram of a Fourier transform infrared (FT–IR) analyzer. (Courtesy of International Sensor Technology.)

Figure 12.20. Temet Gasmet DX-4010 portable FT–IR analyzer ready for field use. (Courtesy of Temet Instruments Oy, Helsinki, Finland.)

position of the mirror is computer controlled; a laser precisely measures the position of the mirror. For each frequency, as the moving mirror moves through its cycle, the interference pattern of the recombined beam will be from totally in phase to totally out of phase to totally in phase again.

Using a beam chopper, the recombined beam alternatively passes through the sample cell and the reference cell to reach the detector. The computer collects the data from the detector, which is a complex measurement of all the phase relationships at a given time, to produce an *interferogram*, which shows the intensity of the infrared radiation as a function of the displacement of the moving mirror. A sophisticated computer program applies a mathematical adjustment, called a Fourier transform, to the interferogram to determine the identity and concentration of the gases that are present. For a multicomponent analysis, the computer retrieves calibration data in the computer's library and tries to calculate a spectrum that is a close match to the actual sample spectrum by mathematically varying the concentration of each component. When it achieves the best possible fit between the calculated spectrum and the sample spectrum, it reports the results.

The GASMET™ DX-4010 from Temet Corporation is a FT–IR device designed for on-site measurements (Figure 12.20). It incorporates advanced hardware and software to provide a portable high-speed identification and quantification of multiple gaseous compounds simultaneously and accurately, with results available in seconds, for applications such as:

• Environmental emissions monitoring
• Quality control
• Workplace air monitoring

The instrument can analyze up to 50 compounds simultaneously. The sample is drawn into the 1.07-L sample cell by an internal pump at a rate of 1–5 L/min. A filter (2 μm) is required on the sample line to protect the optical parts from particulate matter. The sample cell has a multipass, fixed path length of 1.2, 2.5, 5.0, or 9.8 m. After the instrument is powered up and stable, sample analysis time is 120 seconds or less, depending on the gas flow and measurement time. An external personal computer controls the instrument. The proprietary software, called CALCMET™, is used to compute the concentrations and error limits of the different components present in the sample gas. The reference spectra needed for the analysis are stored on the fixed disk of computer and are loaded from the library when used in the analysis of an unknown sample. An RS-232 connection connects the computer to the instrument.[10]

The DX-4010 uses an interferometer called the Temet Carousel interferometer. The Carousel interferometer is rugged and withstands the demanding environmental conditions of nonlaboratory environment. The Carousel interferometer modulates the infrared radiation coming from the infrared source, as described above. The modulated light passes through the temperature-controlled sample cell. The transmitted infrared radiation is detected by a thermoelectrically cooled detector.

It is recommended that the instrument is operated in the following environmental conditions:

- 0–40°C operating temperature in short-term use
- 15–25°C operating temperature in long-term use
- <90% relative humidity at 20°C, noncondensing

The ambient temperature of the use location should be stable. Temperature fluctuations of a few degrees Celsius can in some cases affect the analysis results and the accuracy of the measurements decreases. The influence of the temperature fluctuations can be eliminated by remeasuring the background (zero) spectrum at the existing ambient temperature. Additionally, the use location should be free of strong vibrations. The standard case is not explosion-proof, and so the DX-4010 must not be used to measure explosive gas mixtures or gases that might form an explosive gas mixture with the ambient air.

Advantages and Limitations. The advantage of IR is that units can be purchased that are specific for given chemicals, or tunable units can be purchased and used for many different compounds with a minimum of setup time. In the field they are easy to use and provide stable operation.

Detection limits depend on the absorption coefficient of the compound at a given wavelength or frequency. The limit of detection for many compounds is in the range of 1 ppm to 20 ppm. Maximum concentrations are commonly in the low percent range.

The usefulness of IR analyzers for monitoring complex mixtures is limited because overlapping peaks produce an additive response, making concentrations appear higher than they actually are. A wavelength that is relatively unique to a chemical is selected for monitoring and ideally there should be no others that interfere. However, this is often not the case, and a review of the chemicals likely to be present during monitoring should be done to assure that an interferent will not be a problem. Interferences depend on the contaminant being measured. For example, chlorinated hydrocarbons will all absorb at approximately the same wavelength. This problem can be partially minimized by taking measurements at a secondary absorption wavelength for substance confirmation. A primary interferent is water vapor. The effect of water vapor can be minimized by passing the air sample through silica gel or a similar drying agent, maintaining constant humidity in the sample and calibration gases by refrigeration, saturating the air sample and calibration gases to maintain constant humidity, or using narrowband optical filters in combination with some of the other measures.[11] Carbon dioxide can also be an interference, although its effect at concentrations normally present in ambient air is minimal. According to the American Society for Testing and Materials (ASTM), 750 ppm of carbon dioxide may give a response equivalent to 0.5 ppm carbon monoxide.[11] Some instruments have microprocessors that allow them to correct to some degree for interferences, and the sophisticated mathematical algorithms in a FT–IR device can correct for many interferences.

Calibration. A multipoint calibration is required when an analyzer is first purchased, when the analyzer has had maintenance that could affect its response characteristics, or when the analyzer shows drift in excess of specifications as determined when the zero and single point calibration is performed. A zero and single point calibration is required before and after each sampling period, or if the analyzer is used daily. The flowmeter should be calibrated as well when the analyzer if first purchased, when it is cleaned, and when it shows signs of erratic behavior.[11]

Once calibration plotting curves are prepared for a given compound, air concentrations can be obtained by measuring the absorbance at the analytical wavelength and reading the concentration from the point where the absorbance intersects the curve. Most calibration curves of this type will have some curvature, so it is best to use a plot prepared with three or four data points, rather than using a single-point and a straight-line approximation. If only a single calibration concentration is available, it may be used; the measurements will be most accurate near this concentration point and less accurate at other values due to curvature.

Manual calibration of most long path-length IR instruments is done using a closed-loop system. Small amounts of contaminant (typically microliters) are added to a fixed volume sample chamber (generally 2–6 L) without measurably affecting the pressure of the system, creating a known concentration in the parts per million (ppm) range. User calibrations require considerable practice in using the closed-loop system and becoming familiar with the instrument in order to consistently generate reliable accurate calibrations. Calibration of IRs can also be done using known concentrations in gas cylinders.

Care must be taken during calibration of IRs, because these are nondestructive instruments; thus, any contaminant that enters the instrument will come out essentially unchanged, allowing the user to be exposed to the calibrant gas. The exhaust from the analyzer should be vented to a laboratory hood or other exhaust to remove contaminant, both during calibration and analysis, to prevent buildup in the surrounding environment.

The first step in setting up an IR analyzer to measure a gas is to select the wavelength and path length. Wavelengths are chosen to maximize analytical sensitivity while minimizing interferences from other vapors commonly present in the workplace or atmosphere where this contaminant is typically found. For example, if measuring styrene vapor in a chemical plant where acrylonitrile butadiene styrene (ABS) polymer is manufactured, then 1,3-butadiene is likely to be present. Since this compound absorbs at the most frequently chosen analytical wavelength for styrene (11.10 μm), an alternative wavelength must be chosen. On the other hand, if doing a survey in a fiberglass boat hull construction facility where acetone is commonly used as a cleaning solvent for styrene, the acetone vapors would not be a concern as they are not likely to have any absorption peaks at wavelengths that would interfere with styrene.

Since ambient air always contains water vapor and carbon dioxide, the analytical wavelengths selected must reflect this makeup. Some gases have a limited number of possible analytical wavelengths, and occasionally a wavelength must be chosen where water vapor can interfere. This situation is associated with the 5- to 8-μm region where strong water vapor absorbance occurs. To reduce the interference of any water vapor, the instrument should be "zeroed" with air that has about the same humidity as the sample.

If an IR is overloaded from being exposed to too high a concentration, inaccurate calibrations and negative or suppressed readings at certain concentrations can result. If the absorbances for any of the calibration standards go above one absorbance unit, switch to a shorter path length or use a weaker absorption band.

There are many compounds for which IRs can be used. In the case where the instrument has not already been set up to monitor for a given compound, the manufacturer can often provide information to select appropriate wavelengths and path lengths for setting up IRs to monitor many different gases and vapors. When the path length in a variable path length instrument is changed, the instrument should be recalibrated.

Closed-Loop Calibration for the IR. When using the closed-loop technique, the user introduces pure samples through a septum with a liquid or gas syringe. The sample is circulated through the system by means of the closed-loop pump. Using the closed-loop system for the calibration of organic compounds with low vapor pressures may result in errors due to adsorption on the walls of the closed-loop system. The results of using a calibration curve based on these data would be to overestimate concentration during surveys. Therefore, even though closed-loop calibration systems are not recommended for compounds with vapor pressures less than 15 mm Hg (25°C), from the standpoint of using the instrument the errors may not be significant in certain types of surveys.[12]

Procedure

1. Determine the sample volume required for the desired concentration limit. For gases the volume to be injected is calculated using

$$\text{ppm} = \frac{V_1}{V_2}$$

where V_1 = volume of gas to be injected, V_2 = volume of closed-loop calibration system.

For liquids (at atmospheric pressure and 25°C) the volume to be injected is calculated using

$$\text{ppm} = \frac{V_1(d)(24.45 \times 10^3)}{MV_2}$$

where V_1 is the liquid sample volume in μL, d is the liquid density in g/mL, M is the molecular weight, V_2 is the volume of closed-loop calibration system in L, and 24.45×10^3 is the number of μL of vapor per millimole of analyte at normal temperature and pressure (NTP).

2. Introduce clean air into the sampling chamber using a zero air cylinder or room air drawn through a zeroing cartridge.

3. Connect tubing from the "out" connector of the closed-loop pump to the input port of the analyzer. Connect tubing from the "in" connector of the pump to the "cal port" connector of the analyzer. Turn the sample valve to the "calibrate" position.

4. Turn on the closed-loop pump.

5. Inject the quantity of calibrant gas or liquid needed to fill the sample cell with the desired concentration. Inject equal increments to cover a full range of concentrations. For example, if the maximum concentration is 200 ppm,

and the number of calibration points is to be five, then each injection should increase in concentration by 40 ppm. Frequently replace the silicone rubber disks, called *septa*, used in the injection ports of closed-loop calibration systems. After a number of injections they may not reseal properly.

6. For each concentration, input the level into the instrument's memory or otherwise set the "span" control (depending on the specific instrument). Some instruments automatically fit a calibration curve to the data points.

7. For instruments with a "general purpose" scale (not marked in ppm), prepare a plot of meter reading versus concentration, with meter reading on the y axis and concentration on the x axis. Unknown concentrations can then be found using this curve and the meter reading.

8. Following calibration with each chemical, flush the closed-loop system and cell with clean air or nitrogen.

The microprocessor capability of many instruments has simplified calibration in that certain parameters are programed into the instrument at the factory to allow for an electronic calibration check. The internal electronic calibration system often allows for field calibration without the need for calibration gas, which is sufficiently accurate for performing screening samples. Although this automatic (factory-provided) calibration saves time, it is not as accurate a calibration with a standard.

Field Operation. There are three primary types of sampling modes for IR analyzers: bag sampling, leak or fugitive emissions testing, and continuous monitoring. In bag sampling, discrete breathing zone, area, or process samples can be collected in bags as described in the chapter on sample collection device methods for gases and vapors (Chapter 6) using personal sampling pumps. The flow rate of the sample pump is not critical, as long as it is constant over the duration of the sample. Following sample collection, the bag is transported to the location of the IR analyzer where it is connected to the inlet of the sample cell. An internal or auxiliary pump pulls the sample from the bag, through the sampling chamber, where it displaces the ambient air. When the instrument's response to the sample in the bag reaches a maximum, a reading is taken. The concentration of the contaminant can be determined from comparing the plot of instrument response versus the calibration curve concentrations, or directly from the readout if it is in ppm. A similar application is to collect exhaled breath in bags and analyze it on the IR.

Bags can also be analyzed to identify the presence of unknowns on IRs with scanning capabilities. The scan of the unknown atmosphere is compared with a scan of contaminant(s) expected to be present to see if any unidentified peaks are present. The instrument should generally be connected to a personal computer to manage the data during this analysis.

When sampling in situations where the atmosphere is unknown, one technique is to operate the IR at 3.4 μm (the C–H bond stretching frequency) using the maximum possible pathlength. In the field, the response under these conditions should be reported as equivalents of whatever compounds have been used to calibrate the instrument over the desired concentration range using the closed-loop method.[13]

In survey instruments designed to detect leaks, an extended sample probe of up to 6 feet in length made of inert tubing with a dust filter attached is used to make spot measurements at many different locations, such as door seals, sumps, lines, and fittings. If the filter becomes clogged, the sampling flow rate will decrease, resulting in a slower instrument response. Replace the filter cartridge when the response time has increased noticeably, or after 400 hours of use.

In continuous monitoring, the instrument is stationed to collect samples at a location of interest. A pump continuously draws sample into the sample cell, and the absorbance is continuously displayed on the instrument readout and stored in a data logger for printout later.

If measurements exceed the range for which the instrument was calibrated, it must be remembered that the response of IR analyzers is not linear over the total range of the instrument, particularly at the ends of the scale; therefore, extrapolation past calibration points is not advised. If much higher concentrations than anticipated are identified, monitoring may need to be repeated following calibration at higher concentrations.

If the process being monitored is a constant operation with little variation, readings can be taken with the monitor every 30 minutes. If the process is variable, it may be necessary to take the readings as often as very 5 minutes or continuously as described above.

In compliance sampling with an IR device, calibration should be done using either a premixed gas in a cylinder or the closed-loop calibration system rather than relying solely on the internal calibration system. For personal sampling, Tedlar gas bags capable of collecting a minimum of 6 L and personal sampling pumps equipped with exhaust ports are needed. Care must be taken to assure that sufficient air is collected to adequately purge the sample cell during analysis. In certain environments, a drying tube, Teflon sampling line, particulate filter, or zero gas filter may be necessary.

In general, where ambient air is drawn directly into the instrument all sampling should be done with the sample hose and

particulate filter attached to prevent dust and particulate matter from entering the cell and accumulating on the mirrors and windows, thus interfering with the performance of the instruments by increasing background levels.

IRs should be used outdoors with great care because cooler temperatures and humidity can cause condensation on the mirrors and windows. If the instrument has been in a cold environment, condensation may take place in the cell if the sample is drawn into it before the instrument has warmed up properly. If the mirrors do become fogged, permanently high background readings can result. Some mirror materials are more susceptible to humidity. For example, sodium chloride windows cloud more easily than those made of silver bromide. The mirrors on some instruments can be periodically cleaned. The cell ports in IRs should be covered with protective plastic caps when the IR is not in use to prevent dust and moisture from contaminating the optics.

For all uses of an IR instrument, it is important to prevent corrosive gases and vapors that could damage mirrors, or cell or detector windows, from entering the sample cell.

Zeroing IR analyzers is a critical step. If the sampler zeros in an atmosphere containing the compound of interest or an interference, and then analyzes air that contains less of the same compound, a negative reading or "error" message will occur. This reading can occur when the zero gas filter has become saturated, or if the sampler zeros on humid air and subsequently analyzes air that is drier than the zero. A charcoal canister, sometimes termed a zeroing cartridge or zero gas filter, must be placed over the probe for proper zeroing. Note that charcoal has a limited ability to absorb water vapor. These cartridges should be stored in a sealed plastic bag to prevent exposure to contaminated atmospheres when not in use. Saturation of the cartridge depends on the nature and concentration of the contaminants that are pulled through it. If a new cartridge produces an appreciably lower analyzer reading, the old cartridge should be discarded. If an instrument is zeroed with a contaminated cartridge, a negative value may occur when making measurements, the result being low values for all measurements because the ambient background readings may be lower than the concentration in the cartridge used for zeroing. Whereas some sources recommend using nitrogen for zeroing, others recommend using room air to account for normal humidity conditions.[14]

Bag Sample Collection for Nitrous Oxide Using an IR[15]

1. Set up a portable IR analyzer to the following parameters: λ = 4.48–4.68 μm, path length = 0.5–40 M. Allow the instrument to warm up and equilibrated for 15 minutes.

2. Perform an on-site multipoint calibration at five or more concentrations over the range of 10 ppm to 1000 ppm. *Note:* Because nitrous oxide supports combustion, the shipments of the compressed gas must comply with 49 CFR 171–177 regulations regarding shipment of hazardous materials. Calibration for nitrous oxide can frequently be done using material available on-site when surveys are done in hospitals, dental offices, and veterinary operations, because generally the nitrous oxide in use is of sufficient purity.

3. Zero the instrument while recirculating uncontaminated air through the sample cell. If the area where the instrument is being calibrated is serviced by the same ventilation system as the area to be monitored, it will be necessary to obtain a source of uncontaminated air (or nitrogen or oxygen) for zeroing the instrument.

4. Inject a known volume of nitrous oxide into the sample cell with a gas-tight syringe through tubing, or by using a septum attached to the sample cell. Calculated the concentration of nitrous oxide in the sample cell:

$$C_s(\text{ppm}) = \frac{\text{Volume of N}_2\text{O injected (μl)}}{\text{Volume of cell (L)}}$$

When the instrument reading stabilizes, record meter or display reading.

5. Prepare a calibration graph of ppm (C_s) versus meter or display reading.

6. Select one of the following sampling modes according to the desired form of the data: ambient air or integrated air samples for TWA determinations.

7. Move the instrument to the first area to be sampled. A probe with a sample line can be used to allow more flexibility. Thus, the instrument can be placed nearby and the probe can be used to move, for example, around the doors of a sterilizer, or held over the shoulder of an employee while opening the door of the sterilizer to get a peak measurement.

8. For ambient air, turn on the instrument pump and record the readings on the display. If data are to be expressed as a TWA concentration, an internal datalogging feature or a computer will be needed to store measurements.

9. Pump air to be analyzed through the sample cell to purge the cell. Typically, two to three cell volumes are necessary. When output stabilizes, record the reading as a measurement.

10. Use the calibration graph to determine the concentration by moving along the horizontal axis to the display reading, and moving to the point on the line that corresponds to this concentration. The ppm value for this point on the line can then be read on the vertical axis.

11. During each day's operation, periodically recheck the calibration by repeating measurements with the calibration gas at three or more points on the graph.

12. For integrated air samples for TWA determinations a clean bag of plastic or other material must first be evacuated. It can be done using a personal air sampling pump at the highest airflow available.

13. Attach the plastic bag via tubing to a personal sampling pump and pump air from the contaminated area into the bag at a rate calculated to fill the bag over the sampling period. This rate will be between 20 mL/min and 500 mL/min. Terminate sampling before the bag is 80% full. The pump's flowrate must be within +5% of the initial setting throughout the sampling period.

14. Analyze the bag sample within 2 hours after completion of the sampling to minimize loss of analyte by adsorption and permeation.

Interpreting Measurements with Infrared Analyzers

Most IR instruments express concentration directly in units of parts per million (ppm) or percent. However, some may display results as *absorbance* or *percent transmission*. The percent transmission ($\%T$) readout is used with a calibration curve to determine the concentration of the sample. The absorbance (A) scale is related to the $\%T$ scale by $A = -\log T$ and is not a linear scale as is the $\%T$ scale. It too must be compared to a calibration curve in order to determine the actual concentration of the compound being sampled. The calibration curves must be prepared in advance using the same control settings (path length, wavelength, slit width) which will be used during measurements and the instrument calibrated immediately prior to sampling to assure the curve is still valid.

SUMMARY

This chapter describes direct reading instruments that can identify and measure the concentration of specific compounds in an airborne mixture, which are gas chromatographs (GC), GC/mass spectrometers (GC/MS), and infrared (IR) instruments including fourier transform IR (FT–IR). There are also many IR devices that emit only one wavelength of IR radiation and this are specific to one airborne compound (e.g., CO and CO_2); these are covered in detail in this chapter as part of the overall discussion of IR instruments.

The instruments that can determine the concentrations of specific compounds in an airborne mixture are high-end devices that are expensive and sophisticated and can be complicated to operate. Generally they have these characteristics in common: All are designed to be used with a personal or laptop computer; they are not designed to identify the components of a completely unknown or "mystery" airborne mixture since they must be calibrated with the specific compounds they are measuring; and, because of their size and complexity, they are used for area or source sampling rather than routine personal occupational exposure measurements.

When selecting between a portable GC or IR instrument for field measurements, two important factors to consider are:

- IR devices require less time to complete an analysis than do GCs. Ten minutes for a CG run is typical, while 2 minutes or less is typical for an IR instrument reading.

- GCs have a lower limit of detection for most compounds (in the ppb or ppb range) than do IR instruments (in the ppm range).

Major developments with this category of direct reading instruments has been the increase in functionality due to more powerful microprocessors, and introduction of advanced instruments such as the portable GC/MS and FT–IR. The sampling practitioner should stay current in direct reading technology to ensure that they are aware of the latest advances when selecting instruments for their specific sampling needs.

REFERENCES

1. *Scentograph PLUS II Operating Manual.* Fairfield, NJ: Scentex Systems, Inc.

2. *Voyager Portable Gas Chromatograph Training Manual.* Waltham, MA: Photovac, Inc.

3. *HNU Model GC-311 Product Information.* Walpole, MA: Process Analyzers, LLC.

4. National Institute of Occupational Safety and Health. *NIOSH Manual of Analytical Methods (NMAM®), Method 3702,* 4th ed., M. E. Cassinelli and P. F. O'Connor, eds. DHHS (NIOSH) Publication 94–113, August, 1994.

5. *HAPSITE^{TM} Portable GC/MS Product Information,* Inficon, Inc., East Syracuse, NY.

6. *Radiation and Chemical Hazard Simulation—HAPSITE^{TM}.* Luton, United Kingdom: Argon Electronics.

7. Chou, J. *Hazardous Gas Monitors, A Practical Guide to Selection, Operation and Applications.* New York: McGraw-Hill, 2000.

8. American Society of Heating, Refrigeration and Air Conditioning Engineers, *Standard 110-1995—Method of Testing Performance of Laboratory Fume Hoods,* Atlanta: ASHRAE, 1995.

9. *MIRAN® SapphIRe^{TM} Portable Ambient Air Analyzer Instruction Manual* (MI 611-037). Waltham, MA: Thermo Electron Corporation.

10. *GASMET^{TM} DX-4010 Operating Manual.* Austin, TX: Air Quality Analysis, Inc.

11. American Society for Testing and Materials: Method D3162. Philadelphia: ASTM.

12. Sammi, B. S. Calibration of MIRAN gas analyzers: Extent of vapor loss within a closed loop calibration system. *AIHA J.* **44**(1):40–45, 1983.

13. *Guide to Portable Instruments for Assessing Airborne Contaminants at Hazardous Waste Sites.* Geneva, Switzerland: World Health Organization, 1988.

14. Levine, S. P., et al. Advantages and disadvantages in the use of Fourier transform infrared (FTIR) spectrometers for monitoring airborne gases and vapors of industrial hygiene concern. *AIHA J.* **4**(7):180–187, 1989.

15. National Institute of Occupational Safety and Health. *NIOSH Manual of Analytical Methods (NMAM®), Method 6600,* 4th ed., M. E. Cassinelli and P. F. O'Connor, eds. DHHS (NIOSH) Publication 94-113, August, 1994.

CHAPTER 13

COLORIMETRIC SYSTEMS FOR GAS AND VAPOR SAMPLING

This chapter covers direct-reading devices that rely on color change to measure airborne concentration of chemicals. These devices fall into three main categories:

- Detector tubes and similar products for short-term or grab samples.
- Tubes (either with a sampling pump or passive) and badge-like units or "spot plates" for long-term measurements.
- Electronic instruments that collect air samples on a moving tape or in a liquid and that report the airborne concentration based on the reaction between the target chemical and the reagents in the tape or liquid.

Each type of colorimetric system represents use of appropriate technology to fill a specific application for the sampling practitioner. Detector tubes are probably the oldest direct-reading device still in use and are a convenient way to rapidly sample for 400–500 different gases and vapors. When long-term colorimetric devices were introduced, they filled a gap because they allowed longer-duration measurements than the "grab" sample with standard detector tubes. Although long-term tubes have probably been displaced somewhat by personal-size electronic chemical-specific instruments with data-logging features, they are still a good way to sample for some common contaminants and where chemical-specific electronic sensors do not exist. Colorimetric electronic instruments are mainly used in applications where an electrochemical, solid-state, or other direct-reading sensor is not available.

Colorimetric devices operate on two broad principles: In *length of stain* devices the concentration is related to the amount (length) of reagent that is discolored, whereas with *color intensity* devices the concentration is related to the degree of color change as compared to a standard. Most detector tubes are length of stain design, but badges and other products are often based on color intensity readings. For visual reading of results, length of stain

Air Monitoring for Toxic Exposures, Second Edition. By Henry J. McDermott
ISBN 0-471-45435-4   © 2004 John Wiley & Sons, Inc.

devices are generally considered preferable to color change units since color charts can fade with time or not provide realistic color comparisons, and there is wide variation in human perception of color.[1] Optical reading instruments are provided with some color intensity detectors as a means of improving accuracy.

All colorimetric devices have temperature, relative humidity, and ambient pressure limits as specified by the manufacturer. Often both a *normal range* (which is the range over which no correction from the manufacturer-provided calibration is needed) and a *maximum operating range* (which denotes the outside limits where calibration correction factors can be applied) are specified. Additionally, some colorimetric devices such as tape samplers need some moisture in the air in order to function, and so it is critical to follow the manufacturer's specification; a relative humidity of 40% is a typical requirement for these instruments. Like all monitoring processes, colorimetric systems are subject to interferences from nontarget chemicals that can result in either high or low erroneous readings. In some cases the colorimetric technology has eliminated interferences that prevent the use of other monitoring techniques, thus making colorimetric methods the preferred sampling approach. The sampling practitioner must evaluate possible interferents when selecting the appropriate sampling method for any airborne contaminant.

DETECTOR TUBES

Detector tubes are used for short-term measurements, often termed *grab samples*. They provide the ability to do direct-reading measurements while in the field for a wide variety of gases and vapors. Since no laboratory analysis is required, they provide fast on-site results and are often considered along with direct-reading elec-

Figure 13.1. Piston-style pump for detector tubes. (Courtesy of Nextteq, LLC.)

tronic instruments when immediate results are needed. The primary use for these devices is for occupational sampling, since they are not sensitive enough to detect the low levels of contaminants needed for environmental detection.

They function on the principle that specific sampling media change color when contaminated air is pulled through them, and they have been available for 60 years or longer. A typical sampler is a glass tube filled with a solid granular material, such as silica gel, that has been coated with one or more detection reagents that are especially sensitive to the target substance and quickly produce a distinct layer of color change (Figure 13.1). Table 13.1 lists some typical reagents used inside the tubes for different gases and vapors. A calibration scale is printed on the side of the tube, which is used to read concentrations of the measured substances (gases and vapors) directly when both ends of the tube are broken off and a specified volume of air is drawn through the tube using a hand-powered pump (typically either of "piston" or "bellows" design) or a battery-powered pump. In the case where a tube can be used for different concentration ranges, there may be two scales on a tube, each corresponding to a different number of strokes of the grab sampling pump.

It is important to note that there is a wide variety of tubes available that contain

TABLE 13.1. Color-Change Reactions in Selected Dosimeter Tubes

Gas Measured	Reaction	Color Change
Carbon monoxide	$CO + K_2Pd(SO_3)_2 \rightarrow Pd + CO_2 + K_2SO_3$	Yellow—black-brown
Hydrogen sulfide	$H_2S + Pb(CH_3COO)_2 \rightarrow PbS + 2CH_3COOH$	White—dark brown
Sulfur dioxide	$SO_2 + BaCl_2 + H_2O \rightarrow BaSO_3 + 2HCl$	Green—yellow
Hydrogen cyanide	$HCN + HgCl_2 \rightarrow Hg(CN)_2 + 2HCl$	Yellow—red

Source: Roberson, R. W., et al. Performance testing of Sensidyne/Gastec Dosi Tubes for CO, H_2S, SO_2 and HCN. *AIHA Conf. Presentation*, May 21, 1985.

different chemicals and operate on different reaction principles. The information in this chapter is intended to be a general introduction to this type of air monitoring system. *Always* refer to the instructions for the specific tube(s) being used for accurate operating and safety information. The manufacturer's instructions for the specific detector tubes are included as an insert in the tube box, and they are critical to proper use of the tubes. Some manufacturers also publish a more comprehensive handbook or manual for their tube systems. Read these for information on interferences and relative standard deviations for each tube, as well as the number of strokes, time between strokes, and time necessary for color development, temperature, humidity, and atmospheric pressure effects. The literature also describes the function of each layer in the tube, which is helpful in identifying possible interferences if unanticipated color changes occur. Other valuable information is any special precautions such as: whether the tube may emit smoke or heat up during use; whether the tube requires oxygen or airborne water vapor for proper reactions to take place; whether a tube with no reading after a test can be reused for another measurement; if the tube must be held in a certain position (i.e., vertically upward) for proper operation; and proper disposal procedures for used or out-of-date tubes.

There are several different types of tubes depending on the target chemical (Figure

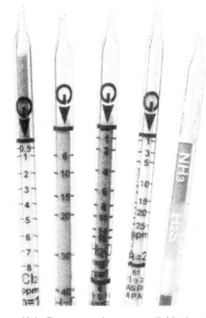

Figure 13.2. Detector tubes are available for a wide variety of toxic chemicals. (Courtesy of Nextteq, LLC.)

13.2). Simple tubes have just one reagent, while others have a mixture of several reagents. For some contaminants the tube contains a *prelayer* before the color development layer to either (a) remove interfering compounds such as moisture or compounds similar to the target compound or (b) otherwise condition the target compound for analysis. For example, the prelayer may react with the gas or vapor of interest to convert it to a different chemical that can be measured by the detecting reagents. The prelayer in tubes detecting benzene is usually designed to remove toluene and xylene because these compounds would also react with the detecting layer due to their similarity in structure to benzene.

Other gases and vapors are not very reactive and thus need to be reacted with very powerful chemical reagents to break down these nonreactive compounds into other more readily detectable substances. These powerful reagents may be contained in a separate tube connected in series or in a liquid-filled ampoule within the main detector tube that is broken just before use to coat the granules with a reactive reagent. An example is the Draeger tube for toluene diisocyanate, which is a long tube with two separate ampoules covered with plastic and an indicating layer. In these situations it is important to follow the specific sequence indicated in the manufacturer's instructions and also practice with the tubes before attempting measurements in the field. The portion of the tube containing the ampoule is bent gently until the ampoule breaks while it is pointing in the right direction so the reagent reaches the correct layer.

Each tube has a specified concentration measuring range that can often be "expanded" by varying the volume of the sample through adjusting the number of hand pump strokes that are drawn. This feature may also allow an *ad hoc* adjustment when the extent of the color change either exceeds or does not reach the calibration scale(s) printed on the tube. For example, when the length of color change does not reach the calibration scale, additional sampling strokes can be taken (up to a specified maximum) until the stain interface extends into the calibration zone. In this case, the true concentration is determined by dividing the tube reading by the ratio of the pump strokes taken to the number of strokes specified in the instructions. Sometimes this information is pre-

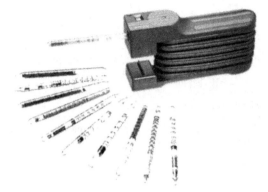

Figure 13.3. Bellows-style pump for detector tubes. (Courtesy of Draeger Safety, Inc.)

sented as "correction factors" in the tube's instruction sheet. When the color change layer exceeds the calibration scale because of higher-than-expected concentration, repeat the test with a fresh tube and sample with half of the standard volume. If the color change layer stays within the calibration scale, the tube reading should be doubled to determine the true concentration.

There are two different types of pumps available: the piston variety (Figure 13.1) and the bellows type (Figure 13.3). Each manufacturer has its own design and manufacturing techniques. For example, some piston-type pumps use orifices to control flow rate while other piston-type pumps use only the resistance of the granules packed in the detector tube to limit flow rate. A pump with multiple sampling orifices will accurately control both the volume of air sampled and the rate of airflow during a test. The specific manufacturer's instructions for each tube will list the required orifice setting and the volume of air required for each tube, as well as the time required to pull that volume through the tube. Bellows-type pumps utilize the tube packing for resistance. Many bellows pump designs have a chain that is taut when the bellows are fully expanded; for these pumps the stroke is finished when the bellows are completely opened and the chain is tight. Usually these pumps are calibrated to pull 100 mL of air per stroke, and 2–10 strokes are required. Stroke counters are very useful for bellows-type pumps, because as noted some measurements require 10 or more strokes and it is easy to get distracted during this period. The stroke counter position must be checked with each stroke, because unless the bellows are fully compressed the counter may not register the stroke. Some pumps have openings to break off the tube ends, which is a convenient feature.

The piston pump handle may be marked to show the volume that the pump will draw when the handle is locked in a certain position. Most piston pumps pull 100-mL volume of air with a full stroke, although at least one has four calibrated volumes: 25, 50, 75, and 100 mL. Some piston pumps also have stroke counters that may be automatic and change with every stroke, or they may be changed manually by the user. Another option is the *flow finish* indicator, which is often a button that pops up when the full volume has been sampled, thus eliminating the need for a stopwatch to time the stroke.

In addition to the basic hand pump and detector tube system, manufacturers offer a variety of accessories to expand the usefulness of detector tubes. These include:

- Battery-powered pumps (Figure 13.4) to use in place of the hand pumps. These are helpful in reducing fatigue and the potential for repetitive strain injuries when making multiple measurements, and in some cases they can extend the measurement range for the tubes through additional strokes.
- Pyrolyzers (Figure 13.5), which are hot-wire instruments operated by batteries that are attached to pumps and used for certain gases and vapors such as Freons™ and other chlorinated compounds. The purpose of the

Figure 13.4. Battery-powered pump to draw air through detector tubes. (Courtesy of Draeger Safety, Inc.)

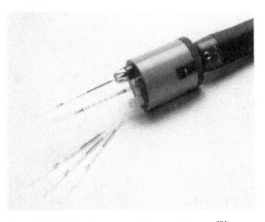

Figure 13.5. Pyrolyzer attachment (Pyrotec™) uses heat to convert some vapors into substances that can be measured using detector tubes. (Courtesy of Nextteq, LLC.)

pyrolyzer is to break down difficult-to-detect compounds into other compounds that are more easily detected. A sample is drawn across the heated filament of the pyrolyzer, which

TABLE 13.2. Draeger Simultaneous Test Kits

Simultaneous Test Set	Compounds Measured
Set I: Inorganic fumes	Acid gases, e.g., hydrochloric acid
	Basic gases, e.g., ammonia
	Carbon monoxide
	Hydrocyanic acid
	Nitrous gases, e.g., nitrogen dioxide
Set II: Inorganic fumes	Carbon dioxide
	Chlorine
	Hydrogen sulfide
	Phosgene
	Sulfur dioxide
Set III: Organic vapors	Alcohols, e.g., methanol
	Aliphatic hydrocarbons, e.g., *n*-hexane
	Aromatics, e.g., toluene
	Chlorinated hydrocarbons, e.g., perchloroethylene
	Ketones, e.g., acetone

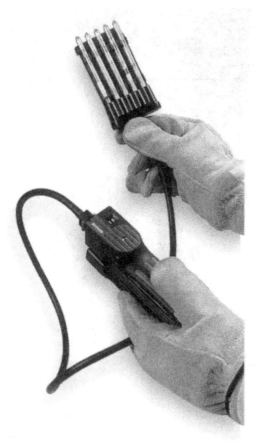

Figure 13.6. Simultaneous test set permits sampling for multiple substances using special tube sets provided by the manufacturer. (Courtesy of Draeger Safety, Inc.)

decomposes the contaminant, thereby releasing a measurable gas. The decomposition products then pass through the detector tube, causing a change in color relative to the concentration. The pyrolyzer is also useful for organic nitrogen compounds, one of the products of breakdown being nitrogen dioxide, which is easily monitored.

• Manifold holders that allow use of several tubes simultaneously (Figure 13.6). For example, Draeger offers

three different sets that are used with a manifold and hand pump for the semiquantitative determination of inorganic gases and organic vapors. Each test set is able to test for five different gases simultaneously (Table 13.2) in less than one minute and are useful in providing on-site information to firefighters, HazMat teams, and environmental agencies. Another special configuration available from several manufacturers consists of a high-pressure regulator, flow controller, and sampling manifold for testing of compressed breathing air cylinders for conformance with the Compressed Gas Association's (CGA) Grade D standard for breathing air. The tubes test for the five substances listed in the standard (carbon monoxide, carbon dioxide, water vapor, oil mist, and oxygen). Note that tests for oxygen are especially important when vendors blend breathing air from indi-

TABLE 13.3. MSA Industry-Specific Detector Tube Sets

Type	Detector Tubes	Type	Detector Tubes
Agricultural set	Phosphine (4)	Synthetic	Toluene (2)
	Sulfur dioxide (2)	Chemicals	Qualitest QL (4)
	Methyl bromide (4)	Manufacturing set	Ethanol (2)
	Hydrogen cyanide (2)		Vinyl chloride (2)
Pulp and paper set	Chlorine dioxide (3)		Trichloroethylene (2)
	Sulfur dioxide (2)		
	Hydrogen sulfide (3)	**Semiconductor Sets**	
	Chlorine (2)	Chemical vapor	Ammonia (3)
	Ozone (2)	Deposition	Carbon dioxide (3)
Pharmaceutical set	Acetic acid (3)	Process	Nitrous fumes (3)
	Aromatic hydrocarbons (3)		Hydrogen sulfide (3)
	Qualitest QL (3)	Etching process	Ammonia (4)
	Hydrogen cyanide (3)		Chlorine (4)
Petroleum set	Hydrogen sulfide (2)		Hydrogen chloride (4)
	Carbon monoxide (2)	Epitaxy process	Carbon monoxide (3)
	Qualitest QL (2)		Hydrogen chloride (3)
	Benzene (4)		Phosphine (3)
	Hexane (2)		Phosgene (3)
Mining set	Carbon monoxide (3)	Crystal growth	Phosgene (6)
	Nitrogen dioxide (2)	Process	Phosphine (6)
	Nitrous fumes (3)		
	Natural gas/methane (4)		

Notes: 1. Number in parentheses indicates the quantity of each tube in the set.
2. Qualitest QL detector tubes are used for screening measurements to identify the presence (but not the concentration) of many common chemicals including acetone, acetylene, benzene, 1,3-butadiene, butanes, butylenes, carbon disulfide, carbon monoxide, cyclohexane, diesel oil, ethyl alcohol, ethylene, formic acid, fuel oil, gasoline, hydrogen chloride, hydrogen sulfide, kerosene, liquefied petroleum gas, methyl ethyl ketone, pentanes and other saturated hydrocarbons, phenol, propane, propyl alcohols, propylene, styrene, perchloroethylene, toluene, town gas (with CO content higher than 1 vol. %), chloroform, vinyl chloride, and xylenes.

vidual gaseous components (oxygen, nitrogen, etc.) rather than by compressing ambient air.

- Special "kits" or sets of detector tubes selected to represent the type of tubes typically used by different user groups. For example, Table 13.3 lists industry-specific sets from MSA; other sets are aimed at HazMat specialists, environmental specialists, and other distinct users.
- Sampling probes, sample gas "coolers" to cool flue gases and other hot gas samples to ambient temperature for accurate analysis, a warmed holder for tubes that are carried in cold environments, and other accessories to extend the uses of detector tube devices.

In addition to the accessories described above, manufacturers have developed innovative colorimetric grab sampling approaches based on detector tube technology. For example, the Draeger CMS™ chip system uses a chip containing 10 measurement capillaries (Figure 13.7) for specific gases in an electronic instrument. A bar code on the chip provides the needed calibration information and other data so the instrument collects the air sample,

Figure 13.7. Draeger CMS™ (chip measurement system) uses an optical sensor to detect color change in reagent-filled chips. (Courtesy of Draeger Safety, Inc.)

senses the color change, and shows the concentration reading on the instrument's LCD display; a data recorder is also available. Currently, chips are available for about 17 different chemicals.

Quality Control

The most important elements of quality control in detector tube manufacturing are the purity of the reagents used, grain size of the gel or adsorbent, method of packing the tubes, moisture content of the gel, uniformity of the tube diameter, and proper storage precautions to preserve the shelf life.[2] Precision, accuracy, and reproducibility vary with the age, conditions of storage, and lot-to-lot manufacturing of these tubes. Grain size of the coated granules in the

tubes may vary among tubes and within tubes. A fine grain size is said to provide a uniform distribution of airflow through the tube along with sharp demarcation lines.[3] Within a given tube, variations in particle size can lead to striations in the stain and cause a poorly defined (fuzzy) stain line.

To ensure a high precision indication, gas detector tubes are carefully manufactured to (a) have key dimensions (such as inner diameter) maintained within strict limits, (b) control the airflow resistance of the filling reagents and packing material, and (c) provide detection reagents with long-term stability. The tubes undergo stringent quality control tests: Individual production lots are tested and calibrated to ensure the highest calibration accuracy for each lot. The certification program described below is an important quality control effort for the limited number of tubes covered by the program.

Detector Tube Accuracy

The typical accuracy specification for detector tube systems is ±25% when the reading is compared to a known gas standard at the occupational exposure concentration. This specification was established by NIOSH in their certification program conducted between 1973 and 1983 and has been continued as an expectation for these devices. While other direct-reading instruments may be available that are more accurate, there are several considerations to keep in mind when judging detector tube accuracy and suitability for use:

- Detector tubes require no user calibration over their listed shelf life, which is usually two years or more. Direct-reading electronic instrument generally require regular calibration, often a daily span gas exposure.
- Detector tubes are also designed to work in a wide array of environmental conditions, from 0 to 40°C and at least

10 to 90% relative humidity. Correction factors are supplied over the specified temperature and relative humidity range. Often a direct-reading electronic instrument has a narrower allowable temperature and humidity range associated with the specified instrument accuracy.

Although the NIOSH certification program was discontinued in 1983, a similar program was implemented in 1986 by the International Safety Equipment Institute (ISEI) and continues today based on American National Standards Institute/International Safety Equipment Association (ANSI/ISEA) Standard 102.[4] The program includes an evaluation of both the detector tubes and sampling pump used by each system, and testing is done for pump volume accuracy and leakage as well as system accuracy. The tubes are exposed to a specified concentration of test gas, and then a six-person team analyzes each tube's corresponding length of stain. Following certification, a manufacturer can use the SEI certification mark on the product. Under both the NIOSH and SEI programs, detector tubes are tested by an independent laboratory and must meet an accuracy level of ±25% at test levels of 1, 2, and 5 times the ACGIH TLV© (*threshold limit value*) and ±35% at half of the TLV©, all at a 95% confidence level. The SEI program is limited to tubes in the TLV range and to about 24 listed substances, so by design the certification does not cover all detector tubes.

In addition to the Safety Equipment Institute (SEI) certification program for detector tubes, international bodies have issued standards regulating the properties of detector tubes for measuring airborne contaminants such as the Japanese Industrial Standard Institution (JIS K0804), British Standard Institution (BS5343), Deutsches Institut für Normung or the German Standard Institution (DIN3382),

and the International Union of Pure and Applied Chemistry (IUPAC—Performance Standard for Detector Tubes). Table 13.4 lists some testing parameters from these standards bodies.

In one study, tests were performed to compare the accuracy of H_2S, CO, CO_2, and toluene detector tubes from various manufacturers. The tests were conducted by the manufacturer sponsoring the study and by three independent laboratories. Each type of tube was tested using the piston or bellows hand pump from the matching manufacturer. All pumps tested passed the specification of <2% leakage after 2 minutes with a sealed tube inserted. Gases were filled into a Tedlar bag from a NIST-traceable standard cylinder (10–50 ppm H_2S, 50–400 ppm CO, 100 ppm toluene, and 0.2–10% CO_2) and a sample volume drawn according to the manufacturer's specification. All tubes gave similar readings within 15% of the standard gas values. The results are well within the industry norm of ±25% accuracy.[5]

Accuracy Versus Precision. When reading the manufacturer's literature for detector tubes, it is important to distinguish between *accuracy* and *precision* information. The *accuracy* of a detector tube system (or any measurement system) is the level of agreement between the system and a known standard, in this case a target gas of known concentration. The *precision* of a measurement value is the level of agreement between it and other measurement values obtained under the same conditions. Many detector tube handbooks and specification sheets lists a *standard deviation* for each tube on the individual data sheet (which often is less than 25%), but the standard deviation is a measure of *precision*. Specifically, standard deviation is an indication of how far a group of repetitive measurements is expected to stray from the average of all the measurements. This is completely independent of accuracy: A box

TABLE 13.4. Comparison of International Standards for Detector Tube Systems (Selected Parameters)

Device	Parameter		Standard				
			JIS K 0804	ANSI/SEA	102 BS 5343	DIN 33882	IUPAC
Gas sampling pump	Volume sampled		±5%	±5%	±5%	±5%	±5%
	Air leakage		≤3%	≤3%	None	≤3 ml/min	≤3%
Detector tube	Channeling[a]		≤2%	≤2%	≤2%	—	≤2%
	Indication accuracy allowance	Higher concentration	±25% for 33% or higher concentration of calibration scale	±25% for 1, 2 or 5 times the standard concentration	−20% to 30% for occupational exposure limit	±30% for 20% or higher of measuring range	±25% for 1, 2, or 5 times the prescribed concentration
		Lower concentration	±35% for 33% or lower concentration of calibration scale	±35% for 50% of standard concentration	−20% to 30% for occupational exposure limit	±50% for 20% or less of measuring range	±35% for 50% of prescribed concentration
	Shelf life		—	—	2 years	2 years	—

[a] Channeling of airflow through detector tube.

Acronyms: JIS, Japanese Industrial Standard; ANSI/SEA, American National Standards Institute/Industrial Safety Equipment Association; BS, British Standard; DIN, Deutsches Institut für Normung; IUPAC, International Union of Pure and Applied Chemistry "Performance Standard for Detector Tubes".

of detector tubes could display a very small standard deviation (i.e., very good precision) if all tubes read about the same value in a test atmosphere even if the value was not close to the actual concentration.[6]

The concept of *standard deviation* can be explained by recalling that two types of errors can occur with devices such as detector tubes:

- Random errors are slight fluctuations in readings compared to the actual concentrations that occur with tube systems meeting manufacturing specifications that are used according to directions. The fluctuations are due to variations in inner diameters of detector tubes, in densities of filling reagents, in sensitivities of reagents, or in the sampling practitioners who read the tubes. To evaluate random errors, the relative standard deviation is used, which shows in percentage how the reading deviates from the mean value. This value is also called the *coefficient of variation* (*CV*):

Relative standard deviation (CV), %

$$= \frac{\text{Standard deviation } (\sigma)}{\text{Mean value}} \times 100$$

- Systematic errors are due to non-random causes such as a leaking sampling pump, incorrectly calibrated detector tubes, improper sampling time, inappropriate storage or usage of detector tubes, or presence of interferents. The concept of *standard deviation* or *CV* does not apply to systematic errors.

Interchangeability of Tubes and Pumps from Different Manufacturers. There has been an ongoing controversy for many years about whether it is acceptable to use one manufacturer's tubes with a hand or other pump from a different vendor. On one hand the pragmatic approach says that once the total sampling volume and flowrate for a given tube is specified, it is unimportant how that air is drawn through the tube. In fact many *ad hoc* sampling arrangements have been developed, such as the multi-tube sampling manifold for hazardous waste sites described later in this chapter, that expand the utility of these tubes and their contribution to protecting workers. The opposing view applies mainly to the tubes certified as part of the SEI or a similar process: It states that the tube and pump are tested and certified as part of a "system," and so it is not permissible to substitute unapproved equipment in the approved system. In particular, flowrate is an important parameter since it determines the absorption rate for the chemical reactions occurring in the detector tubes to produce the color change and length of stain.

The inadvisability of interchanging tubes and pumps was clearly restated in the latest version of ANSI/ISEI Standard 102-1990 (Reaffirmed 1998)[4]:

Since the indicating behavior of a detector tube depends not only on the stroke volume, but also on the suction characteristic of the pump, it must be ensured that each detector tube is used only with the prescribed pump. Pump, tube and other components are designed, manufactured, and calibrated together to form a gas detector tube unit. User interchange of pumps and tubes or components supplied by different manufacturers may provide erroneous and invalid measurements of toxic environments. Accordingly, such interchange is not recommended.

Based on this information, a qualified professional needs to evaluate all relevant information and determine that any proposed deviations from certified systems do, in fact, provide adequate accuracy and precision before using any *ad hoc* detector tube monitoring approaches.

Interferences. Depending on the reaction principles of the tube and the interfering compounds that are present, the interferents may cause higher or lower readings compared to the actual level of the target chemical. Some tubes contain reagents that react directly with the target chemical to produce a color change. If the air contains substances similar to the target substance that also react with the reagents in the detector tube, the result will be a higher reading. For example, many tubes used for chlorine will also react to hydrogen sulfide, ammonia, nitrogen dioxide, ethylene, or halides as well, with the result being an increase in the concentration reading due to these positive interferents. If any of these compounds are present in the air during sampling, the result will not be specific for chlorine. In some cases the color change in a tube is due to a pH indicator that responds to the reaction. With this type of tube, any similar acid or basic compound (depending on the tube) will react as intereferents, giving a higher indication. An example of this interference is hydrogen chloride in the air causing a higher reading on some hydrogen cyanide detector tubes. Negative interferences manifest themselves as less or no color change when an interferent is present. Sulfur dioxide, for example, when present in an atmosphere being measured for hydrogen sulfide, will produce a lower reading than the concentration actually present.

A second type of detector tube involves a two-step reaction in which the target substance is oxidized in the pretreatment layer before reacting with the detecting reagent in the analyzer layer. In this case any nontarget compound that also reacts with the oxidizer (pretreatment reagent) may be an interferent if it causes "consumption" of the oxidizer to a degree that oxidation of the target substance is diminished, thus giving a lower indication. An example of such an interferent is aromatic hydrocarbons to some trichloroethylene detector tubes.

Sometimes manufacturers specify that an interferent will cause a different discoloration of a tube than that specified for the contaminant of interest. For example, chlorine causes a bluish stain in one tube, and nitrogen dioxide will turn the same tube a pale yellow color. Therefore, a discoloration significantly different from what is predicted should be considered suspicious.

The best prevention for erroneous results due to interferences is for the sampling practitioner to evaluate this potential prior to sampling using the information provided by the tube manufacturer.

Temperature Corrections. Chemical reactions occurring in detector tubes are temperature-dependent. With some detector tubes, reaction rates and physical adsorption of reagents are greatly influenced by tube temperatures. Most manufacturers specify the normal operating range for their detector tubes as well as a range of temperatures over which their tubes will work. If the tube is used at an ambient temperature that is outside of its normal operating range, the reading must be adjusted using factors stated in their instructions. The tube should not be used at temperatures outside the specified allowable temperature range. Generally, chemical reaction rates are proportional to the temperature, and varying reaction rates can distort the reading. For example, when the temperature is significantly lower than 20°C the reaction will slow down and some quantity of the sample may not react in the normal reaction zone but instead will partially react further down the tube. As a result, a longer layer of pale color change is produced, giving a higher indication. When the temperature is significantly higher than 20°C, the rate of reaction will be higher than normal and the most or all of the sample may react in a shorter distance than the normal reaction zone for 20°C. As a result, a shorter layer of distinct color change would be produced, giving a

TABLE 13.5. Typical Temperature Correction Factors for Solvent Detector Tube

Tube Reading (ppm)	True Concentration (ppm)				
	0°C (32°F)	10°C (50°F)	20°C (68°F)	30°C (86°F)	40°C (104°F)
100	410	155	100	80	65
80	310	125	80	65	50
60	210	95	60	50	40
40	130	60	40	35	25
20	55	30	20	17	15
10	20	13	10	8	7
5	8	6	5	4	3

lower indication. Temperature also influences physical adsorption of the target substance to the granules and packing in the tube. At lower temperatures, some of the target substance may be physically adsorbed to reagent that has already reacted, and thus it will not react with "fresh" reagent further along in the tube. As a result, a short layer of color change is produced, giving a lower indication. The opposite occurs at higher temperatures: Higher indications may be obtained.

Generally the temperature range refers to the *tube* temperature (not the sample temperature). When air is sampled, its temperature is instantly brought to the temperature of the detector tube. When detector tubes are kept at a certain ambient temperature for a sufficient period, the tube will reach that temperature. Therefore, the detector tubes just taken out of a cool storage location such as a refrigerator have the same temperature as the storage, but it will gradually reach the ambient temperature. Tubes used in hot or cold environments may have to be carried in a temperature-controlled container until used.

Calibration is often performed at 20°C (68°F), so this is considered as the benchmark temperature. If the instruction sheet for the tube indicates that readings are

affected more than ±10% at other ambient temperatures, tube readings should be corrected using the factors provided in the instructions. Temperature correction factors may be a single "multiplying" factor for each temperature, or presented as a table that converts each concentration readings depending on the ambient temperature (Table 13.5).

Correction for Humidity. Detector tubes vary widely in their reaction to relatively humidity (RH). Many are calibrated based on a relative humidity of 50%, and for some there is no effect in the range of 0 to 99% RH as long as there is no condensing moisture in the tube. Any required humidity correction factors are given in the detector tube's instruction sheets.

Correction for Atmospheric Pressure. The gas concentration (number of molecules present) is proportional to the ambient pressure. Detector tubes are calibrated at normal atmospheric pressure (760 mm Hg), and usually their readings will not be affected over the range of ±10% of normal pressure (approximately 685–835 mm Hg). If the pressure at the time of measurement is outside of this range, the tube reading should be corrected as follows:

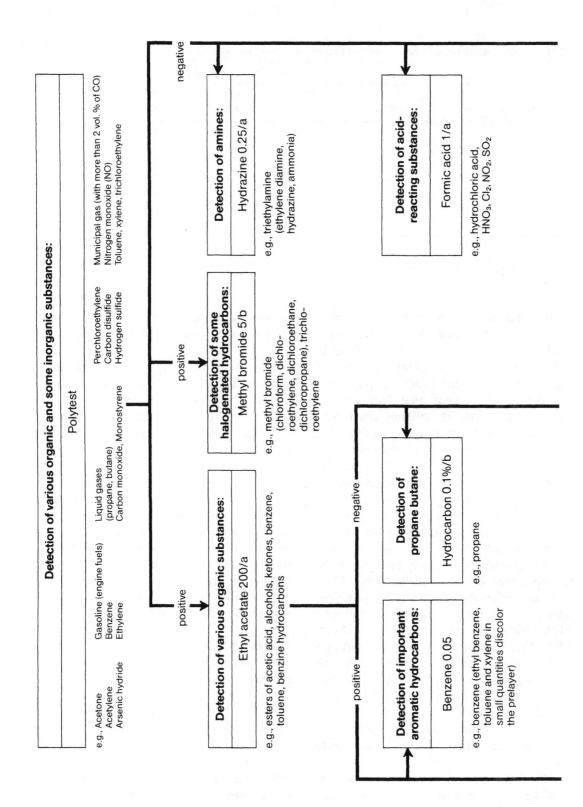

Detection of various organic and some inorganic substances:

Polytest

e.g., Acetone
Acetylene
Arsenic hydride

Gasoline (engine fuels)
Benzene
Ethylene

Liquid gases
(propane, butane)
Carbon monoxide, Monostyrene

Perchloroethylene
Carbon disulfide
Hydrogen sulfide

Municipal gas (with more than 2 vol. % of CO)
Nitrogen monoxide (NO)
Toluene, xylene, trichloroethylene

positive

Detection of various organic substances:

Ethyl acetate 200/a

e.g., esters of acetic acid, alcohols, ketones, benzene, toluene, benzine hydrocarbons

positive

Detection of some halogenated hydrocarbons:

Methyl bromide 5/b

e.g., methyl bromide (chloroform, dichloroethylene, dichloroethane, dichloropropane), trichloroethylene

negative

Detection of amines:

Hydrazine 0.25/a

e.g., triethylamine (ethylene diamine, hydrazine, ammonia)

Detection of acid-reacting substances:

Formic acid 1/a

e.g., hydrochloric acid, HNO_3, Cl_2, NO_2, SO_2

positive

Detection of important aromatic hydrocarbons:

Benzene 0.05

e.g., benzene (ethyl benzene, toluene and xylene in small quantities discolor the prelayer)

negative

Detection of propane butane:

Hydrocarbon 0.1%/b

e.g., propane

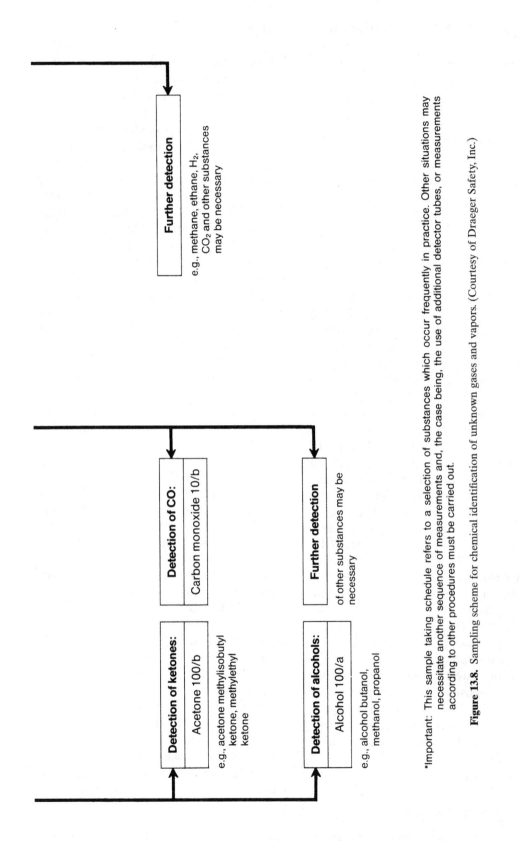

Figure 13.8. Sampling scheme for chemical identification of unknown gases and vapors. (Courtesy of Draeger Safety, Inc.)

*Important: This sample taking schedule refers to a selection of substances which occur frequently in practice. Other situations may necessitate another sequence of measurements and, the case being, the use of additional detector tubes, or measurements according to other procedures must be carried out.

True concentration

$$= \frac{\text{Tube reading (ppm)} \times 760(\text{mm Hg})}{\text{Atmospheric pressure (mm Hg)}}$$

Storage and Disposal

As detector tubes contain chemicals that are reactive, and some reagents might be corrosive, care should be taken for their storage and disposal. They should be stored in a cool (32–50°F), dark location. They should never be exposed to direct sunlight, nor should they be stored above normal room temperature.

Shelf life, represented by the expiration date, is stated on each box of tubes. It is the period of time within which the calibration accuracy of the tubes can be expected to remain within ±25%. It is a common practice to refrigerate detector tubes in order to extend their shelf life, but, because the speed of most chemical reactions is sensitive to temperature, the tubes must be warmed to ambient conditions prior to use if the tubes are going to perform according to the manufacturer's specifications. Refrigeration can ensure that the tubes are not exposed to high temperatures, which might decrease the shelf life.

Prior to use, detector tubes should be examined visually for obvious problems, such as partial discoloration of an indicating layer. If tubes change color during storage, some decomposition has probably occurred, and the tubes should be discarded regardless of whether time remains until the expiration date.

Used or date-expired detector tubes should be disposed properly, and in accordance with local hazardous waste regulations if appropriate. Refer to the tube's instructions or Material Safety Data Sheet (MSDS) for additional handling and disposal guidance.

Use of Detector Tubes

The primary application for detector tubes are grab sampling measurements as a screen to see if contaminants are present as a preliminary to more complex and accurate methods, or in situations where other methods do not exist or the time factor offered by the direct reading tubes is critical. They are also useful for collecting short-term (peak or ceiling) measurements concurrent with long-term sampling using a sorbent tube or other sample collection device. Their use for compliance sampling is limited to situations where a 15-minute short-term exposure limit (STEL) or a "ceiling" standard applies unless a large number of tubes are used to cover longer exposure periods. Other uses include area measurements at hazardous waste sites and for measurements of leaks and spills. A sampling scheme (Figure 13.8) has been developed for use as an aid in identifying unknown chemicals. It is most useful in situations where individuals are already wearing protective gear, or when a spill or the contents of a drum must be identified quickly.

For routine assessments at hazardous waste sites, a manifold device to draw air through 10 detector tubes at a time has been developed in order to allow for rapid identification of compounds (Figure 13.9). It can be assembled using two battery-operated, intrinsically safe personal air sampling pumps and a series of 20 paired-needle valves for sampling and bypass attached to three-way stopcocks. This method overcomes the lengthy time period required to make 10 successive measurements on different tubes. A stopwatch is used to determine when the proper sample volume has been pulled into each tube; at this point the sampling practitioner switches the flow of contaminated air to the blank or bypass valve by turning the stopcock.[7] However, considerable setup time is required in order to ensure that each valve provides the proper flowrate and volume for each individual tube, and the precautions discussed above under "interchangeability" apply.

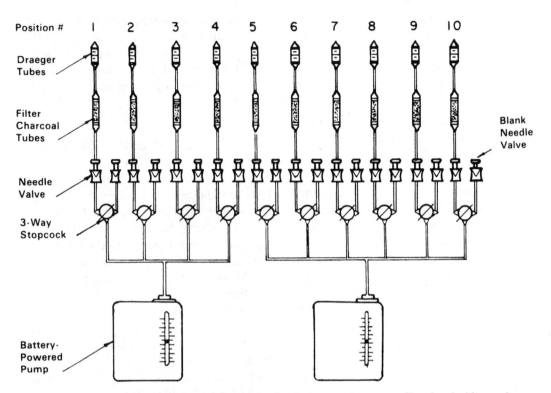

Figure 13.9. Schematic of simultaneous direct-reading indicator tube system. (Reprinted with permission from *Am. Ind. Hyg. Assoc. J.* **44**:617, 1983.)

Inaccessible locations, such as utility vaults, can be sampled by using a flexible tube attached to the pump and putting the detector tube at the other end. The sample should be drawn directly into the detector tube, meaning that the detector tube should be lowered into the hole attached to the end of the tubing, and not with the tubing in front of the detector tube. The "tube in front" configuration avoids the chance of contaminants adsorbing onto the tubing wall, which will result in a low reading.

Detector Tube Pump Maintenance and Performance Checks

Maintenance is important for detector tube pumps. For example, flow orifices may plug from glass fragments from the tube openings, and elastic components such as gaskets may lose their elasticity from

decomposition products of the detection reaction or unreacted contaminants passing through the tube. Pumps should be accurate to ±5% of their stated volume according to the current performance standards (see Table 13.4).[4] There are two types of performance checks that should be conducted on grab sampling pumps to assure accuracy in measurements: leakage tests and flow rate calibrations. The U.S. OSHA *Technical Manual* requires that each pump used by their compliance officers be leak-tested before use, as well as calibrated for proper volume at least quarterly or after 100 tubes using procedures described in the *Manual*.[8]

Leakage Test. A leakage test should be performed on each pump in order to minimize erroneous readings (due to air leaks around the seals) when it is first purchased, after an extended period of non-use, and periodically during use in accordance with the manufacturer's instructions. If a pump has multiple settings, the procedure should be repeated for each setting.

Procedure. Steps 1–4 refer to a piston pump while step 5 describes the test for a bellows pump.

1. If the pump has multiple orifices, turn the rotating head through several revolutions, finally stopping at the highest flowrate position.
2. Insert an unopened detector tube into the tube holder. Line up the guide marks (usually dots on the pump shaft and housing), pull back the handle all the way, and lock the piston in the maximum volume position.
3. Wait 2 minutes and release the pump handle slowly in order to prevent damage as it springs back. When the handle is released, the piston should return completely to the 0-mL mark. If it does not, the pump is not leak-tight and the amount of leakage is indicated by the volume reading as the handle comes to rest. If there is greater than 5 mL of leakage in a 100-mL sample volume within 2 minutes, or whatever time is required for the normal pump sampling stroke, there is excessive leakage which should be repaired to maintain sampling accuracy.
4. If excessive leakage is found, it usually occurs either at the pump inlet or between the piston and cylinder walls. Leakage in the cylinder can usually be corrected by cleaning and relubrication. Leakage at the inlet may result from a poor seal between the detector tube and the rubber inlet tip or between the flange of the rubber inlet tip and the pump body.

5. For a bellows pump, insert an unopened tube and compress the bellows completely. Let it sit for 30 minutes and check for observable expansion. If there is none, the pump passes.

Volume and Flowrate Calibration. Calibrate detector tube pumps for proper volume measurement quarterly. One of the reasons the pumps lose their volume capacity is due to the buildup of particles released during the tubes' reactions. These fine particles are pulled into the interior of the pumps and gradually build up.

The flowrate (mL/min) determines the reaction rate for many chemical reactions that occur in detector tubes, so refer to the manufacturer's literature for flowrate criteria and recommendations for flowrate tests. Some pumps have filters, in which case the filter should be checked for build-up which will reduce flowrate.

Procedure
1. Test the detector tube pump for leakage following the procedures previously described.
2. Connect the inlet of a detector tube pump to the top of a 200-mL bubble burette using Tygon tubing. For procedures on the use of bubble burettes see Chapter 5. For pumps with limiting orifices to control flowrate, use a suitable adapter such as a short piece of glass tubing in place of a detector tube (i.e., do not put a detector tube in line for these pumps). For pumps that rely on the resistance of a detector tube to govern airflow rate, insert a fresh opened tube in line during measurements.

For a piston pump continue with steps 3–6 and 9; for a bellows pump go to step 7.

3. If the pump has flow control orifices, select the orifice to be tested.

4. Review the manufacturer's literature for the pump being used to determine the number of minutes to wait after pulling the pump handle for the full volume of the pump to be drawn.

5. Dip the bubble burette into bubble solution and pull the pump handle slightly to move the bubble front into the calibrated section of the burette. Note the initial position of the soap bubble. Push the pump handle back in, then pull the pump handle all the way out for a full pump stroke and time the bubble front from the start until it comes to rest. Note the final position (reading) of the soap bubble. The difference between the initial and final points is the sample volume. The volume divided by the time is the flowrate.

6. Use the same procedure to test each orifice for pumps with multiple flow controlling orifices.

For a bellows-type pump, follow steps 7 through 9.

7. Squeeze the pump, making sure the bellows are completely and evenly compressed. Allow the pump to open on its own, measuring the volume as described in step 5. Repeat two more times and average the three measurements.

8. Compare the results with the manufacturer's specification for allowable variation in total volume and the time required to draw the total volume.

9. Repair the pump if the volume error is greater than ±5% and repeat the volume calibration test.

Field Measurements with Detector Tubes. The first step is to identify the target chemical(s), their approximate concentration(s), any other compounds (potential interferents) that are likely to be present during sampling, and relevant environmental conditions such as ambient temperature and relative humidity. With this information the proper tube can be selected. The tube should also be selected for the concentration range of interest relative to the allowable exposure limit; usually this is the ACGIH TLV© or OSHA permissible exposure limit (PEL). From the tube instruction sheet, note the number of strokes and the approximate time for the sample to be collected. For example, in order to reach a sensitivity of 1 ppm, at least one tube for benzene requires five strokes, each 2 minutes in length; thus a single measurement requires 10 minutes.

In many cases, the sensitivity of a reading can be improved by taking two to four additional strokes and multiplying the final result by the ratio of the specified number of strokes to the actual number of pump strokes. For example, if one pump stroke results in a reading below the first concentration mark of 10 ppm, two more strokes can be collected. If the reading is now 20 ppm, that value can be divided by 3 to yield a measurement result of 7 ppm, which is more specific than reporting a "<10 ppm" reading.

It is important to keep the tube inlet in the proper location during the sampling phase. If fatigue or awkward position makes holding the proper position difficult, consider using a Tygon tube between the tube and pump. The tube must remain positioned during the entire sampling period in order to make sure all the air comes from the proper point or source.

Procedure for Bellows Pumps

1. Break off the ends of a fresh detector tube by inserting each end in the tube breaker hole usually located in the pump head or by using a separate break-off device. Dispose of the broken glass in a safe place. With a

two-tube system, connect the ends of the two tubes using the rubber tubing supplied after breaking the tube tips. Insert the tube securely into the pump inlet, taking care that the arrow on the tube points toward the pump.

2. Squeeze the bellows pump and release, then wait for the chain straddling the bellows to become taut or for the end of the flow indicator to change color. Bellows-type pumps must be compressed completely flat to get an accurate sample. Both hands may be needed, since equal pressure must be applied to both ends of the pump if the entire bellows does not fully compress when only one hand is used. Repeat the compression process as often as specified in the tube's operating instructions. If using the stroke counter, make sure that it registers each stroke to give an accurate count.

3. When drawing air into a tube, observe the tube to note any color change or stain development. For example, if using a hydrogen sulfide tube with a sensitivity range of 0.5 ppm to 1.5 ppm based on 10 strokes, the tube will fully color with a fewer number of strokes if the concentration is higher than 1.5 ppm. Do not stop until the full number of strokes has been pulled unless the tube saturates as described above. If this happens, put in a new tube and pull one-half the specified number of strokes (in this example 5 strokes) and multiply the reading by a factor of two.

4. Read the tubes immediately after sampling since some changes in the coloration will occur with time, possibly including a complete fading of the stain. Where there is a gradation of color change, the end point should be taken as that point where the slight discoloration can just barely be discerned from the original color. If the end point occurs at an angle, estimate the concentration for each side of the tube and average them.

Procedure for Piston Pumps

1. Make sure the pump handle is all the way in. Align the index marks on the handle and back plate of the pump. Break off the ends of a fresh detector tube; dispose of the broken glass in a safe place. With a two-tube system, connect the ends of the two tubes using the rubber tubing supplied after breaking the tube tips. Insert the tube securely into the pump inlet, taking care that the arrow on the tube points toward the pump.

2. With a multiple orifice pump, select the proper orifice by sliding the locking button on the rotating head forward and turning the head to the designated index number on the flow control plate. When the center line of the locking button is adjacent to the index mark, release the button. The spring-loaded button will then lock the rotating head in this position.

3. Pull the handle straight back (without turning) to the sample volume required (full or partial pump stroke). The piston will automatically lock in this position. On some models the handle can be locked on either 25-, 50-, or 100-mL (full) stroke. Additionally, some pumps feature a "one-hand" valve (Figure 13.10) that allows the sampling to be initiated with only one hand after the piston has been pulled back. The time required for the pump to draw its full volume is important and is specified in the tube's literature; it is necessary to allow this amount of time in order to collect a full sample. If the handle is released before sufficient time has

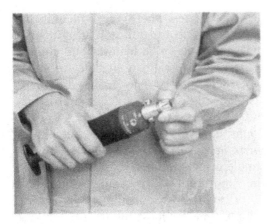

Figure 13.10. "One-hand" valve permits piston pump sampling with only one hand after the piston is pulled back. (Courtesy of Nextteq, LLC.)

passed, the volume will be smaller than is required, with the result being a lower reading than the concentration that is actually present.

4. For subsequent strokes, if needed, unlock the pump handle by making a one-quarter turn and return it to the starting position. Rotate the handle to realign the index marks for the next stroke and pull the handle again. The detector tube should remain in place until all the strokes have been made.

Maximizing the Accuracy of Detector Tube Measurements

Follow these guidelines to maximize the accuracy of detector tube readings[6]:

- Detector tubes are calibrated at 20°C (68°F) and 50% relative humidity. For measurements at significantly different conditions, use the correction factors supplied by the manufacturer.
- The shorter the stain length, the harder it is for a tube to meet the accuracy specification (e.g., ±25% of a 1-mm-long stain is a much smaller window than ±25% of a 10-mm stain). If several range tubes are available, choose the tube that will provide a stain length in the upper two-thirds of the tube's range.
- Ensure that the concentration interval markings on the calibration scale are appropriate for the concentration and setting being evaluated. For example, a sulfur dioxide tube with a 1- to 50-ppm range would not be the best choice for measuring occupational exposures since the permissible exposure limit (PEL) is 2 ppm, while the tube with a 50-ppm range would probably be marked at 5-ppm intervals.
- Regular leak checks of the pump will help to achieve optimum accuracy by ensuring that the proper air volume is passing through the tube.
- Fresh detector tubes will generally be more accurate than tubes beyond their expiration date. Never use detector tubes that are past the posted expiration date.
- Refrigerated storage will prolong the freshness of the tubes and improve accuracy. The recommended storage temperature is 5–10°C (41–50°F). Detector tubes should not be stored frozen.
- For optimum accuracy, allow the tubes to warm up to ambient temperature prior to use.

LONG-TERM COLORIMETRIC TUBES AND BADGES

This category of colorimetric devices includes long-term detector tubes, direct-reading badges, and "spot" plates. Their application is similar to (a) *sample collection device methods* for gases and vapors and (b) personal-size direct-reading instruments with data-logging capability. Although their accuracy is not as high as the accuracy of sorbent tubes analyzed in the laboratory, there are situations where the use of long-term detector tubes can be an advantage. In cases where there is a high degree of employee concern over a potential exposure, these tubes give immediate results that are more representative of employee exposure (integrated) than a series of grab samples using short-term detector tubes. Long-term tubes for a variety of chemicals can be kept on hand for occasional use, whereas maintaining an inventory of direct-reading instruments with fresh sensors for infrequent measurements may not be feasible. Long-term detector tubes can be very useful in situations where a gas or vapor is suspected in that they can be used as a screening device. They can detect much lower levels than other detector tubes because they can collect samples for a longer period of time. In a situation where the concern is what compound (rather than how much) is present, these tubes can be used for 8- to 12- or even 24-hour screening samples. It can be difficult to interpret results when contaminants are present at very low levels due to differences in color changes as compared to the expected color based on a normal sampling time.

It is especially important to remember that these tubes still suffer from the same problem with interferences as the ones used for grab samples. Long-term or long-duration tubes are available for a limited number of compounds, but for some of these chemicals the tubes fill a real need in some monitoring situations if the only other available measurement methods are impingers or expensive real-time instruments.

Long-Term Detector Tubes

Long-term detector tubes fall into two designs: (a) those that are used with battery-operated low-flow personal monitoring pumps for active sampling and (b) passive devices where the contaminants enter the tube via diffusion. These tubes have the same basic features and characteristics as the short-term detector tubes described above. The main difference visible to the sampling practitioner is that the calibration scale is not marked in "ppm" units; instead it is marked in units of *microliters* (μL) or *ppm-hours*, and a simple calculation is needed to determine the exposure level for comparison to standards expressed in units of *ppm*. The sample time for most tubes varies from 1 hour to 4 hours, so several tubes may be needed to measure a full 8-hour time-weighted average exposure.

Long-Term Detector Tubes for Pump Sampling. Long-term tubes for use with a pump usually require very low flowrates (in the range of 10–20 mL/min); not all pumps can achieve these low rates. The setup is essentially the same as that used for gas and vapor sampling with sorbent tubes discussed in Chapters 5 and 6 on *sample collection device methods*. As with sorbent tubes, the tubes should be kept in a vertical position during use. The tubes should also be checked periodically to see if any color change is occurring during the sampling period in order to avoid the potential of exceeding the tube's capacity. The information in the tube's instruction sheet is important since it describes the sampling rate, the total volume, the sampling time, and the response time for the color change to develop as well as temperature, humidity, and pressure correction factors and possible interferents.

While the specified sampling time and flowrate must be followed for accurate quantitative measurements, some flexibility is possible when the tubes are used for screening measurements when it is suspected that the target chemical is present in very low concentrations. For example, very low levels of sulfur dioxide can cause metal to rust, and these extremely low levels can be detected by sampling with a long-term tube for a 24-hour period.

Procedure

1. Prior to use, test each batch of tubes with a known concentration of the air contaminant that is to be measured. Refer to the manufacturer's data for key information and a list of interfering materials. Observe the manufacturer's expiration date, and discard outdated tubes.

2. Break off the ends of a tube. Calibrate a pump at the specified flowrate, usually 10 mL/min to 20 mL/min, with a proper length of tubing and a long-term tube in line using a bubble burette or electronic calibrator.

3. After calibration, replace the tube with a fresh one if there was any possibility of contaminants being present, or soap film or water vapor entering the tube during calibration. Put the pump on a worker or in the area to be sampled. Turn on the pump and record the time or the stroke number if using a stroke volume pump.

4. At the end of the sampling period, turn off the pump and record the time, and also record the stroke reading if appropriate. If the instructions for the tube require that fresh air be pulled through the tube to fully develop the color change, take the tube to an area where clean air is present and pull the required amount of air through it. If only a small amount is required, such as 500 mL, then a suction pump such as a bellows pump can be used; otherwise, the battery-operated pump is used.

5. Read the tube immediately after the last amount of air is pulled through it. Read the length of stain in a well-lighted area. Read the longest length

Figure 13.11. Colorimetric diffusion tubes for long-term measurements. (Courtesy of Nextteq, LLC.)

of stain if the stain development is not sharp or is uneven around the tube in order to give a conservative estimate of exposure. Many tubes will retain their stain for a period of time if the ends are capped.

Passive Detector or Dosimeter Tubes.

Passive detector or dosimeter tubes (Figure 13.11) are available for a variety of chemicals. They operate either based on the length-of-stain or the color intensity that develops following exposure in the atmosphere to be tested. Their primary advantage over the active (pump) detector tube method is their ease of use in that they are simply opened and hung in place for a period of time. They are designed to be exposed and read after a certain number of hours up to a stated maximum. The basic concept is that increasing exposure times results in longer stain lengths. The stain length is thus related to the product of exposure time and ambient concentration.

These devices consist of a glass or plastic tube containing an inert granular material or a strip of paper that has been impregnated with a chemical system that reacts with the gas or vapor of interest. As a result of this reaction, the impregnated chemical changes color. The granular material or paper strip is held in place within the glass tube by a porous plug of a suitable inert material. During use the dosimeter tube is held in a holder that protects the dosimeter and also helps to minimize effects of air currents on performance. The holder has a clip that allows it to be fastened to a collar or pocket during personal sampling, or to some appropriate object during area sampling.

The dose is derived from length-of-stain dosimeters by reading the length of stain from the calibration curve provided with the tubes or printed on the tube. The dose from color intensity units is obtained by comparing the color intensities provided by the manufacturer with the color that develops on the dosimeter. When the reading is in units of *ppm-hours*, this value is divided by the number of hours in the sampling period to obtain a time-weighted average (TWA) exposure result.

Some dosimeters are designed to be used over several days—for example, units placed in homes to measure formaldehyde. These devices often have a calibration for 1 day's exposure and for 7 days' exposure. These dosimeters are usually intended for screening measurements to determine what type, if any, of follow-up measurements should be taken. Results are not considered to be precise.

The sampling rate of passive colorimetric dosimeter tubes is very slow, on the order of 0.1 mL/min. Thus the phenomenon that occurs with some passive devices in low air movement situations where the air layer adjacent to the dosimeter becomes depleted of target chemical molecules is not significant for these devices.

For sampling, the dosimeter is either "broken" open or removed from its protective wrapper, placed vertically on the person's lapel or in a given area, and the time is recorded. At the end of the sample period, the time that the unit is removed is also recorded. The length of stain or the color intensity is compared with the standard or measure provided by the manufacturer, which often is a graph of concentration curves corresponding to stain length at a given time. By finding the

stain length on the "x axis" and drawing a line to intersect with the point on a curve that corresponds to the number of hours for which monitoring was done, the concentration for the sampling period can be determined.

"Spot Plate" Badges. Spot plate badges offer another option for passive colorimetric sampling. Most are designed for visual evaluation and have indicator strips or buttons that change color when a critical accumulation of the target gas is reached. Some have the colors corresponding to various concentrations printed directly on the device or use a colored icon to warn of hazardous levels, while others must be compared to a chart to determine the integrated concentration.

The *SAFEAIR*™ system (Figure 13.12) is a self-contained unit that warns of the threshold concentration of a contaminant because the "exclamation point" icon becomes visible. For higher resolution and wider measurement range, a slip-in color comparator is available for selected chemicals. The color comparator contains a color scale that matches the color developed on the badge at different ppb-hr or ppm-hr exposures. Table 13.6 shows the chemicals for which the *SAFEAIR* badge and color comparator are available.

Some spot plate devices are designed to be used with a sampling pump to increase the minimum detectable level and accuracy. In other respects these units are similar to the passive devices.

An innovative colorimetric badge by Assay Technology relies on an electronic reader to measure ethylene oxide exposures. The ChemChip™ personal monitor (Figure 13.13) is worn as a badge on pocket or lapel. After wearing, the test strips encased in the monitor are removed, developed, and inserted into a calibrated reflectance colorimeter. This electronic reader provides on-site readout of chemical exposure in the parts per million (ppm)

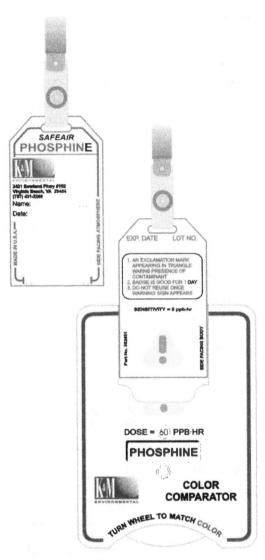

Figure 13.13. ChemChip™ personal monitoring system for determining exposure to ethylene oxide. (Courtesy of Assay Technology.)

Figure 13.12. Colorimetric badges show an immediate color change upon exposure; some devices also use a color comparator for more accurate determination of concentration. (Courtesy of K&M Environmental, Inc.)

COLORIMETRIC ELECTRONIC INSTRUMENTS

There are two types of colorimetric electronic instruments: tape-based monitors and wet chemical instruments.

Tape-Based Monitors

Paper tape-based instruments use chemically impregnated tape to detect toxic gases. The tape changes color when exposed to the target gas; the color change is detected by a photocell, analyzed, and translated into a concentration value (Figure 13.14).[9] These devices have the advantage of providing physical evidence of the gas concentration from a leak or release, and typically are less prone to inferences when compared to electrochemical and solid-state sensors. A disadvantage

range. The operating range is 10–500% of the permissible exposure limit, with optimal accuracy in the range of 50–200% of the PEL. It can reliably measure 0.1 ppm for an 8-hour shift and 1 ppm over a 15-minute STEL period, and it meets OSHA's accuracy requirements for compliance measurements.

TABLE 13.6. Target Chemicals for SAFEAIR™ System Colorimetric Passive Badges

Chemical	Threshold Level	Minimum Detectable Level (8 Hours)	Maximum Sampling Time (Hours)	Minimum Sampling Time (Minutes)
Ammonia	4.0 ppm-hr	0.50 ppm	48	15
Aniline[a]	0.2 ppm-hr	0.025 ppm	48	5
Arsine	18 ppb-hr	2.25 ppb	12	15
Carbon dioxide[a]	8000 ppm-hr	1000 ppm	10	15
Carbon monoxide	7 ppm-hr	1 ppm	10	15
Chlorine[a]	0.18 ppm-hr	0.023 ppm	48	15
Chlorine/Chlorine	Cl_2: 0.18 ppm-hr	0.023 ppm	10	15
Dioxide	ClO_2: 0.2 ppm-hr	0.025 ppm		
Dimethyl amine	5 ppm-hr	0.6 ppm	48	5
1,1-Dimethyl	Front: 30 ppb-hr	3.75 ppb	48	5
hydrazine	Back: 10 ppb-hr	1.25 ppb		
Formaldehyde	0.4 ppm-hr	0.05 ppm	10	15
Hydrazine[a]	8.0 ppb-hr	1.0 ppb	48	5
Hydrogen chloride[a]	2.0 ppm	2.0 ppm STEL only	0.25	15
Hydrogen fluoride[a]	2.8 ppm	2.8 ppm STEL only	0.25	15
Hydrogen sulfide	2 ppm-hr	0.25 ppm	48	15
Mercury	Front: 0.1 mg/m³-hr	0.013 mg/m³	48	15
	Back: 0.2 mg/m³-hr	0.03 mg/m³		
Methyl chloroformate	0.025 ppm-hr	0.008 ppm	48	10
Nitrogen dioxide	1 ppm-hr	0.125 ppm	10	15
Ozone	0.05 ppm-hr	0.006 ppm	48	15
Phosgene[a]	0.015 ppm-hr	0.002 ppm	72	1
Phosphine[a]	5.0 ppb-hr	0.625 ppb	12	15
Sulfur dioxide	0.2 ppm-hr	0.025 ppm	48	15
2,4-Toluene diisocyanate (TDI)[a]	5.0 ppb-hr	0.625 ppb	24	15

[a] Indicates that a color comparator is available for the chemical.

Source: K & M Environmental (www.kandmenvironmental.com).

is that because of cost and complexity, most fixed installations use a central analyzer with sampling tubes run to remote locations. Thus there can be a considerable time delay between a release and when the concentration is sensed by the instrument. These devices are also relatively complex with mechanical, optical, and electronic systems that need preventive maintenance and may be prone to plugging or failure, depending on the ambient environment. Additionally, the tapes usually need some

moisture in the air in order to function; a relative humidity of 40% is a typical specification.[10]

More sophisticated tape-based monitors use advanced optics to measure the light reflected off of the tape before and after the tape is exposed to the sample gas stream. In some cases there are multiple measurement "windows"—some have filters to reduce the instrument's sensitivity and thus increase the measurement range. Sometimes two detection principles are

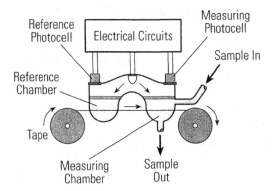

Figure 13.14. Diagram of a tape-based colorimetric direct reading instrument. (Courtesy of International Sensor Technology.)

used in order to monitor over a wide range of concentrations:[10]

- *Density* mode for low concentrations, in which the intensity of the stain is compared to calibration data stored in the instrument.
- *Rate of change* mode, which monitors how fast the stain is developing, for measuring higher concentrations.

Using both principles is necessary to cover a broad concentration range with a high degree of accuracy. With the density mode, there is a flattening of the concentration–response curve as the concentration increases and the stain approaches its maximum intensity. Beyond this point a higher concentration does not produce much increase in stain intensity. The rate of change mode continually monitors how fast the tape is darkening. The higher the concentration the faster the stain develops. The instrument can calculate the concentration based on how fast the tape darkens.

Tape-based monitors are most commonly used for isocyanate compounds because of the difficulty in continuously measuring these compounds. Other compounds for which tape-based monitors are offered include phosgene, hydrazine, ammonia, bromine, chlorine, hydrides (arsine, diborane, disilane, germane, hydrogen selenide, phosphine, silane, and stibine), hydrogen cyanide, hydrogen sulfide, nitrogen dioxide, ozone, sulfur dioxide, and acid gases (hydrogen bromide, hydrogen chloride, hydrogen fluoride, nitric acid, and sulfuric acid). Tape-based monitors are available as small passive badges, personal monitors, stationary monitors, and survey instruments.

As with all direct-reading instruments, a potential disadvantage that must be investigated is the possibility of interference from other airborne chemicals. Some interferents will cause staining of the tape, while other compounds, such as sulfur dioxide, can bleach the tape, thus causing a lower response. In addition to evaluating the possibility of interfering compounds during method selection, it is good practice to routinely inspect the tape following monitoring. In some cases it may be readily apparent that an interferent was present because the color of the stain will be different than expected. For example, whereas TDI commonly stains a tape red-purple, nitrogen dioxide and chlorine will stain the same tape dark brown.

AutoStep Plus™ Portable Paper Tape Toxic Gas Monitor. The Scott/Bacharach Autostep Plus™ is a paper-based colorimetric portable survey instrument for the measurement of chlorine, formaldehyde, hydrazine, hydrides, hydrogen chloride, isocyanate compounds, mono methylhydrazine, phosgene, and TDI (Figure 13.15). Instrument operation is based on the use of variable duration discrete sampling periods. At the start of a period, the AutoStep Plus advances to a new piece of tape and checks the stray light and tape background color. If these are within acceptable limits, the reference light level is measured and stored and the sampling pump is started, pulling air through the

Figure 13.15. Autostep Plus™ tape-based instrument has tape cassettes for different toxic gases and vapors. (Courtesy of Scott Instruments.)

tape. During the sample period the instrument continuously measures the light reflected off of the tape and compares the reading to a preset threshold. If a high concentration of contaminant is present, the reflected light level will eventually exceed the threshold. When this happens, the sampling period is ended at 4 minutes and the stain intensity is used to determine the concentration. The accuracy is whichever is greater between either ±15% of reading or 1 ppb/0.01 ppm (depending on the scale).[11]

Two modes of operation allow the AutoStep Plus to effectively operate as both a portable leak sourcing instrument and a temporary continuous area monitor:

- The "demand sampling mode" provides an indication of toxic gas levels in as little as 15 seconds and is typically used when determining the presence or source of leaks. The user controls the sampling time and the initiation of new samples.
- The "continuous sampling mode" is useful in area monitoring applications where a fixed monitor is not installed or is unavailable. The device can monitor continuously for up to 60

hours when using VAC line voltage or up to 20 hours using the built-in lead acid battery. Relay outputs are also provided for an eternal connection to auxiliary warning devices.

Cassette life depends upon operational conditions and the resolution of the instrument. In standard resolution models where gas concentration is low, cassette life can be as great as 900 samples (60 hours). At the other extreme, when operated in conditions where gas concentrations are very high, cassette life can be reduced to 8 hours. Cassette shelf life for unopened cassettes is four months at 20–30°C, and for opened cassettes it is two weeks.

The AutoStep Plus also functions as a data logger when used with an optional charger/interface unit with software. With this feature all readings are stored in nonvolatile memory for future readout and analysis. This capability allows results of surveys to be analyzed and permanently stored. Data points are stored in memory at the end of each individual sample period. The duration of these periods range from 20 seconds to 4 minutes, depending on gas concentration. Data can be downloaded via an RS-232 serial port, and the software controls data downloading, analysis, storage, and a wide range of displays and printout options.

The instrument's LCD panel shows concentration and user-set alarm levels. It has audible and visual alarms, and an acoustic chamber built into the instrument's handle also provides a tactile warning alarm to the user. The acceptable operating range is 32–104°F temperature and 5–95% relative humidity (noncondensing). Battery service life of the rechargeable lead acid is 16–20 hours, and it can be operated continuously from line voltage with the charging unit or interface. The AutoStep Plus is UL approved as intrinsically safe for Class I, Division I operation. It weighs 4.75 pounds.

Wet Chemical Colorimetric Techniques

A unique niche in air monitoring techniques is filled by wet chemical colorimetric instruments. These devices collect an air sample in a manner that removes the target compound from the air stream and transfers it to a liquid chemical reagent where it undergoes a color change that is proportional to its concentration. These instruments are selected when there is not a more feasible or convenient direct reading technique for the application. The most common reasons for using wet chemical instruments are the need to monitor very low levels of acutely toxic compounds, or the presence of interfering compounds that render other techniques unsatisfactory.

For example, one manufacturer makes a wet chemical analyzer that is capable of continuously monitoring for compounds such as ammonia, chlorine, formaldehyde, hydrazine, hydrogen chloride, hydrogen cyanide, hydrogen fluoride, hydrogen peroxide, mercaptans, nitrogen dioxide, NO_x, phenol, phosgene, sulfur dioxide, titanium oxides, nicotine, and total aldehydes. The target compound can be changed by changing reagent(s) and, in some cases, by changing the analytical module tray which contains the necessary glassware. The unit is extremely sensitive and for many of the analyses can operate over a range of maximum sensitivity of 0- to 0.25-ppm full scale, detecting as minute a quantity as 0.005 ppm or 5 ppb of the gas to be measured, adjustable to a full scale range of more than 10 ppm. As an illustration, to measure formaldehyde with this instrument, an air sample is continuously drawn into the unit and any formaldehyde that is present is scrubbed with a sodium tetrachloromercurate solution that contains a fixed quantity of sodium sulfite. Acid-bleached pararosaniline is added, and the intensity of the resultant color is measured at 550-nm wavelength by a colorimeter and is displayed on a digital readout and stored in memory. Using the standard range, concentrations of 0–2 ppm can be measured; a low-level option will measure 0–500 ppb. The minimum detection level is 0.005 ppm (5 ppb) using the 0- to 0.5-ppm full-scale setting or 1% of other scales. There is a lag time of several minutes between collection of the sample and when the results are available, although 90% of the value is displayed within less than 5 minutes. The instrument is usable at temperatures of 40–120°F, with an optimum range of 60–80°F and 5–95% RH.

SUMMARY

Colorimetric direct-reading devices rely on color change to measure airborne concentration of chemicals and include: detector tubes and similar products for short-term or grab samples; tubes (either with a sampling pump or passive) and badge-like units or "spot plates" for long-term measurements; and electronic instruments that collect air samples on a moving tape or in a liquid and that report the airborne concentration based on the reaction between the target chemical and the reagents in the tape or liquid.

Colorimetric devices operate on two broad principles: In *length of stain* devices the concentration is related to the amount (length) of reagent that is discolored, whereas with *color intensity* devices the concentration is related to the degree of color change as compared to a standard.

Like all monitoring processes, colorimetric systems are subject to interferences from nontarget chemicals that can result in either high or low erroneous readings. In some case the colorimetric technology has eliminated interferences that prevent the use of other monitoring techniques. All colorimetric devices have temperature, relative humidity, and ambient pressure limits as specified by the manufacturer. Some colorimetric devices such as tape samplers need some moisture in the air in order to function. The sampling practitioner must evaluate possible interferents and whether the operating conditions are suitable when determining if colorimetric systems are a good option for their sampling needs.

REFERENCES

1. Cohen, B. S., and C. S. McCammon, Jr., eds. *Air Sampling Instruments for Evaluation of Atmospheric Contaminants*, 9th ed. Cincinnati, OH: ACGIH, 2000.

2. Plog, B. A., and P. J. Quinlan, eds. *Fundamentals of Industrial Hygiene*, 5th ed. Itasca, IL: National Safety Council, 2002.

3. Samer, S. F., and N. W. Henry. The use of detector tubes Following ASTM Method F-739-85 for Measuring Permeation Resistance of Clothing. *AIHA J.* **50**(6):298–302, 1989.

4. American National Standards Institute/International Safety Equipment Association Standard 102-1990 (Reaffirmed1998). *American National Standard for Gas Detector Tubes—Short Term Type for Toxic Gases and Vapors in Working Environments.* ANSI: New York, 1998.

5. Technical Note (TN)-143, Accuracy Comparisons of RAE Systems Gas Detection Tubes. RAE Systems, Sunnyvale, CA, 1999.

6. Roberson, R. *Tech Corner—Common Questions Revolving Around Detector Tube Accuracy.* Clearwater, FL: Sensidyne, 2003.

7. King, M. V., P. M. Eller, and R. J. Costello. A qualitative sampling device for use at hazardous waste sites. *AIHA J.* **44**(8):615–618, 1983.

8. Occupational Safety and Health Administration. *OSHA Technical Manual.* Washington, D.C.: U.S. Department of Labor, 2003 (http://www.osha-slc.gov/dts/osta/otm/otm_extended_toc.html).

9. Chou, J. *Hazardous Gas Monitors, A Practical Guide to Selection, Operation and Applications.* New York: McGraw-Hill, 2000.

10. *What You Should Know About Gas Detection.* Exton, PA: Scott/Bacharach Gas Detection Products, 2003 (http://www.scottbacharach.com).

11. *Scott/Bacharach AutoStep Plus^{TM} Product Information*, Exton, PA: Scott/Bacharach, 2003 (http://www.scottbacharach.com).

CHAPTER 14

REAL-TIME SAMPLING METHODS FOR AEROSOLS

Real-time field instruments for measuring aerosols can determine parameters such as mass, respirable mass, total particle count, or particle size distribution. Most instruments are too large to be worn by the worker; thus they are used mainly for area or spot readings to measure short-term exposures such as occur when a worker dumps bags, or to compare one operation with another or to break an operation down into specific exposures with various sequential tasks. However, a few devices are small enough to be worn by the worker. These instruments have data-logging capabilities, so they can be used to provide a time-dependent profile of exposure over a work period, as well as integrated TWA exposures.

These instruments are very useful as screening devices during these situations:

- During walk-around surveys to estimate total and respirable dust levels.
- Developing a sampling strategy by identifying the highest potential exposures.

- Measuring the effectiveness of dust control measures.
- Identifying dust sources including fugitive emission sources.
- Evaluating work practices.
- Increasing awareness among workers of the aerosol concentrations generated during their tasks, to encourage proper use of controls measures.

Some devices can be fitted with cyclones or other size-selective separators on their inlet. Additionally, a filter can often be placed at the exhaust of the instrument to collect a sample for subsequent laboratory analysis. The filter sample can be used for a field calibration check (by comparing its results to the instrument readings) or to determine the chemical composition of the sampled dust.

Real-time aerosol monitors do not replace integrated filter sampling methods to the same extent as do direct-reading instruments for gases and vapors. The main reason for this is that aerosol monitors

Air Monitoring for Toxic Exposures, Second Edition. By Henry J. McDermott
ISBN 0-471-45435-4 © 2004 John Wiley & Sons, Inc.

cannot determine chemical composition. This factor can be a problem in situations where highly toxic dusts, such as lead, cadmium, and cobalt, may be present. Nonspecificity also impacts concentration measurements when an instrument has been calibrated with one type of aerosol and is used to measure another. In some applications, such as hazardous waste sites, direct-reading instruments are preferred to traditional filter methods because the results are available immediately.

The accurate measurement of aerosols is more complex and difficult than the measurement of gas and vapor molecules. Aerosols are subject to many factors such as size, shape, settling velocity, and other influences relating to the dynamics of particle collection as discussed in Chapter 7.

Direct-reading instruments for aerosols are calibrated by the manufacturer with one or more "standard" aerosols with known size distributions. Although there are several standard dusts in common use, Arizona road dust is typical. This dust is available in four different grades, which are designated in ISO 12103-1 (Road Vehicles—Test Dust for Filter Evaluation):

- A1 (ultrafine test dust) has a nominal 0- to 10-μm size.
- A2 (fine test dust) has a nominal 0- to 80-μm size.
- A3 (medium test dust) has a nominal 0- to 80-μm size, but with a lower 0- to 5-μm content than A2.
- A4 (coarse test dust) has a nominal 0- to 180-μm size.

Table 14.1 shows the particle size distribution for each grade of Arizona road dust.

Almost all instruments for field measurement are based on the light-scattering principle that analyze the particles as they interact with a beam of light as they move through a counting section; the deflected light is measured by a sensor and converted to airborne concentration based on calibration values programmed into the instrument. A second principle, the Piezobalance [also called the *Quartz Crystal Microbalance* (QCM)], is used for some specialized applications. The Piezobalance contains two oscillating crystals and electronic circuitry that monitors changes in frequency between the two crystals. Airborne dust

TABLE 14.1. Particle Size Distribution for Arizona Road Dust

Particle Size (μm)	Cumulative Volume (Percent) Smaller than Stated Particle Size			
	A1 (Ultrafine)	A2 (Fine)	A3 (Medium)	A4 (Coarse)
1	1.0–3.0	2.5–3.5	1.0–2.0	0.6–1.0
2	9.0–13.0	10.5–12.5	4.0–5.5	2.2–3.7
3	21.0–27.0	18.5–22.0	7.5–9.5	4.2–6.0
4	36.0–44.0	25.5–29.5	10.5–13.0	6.2–8.2
5	56.0–64.0	31.0–36.0	15.0–19.0	8.0–10.5
7	83.0–88.0	41.0–46.0	28.0–33.0	12.0–14.5
10	97.0–100	50.0–54.0	40.0–45.0	17.0–22.0
20	100	70.0–74.0	65.0–69.0	32.0–36.0
40	—	88.0–91.0	84.0–88.0	57.0–61.0
80	—	99.5–100	99.0–100	87.5–89.5
120	—	100	100	97.0–98.0
180	—	—	—	99.5–100
200	—	—	—	100

deposits on one crystal, which changes its oscillating frequency slightly as compared to the other (reference) crystal; the change in frequency is expressed as particulate concentration. These two types of instruments are described in this chapter.

Another type of instrument is the beta gauge, which operates on the principle that a near-exponential decrease in the number of beta particles transmitted through a thin sample occurs as the unit area density of the sample increases. They are used for continuous ambient particulate sampling for total suspended particulate (TSP) matter, PM_{10}, $PM_{2.5}$, and $PM_{1.0}$ to meet U.S. EPA and other regulatory requirements. A typical beta instrument contains the appropriate inlet sizing cyclone or other device, a sampling pump, and the instrument section, which consists of a moving filter strip mechanism, a beta source (usually $100\,\mu$Ci carbon-14), and a radiation measuring chamber, along with associated electronic circuitry, a control keypad, and an LCD display. Calibration is performed in the field with a set of calibration control foils. The measurement range for beta gauge instruments is typically up to 5.0 mg/ m^3 (5000 $\mu g/m^3$). Since these instruments have such specialized environmental application, they are not further described.

Another consideration is whether an instrument measures total or respirable dust. When set up to measure respirable dust, U.S.-made instruments follow the criteria of the American Conference of Governmental Industrial Hygienists (ACGIH) previously described in the chapter on sample collection device methods for aerosols (Chapter 7). Some monitors (passive) do not use pumps to pull the aerosol into the sensing chamber, whereas others (active) depend on them. Studies have shown that when the same aerosol is measured using both active and passive methods, the passive instrument will show more deviations in the results.[1] One explanation may be homogenization of the aerosol during active sampling, and when an aerosol is collected as a result of natural air convections, concentration variations may show up more readily.

The aerosol monitoring instruments described within this chapter were selected because they are easily obtainable for use in the field. However, there are other more sophisticated and expensive units available, and the instrument of choice will depend on the specifics of the sampling situation.[2]

LIGHT-SCATTERING MONITORS

Light-scattering aerosol monitors are also called nephelometers or aerosol photometers. In these devices, the instrument continuously senses the intensity of light from the combined scattering of the population of particles present within the sensing volume at a given angle relative to the incident beam as they pass through a sensor cell of defined volume (Figure 14.1). As the number of particles increases, the light reaching the sensor increases. The scattered light detected by the photodetector is transformed into a voltage proportional to the light intensity.

The light source can be monochromatic, such as light-emitting diode or laser, or a broadwavelength light source, such as a tungsten filament lamp. The choice of light source generally has more to do with the ability to control the light output level than with the wavelength of the output. The detector is generally a solid-state photodiode but can be a photomultiplier tube. The

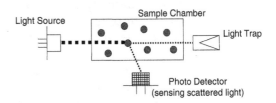

Figure 14.1. Diagram of a light-scattering particle analyzer.

scattered light received from a single particle depends on the size and shape of the particle, the refractive index of the particle, the wavelength of the light, and the angle of scatter.[2] Thus, the amount of scattered light received from the multiple particles in a cloud when the wavelength and angle of scatter are fixed will vary with the concentration, the particle size distribution, and changes in refractive index associated with the aerosol's composition. The relationship between the measured light scattered by a dust and its mass also depends on dust density.[3] The response to light-scattering measurements for larger particles is influenced more by their surface area than their mass.[4]

The components of nephelometers that have the greatest influence on their readings are the light source, scattering angle, and particle size preselector. Of these, the scattering angle may be the most important. The angle of scattering is defined with respect to the beam of light passing through the aerosol in the detection volume. The smaller the value of this angle, the more the detection is weighted toward larger particles. A scattering angle of 90° provides the greatest sensitivity for small particles. Some instruments use near forward light scattering with scattering angles of 12° to 20°. This results in smaller variations based on the optical properties of the aerosol (refractive index and absorption coefficient) than with the other photometers.

Other factors influencing the sensitivity, or response of the light-scattering aerosol monitors are the geometry of the optical system, wavelength (distribution) of the light source, spectral sensitivity of the light detector, particle-size distribution, physical properties of the particles to be measured, and humidity.

Light-scattering methods have distinct advantages: measurements can be made without interfering with the particles so the particles can be collected on a filter at the outlet of the instrument for laboratory analysis. Because they actually measure particle count and not mass, aerosol photometers should be used to determine mass concentration only under conditions in which the optical properties, density, and size distribution of the particles are constant.[5]

Most instruments will have some detection losses, as some dust particles may fail to reach the detector. However, these losses are generally small, unless there are strong air currents blowing across the sampling path or the particles are highly charged. Some aerosols can be difficult to measure with light-scattering devices, such as welding fume particles that have a tendency to stick together, forming agglomerates over time.[4]

A *condensation nuclei counter* is a special application of scattered light devices that is used for measurement of extremely small particles (in the nanometer size range). The sample passes through a heated chamber where alcohol or another liquid evaporates into the sample stream. The sample then moves to a cooled section where the alcohol vapor becomes supersaturated and condenses onto any particle present. Droplets then pass into a light-scattering particle counting section. These devices are used for respirator fit testing and also to test the removal efficiency of high efficient particulate air (HEPA) filters.

Nephelometer Instruments

This section describes some typical nephelometers designed for personal monitoring, area sampling, and fixed use. Key features of each device are described to illustrate the options available. Refer to the current product literature for these and comparable instruments from other manufacturers for a full understanding of the real-time aerosol measurement technology that is available.

Figure 14.2. TSI DUSTTRAK™ aerosol monitor with a measurement range from 0.001 to 100 mg/m³. (Courtesy of TSI, Inc.)

DUSTTRAK™ Aerosol Monitor.

The TSI Corporation's Model 8520 DUSTTRAK™ aerosol monitor (Figure 14.2) is a 90° laser diode nephelometer with a measurement range from 0.001 to 100 mg/m³; its stated resolution is ±1% of reading or ±0.001 mg/m³ (whichever is greater) with a long-term zero stability of ±0.001 mg/m³ (10-second averaging) over 24 hours. It responds to a particle size range of 0.1 to approximately 10 μm (depending on flowrate). A pump draws the sample aerosol through an optics chamber where it is measured. A sheath air system isolates the aerosol in the chamber to keep the optics clean for improved reliability and low maintenance. It provides reliable exposure assessment by measuring particle concentrations corresponding to PM_{10}, $PM_{2.5}$, $PM_{1.0}$, or respirable size fractions. Applications include site perimeter monitoring, ambient monitoring, process area monitoring, and other remote uses.[6]

The internal data-logging capability is >31,000 data points, which allows 21 days of logging with a sampling interval of 1 second. Data-logging averaging periods range from 1 second to 1 hour. The device is programmed from a personal computer using the provided software with the operating parameters such as logging period, display averaging time, alarm level, and sampling mode. It also has an optional portable printer for use without a computer. Additionally, newer models contain an analog output feature. This means that the instrument is capable of providing an analog voltage signal that is proportional to mass concentration. There are five ranges for the 0- to 5-V output signal (0–0.10 mg/m³, 0–1.0 mg/m³, 0–10 mg/m³, and 0–100 mg/m³).

The DUSTTRAK monitor is calibrated to the respirable fraction of ISO 12103-1, "A1" (formerly called ultrafine Arizona test dust or SAE ultrafine). The calibration data are stored internally and cannot be accessed or modified by the sampling practitioner. This standard test dust is used because of its wide particle size distribution, which makes the internal calibration representative of an average of most types of ambient aerosols. Because optical mass measurements are dependent upon particle size and material properties, there may be times in which a custom calibration improves the accuracy for a specific aerosol. The monitor can be calibrated to any arbitrary aerosol by adjusting the custom calibration factor.

A 10-mm nylon Dorr–Oliver cyclone with a 4-μm 50% cutoff size is provided to discriminate between the respirable fraction and other size portions of an ambient aerosol. Since the cutoff size for any cyclone is dependent on flowrate, the sample flowrate must be set at 1.7 L/min to maintain the cutoff size at 4 μm.

An available option is an environmental enclosure for outdoor unattended use, which is a weatherproof cabinet plus additional battery and inlet configurations. The enclosure should be set up in a location that is away from obstructions which may affect wind currents. For example, it should not be placed at the corner of a building which would cause swirling wind currents and result in poor particle sampling. It should be used in wind conditions with speeds of 22 mph or less to obtain the most accurate readings. An increase in wind speed over 22 mph can decrease the sampling efficiency of the inlet to under the efficiency specified by PM_{10} standards. If wind gusts of over 22 mph are present, the data collected are still valid but the readings will be slightly lower than the actual mass concentration of aerosol present.

The instrument has an alarm whenever a user-defined level is exceeded; the alarm also has an external output to allow programming of a switch closure (fan, area alarm, etc.). The manufacturer recommends that the monitor be returned to the factory for annual calibration.

The personalDataRAM™. The *personal*-DataRAM™ hand-held real-time aerosol monitor/data logger from Thermo Electron Corporation (Figure 14.3) measures mass

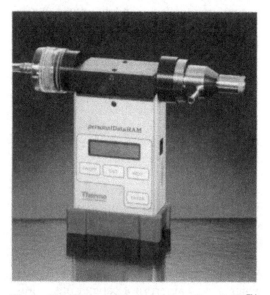

Figure 14.3. Thermo Electron *personal*DataRAM™ hand-held real-time aerosol monitor/data logger measures mass concentrations of dust, smoke, mists, and fumes in real time. (Courtesy of Thermo Electron Corporation.)

concentrations of dust, smoke, mists, and fumes in real time. It is designed for personal/breathing zone monitoring, plant walk-through surveys, remediation site worker exposure monitoring, and indoor air quality monitoring.[7]

The *personal*DataRAM is a nephelometer that incorporates three components: a pulsed, high-output, near-infrared light-emitting diode source; collimating optics; and a silicon detector with amplifying circuitry. The intensity of the light scattered over the forward angle of 45–90° by the airborne particles passing through the sensing chamber is linearly proportional to their concentration. The optical configuration produces optimal volume response to particles in the size range of 0.1 to 10 μm, achieving high correlation with standard gravimetric measurements of the respirable and thoracic fractions.

It has a measurement range from 0.001 to 400 mg/m³; its stated accuracy is ±5% of reading with a long-term zero stability of ±0.003 mg/m³ (1-second averaging) over at least 180 days. It sounds an alarm whenever a user-defined level is exceeded, weighs 18 ounces, and has several power options including line power and a battery pack that provides up to 72 hours of operation between charges.

Integral large-capacity data-logging capability permits storage of 13,000 data points in up to 99 discrete data sets. Each data point shows average concentration, time/date, and data point number. The run summary shows overall average and maximum concentrations, time/date of maximum, total number of logged points, start time/date, total elapsed time (run duration), STEL concentration and time/date of STEL, averaging (logging) period, calibration factor, and data set number. Data-logging averaging periods range from 1 second to 4 hours. At the beginning of each sampling run, the instrument automatically tags and time stamps the data collected. The device is programmed from a personal computer using the provided software with the operating parameters such as logging period, display averaging time, alarm level, mode, and so on. It also has a two-line LCD readout which is updated every second to display both real-time and time-averaged (TWA) concentration values; other screens are selected by scrolling through a menu.

Each *personal*DataRAM is gravimetrically factory-calibrated (NIST-traceable) in mg/m^3 using standard SAE fine (ISO Fine) test dust. Zeroing with particle-free air is performed in the field using a zeroing kit provided with the instrument. An internal calibration check automatically references to the optical background set at the factory. Gravimetric field calibration can be performed by comparison with a filter sampler and then programming the internal calibration constant.

The instrument is available in two configurations:

- A passive air sampling model. Air surrounding the monitor circulates freely through the open sensing chamber by natural convection, diffusion, and background air motion. With this passive nephelometer sampling method, concentration measurements do not depend on the air velocity through the sensing chamber.
- An active sampling model that uses a pump module or other sampling pump at a flowrate of 1–10 L/min to perform particle-size-selective measurements. By operating the pump at specific sampling flowrates, the cyclone separator provides precisely defined particle size cuts. For example, at 4 L/min, the cut point of the cyclone is 2.5 µm as required for PM$_{2.5}$ monitoring. With other flowrates the particle cut points can be varied between 1.0 and 10.0 µm. It is suitable for respirable, thoracic, and PM$_{2.5}$ monitoring and has an isokinetic sampling option for stack and

duct particulate sampling. It also can be equipped with special inlet accessories for ambient air measurements under variable wind and high humidity conditions. An integral filter holder directly downstream of the photometric sensing stage accepts 37-mm filters. Membrane filters of various pore sizes can be used for chemical analysis or for concurrent gravimetric measurements. Primary gravimetric calibration of the instrument concentration readout can be performed under actual field conditions by means of this integral filter. The calibration constant can be adjusted to match the filter-determined concentration.

Casella Microdust pro™. The Casella Microdust pro™ (Figure 14.4) is a hand-held real-time monitor for measuring the concentration of suspended particulate matter with a range of 0.001 mg/m^3 to 2500 mg/m^3. It has a built-in data-logging capability of over 15,000 points to give a time history showing how the concentra-

Figure 14.4. Casella Microdust-pro™ measures the concentration of suspended particulate matter with a range of 0.001 mg/m^3 to 2500 mg/m^3. (Courtesy of Casella USA.)

tion levels changed during the measurement period. A proprietary software package is supplied with the instrument that allows download of run results and the viewing of real-time plots of particulate concentration on the screen.[8]

The unit can be calibrated for zero and span and exhibits good linearity throughout the measurement range. It can be operated in conjunction with conventional sampling pumps and any correction factors necessary for comparison with the standard measurement method can be stored and applied to future run data. The individual calibrations for up to four user-selectable dust types can be stored in the memory and recalled for specific measurements.

For area sampling the Microdust pro™ can be specified with a lockable enclosure and a special kit containing a Casella Vortex sampling pump, rechargeable battery pack, size-selective adaptor (to suit individual applications), filter support assembly, and an external omnidirectional air sampling inlet (Figure 14.5).

DataRAM-4™ Portable Particle Sizing Aerosol Monitor.

The DataRAM-4™ real-time aerosol monitor from Thermo Electron Corporation (Figure 14.6) measures mass concentrations of airborne dust, smoke, mist, haze, and fumes, while providing continuous real-time readouts of both air temperature and humidity. It is a high-sensitivity nephelometric monitor that samples the air at a constant, regulated flow rate by means of a built-in diaphragm pump. The device's optical configuration is optimized for the measurement of airborne particle concentrations in the range $0.0001\,mg/m^3$ $(0.1\,\mu g/m^3)$ to $400\,mg/m^3$ and is immune to particle coincidence errors, even at the highest concentrations. With optional accessories, the device can also provide respirable, $PM_{2.5}$, or PM_{10} correlated measurements.[9]

The temperature operating range is −15 to +50°C, with a relative humidity operating range (at 25°C) of 10–95% (noncondensing). For zeroing of the instrument, a solenoid valve diverts the entire filtered air stream through the optical sensing stage in order to achieve "zero" air reference. In addition, instrument secondary calibration can be performed by turning a knob on the device's back panel, which inserts a built-in optical scattering/diffusing element into the filtered air stream. After passing

Figure 14.5. Environmental enclosure for the Casella Microdust-pro™ permits outdoor environmental monitoring with a variety of inlet size selectors. (Courtesy of Casella USA.)

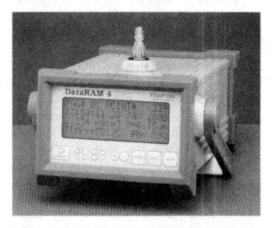

Figure 14.6. Thermo Electron DataRAM-4™ real-time aerosol monitor measures mass concentration while providing continuous real-time readout of both air temperature and humidity. (Courtesy of Thermo Electron Corporation.)

through the optical sensing stage, all the particles are retained on a HEPA filter. Part of the filtered air stream is continuously diverted through and over all optically sensitive areas (lens, light traps, etc.) to form a continuous air curtain that protects against particle deposition. A membrane filter can be substituted for the HEPA cartridge for gravimetric and/or chemical analysis of the particles collected downstream of the sensing stage; the sample flowrate is adjustable from 1 to 3 L/min.

The DataRAM-4 has a built-in data logging with a capability of 10,000 data points (each point contains average, minimum, and maximum concentrations) in up to 99 data groups. Stored information includes time and date, average concentrations, maximum and minimum values over selected periods, STEL concentration, and tagging codes. Data-logging averaging periods range from 1 second to 4 hours. Logged information can be retrieved either by scrolling through the LCD display or by downloading to an external device such as a personal computer or printer. Any standard serial communications software can be used to download data, and standard spreadsheet packages (e.g., Microsoft Excel™ and Lotus™) can be used to access and analyze data log files. The DataRAM-4 provides a continuous digital output (by means of an RS-232C data port) as well as an analog output, along with a switched output for selectable high-level alarm with a built-in audible signal. Additionally, real-time and date, time-weighted average concentrations, elapsed run times, and other information are easily viewed on the 8-line LCD screen using a scroll-through menu.

Several optional accessories are available for a wide range of sampling applications:

- A cyclone precollector allows respirable particle measurements.
- An omnidirectional air sampling inlet

(with or without a $PM_{10/2.5}$ head) is available for ambient monitoring.
- Isokinetic inlet nozzles for duct sampling.
- An in-line heater module for accurate monitoring of solid particles in high humidity/fog conditions.
- A sample dilution accessory permits elevated temperature and/or very high concentration monitoring.
- A portable battery-powered printer and cabling accessories.

The alarm level can be selected over entire measurement range, with an alarm averaging time of 1 to 10 seconds, or 15 minutes for a short-term exposure limit (STEL) criterion. It weighs 11.7 pounds, and it can operate for about 24 hours using the rechargeable sealed lead-acid battery or continuously with the battery charger plugged into to line current.

Met One Hand Held Particle Counter (HHPC-6). The Met One HHPC-6 Handheld Airborne Particle Counter is a light-scattering device that can measure size distribution of airborne particles as well as their concentration (Figure 14.7). It is used to: monitor the air quality in clean rooms, manufacturing processes, pharmaceutical production, and hospital surgical suites; test filter seals and filter efficiency; locate particle contamination sources; monitor particle size distributions; and monitor for mold during remediation and indoor air quality (IAQ) investigations.[10]

It can measure particle concentrations in six different size ranges and is most often configured for either:

- Standard measurements: 0.3, 0.5, 0.7, 1.0, 2.0, 5.0 μm
- Indoor air quality investigations: 0.5, 0.7, 1.0, 2.0, 5.0, 10.0 μm

Additionally, the manufacturer can provide custom configurations to measure

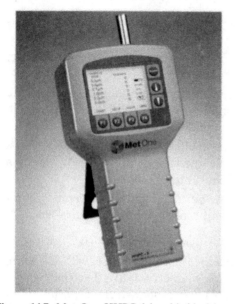

Figure 14.7. Met One HHPC-6 hand-held airborne particle counter measures size distribution of airborne particles as well as their concentration. (Courtesy of Hach Ultra Analytics.)

other size distributions in the 0.3- to 20-μm size range.

The instrument measures particle size by passing a collimated beam from a laser diode through a very small sensing chamber (Figure 14.8). The chamber volume is so small that only one particle is normally present at a time, and so at the proper airflow rate the output signal from the photodetector is proportional to the particle diameter (Figure 14.9). Careful calibration and operation of the instrument is important to avoid situations where sizing or counting errors might occur. These include high particle concentrations causing multiple particles to be in the sensing chamber at one time, or incorrect residence time for the particle in the chamber since too short a time will cause an underestimation of size while too long a

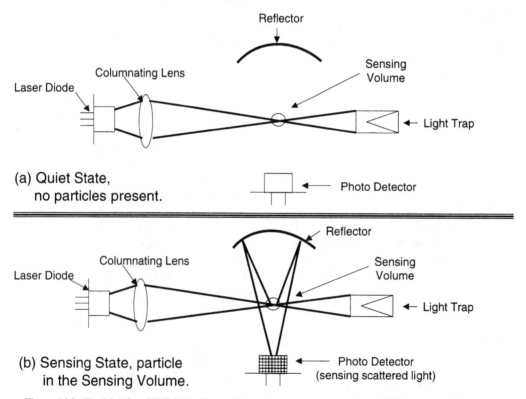

Figure 14.8. The Met One HHPC-6 collimated laser beam sensing chamber: (a) With no particle present and (b) with a particle in the sensing volume. (Courtesy of Hach Ultra Analytics.)

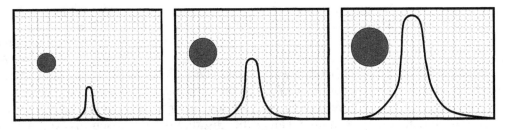

Figure 14.9. The signal pulse is proportional to the size of particle passing through the HHPC-6's sensing volume. This allows particle sizing as well as concentration measurements. (Courtesy of Hach Ultra Analytics.)

residence time will overestimate particle size. It has a counting accuracy of 50% for 0.3-µm particles and 100% for particles >0.45 µm.

Since the device measures particle size, it must be used with an isokinetic sampling probe (either attached directly to the instrument or attached to a sampling tube) to ensure that the particulate size distribution in the sampled air matches that in the ambient air.

Samples are usually collected using a preset sampling volume, which is achieved by controlling the sample duration using the internal 2.8-L/min sampling pump. Typical sampling durations are 21 seconds (1.0 L), 60 seconds (2.83 L), 3.53 minutes (10.0 L), or 10 minutes (28.3 L).

Before samples are collected, the unit's operation is verified using the zero count filter. When used in a very clean environment such as a clean room application, the zero count filter verifies that the unit is not counting electrical noise signals from internal components and is not subject to external interference. In indoor air quality and other applications, the zero count filter cleans the sensor immediately after a high concentration sampling. To perform this verification, a zero count filter is attached to the intake nozzle and the unit is operated for approximately 15 minutes. Satisfactory performance is verified if not more than one particle greater than 0.3 µm in 5 minutes on the average or not more than one particle per 0.5 cubic foot is counted.

The data-logging feature holds 500 samples (2000 with an optional expansion) in memory; each record includes date, time, counts, sample volume, temperature, and relative humidity. The data can be downloaded to a computer or printer using an RS-232/RS-485 interface cable and proprietary software.

Calibration of Light-Scattering Monitors

As stated earlier, light-scattering monitors are calibrated both in the factory and in the field. The purpose of the factory calibration is to ensure that an instrument is operating properly and responding correctly when exposed to the "calibration dust" while the field calibration is to improve the specificity of the instrument for dust generated by a given process. Most light-scattering monitors are factory calibrated by comparing the instrument response in a well-defined aerosol to measurements by the gravimetric method. During factory calibration, the rate at which particles are detected, expressed in a unit such as "counts per minute" (cpm), is related to one or more concentrations of a particular calibration aerosol. For a given aerosol, a certain value of cpm's will be equal to a given mg/M^3. In most cases, the instrument's response is modified to read directly in mg/M^3. The calibration aerosol is the manufacturer's choice; typical choices are Arizona road dust, oil mist, or a cement dust.

Scattered-light instruments provide a reliable indication of aerosol mass concentration if the instruments are calibrated with the same aerosol to be measured and the size distribution of this aerosol stays constant. This is necessary because aerosols from different sources, of different composition, or different size distribution can produce photometer responses that differ widely.

Field calibration for a given process is carried out by side-by-side sampling to compare the TWA instrument readings directly with field measurements using a cyclone for respirable measurements or filter cassette for total dust measurements (NIOSH Methods 0600 and 0500) to establish an instrument-specific ratio. The object of side-by-side sampling with direct reading dust monitors and gravimetric methods is to generate calibration factors that adjust for differences in the type of dust being measured compared to that being used to calibrate the instrument. To do this, the average mass concentration determined on filters is compared to the average mass concentration determined by the aerosol photometer. When the methods are in agreement, the ratio of these measurements is 1:1. However, when measurements differ in the field, this ratio will change. It is important to note that since any calibration factor determined this way will be an average calibration factor, because no industrial source produces dusts with constant particle-size distributions or optical properties and so, actual results may differ somewhat.

Adjusted levels are then obtained by multiplying the instrument readings by the appropriate calibration factors for each type of dust. Although differences between gravimetric measurements and light-scattering instrument results are usually attributed to differences in dust type, sampling factors such as inlet cyclone orientation have been found to be significant. Orientation also impacts the velocity of particles as they enter the instrument's inlet.

Some instruments are equipped with an internal light-scattering reference with which the instrument can be checked and readjusted to the original calibration. A field check to see if the instrument is still within the manufacturer's calibration can be done using a reference scatterer that can be inserted into the sensing volume in order to check the calibration and overall operation of the instrument. For one instrument this is done by inserting the reference element and adjusting the instrument to achieve a known and traceable reading on the digital display. The mg/M^3 value of this reference element is specific to each instrument. This check does not compensate for differences between the calibration aerosol and that being measured, so the field calibration against NIOSH methods is still needed.

Use of Light-Scattering Instruments

Light-scattering devices cannot be used to discriminate between different types of aerosols, since these instruments will respond to any and all aerosols present in the sensor cell. For example, if lead pigment is to be measured but there are dusts present, the nephelometer reading might overstate the concentration of lead dust. However, if the sampling practitioner is relatively sure that the contaminant of interest is the major component of a dust, the instrument will give results that are representative of this compound.

Care should be taken when measuring aerosols in high-humidity situations. An example is in cotton mills and carpet manufacturing, where water mists are deliberately introduced to decrease airborne dusts, mostly for the purpose of minimizing the potential for dust explosions. Water droplets can exist in the air for extended periods of time and will be detected by a photometer.

Should a light-scattering analyzer become contaminated by overloading or sampling too high a concentration of oil or

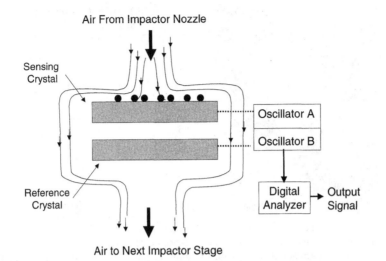

Air From Impactor Nozzle

Air to Next Impactor Stage

Figure 14.10. One stage in a quartz crystal microbalance impactor showing how particles collect on the *sensing* crystal. This causes a decrease in its oscillating frequency relative to the *reference* crystal that is proportional to the mass of the deposited particles.

other particulates, the critical optical surfaces should be cleaned. Also, if sensor calibration and zeroing procedures result in settings far from the factory-set values, contamination should be suspected. Erratic behavior in mass concentration readings can also be caused if fibrous or other foreign material is deposited in the particle sensing chamber and interacts with the light beam. Use a small flashlight and check for contamination of the chamber by shining it through the sampling ports. For some instruments cleaning can be done by the user. When cleaning optical surfaces, do not rub or attempt to polish them. Dip a cotton tip applicator in freon or alcohol and clean the surfaces with a single very lightly applied motion.

PARTICLE MASS MEASUREMENTS WITH THE PIEZOBALANCE

The Piezobalance or Quartz Crystal Microbalance (QCM) utilizes a different method for particle measurements than the optical devices described previously. The Piezobalance measures airborne particle mass by depositing the particles on a piezo-electric sensor via impaction or electrostatic precipitation.

The sensing portion of the device consists of two quartz crystals that oscillate at a fixed frequency (Figure 14.10). One of the crystals, the *sensing* crystal, collects particles while the other crystal, the *reference* crystal, does not collect particles but is there to balance out temperature or other effects on both crystals. As particles impact on the *sensing* crystal, its frequency decreases while the *reference* crystal frequency remains unchanged. The amount of frequency change is directly proportion to the particle mass deposited on the *sensing* crystal. Knowing the air volume flowrate and the length of time of sample collection, the mass concentration of particles can be readily determined. The concentration is calculated from

$$C = \frac{K(\Delta F)}{\Delta T}$$

where C is the concentration ($\mu g/m^3$), K is the sensitivity constant (unique for each crystal sensor), ΔF is the frequency change (hertz), and ΔT is the sampling time (minutes).

Figure 14.11. The 10-stage QCM cascade impactor air particle analyzer provides real-time aerosol size and concentration data, plus the particles deposited on each *sensing* crystal can be analyzed to identify their chemical composition. (Courtesy of California Measurements, Inc.)

TABLE 14.2. QCM Cascade Impactor Air Particle Analyzer Cut-Points

Stage	D_{50} (µm)	D_{EAD} (µm)
1	>25	>35
2	12.5	18
3	6.4	9
4	3.2	4.5
5	1.6	2.3
6	0.8	1.1
7	0.4	0.56
8	0.2	0.30
9	0.1	0.16
10	0.05	0.10

Notes: D_{50} is the cutoff diameter for particles with 2-g/cc density, which is close to the density of typical ambient particles. D_{EAD} is the cutoff equivalent aerodynamic diameter for particles with a density of 1 g/cc.

QCM Cascade Impactor Air Particle Analyzer. One of the few Piezobalance devices commercially available is the QCM cascade impactor air particle analyzer from California Measurements, Inc. (Figure 14.11). This instrument is a 10-stage inertial impactor similar to the impactors described in Chapter 7 except that each stage consists of a pair of quartz crystals that function as a very sensitive microbalance. It provides real-time aerosol size and concentration data, and the crystal containing the impacted particles can be further analyzed using standard laboratory techniques to identify the chemical composition of the particles.[11] Table 14.2 shows the cut-point for each stage in the standard model with a range of 0.05–5.0 mg/m^3 that is typically used for industrial hygiene sampling. There are several other models available with lower range and different size separation capability, including a six-stage model tai-

lored for ambient air monitoring for environmental purposes to cover the PM_{10} and $PM_{2.5}$ size ranges.

Like all cascade impactors, this instrument is a bulk sampling instrument that requires the collection of a certain sample volume of particle-laden air to determine the size composition and mass concentration of particles suspended in it. Unlike conventional cascade impactors, this system utilizes vibrating quartz crystals as sample collection plates in its impactor stages. As particles impact on the oscillating crystals, their mass loadings change the crystals' frequencies of oscillation in proportion to the amount of the mass collected. By monitoring and recording the frequency changes in each impactor stage before and after a sampling run, the instrument electronically determines the mass concentration and size distribution of the particles in the air sample and automatically provides a printout of the data soon after a sample run (Figure 14.12).

Figure 14.12. Printout of data from the QCM showing readings at a brush fire. During the first sampling period the concentration was 0.178 mg/m^3, with particle size in the 23-µm range most prevalent. Later that day the concentration had dropped to 0.075 mg/m^3 with a more widespread size distribution. (Courtesy of California Measurements, Inc.)

DATE OCTOBER 28, 1993
          11:15 am

LOT AMBIENT AIR BRUSHFIRE
         SMOKE, SIERRA MADRE, CA

CH10

```
.14 |▄▄
.23 |▄▄▄▄▄▄▄
.40 |▄
.62 |▄
1.3 |▄
2.3 |
3.2 |
4.6 |▄
6.5 |▄
9.2 |▄
```

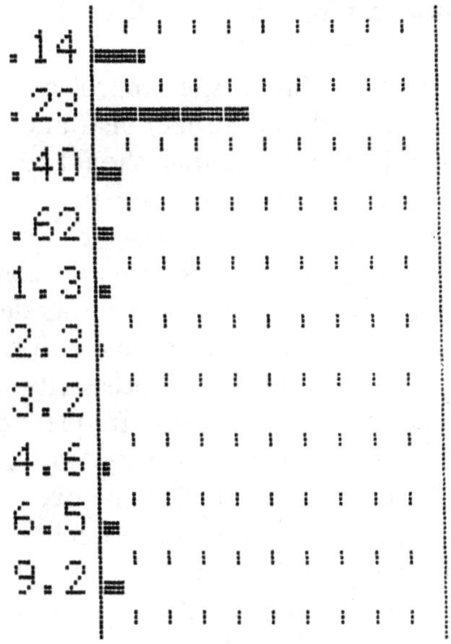

CH 1

△T=60.0

RUN 03

TOTAL   0.17794

CH	△F	m₃/m³

CH	ΔF	mɡ/m³
1	0004	0.01256
2	0003	0.00894
3	0002	0.00544
4	0001	0.00260
5	0002	0.00508
6	0003	0.00756
7	0004	0.01000
8	0006	0.01476
9	0027	0.08100
10	0010	0.03000

DATE OCTOBER 28, 1993
          3:15 pm

LOT AMBIENT AIR BRUSHFIRE
         SMOKE, SIERRA MADRE, CA

CH10

```
.14 |▄
.23 |▄▄▄▄
.40 |▄
.62 |▄
1.3 |▄
2.3 |▄
3.2 |▄
4.6 |▄▄
6.5 |▄
9.2 |▄
```

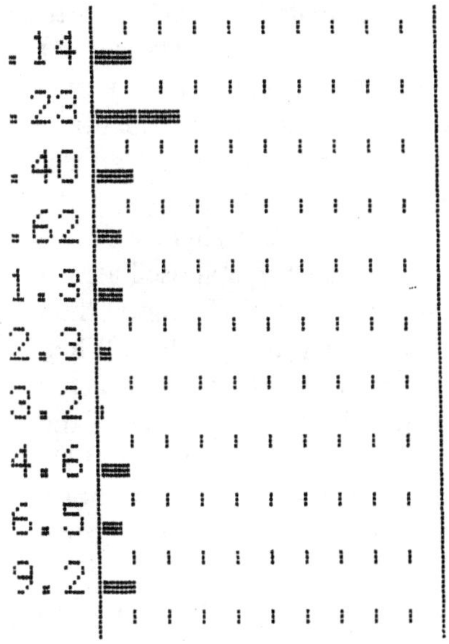

CH 1

△T=96.0

RUN 03

TOTAL   0.07491

CH	ΔF	mɡ/m³
1	0004	0.00785
2	0003	0.00558
3	0004	0.00680
4	0001	0.00162
5	0002	0.00317
6	0004	0.00630
7	0004	0.00625
8	0006	0.00922
9	0010	0.01875
10	0005	0.00937

A typical QCM impactor stage is shown in Figure 14.10. It contains a nozzle, two open-face quartz crystals, and electronic circuits for the crystals. The nozzle accelerates the air passing through it, and its jet diameter controls the size of particles that will be collected through impaction. One of the crystals, the *sensing* crystal, is placed close to the nozzle exit and collects particles while the *reference* crystal balances out temperature or other effects on both crystals. The two crystals are almost identical and have a resonant frequency of approximately 10 MHz. But the circuits that control their oscillation are slightly different so that the *reference* crystal oscillates at a frequency that is about 3 kHz higher than the sensing crystal before sampling. When the oscillator frequencies are subtracted from each other in a mixer circuit, the difference in frequency of about 3 kHz becomes the output signal that is monitored by a microprocessor in the control unit.

For sampling, a vacuum pump is used to draw particle-laden air through the nozzle. As described in Chapter 7, in each stage particles larger than a certain size impact on the crystal and are retained, whereas smaller particles negotiate the turn around the crystal and follow the air stream to the next stage. In the next stage, the nozzle jet diameter is smaller and the velocity of the air passing through it becomes higher than the previous stage. In like manner, a portion of the smaller particles will impact on that crystal. The process repeats itself as the air travels down the impactor stages, and progressively smaller particles are impacted and collected.

During sampling, a ball valve on the top of the device is used to control sampling time. It allows either clean filtered air or particle-laden air into the cascade impactor. Before a measurement, clean air is used to purge the instrument to keep it in equilibrium while on standby. The goal during a measurement is to collect just

enough particles in the peak size range to get a signal frequency change of 30–50 Hz. At this signal level, the signal-to-noise ratio is sufficient to overcome background "noise" and yield an accurate reading. Collecting more particles than necessary will cause premature overloading of the crystals, which will require cleaning before more samples can be collected.

Sampling Time. The control of sampling time is critical to proper instrument operation. The sampling time should be just long enough to yield a frequency change of 30–50 Hz in the peak particle size range. If the sampling time is properly controlled, many measurements can be made before crystals need to be cleaned. Too long a sampling time will unnecessarily load the crystal and require more frequent cleaning than is necessary. If it is not controlled at all, the crystals may become overloaded in a single measurement. Sampling time is inversely proportional to aerosol concentration; the higher the concentration, the shorter the sampling time, and vice versa. If the aerosol concentration is unknown, often a trial run is made depending on the estimated concentration using the following guidelines:

- For levels of 10–100 $\mu g/m^3$, start with 100 seconds.
- For 100–500 $\mu g/m^3$, use 30 seconds.
- For concentrations over 1 mg/m^3 (1000 $\mu g/m^3$), start with 1 or 2 seconds.

Crystal Saturation Limits. The QCM mass to frequency conversion relationship is linear within a certain range. For general applications, such as sampling particles in ambient air indoors or outdoors, the linear relationship holds for up to a cumulative mass loading on the crystal that changes the crystal oscillation by 500–1000 Hz. If optimum sampling time as described above is used, about 10–15 sampling runs can be made before the saturation limit is reached and the crystal requires cleaning.

The mass change to frequency change linear relationship varies with different types of aerosols. If they are wet and sticky (e.g., cigarette smoke or oil vapor particles), the linear region extends to a maximum cumulative total of up to 2000 Hz. On the other hand, if the particles are dry and fluffy, the linear relationship may only apply for a range of 200–300 Hz. The sampling practitioner must make some judgment on the nature of the aerosol being measured and adjust the sampling plan accordingly.

Post-Sampling Analysis. The particles collected during a measurement can be retained for analysis to determine their morphology and elemental composition using scanning electron microscopy (SEM), energy dispersive X-ray spectroscopy (EDXS), or another technique. If this type of analysis is to be performed, the *sensing* crystal is removed, capped with a metal cover, and shipped to the laboratory. The laboratory analyst then mounts the crystal in the analytic instrument; the particle sample is not disturbed after it is collected in the impactor. Since the metal electrode on the crystal is nickel, a prominent nickel energy peak will appear in the X-ray spectrum. Under the nickel coating is a thin layer of silver, and below that there is a thin layer of chrome. Energy peaks of these metals may also appear in the spectrum. If the crystal is used for SEM analysis, it may be damaged to a degree that it cannot be reused.

Crystal Cleaning. After a sensing crystal reaches saturation, it needs to be removed and cleaned. The cleaning process involves wiping off the particles with a cotton swab soaked with clean hexane and then coating the crystal with a special grease. If the device is being used in the field and time is of the essence, it is possible to flip a saturated crystal over and use the other side

rather than cleaning it. Usually only a few stages need to be serviced often since in most applications the collected particles are concentrated in a few size ranges.

The crystals are reusable. The nickel electrodes can withstand exposure to corrosive aerosols, although over time the electrode may become pitted or stained. The pitting results in the loss of surface area and will cause the electrode's frequency to increase. Within certain limits, more grease can be applied to lower its frequency.

QCM Applications. In addition to occupational exposure sampling, the QCM particle analyzer can be used for a variety of applications:

- Inhalation toxicology research
- Pharmaceutical aerosol size analysis
- Industrial process emission studies
- Clean room contamination analysis
- Air cleaner and filter efficiency measurements
- Environmental particle measurements

For example, Figure 14.13 shows the device mounted on an aircraft for sampling

Figure 14.13. QCM cascade impactor air particle analyzer mounted in a special module for fixed-wing aircraft airborne sampling. (Courtesy of California Measurements, Inc.)

of volcanic dust emissions. It has also been mounted on a remotely piloted aircraft for low-altitude environmental sampling in a heavily polluted industrial area.

SUMMARY

This chapter describes direct-reading instruments that can measure the concentration of airborne aerosols and that in a few cases also determine the size of particles. Almost all of the commercially available instruments operate on the principle of scattered light, while the use of the Piezobalance (or *Quartz Crystal Microbalance*) is much more limited.

Accurate real-time measurement of aerosols is complex due to factors such as size, shape and settling velocity of the particles. Calibration is another consideration since these devices are factory-calibrated using standard dusts. It is generally possible to conduct field calibration checks using side-by-side pump sampling to measure actual concentrations for comparison to the device's internal calibration setting.

Since these instruments measure "all" airborne particles, they cannot differentiate between different compounds if more than one chemical dust is present. This can be a problem if the exposure involves dusts of mixed composition where the composition can change over time.

REFERENCES

1. Willeke, K., and S. J. Degarmo. Passive versus active aerosol monitoring. *Appl. Ind. Hyg.* **3**(9):263–266, 1988.
2. Cohen, B. S., and C. S. McCammon, Jr., eds. *Air Sampling Instruments for Evaluation of Atmospheric Contaminants*, 9th Ed. Cincinnati, OH: ACGIH, 2000.
3. Schnakenberg, G., and B. Chilton. Direct reading systems. *ACGIH Ann.* **1**:272–290, 1981.
4. Kusisto, P. Evaluation of the direct reading instruments for the measurement of aerosols. *AIHA J.* **44**(11):863–874, 1983.
5. Lehtimaki, M., and J. Keskinen. A method of modifying the sensitivity function of an aerosol photometer. *AIHA J.* **49**(8): 396–400, 1988.
6. *DUSTTRAK*™ *Model 8520 Aerosol Monitor Operation and Service Manual*, Shoreview, MN: TSI Inc.
7. *personalDataRAM*™ *Series Hand-Held Real-Time Aerosol Monitor/Data Logger Product Literature*. Waltham, MA: Thermo Electron Corporation.
8. *Casella Microdust pro*™ *Product Literature*. Amherst, NH: Casella USA.
9. *DataRAM 4*™ *Portable Particle Sizing Monitor Product Literature*. Waltham, MA: Thermo Electron Corporation.
10. *Met One HHPC-6 Handheld Airborne Particle Counter Operator's Manual*. Grants Pass, OR: Met One.
11. *Air Particle Analyzer: QCM Cascade Impactor Operating Manual*. Sierra Madre, CA: California Measurements, Inc.

PART IV

MONITORING FOR AIRBORNE AGENTS OTHER THAN CHEMICALS

CHAPTER 15

RADON MEASUREMENTS

Radon (Rn) comes from the radioactive decay of uranium. There are several isotopes of radon, but ^{222}Rn is the most common. Radon-222 is widely distributed through the environment due to its chemical inertness, gaseous state, and relatively long half-life. Radon can be found in high concentrations in soils and rocks containing uranium, granite, shale, phosphate, and pitchblende. Radon may also be found in soils contaminated with certain types of industrial wastes, such as the byproducts from uranium or phosphate mining.

Radon gas has a half-life of 3.8 days and decays in several steps to radioactive particles, called radon progenies. These are polonium (Po), bismuth (Bi), and lead (Pb). Table 15.1 lists the isotopes, along with the half-life and mode of decay for each.

Radon can be immobilized in rock. Radon atoms originating from this source may migrate within the rock and subsequently dissolve into ground or surface waters.[1] Consequently, radon can occur in the soil and in water. Direct sources of radon include tap water and seepage of radon gas and natural gas.

Because radon is a gas, it can move through small spaces in the soil and rock on which dwellings are built (Figure 15.1). Radon in the soil can enter buildings in two ways: (1) by passive diffusion through the air-pore system in the soil, through dirt floors, cracks in concrete floors and walls, floor drains, sumps, joints, and tiny cracks or pores in hollow-block walls; and (2) by pressure-driven flow, which is created by thermal effects. The thermal effects are due to the pressure differential that builds up whenever there are air masses of different temperatures. During the heating season, this buildup gives rise to a pressure gradient, with the net effect that the radon gas is actually pulled into the building as the hot air rises.

Loose, sandy soil promotes radon transfer, whereas clayey, compacted, wet, or frozen soil inhibits gas flow. The soil properties that most influence radon diffusion are soil moisture regime, depth to bedrock,

Air Monitoring for Toxic Exposures, Second Edition. By Henry J. McDermott
ISBN 0-471-45435-4 © 2004 John Wiley & Sons, Inc.

Figure 15.1. Major routes of radon entry into buildings.

TABLE 15.1. Decay Products for Radon-222

Isotope	Half-Life	Radiation Emitted[a]
Radon-222	3.8 days	Alpha—5.45 MeV
Polonium-218	3 min	Alpha—6.00 MeV
Lead-214	27 min	Beta
Bismuth-214	20 min	Beta
Polonium-214	180 μsec	Alpha—7.69 MeV
Lead-210	22 years	Beta
Bismuth-210	5 days	Beta
Polonium-210	138 days	Alpha—5.50 MeV
Lead-206	NA	None (stable)

[a] MeV, million electron volts.

and porosity. Although soil water decreases radon permeability, it can increase chemical weathering; facilitate the transport of the mobile, oxidized form of uranium (hexavalent); and allow more radon isotopes to be emanated in the interstitial pore space of rock instead of remaining embedded into the solid matrix during decay. A characteristic of the upward flow of radon, as well as other gases, is that it is highly variable because geological material tends to be in horizontal formations due to anisotropic development of soil horizons.[2] Table 15.2 lists risk classification guidelines for radon in soil gas.[3]

In some situations, radon may be released from construction materials, for example, a large stone fireplace or a solar heating system in which heat is stored in large stone beds. The potential for radon exposure in schools has also become an issue.[4]

Radon can also enter water in private wells and be released into the air when the water tap is on. Activities and appliances that spray or agitate heated water, such as showers, dishwashers, and clothes washers, create the largest releases of waterborne radon. The airborne contribution of radon from water is usually small compared to that from soil gas. Usually radon is not a problem with large community water supplies.

TABLE 15.2. Risk Classification Guidelines for Radon in Soil Gas

Classification	Soil Type and Radon Concentration
High-risk	Uranium-rich granites, pegmatites, phosphates, and alum shale Highly permeable soils, such as gravel and coarse sand Radon concentration in soil gas >1350 pC$_i$/L
Medium-risk	Rocks and soils with low or normal uranium content Soils with average permeability Radon concentration in soil gas from 270 pCi/L to 1350 pC$_i$/L
Low-risk	Rocks with very low uranium content, such as limestone, sandstone, and basic igneous and volcanic rocks Soils with very low permeability, such as clay and silt Radon concentration in soil gas <270 pCi/L

The primary populations exposed to radon are individuals living in residential housing and uranium and phosphate miners. Another potential source of occupational exposure to radon is heating gas. During fractionation at processing plants, radon may be concentrated in the liquefied petroleum gas (LPG) product stream. Radon can contaminate the interior surfaces of plant machinery and when equipment is repaired, exposures can occur.[5] Safe levels for occupational exposures are higher than safe levels for the community where daily exposures can be 24 hours.[6]

The primary adverse health effect associated with radon exposure is lung cancer from inhalation of the radon decay products. Radioactive substances emit three principal types of radiation: alpha, beta, and gamma. The primary hazards associated with radon decay are alpha particles. They act at very short range; and although they cannot penetrate infact skin, if they are released in the lungs they can do significant cellular damage, including causing lung cancer. Table 15.3 is a radon risk chart.[3]

The concentration of radon in the air is measured in units of picocuries per liter (pCi/L) or becquerels per cubic meter (Bq/m^3). One Bq corresponds to one disintegration per second. One pCi/L is equivalent to 37 Bq/m^3.

The concentration of radon progeny is expressed in units of working level (WL) to account for the different decay rates and energies of the various progeny. The WL is a measure of the potential alpha particles energy per liter of air. One WL of radon progencies corresponds to approximately 200 pCi/L of radon in a typical indoor environment. However, the relative concentration of radon and radon progencies may vary from one building to another. In the extreme case, 1 WL corresponds to 100 pCi/L of radon. This situation is called full equilibrium and is extremely unlikely to occur. Occupational exposure to radon

progencies is expressed in working level months (WLM), and a working level month is equivalent to the exposure at an average concentration of 1 WL for 170 working hours. Measurement data are reported in either of the above units. For making comparisons between the data from different sources, the following conversion chart may be useful:

$$1\,pCi/L = 37\,Bq/m^3$$
$$1\,m^3 = 1000\,L$$
$$0.01\,WL = 74\,Bq/m^3 = 2\,pCi/L$$
$$0.02\,WL = 148\,Bq/m^3 = 4\,pCi/L$$
$$0.1\,WL = 800\,Bq/m^3 = 20\,pCi/L$$

COLLECTION METHODS FOR RADON AND ITS PROGENY IN AIR

Several different measurement methods can be used to determine radon concentrations. In practice, the choice of a method is often dictated by availability. If alternate methods are available, then the cost or the duration of the measurement may become the deciding factor.

These sampling methods are designed to detect either radon or its decay products, radon progeny, and the basis for the measurements is the detection of alpha particles from radon decay.

The techniques for measuring radon decay are different from those used to measure its decay products. There are three basic types of samples: short-term, continuous, and grab. Short-term sampling is the most common, and involves the use of charcoal canisters and alphatrack detectors. These samples are generally used for screening measurements. Continuous sampling is done with real time instruments and is often used to determine the effectiveness of controls or for very long-term monitoring. Grab sampling, the most involved method, is generally reserved for calibrating other methods and collecting occupational samples. It is sometimes

TABLE 15.3. U.S. EPA Radon Risk Evaluation Chart (2003)

Part A. RADON RISK IF YOU SMOKE			
Radon Level	If 1000 people who smoked were exposed to this level over a lifetime . . .	The risk of cancer from radon exposure compares to . . .	WHAT TO DO: Stop smoking and . . .
20 pCi/L	About 135 people could get lung cancer	100 times the risk of drowning	Fix your home
10 pCi/L	About 71 people could get lung cancer	100 times the risk of dying in a home fire	Fix your home
8 pCi/L	About 57 people could get lung cancer		Fix your home
4 pCi/L	About 29 people could get lung cancer	100 times the risk of dying in an airplane crash	Fix your home
2 pCi/L	About 15 people could get lung cancer	2 times the risk of dying in a car crash	Consider fixing between 2 and 4 pCi/L
1.3 pCi/L	About 9 people could get lung cancer	(Average indoor radon level)	(Reducing radon levels below 2 pCi/L is difficult.)
0.4 pCi/L	About 3 people could get lung cancer	(Average outdoor radon level)	(Reducing radon levels below 2 pCi/L is difficult.)

Note: If you are a former smoker, your risk may be lower than indicated above.

Part B. RADON RISK IF YOU HAVE NEVER SMOKED			
Radon Level	If 1000 people who never smoked were exposed to this level over a lifetime . . .	The risk of cancer from radon exposure compares to . . .	WHAT TO DO:
20 pCi/L	About 8 people could get lung cancer	The risk of being killed in a violent crime	Fix your home
10 pCi/L	About 4 people could get lung cancer		Fix your home
8 pCi/L	About 3 people could get lung cancer	10 times the risk of dying in an airplane crash	Fix your home
4 pCi/L	About 2 people could get lung cancer	The risk of drowning	Fix your home
2 pCi/L	About 1 person could get lung cancer	The risk of dying in a home fire	Consider fixing between 2 and 4 pCi/L
1.3 pCi/L	Less than 1 person could get lung cancer	(Average indoor radon level)	(Reducing radon levels below 2 pCi/L is difficult.)
0.4 pCi/L	Less than 1 person could get lung cancer	(Average outdoor radon level)	(Reducing radon levels below 2 pCi/L is difficult.)

Note: If you are a former smoker, your risk may be higher than indicated in Part B.

TABLE 15.4. Overview of Radon Measurement Methods

Instrument	Sampling Times
Charcoal canister	2 to 7 days
Alpha-track detector	3 months (or less if laboratory uses adequate lower limit of detection)
Radon progeny integrating sampling unit (RPISU)	100 hours minimum, 7 days preferred
Continuous working level monitor	6 hours minimum, 24 hours or longer preferred
Continuous radon monitor	6 hours minimum, 24 hours or longer preferred
Grab working level	5 minutes
Grab radon	5 minutes

called Kusnetz or modified Tsivoglou methods. Table 15.4 provides an overview of sampling methods.

Short-term screening is useful for comparative measurements in various locations in the same building or in different buildings. Results of remediation activities can be verified by doing short-term measurements. Long-term measurements will give a more valid measurement of actual radon concentrations in that seasonal and other variations will be averaged out.

The equilibrium factor (EF) is an index of the extent to which radon gas has reached equilibrium with its progeny. This index is useful to describe the efficiency of air exchange, especially in confined spaces such as caissons or basements. The EF is defined as

$$EF = \frac{100\,WL}{Rn}$$

where WL is the working level and Rn is the radon gas concentration. A high EF reflects a lack of air exchange, leading to rapid equilibrium between radon gas and its progencies. The maximum EF value of 1 could be obtained in an unventilated area.

Problems encountered when measuring indoor radon and radon decay product concentrations include variability due to nonstandardized procedures, different house conditions prior to and during measurement, seasonal and other weather conditions, and different interpretations of the results. For example, in the case of a house next to a lake, the rising water level in early summer could help drive radon into the basement if ground-water levels also rise. In the winter the lake is frozen and the water table is considerably lower. Because of seasonal trends several measurements are required over a year to determine the actual radon levels in a building. It has been suggested that a 24-hour measurement once a month would be sufficient to determine the seasonal trend and annual average for most buildings.[4]

Measurements should be made under "closed-house" conditions. To a reasonable extent, windows and external doors should be closed for 12 hours prior to and during the test. Normal entrances and exits are permitted, but should be limited to brief opening and closing of doors. In addition, external–internal air-exchange systems, such as high-volume attic and window fans, should not be operating.[7] Air-conditioning systems that recycle interior air can be operated during the closed-house conditions.

In northern climate where the average daily temperature is less than 40°F, measurements should be made during the coldest months of the winter season. In southern areas that do not experience extended periods of cold weather, an attempt should be made to identify if there are time periods when closed-house conditions normally exist.

Measurements of 3 days or less should not be conducted if severe storms with high winds are predicted. Severe weather will affect the measurement results in a variety of ways. A high wind will increase the variability of radon concentration due to wind-induced differences in air pressure between the house interior and exterior. Rapid changes in barometric pressure increase the chance of a large difference in the interior and exterior air pressures, therefore changing the rate of radon influx. Measurements should also not be made if remodeling, changes in the heating, ventilating, and air-conditioning (HVAC) system, or other modifications that may influence the radon concentration during the measurement period are planned.

Schools and other public buildings should conduct measurements on the weekends, so that closed conditions can be more easily satisfied. Ventilation systems should not be shut down or operated at a reduced rate during the time when the measurement is made.[8]

A sampling site must be selected for stationary monitors where they will not be disturbed during the measurement period. The location must be at least 20 inches above the floor with the detector's top face at least 4 inches from other objects and well away from exterior house walls. The unit must sit in open air that people might breathe rather than in a closet or a drawer. All measurements should be made away from drafts caused by HVAC vents, doors, windows, and fireplaces in order to reduce the influence of changes in ventilation and on radon and radon decay product concentrations.

Samples should be collected in areas permanently occupied (office space, break areas) and areas routinely visited (mechanical areas, storage areas, work shops, garage areas) by employees of the building on floors directly in contact or directly over (crawl spaces) the ground. Rooms should be selected that are expected to have the lowest ventilation rates, such as interior rooms with no windows and tight doors. Avoid locations near excessive heat or in direct, strong sunlight and areas of high humidity. Detectors can be placed close to, but not directly in, inaccessible areas, such as closets, sumps, crawl spaces, or nooks within the foundation. Detectors should be placed inside the mixing chambers in HVAC systems where the return air from the ground level mixes with outside air before recirculation to other areas of the building. In gymnasiums or buildings with large open rooms, detectors can be placed every 2000 square feet. A knowledge of the building's profile (structure, mechanical areas, utilities) will assist in identifying the sampling areas. Review below-grade (basement) floor plans to help determine sampling positions.

Documentation is critical so that data interpretations and comparisons can be made. It includes:

- Start and stop times, and the date(s) of the measurement
- Information about how the standardized conditions have been satisfied
- Exact location of the monitor, including a floor plan of the building
- Any other useful information, including type of building, heating system, existence of a crawl space, smoking habits of the occupants, operation of humidifiers, air filters, electrostatic precipitators, or clothes dryers

The primary types of short term passive monitors for radon are charcoal canisters and alpha track detectors. There are a number of variations on these. Some are designed for shorter sampling periods.

Figure 15.2. Charcoal canister monitor for radon. (Courtesy of F&J Specialty Products, Inc.)

Charcoal Canisters

Charcoal canisters are passive integrating devices that measure radon gas directly. They consist of containers most often shaped like a respirator cartridge or round canister with a screw top filled with a measured amount of activated charcoal (Figure 15.2). In one version, one side of the container is fitted with a screen that keeps the charcoal in but allows air to diffuse into the charcoal when the cover is removed. When the canister is prepared by the supplier, it is sealed until it is ready to be used. Another type of charcoal detector is shaped like a large pouch into which room air can diffuse. The top of the container is usually perforated to allow collection without loss of charcoal. Some have a filter to keep out radon-decay products. All charcoal adsorbers should be stored in airtight containers when not being used for sampling.

Charcoal canisters are generally set out for 4 to 7 days, although some EPA guidelines recommend only 2 days.[4] Sealed environmental conditions in the testing area (or building) for 12 hours prior to and following the test are required for accurate results. Laboratory analysis is performed using a sodium iodide gamma-scintillation detector to count the gamma rays emitted

by the radon-decay products on the charcoal. Gamma rays of energies between 0.25 MeV and 0.61 MeV are counted. The detector may be used in conjunction with a multichannel gamma spectrometer or with a single-channel analyzer with the window set to cover the appropriate gamma-energy window.

A correction must be made for the reduced sensitivity of the charcoal due to adsorbed water. It may be done by weighing each canister when it is prepared, and then reweighing it when it is returned to the laboratory for analysis. Any weight increase is attributed to water adsorbed on the charcoal. The weight of water gained is correlated to a correction factor that should be empirically derived. The correction is unnecessary if the charcoal canister configuration is modified to significantly reduce the adsorption of water, and if the user has experimentally demonstrated that, over a wide range of humidities, there is negligible change in the collection efficiency of the charcoal.

Advantages of charcoal canisters include:

- Low cost per canister
- No special skills required for use
- Convenient to handle and install
- Unobtrusive when installed
- No external source of power needed
- Precise results with proper analytical techniques.

Disadvantages include:

- Some charcoal adsorbers are more sensitive than others to temperature and humidity.
- Canisters can only be used for short-term testing, that is, 7 days or less.

The primary use of canisters is as screening devices; due to the short sampling period their accuracy is lower than other passive methods. The downside of short-term monitoring is that it may detect an unrepresentative peak or a valley of radon concentration, causing a false sense of alarm or security depending on the situation.

Calibration. For larger sampling projects, duplicate canisters should be placed in enough rooms or buildings to monitor the precision of the canisters, that is, approximately 10% of the number of sites monitored in a month, or 50 duplicates, whichever is less.[4] The duplicate canisters should be shipped, stored, exposed, and analyzed under the same conditions.

Procedure[4]. This procedure describes use of one or several canisters for a residential screening study.

1. Prior to sampling, make sure for 12 hours before and during the 4- to 7-day measurement period that windows and external doors are kept closed, except for normal entry and exit. Fans or ventilation systems that use outside air, such as attic fans, must not be operated.

2. Put a unique identifying number on the canister.

3. Record on the sampling data sheet the type of building in which the measurement is being made: single-family home, multifamily dwelling, business, school, other.

4. Generally, if the building being sampled has a basement, this place is the best to put the canister. Garages, root cellars, and crawl spaces have too much ventilation and are undesirable. If there is no basement, place the canister in any room on the lowest floor of the house, except in a bathroom, kitchen, or porch. Record the floor of the building where the measurement is being made.

5. Within the selected room, the canister should not be in a location frequently exposed to noticeable drafts of an open door, window, fireplace, and so forth. The canister should be exposed to the same air that the building's occupants are breathing. It should be placed on a table or shelf at least 2 feet above the floor, and should be in the open air, not in a closet, drawer, or cupboard. Record the room and the exact location in which the measurement is to be made: bedroom, family room, office, living room, unfinished basement, classroom, or other.

6. Record the date and time, and unscrew the cap or lift off the tape to start the measurement. Some canisters are designed to be exposed for 48 hours (2 days), and others for 72 hours (3 days), 96 hours (4 days), or 168 hours (7 days). A deviation from the schedule of up to 6 hours is acceptable as long as the actual time of termination is documented.

7. At the end of the sampling time, replace the cap or the tape on the canister. Record the date and time and immediately send the canister to the laboratory.

Alpha-Track Detectors

The alpha-track (AT) method measures radon. An AT detector is a small sheet of special plastic material (usually polycarbonate) enclosed in a container with filter-covered opening. Some AT devices look like large pill boxes; others look like clear plastic drinking glasses (Figure 15.3). Alpha particles that strike this plastic cause microscopic markings while other types of radiation pass through without causing any changes. At the end of the measurement period, the detectors are returned to an analytical laboratory for processing. The plastic is placed in a caustic solution that

Figure 15.3. Alpha-track monitor for radon. (Courtesy of RSSI.)

TABLE 15.5. Precision of Alpha-Track Detectors

Number of Net Tracks Counted	2 Sigma Error (%)[a]
4	100
6	82
10	63
15	52
20	45
50	28
75	23
100	20

[a] This is the minimum error for the number of net tracks indicated; the absolute error is dependent on the actual number of background tracks counted.

accentuates the damage tracks so they can be counted using a microscope or an automated counting system. The number of tracks per unit area is converted to an average radon gas concentration in air, using a conversion factor derived from data generated at a calibration facility. When these detectors are used according to guidelines in the EPA protocol, they are left for periods of up to 3 months for screening and 12 months for follow-up measurements.[4]

These detectors require no special skills to use, and can be used to measure the long-term average concentrations over a 12-month period, which is the optimal measurement period for long-term concentrations. During this period, closed-house conditions do not have to be satisfied. The long measurement period, with a 3-month minimum, limits the situations for which these devices can be used.

Many factors contribute to the variability of AT detector results, including differences in the detector response within and between batches of plastic, nonuniform plateout of decay products inside the detector holder, differences in the number of tracks used as background, variations in etching conditions, and differences in readout. The variability in AT detector results decreases with the number of net tracks counted, so counting more tracks over a larger area of the detector will reduce the uncertainty of the result. In addition, use of duplicate AT detectors will reduce error.

The sensitivity of an AT detector system is dependent on the area of the detector that is counted for alpha tracks (Table 15.5).[4] At a minimum, the detector should provide an adequate sensitivity at 4 pCi/L.

Advantages of AT detectors include:

- Low cost per detector
- Easy to handle and install
- Unobtrusive when installed
- No training required for use
- No external power source needed
- Ability to measure the integrated average concentration over a 12-month period, which is the optimal measure of long-term concentration

Disadvantages include:

- Relatively long measurement period needed, minimum of 3 months
- Large inherent variability (precision errors), especially at low concentra-

tions, if the area of the detector that is counted is small

Calibration. Determination of a calibration factor is performed by the manufacturer. It requires exposure of AT detectors to a known radon concentration in a radon exposure chamber. These calibration exposures are to be used to obtain or verify the conversion factor between net tracks per unit area and radon concentration. AT detectors should be exposed in a radon chamber at several different radon concentrations, or exposure levels similar to those found in at least three tested buildings. A minimum of 10 detectors should be exposed at each level. The period of exposure should be sufficient to allow the detectors to achieve equilibrium with the chamber atmosphere. A calibration factor should be determined for each batch of detector material received from the material supplier.

Use. AT detectors should be used as soon as possible after delivery from the supplier. If the storage time exceeds more than a few months, the background exposures from a sample of the stored detectors should be assessed.

Always review the manufacturer's instructions furnished with the sampler. As with all passive detectors, measurements should be collected in an area away from drafts that can be caused by heating and air conditioning systems, openings such as doors and windows, or ventilation from cracks and openings in exterior walls. A rule of thumb is the placement of a detector for each 2000 square feet of floor space.

Since the device will be used for an extended period chose a location that will discourage tampering or inadvertent interference with the test.

Procedure[4]
1. Remove the protective covering by cutting the edge of the bag, or remove

it so that it can be reused to reseal the detector at the end of the exposure period. The monitoring period begins when the detector is exposed.

2. Inspect the detector to make sure the front is intact and has not been physically damaged in shipment or handling.

3. It is usually convenient to suspend the detector from the ceiling or on a joist. Some have adhesive strips for this purpose. A thumb tack can also be used. The detector should be positioned at least 8 inches below the ceiling. However, it can also be placed upright with holes showing at least 2–3 feet above the floor on an exposed surface.

4. Record the serial number of the detector in a log book along with a diagram showing the location in the room where the detector was placed. If during the exposure period it is necessary to relocate the detector, make certain it is noted in the log book, along with the date it was relocated and recorded on the diagram.

5. At the end of the measurement period, the detector should be inspected for damage or deviation from the conditions entered in the log book at the time of installation. Any changes should be noted. The date of removal is entered on the data form for the detector and in the log book. The detector is then resealed using the protective cover provided with the detector, sometimes an adhesive seal, or a bag with the correct serial number for that detector. If a bag is used, the open edge of the bag is folded several times and resealed with tape. If the bag or cover has been destroyed or misplaced, the detector should be wrapped in several layers of aluminum foil and taped shut. After retrieval, the detectors should

be returned as soon as possible to the analytical laboratory for processing.

Comparison of Charcoal and Alpha-Track Detectors

Measurements made with AT detectors give a better estimate of average radon concentration than charcoal canisters. AT detectors are better integrating devices, and are not as affected by fluctuating radon levels. AT detectors do not require immediate analysis. Unlike charcoal canisters, no radon decay occurs in the AT detector once the measurement is taken. Therefore, the time between when the radon measurement is completed and when the devices are shipped is not as critical, allowing large numbers of detectors to be handled.

Grab Sampling Methods

The major differences in various grab sampling methods are the total time period required for sampling and analysis, the capability of determining exposure concentration at the work site, and the amount of routine maintenance and calibration requirement of an instrument. Grab sampling methods measure concentrations of radon gas or radon decay product concentrations.

Some detector systems can sample both radon and radon decay products simultaneously. Several samples can be collected and analyzed each day. Due to the short measurement periods, the samples may not be representative of the typical long-term concentrations to which people may be exposed. Careful control of the environment to assure closed conditions is required for 12 hours prior to and during the test. Because of the highly reactive nature of radon decay products, WL measurements are more susceptible to sampling error than radon gas measurements.

Grab sampling is useful for sites where high concentrations are suspected because of the need to implement control measurements as soon as possible. They can also be used as diagnostic tools to trace the probable cause of elevated levels in a building. Multiple grab samples are preferred to single grab samples. Grab samples alone are generally not recommended for measurements made to determine if remedial steps are needed. Grab samples are also not recommended for follow-up measurements because of their poor correlation with long-term averages. Advantages of grab sampling include:

- Results are quickly obtained.
- Equipment is portable.
- Both radon and its decay products can be measured at the same time.
- Several samples can be run per day.
- Conditions during measurement are known to the sampling professional.

Disadvantages include:

- Since the measurement period is very short, the results may not be representative of the long-term average.
- Sampling requires considerable skill.
- House conditions must be under careful control for 12 hours prior to measurements.
- Cost of a system is relatively expensive.

Critical factors in assuring accuracy when collecting grab samples include the proper calibration of radiation detectors and pumps, filters that precisely fit the equipment, and accurate maintenance of the flow rate during the sampling period. It is also important to prevent contamination of the pump, counting equipment, and filters.

Grab Sampling Methods for Radon.
Grab samples for radon are very quick determinations of the concentration col-

lected in scintillation cells holding 100 to 2000 cubic centimeters (cc) of air. In this method a sample of air is drawn into these cells that have a zinc sulfide phosphor coating on the inner surfaces with a clear window at one end. To take a measurement, the cell is evacuated, taken to the sampling location, opened to allow room air to rush in, and then sealed to trap the radon inside.

The cell is counted within 4 hours after filling to allow the short-lived radon progeny products to reach equilibrium with the radon. In the laboratory, the cell is put into contact with a photomultiplier tube and tiny light flashes (scintillations) produced by radon decay products striking the cell's zinc-sulfide-coated interior are electronically detected. The number of light flashes is proportional to the radon in air concentration. Correction factors are applied to the counting results to compensate for decay during the time between collection and counting to account for decay during counting.

For accurate measurements it is necessary to standardize cell pressure prior to counting, because the path lengths of alpha particles are a function of air density. The counting system, consisting of the scaler, detector, and high-voltage supply, must be calibrated. The correct high voltage is determined via a plateau. Each counting system should be calibrated before being put into service, after any repair, or at least once per year. Also, a check source or calibration cell should be counted in each system each day to demonstrate proper operation prior to counting any samples.

An accurate calibration factor must be obtained for each counting cell. This is done by filling each cell with radon of a known concentration and counting the cell to determine the conversion factor of counts per minute per picocurie. The known concentration of radon may be obtained from a radon calibration chamber or estimated from a bubbler tube containing a known concentration.[7]

Procedure[7]

1. Prior to collection of the sample, counter efficiency must be verified and a background measurement must be taken.

2. Evacuate the cell (Lucas type) to at least 25 inches of mercury, attach the filter to the cell, and open the valve, allowing the cell to fill completely with air. Allow at least 10 seconds for the cell to completely fill. To ensure a good vacuum at the time of sampling, the cell may be evacuated using a small hand-operated pump in the room being sampled. It is a good practice to evacuate the cell at least five times, allowing it to fill completely with room air each time. Make sure the air to be sampled flows through the filter each time. If it can be demonstrated that the cells and valves do not leak, the cells can be evacuated in a laboratory.

3. With a double-valve, flow-through-type cell, attach the filter to the inlet valve and a suitable vacuum pump to the other valve. The pump may be motor-driven or hand-operated. Open both valves and operate the pump to flow at least 10 complete air exchanges through the cell. Stop the pump and close both valves.

4. Record all pertinent sampling information after taking the sample, including the date and time, cell number, name of person collecting the sample, and any other significant conditions within the building or notes regarding weather conditions.

5. After the cells have been counted and the data recorded, the cells must be flushed with nitrogen to remove the sample. Flow-through cells should be flushed with at least 10 volume

exchanges at a flow of about 2 Lpm. Cells with single valves are evacuated and refilled with nitrogen at least five times. The cells are left filled with nitrogen and allowed to sit overnight before being counted for background. If an acceptable background is obtained, the cell is ready for reuse.

Grab Sampling for Radon Decay Product Measurements. Grab sampling methods for radon progeny involve drawing a known volume of air through a filter and counting the alpha activity during or following sampling. They are collected by drawing a known volume of air through a filter using an air sampling pump for very short sampling periods, usually 5 minutes. The radon decay products, if present in the air, are collected on the filter. Filters are counted for alpha particle emissions during mathematically determined periods after the sample is collected. There are two main methods: the Kusnetz method, where the filter is counted once, and the modified Tsivoglou method, where the filter is counted three times to measure the decay that has occurred.

For a 5-minute sampling period (10–20 L of air) on a 25-mm filter, the sensitivity using either method is approximately 0.005 working level.

Kusnetz Procedure[7]

1. Place a new filter in the holder prior to entry to the building where sampling will be done. Care should be taken to avoid puncturing the filter and to avoid leaks.

2. Start the pump and the clock simultaneously. Note the flow rate and record it on the sampling data sheet. Collect up to 100 L of air on a filter over a 5-minute sampling period. The sampling time should be carefully monitored and recorded.

3. Remove the filter from the holder using forceps and carefully place it facing the scintillation phosphor. The side of the filter on which the decay products were collected must face the phosphor disc. The chamber containing the filter and disc should be closed and allowed to adapt to the dark prior to starting the count.

4. The total alpha activity on the filter must be counted at some time between 40 and 90 minutes after the end of sampling. Counting can be done using a scintillation counter to obtain gross alpha counts for the selected period. Counts from the filter are then converted to disintegrations using the appropriate counter efficiency.

5. The disintegrations from the decay products collected from the known volume of air can then be converted into working levels using the appropriate Kusnetz factor (Table 15.6) for the counting time utilized and the following calculation:

$$\mathrm{WL} = \frac{C}{K(t)VE}$$

TABLE 15.6. Kusnetz Factors

Time (min)	$K(t)$	Time (min)	$K(t)$
40	150	66	98
42	146	68	94
44	142	70	90
46	138	72	87
48	134	74	84
50	130	76	81
52	126	78	78
54	122	80	75
56	118	82	72
58	114	84	69
60	110	86	66
62	106	88	63
64	102	90	60

where WL is the working level, C is the sample cpm—background cpm, $K(t)$ is the factor determined from Table 15.6 for the time from the end of collection to the mid-point of counting, V is the total sample air volume in liters (flowrate × time), and E is the counter efficiency in counts per minute (cpm) or disintegrations per minute (dpm).

Sample Problem for the Kusnetz Procedure

Background count = 3 counts in 5 min, or 0.6 cpm

Standard count = 5985 counts in 5 min, or 1197 cpm

Efficiency = 1197 cpm − 0.6 cpm/2430 dpm = 0.49 (from known source of 2439 dpm)

Sample volume = 4.4 L/min × 5 min = 22 L

Sample count at 45 min (time from end of sampling period to start of counting period) = 560 counts in 10 minutes, or 56 cpm

K_t at 50 minutes (from Table 15.6) = 130

WL = (56 cpm − 0.6 cpm)/130 × 22 L × 0.49

WL = 0.04

Modified Tsivoglou Procedure[7]

1. Collect a sample on a filter using the same method as described in the Kusnetz procedure.

2. Remove the filter using forceps and carefully place it facing the scintillation phosphor. The side of the filter on which the decay products were collected must face the phosphor disk. The chamber containing the filter and disk should be closed and allowed to adapt to the dark prior to starting counting. If the counter used is slow to dark adapt, the counting procedure should be done in a darkened environment.

3. The filter must be placed into the counting position very quickly, since the first of the three counts must begin 2 minutes following sampling. Count the filter at the following intervals following collection: 2–5, 6–20, and 21–30 minutes.

4. The concentration, in picocuries per liter (pCi/L), of each of the radon-decay products (Po-218, Pb-214, and Po-214) can be determined by using the following calculations:

$$C_2 = [1/FE](0.16921G_1 - 0.08213G_2 + 0.07765G_3 - 0.5608R)$$

$$C_3 = [1/FE](0.001108G_1 - 0.02052G_2 + 0.04904G_3 - 0.1577R)$$

$$C_4 = [1/FE](-0.02236G_1 + 0.03310G_2 - 0.03765G_3 - 0.05720R)$$

It is important to note that the constants in these equations are based on a 3.04-minute half-life of Po-218. The working level (WL) associated with these concentrations can then be calculated using the following relationship:

$$\begin{aligned} WL = &(1.028 \times 10^{-3} \times C_2) \\ &+ (5.07 \times 10^{-3} \times C_3) \\ &+ (3.728 \times 10^{-3} \times C_4) \end{aligned}$$

where C_2 is the concentration of Po-218 in pCi/L; C_3 is the concentration of Pb-214 in pCi/L; C_4 is the concentration of Po-214 in pCi/L; F is the sampling flowrate in liters per minute (Lpm); E is the counter efficiency in counts per minute/disintegrations per minute (cpm/dpm); G_1 represents the gross alpha counts for the time interval of 2–5 minutes; G_2 represents the gross alpha counts for the time interval of 6–20 minutes; G_3 represents gross alpha counts for the time interval of 21–30 minutes; and R is the background counting rate in cpm.

Sample Problem for the Modified Tsivoglou Procedure

Given:

F = sampling flowrate = 3.5 Lpm
E = counting efficiency = 0.47 cpm/dpm
$G_1 = 880$
$G_2 = 2660$
$G_3 = 1460$
$R = 0.5$

Calculations:

$$C_2 = [1/3.5 \times 0.47](0.16921 \times 880)$$
$$- (0.08213 \times 2660) + (0.07765 \times 1460)$$
$$- (0.05608 \times 0.5)$$
$$C_2 = 26.8 \text{ pCi/L}$$
$$C_3 = [1/3.5 \times 0.47](0.001108 \times 880)$$
$$- (0.02052 \times 2660) + (0.04904 \times 1460)$$
$$- (0.1577 \times 0.5)$$
$$C_3 = 10.9 \text{ pCi/L}$$
$$C_4 = [1/3.5 \times 0.47](-0.02236 \times 880)$$
$$- (0.03310 \times 2660) + (0.03766 \times 1460)$$
$$- (0.05720 \times 0.5)$$
$$C_4 = 8.1 \text{ pCi/L}$$
$$\text{WL} = (1.028 \times 10^{-3} \times 26.8)$$
$$+ (5.07 \times 10^{-3} \times 10.9)$$
$$+ (3.78 \times 10^{-3} \times 8.1)$$
$$\text{WL} = 0.11$$

Continuous Monitors

Continuous radon and continuous WL monitoring measurement methods are similar in that they both use an electronic detector to accumulate and store information related to the periodic (usually hourly) average concentration of radon gas or radon-decay products. The primary difference between methods is that a continuous WL monitor samples the ambient air collected on a filter cartridge at a flow rate of about 0.1–1.0 Lpm, while a continuous radon instrument samples the air after it has passed through a filter. Radon WL instruments are generally faster than radon instruments (0.5 hours versus 3-hour response time).

A continuous monitor for radon consists of three parts: a pump, a filter to remove dust and radon-decay products, and a scintillation cell. These monitors sample the radon in the ambient air that is collected in a scintillation cell after passing through a filter that removes dust and radon-decay products. As the radon in the cell decays, ionized radon-decay products plate out on the interior surface of the scintillation cell. The alpha particles that are produced strike the coating on the inside of the scintillation cell, causing scintillations to occur that are detected by a photomultiplier tube. The resultant electronic signals are processed and the data are either stored in the memory of the continuous monitor or printed on paper tape by the printer. These units can be the flow-through cell type or the periodic-fill type. In the flow-through cell type, air continuously flows into and through the scintillation cell. The periodic-fill type fills the cells during preselected time interval, counts the scintillations, then begins the cycle again.

Continuous Radon Measurements

The Sun Nuclear Model 1027 Professional Continuous Radon Monitor (Figure 15.4) is an example of a portable continuous monitor using a diffused-junction photodiode sensor that provides short-term and long-term measurements of radon and radon progeny concentration. Ambient room air, laden with radon, diffuses into the monitor's detection chamber. Radon decay byproducts emit alpha particles that are detected by the photodiode. A microcomputer stores the pulses and computes

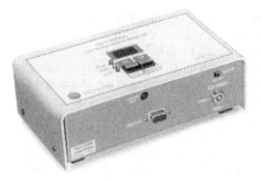

Figure 15.4. Sun Nuclear Model 1027 Professional Continuous Radon Monitor for short-term and long-term measurements of radon and radon progeny concentration. (Courtesy of Sun Nuclear Corporation.)

the radon concentration from an internal calibration factor. The measurement range is from 0.1 to 999 pCi/L with an accuracy of ±25% or 1 pCi/L, whichever is greater after 24 hours. The usual measurement interval is 1 hour; 4, 8, or 24-hour intervals are available by special order. The lower limit of sensitivity as stated by the manufacturer is 2.5 counts per hour per pCi/L. It has been evaluated and accepted by the U.S. EPA for use in real estate transaction testing.[9]

The device is operated using line power. The instrument contains a 9-volt battery that provides backup power for about 20 hours in the event of a power outage during a test. It also employs internal sensors to detect and report movement during a test. The testing is not interrupted if tampering is detected, but the tampering is noted on the final report. The instrument has no consumable parts; it is ready for a new test after the memory is cleared.

In a typical test, the Model 1027 device is placed in the home to be monitored, connected to line power, cleared and set to operate, and left unattended for a period of two or more days. The data from the test include the average radon concentration in pCi/l over the total monitoring period and the current radon concentration. The average and current display values continue to update during the entire test, even

if the test period exceeds the maximum storage of 90 intervals. The interval storage stops updating after 90 intervals until the memory is cleared. At the end of the monitoring period, the results are available in any of three forms: numerical display on the control panel, printed report on an accessory dedicated dot-matrix printer that uses ordinary adding machine tape paper, or downloaded to a personal computer using an RS-232 interface. Data can be displayed at any time during the test on the instrument's LED display. Once the data are read or downloaded to the computer, the memory can be cleared and another test started. Typical reports are displays of the reading by interval period as well as average values.

The tamper sensor is active except for a 15-second period immediately following the "Clear" command. A "tamper" notation appears with any interval readings during which the device was moved or jarred. To avoid a tamper notation on the first reading, the monitor cannot be moved or jarred after the tamper sensor becomes active. As an additional security feature, the keyboard and operating controls can be locked during unattended periods.

With instruments like this it is important to read the operating instructions carefully to yield valid tests. For example, in some instruments both the line power and backup battery power must be shut off to stop readings; if only the line power is shut off, the instrument might continue to record data using the backup battery source. Also, instruments vary in how they store data. With some instruments, if the average reading is recorded, then power is completely shut off, then power is resumed and another average reading displayed, the two values might be different since some partial interval data were lost during the time when power was off.

Continuous radon monitors can be used to look at changes in radon concentrations over periods of minutes to hours to days.

Errors that may occur with continuous monitors include inherent errors caused by measuring events instead of alpha energy, absorption errors that affect the system's sensitivity and inherent error, statistical errors from counting random events, calibration errors, airflow rate errors, radon progeny plate out error, instrument instability and nonlinearity errors, and external background activity errors.[10]

Maintenance and Calibration. These monitors must be calibrated in a known radon environment to obtain the conversion factor used by the electronics to convert count rate to radon concentration. After every 1000 hours of operation, units should be examined to check the background rate by purging with clean, aged air or nitrogen. A second check for background count rate is done by operating the instrument in an outdoor or other low-radon environment. Twice a year the unit's response to a known radon concentration should be measured, and the flowrate of the pump should be measured.

Use. The continuous monitor should be programmed to run continuously, recording the hourly integrated radon concentration measured and, if applicable, the total integrated average radon concentration. The sampling period should not be less than 24 hours.

Prior to and after each measurement, test the monitor to verify that the correct input parameters and the unit's clock are set properly and to verify that the pump is operating properly.[7]

Continuous Working-Level Monitors.
As noted, the continuous WL monitor samples the radon progeny in ambient air by filtering airborne particles as the air is drawn through a filter cartridge at a flowrate between 0.1 Lpm and 1 Lpm. An alpha detector, such as a diffused-junction or surface-barrier detector, counts the

alpha particles produced by ^{218}Po and ^{214}Po as they decay on the filter. The detector is normally set to detect alpha particles with energies between 2 MeV and 8 MeV. The event count is directly proportional to the number of alpha particles emitted by the radon-decay products on the filter. Total counts over a specific time period can also be measured.[7]

An example of this type of instrument is the Thomson Nielsen Radon WL Meter (Figure 15.5). The Radon WL Meter operates on the same basic principles as most other active WL meters. Air containing radionuclides is sampled through a slot in the top of the instrument via a small continuously operating pump and is deposited on a standard 0.8-µm MCE filter. Alpha particles are detected by a semicustom digital-integrated circuit (silicon detector) and counted with their total displayed as alpha counts. The air sampling and alpha counting are controlled by a timer that can be preset to operate for 0.5–8 hours or, optionally, 24 hours. Background gamma

Figure 15.5. Thomson Nielsen Radon WL Meter collects airborne radionuclides on an internal filter, which is counted using a semiconductor detection system. (Courtesy of Thomson Nielsen.)

radiation does not interfere as the instrument counts alpha particles only. At the end of the air sampling period, the total alpha count is displayed on the readout and the WL is manually calculated using the alpha counts and the instrument's calibration factor. The instrument can be operated with AC for long periods or with a battery for shorter periods.

These instruments require a relatively short measurement period (24 hours). Hourly results can track the variation of concentrations present. Most models are very precise, and results are available on-site following the measurements.

Advantages of continuous radon monitors and continuous WL monitors include:

- Relatively short measurement durations—a minimum of 6 hours for screening, 24 hours for follow-up measurements
- Hourly results that allow identification of variations of building concentrations
- Small precision error
- Result available on site

Disadvantages of these monitors include:

- Higher cost than most other methods
- Heavy and awkward to move some models
- Extensive calibration requirements utilizing a radon calibration chamber
- Trained operators necessary

Maintenance and Calibration. Calibration is performed by the manufacturer in a calibration facility and a calibration factor is printed on a label on the back of the instrument. The EPA recommends recalibration by participation in a semiannual laboratory intercomparison program in which the device's response is compared to a known radon-decay product concentration. Recal-

ibration is also necessary after repair or modification of the instrument. A checkout should be performed prior to use to check the detector calibration and the pump performance as follows:

1. Turn over the instrument and open the filter holder compartment. Remove the cap from the filter holder and replace the filter with a capped alpha-emitting check source using tweezers so as to prevent damage from occurring to the active surface of the source and to avoid contact of the source with the skin. The check source used must be the one supplied by the manufacturer of the instrument.
2. Set the timer to 30 minutes and turn on the instrument but not the instrument's pump.
3. Note the alpha counts displayed on the readout at the end of the 3-minute period and divide this count by 30 to determine the calibration check factor in cpm. Compare this factor with that provided by the manufacturer of the instrument. Generally they should agree within 10%.
4. The flowrate is checked using the methods discussed in Chapter 5. The calibration factor is used to convert the number of counts into average milli-working-level (mWL) using the following calculation:

$$mWL = \frac{N}{(T - 0.5)(CF)}$$

where N is the total number of counts at the end of the sampling period, T is the sampling time in hours, CF is the calibration factor in counts per hour per mWL, and $1\,WL = 10^3\,mWL$.

After every 100 hours of operation, the unit should be checked to measure the background count rate using the procedures that may be identified in the operating manual for the instrument. Twice annually the unit's response should be compared to a known radon-decay product concentration using a calibration chamber and the flow rate of the pump.

Use. The continuous WL monitor should be programmed to run continuously, recording the hourly integrated WL measured and, when possible, the total integrated average WL. The sampling period should not be less than 24 hours for most purposes. The longer the operating time, the smaller the uncertainty associated with the measurement result. The integrated average WL over the measurement period should be used as the measurement result.[7]

1. Prior to use the continuous WL monitor should be tested to verify that a new filter has been installed and the input parameters and clock are set properly. The detector performance should be checked daily using the procedure described earlier.

2. The following information should be recorded during measurements: date and time of the start and finish of the measurement period; whether the conditions during the measurement period were standard or if exceptions occurred; the exact location of the measurement on a floor plan of the building; type of building and number of stories; type of heating system; existence of a crawl space or basement; occupants' smoking habits; operation of humidifiers, dehumidifiers, air filters or electrostatic precipitators, and clothes dryers.

3. As with all radon measurements, closed-house conditions must prevail. Measurements should not be taken near drafts caused by heating, ventilation and air conditioning vents, doors, windows, and fireplaces. The measurement location should also not be close to the outside walls of the building. The instrument should be placed at least 20 inches from the floor.

4. For long-time averaging measurement, the instrument should be set to run for a minimum of 8 hours, preferably 24 hours. Converting the number of counts into "average mWL" by performing the long-term averaging calculation is shown in the preceding section on maintenance and calibration of continuous WL monitors.

5. For short-time screening, the timer can be set to between 0.5 and 4 hours, although 1 hour is most commonly used. This mode of operation can be used to do comparative measurements in various locations in the same building or in different buildings. The equation used for long-time averaging is not accurate when used for short-term screening; instead the best use of the results is to compare total counts from various areas in order to target the best areas to perform long-term measurements.

6. Turn over the instrument and remove the filter holder. Remove the cap holding the filter and put a clean 25-mm filter inside. Do not touch the alpha particle detector located below the filter holder. It has a thin Mylar-film window that is easily penetrated. Any contact with the detector may destroy it.

7. Set the timer to the desired sampling time.

8. Turn on the power and check the display for the "000000" readout. Turn on the pump.

9. At the end of the sampling period, note the alpha counts and calculate the WL if appropriate.

Radon Progeny Integrating Sampling Units

There are several different electronic instruments available for measuring concentrations of radon progeny over a period of about two days to seven weeks. They all employ a sampling pump to draw ambient air into the device and a filter to capture particulate matter. These devices are true integrating instruments if the pump flowrate is uniform throughout the sampling period. The instruments differ in how the radioactive particles on the filter are counted:

- *Thermoluminescent Dosimeter (TLD):* TLDs are special crystal devices that store energy when impacted by ionizing radiation, and then when placed in a "reader" emit an amount of light that is proportional to the radiation dose. The TLD-type instrument contains at least two TLDs. One TLD measures the radiation emitted from radon decay products collected on the filter, and the other TLD is used for a background gamma correction. Analysis of the detector TLDs is performed in a laboratory using a TLD reader. Interpretation of the results of this measurement requires a calibration for the detector and the analysis system based on exposures to known concentrations of radon decay products.

- *Alpha Track Detector:* This type has a cylinder with three collimating cylindrical holes opposite the filter. Alpha particles emitted from the radon decay products on the filter pass through the collimating holes and through different thicknesses of energy-absorbing film before impinging on a disk of alpha track detecting plastic film. Analysis of the number of alpha particle tracks in each of the three sectors of the film allows the determination of the number of alpha particles derived from ^{218}Po and ^{214}Po. This feature allows the determination of the equilibrium factor for the radon decay products. Etching and counting of the alpha track assembly is carried out by mailing the detector film to the analysis laboratory. Interpretation of the results of this measurement requires a calibration for the detector and the analysis system based on exposure to known concentrations of radon decay products.

- *Electret Detector:* An *electret* is an electrostatically charged disk detector situated within a small container (ion chamber). The electret instrument is similar in operation to the TLD-type unit, except that the TLD is replaced with an electret. As the radon decay products that are collected on the filter decay, negatively charged ions generated by alpha particle radiation are collected on a positively charged electret, thereby reducing its surface voltage. This reduction has been demonstrated to be proportional to the radon decay product concentration; a calibration factor relates the measured drop in voltage to the radon concentration. The electret must be removed from the chamber, and its voltage must be measured with a special surface voltmeter both before and after exposure. To determine the average radon concentration during the exposure period, the difference between the initial and final voltages is divided first by a calibration factor and then by the number of exposure days. A background radon concentration equivalent of ambient gamma radiation is subtracted to compute radon concentration. Electret voltage measurements can be made in a laboratory or in the field. Variations in electret design determine whether detectors are appropriate for making long-term or short-term measurements.

Occupational Air Sampling for Radon.
The primary situations where occupational exposure to radon is measured are in uranium and phosphate mines. Other hard rock mines where radon progeny exposure may be a problem are iron, zinc, fluorspar, and bauxite. In uranium mines the exposure usually depends on the quality and amount of uranium ore present and the effectiveness of the ventilation system.[11] The primary methods used to measure occupational exposure to radon progeny in mines are grab samples collected in various areas rather than a continuous integrated personal sample.

Radon samples in mines are initially collected on the exhaust air using procedures specified by the MSHA standards.[12] If greater than 0.1 WL is detected in the exhaust air of a uranium mine, sampling representative of the breathing zone of miners must be conducted every 2 weeks at random times in all active working areas. This includes stopes, drift headings, travelways, haulage ways, shops, stations, lunchrooms, magazines, and any other place or location where people work, travel, or congregate. Higher levels increase the sampling period to weekly. The monitoring schedule is required until levels decrease to less than 0.1 WL. In mines other than uranium mines, radon also has to be sampled in the exhaust air, and if levels exceed 0.1 WL, samples representative of the breathing zone of miners must be taken every 3 months.

When relying on grab samples, a sampling strategy must take into account the typical variability in radon progeny concentrations that can occur in any given area. Furthermore, once the samples are collected a correlation must be made for individual occupations to account for the times spent in various areas often within a single work shift. A statistical approach is used to attempt to randomize the results as much as possible. Work stations in close proximity are clustered together so that they can all be sampled on the same day. Grab samples are collected at each station in a single day, but the sequence in which the samples are collected is rotated.

Sampling is done on 2 different days within a given 2-week period with a provision for additional repeat next-day sampling whenever results exceed 0.14 WL. Calculations of results include average work shift concentrations for a given day or 2 sequential days. By apportioning the length of time an individual spends in each location with the measured concentration in that area, an estimated value of the working-level month (WLM) is determined. A WLM is the exposure that takes place to radon progencies for 170 hours in a given month. The time spent by each worker at a given workplace can be determined from the time sheets and can be used to calculate individual monthly exposure to alpha. Gamma radiation measurements using Geiger counters are often done simultaneously to see if this is a significant contribution to the exposure.[11]

Occupational grab sampling techniques use the Kusnetz count method or the instant WL monitor. Grab sampling methods for radon progeny are described in the section on grab sampling methods in this chapter.

Continuous monitoring systems are rarely used for occupational sampling, since they are usually stationary systems and their primary use is for indicating problems with the ventilation system and identifying exposure sources. While personal sampling dosimeters for radon progencies exist, they are not commonly used in occupational measurements due to problems with calibration and precision of results. Where there are high thorium concentrations, special sampling techniques may be required in order to accurately monitor the contribution from radon progeny.[13]

COLLECTION METHOD FOR RADON IN WATER

Since radon is a dissolved gas when it is in water, it can escape if sampling is not done carefully. Other types of contamination that can release radiation in water are iodine, strontium, radium, and uranium.

Procedure

1. For sampling, 40-mL glass vials with Teflon liners must be used. Bladder pumps and tubing, a bailer with dual-check valves, a bottom emptying device, or a positive displacement pump can be used if a sump or well is being sampled. Do not use suction, airlift, or peristaltic pumps to collect samples of water suspected of containing radon.

2. Prior to sampling each point, label two vials with a unique sampling number and record the number on the sampling data sheet. For duplicate samples, and "A" and "B" designation added to the number can be used.

3. Make sure the liner is in the correct position in the cap with the Teflon-coated septum, which is usually white on the outside; when the lid is screwed on, it should point toward the sample.

4. For tap water samples, turn on the tap and overfill the vial. Screw the lid down while the water is still flowing.

5. If pumps or bailers are used, minimize air contact with sampled water while filling two vials almost to overflowing. While filling the vial, try to minimize turbulence and air—water contact. Cap the vial immediately and screw the cap on snug (Teflon side down). There should be no air space at the top of the vial.

6. Invert the vial and tap it on a hard surface to determine if any air bubbles remain inside. If there are any, refill the vial(s). If samples without bubbles are not possible to collect, open the vial and add more sample. Reclose the vial and again inspect it for air bubbles. Continue to work with the sample until no bubbles are present.

7. Fill two vials from each sampling point. Do not filter samples for radon even if the water is silty.

8. Store samples in a cooler at 4°C upside down, but do not allow them to freeze. Keep a thermometer in the cooler with the samples. Keep the trip blank and any field blanks with the samples.

9. Send samples to a laboratory and request an analysis as a gross alpha count.

INTERPRETATION OF RADON MEASUREMENTS

Radon measurements may be reported in two ways. One is to report the concentration of the radon gas itself. In this case, the unit used is pCi/L. The other method is to report in units that represent the energy released by the chain of radon decay products. This unit is the WL. One WL is any combination of radon progeny in one liter of air whose ultimate decay through ^{214}Po will produce 1.3×10^5 MeV of alpha energy. It is also the amount of alpha energy delivered by short-lived radon progeny in equilibrium with 100 pCi of ^{222}Ra. The most common area monitoring devices read-out in pCi/L while the unit most directly related to dose and health effects is the WL.

Because exposure recommendations to radon and its progeny are under constant review, only the latest guidelines and recommendations should be used for interpreting measurements. Measurements generally relate to the following exposures:

- *Residential and Other Building:* Tables 15.3 and 15.7[14] summarize some recommendations from the U.S. EPA. This agency as well as the ASTDR and other federal bodies issue guidance from time to time. One specific reference is the The National Research Council's *Health Effects of Exposure to Radon: BEIR VI,*[15] part of their continuing study on the Health Effects of Exposure to Ionizing Radiation. State air and radiation safety agencies may also issue regulations and guidance documents.

- *Occupational Exposures:* The Mine Safety and Health Administration issues standards that cover miners. Their current standard is 4 WL-months per year over a 30-year exposure.[12] The National Institute for Occupational Safety and Health has a recommended exposure limit of 1.0 WL-months per year and an average work shift concentration of one-twelfth of 1 working level. NIOSH considers these limits as the upper boundaries of exposure, and it recommends that every effort be made to reduce exposures to the lowest concentrations possible. In addition to the REL, NIOSH recommends specific provisions for medical monitoring, recordkeeping, respiratory protection, worker education, and sampling and analytical methods.[11]

- *Water:* Consult U.S. EPA and other agency recommendations for water sample results. As a rule of thumb, 10,000 pCi/L in incoming tap water is equivalent to 1 pCi/L of radon in indoor air. If the gross alpha count is high, samples should be collected for radium and uranium analysis.

PERFORMING FOLLOW-UP MEASUREMENTS (AFTER SCREENING)

The purpose of follow-up measurements is to estimate the long-term average radon or radon-decay product concentrations in general-use/living areas with sufficient confidence so that the need for remedial action can be estimated. Table 15.7 lists recom-

TABLE 15.7. Recommended Actions Based on Results of Residential Screening Measurements

Step	U.S. EPA Recommended Testing Steps:
1	Take a short-term test. If result is 4 pCi/L or higher (0.02 working levels [WL] or higher), take a follow-up test (Step 2) to be sure.
2	Follow up with either a long-term test or a second short-term test: • For a better understanding of your year-round average radon level, take a long-term test. • If you need results quickly, take a second short-term test. The higher the initial short-term test result, the more certain it is that a short-term rather than a long-term follow-up test should be taken. If the first short-term test result is more than twice EPA's 4 pCi/L action level, a second short-term test should be performed immediately.
3	If the follow-up was a long-term test: Fix the home if the long-term test result is 4 pCi/L or more (0.02 working levels [WL] or higher). If the follow-up test was a second short-term test: The higher the short-term results, the more certain it is that the home should be fixed. Consider fixing the home if the average of the first and second test is 4 pCi/L or higher (0.02 WL or higher).

TABLE 15.8. Follow-up Measurement Periods to Estimate Annual Averages

Instrument	Result >20 pCi/L[a]	Result <20 pCi/L[b]
Alpha-track detector	3-month intervals	12-month intervals
Charcoal canister	2–7 days	Four measurements every 3 months
Continuous working level	24 hours	Four 24-hour measurements every 3 months
Continuous radon	24 hours	Four 24-hour measurements every 3 months

[a] Made under closed-house conditions.
[b] Made under normal living conditions.

mended actions based in results of residential screening measurements.

Average indoor concentrations are about 1.3 pCi/L while typical outdoor levels are 0.4 pCi/L. Usually 2.0 pCi/L or lower is achievable in residential property with remediation.

These measurement guidelines must be followed:

1. Measurements, should be made in each level of the building that is frequently used as a living, classroom, or work area.
2. Measurements should be made in the most frequently occupied room of each level. For example, in residences a bedroom is a good choice, because most people spend more time in their bedrooms than in any other room of a house.
3. If children use or live in the building, also perform measurements in the areas where they spend most of their time.
4. Do not make measurements in kitchens, because they often contain stove exhaust systems and cooking contributes to particle levels. Bathroom air measurements are also not useful because of the high humidity often present and the small amount of time usually spent in these rooms. Bathrooms can be used for sampling radon in water.

Measurements should also be made after control measures have been installed to determine the effectiveness of these mitigation techniques. Once satisfactory levels have been obtained, follow-up measurements should only be done if physical changes occur, such as new construction, repair and alterations to the building, or settling of the foundation. Table 15.8 lists follow-up measurement periods to estimate annual averages.

SUMMARY

Sampling practitioners should be knowledgeable about sampling for radon and its progeny because of continuing concern over the issue and the growing use of measurements as part of real estate transactions. There are a variety of sampling approaches and instruments available to satisfy various applications. These will yield results expressed as pCi/L for radon, or as the working level (WL) that accounts for the energy emitted by radon and its decay products.

When planning a sampling project it is important to be aware of the relevant exposure standards and guidelines that apply to the residential, commercial, occupational or other environments under evaluation. This will help to ensure that the measured values can be directly compared to the exposure standards and guidelines.

REFERENCES

1. McManus, T. N., and J. W. Smith. A simple compact system for the extraction of radon from water samples. *AIHA J.* **48**(3):276–286, 1987.

2. Boyle, M. Radon testing of soils. *Environ. Sci. Technol.* **22**(12):1397–1399, 1988.

3. USEPA. Interim Guide to Radon Reduction in New Construction. Draft, January 1987.

4. USEPA. Radon Measurements in Schools: An Interim Report. 520/1-89-010, 1989.

5. Summerlin, J., and H. M. Prichard. Radiological health implications of lead-210 and polonium-210 accumulations in LPG refineries. *AIHA J.* **46**:202–205, 1985.

6. Gesell, T. F. Occupational radiation exposure due to ^{222}Rn in natural gas and natural gas products. *Health Phys.* **29**:681–687, 1985.

7. Ronca-Battista, M., et al. Interim Indoor Radon and Radon Decay Product Measurement Protocols. EPA 520/1-86-04, February 1986.

8. Thomson, I., and T. K. Nielson. A New Portable WL Meter for Indoor Radon. Presented at the 32nd Annual Meeting of the Health Physics Society, July 5–9, Salt Lake City, Utah.

9. USEPA, Office of Radiation Programs, Las Vegas Facility. *Operational Evaluation of the AT EASE Model 1020 Radon Monitor.*

10. Droullard, R. F. Instrumentation for Measuring Uranium Miner Exposure to Radon Daughters.

11. National Institute of Occupational Safety and Health. *Criteria for a Recommended Standard to Radon Progeny in Underground Mines.* NIOSH, Cincinnati, 1987.

12. 30 *CFR.* Chapter 1, MSHA, Radiation—Underground Only.

13. Phillips, C. R., and H. Leung. Working Level Measurement of Radon Daughters and Thoron Daughters by Personal Dosimetry and Continuous Monitoring.

14. USEPA, Office of Air and Radiation. *A Citizen's Guide to Radon: The Guide to Protecting Yourself and Your Family from Radon,* 4th ed. EPA Document 402/KOZ006, 2002.

15. National Research Council. *Health Effects of Exposure to Radon: BEIR VI.* Washington, D.C.: National Academy Press, 1999.

CHAPTER 16

SAMPLING FOR BIOAEROSOLS

Bioaerosols, meaning airborne particles derived from microbial, viral, and related agents, exist in a wide variety of sizes, shapes, and microbiological classifications. They are currently of extreme interest to the sampling practitioner because of (a) mold problems in buildings and (b) the potential use of bioaerosols as bioterrorism agents. The term *mold* is a common term for certain types of fungi. For specific information on potential bioaerosol bioterrorism agents, refer to Chapter 4.

Although sampling techniques for bioaerosols follow the same principles as for other airborne particles, three significant differences that must be considered are the need to:

- Collect most bioaerosols intact so they can be grown in a culture media, or be identified or counted using direct microscope observation. Thus sampling techniques that could damage or destroy the structure of the particles may not be useable.

- Collect some bioaerosols directly onto a growth media such as an agar plate or nutrient-treated filter.

- Avoid or account for the possible deposition of multiple bioaerosols directly on top of each other for analytical techniques that involve counting the resulting colony forming units (CFUs). This may result in very short sampling times (possibly less than one minute for heavy concentrations) or require statistical adjustments to counting results.

This chapter deals primarily with air sampling for bioaerosols. Most of these techniques involve *Sample Collection Devices* although there are limited *Direct Reading Instruments* available. Air sampling is only a small part of identifying and dealing with potential bioaerosol hazards; and, in many cases, air sampling is not the preferred method for identifying bioaerosols and assessing potential hazards. A comprehensive reference for the entire topic is

Air Monitoring for Toxic Exposures, Second Edition. By Henry J. McDermott
ISBN 0-471-45435-4   © 2004 John Wiley & Sons, Inc.

Bioaerosols: Assessment and Control published by the ACGIH, which should be consulted for more information.[1]

Bioaerosols can cause two basic conditions: infections and allergies. Infections are generally the result of multiplication and growth of microbes inside humans while allergies are the result of exposures to antigens. Not all infectious organisms cause pathogenic diseases in humans, but those that can are of concern. Well-known diseases associated with occupational exposures include anthrax, Q fever, and brucellosis. Diseases for which concerns are increasing are those associated with health workers and include AIDS and hepatitis.

Antigens are capable of stimulating the production of antibodies that produce allergic diseases. Allergic reactions are the result of an antigen producing a response from the immune system. *Hypersensitivity disease* is another term for the allergic reactions produced by these agents. These include hypersensitivity pneumonitis, allergic asthma, and allergic rhinitis. Sources of airborne antigens include bacteria, fungi, pollen, insect body parts, and skin scales (dander) and saliva of mammals.[1] In these situations antibody assays on blood from affected individuals may be performed in conjunction with monitoring for bioaerosols.

Sometimes it is not the microbe itself that produces the harmful effect but the fact that it produces a toxin. Botulism is an example wherein the botulinum toxin is the responsible agent. When release of a toxin is involved, the organism can produce a disease without extensive multiplication or dissemination throughout the body.

Just as there are factors that can predispose individuals to the health effects caused by chemicals, certain persons are also at increased risk when exposed to bioaerosols. Risk factors include age (both the very young and those who over 50 years old), drinking alcohol excessively, smoking, an impaired immune systems, or preexisting respiratory disease or other illnesses such as diabetes or kidney disease.

Bioaerosols can exist in both viable (living) and nonviable states. Viable microorganisms such as bacteria, fungi, yeasts, and molds originate from sprays or splashes of media, from the agitations of dusts, and from sneezes and coughs of which only the small particles ($<10\,\mu m$) remain in the air long enough to travel any distance. Examples of nonviable agents that are occasionally sampled include pollens and insect parts. Grains, clusters of cells, and skin scales are much larger-sized particles (10–$50\,\mu m$) than bacteria and viruses[2] (Figure 16.1). Spores, which can be formed by fungi and certain bacteria, can be both viable and nonviable and are capable of causing disease in both forms. Most techniques attempt to sample for only viable particles as these can be cultured so that they multiply, making identification easier.

The specialized characteristics of viable agents require specialized sampling instruments in order to preserve the organisms for laboratory culture, which is the primary means of identification. Their fragility and temperature, moisture, and nutrient needs are the primary considerations when selecting a sampling device. While passive air sampling is simple and can be done by setting out plates containing culture media, it is not as effective as the use of active techniques involving the use of pumps. There are two basic methods for collecting these air samples: (1) specialized bioaerosol sampling instruments and (2) air sampling trains incorporating a personal air sampling pump, rotameter and media, such as is used for sample collection device (SCD) chemical sampling. Area air samples are often collected for bioaerosols rather than personal samples, regardless of the type of situation being monitored, due to the need to house culture media inside of instruments specialized for sampling viable bioaerosols.

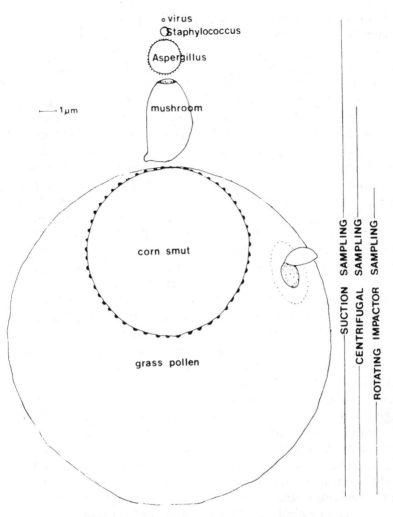

Figure 16.1. Relative sizes of microbial agents. (From reference 9.)

Typically, sampling for bioaerosols related to occupational exposures is performed in hospitals, laboratories, and research facilities; certain industrial operations, such as brewery fermentation, cotton preparation and ginning, wool sorting, hemp handling, and sawmills; and agricultural operations, including hay preparation and the use of biological insecticides and wastewater and sewer treatment facilities. One application is illustrated by a survey for a biological insecticide, *Bacillus thuringiensis*, to characterize exposures to personnel during a large-scale insecticide spraying application.[3] Workers in some sectors of the food industry where fruits and vegetables are processed may also be affected. As an example, slicing sugar beets was found to generate exposures to bacteria originating in soil during beet growth.[4] Other examples of occupational operations where bioaerosol sampling might be done include pharmaceutical manufacturing plants, clean rooms in semiconductor manufacturing plants, animal laboratories, and food processing plants.

Historically most environmental monitoring has consisted of collecting bulk samples of water, especially drinking water and wastewater. Indoor air surveys in buildings incorporate both indoor and outdoor air samples for various agents.

Sampling usually attempts to determine whether the agents are being generated from a source within a building rather than from an outside source where they naturally occur. It has been noted that the majority of fungal spores found indoors are derived from outdoor sources, such as dead or dying plant and animal materials, while the primary source indoors is human bacteria shedding. Sources of microbes include organic materials; humidifiers; vaporizers; heating, ventilating, and air conditioning systems (HVAC); as well as their associated equipment, such as cooling towers.

Another situation of growing interest that incorporates both occupational and environmental exposures is bioremediation of contaminated soils and waters using specially engineered "super bugs." These techniques have proved very successful for treating petroleum compounds. In these situations, the release of volatile organic compounds (VOCs) is also a concern, so an environmental monitoring strategy would need to address both biological and chemical agents.

Air sampling for bioaerosols may also be combined with sampling for chemical agents in situations such as indoor air surveys and exposures to wood dust or bark. In some situations, monitoring is performed for chemicals where there are metabolic byproducts of the organism, such as endotoxin, released by certain bacteria. Biological monitoring of exposed individuals may also need to be performed. For example, blood and urine samples have been collected in cases of suspected Legionnaires' disease, since *Legionella pneumophila* infections cause the release of antigens to the urine.[5] Surface contamination may also be a concern.

BACTERIA

Bacterial infections are the most common type of infection seen in humans, such as those that occur in minor wounds and scratches. Diseases related to bacteria that are found in occupational exposures include those caused by anthrax, also called wool sorter's disease, transmitted by handling imported goat hair, wood, and hides; *Brucella canis* infections (brucellosis) from the contaminated blood of slaughtered animals; and Leptospira-induced disease (leptospirosis), associated with farm animals, dogs, and rodents, which is spread through contact with infected urine, animal tissue, or water.[6]

Other bacteria of concern in sampling within the occupational environment include staphylococcus and streptococcus, which are carried by humans and cause infections under the right conditions; pseudomonas, which cause pneumonia; and bacillus, which is associated with hypersensitivity pneumonitis. Thermophilic bacteria of concern include Thermoactinomyces, known to cause hypersensitivity; and Micropolyspora, Thermomonospora, and Saccharomonospora.[7] Another exposure of concern is the salmonella bacteria responsible for food poisoning. In this case, ingestion rather than inhalation is the route of exposure. While not strictly occupational in nature, this may be a concern in indoor air investigations.

Rickettsiae are intracellular parasites in fleas, ticks, and lice that are considered to be bacteria. The tick is the most common reservoir and tick bites are the primary route of transmission. Rickettsiae do not appear to produce symptoms of disease in their hosts, but if they are transmitted to humans, a severe and often fatal infection may result. The major rickettsial disease of humans is epidemic typhus. Other human diseases are Rocky Mountain spotted fever and Q fever, both transmitted by ticks, and scrub typhus, normally transmitted by mites to field mice, but also transmissible to humans. The chlamydias, other specialized bacteria, are carried in birds and the primary disease they are associated with, ornithosis, is transmitted by inhalation of dried discharges and droppings of birds.

Most bacteria are 1 to 5 μm in size and are classified through their nutritional and growth requirements, shape (round or rod), ability to utilize certain compounds, oxygen needs, metabolites produced, and the type of stain they take in the Gram stain test (positive or negative). There are four main factors that influence bacterial growth: temperature, nutrient supply, availability of stagnant water, and presence of particulate matter to bind to. Temperature is the most important factor for many bacteria. While some organisms, called thermophilic, can exist at elevated temperatures, most bacteria exist in a range of 20°C to 45°C. Nutrients can include nitrites; by-products of other microbials since the growth products of one organism may support another; certain chemical additives often used for water treatment; nonmetallic materials such as insulation, rubber parts, gaskets, and sealants; and atmospheric contamination by dusts and insects. In order to select the proper sampling medium, it will be necessary to know what the specific characteristics are of the bacteria suspected of being present. It has been suggested that gram-positive bacteria are more likely to survive during air sampling than gram-negative. Gram-negative bacteria produce endotoxins that are pyrogenic and induce local inflammatory responses.[8] It is currently thought that endotoxins may be responsible for a number of diseases in workers.[1] Some bacteria, like fungi, can produce spores whose characteristics are discussed further in the section on fungi.

Bacteria are widely present in the environment and the presence of certain odors, slime, and foam on a surface are often an indication of bacterial growth. Certain types live in the human body while others live outdoors on vegetation; therefore, when sampling indoors, those bacteria associated with humans will predominate while outdoor samples will contain mostly the other type. Bacteria are commonly spread indoors from the human respiratory tract, although bacteria that are very small are often dispersed on skin scales.[9] It has been estimated that 7 million skin scales are shed per minute by humans, each containing an average of 4 viable bacteria.[10] Bioaerosols are usually associated with water, and an undisturbed source of water or a humid environment is highly conducive to growth. When water is aerosolized, droplets range in size, but larger droplets can evaporate and become smaller, thus increasing the potential for inhalation. Humidifiers, taps, showers, whirlpool spas, and cooling towers are all source. When bacteria attach to particles, they are often protected against the environment. Table 16.1 lists the factors that affect the survival and dispersion of bacteria in wastewater aerosols.

Bacteria vary in their ability to survive the stresses of collection and subsequently grow under laboratory conditions. As a result, knowledge of the specific bacterium of interest's structure and physiological properties is needed to select the proper air sampling method. The method most commonly used is impaction onto an agar collection surface for airborne concentrations that will not overload the agar surface (typically <10^4 organisms/m³).[1] At higher concentrations, or where the sample is to be diluted as part of the analytical procedure, collection into a liquid using an impinger or wetted cyclone is used. Bacterial endotoxins are collected on filters.

Coliforms

Coliform bacteria are a concern in potable water because their presence indicates possible contamination by sewage or animal wastes. Fecal coliforms and *Escherichia coli*

TABLE 16.1. Factors that Affect the Survival and Dispersion of Bacteria and Viruses in Wastewater Aerosols

Factor	Comment
Relative humidity	Bacteria and most enteric viruses survive longer at high relative humidities, such as those occurring during the night. High RH delays droplet evaporation and retards organism die-off.
Wind speed	Low wind speeds reduce biological aerosol transmission.
Sunlight	Sunlight, through UV radiation, is deleterious to microorganisms. The greatest concentration of organisms in aerosols from wastewater occurs at night.
Temperature	Increased temperature can also reduce the viability of organisms in aerosols, mainly by accentuating the effects of RH. Pronounced temperature effects do not appear until a temperature of 80°F (26°C) is reached.
Open air	It has been observed that bacteria and viruses are inactivated more rapidly when aerosolized and when the captive aerosols are exposed to the open air than when held in the laboratory.

are not considered harmful in themselves, but are analyzed for instead of more difficult to isolate pathogenic organisms. Potable water samples are periodically tested for coliforms, and standing water or other sources may be tested during an indoor air quality (IAQ) investigation. Finding a few of these bacteria during an IAQ survey along with normal human-source bacteria is not a concern. However, finding a significant number of coliforms in an air sample may indicate that standing water has been aerosolized.

Legionella Pneumophila

Legionnaires' disease became a concern when it became apparent that aerosol from contaminated cooling towers could enter fresh air vents and spread this agent to the HVAC system in buildings. *Legionella pneumophila*, the agent in this case, is a rod-shaped, slow-growing, gram-negative bacterium. Ideal growth conditions require a temperature of 35–45°C and a pH of 6.9–7.0. The bacterium is ubiquitous in the environment. It can coexist with amoebae and can survive and grow on blue-green algae. Prevention of Legionella outbreaks

requires careful management of possible sources.

The vast majority of outbreaks have been associated with Legionella sero group 1, but other sero groups, if detected, are also of concern, and control measures should be initiated if they are detected. Legionella sero group 1 has been isolated from a variety of surface and potable aquatic habitats. It is viable in tap water for more than a year. Hot-water tanks, and in particular their bottom sediments, are excellent media for its survival and proliferation.

Cooling towers are especially susceptible to contamination, since their primary function involves inducing large volumes of air into large amounts of flowing water; thus, they act as air scrubbers, washing out dust, debris, pollen, insects, and plant materials. These bacteria have also been known to build up in water softeners.

Most sampling for Legionella is done by collecting bulk samples of suspected sources. One source considers heavy contamination by Legionella to be counts greater than 10 colony-forming units per liter in a bulk sample.[11] Sampling should be done in both suspected and background areas. Samples should not be collected following cleaning or immediately after startup. It is best to sample during a period of normal operation. The best places are system water dead spots or slow-flowing areas; however, these should be selected so they are not near incoming fresh water or biocide treatment points. Testing should precede but not follow slug feed of biocide.

In the course of evaluating the potential for this microorganism to be the source of an outbreak, the operational procedures and facilities are considered as important as the sources of microbial contamination and dissemination. An investigation for Legionella in cooling towers would focus on the following areas[11]:

- *Temperature:* If the normal operating temperature of the water is greater than 30°C, there is a higher risk of bacterial growth.
- *Contamination:* Nonmetallic materials such as washers, coating, gaskets, linings, and sealants in the system can harbor bacterial growth.
- *Stagnation:* If the water is left standing for more than 5 days at a time, there is a higher risk of bacterial growth.
- *Particulate Matter:* The sump can accumulate sludge, debris, scale and bacterial growth.
- *Aerosol Generation:* If there is a significant amount of spray the likelihood of spread is increased, especially if there is any possibility of aerosol escaping from the cooling tower.
- *Susceptible Populations:* If the cooling tower is associated with any buildings or sites occupied by susceptible people, such as hospitals or schools, the risk is increased.

Other factors of importance include whether a competent person is in charge of the cooling system, the type of training provided to the staff responsible for its upkeep, and the availability of adequate record keeping.

If a routine test for Legionella is positive, the cooling tower should be cleaned immediately. Following startup, the system should be resampled within seven days. As Legionella require soluble iron for growth, control of corrosion is an important factor. Often it involves adjusting the pH of the water. Biocides are often used to kill Legionella and other waterborne microbes, but their effectiveness depends on controlling water chemistry including pH, alkalinity, and cycles of concentration (i.e., dose intervals).

Endotoxin

Endotoxin is a distinct lipopolysaccharide (LPS) found in the outer membrane of gram-negative bacteria, and it varies among bacterial types. Endotoxin can be present in several forms: the bacteria, fragments of bacteria membranes incorporated in dust, as well as the endotoxin molecule itself. Endotoxin is considered highly toxic and is suspected of causing pulmonary impairment in humans. Endotoxin has been implicated as having a significant role in the development of byssinosis from cotton dust exposures.[12] It has been found in agricultural, industrial, and office environments.

Endotoxin in air has been sampled using filters attached to personal sampling pumps. Bulk samples are useful when endotoxin is suspected.

Source Versus Air Samples

Typical source samples are liquids from water reservoirs, portions of water-damaged building materials and furnishings, and samples of sludges or biofilms. Clinical samples (e.g., sputum from contagious disease patients) also help to show if the patient is a potential source of airborne bacteria. Source samples for bacteria are valuable for:

- Identifying the presence and type of organisms before undertaking air sampling.
- Collecting fragile or otherwise difficult to collect organisms that probably would not survive the stresses of air sampling.
- Samples of organisms where the likely airborne concentration is very low, and thus identification through air sampling would be difficult.

FUNGUS AND MOLDS

The fungi class includes yeasts, mold, mildew, and mushrooms. *Mold* is a common term for visible fungal growth while *mildew* is a lay term to describe fungi growing on fabrics or bathroom tiles. Soil is the most common habitat of the fungi, although many of the primitive fungal groups are aquatic. Fungi occur on the surface of decaying plant or animal materials in ponds and streams or grow on top of aqueous industrial fluids such as metal-working coolants. They are common in grain-handling facilities, paper mills, fruit ware-houses, and agricultural environments as well as indoor air environments. Their presence in residential and commercial buildings, along with resulting indoor air quality and health concerns is a current important issue. However, fungi have been implicated in occupational health outbreaks for many years.

Fungal species commonly encountered include Aspergillus, which is ubiquitous in the soil and air, especially in agricultural products and in standing water, whose spores are known to cause a variety of pulmonary effects; and Histoplasma and Cryptococcus, found in bird droppings. Penicillium is a mold that grows on damp

organic materials and standing water and is associated with hypersensitivity pneumonitis. *Candida albicans* is a yeast that is ubiquitous and known to cause Candidiasis, a disease of the skin and mouth that occurs in dishwashers, cooks, cannery workers, and others who frequently have their hands in food-contaminated water. In immuno-suppressed individuals it can have systemic effects. Other fungi of concern are Alternaria, Aureobasidium, Chaetomium, Cladosporium, and Mucur.[7]

Fungal-related diseases can be divided into four types:

- Mycosis represents a variety of toxic effects, including dermatitis, hypersensitivity pneumonitis, and some systemic diseases that result from an infection by the organisms themselves.
- Mycotoxicosis is produced by metabolites of various fungi and causes diseases such as toxic aleukia and yellow rice disease.[13]
- Glucans comprise most of the cell walls of most fungi and can have irritant effects similar to, although less potent than, bacterial endotoxin.
- Volatile organic compounds produced by growing fungi may have distinctive odors with low odor thresholds. As a minimum, some people find these VOCs offensive.

Occupations associated with exposure to fungi include sawmill, sugarcane, and cork workers as well as jobs where seeds and textile fibers are handled. Other work environments conducive for the growth and sporulation of fungi are farming, grain handling, mushroom cultivation, insect rearing, and pharmaceutical manufacturing.[13]

Like other microbes, fungi have specific nutritional requirements that vary among the species and produce metabolic products, a classic example being penicillin produced from the mold penicillium. Fungi are also dependent on having water present. The presence of a moldy odor is suggestive that fungi are growing.[1]

A unique stage of some fungus' life cycle is the spore stage consisting of a wide variety of shapes in a very broad size range (<2 to >100 μm). In this stage they form a durable coating over the exterior and become dormant. Spores can be classified by size, morphology, and color, allowing them to be categorized into different taxonomic groups. Since spores are relatively hardy structures, they can survive in dry environments and become airborne when disturbed. Fungal spores are released into the air either by mechanical means, such as wind or other agitation, or biologically by specialized (active) spore discharge mechanisms usually occurring during periods of high relative humidity.[14] When airborne, fungal spores tend to travel as single units.[9]

Certain foods such as peanuts and animal feed contain fungal spores that begin growing and producing aflatoxins when environmental conditions (time, temperature, moisture, nutrients, and pH) are favorable. Aflatoxins are a group of chemically similar compounds known to be acutely toxic and carcinogenic at low doses, and are metabolites of two common fungi: *Aspergillus flavus* and *Aspergillus parasiticus*. If fungal spores are suspected, water reservoirs should be identified and bulk samples should be collected at all suspected sources. Suitable niches for growth and sporulation include stored food, house plants, air conditioners, humidifiers, cold air vaporizers, books and papers, carpets, and damp areas.

The primary air sampling tool used for fungi spores is impaction onto agar to be cultured or an adhesive-coated transparent surface for direct counting (spore trapping). Cultured samples may underestimate actual concentrations due to the spores that do not or cannot grow on the agar. Spore trapping provides an accurate counting of spores and visual identification of some spores. Screening samples can be collected with a centrifugal impactors.

VIRUSES

Viruses represent a unique class of agent and are different from cellular organisms. A virus alternates in its life cycle between two phases: one extracellular and the other intracellular. In its extracellular phase, a virus exists as an inert, infectious particle, or virion. A virion consists of one or more molecules of nucleic acid, either DNA or RNA, contained within a protein coat, or capsid. In its intracellular phase, a virus exists in the form of replicating nucleic acid, either DNA or RNA. Viruses utilize the host cell for replication (reproduction) and thus are intracellular parasites. In the extracellular phase, some viruses are quite stable and resistant to heat and light.

Viral infections may be acquired from vectors such as needles or from handling of animals or animal products and from humans. Laboratory-acquired infections may result from needle sticks; animals; clinical or autopsy specimens; or contaminated glassware. Diseases include rabies, cat scratch disease, and viral hepatitis (both serum and infectious).[6]

Viruses survive best in situations where high humidity and moderate temperatures are present. Situations where water containing a virus is being aerosolized are especially conducive to viral multiplication. Table 16.1 describes factors affecting survival and dispersion of bacteria and viruses in wastewater aerosols.

Collection of viruses often requires very specific techniques, although some have been collected on filters.[15] Viruses have also been collected with the slit sampler and the multistage cascade impactor.

Viruses are usually measured as either infectious units or total particle numbers.[1] While it is important to be aware that viruses may be a cause of an outbreak of illness, it is unlikely that air sampling would be useful in most situations, due to complicated analytical techniques usually requiring that a live species be injected and the degree of specialization necessary to perform these analyses. Instead, a more common technique is to identify the symptoms associated with a suspect virus and determine if the disease is present through a physician's clinical evaluation.[1]

OTHER MICROORGANISMS

Spirochetes have a unique cell structure relative to other bacteria in that they have very long, wormlike bodies. Thus, they can swim in liquid media and are found in mud and water.

Mycoplasmas, the smallest known cellular organisms, are a large and widespread group. The first member of this group to be identified was the agent of bovine pleuropneumonia. Many other species have been isolated from humans and other vertebrates, where they occur as parasites on moist mucosal surfaces. These organisms contaminate tissue cultures, and mycoplasmas have been found in hot springs and other thermal environments, in which case a bulk sample is useful.

Protozoa are large in size compared to other microorganisms and include amoeba. Most protozoa are encountered as a result of contaminated water reservoirs. Naegleria is a protozoan that can infect humans and Acanthamoeba releases antigens that cause hypersensitivity pneumonitis.[6] Protozoa may pose an exposure in wastewater treatment (sewage) workers. Sampling is labor intensive and problematic, so generally a bulk sample of a suspected source is more useful than an air sample for these microorganisms.[1]

SAMPLING METHODS AND STRATEGIES

Although the overall sampling strategic considerations for bioaerosols is similar to other airborne contaminants as discussed in Chapter 3, there are several specific factors that must be considered when developing a sampling strategy and plan:

- Sampling is usually designed to collect *culturable* samples—samples that are collected on or transferred to a specific growth medium that will allow or even foster the growth of some microorganisms while either not supporting or inhibiting the growth of others. The result is likely to be growth patterns that are different from the patterns of viable organisms in the sample under normal environmental conditions, so the sampling and culturing methods must be selected to provide information about the organisms of interest.

- Sampling times for many bioaerosol samples are very short compared to the sampling period for chemical contaminants in order to prevent overloading the sample device. Sampling periods of from one to 30 minutes is typical. With these short duration samples there is a greater chance that variable concentrations either between locations in the occupied space or over time will be missed unless a sufficient number of samples are collected. Another option may be to collect samples in a medium that can be serially diluted such as a liquid-filled impinger or a membrane filter.

- For most bioaerosols there are not specific airborne standards (or specified sample collection methods). This makes it imperative for the sampling practitioner to have a clear understanding of the reason for the sampling and also to work closely with the lab-

TABLE 16.2. Summary of Bioaerosol Air Sampling Parameters

	Air Sampling Method						Typical Analysis					Data Obtained				
	Filter	Impactor	Impinger	Wetted cyclone	Cyclone	Sorbent Tube	Culture	Microscopy	Bioassay	Chemical Analysis	Biochemical Assay	Concentration	Species Identification	Species Confirmation	Biological Activity	Toxic Activity
Amoebae		✓	✓					✓	✓			✓	✓	✓		
Bacteria	✓	✓	✓	✓			✓	✓	✓			✓	✓			
Endotoxins	✓									✓		✓				
Whole-cell lipids, phospholips	✓		✓							✓		✓			✓	
Fungi	✓	✓	✓	✓				✓				✓	✓			
Yeasts		✓	✓	✓				✓			✓		✓			
Cell-wall components	✓									✓	✓	✓			✓	
Fungal toxins	✓	✓	✓	✓						✓	✓	✓				✓
Viruses	✓	✓	✓		✓			✓	✓		✓	✓	✓	✓		
MVOCs[a]						✓				✓		✓	✓	✓		

[a] MVOCs, microbial volatile organic compounds.

Source: Excerpted from reference 1.

oratory that will be analyzing or counting the samples.

Often bioaerosol sampling is designed to meet one or more of these goals:

- Determine whether the relative concentrations of certain microorganisms are higher within the building or area under study as compared to outdoor concentrations.
- Identify whether certain microorganisms are present at all.
- Determine if potential "reservoirs" or "amplifiers" of microorganisms, as identified through visual inspection, patient cultures, or other evaluation techniques, are actually resulting in increased airborne concentrations.

New collection and analytical techniques are continually being developed for bioaerosols. Table 16.2 outlines typical sampling methods, analytical techniques, and data output for bioaerosol sampling. The best way to select a monitoring technique is to review the literature and discuss the specific situation under evaluation with knowledgeable staff at qualified analytical laboratories.

Selection should consider sampling location(s), type of agent, expected concentration, analytical methods, and organism characteristics. The type of agent will dictate whether results will be expressed as the total number of viable particles (colony-forming units) or the total number of individual viable cells.

Selection of an air sampling method will depend on the type of organism being sampled as well as on the expected concentration. Viable bacteria and fungi including yeasts and molds are most commonly sampled using agar culture media contained in a variety of different sampling instruments or in impingers containing nutrient solutions. Nonviable agents such as spores are usually collected on adhesive surfaces or filters. A rule of thumb is to limit the use of culture plate samplers to environments where concentrations are expected to be less than 10,000 organisms/M[3].[1] Filters have been recommended for situations where high concentrations of microorganisms are suspected[16] since the microbes can be rinsed from the filter and serially deleted for analysis. The selection of culture media and conditions such as temperature and humidity can determine which organisms grow and which do not, because organisms vary significantly in culture requirements. Media are usually selected for their ability to culture virtually every organism of a certain type that might be present for screening samples or to selectively grow only certain species for identification samples.

Malt extract agar is recommended for the general detection of fungi while agar containing casein peptone, soy peptone, and sodium chloride is used similarly for bacteria.[7] Trypticase soy agar has been used to collect and support the growth of both fungi and bacteria.[17] In order to make a culture media more specific, antibiotics and other compounds may be added to inhibit the growth of certain microbes. For example, inhibitory mold agar will promote fungi growth. Rose-Bengal-Streptomycin has also been recommended as a useful medium for fungal sampling as it yields high colony counts and impedes colony spreading; however, it is sensitive to direct sunlight and must be shielded when sampling outdoors.[13] All media should be prepared within days of use. Poured plates should be stored upside down and any condensate that accumulates in the lid should be shaken out before use.

Ideally samplers to be used for viable microorganisms should be sterilized before each use. Most samplers commonly used with culture media can be swabbed with isopropyl alcohol or bleach solution before each use. All glass impinger fluid should be sterile and the impinger should be rinsed with sterile fluid before use. Table 16.3 describes samplers recommended for viable bioaerosols.[7]

Analytical methods are direct, meaning the organism or its toxin is identified, or they are indirect in that a skin bioassay or immunological assay on blood for an allergic reaction to that organism is performed and considered evidence that an individual has been exposed. Direct methods include microscopy that may be optical, scanning electron microscopy (SEM), or epifluorescence (requires that the sample be stained) and culturing followed by a variety of chemical tests for specific properties of microbes. Biochemical tests can be performed to identify metabolites, especially those considered toxic.

Prior to sampling, an appropriate laboratory must be selected for analyses. Many clinical laboratories specialize in medical samples such as throat cultures and other human cultures and do not have experience in analyzing environmental air samples. Some laboratories specialize in mycology, the study of fungi, while antigen, endotoxin, and virus sampling often are still research techniques, and arrangements for analysis must be made with laboratories conducting research in a given area. Often a review of the literature will provide information on laboratories specializing in various techniques.

TABLE 16.3. Samplers Recommended for Viable Bioaerosols

Sampler	Principle of Operation	Sampling Rate (Lpm)	Rec. Sample Time (min)	Min. CFU Detected
Slit impactor	Impaction onto adhesive-coated slide	10	15	N/A
N-6 Single-stage impactor	Impaction onto agar in 100-mm culture plate	28	1	35
2-Stage impactor	Impaction onto agar in 2–100-mm culture plates	28	1	35
Filter cassettes	Filtration	1–2	15–60	8–33
Glass impingers	Impingement into liquid	1.5–2.5	30–60	5–25
Centrifugal impactor	Impaction onto agar strips	40	$\frac{1}{2}$	50

Source: ACGIH BioAerosols Committee. Guidelines for assessment and sampling of saprophytic bioaerosols in the indoor environment. *Appl. Ind. Hyg.* **2**(5):R-10–R-16, 1987.

Colony counts should be done according to the manufacturer's specifications for the sampling instrument used. If using a new laboratory, these specifications should be provided along with the samples. Samples for bioaerosols associated with allergic illnesses, such as hypersensitivity pneumonitis and asthma, must be sent to a laboratory proficient in growing and identifying environmental molds. Clinical hospital laboratories typically have little experience in identifying environmental fungi and therefore may not be the best analytical choice.

Sampling practitioners should be aware that not all microbes can be cultured, and some are extremely small and difficult to see with an optical microscope. In the case where sampling is being done for a specific strain, a nonspecific method may lead to confusion in the situation where other interfering strains may also be present. Identification of specific strains can be time consuming and difficult for laboratories and is not always necessary to identify what controls are needed. In some cases, rather than conducting extensive sampling for bioaerosols, it will be better to identify possible sources of microorganisms such as stagnant water and continuously damp organic materials, and correct the source of

moisture or humidity. Any organic material may support mold growth when wet, including leather, cotton, paper, carpets, furniture, and furniture stuffing.

Sampling strategies for bioaerosols depend on the situations being sampled. When identifying the source of disease outbreak, sampling must be done as soon as possible following the outbreak. In this case, the best strategy may be to have affected individuals examined by a physician and specific sampling should be performed to identify the source of exposure once likely pathogens have been identified.

Viable sampling indoors will be affected by HVAC filtration units, and seasonal variations in loads and types of microorganisms entrained from the outdoor air.[18] More variations will occur when samples are collected outdoors rather than indoors, due to differing conditions, such as climate changes and unpredictability of the wind. Other variations can also affect sample results. For example, in one study on bacteria in grain dust, it was found that *Enterobacter agglomerans* was the predominant species in warm months, but in winter it was Pseudomonas and Klebsiella species.[8]

One strategy suggested for situations where a bioaerosol is suspected but sampling must be performed for a broad range of organisms is as follows[9]: Collect screening samples with a slit impactor, followed by sampling with at least three cascade impactors, each containing a different type of culture medium for bacteria, and filters for spore collection and examination. Identify specific bioassays or biochemical tests for any suspect toxins and allergic reactions, and have exposed individuals tested.

For indoor air studies, sampling should be performed during work periods when occupants are present and also when the building is empty. A concern is that during off-periods, such as evenings, ventilation systems are often set to 100% recirculation air whereas during the workday a certain percent of outdoor air is utilized. In the indoor environment, samples should be collected in the supply and return air of rooms housing the affected occupants, as well as in other rooms where occupants have no complaints. Outdoor samples should be taken close to the intakes of HVAC systems when they are open and, if applicable, in the vicinity of potential bioaerosol sources such as cooling towers, stored organic material, and dense vegetation. Air samples should also be collected at an outdoor site remote from obvious sources. It is essential to sample at different times during the day and, if at all possible, on different days. Seasonal sampling may be required if the sampler is dealing with a past event. The following sequence is one option for collecting samples for indoor air situations[19]:

1. Sample before the HVAC system is turned on and before occupants arrive for work.
2. Sample before occupants arrive for work at a time when the HVAC system is operating.
3. Sample at the time of maximal occupancy with the HVAC system on.
4. Sample after occupants leave with the HVAC system off.

5. Sample in relation to normal changes in energy conservation that affect the HVAC system.

Humidity or the lack of it is important to measure when sampling microbials for indoor air exposures, since virtually all microbes require a consistent source of moisture for growth. While growth can be supported at between 25% and 75% relative humidity (RH), greater than 70% is considered optimal for growth.[10] Since buildings are often equipped with humidifying facilities along with heating and cooling systems, these can become infested with a variety of molds and other microbials that can release spores to the ventilation air when the systems cycle from wet to dry operation, so bulk samples should also be collected.[20] Indoor air surveys are discussed in greater detail in the chapter on specific sampling situations (Chapter 17).

Guidelines for Indoor Mold Evaluations

Because of the current interest in indoor mold problems, this section provides more detailed information on this topic. As with any dynamic issue, the sample practitioner should obtain the latest information before planning and carrying out a sampling project.

Most building problems are due to molds proliferating within the building rather than molds from outdoor sources. These molds usually grow in locations where specific conditions such as moisture and substrate materials (wallboard, carpet, insulation, debris, etc.) favor their proliferation. In many cases, any air sampling is performed to verify that visually identified mold sources are actually causing increased airborne levels. However, where the source is diffuse and visual inspection is difficult (such as possible wet insulation within ventilation system ducts), air sampling may be the most practical way to first identify the existence of a mold reservoir.

Air monitoring generally should be aimed at collecting *culturable* samples that will permit differentiation to the fungal species level. Any of the sampling devices described later in this chapter may be used for mold monitoring; impaction is probably the most common technique in use today. Remember that absolute accuracy in measuring airborne concentrations is not always required when the air monitoring is looking for relative levels in different locations or whether a specific fungus is present at all. Typical indoor fungi to be identified in air samples include:

- *Cladosprium* spp.
- *Alternaria* spp.
- *Aureoblasidium* spp.
- *Pencillium* spp.
- *Mucor* spp.
- *Stachybotrus* spp.
- *Aspergillus* spp.
- Other fungi present in high concentrations or present in problems areas but not in nonproblem areas of the building or outdoors.

Not all of these fungi can be cultured on the same growth media, so careful planning is necessary to ensure the sampling will yield the desired information.

Surface swab, contact plate, vacuumed and actual material samples from carpets, upholstery, duct insulation, and similar materials will not reflect airborne levels, but usually contain a long-term record of fungal growth deposition and can overcome the variability problems of short-term air samples.

In addition to culturable samples, special air samples for bioassays or to identify volatile organic compounds released from fungi may be appropriate in some situations. Refer to a current reference on bioaerosols[1] for more specific guidance.

Settled Dust Samples

Passive methods were the first attempt to collect airborne microorganisms and are still used. In setting out culture plates, otherwise known as settling plates or gravity sampling, dishes of culture media or adhesive coated glass slides are placed in various areas and left open for air exposure for a number of hours. Then they are incubated and the number of colonies is counted. These samples are greatly influenced by particle size and air movement. Because of these factors settled dust samples cannot be considered to reflect airborne particles.

If settled dust samples are collected, extreme care must be taken to locate them away from supply and exhaust (return) vents, windows, sitting areas, and heaters. The best application is as a screening technique when instruments for active air sampling are unavailable. Figure 16.2 demonstrates problems with settling methods.

If setting plates are left near HVAC air vents, open windows, or other sources of air currents, the large : small particle ratio will change even more. Also they are susceptible to outdoor contamination if left near open windows and to human contamination if too close to areas where persons are present.

Active Air Sampling

Active methods use a pump or other air-moving equipment to draw ambient air into the sample collection device. These methods are used for almost all bioaerosol sampling. Most of the equipment and techniques follow the *sample collection device* methods for particulates described in Chapter 7 except that modifications are needed to collect organisms in a manner that allows them to be cultured in the laboratory or counted microscopically.

Although generally not a problem for active sampling indoors, wind speed and

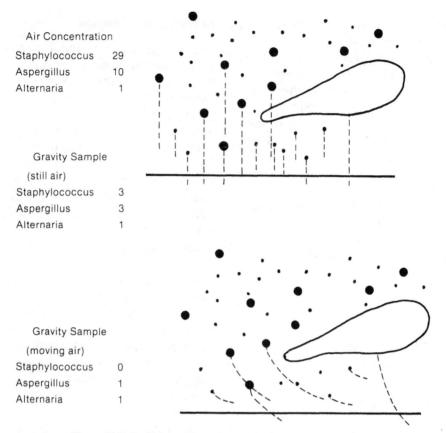

Air Concentration

Staphylococcus 29
Aspergillus 10
Alternaria 1

Gravity Sample

(still air)

Staphylococcus 3
Aspergillus 3
Alternaria 1

Gravity Sample

(moving air)

Staphylococcus 0
Aspergillus 1
Alternaria 1

Figure 16.2. Problems with settling methods. (From reference 9.)

direction must be considered when sampling outdoors. Incident airflow should be parallel and going in the same direction as the suction flow of the sampler.[9] All sample volumes should be adjusted to provide a sensitivity near 50 CFU. Sample times at each site should be varied from this maximum time/volume to prevent overloading, especially when culture plate samplers are used, and to allow for the logarithmic variability that these organisms display.[7] Compared to the collection of chemical samples, sampling periods for bioaerosols are much shorter, especially when viable organisms are being collected. Table 16.3 details sampling characteristics for some common types of bioaerosol samplers.

Inertial Impactors. Inertial impactors, such as the Andersen-type cascade impactor, can be used to separate viable particles into respirable and nonrespirable particles. Impactors for viable microbe sampling are available from several manufacturers. The Andersen-type viable sampler was first recommended as a standard microbial sampler around 1970. Inertial impactors, like all samplers relying on collection on culture plates, are used for situations where airborne counts are expected to be low, such as for indoor sampling.

There are three sizes of impactors designed for collecting microbials: 6-stage, 2-stage, and single-stage impactors (Figures 16.3 and 16.4). When impactors other than the Andersen are used, their collection effi-

Figure 16.3. Six-stage inertial impactor viable sampler for microbials. (Courtesy of Thermo Electron Corporation.)

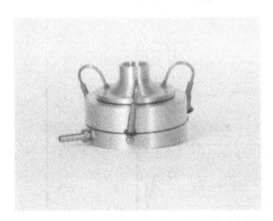

Figure 16.4. Single-stage Bio-Aire™ B6 inertial impactor viable sampler for microbials. (Courtesy of A.P. Buck, Inc.)

ciency relative to this unit should be known, because the Andersen instrument has been in use for so long. This can sometimes be done by consulting the literature. If the efficiency is significantly different, the data should be adjusted. A personal-size impactor has also been used for personal microbial sampling. The principle of collection is the same as it is for other impactors. For more information on the use of this device, see Chapter 7.

The basic mechanism for collection by single and multistage inertial impactors is as follows (Figure 16.5): As air is drawn through the sampler, it passes through the orifices where the velocity of the air increases in a manner that is inversely proportional to the area of the orifice: the smaller the orifice, the higher the velocity of the air. When the velocity imparted to a particle entrained in the airstream is sufficiently great, its inertia will overcome its aerodynamic drag, and the particle will leave the turning airstream to be impacted on the agar surface. If the particle does not achieve a velocity sufficient to cause impaction, it remains entrained in the airstream and proceeds to the next stage.

Each stage in the 6-stage Andersen impactor contains a plate with 400 orifices. The orifice size is constant for a given stage, but decreases for each following stage. The 6-stage sampler separates particles into six aerodynamic ranges (Table 16.4). Directly below each stage is a glass petri dish containing a nutrient agar medium. The stages are held together by three spring clamps. Air is pulled through at a rate of 28.3 L/min (1 ft³/min).

Each stage of the two-stage unit contains 200 tapered orifices. The diameter of the orifices on the first stage is 1.5 mm, and on the second stage it is 0.4 mm. Directly below each stage is a standard 100-mm by 15-mm disposable plastic petri dish that will hold various culture mediums. The stages are held together by three dowel pins with Teflon caps. The particles on the first stage are considered nonrespirable, and those on the second stage are respirable. This sampler also requires a flow rate of 28.3 L/min.

When compared with the 6-stage unit, the 2-stage model has been determined to have decreased counts.[21] It has been suggested that smaller particles of 0.65 μm to 1.1 μm may not be collected by the 2-stage model due to its larger orifice sizes, whereas they are collected by the fifth and sixth

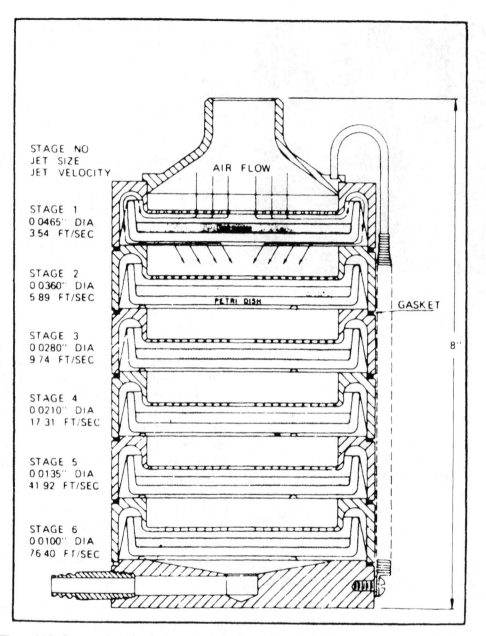

STAGE NO
JET SIZE
JET VELOCITY

AIR FLOW

STAGE 1
0 0465" DIA
3 54 FT/SEC

STAGE 2
0 0360" DIA
5 89 FT/SEC

PETRI DISH

GASKET

STAGE 3
0 0280" DIA
9 74 FT/SEC

STAGE 4
0 0210" DIA
17 31 FT/SEC

8"

STAGE 5
0 0135" DIA
41 92 FT/SEC

STAGE 6
0 0100" DIA
76 40 FT/SEC

Figure 16.5. Cross section of a six-stage viable impactor showing air flow pattern through the device. (Courtesy of Thermo Electron Corporation.)

stages of the 6-stage model. It may only be a problem for certain types of microbes of very small size. Another observation is that the 2-stage sampler has one half the number of orifices on each stage as the 6-stage sampler. Therefore, the probability of multi-particle impaction through the same orifice is greater than for the 6-stage impactor. Since each orifice point is only counted as one CFU, the result in the 2-stage readings would be expected to be consistently lower than the 6-stage readings.

TABLE 16.4. Particle-Size Ranges for the Six-Stage Andersen Sampler

Stage	Orifice Diameter (mm)	Range of Particle Sizes (µm)
1 (top)	1.18	7.0 and above
2	0.91	4.7–7.0
3	0.71	3.3–4.7
4	0.53	2.1–3.3
5	0.34	1.1–2.1
6 (bottom)	0.25	0.65–1.1

The single-stage sampler (Figure 16.4) is useful in situations such as large office buildings where it is important to sample simultaneously in many areas of a building in order to determine the relationship of the ventilation system and its variables with levels of viable microbes in various areas. Two advantages of the N-6 method are that shorter sampling times and less collection media are needed. There is a tendency for the very small holes of the lower stage to become plugged, however, and for the single collection plate to be overloaded.

In one study involving 6-stage Andersen samplers to collect airborne bacteria and fungi, 4 minutes was the ideal sampling time for a satisfactory count for total microorganisms to assure that a sufficient number of microorganisms had reached all six stages.[22] For respirable samples, defined as the amount of microorganisms that deposit on stages 3 through 6, 20 minutes was required. The average flowrate was the 28.3 L/min rate. One method for decontamination of Andersen samplers between uses is to wash with 70% ethanol and follow up with acetone to dry.[22] Excess cleaning solution should be collected for proper disposal.

Most inertial impactors for bioaerosol sampling are constructed of aluminum. Stainless steel substrates (plates) are often used for viable particles because they can be sterilized more easily (wrapped in paper and autoclaved) or placed in boiling water for 10 minutes and stored in sterile petri dishes.

For Andersen-type inertia impactors, corrections must be made for multiple impactions at single holes. No more than one fourth of the holes should have multiple impactions, especially for fungus counts, to avoid inhibition effects resulting from overcrowding. Statistical charts are available that show these corrections.[1]

Procedure for N-6 (Single-Stage) Sampler

1. Remove the inlet cone and jet classification stage by gently lifting and pulling the three spring clips out of the indentations surrounding the cone.

2. Clean all surfaces of the sampler by wiping all surfaces with isopropyl alcohol on a sterile gauze pad.

3. Remove the lid from an agar plate. Typical agars include (a) Tryptic Soy Agar (TSA) or Tryptic Soy Agar with 5% Sheep Blood for bacteria and (b) Potato Dextrose Agar (PDA), Malt Extract Agar (MEA), or Dichloran Glycerol 18 Agar (DG-18) for fungi.

4. Place the agar plate on the base plate of the sampler so that the agar plate rests on the three raised metal pins.

5. Replace the jet classification stage and inlet cone. Secure by placing the three spring clips into the indentations surrounding the inlet cone. Visually inspect for a proper seal.

6. Calibrate the sampler by connecting a rotameter or other flow meter inline between the sampler outlet and the pump. Turn on the pump and adjust flow until the rotameter reads 28 L/min (see Chapter 5 for

calibration instructions). Turn off the pump.

7. After calibrating the pump, place a fresh agar plate in the sampler.

8. Turn on the vacuum pump and sample for an accurately known time (e.g., 10 minutes).

9. After the sampling time is complete, turn off the pump and disconnect the flexible tubing.

10. Unhook the three spring clips and remove the jet classification stage and inlet cone. Remove the agar plate and replace its cover immediately.

11. Label the bottom of the plate with all pertinent information and place in a sealable plastic bag. Store the collection plate in an ice chest with coolant. Send the sample and a blank unexposed plate to a laboratory to be analyzed within 24 hours.

12. Outdoor samples should be collected for comparison to indoor samples. An indoor control sample should also be taken for noncomplaint areas. Clearly mark each plate.

13. In between sampling, sanitize hands and the impactor. Also, never use sampling media that has expired, displays visible cracks, or has been contaminated.

14. Air should not be drawn through the sampler unless the petri dishes are inplace; otherwise, dirt may lodge in the small holes of the lower stage.

If it is desired to know the number of viable cells in each particle, as well as the number and size of particles collected by the sampler, duplicate samples should be collected. One set of plates is used for particle counts and the other set for cell counts. The cell count may be obtained by immediately washing the collected mater-

ial from each plate into a flask, shaking the flask vigorously, and pouring the contents through a membrane filter. The plates are counted by selecting a number of fields, and then counting the total number of colonies in these fields. In stages three through six, the colonies conform to the pattern of jets, and are counted by either the positive-hole method or by the microscope method. The positive-hole method is essentially a count of the jets that delivered viable particles to the petri plates, and the conversion of this count to a particle count by the use of a positive-hole conversion table.

CFUs are calculated by the following:

$$\frac{\text{CFU}}{\text{M}^3} = \frac{\text{adjusted number of colonies on plate}}{\text{total volume of air sampled in M}^3}$$

Slit Samplers. The Burkard personal volumetric sampler is an example of a slit sampler, also considered a type of inertial impactor for collecting screening samples of bioaerosols (Figure 16.6). The unit is a battery-powered air sampler that can collect area samples directly on microscopic slides. An impaction orifice is located on top of the unit. A slide is slipped in through the side of the sampler with its collection surface facing up and air is pulled into the unit at a rate of 10 L/min. The batteries must be fully charged to ensure this flow rate.

The Burkard sampler is useful for identification of specific sources of microbes. As an example, in an indoor air situation where occupants on only one side of a building are complaining, samples can be collected in several representative rooms in a number of different areas for comparison over the same day. Each sample must be sealed in a container immediately following collection to avoid contamination.

Clean glass slides can be used without adhesive if the final sample will not be preserved. However, an adhesive is often used.

Figure 16.6. The Burkard Personal Volumetric Air Sampler is a slit impactor. (Courtesy of Burkard Manufacturing Company, Ltd.)

the glass slide through the aperture, adhesive side up, until the end of the slide is firmly against the stop inside the unit. Close the sampling chamber by rotating the upper knurled ring by at least 1 inch. Make sure the slots are fully covered.

3. Turn on the unit. When warmed up, it draws 10 L/min. Normal sampling time is 15 minutes.

4. At the end of the sampling period, open the ring and push the slide out.

Burhand also makes slit impactors that utilize adhesive tape strips to collect samples over longer time periods.

The SKC Air-O-Cell Cassette™ (Figure 16.7) is a slit impactor contained in a small "filter cassette" unit. It can collect particles such as mold spores, pollen, insects, skin cell fragments, and plant fragments as well as general dusts and fibers. The small, sealed filter cassette contains a special high-tack sampling medium on a slide that permanently fixes particles for easy sampling, transportation, and analysis.

Particles in the ambient air entering the device accelerate to a minimum velocity of 40 ft/second through a tapered slit. Below the slit is a slide coated with a sticky sampling medium. The airflow is aimed directly at the sampling medium and then forced to make a 90° turn at the surface of the slide. The velocity, combined with the mass of the particles in the air stream, causes the particles (up to 2 μm) to impact and adhere to the sampling medium instead of following the air stream to the exit at the back of the cassette.

The Air-O-Cell Cassette operates at flowrates from 5 to 30 L/min, with an optimum flowrate of 15 L/min. Any air sample pump capable of producing a constant 15 L/min flow can be used with the Air-O-Cell Cassette. Flows less than 15 L/min may result in a loss of particles since impaction will be lessened. Conversely, higher flows may damage the morphology

A water-soluble plastic material is often used as an adhesive based on the manufacturer's or laboratory's recommendations. A number of factors are involved in the selection of an adhesive for the slides: (1) stickiness, since the surface should be as wet as possible, especially for sampling spores; (2) weather resistance, since fog or high humidity will cause adhesives to slough off or to be emulsified; (3) the type of microscopy to be used on the samples including stains and mounting agents.

Maintenance of the device involves cleaning the sampling orifice to remove contaminants and obstructions to airflow.

Procedure

1. Prepare the glass slide with a suitable adhesive.

2. Rotate the upper ring assembly until the red dots are in line; this exposes the aperture for the glass slide. Insert

Figure 16.7. The SKC Air-O-Cell Cassette™ is a slit impactor contained in a small "filter cassette" unit for collection of mold spores, pollen, insects, skin cell fragments, and plant fragments as well as general dusts and fibers. (Courtesy of SKC, Inc.)

of some materials (e.g., pollen, mold spores). Typical sampling times at 15 L/min are 10 minutes for "clean" buildings or outdoor locations, 5 minutes for indoor environments with high activity or number of people, and 1 minute where airborne levels are expected to be high such as drywall renovation. A special configuration can be used for nondestructive testing of air in wall cavities to assess for mold growth.

Rotating Impactors. The Biotest RCS sampler (Figure 16.8) is an example of a rotating impactor. A multibladed impeller draws air into the sampler, depositing microbes on a plastic strip containing agar culture media lining the inside of the impeller housing. After the samples are collected, the strips are incubated and the colonies are counted. This device is a compact field sampler that is battery-operated, and samples at 40 L/min. There are 5 preset sampling times: 30 seconds and 1, 2, 4, and 8 minutes.

An advantage of this sampler is its ability to collect samples to screen for a number of different organisms by collecting sequential samples on strips that contain different types of agar. The follow-

Figure 16.8. The Biotest RCS sampler is a rotating impactor. (Courtesy of Biotest Diagnostics Corporation.)

ing types of agar are recommended by the manufacturer:

- *TC (Tryptic Soy Agar).* A standard USP formulation designed to promote the growth of bacteria, yeast and mold. Incubation is typically at 30–35°C for 1–3 days.
- *TC-γ (γ-Irradiated Modified TSA).* A modified TSA formulation that is sterile, double wrapped and gamma-irradiated, for use in aseptic and contamination-controlled environments. Incubation is typically at 30–35°C for 1–3 days.

- *TSM (Modified TSA with Neutralizers).* A TSA formulation with lecithin, Tween 80, and supplements to support growth of sublethal damaged airborne microorganisms from ultraclean environments. Incubation is typically at 30–35°C for 1–3 days.

- *TCI-γ (γ-Irradiated Modified TSA).* A modified TSA formulation with H_2O_2 neutralizers. TCI-γ is sterile, double-wrapped, and gamma-irradiated, for use in isolator monitoring and/or cleanrooms. Incubation is typically at 30–35°C for 1–3 days.

- *YM (Rose Bengal Agar).* A medium designed for selective enumeration of yeasts and molds. Rose Bengal is present to restrict (not inhibit) the growth of Rhizopus and Mucor that can overgrow the strip. Streptomycin is present to inhibit bacterial growth. Incubation is typically at 20–25°C for 3–7 days.

- *SDX and SDX-γ (Sabouraud-Dextrose Agar, Modified).* A medium used for the detection of fungi. Sabouraud-Dextrose Agar is an acceptable media for the detection and enumeration of yeasts and molds as indicated in the United States Pharmacopoeia. SDX should be used routinely for the recovery of yeasts and molds. Incubation is typically at 20–25°C for 3–7 days.

- *C (MacConkey Agar).* A medium for the isolation and differentiation of lactose fermenting gram-negative organisms (coliforms) from lactose nonfermenting gram-negative enteric bacteria. Fermenters will appear red while nonfermenters will remain colorless and translucent. Gram-positive organisms are largely inhibited due to the presence of crystal violet and bile salts. Incubation is typically at 30–35°C for 1–3 days.

- *S (Mannitol Salt Agar).* A medium for the selective isolation of pathogenic staphylococci. Pathogenic staphylococci that coagulate rabbit plasma produce colonies with yellow zones. Nonpathogenic staphylococci produce small colonies with no color change of the medium. Other bacteria are generally inhibited by a concentration of 7.5% sodium chloride. Incubation is typically at 30–35°C for 1–3 days.

- *PEN (Tryptic Soy Agar with Penase).* A standard USP formula supplemented with 160,000 L.U. of penicillinase. This medium is used by penicillin manufacturers for the purpose of neutralizing any airborne penicillin captured during air sampling. This allows for a more accurate assessment of the air quality. Incubation is at 30–35°C for 1–3 days.

- *DG18 (DG-18 Agar).* A specialty media containing 18% glycerol. It is used for the recovery of xerophilic yeasts and molds. DG-18 should be used in very dry locations such as dry powder storage and manufacturing areas. DG-18 will result in a higher yield compared with YM. Incubation is typically at 20–25°C for 3–7 days.

- *Blank Strips.* Sterile, blank strips that can be filled with a prepared custom agar. Each strip requires a volume of 9 mL of medium.

The sampling time should be measured monthly. The 8-minute sampling time is the one most often checked. The manufacturer recommends that it not exceed or be less than 480 ± 9.6 seconds, which is an accuracy of ±2%. If the result is not within this limit, the batteries should be changed. If the result is still not within this limit following this change, the unit must be sent in for electronic calibration.

The primary application of the Biotest is as a probe for hot spots, in order to screen areas to determine those with the highest concentrations. Like all samplers relying on collection on a culture plate, the Biotest is limited to situations where airborne counts are likely to be low, which is generally for indoor sampling.

Procedure for Use of the Biotest Sampler

1. Prior to use, the impeller drum should be removed from the unit and sterilized. It can be done by boiling the unit in water or soaking it in isopropyl alcohol.

2. Select the sampling time. The instrument provides five selections for sampling times that yield a sample volume of 20, 40, 80, 160, or 320 L. The sampler should not be used more than 8 minutes (320 L) for any one agar strip as excessive desiccation and microbial death will result. Sampling time is selected by setting the switches in a variety of configurations.

3. Remove the agar strip from the wrapper and inspect it for contamination. Do not use a strip with growths on it. Handle it by the edges only.

4. Insert the strip into the slot in the open-end drum with the agar surface facing toward the impeller blades (inside). Continue inserting until the slot is closed by the strip and the tab protrudes about 2 cm (Figure 16.9).

5. Switch on the instrument and check that the control light is on.

6. Press the start button to begin sampling. Depending on what is being sampled, such as the vent of an HVAC system, it is important to keep the unit placed correctly. Even though it is light weight, after 8 minutes it can be difficult to hold in place. After the drum stops turning, move the main switch to off.

7. Remove the agar strip by holding its tab and gently pulling it out of the impeller drum (Figure 16.10).

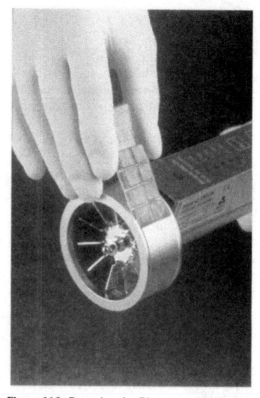

Figure 16.9. Preparing the Biotest sampler for use. (Courtesy of Biotest Diagnostics Corporation.)

Replace it in its original wrapper with the agar surface facing away from the lid (face down). Seal the wrapper by sliding the plastic seal over the top to prevent the media from drying out during incubation. Label the holder immediately using an indelible felt tip pen or grease pencil with a unique number.

8. Calculations are done following the count.

$$CFU/L = \frac{\text{colonies on agar strip}}{40 \times \text{sampling time in minutes}}$$

$$CFU/M^3 = \frac{\text{colonies on agar strip} \times 25}{\text{sampling time in minutes}}$$

$$CFU/ft^3 = \frac{\text{colonies on agar strip} \times 0.708}{\text{sampling time in minutes}}$$

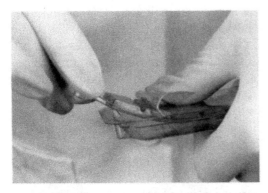

Figure 16.10. The agar strip sampling media for the Biotest sampler. (Courtesy of Biotest Diagnostics Corporation.)

Impingers. Liquid impingment collects microbes directly into a liquid, providing some protection for the microorganisms, and allows immediate initiation of repair of any damage that may have been caused by the collection process itself. Impingers are described in Chapters 5 and 7. The sample liquid can be processed to detect a wide range of aerosol concentrations. It can be sampled directly or filtered with all particles removed from the total volume. Impingers can be used in situations where high concentrations of microorganisms are suspected as the solutions can be diluted prior to culturing.[1] An advantage of having the sample in a liquid is that several types of media can be inoculated from the suspension or several plates of the same medium can be inoculated and incubated under different conditions.[23] Aggregates of cells that would grow as single colonies on an impactor sampler are broken up in the impinger liquid, allowing each cell to develop into a separate colony. Calculation of inhaled doses from this measurement of air concentration may be preferred for some disease agents, especially those for which a small number of organisms constitutes an infective dose. A spillproof impinger has been used successful to collect microbials with the same efficiency as the standard all-glass impinger and

showed more efficient recovery of hardy spores.[2]

Impingers are recommended for bacteria collection as long as appropriate media are used. In general, impingers are useful for the recovery of soluble materials, such as mycotoxins, antigens, and endotoxins, and for sampling aerosols of bacteria and viruses that require gentle handling.[9] All-glass impinger fluid should be sterile and the impinger should be rinsed with the sterile fluid prior to use.

Procedure[18]

1. Prepare a sampling train with a portable pump and a sterile glass midget impinger containing 10 mL of sterile phosphate buffer solution.

2. Collect a sample at a flow rate of 2.0 Lpm ± 0.5. Sampling time will be dependent on an estimated bacterial concentration.

3. After sampling, cap the impinger openings and return to the laboratory within 30 minutes.

4. Aspirate the sample through an opened Millipore sterile clinical field monitor (catalog no. MHWG037HO) with a 0.45-µm pore-size gridded filter inside.

5. Rinse the empty impinger with 10 mL of sterile phosphate buffer solution, and aspirate this washing through the filter.

6. Add 1 ampoule of bacterial media to the filter by aspiration.

7. Cap the filter cassette, and incubate the cassette sample in a 30°C incubator for 2–3 days.

8. Count each colony using a stereoscopic microscope.

9. Calculate the volume of air sampled, and report as CFU/M^3.

Special impinger-like devices have been developed for bioaerosol sampling that

Figure 16.11. The SKC BioSampler™ is a special impinger-like device featuring three tangential nozzles that collect airborne particles that would pass through the human nose. Its design reduces particle bounce off of the inner wall helping to ensure bioaerosol viability. (Courtesy of SKC, Inc.)

overcome some limitations of the standard impinger design. For example, the SKC BioSampler™ (Figure 16.11) features three tangential nozzles that act as critical orifices, each permitting 4.2 L/min of ambient

air to pass through for a total of approximately 12.5 L/min. The design limits the collection of airborne particles to those that would pass through the human nose and also reduces particle bounce off the inner wall helping to ensure bioaerosol viability. The sampler is normally used with a liquid that swirls upward on the sampler's inner wall and removes collected particles. This gentle swirling motion generates very few bubbles, thus minimizing re-aerosolization of collected particles.

The BioSampler also provides greater efficiency over longer sampling times because it can be used with nonevaporating liquids that have viscosities higher than water such as a special mineral oil for sampling bioaerosols. When used with this oil, the device's collection efficiency remains constant over an 8-hour sampling period. Typical sampling times with standard impingers and water-based impinger liquids are only 1 to 1.5 hours. The BioSampler's longer sampling times also provide increased sample volumes for detecting organisms at lower concentration levels.

Filtration. Filters in plastic cassettes attached to a personal air sampling pump can be used to collect nonviable dusts such as spores, but viable cells will most likely be dehydrated and killed by the large volumes of air that pass over the filter unless the sample duration is short. Filters are also useful for sampling highly contaminated environments. Examples include sawmills during wood chip handling, grain elevators, and straw handling in pig houses.[16] Spores that have a tougher cell wall and are not actively metabolizing can be collected on gelatin and cellulose ester and polycarbonate filters. The analytical method often dictates what type of filter can be used. One study evaluated three different analytical methods for fungi collection on filters.[16] Filters were analyzed by SEM and light microscopy for direct counts and epifluorescence microscopy, a staining technique. The material collected on filters can also be resuspended and cultured on different agar media, and analyzed for mycotoxins, endotoxins, or specific allergens. Polycarbonate 0.4-μm filters in 37-mm cassettes were used to collect spores to be examined by SEM or epifluorescence microscopy. Cellulose ester 0.8-μm membrane filters were used for samples for light microscopy and for culturing. The flow rate for all samples was 1.5 Lpm and samples were collected closed-face.

An advantage of gelatin filters is that they can be used to provide an estimate of both total number of bacteria-laden particles and total number of bacteria present. A gelatin filter placed on warm, moist agar readily dissolves and is absorbed. Intact particles are deposited on the agar, and if viable, they grow into visible colonies, one colony per particle, even if a particle carries more than one bacterium. A gelatin filter dissolved in water releases collected particles, and clusters of bacteria adhering to particles are dispersed by vigorous shaking. Dilution of the liquid avoids the problem of an overloaded filter, resulting in colonies too numerous to count (TNTC).[2] When polycarbonate filters are used, spores can be washed from the filter and centrifuged to form a pellet that is then examined in a haemocytometer under a microscope.[7]

Culturing filters allows for identification of both bacteria and fungi, although different types of agar must be used. Since counts usually involve inspection of only a portion of the filter, the loading characteristics are important. Accuracy is affected by uneven distributions of particles on filters and the difficulty of counting small spores close to the limits of microscopy resolution.

Filter counting procedures must deal with the fact that a typical sample may contain single spores, aggregates of spores, and other particles carrying microorganisms.[16] If a filter is washed so the sample can be resuspended, spores may be lost.

Plastic cassettes used to hold filters should be sterilized prior to use.

Procedure[18]

1. Prepare a sampling train with a personal air sampling pump and a sterile Millipore clinical field monitor (catalog no. MHWG037HO) containing a 0.45-μm gridded filter.

2. Collect a sample at a flow rate of 2.0 Lpm ± 0.5. Sampling time will depend on estimated fungal concentration.

3. Return the sample to the laboratory within 30 minutes.

4. Open the cassette and aspirate 10 mL of sterile phosphate buffer solution through the filter. This wetting step helps to distribute the media added next.

5. Add fungal media, by aspiration, from one Millipore ampoule, to the opened filter cassette.

6. Cap the cassette and place the sample in a drawer for 3–5 days to incubate at room temperature.

7. Count each individual colony that has grown during the incubation period. The use of a magnifier or stereoscopic microscope is helpful. Ideally, if the sampling time has been estimated correctly, the colonies should be easy to read with no overcrowding.

8. Calculate the volume of air sampled and report as CFU/M^3.

Size-selective personal sampling devices such as the IOM sampler (Figure 7.9) and the SKE Button Aerosol™ inhalable sampler (Figure 7.11) can also be used for bioaerosol sampling. Typical filters include gelatin, endotoxin-free polycarbonate, and mixed cellulose acetate (MCE). All filters as available presterilized for bioaerosol sampling.

Storage and Shipment of Samples

Prompt shipment and proper storage of samples following collection is critical for viable organisms. A rule of thumb is to get viable samples to a laboratory within 24 hours.[1] During storage the sample temperature should be maintained as close as possible to that of the source.[24] Samples taken on solid or liquid media will begin to grow almost immediately after collection. To prevent growth of bacteria that grow best at temperatures between 20°C and 55°C on media designed for high-temperature bacteria, refrigerate samples or begin incubation at the proper temperature immediately.[7] Blanks should be included of sampler and/or culture medium. Containers should be sterilized prior to use and shipping packages should insulate samples from environmental stresses, especially heat and freezing cold.

Impactor plates and filters should be protected from bright light and maintained at refrigerated temperatures for 12 hours to 36 hours until transported to the laboratory where they can be incubated.

DIRECT-READING INSTRUMENTS FOR BIOAEROSOLS

There are limited direct reading instruments available for measuring bioaerosols.

One is the TSI Model 3312A Ultraviolet Aerodynamic Particle Sizer® (UV-APS) spectrometer (Figure 16.12). It measures aerodynamic diameter, light-scattering intensity, and fluorescence intensity in real time. Therefore, it provides extremely rapid measurements of aerodynamic size and scattered light for particles from 0.5 to 15 µm, and it also measures fluorescence characteristics of individual particles allowing the user to distinguish airborne biological particles from most inanimate material.

The UV-APS uses two detection methods:

- An optical system for aerodynamic and scattered-light measurements.
- An ultraviolet laser beam produces biological-related fluorescence by exciting particles at a specific wavelength, allowing the instrument to measure fluorescence intensity.

Typical uses for this instrument are biohazard detection, inhalation toxicology, indoor air-quality monitoring, and filter and air-cleaner testing. It is a sophisticated device that works in conjunction with a computer and dedicated software for data management, and it has not been routinely applied to workplace exposures or problem building situations.

Figure 16.12. The TSI Model 3312A Ultraviolet Aerodynamic Particle Sizer® (UV-APS) spectrometer measures aerodynamic diameter, light-scattering intensity, and fluorescence intensity in real time. (Courtesy of TSI, Inc.)

INTERPRETATION OF RESULTS

No Threshold Limit Values (TLVs) or other official standards exist for levels of microbials in an air sample. Results of bioaerosol sampling will be presented differently depending on how samples were collected. Air samples are reported as CFU/M^3. Specialized sampling is often reported in terms of the entity collected, for example, if only spores were sampled, results would be reported as $spores/M^3$, or if pollen was sampled, results would be reported as $grains/M^3$. Impinger samples are reported as CFU/ml while bulk samples are reported as CUF/g in the case of dusts and CFU/cm^2 for carpet. Viruses are reported as infections units describing the ability of the organisms to infect, multiply, and produce new virus. CFU data must be interpreted with care because many fungus spores travel in chains and many bacteria are carried on other larger particles that produce only 1 colony for 5 to 100 cells or spores. The designation TNTC (too numerous to count) on a report indicates that sample results were too high to be effectively counted. The EPA defines TNTC as a situation where the total number of bacterial colonies exceeds 200 on a 47-mm diameter membrane filter for coliform detection.[25] If blanks are positive, sampling must be repeated.

When using culture techniques for collecting air samples, most results will be less than the actual number present due to losses in viable organisms during sampling as well as the fact that other organisms will not grow.[1] Some organisms have specialized growth needs: Growth inhibitors produced by one microbe may reduce levels of other organisms; overcrowding can reduce recoveries due to both soluble inhibition factors and "contact" suppression by adjacent growth points; filters can allow microbes to dry out and die; and there are limited periods when organisms are viable when airborne.[9] Results of cultures can be affected by the viability of the organisms; choice of nutrient medium; conditions for growth, especially temperature; and interactions between different organisms.[16]

In indoor air investigations, the best standard is the comparison of the outside air at the intakes of a building to the indoor air to determine whether there is indoor amplification of particular organisms. The percent of outdoor air allowed into the building and the percent of air that is recirculated must also be considered. For fungi, a rule of thumb is that indoor counts should be less than one half of outdoor levels present when HVAC systems are on.[1] The presence of any one fungus in levels exceeding $500/M^3$ indoors when comparison samples collected outdoors have no detectable levels of this strain or low levels of fungus can lead to a presumption of an indoor source. Levels of fungal spores indoors should be less than one third of those collected on simultaneous outdoor samples or a significant amount of contamination may exist.[7]

Overall, it is the quality of data collected during all sampling and other evaluations that governs the robustness and accuracy of interpretation. Where data are adequate, rigorous statistical analysis may be applied to evaluate possible causal relationships.[1] However, when data are limited, less formal guidelines must be applied when interpretating data.

According to one source, if the total count of microorganisms at an affected person's area equals or exceeds $10,000/M^3$, remedial action is necessary.[19] Another source suggests it is a good practice to attempt to identify any viable microorganism recovered from air in levels greater than $75 CFU/M^3$.[7] It has been suggested that a level of viable microorganisms in excess of 1×10^3 viable particles per M^3 indicates that the indoor environment may be in need of investigation and improvement.[7] High levels of human source bacteria (gram-positive) in an indoor environment indicate overcrowding and/ or inadequate ventilation. High levels ($>500\,CFU/M^3$) of gram-negative bacteria or Bacillus species in an indoor air situation suggest contamination or a source, since these organisms are not usually present indoors.[1] Actinomycetes bacteria are usually associated with agriculture and their presence in other situations is suspect.[1] The presence of thermophilic actinomycetes in indoor environments at levels above $500/M^3$ has been associated with outbreaks of hypersensitivity lung illness.[19]

One method for evaluating the contribution of outdoor air to indoor bioaerosols and wipe samples is rank-order asessment.[19] Identify taxa at experimental sites and at control (outdoor) sites. Calculate relative abundance of each taxon. If locations have the same flora, the taxa should fall into a similar order of abundance. Individual taxa are listed in descending order of abundance for indoor sites and outdoor controls. For example if one fungus is much higher indoors than outdoors, even though others are found, it is likely that it may be the source of the problem.

As a precaution, it should be noted that in some ventilation systems filtration may remove only large spores from the outside air, and thus change the rank order of smaller spores in the interior. A combination of quantitation and rank order assessment is usually necessary.

The type of organisms detected often depends on the type of operation being sampled. For example, bacteria are more prevalent than fungi in straw handling operations, grain elevators, and pig houses whereas fungi are more prevalent than bacteria in wood-related operations such as wood chip handling and sawmills.[16] Fluctuations in results from sampling situations where animals are suspected as the source can be due to variations in feeding activities, maintenance, and cleaning of animal housing facilities.[22]

The presence of a microbe in a bulk sample of water or other material does not mean it is airborne unless a means of aerosolizing the microbe can be identified. If a growth is positive, it may or may not be the source of disease. There are often several different strains of a given organism and some are pathogenic to humans whereas others are not.

SUMMARY

This chapter covered the details of bioaerosol monitoring, building on more general information in Chapters 5 and 7. When sampling for bioaerosols, the concepts of general aerosol sampling apply along with special considerations such as sampling is usually designed to collect *culturable* samples for the organisms of interest, sampling times for many bioaerosol samples are very short compared to the sampling period for chemical contaminants in order to prevent overloading the sample device, and for most bioaerosols there are not specific airborne standards (or specified sample collection methods).

Because of the complexities involved with bioaerosol sampling, it is imperative for the sampling practitioner to have a clear understanding of the reason for the sampling and also to work closely with the laboratory that will be analyzing or counting the samples.

REFERENCES

1. Macher, J. M., ed. *Bioaerosols: Assessment and Control.* Cincinnati, OH: ACGIH, 1999.

2. Macher, J. M., and M. W. First. Personal air samplers for measuring occupational exposures to biological hazards. *AIHA J.* **45**(2):76–83, 1984.

3. Elliott, L. J., R. Sokolow, M. Heumann, and S. L. Elefant. An exposure characterization of a largescale application of a biological

insecticide, *Bacillus thuringiensis*. *Appl. Ind. Hyg.* **3**(4):119–122, 1988.

4. Forster, H. W., et al. Investigation of organic aerosols generated during sugar beet slicing. *AIHA J.* **50**(1):44–50, 1989.

5. Muraca, P. W., et al. Legionnaires' disease in the work environment: Implications for environmental health. *AIHA J.* **49**(11): 584–590, 1988.

6. National Institute of Occupational Safety and Health: *Occupational Diseases: A Guide to Their Recognition.* Cincinnati, OH: NIOSH, 1977.

7. ACGIH Bioaerosols Committee. Guidelines for assessment and sampling of saprophytic bioaerosols in the indoor environment. *Appl. Ind. Hyg.* **2**(5): R-10–R-16, 1987.

8. DeLucca, A. J., and M. S. Palmgren. Seasonal variation in aerobic bacterial populations and endotoxin concentrations in grain dusts. *AIHA J.* **48**(2):106–110, 1987.

9. Burge, H. A., and W. R. Solomon. Sampling and analysis of biologic aerosols. *Atmos. Environ.* **21**(2):451–456, 1987.

10. Burge, H. A. Indoor sources for airborne microbes. In *Indoor Air and Human Health*, H. B. Gammage and S. V. Kaye, eds. Chelsea, MI: Lewis Publishers, 1985.

11. DuBois Chemicals, Ltd. *Reducing the Risk of Legionnaires' Disease.* 1990 management report, England, 1990.

12. Fischer, J., et al. Environment influences levels of gram-negative bacteria and endotoxin on cotton bracts. *AIHA J.* **43**:290–292, 1982.

13. Morring, K. L., W. G. Sorenson, and M. D. Aitfield. Sampling for airborne fungi: A statistical comparison of media. *AIHA J.* **44** (9):662–664, 1983.

14. Burge, H. A. Fungus allergens. *Clin. Rev. Allergy* **3**:319–329, 1985.

15. Chatigny, M. A. Sampling airborne microorganisms. In *Air Sampling Instruments*, 3rd ed. Cincinnati, OH: ACGIH, pp. E-1–E-9.

16. Eduard, W., et al. Evaluation of methods for enumerating microorganisms in filter samples from highly contaminated occupational environments. *AIHA J.* **51**(1): 427–428, 1990.

17. McJilton, C. E., et al. Bacteria and indoor odor problems—three case studies. *AIHA J.* **51**(10):545–549, 1990.

18. Morey, P. R., et al. Environmental studies in moldy office buildings: Biological agents, sources and preventive measures. *Am. Conf. Gov. Ind. Hyg. Ann.* **10**:21–35, 1984.

19. ACGIH BioAerosols Committee. Airborne viable microorganisms in office environments: Sampling protocol and analytical procedures. *Appl. Ind. Hyg.* **1**(4):R-19–R-23, 1986.

20. Furst, M. W. Air sampling. *Appl. Ind. Hyg.* **3**(12):F-20, 1988.

21. Macher, J. M. Positive-hole correction of multiple-jet impactors for collecting viable microorganisms. *AIHA J.* **50**:561–568, 1989.

22. Cormier, Y., et al. Airborne microbial contents in two types of swine confinement buildings in Quebec. *AIHA J.* **51**(6): 304–309, 1990.

23. Macher, J. M., and H. C. Hansson. Personal size-separating impactor for sampling microbiological aerosols. *AIHA J.* **48**(7): 652–655, 1987.

24. American Public Health Association. *Standard Methods for the Examination of Water and Waste Water*, 16th ed. New York: APHA, et al., 1985.

25. Environmental Protection Agency. *National Drinking Water Regulations.* CFR 40. Chap. 1, Subchapter D, Part 141.

PART V

SPECIFIC SAMPLING APPLICATIONS AND SUPPLEMENTARY INFORMATION

CHAPTER 17

SPECIFIC SAMPLING SITUATIONS

This chapter is intended to provide an overview to the sampling practitioner on typical approaches and considerations when performing air monitoring for toxic contaminants. These examples were selected to cover a variety of sampling situations:

- Confined space measurements
- Indoor air quality investigations
- Leak testing for fugitive emissions
- Welding fumes
- Carbon monoxide from forklifts
- Multiple solvents in printing ink manufacturer

Although air sampling is the main topic in this book, it is important to note that for confined space entry many other safety and health considerations besides air monitoring are critical to protecting people, and for indoor air surveys other investigative approaches besides air sampling are gener-

ally required to diagnose and resolve problems.

CONFINED SPACES

Confined spaces can present many serious, even life-threatening, hazards to those who must enter and work within these spaces. Potential hazards include oxygen deficiency, toxic contaminants, fire and explosion, drowning, risk of engulfment by solid materials, risk of being trapped by small passageways, and other hazards. The difficulty in rescue or escape in case of injury or medical problems must also be considered an additional risk from confined spaces. Tanks, pits, utility vaults, silos, tunnels, trenches, and crawl spaces are well-known examples of confined spaces; all sampling practitioners should be on the alert for spaces that are less well known but still fall under the definition of a confined space.

There are numerous sources of toxic or

Air Monitoring for Toxic Exposures, Second Edition. By Henry J. McDermott
ISBN 0-471-45435-4 © 2004 John Wiley & Sons, Inc.

dangerous atmospheres within confined spaces including:

- Residues of prior contents may evaporate and form hazardous levels or may displace oxygen in the space.
- Harmful contaminants may be released when sludge or surface deposits are agitated or disturbed during cleaning operations.
- Leaks from pipelines, flanges, and other equipment may contaminate the space or displace oxygen.
- Decomposing organic matter from vegetation or other sources can produce methane, carbon monoxide (CO), and hydrogen sulfide and can consume oxygen.
- Welding, arc, or torch cutting or brazing on surfaces coated with paint or another surface coating, or contaminated with chemicals or oil, can produce CO and other contaminants and deplete the oxygen.
- Combustion products (e.g., CO) from nearby gasoline or diesel engines or from welding, cutting, or brazing can contaminate the space.
- Oxygen-deficient atmospheres can be formed by a variety of often-overlooked "processes" such as rusting metal and adsorption of O_2 by activated carbon beds.
- Oxygen above the normal level of 21% increases the flammability range of combustible materials and causes them to burn more violently than usual.

These hazardous conditions may exist at time of entry or develop after initial air monitoring has determined that the atmosphere is "safe" for entry. Explosive and toxic gases, such as hydrogen sulfide and carbon monoxide, along with a lack of oxygen, cause the majority of confined space injuries and fatalities. More than half of the fatalities are among would-be rescuers of the initial victims.

OSHA Standard for Permit-Required Confined Spaces

The OSHA standard for permit-required confined spaces[1] contains requirements for practices and procedures to protect employees from the hazards of entry into confined spaces where hazardous conditions exist or could develop. The discussion below only covers the requirements relating to air monitoring. The standard sets other requirements covering topics such as an entry permit, personnel qualifications, medical and rescue resources, respirator use, labeling or posting of signs, ventilation, and maximum storage limits for flammable materials. Since this chapter is intended to help explain air monitoring requirements, it does not cover the other compliance requirements that apply. Additionally, it does not cover state OSHA standards provisions that differ from federal OSHA requirements.

A *confined space* is defined by OSHA as a space that is (a) large enough and so configured that an employee can bodily enter and perform assigned work, (b) has limited or restricted means for entry or exit (for example, tanks, vessels, silos, storage bins, hoppers, vaults, and pits are spaces that may have limited means of entry), and (c) is not designed for continuous employee occupancy (Figure 17.1).

Recognizing the different potential hazards that can occur in confined spaces, the standard covers two broad situations:

- Circumstances where the employer can demonstrate that continuous forced air ventilation alone is sufficient to maintain the space safe for entry. Entry is allowed if (a) the atmosphere is tested and found safe, (b) continuous forced air has eliminated any hazard and is continued for the dura-

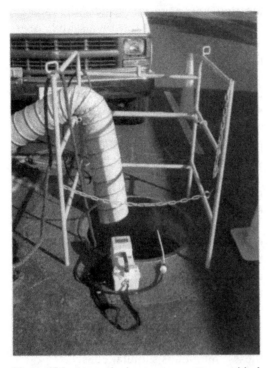

Figure 17.1. Atmospheric measurements are critical for safety before any confined space entry. (Courtesy of RKI Instruments, Inc.)

tion of the work inside the space, (c) there is a periodic testing of the air inside the space, and (d) there is a signed tag (or other certification) completed before entry is allowed that describes how the requirements of this section of the standard are being met. There may be no hazardous atmosphere within the space while employees are inside; if any develop, the employees must be removed immediately.

• For confined spaces that have potential hazards beyond the conditions described in the previous paragraph, detailed requirements apply before entry is allowed. These include preventing unauthorized entry, identifying and controlling the potential hazards, providing rescue equipment and personnel, and training employ-

ees. As part of these extensive requirements, air monitoring and ventilating equipment must be provided if needed.

The standard requires atmospheric testing for two distinct purposes:

• *Evaluation Testing.* Where the atmosphere of a confined space is analyzed using equipment of sufficient sensitivity and specificity to identify and evaluate any hazardous atmospheres that may exist or arise, so that appropriate permit entry procedures can be developed and acceptable entry conditions stipulated for that space. Evaluation and interpretation of these data and development of the entry procedure are to be performed or reviewed by a technically qualified professional based on evaluation of all serious hazards. OSHA provides these examples of qualified professionals: the OSHA consultation service, or a certified industrial hygienist, registered safety engineer, certified safety professional, or certified marine chemist, and so on.

• *Verification Testing.* Where the atmosphere of a permit space which may contain a hazardous atmosphere is tested for residues of all contaminants identified by evaluation testing using equipment that is specified in the permit to determine that residual concentrations at the time of testing and entry are within the range of acceptable entry conditions. According to OSHA, results of testing (i.e., actual concentration, etc.) should be recorded on the entry permit in the space provided adjacent to the stipulated acceptable entry condition.

Measurement of values for each atmospheric parameter should be made for at least the minimum response time of the test instrument specified by the manufacturer; and if a person has to enter a space in order to monitor by descending into atmospheres that may be stratified, the atmospheric envelope should be tested a distance of approximately 4 feet in the direction of travel and to each side. If an extension hose or sampling probe is used, the entrant's rate of progress should be slowed to accommodate (a) the effective sampling rate with the probe and extension hose in place and (b) the detector response.

A test for oxygen is to be performed first because most combustible gas meters are oxygen-dependent and will not provide reliable readings in an oxygen-deficient atmosphere. Combustible gases are tested for next because, according to OSHA, in most cases the threat of fire or explosion is both more immediate and more life-threatening than exposure to toxic gases and vapors. If tests for toxic gases and vapors are necessary, they are performed last.

Confined Space Measurements

In addition to air monitoring, if a tank or other container held an unknown compound, a bulk sample of the material should be collected and analyzed prior to entry. The potential for dermal exposure should always be evaluated whenever individuals enter a process vessel that recently contained hazardous materials, since flushing vessels may be difficult and often sludge and other residual material is left behind on the bottom and on various structures in the vessel. If the purpose of entry is cleaning, sometimes wipe samples are done to identify material adhering to walls and other surfaces.

When sampling in a confined space, do not enter before taking measurements of contaminants. Instead use either the tubing provided with pump-drawn samplers or the remote sampling cord attached to a passive sensor that allows it to be lowered or

inserted on a probe inside the space. (Figure 17.2) Do not rely on a sample near an entry way to represent the atmosphere throughout the vessel. Also, sampling should be done periodically and not just prior to the start of the job, especially when work is done that disturbs the interior surface. Even after an empty tank has been purged, gases can desorb from porous walls or be liberated from sludge or from cleaning solvents, or be produced by chemical reactions with cleaning solvents and other materials.

The primary confined-space measurements are for oxygen deficiency and combustible gases. However, other gases should also be tested for if there is any possibility they may be present, such as hydrogen sulfide and carbon monoxide. There are a number of instruments for this application that can continuously monitor for all of these compounds simultaneously.

Be very careful when suspending passive detectors into tanks where there may be liquids or in which there is uncertainty of the contents. Submersing the detector in

Figure 17.2. A portable gas monitor equipped with a diffusion head that can be located up to 50 feet from the instrument (using a remote cable) to provide rapid readings. (Courtesy of RKI Instruments, Inc.)

liquids will require repair or replacement. Similarly, drawing liquids into a pump-driven instrument may damage the pump.

The type of work to be done inside the confined space will determine what is an acceptable measurement reading. For example, if welding is to be done and there is any possibility that chlorinated materials may have been inside of a tank at any point, the atmosphere should also be tested for very low levels of halogenated hydrocarbons in addition to combustibles and oxygen deficiency. This is good practice since the welding arc as well as other high heat sources can convert chlorinated compounds to phosgene, which is highly toxic at very low levels.

Procedure

1. All instruments should first be checked for a proper zero indication for combustible and toxic gases and for 20.9% oxygen in fresh air. Check alarm response using a test source.
2. For spaces such as a utility vault, first sample through a pick hole or open the cover slightly on the downwind side before opening the cover completely.
3. If required, hook the extension tube onto the instrument. Note that the addition of the extension tube increases the response time of the instrument. Check the hose for leaks by placing a finger over the end of the hose inlet while listening to the pump. If leakage is occurring, the pump will not labor.
4. Sample at several levels and take multiple samples. In general, levels of gases will form gradients based on their densities relative to air. The lack of normal ventilation allows these gases to stratify because normal air currents that cause mixing an dilution are not present.
5. Check any areas where gases can

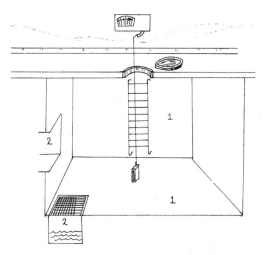

Figure 17.3. Diagram of recommended testing points (1 = test before entry; 2 = test after entry) for confined spaces. (Courtesy of Enmet.)

"pocket," such as below gratings, between rafters or in small chambers attached to the main space (Figure 17.3).
6. Once work begins, sample frequently or continuously, because conditions can change. As work progresses, a once-safe atmosphere can become hazardous due to leaks, combustion, cleaning processes, or other operations such as welding.

There are many additional requirements for making a confined space entry. For this reason, it is important to be aware of all the actions needed to do a safe confined space entry even if the sampler's responsibilities are limited to only making the measurements.

INDOOR AIR QUALITY INVESTIGATIONS

The quality of air in residential, commercial, and office buildings is an important issue for building managers, engineers, and health professionals. Complaints about air quality or possible illness caused by the indoor environment trigger investigations and surveys in which air monitoring can play a significant role.

The heating, ventilating, and air conditioning (HVAC) systems in a building play a vital role by providing the fresh air and circulation to maintain good air quality and comfort. Many ventilation-related factors such as total airflow, amount of outdoor air provided, temperature, humidity, air motion in the occupied space, and presence of odors or other contaminants in the outdoor and recirculated air all affect indoor air quality (IAQ) and the potential for indoor air problems.

HVAC systems are designed to meet three objectives: good air quality; thermal comfort; and, reasonable energy costs. These three objectives provide clues to how building problems can develop. In particular, soaring energy costs in the mid-1970s caused a reexamination of (a) established ventilation airflow standards and (b) the ways the HVAC systems (and all aspects of buildings) were designed and operated. Experience now shows that some of the HVAC systems designed in this period either provided too little outdoor air or did not deliver the air to the "occupied zone" within 6–10 feet above the floor. Sometimes the inability to delivery air properly was due to insufficient flexibility as building use or layout changed. Also during this period, building materials and furnishings in new or renovated buildings were selected without a full understanding of indoor-generated contaminants. What followed were increased complaints of poor air quality and symptoms of eye and respiratory tract irritation, headache, or allergic responses. Thermal comfort also suffered in some cases, since lower air circulation rates made it difficult to maintain adequate air motion. As a result, there is an "inventory" of buildings designed during this period with low outdoor air ventilation rates that are more susceptible to IAQ

problems unless the HVAC system has been modified or other steps taken to ensure adequate indoor air quality.

Since that time period, building and ventilation standards have been revised based on experience and new research. Standards now recommend higher outdoor air quantities. Indoor air quality considerations are now usually a selection criterion for construction materials and furnishings for new or remodeled buildings. Building managers are more aware of the impact of maintenance, landscaping, and other activities on IAQ. The importance of HVAC system preventive maintenance is better understood. The reduction or elimination of smoking in many occupancies has also reduced IAQ complaints. Overall these factors have resulted in (a) more awareness of indoor air quality issues and (b) the need to deal with complaints or episodes of possible building-related illness promptly and professionally.

Regardless of building age or HVAC system design, microbial growth including mold is an IAQ consideration wherever there is standing water or water damage has occurred. In some cases, microbial buildup in ducts and other system components can be the cause or a contributing factor in indoor air quality complaints. See Chapter 16 for more details on this topic, including air monitoring information.

Of particular concern to an air sampling practitioner involved in an IAQ investigation is the fact that airborne levels of specific chemical contaminants are generally very low as compared to published allowable levels for the occupational setting. Often no single contaminant can be identified as the cause of the complaints. In those cases where the level of a specific contaminant is elevated above "background," it may serve as a marker for the presence of the problem rather than be the cause of the problem itself. For example, *slightly* elevated carbon monoxide (CO) levels may indicate that vehicle exhaust (consisting of numerous irritating or odorous compounds) is entering the building even though the CO levels are not hazardous. Low levels of volatile organic compounds (VOCs) from microbial growth may also be detected; although not at "hazardous" levels, these VOCs do indicate the presence of mold which may be causing the IAQ problems.

Additional factors that often make IAQ problems difficult to diagnose include[2]:

- Often the complaints or symptoms (e.g., odors, headache, nose and throat irritation, etc.) begin with a few individuals and are easy to overlook in the early stages.
- Individual health problems (flu, allergies, etc.)—as well as general stress from work pressures, management–employee relations, and other factors—may be involved.
- Widespread episodes may be triggered by an accumulation of smaller problems that remain unsolved. Thus a direct cause–effect relationship, valuable in diagnosing problems, may be absent. This makes it difficult to identify the extent and cause of the complaints.
- Where the existence of a problem was not recognized and dealt with promptly, employees' health concerns and frustration at being ignored may add an emotional component to the problem that makes a purely technical solution difficult to achieve.

Because of these factors, diagnosis and resolution of IAQ problems should follow well-established protocols.[3] These protocols usually include a questionnaire or interviews with employees to determine the type and extent of symptoms and their commonality throughout the building. It is important to identify when symptoms occur (time, day of week, or season), and what other factors employees feel may contribute to the problem in order to reach a final resolution.

Typical Contaminants

The following contaminants can be measured during IAQ investigations if there is any apparent source of the contaminant. In some cases, measurements are made to "rule out" the possibility that the chemicals are present. These substances are also listed in Table 17.1 along with a summary of typical indoor levels, air monitoring techniques, and other comments.[2]

Carbon Monoxide. Carbon monoxide (CO) from faulty furnaces or automobile exhaust is a well-known hazard. Simple detector tubes or a direct-reading monitoring instrument are used to evaluate any situation where CO could be present and acute symptoms resemble CO over exposure. As described above, even slightly elevated CO levels from adjacent parking structures or vehicles operated or idling near building air intakes need to be addressed to resolve complaints even though the CO levels are not harmful. Studies of CO levels in buildings near freeways show that higher CO concentrations can occur if HVAC systems are set to draw outdoor air into the building during the morning or evening commute period.

In buildings where heavy smoking occurs, indoor CO levels may rise above ambient levels. Current ventilation standards recognize that significantly more airflow is needed in smoking areas than in nonsmoking areas to achieve acceptable odor levels rather than controlling CO levels.

Organic Vapors. Organic vapors are discussed as a general category because individual organic compounds tend to be found at levels far below concentrations that would be expected to cause health problems. There are also a large number of

TABLE 17.1. Typical Indoor Contaminants in Office and Commercial Buildings

Contaminant	Indoor Source	Indoor Levels	Implicated in Acute Symptoms?	Air Monitoring Techniques	Comments
Carbon monoxide	Faulty furnaces, cigarette smoking, vehicle exhaust	See *Comments* column	Yes — Faulty furnaces, vehicle exhaust —headache, nausea, etc. No — Cigarette smoking	Detector tubes Direct-reading instruments	CO levels up to 25 ppm have been found in heavy smoking areas. Above this level other sources should be suspected.
Organic vapors	Solvents, resins, aerosol sprays, some building materials	0.05–1.0 ppm	Yes — Odors, eye and respiratory tract irritation, headache, nausea	Activated carbon or other adsorbent with sampling pump Passive dosimeters Detector tubes	Levels may be too low to detect with routine techniques.
Respirable particulate matter	Smoking, aerosol sprays, dust resuspension	100–300 μg/m^3	Yes — Eye and respiratory irritation	Respirable dust cyclone or impactor with pump	None
Airborne microorganisms	Infected occupants, contaminated HVAC systems, molds and fungi	Not applicable	Yes — Allergic response, infection	Air sampling with laboratory culture to identify agents	Often difficult to identify organism causing problem.
Radon and its progeny	Building materials, soil and groundwater	0.1–30 nCi/m^3	No	Passive dosimeter	None
Carbon dioxide	Metabolic activity of occupants	Up to 3000 ppm	No	Infrared or other direct-reading instrument Detector tubes	About 400 ppm is typical for ambient air. Elevated indoor levels may indicate too many occupants for the amount of fresh air supplied.

separate compounds present; identifying and quantifying each one is extremely difficult if not impossible, and often the effort and expense are not warranted during IAQ investigations. Some of these compounds may be lacrimating (tearing) agents or particularly irritating, even at extremely low levels. Major sources of organic contaminants are as follows:

- Bioeffluents—humans emit organic substances from natural biologic processes. Compounds such as methanol, ethanol, acetone, and butyric acid were identified in a school auditorium seating 400 people.[4] An adequate HVAC system should keep bioeffluents at acceptable levels.
- Building materials—adhesives, paints, sealants, carpeting, and other materials release organic vapors, especially when new. Emission rates generally decline as the building or product "cures" or ages.
- Consumer products—spray cleaners, pesticides, photocopy machine toners, and other materials all add to the overall level of organic vapors. Where feasible, these can be controlled by restricting use, substituting more IAQ-friendly products, or local exhaust ventilation.
- Tobacco smoke contains a variety of organic vapors in addition to the CO discussed earlier.

Respirable Particulate Matter. Respirable particulate matter (RPM) is another broad category of indoor pollutants that includes particles, such as dusts and pollen, and aerosols in the size range that can reach the lungs when inhaled. Although these pollutants vary in chemical composition, they are grouped together because air sampling devices collect particulates according to size distribution. High RPM levels can cause eye and respiratory tract irritation.

For many people the main exposure to RPM is from tobacco smoke.[4] Smokers receive a higher total dose than do nonsmokers, but side stream smoke is produced as long as the tobacco burns, and many nonsmokers are more sensitive to these substances than are smokers. Smoking can cause an increase by a factor of 3–40 in RPM over background levels.[5] Other sources of RPM are aerosol sprays and resuspension of dust.

Practically all HVAC systems have a filter in the system to control particulates. These filters may not be adequate when RPM levels are very high. If the source of RPM is confined to a relatively small area, local exhaust systems or local air cleaners such as portable electronic air cleaners may be helpful. Limitations on smoking are a solution when tobacco smoke is a major source of RPM.

Airborne Microorganisms. As described earlier, microbial growth is a consideration wherever there is standing water or water damage has occurred. Microbes drawn into the building (such as *Legionella* bacteria from nearby cooling towers) or disease organisms from building occupants may be an issue in some IAQ evaluations. Air monitoring and other sampling techniques for microorganisms or the VOCs they emit are covered in Chapter 16.

Radon and Its Progeny. Building materials, soil around buildings, and water may release radon gas through radioactive decay of naturally occurring radium-226. The radon, in turn, decays to form short-lived radioactive decay products (called radon daughters or progeny) that include polonium, lead, and bismuth. These materials exist in the air and also attach to airborne particles. They can be deposited in the lungs when inhaled, which may pose a risk of increased lung cancer.

The potential for higher radon concentrations varies widely by geographical area, depending on the level in soil, water, and building materials. However, since there are no acute symptoms, radon and its progeny are not factors in tight building syndrome outbreaks. See Chapter 15 for more details.

Carbon Dioxide. Carbon dioxide (CO_2) is present in the ambient environment at levels averaging about 400 ppm. The main indoor source of CO_2 is the metabolic activity of occupants; smoking and wood-burning fireplaces do not increase levels significantly.

CO_2 levels are now recognized as an important factor in setting adequate ventilation standards for energy efficient buildings since IAQ complaints occur more often in buildings with CO_2 levels in the range of 1000 ppm or greater. CO_2 levels are often measured during IAQ surveys.

Humidity. Water vapor is not usually considered as a contaminant (and is not listed in Table 17.1), but it affects indoor air quality in several ways:

- Low relative humidity can cause particles and vapors in the air to be more irritating than at high levels of humidity.
- High humidity aggravates odor problems and fosters mold growth. High humidity also leads to complaints of stuffiness and thermal discomfort.

As a guideline, relative humidity should be maintained in the 20–30% range during the heating season and maintained at <60% during the cooling season.[6] Many direct reading instruments specifically designed for IAQ surveys measure and record both humidity and temperature.

Determining Outdoor Air Ventilation Rates Using Air Monitoring

HVAC standards such as ASHRAE-62, "Ventilation for Acceptable Indoor Air Quality,"[7] usually specify the amount of outdoor air to be provided based on the number of occupants in the area. There are several techniques for estimating the quantity of outdoor air; a few involve air monitoring or similar measurements and thus are described below.

Two methods involve measuring either CO_2 levels or temperature at these three specific locations in the HVAC system and then applying a simple formula[6]:

- Return air—the air coming back from the occupied space that will be recirculated to the building after mixing with incoming outdoor air.
- Outdoor air—fresh air being introduced into the HVAC system.
- Mixed air—the mixture of recirculated return air and outdoor air.

Using the Temperature Method to estimate the amount of outdoor air:

$$\text{Outdoor air } (\%) = \frac{T_{\text{return}} - T_{\text{mixed}}}{T_{\text{return}} - T_{\text{outdoor}}} \times 100$$

where outdoor air (%) is the percent of outdoor air in total airflow, T_{return} is the temperature in return air (°F), T_{mixed} is the temperature in mixed air, (°F), and T_{outdoor} is the temperature in outdoor air (°F).

Using the Carbon Dioxide Method to estimate the amount of outdoor air, CO_2 concentrations can readily be measured in the field using direct-reading instruments:

$$\text{Outdoor air } (\%)$$
$$= \frac{CO_{2(\text{return})} - CO_{2(\text{mixed})}}{CO_{2(\text{return})} - CO_{2(\text{outdoor})}} \times 100$$

where outdoor air (%) is the percent of outdoor air in total airflow, $CO_{2(\text{return})}$ is the carbon dioxide concentration in return air (ppm), $CO_{2(\text{mixed})}$ is the carbon dioxide concentration in mixed air (ppm), and $CO_{2(\text{outdoor})}$ is the carbon dioxide concentration in outdoor air (ppm).

The building engineering or maintenance staff can help to identify the best locations for these measurements.

Example: Velocity measurements inside a 16in. × 10in.-rectangular duct supplying one work area show the average velocity to be 820ft/min. CO_2 readings at the three specified locations are

$$CO_{2(return)} = 535\,ppm$$

$$CO_{2(mixed)} = 425\,ppm$$

$$CO_{2(outdoor)} = 300\,ppm$$

If there are 12 people in the work area, calculate the amount of outdoor air per person being supplied to the area.

Answer: First calculate total flowrate (Q) using

$$Q = V \times A$$

where

$$A = 16\,in. \times 10\,in. = 160\,in.^2$$

$$A = 160\,in.^2 \left(\frac{ft^2}{144\,in.^2} \right) = 1.11\,ft^2$$

$$Q = 820\,ft/min \times 1.11\,ft^2 = 910\,ft^3/min$$

The percent of outdoor air in the total airflow is calculated as follows:

Outdoor air (%)

$$= \frac{CO_{2(return)} - CO_{2(mixed)}}{CO_{2(return)} - CO_{2(outdoor)}} \times 100$$

$$= \frac{(535 - 425)}{(535 - 300)} \times 100 = \frac{110}{235} \times 100$$

Outdoor air (%) = 46.8%

$$Q_{outdoor} = 910\,ft^3/min\,(0.468)$$

$$= 426\,ft^3/min$$

$$Q_{outdoor}\text{ per person} = \frac{426\,ft^3/min}{12\text{ people}}$$

$$= 35.5\,ft^3/min\text{ per person}$$

To determine if this value meets current guidelines, refer to the latest version of ASHRAE Standard 62, "Ventilation for Acceptable Indoor Air Quality" for the type of occupancy (office, school, etc.). Not meeting current criteria does not necessarily indicate a problem, especially if the building was designed while an earlier version of the standard was in effect. However, the building should be further evaluated to determine if the outdoor air rates are adequate to prevent IAQ problems.

Tracer Techniques. In some situations, tracer gas studies are used to estimate outdoor air rates, especially with complex HVAC systems, rooms with very high ceilings where duct or outlet air volume measurements are not feasible, and in buildings where a significant amount of outdoor air infiltrates through doors, windows, or other openings.

The tracer technique involves use of a stable, nontoxic, easily measured gas with a density near that of air. The tracer is distributed throughout the work area, and then its level is measured periodically. The airborne level decreases with time due to dilution with supply air or infiltration of outdoor air through open doors or cracks. A simple equation is used to estimate the air exchange rate in the space based on tracer concentration decrease over time.

Where the occupant density is sufficient to generate increased CO_2 levels, this can be used as a tracer. To be a successful test agent, the indoor CO_2 levels must be higher than ambient levels so that the dilution with outside air can be measured over time. These tracer tests are conducted immediately after workers leave for the day to avoid interference caused by additional CO_2 that is contributed by the occupants during the test. If indoor CO_2 levels are not high enough, additional CO_2 (using "dry ice") or another nontoxic tracer gas [e.g., sulfur hexafluoride (SF_6)] may be released when the building is vacant if permission can be obtained.

To perform a tracer study, obtain the gas, a properly calibrated direct-reading analytical instrument, and mixing fans, if needed. Then follow this general procedure[2]:

- Make sure that the HVAC system controls are set to operate in a manner typical of the time period being evaluated. For example, in studies conducted after the close of business, be certain that automatic setbacks will not reduce airflow rates if the tests are to represent normal daytime operations.
- Measure background levels of the tracer gas both indoors and outside.
- Release the gas and, when it is thoroughly mixed, make the initial reading and record it along with the time. Temperature, humidity, and system operating parameters should also be recorded so that the factors affecting ventilation rates can be reconstructed later.
- Periodically repeat the concentration measurements and record the level and time until the indoor level of the tracer drops to near ambient levels.
- Calculate the outdoor air rate using this equation:

$$C_{final} = C_{original}e^{[-Q' \times \Delta t/V_r]}$$

where Q' is the effective air ventilation (including infiltration) rate (ft^3/min), $C_{original}$ is the indoor concentration of tracer gas at start of test (minus any level in outdoor air) (ppm), C_{final} is the indoor concentration at end of test (minus any level in outdoor air) (ppm), Δt is the elapsed time of test (minutes), and V_r is the room volume (ft^3).

Example: A tracer study using occupant-generated CO_2 is to be performed in one large room in an office building. The test will begin just after quitting time (5:15 P.M.). CO_2 levels will be measured with an infrared instrument. Room dimensions are $100\,ft \times 50\,ft \times 10\,ft$; room occupancy is normally 45 people doing sedentary office work. Outdoor CO_2 level is 325 ppm.

Answer: The following CO_2 levels are measured during the test:

Time	Indoor CO_2 Level (ppm)	Indoor CO_2 Level – Outdoor Level (ppm)
5:20 P.M.	905	580
6:00 P.M.	750	425
6:40 P.M.	610	275
7:20 P.M.	530	205
8:00 P.M.	425	100

From these data and the information given, the parameters are

$$C_{original} = 580\,ppm$$

$$C_{final} = 100\,ppm$$

$$\Delta t = 2\,hr\,40\,min = 160\,minutes$$

$$V_r = 100\,ft \times 50\,ft \times 10\,ft = 50{,}000\,ft^3$$

$$C_{final} = C_{original}e^{[-Q' \times \Delta t/V_r]}$$

$$100\,ppm = (580\,ppm)e^{[-Q \times 160/(50{,}000)]}$$

$$\frac{100\,ppm}{580\,ppm} = e^{[-Q' \times 160/(50{,}000)]}$$

$$0.17 = e^{[-Q' \times 160/(50{,}000)]}$$

$$\ln 0.17 = \frac{-Q' \times 160}{50{,}000}$$

$$-1.77 = \frac{-Q' \times 160}{50{,}000}$$

$$Q' = 552.4\,ft^3/min$$

On a per-person basis, this ventilation rate is

$$\frac{552.4\,ft^3/min}{45\,people} = 12.3\,ft^3/min \text{ per person}$$

This rate can be compared to the current ASHRAE Standard 62 to see if it meets the airflow guidelines for a system being designed today.

The remainder of this section covers data gathering and other aspects of an IAQ investigation for use by the sampling practitioner in developing an approach and performing the investigation.

Guidelines for IAQ Investigations

Indoor air investigations are often most effective when performed by a team involving sampling practitioners, industrial hygienists, ventilation engineers, and physicians. Develop a questionnaire for building occupants regarding their symptoms. Map the reported symptoms on a building floor plan. Generally there will be a wide spectrum of response, from individuals with no symptoms to those with pronounced symptoms. Not everyone will experience the same symptoms. The areas where the highest incidence of symptoms reside can be targeted for priority inspections and monitoring.

The building inspection involves collecting information on potential sources of contaminants inside and outside of the building. Find out if any changes occurred in the building since the symptoms were noted. For example, a carpet may have been installed. A number of different glues are used to install carpets and the carpets themselves contain various chemicals. Bulk samples are often collected of materials such as carpet that are suspected of emitting contaminants.

Identify outdoor sources of contamination that might enter the building. Check for local exhaust system discharge stacks on the roof of the problem building or adjacent buildings. If there are factories nearby or specific sources of a chemical such as sulfur dioxide or ammonia, these should be taken into consideration. A walk-through of the neighborhood may be needed in some cases to identify sources of pollutants such as factories within a 1-mile radius. It is also useful in the case of a new building to research the history of the property. For example, sanitary landfills release a number of different gases including vinyl chloride and methane. Leaking underground storage tanks can contaminate neighboring properties, causing significant odors inside basements. Combustion products are especially important due to their ubiquitous presence and the fact that often they migrate through vents as a result of discharge sources that are located too near to building air intakes. Building exhaust and intake vents should be identified including their proximity to each other. A poorly located flue pipe for a hot-water heater could result in nitrogen oxides being pulled inside the building. Figure 17.4 contains information that can be used to identify sources of indoor and outdoor contamination.

There are many areas indoors where growing conditions are favorable to microorganisms, such as stagnant pools of water in basements and sumps, air-conditioning reservoirs and ducts, and cooling towers. Microorganisms can also enter from outdoors. If individuals are symptomatic, air sampling may be needed to at least rule out microorganisms as a source of disease. If a reservoir of a contaminant is suspected, a bulk sample should be collected.

Where strong odors are noted, a general survey instrument such as a PID analyzer can be useful. The area where odors are the strongest should be monitored first. If a basement or lower level has the highest levels and no intakes for outdoor air are present, then an underground source should be suspected. Since odors fluctuate, every effort should be made to perform surveys on days when odors may be present.

A good practice is to document the background levels of outdoor contaminants in

I. General Area Description

1. Building Address:

    Provide location on sketch.

2. Describe geographic characteristics of the immediate (½-mile radius) neighborhood. Include:

    a. Percent open land   _____ %
    b. Is a stream located in the area?   ☐ yes ☐ no
    c. Is the area hilly?   ☐ yes ☐ no
    d. Are there tall structures that affect wind flow?   ☐ yes ☐ no
    e. Are there any condemned or demolished structures in the area? ☐ yes ☐ no

3. Describe each of the neighboring buildings or open areas. Indicate location of each on sketch.

Relative Location	*Type of Structure/Open Area*	*Approx. Age of Structure*	*Known or Observed Chemical Sources*

Figure 17.4. Indoor air screening data. (From Environmental Protection Agency 600/6-88009A, August 1988.)

4. Describe any potential point source for the chemicals or pollutants of interest within an area described as follows. Establish a ½-mile radius centered on the structure of interest. Extend the area an additional ½ mile upwind, creating an oval shaped area. Indicate the location of each point source on the area sketch.

Relative Location	*Distance*	*Type of Chemical*	*Comments*

5. Are there any major freeways within the area created in Section 4? If so

   a. Draw freeway on sketch
      Average traffic levels during:

      Rush hours   _____ vehicles/hour

      Other times   _____ vehicles/hour

   b. Indicate any major city streets
      Average traffic levels during:

      Rush hours   _____ vehicles/hour

      Other times   _____ vehicles/hour

   c. Average traffic flow on nearest street: _____ vehicles/hour

6. Has there been any exterior pesticide application at any neighboring buildings or areas within the last 30 days?

7. a. Describe prevailing winds or attach a wind rose.

   b. Outside temperature: high _____ °F low _____ °F

Figure 17.4. *Continued*

    c. Outside RH:

    d. Wind speed and direction:

    e. Barometric pressure and tendency:

II. Building Characteristics

1. Describe the site in terms of usage and surroundings.

2. Age of building. Include information on additions and major renovations.

3. a. Approximate square footage of the building, including each floor if multistory.

    b. What is the approximate ceiling height in the majority of the structure?
    _____ feet

4. Number of floors above the substructure: _____ floors

5. What are the structural materials of the exterior of the building?

6. Does the building have an attached or enclosed garage or a loading dock?

Figure 17.4. *Continued*

7. a. What is the source of water for the structure?

    b. What is the primary source of public supply?

8. Describe the internal construction characteristics of the building:

    a. Are there:

False walls?	☐ yes	☐ no
Movable walls?	☐ yes	☐ no
Movable partitions?	☐ yes	☐ no
False ceilings?	☐ yes	☐ no
Interfloor spacing?	☐ yes	☐ no

    b. What are the surface materials of:

    Walls   _____

    Floors   _____

    Ceilings   _____

    Nonfixed structures _____

9. a. Main type of heating:

    b. Secondary sources of heating:

Figure 17.4. *Continued*

c. Main type of heating fuel:

d. Type of fuel used for cooking in the main kitchen:

10. What is the source of hot water for the site?

11. What type of air-conditioning system(s) are present?

12. a. Are there any fixed ventilation systems, other than the heating and air conditioning, present in the structure?   □ yes   □ no
    Specify type, location, control mechanism, filters, and particle collectors.

    b. Describe all secondary ventilation devices, including portable fans, and kitchen and bathroom exhausts.

13. Describe openings in walls and ceilings.
    a. How many windows are present?   _____ windows
    b. What percent can be opened?   _____ %

Figure 17.4. *Continued*

c. Describe any windows of unusual type or construction.

d. How many doors and other penetrations of walls are present?

_____ doors

_____ other penetrations

e. How many penetrations through the ceilings and floors are present?

14. If known, what is the general rate of air exchange for this building?

15. What are the normal temperature and RH settings for the common areas of the building?

	Winter		*Summer*	
	Day	*Night*	*Day*	*Night*
Temperature				
RH				

16. Changes in the building in the past six months:

a. Describe any major contruction, renovations, or modifications.

b. Describe any specific weatherization, or building tightening actions taken during the same period.

Figure 17.4. *Continued*

c. Have any changes been made in this building that include the addition of:

Furniture, drapes	☐ yes	☐ no
Synthetic carpet	☐ yes	☐ no
Wallpaper	☐ yes	☐ no
Interior paint	☐ yes	☐ no
Ceiling, floor finishes	☐ yes	☐ no
Plywood, particleboard	☐ yes	☐ no
Foam insulation	☐ yes	☐ no
Interior paneling	☐ yes	☐ no

17. a. How often is the interior of this building treated with pesticides?

b. When was the last time this building was treated?

c. By whom was the building last treated?

d. What product(s) were used and in what quantities?

e. Describe formulation, application instructions, and other details of application.

Figure 17.4. *Continued*

18. Describe cleaning products ge erally used in this building. Include soaps, waxes, deodorants, disinfectants, polis es, etc.

Product	Quantity Us d in Average Mo th	Frequency of Application	Last Application

19. In the location, is there:

   a. Foam insulation   ☐ yes ☐ no

   b. Polyurethane products   ☐ yes ☐ no

   c. Plywood subflooring   ☐ yes ☐ no

   d. Wall paneling   ☐ yes ☐ no

   e. Composition board   ☐ yes ☐ no

   f. Particleboard   ☐ yes ☐ no

20. Identify and give the age of fur iture in this location.

Item	Age	Co struction Material	Approx. Surface Area

Fi ure 17.4. *Continued*

21. Describe the type, surface area, and age of any wall coverings in the location including backing material and glue/paste.

22. Describe any carpeting:
    a. Age:
    b. Backing:
    c. Glues:
    d. Surface Area:

23. Describe any pressure gradients present in this structure, especially those near potential sources of the pollutants of interest.

24. Average number of occupants in the location.
    Day:
    Night:

25. Number of smokers present during the monitoring period:

26. Were there any pets present during the monitoring period and were flea collars or flea powders in use?

27. Did anyone engage in any of the following hobbies in the location during the monitoring period?
    a. Woodworking   ☐ yes   ☐ no
    b. Painting   ☐ yes   ☐ no
    c. Ceramics/Pottery   ☐ yes   ☐ no
    d. Photographic developing ☐ yes ☐ no
    e. Other   Type _____

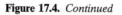

Figure 17.4. *Continued*

28. a. Describe the number and type of pest strips in use.

    b. Identify any other pesticides in use and time of use during monitoring.

29. Were any windows in the location open during the monitoring period?
    Percent of time open _____ %

30. a. Is there a gas cooking stove at the location?
    b. Is it vented?
    c. Does it have a pilot light?
    d. What percent of time was it in use during monitoring?

31. a. Is there a gas or kerosene space heater in the location?
    b. Is it vented?
    c. Does it have a pilot light?
    d. What percent of the time was it in use?

32. a. Was a free-standing stove or fireplace in the location in use during the monitoring period?
    b. Percent of time in use.
    c. Energy source.
    d. Is it vented?

33. a. Was a clothes dryer in use during the monitoring period?
    b. Percent of time in use.
    c. Energy source.

Figure 17.4. *Continued*

34. a. Was a humidifier in use in the location during the monitoring period?

b. Percent of time in use.

35. a. When was the area last vacuumed?

b. When was the area last dusted?

36. a. What is the rate of air exchange in the kitchen?

b. What is the rate of air exchange in the rest of the location?

37. Household products

Product	Number of Hours Since Last Use	Amounts Used	Frequency of Use	Surface Area Covered
a. Cleaning				
b. Aerosol				
Drugs/Bronchodilator				
Vaporizer				
Hair Products				
Personal Hygiene				
Deodorant				
Foot Spray				
c. Housekeeping				
Disinfectants				
Waxes				
Bathroom/Kitchen Deodorants				

Figure 17.4. *Continued*

addition to the levels of contaminants inside in order to determine whether these contaminants are being entrained from the outside by the ventilation system or being generated by sources within. Therefore, both indoor and outdoor samples should be collected for contaminants such as carbon monoxide, nitrogen oxides, total particulate, ozone, and volatile organic hydrocarbons. Portable GCs have also been used to take samples of indoor and outdoor air to identify whether a source was inside or outside the building. Sampling results should be entered on a floor diagram of the building. Radon and asbestos do not normally cause symptoms in building occupants; however, they are commonly sampled due to concerns for their carcinogenicity.

Initiate a discussion of the building heating, ventilating, and air-conditioning (HVAC) system with maintenance personnel to identify how often and what percent of outside air is allowed into the building. The amount of outdoor air is generally changed during the day in order to control heating and cooling costs while maintaining adequate indoor air quality. It should be noted that while most often indoor air quality problems are remedied by increasing the ventilation rate in the affected area, this remedy is in conflict with strategies for reducing the energy consumption by lowering ventilation rates; therefore, sources and chemical characteristics of indoor air pollutants should be identified so that energy-efficient control strategies can be implemented to eliminate them.[8]

Carbon dioxide is used as a surrogate measure for the lack of ventilation in a building and the potential that other contaminants could be building up. An increase of this byproduct of human respiration over the course of a day is an indicator. Carbon dioxide is measured with an infrared spectrometer. As mentioned, temperature and humidity should also be monitored, since they can have a significant

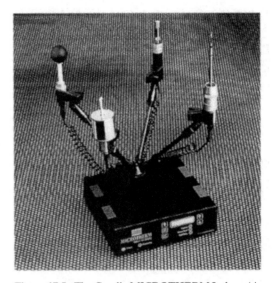

Figure 17.5. The Casella MICROTHERM Indoor Air Quality instrument consists of a central control and data-logging unit to which different sensors are connected. (Courtesy of Casella USA.)

impact on the discomfort of building inhabitants. One way is through the use of an instrument that will measure all of these factors simultaneously and log the data for printout and analysis.

The MICROTHERM Indoor Air Quality instrument from Casella (Figure 17.5) illustrates a typical device designed for IAQ measurements. It consists of a central control and data logging unit to which different sensors are connected. It measures temperature, relative humidity, air velocity, globe temperature, wet bulb temperature, carbon dioxide, carbon monoxide, and ozone. It can be connected to other devices to simultaneously record noise levels, light intensity, and particulate concentrations.

It may be configured as a single station monitor with all of the sensors grouped together, or with sensors located remotely throughout the work area and connected to the unit by cable. The internal memory can hold up to 13,000 data points with an expansion memory available. Results can be downloaded to a computer or viewed on the instrument's LCD output.

TABLE 17.2. Formaldehyde Guidelines for Indoor Air Situations

Level (ppm)	Interpretation
0.01–0.05	Background (outdoor) levels
0.04–0.09	Average conventional home
0.09–0.62	Average mobile or prefab home
≤0.10	ASHRAE guideline for indoor air
≤0.40	HUD standard for prefab homes
>0.1	Mild irritation or allergic sensitization in some people
>0.5	Irritation to eyes and mucous membranes
>1.0	Possible risk of nasopharyngeal cancer
>3.0	Respiratory impairment and damage

Source: Office of Pesticides and Toxic Substances, US EPA, *Assessment of Health Risks to Garment Workers and Certain Home Residents from Exposure to Formaldehyde.* Washington, D.C., 1987.

TABLE 17.3. Comparison of Daily Outdoor Levels with Indoor Levels for Selected Chemicals

Compound	Average Daily Concentration (ppb)	
	Indoor	Outdoor
Acetone	7.955	6.927
Benzene	5.162	2.800
Carbon tetrachloride	0.40	0.168
Chloroform	0.832	0.630
1,3-Dichlorobenzene	3.988	0.889
1,4-Dichlorobenzene	0.932	0.996
1,4-Dioxane	1.029	0.107
Formaldehyde	49.4	8.293
α-Pinene	0.547	0.484
1,1,2,2-Tetrachloroethane	0.014	0.101
Tetrachloroethene	3.056	0.853
Toluene	7.615	0.775
Trichlorobenzene	0.065	0.016
1,1,1-Trichloroethane	48.90	0.911
Tricholoroethene	1.347	0.495

Source: Shah, J. J., and H. B. Singh. Distribution of volatile chemicals in outdoor and indoor air. *Environ. Sci. Technol.* **22**(12):1381–1388, 1988.

The American Society of Heating, Refrigeration, and Air Conditioning Engineers (ASHRAE) has developed a number of guidelines for indoor air contaminants, in most cases based on state or EPA standards or guidelines. ASHRAE standards are useful for interpreting results in indoor air. In some cases, a level can be compared to a known health effect occurring at that level. An example of a compound typically measured in residences and indoor air situations is formaldehyde. Table 17.2 describes guidelines that have been developed for formaldehyde levels in indoor air.[9] In these situations, workplace levels are unlikely to be present and therefore those standards would not be applicable. When adapting NIOSH methods to collect samples to compare to these low-level standards, modifications such as extended sample times are required.

Outdoor samples (Table 17.3)[10] should be compared to a relevant EPA standard if one exists. In addition, whenever collecting outdoor samples, an estimation of typical levels present in the background should be made. This information is available from a number of sources, including a data base established by the EPA. The ratio of the results of indoor:outdoor sampling can be used to provide insight as to the source of contamination. For example, for volatile organic hydrocarbons such as benzene and xylene, ratios close to one suggest an outdoor source.[11]

LEAK TESTING: FUGITIVE EMISSIONS MONITORING

Fugitive emissions monitoring is performed with general survey instruments such as the flame ionization detector (FID) and photoionization detector (PID) described in Chapter 11. Fugitive emissions are generally defined as leaks from process components, such as valves, pumps, compressors, pressure relief devices, sampling systems, process drains, storage tanks, and open-ended lines in volatile organic compounds (VOCs) service. As defined by the Environmental Protection Agency (EPA), these are limited to emissions that do not occur as part of the normal operation of plants, but are the result of the effects of age, lack of maintenance, improper equipment specifications, or externally caused damaged.[12] Other types of emissions which are monitored include breakthrough from carbon canisters, and emissions from stacks, vents, and roof monitors. Emissions sampling can involve a determination of emission factors, the emission rate for a specific piece of equipment, or the concentration and composition of the ambient plant air. Identifying leak sources, which often are not repaired until a process can be shut down for maintenance, is also important for protecting plant personnel when working in these areas.

In some cases, leaks can be detected using a bubble solution that is sprayed on a potential leak source and then observing the potential leak sites to determine if any bubbles are formed. If no bubbles are observed, the source is presumed to have no detectable emissions or leaks. This method is still used, but only for sources that do not have continuously moving parts, that do not have surface temperatures greater than the boiling point or less than the freezing point of the soap solution, that do not have open areas to the atmosphere that the soap solution cannot bridge, or that do not exhibit evidence of liquid leakage. In general, the soap solution is best reserved as a gross screening technique, and sources passing this test should be followed up with a monitoring instrument.

The first portion of the sampling program involves determining the actual number of potential leak sources, such as valves in a plant. The second portion involves testing these sites. Regardless of the source being tested, it should be remembered that a regular, scheduled inspection and maintenance program is the key to reducing fugitive emissions.

There are two types of standards used to evaluate the results of fugitive emissions surveys. The first involves classifying leaks compared to a concentration defined in a particular standard. The second involves classifying leaks based on the difference between this measurement and a previous background measurement. A response factor must be determined for each compound that is to be measured. The EPA has specified that a VOC analyzer, such as an FID-based instrument, used for fugitive emissions is to be calibrated at 10,000 ppm for methane or hexane, and that daily recalibration checks must be performed.

While the procedure below describes a method for identifying leaks, it should be remembered that the leak rate is more important. However, determining this is more difficult than just using a spot measurement. One technique that has been used is to enclose the leak source within an air bag of known volume and to measure the increase in concentration inside the bag at regular intervals.[13] The leak rate can then be calculated.

Procedure for Leaks Based on Defined Concentrations in Standards[12]

1. The sampling practitioner should be alert for audible indications of leaks and also odors during the survey. Several problems can occur when sampling gases that have too high a concentration for an instrument. These include flame-out of FIDs and condensation on the lamp of PIDs. For these reasons, the user should monitor the hydrocarbon concentration while slowly approaching the valve stem, pump shaft seal, or other source. If the concentration exceeds the leak definition before the probe is placed close to the leak site, there is no reason to continue monitoring at that spot.

2. When monitoring sources such as valves and pumps that handle heavy liquid streams at high temperatures, relatively nonvolatile liquids can condense in the probe and the detector. This can slow down the instrument's response.

3. Place the probe inlet at the surface of the interface where leakage could occur. Move the probe along the interface periphery while observing the device's display. If an increased reading is observed, slowly sample the interface where leakage is indicated until the maximum meter reading is obtained. Here are guidelines for specific types of equipment:

   • *Valves:* The most common source of leaks from valves in closed systems is at the seal between the stem and housing (valve stem packing gland). Place the probe at the point where the valve stem leaves the packing gland. The normal procedure is to circumscribe this location within 1 cm of the valve stem. Also place the probe at the interface of the packing gland take-up flange seat and sample the periphery. Valves used at the end of drains or sample lines have two sources of leakage: the valve stem and the valve seat. The probe is usually placed at the center of the discharge pipe for monitoring any valve seat leakage.

   • *Flanges and Other Connections:* For welding flanges, place the probe at the outer edge of the flange–gasket interface and sample the circumference of the flange.

   • *Pumps and Compressors:* Fugitive emissions from pumps occur from the pump shaft seal used to isolate the process fluid from the atmosphere. The most commonly used seals are single mechanical seals, double mechanical seals, and packed seals. Conduct a circumferential traverse at the outer surface of the pump or compressor shaft and seal interface. If the source is a rotating shaft, position the probe inlet within 1 cm of the shaft–seal interface for the survey. In addition, sample all other joints on the pump or compressor housing where leaks could occur.

   • *Pressure Relief Devices:* The configuration of most pressure relief devices prevents sampling at the sealing–seat interface. For those devices equipped with an enclosed extension, or horn, place the probe inlet at approximately the center of the exhaust discharge to the atmosphere.

   • *Process Drains:* For open drains, place the probe inlet at approximately the center of the area open to the atmosphere. For covered drains, place the probe at the surface of the cover interface and conduct a peripheral traverse.

   • *Open-Ended Lines or Valves:* Place the probe inlet at approximately the center of the opening to the atmosphere.

   • *Seal System Degassing Vents and Accumulator Vents:* Place the probe inlet at approximately the center of the opening to the atmosphere.

   • *Access Door Seals:* Place the probe inlet at the surface of the door–seal interface and conduct a peripheral traverse.

4. Leave the probe inlet at each maximum reading location for approximately two times the instrument response time.

5. If the maximum observed meter reading is greater than the leak definition in the applicable regulation, record and report the results as specified in the appropriate regulation.

1. Determine the local ambient concentration by moving the probe inlet randomly upwind and downwind at a distance of 1 meter to 2 meters from the source. Use the most sensitive scale for this reading.
2. Next move the probe inlet to the surface of the source and determine the concentration by using the procedures described in the previous section for the various sources.
3. The difference between the ambient concentration and those detected around the leak sources determines whether there are any detectable emissions.

WELDING FUMES

Welding fumes cannot be classified simply. The composition as well as the amount of fume released depends on the base metal being welded and type of welding process, including the type of electrodes used. Stick (i.e., shielded metal arc) welding produces high levels of fume. Gas metal shielded welding tends to produce lower levels of fume and higher levels of gases, such as ozone. Fumes often contain iron oxide, manganese, silicon dioxide, and other metals such as chromium and nickel depending on the materials involved. For example, welding on stainless steel will produce chromium and nickel oxide. Some coated and flux-cored electrodes are formulated with fluorides, and the fumes associated with them may contain significant amounts of fluorides.

In most fusion welding operations, the base metal does not contribute significantly to the total welding fume. Surface coatings, such as primer paints or galvanizing, can contribute significantly to the fume. Therefore, the coating constituents should be analyzed in the fume when they are present on the base metal; otherwise, the composition of the constituents depends largely on the filler metal rod being used and the process conditions. Metals are predominantly present as mixed metal oxides together with some silicates and unoxidized metal. Fume from welding of stainless steels may contain about 10% chromium predominantly in the hexavalent state.[14] One way of determining the constituents of the rods, electrodes, or sticks is a review of the appropriate material safety data sheet (MSDS).

Welding fume particulate contains a wide range of airborne particulates from fine chains to spherical particles.[13] Particle size sampling of welding fume is complicated by the fact that the particles form agglomerates so that smaller particles decrease in number, but the number of larger-size particles increases along with a change in the overall dispersion of particles.[15] The type of welding process will also affect particle size. For example, particles from the shielded metal arc welding process are somewhat larger in size than those from other welding processes, ranging up to $2\,\mu m$ in size.[16]

Therefore, it is important to measure the respirable fraction of the fume as well as to determine its individual constituents. Stick welding fume consists of chains and clusters of submicron particles and glassy coated spheres up to $10\,\mu m$ in diameter.[14] Larger particles originating from electrode spatter are also present, including unoxidized metal. Spatter particles are typically above the respirable range, but as they oxidize during passage through the air outside the arc, they may contribute to the fume.

When selecting which welders to sample, a sampling priority can be developed based on expected exposure levels. Welding in confined spaces presents the greatest hazard of exposure and is the first priority to sample. Persons using filler or base metals that contain highly hazardous components, such as lead or cadmium, are next. If highly visible fumes are present, they are an indicator that high exposures may be occurring. Another way of categorizing operations is based on current settings and arc times. In general, the higher the current settings and the longer the arc times, the greater priority for sampling.[17] Following is a list of observations to make when doing sampling for welding fumes in order to prioritize sampling[18]:

- Identify and diagram the welding area and general work area.
- Determine if the work is being done in a confined space.
- Measure the size of the work area or room.
- Determine the number of welders in the work area.
- Identify any barriers in the work space that will obstruct general dilution ventilation, such as curtains, partitions, trucks, or other machinery.
- Review chemicals to determine if there are any solvents that are chlorinated or flammable in use in the area.
- Identify which welding processes are in use, including the specific types of rods, size of wire, and grades of gases.
- Determine the welding conditions: amperage, voltage, and polarity.
- Identify the base metal and any surface coatings such as paints, rust inhibitors, or plating.
- Review schedules to determine if the job being sampled is typical or nonroutine.
- Determine if the rate of production is light, heavy, or normal.
- If the work is being done at a constant rate, identify the longest arc time.
- Review the work practices of the welders to determine if their posture causes them to work in the welding plume, lean away from the plume, or if

it varies with the location of the arc and the plume.

- Identify if any personal protective equipment is in use, including helmets, face shields, and respirators.
- Determine what type of ventilation is in use.
- For local exhaust, determine the type, location with respect to the arc and distance from the arc, and the size of the hood and its flowrate.
- See if the positions of local exhaust hoods are fixed or mounted on flexible ducting.
- Identify any area fans and their placement with respect to the local exhaust and the welder.
- For air makeup units, determine their location relative to the welder and local exhaust hoods, as well as the air movement direction relative to the welders.

Welding fumes are collected on filters contained in cassettes. The cassette should be placed inside the welding hood. Welding hoods have been reported to provide protection factors ranging from 3.3 to 15.[19] The extent of reduction of fume concentrations at the breathing zone due to the use of welding helmets ranged from 29% to 64% in one study, depending on the type of welding and the welders' postures.[20] This illustrates why it is important to collect samples under the welders' hood.

The American Welding Society has defined the welder's breathing zone to be the area immediately adjacent to the welder's nose and mouth, inside the welder's helmet when worn, or within 9 inches of the nose and mouth when a helmet is not worn.[21]

In addition to particulate exposures, it should be remembered that many gases (Table 17.4) are generated during welding operations as well.[18] For more information

TABLE 17.4. Gases Associated with Specific Arc Welding Processes

Welding Process	CO	Fluorine	NO$_x$	Ozone
Shielded metal arc welding	X		X	
Flux cored arc welding	X	X	X	
Argon helium shielding			X	
Carbon dioxide shielding	X		X	
Gas tungsten arc welding			X	X

on how to sample these, see Chapters 6 and 10.

Procedure

1. Use a three-stage cassette with a 0.8-μm MCEF filter inside. The pump should have been calibrated using the technique for open face filter sampling. The top should be taken off the cassette for sampling.

2. Tape or otherwise attach the filter cassette under the welder's hood.

3. Analysis will depend on whether any toxic elements are present in the welding rods, base metal, or if a metal coating exists. Sampling for fumes does not eliminate the need to evaluate the potential for hazardous gases to be present.

Total welding fume samples are analyzed gravimetrically, that is, they are weighed. This type of sample provides baseline information and can be compared to the TLV for total welding fume, but does not provide information on metals of concern such as Ni or Cr^{6+}. If analyses for specific contaminants is required, coordinate with the analytical laboratory in advance to ensure that the proper filters are used for sampling. Results should be converted to the relevant compounds, for example, iron oxide (Fe$_2$O$_3$) is the constituent generated during welding whereas iron (Fe) is what is measured during the analysis.

CARBON MONOXIDE FROM FORKLIFTS

A typical situation involving the use of propane-fueled forklifts is moving pallets of raw materials or products in a warehouse. The exhaust from propane can contain significant amounts of carbon monoxide. These situations often are aggravated by the fact that doors are blocked with plastic strip curtains to minimize leakage of cold air into the buildings. The forklifts usually move in and out of various storage rooms, in and out of truck trailers or railroad boxcars, as well as going outside occasionally. Often the survey is a result of employee complaints of symptoms such as headaches while on the job (Table 17.5).

There are several options for CO sampling for this exposure: A direct-reading CO instrument with data-logging capability or long-term colorimetric tubes are the most common techniques. Personal monitoring data are preferred when judging exposures, so area monitoring instruments would not be a good choice. Field notes should record peak traffic periods and time spent inside enclosed spaces such as truck trailers if possible to allow a better understanding of exposure situations. Samples should be collected according to a *monitoring plan* based on a sound *exposure assessment strategy* (Chapter 3).

Typical levels that might be measured in an uncontrolled exposure situation are shown as follows:

Time	CO level (ppm)
8 A.M.	17
9 A.M.	25
10 A.M.	37
11 A.M.	25
Noon	48
1 P.M.	52
2 P.M.	70
3 P.M.	53
4 P.M.	41

When compared to the current ACGIH TLV® of 25 ppm expressed as an 8-hour time-weighted average exposure, it is clear that employees are being overexposed.[22] Controls such as replacement of the propane-powered forklift trucks with battery-powered models or increased ventilation are required.

In addition, if employees had been complaining of symptoms such as headaches and nausea, biological monitoring as described in Chapter 18 might be appropriate.[23] This could be blood sampling for carboxyhemoglobin, or carbon monoxide in exhaled air. Table 17.5 shows symptoms caused by different carboxyhemoglobin levels in blood. Smokers should be identified if biological samples are collected so the influence of smoking will not be attributed to occupational exposures.

MULTIPLE SOLVENTS IN PRINTING INK MANUFACTURE

Printing ink manufacture is a typical situation where many different solvents are in use in one operation. For example, isopropanol, 1,1,1-trichloroethane, methyl ethyl ketone, and toluene might be used in various blending operations being performed in the same room along with other solvents.

In this type of survey, both air samples and biological monitoring may be needed. In addition, several OSHA standards and ACGIH TLVs® would be of concern affecting the way that samples are collected. All of these compounds have 15-minute short-term exposure limits (STELs), as well as 8-hour TWA standards.

Integrated air samples would be collected using multiple-tube holders, since methyl ethyl ketone requires a different type of sorbent tube than the other solvents and isopropanol should be analyzed separately from toluene and 1,1,1-trichloroethane. A large charcoal tube

TABLE 17.5. Symptoms Caused by Various Amounts of Carbon Monoxide Hemoglobin in the Blood

Blood Saturation (% COHB)	Symptoms
0–10	None
10–20	Tightness across forehead, possible slight headache, dilation of cutaneous blood vessels
20–40	Headache and throbbing in temples, severe headache, weakness, dizziness, dimness of vision, nausea, vomiting, collapse
40–50	Same as previously noted with increased likelihood of collapse and syncope, increased respiration and pulse
50–60	Syncope, increased respiration and pulse, coma with intermittent convulsions and Cheyne–Stokes respiration
60–70	Coma with intermittent convulsions, depressed heart action and respiration and possible death
70–80	Weak pulse and slow respiration, respiratory failure and death

would be used for the other compounds. Tubes would be changed approximately every 1 to 2 hours, to avoid potentially saturating the tubes. For sampling periods designed to measure STELs, tubes would be changed after 15 minutes.

Biological monitoring would be done if there was a concern that employees were overexposed. For example 1,1,1-trichloroethane can be sampled in the exhaled air, or its metabolite trichloroethanol can be measured in the urine. Toluene can be monitored as its metabolite, hippuric acid, in urine, in venous blood, or in exhaled air. While ethanol is commonly monitored in the breath of drivers suspected of intoxication, the levels that can be achieved by drinking are unlikely to occur from inhalation while working around this material.

When results are received, in addition to comparisons with these standards, the potential for additive effects must be considered. It is not uncommon for individual exposures to be less than the PEL, but for the additive exposure—the ratio of the levels of each compound to its PEL added together—to exceed one and represent an overexposure.

SUMMARY

This chapter describes possible monitoring approaches and considerations for typical sampling situations. It is important to note that for some conditions (e.g., confined space entry) many other safety and health considerations besides air monitoring are critical to protecting people, and for other applications (such as indoor air quality problems) air sampling is performed but other investigative approaches besides air sampling are generally required to diagnose and resolve problems.

Exposure monitoring should always be performed according to a *monitoring plan* developed from a sound *exposure assessment strategy* as described in Chapter 3.

REFERENCES

1. OSHA General Industry Safety and Health Regulations, U.S. Code of Federal Regulations, Title 29, Chapter XVIII, Part 1910.146.

2. McDermott, H. J. *Handbook of Ventilation for Contaminant Control*, 3rd ed. Cincinnati: ACGIH, 2001.

3. American Industrial Hygiene Association. *The Industrial Hygienist's Guide to Indoor Air Quality Investigations.* Fairfax, VA: AIHA, 1993.

4. National Research Council—Committee on Indoor Pollutants. *Indoor Pollutants* Washington, D.C.: National Academy Press, 1981.

5. Cain, W. S., et al. Ventilation Requirements in Buildings—I. Control of occupancy odor and tobacco smoke odor. *Atm. Environ.* **17**(6): 1183–1197, 1983.

6. American Conference of Governmental Industrial Hygienists. *Industrial Ventilation—A Manual of Recommended Practice*, 24th ed. Cincinnati: ACGIH, 2001.

7. American Society of Heating, Refrigerating and Air Conditioning Engineers. *ASHRAE Standard 62. Ventilation for Acceptable Indoor Air Quality.* New York: ASHRAE, current edition.

8. Hollowell, C. D. *Indoor Air Quality.* Lawrence Berkeley Laboratory, Energy and Environment Division, June 1981. US DOE Contract W-7405-ENG-48.

9. Office of Pesticides and Toxic Substances, US EPA. *Assessment of Health Risks to Garment Workers and Certain Home Residents from Exposure to Formaldehyde.* Washington, DC, 1987.

10. Shah, J. J., and H. B. Singh. Distribution of volatile chemicals in outdoor and indoor air. *Environ. Sci. Technol.* **22**(12): 1381–1388, 1988.

11. Montgomery, D. D., and D. A. Kalman. Indoor/outdoor air quality: Reference pollutant concentrations in complaint-free residences. *Appl. Ind. Hyg.* **4**(1): 17–20, 1989.

12. EPA Reference Method 21. *Determination of Volatile Organic Compound Leaks.*

13. Jones, A. L. The measurement and importance of fugitive emissions. *Ann. Occup. Hyg.* **28**(2): 211–215, 1984.

14. Hewitt, P. J., and C. N. Gray. Some difficulties in the assessment of electric arc welding fume. *AIHA J.* **44**(10): 727–732, 1982.

15. Glinsmann, P., and F. S. Rosenthal. Evaluation of an aerosol photometer for monitoring welding fume levels in a shipyard. *AIHA J.* **46**(7): 391–395, 1985.

16. American Welding Society. *Effects of Welding on Health.* AWS, Miami, FL, 1979.

17. American Welding Society. *Evaluating Contaminants in the Welding Environment: A Sampling Strategy Guide.* AWS F1.3-83, Miami, FL: AWS, 1982.

18. American Welding Society. *The Welding Environment.* Miami, FL: AWS, 1979.

19. Goller, J. W., and N. W. Paik. A comparison of iron oxide fume inside and outside of welding helmets. *AIHA J.* **46**(2): 89–93, 1985.

20. American National Standards Institute/ American Welding Society. *Method for Sampling Airborne Particulates Generated by Welding and Allied Processes.* Method No. F1.1-1978.

21. National Institute of Occupational Safety and Health. *A Guide to the Work-Relatedness of Disease.* NIOSH Pub. No. 79-116. Cincinnati: NIOSH, 1979.

22. American Conference of Governmental Industrial Hygienists. *Threshold Limit Values and Biological Exposure Indices for 2002.* Cincinnati: ACGIH, 2002.

23. Plog, B. A., and P. J. Quinlan, eds. *Fundamentals of Industrial Hygiene*, 5th ed. Itasca, IL: National Safety Council, 2002.

CHAPTER 18

BIOLOGICAL MONITORING

Biological monitoring (also called "bio-monitoring") measurements of bodily fluids or breath are considered better measurements of exposure than the values measured in the air since they represent an individual's actual absorbed dose. Biological monitoring techniques continue to be developed, and the number of compounds for which methods are available can be expected to increase.[1] Currently it is almost exclusively applied to occupational exposures.

There are many specific purposes for conducting biological monitoring. Many of the Occupational Safety and Health Administration's (OSHA) substance-specific standards require monitoring of urine or blood for compliance if certain airborne levels are exceeded. Biological monitoring can assist in evaluating the contribution from dermal and oral exposures and in the absence of hazardous airborne levels, whether skin or oral exposures are occurring. In the case of an emergency release, an assessment of the degree of exposure can be performed (but not detection of whether an acute exposure has occurred as discussed in the section on limitations later in this chapter). It can also be used to monitor the effectiveness of personal protective equipment (PPE). For example, in one study it was determined that a respirator cartridge was overloaded due to high mandelic acid levels in the urine of a worker exposed to styrene in a fiberglassing operation.[2] Another use of biological monitoring is to assess the effectiveness of engineering controls or work practices to limit uptake of environmental chemicals.

As with air sampling results, the presence of a biological marker of exposure in an individual does not establish that the exposure caused a toxic effect.

Sometimes biological monitoring is termed medical monitoring or biomedical surveillance; however, this should not be confused with health or medical surveillance that consists of periodic (often annual) exams and tests since its purpose is

Air Monitoring for Toxic Exposures, Second Edition. By Henry J. McDermott
ISBN 0-471-45435-4   © 2004 John Wiley & Sons, Inc.

to identify whether an adverse effect has occurred such as increased protein in urine from exposure to hepatotoxic chemicals. Biological monitoring is meant to be preventive in that the tests are selected to identify whether an overexposure has occurred but not an adverse effect. These changes and concentrations that are measured are temporary in nature and should be representative of events that occur long before those changes measured in conventional health surveillance tests such as liver function, blood counts, and X rays. Clinical pathology tests can be difficult to interpret as normals fall into a wide range. Therefore, if a result is near the end of a range it is difficult to tell if damage has occurred or not. Changes are easiest to detect when monitoring has been conducted on the same individual over a period of time, with an initial "base line" value for comparison to later measurements.

Another way to view the difference is to consider that the aim of biological monitoring is to assess whether a significant dose has been absorbed, whereas many medical monitoring tests are aimed at determining whether a disturbance in physiology as a result of the absorbed does has occurred. For example, liver function test are used to indicate abnormalities in function rather than the mere presence of a compound. While an abnormality in such a test does not mean the organ is compromised, it does indicate some change may be taking place. Also, there are often two levels of concern when biological monitoring is performed: one being a statistical significant level, meaning that there is statistical confidence that the absorbed dose is higher than background, and the other of clinical concern, because at this point the individual may develop an exposure related illness or medical condition.

Finally, the types of tests appropriate for biological monitoring can vary including those that provide useful information for medical surveillance. For example, an increase in cadmium levels in urine may be indicative of an adverse effect on the kidneys (saturation of the detoxification mechanism), whereas cadmium in blood correlates much better with recent exposure; therefore, it is the preferred test for biological monitoring.[3] In some cases, biological monitoring is preferred to air sampling. For example, studies of occupational exposures to agricultural (primarily organophosphate pesticide-related) workers have found that inhalation exposures contribute much less to the absorbed body burden than dermal exposures.[4] Other situations include that assessment of individual work practices, determination of the effectiveness of PPE, and assessment of the contribution of oral and dermal exposures.

Blood and urine are sampled by collecting the bulk fluid. Breath sampling is done using basic air sampling techniques such as Tedlar bags, sorbent tubes, and a specialized glass pipet. Analysis of blood and urine is always done in the laboratory and real time instruments are sometimes used for breath samples. Respirators with specially adapted cartridges have also been used for breath collection, although as of now there are no biological standards based on this technique.

Biological monitoring has often been called the ultimate personal sampler, because when properly used it can assess worker exposure to industrial chemicals by all routes, including skin absorption and oral indigestion. Specifically, biological monitoring includes the measurement of the absorption of an environmental chemical in an individual. There are two basic types of biological monitoring used to develop methods: direct measurement of a chemical or a metabolite in biological media, or indirect measurement by quantifying a nonadverse biological effect related to the exposure.

Nonadverse or subcritical effects have been defined as effects that do not impair

TABLE 18.1. Sources of Pharmacokinetic Variability

Absorption	Distribution	Metabolism
Route	Body size	Genetic factors
Physical form	Body composition	Age and gender
Solubility	Protein binding	Environment (pollution, diet, habits)
Physical work load	Physical work load	Chemical intake (alcohol, medications)
Exposure concentration	Exposure concentration	Physical activity (pulmonary ventilation, blood flow)
Exposure duration	Exposure duration	Protein binding
Skin characteristics		Life-style (smoking)
		Exposure level

Source: Reference 7.

cellular function, but are still evident by means of biochemical or other tests, and as a precursor of a critical effect, a subcritical effect may be a more useful measurement than others in preventing harmful exposures.[5] Direct measurements are done more often than indirect ones and include such tests as blood lead as an assessment of chronic lead exposure, measurement of urinary phenol for assessment of exposure to benzene, and measurement of carboxy-hemoglobin in blood as an index of carbon monoxide or methylene chloride exposure. Examples of indirect measurements include cholinesterase inhibition by organophosphate compounds and inhibition of a delta-amino levulinic acid dehydratase (ALAD) (a red blood cell enzyme) by inorganic lead.

Another type of test is the measurement of a specific antibody to determine if exposure to an allergen has taken place. In one study individuals who had become sensitized to pigeons were identified by testing for a reaction of a specific IgG antibody in their blood with pigeon serum proteins.[6] The limitation of direct and indirect measurement methods is that although they measure exposure, they do not quantify the amount that has reached a target organ, such as the liver in the case of exposures to chlorinated solvents.

The basis for biological monitoring is pharmakinetics (also sometimes called toxicokinetics when used to describe the effects of nondrugs or toxic compounds), which is the study of the process of uptake, distribution, and elimination of substances from the body. Once in the body, chemicals are metabolized and excreted or stored in an organ, fat, or bone. The liver is the primary site for metabolism. Some of the chemical may be excreted unchanged, such as occurs when solvents are inhaled and partially eliminated in the breath.

Accumulation of a chemical such as lead in the blood or in the hepatobiliary intestinal loop can also occur. Once in the blood, chemicals can bind to plasma or other proteins or circulate unbound. The unbound chemical is usually the entity that is responsible for toxic effects unless the target organ is the blood.

Each individual represents a unique biological system (Table 18.1). Workers differ in age, body build, weight, fitness, physiological and nutritional status, and habits such as smoking and alcohol consumption, all of which can affect their intake. Work rate has also been shown to affect intake in a given exposure situation. As an example, concentrations of mandelic acid in urine due to styrene exposure decrease when light work rather than heavy work is performed.[2] Metabolizing enzymes can be induced by smoking, medication, and absorption of other chemicals present in the environment or diet. The simultaneous absorption of chemicals in relatively high dosages, as in alcohol consumption, drug intake, or simultaneous occupational exposure to other chemicals, can slow down the metabolic rate by inhibition of the metabolizing enzyme system.[7] For example, alcohol (ethanol) consumption during or shortly following an exposure period to toluene will increase toluene levels in the blood.[8]

As a result, when the same group of workers is exposed to the same airborne concentration of a given chemical, each individual will have differences in the uptake, absorption, biotransformation, and elimination of that compound. This will produce differences in the amount of active metabolite(s) reaching their target organs.[9]

Some types of organic compounds can undergo extensive biochemical transformations during metabolism. In these situations, it is a metabolite that is usually monitored rather than the individual compound. Since there is often more than one metabolite, selection of the proper one to monitor is important. The metabolite must be specific to the exposure and must accumulate in biological fluids in quantities proportional to the dose of the original compound in order for monitoring to be effective. An example is the determination of 2,5-hexanedione in urine following exposure to *n*-hexane. Often compromises are necessary and the metabolite does not always meet this criterion perfectly. In the case of *n*-hexane, while there are no natural (endogenous or diet) sources of 2,5-hexanedione, it is also a metabolite product of methyl *n*-butyl ketone (MBK).[8]

The biologic half-life (BHL) determines whether a chemical is rapidly excreted or stored in the body (Table 18.2). This value is specific for a given chemical and can be determined for a chemical's residence time in the entire body or in a specific tissue, in the case where a chemical is stored.

TABLE 18.2. Biologic Half-Lives

Chemical	Half-Life (hours)
Acetone	4
Aniline	2.9
Benzene	3
Carbon disulfide	0.9
Dieldrin	365
Ethylbenzene	5
Hexane	3
Lead	840
Pentachlorophenol	33
Phenol	3.5
Toluene	1.5
Xylene	3.6

BIOLOGICAL EXPOSURE INDICES (BEIs®)

The value of any sample results are enhanced when they can be compared to existing standards or guidelines. This permits the sample practitioner to evaluate whether the results are within expected limits or whether additional follow-up or exposure controls are required.

Biological exposure indices (BEIs®)[10] are developed by the ACGIH Biological Exposure Indices Committee as one means to assess exposure and health risk to workers. They currently cover about 38 different substances. BEIs® generally represent the biological monitoring results that would be typical of healthy workers exposed to the chemical via inhalation at the TLV®. They also cover some chemicals where significant exposures can occur through dermal or another noninhalation route and so biological monitoring of total absorbed dose is appropriate. Thus the BEIs® indirectly reflect the total dose to the worker from the chemical(s) of interest, and they represent a concentration below which healthy workers should not experience adverse health effects.

The BEIs® are expressed as the level of the chemical itself, one or more metabolites of the chemical, or a characteristic reversible biochemical change caused by the chemical. Most BEIs® are based on urine, blood or exhaled air measurements. They are not intended to measure adverse effects or as a diagnosis of occupational illness. They are based on a variety of information: mechanisms of absorption and elimination; all possible toxic effects; laboratory studies; and extrapolation of data.[11]

BEIs® should be applied in conjunction with air monitoring results and on-site evaluations of working conditions to identify skin contact and other routes of potential exposure. In some cases a BEI® cannot be considered as a valid exposure indication if there is not a certain airborne exposure level. For example, the BEI® for monitoring toluene exposure via the concentration of hippuric acid in urine is not considered useful below a certain airborne concentration of toluene (about 50 ppm) due to interferences from nonoccupational sources of hippuric acid including sodium benzoate, a food preservative, and many fruits.[8]

In addition to BEIs®, some OSHA standards covering specific chemicals contain acceptable biological monitoring levels.

ADVANTAGES AND DISADVANTAGES OF BIOMONITORING

Biomonitoring also has its limitations. While it is useful for preventing adverse effects that might result from chronic exposures, it usually will not help identify acute (peak) exposures since it is not done frequently enough, for example, daily. The same tests can be used, however, to measure the extent of exposure as an aid to identifying proper medical treatment following acute or emergency exposures that can be identified by other means. Biomonitoring is not useful for assessing exposures to substances that exhibit toxic effects at the site of first contact where these substances are poorly absorbed. Such is the

case for primary lung irritants.[9] Biomonitoring is only useful when the relationship between airborne (and in some cases, dermal) levels, internal dose, and adverse effect is known for a specific chemical.

A major limitation to widespread use of biological monitoring is the lack of detailed information on the fate of industrial chemicals in humans. Most of the toxicokinetic data available are from experimental animal studies and must be extrapolated to humans. Attempts are being made to use modeling to develop more data.[12]

Biological monitoring, unlike environmental monitoring, should be considered a medical procedure since by definition the specimen comes directly from a human. Permission, as well as trained and in some cases (blood) licensed personnel, may be required for collection of biological samples. These requirements can be an obstacle in some situations, for example, when workers are spread out over a wide geographical area.

As air monitoring standards become more stringent, biological monitoring methods may need to change, since metabolites must be detected at increased levels of instrument sensitivity. Sometimes this requires a change in metabolite. As an example, a standard of 50 mg/L of phenol in urine can be reliably used to detect benzene overexposures associated with an airborne standard of 10 ppm, but for airborne standards of 1 ppm and less the background levels of urinary phenol that range up to at least 20 mg/L in nonoccupationally exposed individuals may mask any changes due to occupational exposures.[11,13]

There are situations in which air monitoring is a better measure of exposure than biological monitoring. In atmospheres where the airborne concentration is rapidly fluctuating, the amount of material actually absorbed by an individual will be less than that measured by a continuous monitoring sampler in the breathing zone. The number of available biological monitoring methods is limited, and for many substances only air monitoring methods exist. However, given the limitation involved in collecting, analyzing, and interpreting air and biological samples, the best approach often is to use air sampling, wipe sampling, and bulk sampling in conjunction with biological monitoring to assess not only the true dose a worker is receiving but also the sources of exposure.

METHOD SELECTION

Considerable advance planning is necessary when biological monitoring is contemplated to aid in making sure that the appropriate test, time for monitoring, and properly trained, and in some states, licensed, personnel are available for sample collection. The type of exposure (chronic, acute, intermittent), the major route of exposure, and a knowledge of the process and all chemicals is use are also important considerations when contemplating biological monitoring. The proper medium to sample, the frequency of sampling, the parameters to measure, additional data to collect, and the need for air sampling will be determined by the goals and objectives.

In general, blood serum, urine, and breath sampling require different sample preparation and involve completely different collection methodologies. The limit of detection for the test should be sensitive (low) enough to differentiate exposed from nonexposed workers, and if exposure limits

have recently been lowered, existing methods should be reviewed for adequacy. Also, when selecting a test it must be remembered that the fate of any chemical in the body will depend on its volatility, polarity, and chemical and biological stability.

Labeled containers with the proper preservative or anticoagulant must be used, and immediate shipping procedures must be identified. All laboratories that analyze biological specimens must be licensed as clinical laboratories under either a federal or state program. Several professional organizations have established accreditation programs for clinical laboratories, including the College of American Pathologists and the American Association of Blood Banks.

Selection of the test method is important since different tests can represent different types of exposure, for example, the ALAD test is useful for detecting early lead exposure, and the free erythrocyte protoporphyrin (FEP) or zinc protoporphyrin (ZPP) tests indicate chronic (long-term) exposure to lead.

While standards such as BEIs often specify when samples are to be collected (Table 18.3); in some cases it is up to the biomedical professional. Timing is important because levels of some rapidly metabolized compounds decrease significantly within hours. Therefore, a measurement immediately following the shift is an indicator of the most recent exposure, and samples collected prior to a shift, 16 hours

TABLE 18.3. Sample Time Guidelines for ACGIH BEIs®

Specimen Collection Time[a]	Guidelines
Prior to shift	16 hours after exposure ceases.
During shift	Anytime after 2 hours of exposure.
End of shift	As soon as possible after exposure ceases.
End of workweek	After 4 or 5 consecutive working days with exposure.
Discretionary	At any time.

Source: Reference 10.

away from work, reflect the average exposure of the prior workday.[11]

Sample collection times often depend on the BHL of a compound. For example, toluene if sampled in breath must be collected immediately following the end of exposure due to its short half-life and rapid clearance from the lung.[8] The nature of the operation and availability of workers for sampling will also affect the timing.

An understanding of the chemicals used in the process can help interpret results. For example, the same metabolite can be a product of more than one substance. Mandelic acid is the metabolite of styrene and ethyl benzene, while 2,5-hexanedione can be produced by *n*-hexane and methyl *n*-butyl ketone. The same biochemical change can be induced by different compounds, for example, methylene chloride is metabolized to carbon monoxide in the body, thus creating carboxyhemoglobin just as a carbon monoxide exposure does. As described, interferences can occur as a result of diet, drugs, alcohol, disease, or other workplace chemicals.

Blood Sampling

Most chemicals capable of causing systemic effects are generally transported by the blood. The blood is often considered the most useful biological medium to monitor, since it generally provides an accurate, although indirect, measurement of the level of most toxic agents in target organs or tissues. Blood samples are usually collected if a urinary or other noninvasive test (breath) is not applicable. This can occur when airborne exposure concentrations will not produce a sufficient amount of a key metabolite in urine to be detected, when a compound does not undergo significant biotransformation, or when urinary metabolites are not specific to that compound.

Chemicals that enter the blood can be found at different concentrations depending on whether the site of entry is measured or what blood vessel is sampled. Concentrations in venous and arterial blood differ with the same exposure, and capillary blood concentrations will resemble those found in arterial blood. Therefore, blood collected from an area where dermal exposure has occurred may have higher levels of a contaminant than blood from another location.

Measurements on blood can be done on whole blood, plasma, or serum, depending on the contaminant. Chemicals can concentrate in red blood cells or plasma, or can be found in equal concentrations in both. For example, when measuring a cholinesterase in organophosphate-exposed workers, both red blood cell and plasma cholinesterase—two different enzymes—are determined, because with some pesticides, such as Demeton, plasma cholinesterase is affected by exposure long before red cell cholinesterase.[14]

Blood samples also require an anticoagulant, usually heparin. Vacuum tubes, the most common method of collection, are coated on the inside with the anticoagulant. The color of the cap indicates what has been added to the tube and the color coding is standardized. Blood samples cannot be frozen but can be stored at 4°C for 5 to 7 days.[8] Preservatives such as sodium fluoride can also be added and storage in the dark at 4°C is recommended for some samples, such as those collected for carboxyhemoglobin analysis.[15]

Blood is usually collected from the cubital vein or finger or earlobe capillaries. Cubital vein blood contains the same concentration of contaminant as the muscles. Capillary blood resembles the concentrations found in arterial blood.[9]

The primary advantage of sampling blood is the relatively small amount of variation in its composition relative to its effect on concentrations of chemicals as well as the fact that the sampling technique is simple and straightforward. The measurement of compounds in blood is much more specific than that of metabolites in urine and is subject to much less interference. However, sampling blood is also an invasive technique that causes discomfort among the subjects and thus most individuals are reticent to regularly provide samples. In particular, routine blood sampling in the field is impractical. Venipuncture requires trained personnel who are not usually available under these conditions. Also, the importance of cleanliness to prevent sample contamination makes field collection difficult. There is a concern among both analytical personnel and those qualified to collect blood samples about the potential for blood-borne diseases. Also, the samples must be carefully stored and handled or deterioration will occur.

General Method for Sampling Blood

1. If more than 0.5 mL of blood is needed for the test, a venous blood sample is required.
2. For less than 10 mL of unclotted blood, one of the following anticoagulants should be used: 20 mg of potassium or sodium oxalate, 50 mg of sodium citrate, 15 mg of disodium ethylene diamine tetraacetic acid (EDTA), or 2 mg of heparin. The anticoagulant should be dispersed in a concentrated solution along the bottom wall of the tube, and then desiccated. Care must be taken with vacutainers, because the rubber stoppers may contaminate the blood with low levels of organics.[14] If the sample is to be analyzed for metals, acidwashed, metal-free glass containers should be used for collection.
3. The skin should be washed prior to sampling, first with soap and water, and then with isopropanol.
4. Gently rotate the specimen container to mix.

5. Do not freeze whole blood or hemolysis will occur.

Blood Sampling for Polychlorinated Biphenyls in Serum[16]

1. Collect 20 mL to 25 mL of whole blood by venipuncture using a 30-mL glass syringe.
2. After the blood has clotted, centrifuge for 10 minutes at 2000 rpm. Transfer the serum to a 16-mm by 150-mm culture tube with a Teflon-lined screw cap using a sterile, disposable pipet.
3. Ship the serum in an insulated container with ice to keep the samples at 4°C.
4. Freeze samples upon arrival at the laboratory until analysis.

Blood Sampling for 2-Butanone, Toluene, or Ethanol[17]

1. Collect 5 mL of venous whole blood in a vacuum tube containing heparin. Invert the tube several times to mix.
2. Ship samples in polyfoam packs containing bagged ice or refrigerant.

Urine Sampling

Urine is suitable for monitoring hydrophilic chemicals, metals, and metabolites. It is noninvasive and relatively easy to analyze. Urine consists of 90% to 98% water, and the balance is solids consisting of many inorganic and organic compounds. Total daily urine output of an average adult varies between 600 mL and 2500 mL. The volume at any given time depends on the time of day, diet, temperature, and humidity.

There are three types of urine samples: collection of all urine voided during a 24-hour period; collection of a single sample, often called "spot" collection; and pooling of several spot samples over the course of a day. The 24-hour sample is the most desirable, since the levels best represent actual exposures; however, this sample is the most difficult to obtain. Pooled specimens, representing several collections during the day on the same individual, are also representative of exposure but are more likely to be contaminated. A single spot sample is the easiest to obtain, but the least representative of exposures.[18]

Regardless of the urine sample type, care in collecting the sample is needed since fallout from work clothing or contaminated hands can contaminate a urine sample.

In the case of 24-hour urine samples, the volume of the urine is measured using a graduated cylinder, followed by saving some of the sample in sealed vials for analysis. Other measurements such as specific gravity and pH are often done.

Most exposure measurements are based on a single sample. Due to the variation in urine volume over the day, urine samples must be corrected for dilution. Since the excretion rate of the solids is relatively constant, whereas that of the water is not, measurements are generally adjusted to the specific gravity of the solids or creatinine, a metabolic product of skeletal muscles, whose excretion rate is relatively steady. Specific gravity can vary up to a factor of 10 over a day while creatinine excretion varies much less.[19] Results are expressed as mg/L when measurements are adjusted for specific gravity or in grams per gram of creatinine. NIOSH recommends correcting urine to a specific gravity of 1.024.[20] Since other factors have also been used for specific gravity corrections, when comparing results of one sample to another or to a standard, or samples from different laboratories, it is important to make sure that the same correction factor was used. Sampling professionals should also be aware that creatinine excretion is affected by kidney function. The concentration of creatinine is also dependent on age, gender, and muscle mass.[21]

Urine samples are often preserved using thymol or acid to prevent bacterial degradation and then stored at 4°C until analysis. Frozen specimens are stable for greater than a month while specimens at 4°C are stable for 3 to 4 weeks.[8]

One strategy in situations where the metabolite selected for analysis also occurs in nonexposed individuals is to collect samples of urine after a period of nonexposure such as a weekend to use as a background for those collected following an exposure period.[21]

General Method for Sampling Urine[22]

1. The subject is asked to empty the bladder and record the time. For end-of-shift sampling, this procedure should be done 3 hours before the end of the shift.

2. The subject is provided with a 500-mL (or 200- to 300-mL) glass container with a wide mouth and a Teflon-lined screw-top lid inside a securely sealed bag, and asked to collect the next void. Do not discard any sample; leave this decision to the laboratory. If volatiles must be collected, a 50-mL container with a screw-top lid should be used. Ask the subject to fill the container completely. It must be immediately sealed following collection to prevent losses. These containers should be able to endure changes of temperature during transportation and storage that might result in overpressure. The time of this next collection should be recorded. The container should be labeled with a unique number.

3. For 24-hour collections, subjects void into a 4-L plastic jug with a large screw cap. In between collections, the jug is stored in an ice-pack, cooled Styrofoam chest.

4. For storage periods of 5 days or less, refrigerate; for longer periods, freezing is necessary.

5. The total volume of urine collected divided by the total elapsed time between voids represents the urine output. Compensate for dilution by correcting for either specific gravity or creatinine. Samples with a specific gravity <1.01 are too dilute and sampling should be repeated. If creatinine is to be measured, it should be done on the same sample as the chemical measurement.

Collection of Urine Samples for Benzene Exposure[23]

1. Collect 50 mL to 100 mL of urine in a 125-mL polyethylene bottle containing a few crystals of thymol. Close the bottle immediately after sample collection and swirl gently to mix.

2. Collect two urine samples for each worker, one prior to exposure and one after. A representative number of workers should be sampled. Submit individual samples as a group.

3. Collect and pool urine samples from nonexposed workers to use as background phenol levels.

4. Freeze the urine and ship in dry ice in an insulated container. Submit for analysis for phenol.

Exhaled Air (Breath) Sampling

Generally, when doing measurements on exhaled breath the compound representing the exposure rather than its metabolite(s) is monitored because metabolites are usually not volatile and therefore are not excreted through the lung.[24] Breath analysis is suitable for monitoring volatile solvents, carbon monoxide, and other gases excreted through the lungs. Once inhaled, volatile compounds can pass through the alveoli in the lungs to the bloodstream very rapidly. Gas (air) in the alveolar region of the lungs is almost in equilibrium with the arterial blood gases. Compounds that are poorly soluble in water and fat, compounds that are poorly metabolized, and compounds that have a high vapor pressure are poorly retained in the lungs.[25] For other compounds factors that influence the concentration passing from the alveoli to the blood are solubility in blood, breathing rate, solubility in fat, duration of exposure, and biotransformation. The less soluble a compound is in the blood, the higher the concentration in the alveolar air. As a general rule, the ratio of alveolar air to outside air provides some guidance in predicting the degree of solubility. Compounds such as hexane with a ratio greater than 0.5 (0.8–1.0 is best) are poorly soluble in blood while the opposite is true when this ratio is less than 0.5, as is the case for methyl ethyl ketone and toluene. Breath analysis works best on compounds with low blood solubility; these compounds are eliminated into expired air unmetabolized.[9]

The concentration of a volatile compound in the exhaled breath is directly related to the blood concentration, and is dependent on the total amount absorbed, the time passed since absorption, and the rate of elimination from the body. An example of the typical decrease in breath concentration as time passes following exposure can be determined from a study done using trichloroethylene (TCE) during which subjects were exposed to 100 ppm for a 7.5-hour period (Figure 18.1).[26]

Solubility can also affect BHL. In one study concentrations of methylene chloride, a poorly soluble compound, leveled off in alveolar air during the work week, staying relatively constant, while concentrations of toluene, which is highly soluble (in blood), increased throughout the week.[12]

Breath sampling is not as simple as it might appear. When a breath of an exposed person is exhaled it contains some plain air

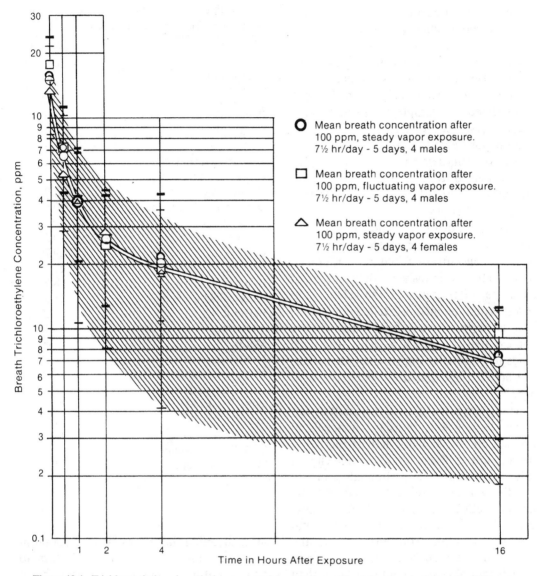

Figure 18.1. Trichloroethylene breath-decay curve: Mean and range breath concentrations after 100 ppm vapor exposure for 7.5 hours/day for 5 days with 12 subjects. The shaded area encloses the mean ±25 standard deviation. (From reference 26).

and some contaminated air. This is because the total exhalation, or mixed expired air, as it is called, is not the same as air from the alveolar region. After inhalation the contaminant is concentrated in this alveolar region while the air from the upper respiratory tract, sometimes called the "dead space," does not contain contaminant. In quiet breathing, a healthy human inhales and exhales about 500 mL of air 15 times per minute. When the total air exhaled during normal breathing is collected, the concentration in the sample represents the gas mixture displaced from the dead space (150 mL) and from the alveoli (350 mL).[18] When the air is collected only at the end of an expiration phase, the sample is called end-exhaled air.

Ideally, the concentration of vapors and gases in end-exhaled air equals the concentrations in alveolar air as this space is in equilibrium with concentrations in arterial blood.[27] Benzene provides an example of the difference in concentrations that can be expected in a sample of mixed-exhaled breath versus end-exhaled breath for the same exposure. If the airborne concentration is 10 ppm, benzene in mixed-exhaled air is expected to be 0.08 ppm while for an end-exhaled breath the concentration would be 0.12 ppm.[13]

During exposure, the end-exhaled concentration is smaller than the mixed-exhaled concentration, but the difference between them diminishes with the length of exposure. After exposure, the relationship reverses and the mixed-exhaled concentration is approximately two-thirds of the end-exhaled concentration. The difference tends to be smaller during increased physical activity, and generally increases with age.[24]

There are a number of factors that can affect the concentrations of chemicals in exhaled breath. As work activity increases, the amount of highly blood-soluble chemicals (such as toluene) eliminated in the breath will increase, but there will be minimal effects on poorly blood-soluble compounds.[12] Postexposure activity and obesity also affect concentrations in the breath. In the case of obesity, these individuals may develop higher exposures than others. Increased activity after work has decreased levels when compared to individuals who rested.[12] The breathing rate of subjects is also a source of fluctuations: Some individuals breath normally, others hyperventilate, and still others breathe very slowly.

In addition to the physiologic variations, there are other considerations. Storage of breath samples after collection can be a problem, especially if there is any possibility of adsorption onto the walls of the bag. Samples must be collected in a "clean"

TABLE 18.4. Gases and Vapors for Which Breath Screening Is Currently Recommended

Compound	Infrared Wavelength (μm)
Ammonia	10.77
Benzene	9.62
Carbon dioxide	4.27
Carbon monoxide	4.58
Carbon tetrachloride	12.60
Chlorobenzene	9.16
Chloroform	12.95
Dichlorodifluoromethane (freon)	10.85
Ethanol	9.37
Ethyl benzene	9.70
Ethylene oxide	11.48
Ethyl ether	8.75
Fluorotrichloromethane	11.82
Furfural	13.27
n-Hexane	3.40
Methanol	9.45
Methyl chloride	3.35
Methylene chloride	13.10
Perchloroethylene	10.92
n-Propyl alcohol	10.47
Styrene	12.90
Toluene	13.75
1,1,1-Trichloroethane	9.20
Trichloroethylene	11.78

area, meaning that no workplace contaminants are present. With these factors to consider, it may come as no surprise that it is more difficult to obtain good reproducible samples with breath than with blood or urine. However, since the technique is noninvasive and can be repeated without causing significant discomfort to the subject, it may be useful for screening exposures, accident investigations, and providing supporting data for other measurements. Table 18.4 lists gases and vapors for which breath screening is currently recommended.

Breath samples represent different aspects of exposure depending on how the sample is collected. Therefore, there are a number of approaches to take when it comes to breath sampling. Depending on the sampling technique, breath measurements reflect either the instantaneous blood levels of the contaminant in the body (a single breath) or the average blood level (multiple breaths) during the sampling period. Multiple breath samples are generally considered more representative of actual body concentrations than single breaths.

One approach is to use the breath-holding technique that consists of having the subjects hold their breath for 5 seconds to 30 seconds before exhaling into the sampling device. The content of the air can differ here as well, depending on whether the lungs were full of air or partially empty before starting.[18] Other aspects that have been considered include the use of end tidal volume where the breath is sampled only at the end of an expiration. Depending on whether the expiration is normal or forced, the content of the breath may differ.

A variation involves using a mouthpiece with a Y tube in which one tube of the Y functions as an exit bypass for that portion of the breath that would dilute the sample, and the other tube is attached to the inlet valve on the bag. During collection the valve on the bag is open, but the exit bypass prevents sample from entering the bag until the subject has exhaled 60% of lung capacity. At this point, a thumb is placed over the fixed end of the Y to shunt the balance of the exhalation into the bag. This is repeated until 100 mL to 1 L has been collected. The bag is stored in a shaded area for analysis.[28]

End-Exhaled Air Sampling Using a Bag

1. Select a direct reading instrument that is specific for the compound to be measured. Often these are electrochemical, solid-state infrared, or GC-based units. For more information on their operation, see the chapters on direct-reading instruments. The instrument should be hooked up to a data recording or output device, such as a printer or data logger. Prior to sampling, take a baseline measurement in the area where the instrument is stationed. If several samples are being processed, take a background after every 10 runs.

2. Explain the technique to the employees and have them practice during a briefing session prior to sample collection. Attach a fresh piece of Tygon tubing to the valve of a Tedlar bag.

3. Timing of sampling is important. Preshift samples should be collected 16 hours after the previous shift in clean air and prior to the next shift. Postshift samples should be collected immediately following the shift. During-shift samples should be collected as soon as the employee leaves the work area. See Table 18.3.

4. Have the employee take several deep breaths, then hold a deep breath for 25 to 30 seconds.

5. The employee should then exhale half of this breath and blow the remaining half of the breath into a Tedlar bag. Close the valve on the bag immediately.

6. If using an analyzer with a pump, attach the bag to the inlet and record the reading. If using a GC, use a syringe to withdraw a sample from the septum on the bag and inject it directly into a properly set-up GC.

7. Subtract the baseline reading from the final level. The result represents the change in gas concentration due to the exposure.

Carbon Monoxide Breath Sampling in Bags[29]

1. The smoking habits of the person being sampled should be obtained

(especially for that day). The normal carboxyhemoglobin level of the blood ranges from 0.5% to 2%. A person who smokes one pack of cigarettes per day can be expected to have an average level of 5% carboxyhemoglobin, and a two-pack-per-day smoker can be expected to have an 8% to 9% level. In terms of impacting the results, one pack a day yields 30 to 35 ppm while two to three packs per day will produce as much as 45 to 50 ppm background CO in exhaled breath.[15]

2. Have subjects hold their breath for 20 seconds, and then discard the first portion of the expired breath and collect the last portion in a 5-L Saran bag.

3. Bags are immediately hooked up to a direct reading carbon monoxide monitor.

4. Correlate exposure duration, carboxyhemoglobin, and carbon monoxide concentration.

Breath Sampling with Sorbent Tubes[18]

1. Breath sampling should be carried out in an uncontaminated area. Set up a sampling train that is connected in the following order: expired air bag, silica gel or charcoal tube, ascarite tube, personal air sampling pump calibrated to 500 mL/min. The choice of tube depends on the contaminant being collected. For guidance, consult the NIOSH air sampling method.

2. Heat the expired air bag until it is warm to the touch.

3. Instruct the worker to take a normal inspiration, expel a small amount, and then direct the rest of the expiration into a heated bag through a piece of tygon tubing and close the valve on the bag.

4. Attach the tubing on the bag of expired air to the inlet end of the sampling tube. While keeping the bag warm, 1000 mL of air (2-minute sample) from the bag is collected using the pump.

5. The time of sampling, the time of the end of work, the ambient pressure, and the temperature of the sampling bag should be recorded.

6. After sampling, the tubes are again separated and capped. Care should be taken to use the same plugs for the ascarite tube if it was preweighed with the plugs on.

7. One sorbent tube and one ascarite tube should be handled the same way, but without drawing air through, for laboratory blanks.

8. Label each tube with a unique number and send to the laboratory in a refrigerated container.

Pipet Breath Sampling[26]. The glass pipet method is best for those applications in which highly sensitive detection methods will be used. A 50-mL glass tube with screw caps on each end is used. This technique is useful for situations where a large number of repetitive samples must be collected.[30]

1. Alveolar breath samples are obtained from each subject prior to exposure and immediately following exposure. Duplicate background samples are also collected by unscrewing the caps and allowing the tube to lie for several minutes.

2. The samples are collected in duplicate for each subject. One end of the pipet (Figure 18.2) is sealed with the lips, flushing the chamber with three exhaled breaths, and then, after holding a fourth breath for 30 seconds, exhaling through the pipet chamber so that the end-tidal portion can be collected.

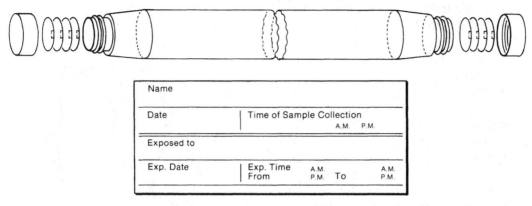

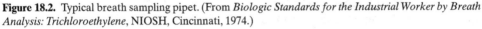

Figure 18.2. Typical breath sampling pipet. (From *Biologic Standards for the Industrial Worker by Breath Analysis: Trichloroethylene*, NIOSH, Cincinnati, 1974.)

3. The distal cap is secured while the subject is still expiring through the pipet chamber. After the distal cap is secure, the proximal end of the pipet is removed from the mouth and quickly sealed with a fingertip.

4. The proximal cap is then screwed shut. Screw the caps tightly; leaks in caps are the most common source of sample loss with this technique.

5. While it is desirable to construct individualized breath decay curves for each subject, more often these results are compared to standard curves.

Hair and Nail Sampling

Hair and nails are usually not suitable for use as biologic indicators, because they are frequently contaminated.[11] However, hair has been used for monitoring exposures to such heavy metals as arsenic, lead, cadmium, and mercury. One problem is the wide spread of data when attempting to come up with average hair values for the normal individual from exposures to typical environmental contaminants in food and water (Table 18.5).[31] Another concern is the likelihood of concentration gradients in a single hair.

Human hair consists of approximately 80% protein and 15% water, with smaller

TABLE 18.5. Normal Ranges for Trace and Minor Elements in Hair

Element	Concentration Range (µg)
Aluminum	20–40
Arsenic	2.0–3.0
Cadmium	1.0–2.0
Calcium	200–600
Chromium	0.50–1.50
Cobalt	0.2–1.0
Copper	12–35
Iron	20–50
Lead	20–30
Lithium	0.10–0.80
Magnesium	25–75
Manganese	1.0–10
Mercury	2.5–5.0
Molybdenum	0.10–1
Nickel	1.0–2.0
Phosphorus	100–170
Potassium	75–180
Selenium	3.0–6.0
Sodium	150–350
Vanadium	0.5–1.0
Zinc	160–240

amounts of lipid and inorganic materials. The water content of hair varies directly with the ambient RH. One consideration with selecting hair as a sampling medium is that hair does not grow continuously, but over periods of alternating activity. These periods include active growth, a resting stage, and a transitional phase. During the active growth phase, the follicles of the scalp produce hair at a rate ranging from 0.2 mm/day to 0.5 mm/day. Growth rates and the duration of activity are highly variable, and are dependent on such factors as the individual's age, race and gender, and the anatomical location of the hair and the season. However, it is generally assumed that human hair grows about 1 cm/month. Levels of various trace elements in hair are affected by various disease states, such as anemia, cirrhosis, and epilepsy.[32] Nonoccupational sources that can contribute to the types and concentrations of metals found in hair are the diet, especially fruits, vegetables, and grains, and hair dyes.[19]

It has been suggested that hair analysis would be most useful when a regular testing program is conducted on an individual over several years.[19] Test data used for comparison should be based on the same cleaning procedures. Detergent wash followed by a distilled water rinse is the most common method for cleaning hair, although other means, such as solvents, have also been used. The biggest concern in sampling hair is an inability to identify when the exposure occurred and the potential for exterior and nonoccupational contamination.

The International Atomic Energy Agency protocol recommends that at least 100 individual hairs be taken for an analysis: 5 to 10 strands from 10 to 20 different sites on the scalp. If analyses are to be compared between individuals, the lengths of hair sampled must be the same. Scalp hair can be divided into five regions: frontal, temporal, vertex anterior, vertex posterior, and nape (Figure 18.3).[32]

Collection Method for Hair[18,31]

1. Collect from the back of the head at the nape of the neck. Use clean stainless steel or plastic scissors and have clean, washed and dried, hands.

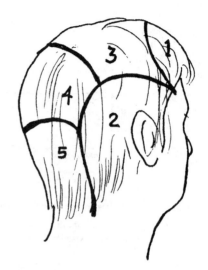

Figure 18.3. Regions of the scalp: 1, Frontal; 2, temporal; 3, vertex anterior; 4, vertex posterior; 5, nape. (Reprinted by permission of VCH Publishers, Inc., 220 East 23rd St., New York, NY 10010, from reference 32, p. 75.)

2. Measure lengths of 1–2 inches (5–10 cm) from the area closest to the scalp outward, and cut as close to the scalp as possible. This hair represents the most recent growth.

3. Collect 0.5–1 gram of hair (1–2 tablespoons), and place it in a plastic bag for shipment.

4. Interview the subject regarding the type of shampoo, hair colorants, and other hair cosmetics in use, such as mousse, sprays, and conditioners, including brand names.[17]

5. When submitting the sample to the laboratory, indicate from where the hair was collected (portion of the scalp) and include an empty plastic bag as a blank in the event contamination is suspected.

INTERPRETATION OF RESULTS

Interpretation of data is difficult, because there are very few standards relative to the types of biomonitoring that can be done for chemical exposures. When ACGIH BEIs® or governmental standards are available, they should be used.[10] The aceptable daily intake (ADI) may be a useful standard where no BEIs exist.

Results obtained are often dependent on time of collection, specimen integrity, drug or chemical interference, and methodologies. Concentrations in blood and breath reflect the most recent exposure if the sample is collected during the work shift or represent an integrated exposure (for an entire period) if performed 16 hours after exposure stops. For cumulative compounds such as hexachlorobenzene, concentrations represent the body burden.[9]

When attempting to compare the results of blood and breath samples, it is important to note that alveolar air represents arterial blood concentrations while blood samples are collected from veins. During exposure, tissue uptake of a contaminant decreases its concentration in the blood so the concentration in veins is less than that in the arteries.[9]

An analysis of very dilute urine samples (specific gravity less than 1.010 or creatinine concentration less than 0.3 g/L) is not likely to be accurate and should be repeated.[9] It is important to review test data carefully and make sure that the proper analysis has been requested and performed. For example, in the case of styrene, results may be given as either mandelic acid or total mandelic acid. In the latter case, they represent the total concentration of both mandelic and phenylglyoxylic acids, both the metabolites of styrene.[21]

Results of biological monitoring of chemicals with a long biological half life are indicators of long-term exposure, and do not correlate well with the current time-weighted average (TWA) inhalation exposure. Examples are metals, such as lead, and cadmium or chlorinated compounds, such as DDT and PCBs. On the other hand, bio-

logical "determinants" with a short half life are indicators of the most recent exposure, and do not correlate well with the TWA exposure if the air concentration fluctuates widely. Examples are blood and expired air measurements in samples taken at the end of a shift. Factors such as the chemical's half life and the volume of body fat, and variability of metabolism among individuals as well as differences in workloads are also important to consider when making correlations between airborne exposure and biological results.

Nonoccupational exposures must also be taken into account. For example, smokers have significantly higher carboxyhemoglobin than nonsmokers, which must be considered when interpreting results of measurements for employees exposed to carbon monoxide. Correlations must also be made with the observed levels of "natural" pollutants in the environment, as bioaccumulation of these pollutants can occur, resulting from long-term exposure to low levels in the environment. One approach to this problem is to collect specimens from nonexposed control subjects at the workplace, and to use this group mean as a baseline to interpret the values from an exposed population. If a nonexposed control group cannot be obtained from the workplace, the use of an outside control group should be considered.

SUMMARY

Biological monitoring of bodily fluids and exhaled air can be a valuable indication of exposure since they measure the individual's absorbed dose of the substance. As such, the measurements can complement or even replace traditional air sampling.

BEIs® cover about 38 different chemicals and represent the most comprehensive set of "allowable levels" for comparing biological monitoring results.

Even when technically feasible, biological monitoring programs must be carefully developed and communicated to the workforce to avoid resistance due to:

- The discomfort or inconvenience of collecting certain biological samples.
- The concern that samples may be used to screen for alcohol or illegal drug use in addition to the chemical contaminants.

REFERENCES

1. Corn, M. Strategies for prospective surveillance. In *Advances in Air Sampling*. Cincinnati: ACGIH, 1988.

2. Bowman, J. A., J. L. Held, and D. R. Factor. A field evaluation of mandelic acid in urine as a compliance monitor for styrene exposures. *Appl. Occup. Environ. Hyg.* **5**(8): 526–535, 1990.

3. Thomas, V. Five years of the biological exposure indices committee. *Appl. Ind. Hyg.* **3**(10):F-26–F-28, 1988.

4. Yeary, R. A. Urinary excretion of 2,4-D in commercial lawn specialists. *Appl. Ind. Hyg.* **1**(3):119–121, 1986.

5. Friberg, L. T. The rationale of biological monitoring of chemicals—with special reference to metals. *AIHA J* **46**(11):633–642, 1985.

6. Mc Sharry, C., et al. Seasonal variation of antibody levels among pigeon fanciers. *Clin. Allergy* **13**:293–299, 1983.

7. Droz, P. O. Biological monitoring I: Sources of variability in human response to chemical exposure. *Appl. Ind. Hyg.* **4**(1):F-20–F-24, 1989.

8. Lowry, L. K. Review of biological monitoring tests for toluene. In *Biological Monitoring of Exposure to Chemicals. Organic Compounds*, M. K. Ho and H. K. Dillon, eds. New York: Wiley, 1987, pp. 99–109.

9. Bernard, A., and R. Lauwerys. General principles for biological monitoring of exposures to chemicals. In *Biological Monitoring of Exposure to Chemicals. Organic Compounds*, M. K. Ho and H. K. Dillon, eds. New York: Wiley, 1987, pp. 1–16.

10. American Conference of Government Industrial Hygienists. *TLVs® and BEIs®*. Cincinnati: ACGIH, 2003.

11. Fiserova-Bergerova, Thomas V. Development of biological exposure indices (BEIs) and their implementation. *Appl. Ind. Hyg.* **2**(2):87–92, 1987.

12. Fiserova-Bergerova, V. Simulation model as a tool for adjustment of BEIs to exposure conditions. In *Biological Monitoring of Exposure to Chemicals. Organic Compounds*, M. K. Ho and H. K. Dillon, eds. New York: Wiley, 1987, pp. 27–57.

13. American Conference of Governmental Industrial Hygienists. Benzene—recommended BEI. In *Documentation of the Biological Exposure Limits*. Cincinnati: ACGIH, 2001.

14. Lowry, L. K., et al. Biological monitoring III: Measurements in blood. *Appl. Ind. Hyg.* **4**(3):F-11–F-13, 1989.

15. American Conference of Governmental Industrial Hygienists. Carbon monoxide—recommended BEI. In *Documentation of the Biological Exposure Limits*. Cincinnati: ACGIH, 1986.

16. NIOSH. *Manual of Analytical Methods, Method 8004*, 3rd ed. Cincinnati: NIOSH, 1994.

17. NIOSH. *Manual of Analytical Methods, Method 8002*, 3rd ed. Cincinnati: NIOSH, 1994.

18. Hill, R. H., et al. Sample collection. In *Methods for Biological Monitoring: A Manual for Assessing Human Exposure to Hazardous Substances*, T. J. Kneip and J. V. Crable, eds. Washington, D.C.: American Public Health Association, 1988.

19. Waritz, R. S. Biological indicators of chemical dosage and burden. In *Patty's Industrial Hygiene and Toxicology. Vol. III: Theory and Rationale of Industrial Hygiene Practice. 3B: Biological Responses*, L. G. Cralley and L. V. Cralley, eds. New York: Wiley, 1985, pp. 175–312.

20. NIOSH. *Criteria for a Recommended Standard ... Occupational Exposure to Benzene.* Cincinnati: NIOSH, 1974.

21. van Hemmen, J., and G. de Mik. Biological monitoring of solvents, no panacea. In *Bio-*

logical Monitoring of Exposure to Chemicals. Organic Compounds, M. K. Ho and H. K. Dillon, eds. New York: Wiley, 1987, pp. 73–84.

22. Rosenberg, J., et al. Biological monitoring IV: Measurements in urine. *Appl. Ind. Hyg.* **4**(4):F-16–F-21, 1989.

23. NIOSH. *Manual of Analytical Methods, Method 8305*, 3rd ed. Cincinnati: NIOSH, 1994.

24. Fiserova-Bergerova, Thomas V., et al. Biological monitoring II: Measurements in exhaled air. *Appl. Ind. Hyg.* **4**(2):F-10–F-13, 1989.

25. Bardodej, Z., J. Urban, and H. Malonova. Important considerations in the development of biological monitoring methods to determine occupational exposures to chemicals. In *Biological Monitoring of Exposure to Chemicals. Organic Compounds*, M. K. Ho and H. K. Dillon, eds. New York: Wiley, 1987, pp. 17–28.

26. NIOSH. *Biologic Standards for the Industrial Worker by Breath Analysis: Trichloroethylene.* Cincinnati: NIOSH, 1974.

27. West, J. B. Ventilation. In *Respiratory Physiology—the Essentials*. Baltimore: Williams and Wilkins, 1974.

28. Prevost, R. J., et al. Biological monitoring of exposures to chemical vapors released in marine operations. In *Biological Monitoring of Exposure to Chemicals. Organic Compounds*, M. K. Ho and H. K. Dillon, eds. New York: Wiley, 1987, pp. 179–195.

29. Stewart, Richard, et al. Rapid estimation of carboxyhemoglobin level in fire fighters. *JAMA* **235**(4):390–392, 1976.

30. Linch, A. L. *Biological Monitoring for Industrial Chemical Exposure Control.* Cleveland: CRC Press, 1974.

31. Sheldon, L., et al. (Research Triangle Institute, EPA, NC). Chemicals identified in human biological medium. In *Biological Monitoring Techniques*. Park Ridge, NY: Noyes Publishers, 1986 (adapted from MineraLab, Inc., 22455 Maple Court, Hayward, CA 94541).

32. Katz, S. A., and A. Chatt. *Hair Analysis: Applications in the Biomedical and Environmental Sciences.* New York: VCH, 1988.

CHAPTER 19

SURFACE SAMPLING METHODS

There are several reasons why surface contamination, especially easily removable contamination, is sampled:

- Many toxic materials can enter the body through ingestion, and some can penetrate intact skin. Contact with contaminated surfaces or direct fallout onto skin and other surfaces are key factors governing exposure via these routes. Any ingestion or dermal dose must be considered in addition to inhalation exposures. Air sampling will not detect the surface contamination involved in these exposures; their presence must be established by surface sampling.

- Loose particulate matter on surfaces can be resuspended into the air due to air movement or agitation. Surface sampling can help evaluate the extent of loose surface contamination, so cleanup or other steps can be taken to prevent harmful exposures from these sources.

- Surface contamination can serve as an indicator of the adequacy of contamination control programs. For example, many facilities handling radioactive materials routinely collect surface wipe samples in both radioactive handling areas and public areas (cafeteria, main gate, etc.) to ensure that contamination is not inadvertently tracked out of regulated areas.

Once the presence of surface contamination or the potential for ingestion or dermal hazards is documented, dermal monitoring, biological monitoring, or air sampling may be required to further define the extent of the hazard and the adequacy of controls.

The control of ingestion hazards is largely based on good hygiene practices such as cleaning up the contamination, hand washing before eating, drinking or smoking, segregating eating facilities from chemical handling areas, and ensuring the contaminated work clothing is handled

Air Monitoring for Toxic Exposures, Second Edition. By Henry J. McDermott
ISBN 0-471-45435-4   © 2004 John Wiley & Sons, Inc.

TABLE 19.1. Comparison of Respirable and Dermal Exposures to Parathion During Citrus Harvesting

Exposure	Week 1 (mg)	Week 2 (mg)	Week 3 (mg)
Total dermal	0.382	0.246	1.030
Total respiratory	0.008	0.009	0.011

Source: Popendorf, W. J., and J. T. Leffingwell. Regulating OP pesticide residues for farmworker protection. *Res. Rev.* **82**:125–201, 1982.

properly. Evaluation and control of chemicals capable of penetrating intact skin is often more difficult since for these chemicals skin contact must be prevented to the degree possible.

In some cases, dermal exposure is greater than inhalation exposures. Field studies on agricultural workers handling certain pesticides reinforce this difference. Table 19.1 compares respirable and dermal exposures to parathion during citrus harvesting, and it shows that in two of the three weeks of evaluation the dermal exposure was approximately 100 times higher than the respiratory exposure.

It is important to understand that not all chemicals can penetrate intact skin and result in a significant exposure. The ACGIH Threshold Limit Values® (TLVs®) for chemical substances uses a designation called a "skin notation" to identify chemicals with the potential for a significant contribution to the overall exposure by the cutaneous route, including skin, mucous membranes and eyes, either through contact with vapors or, of probable greater significance, by direct skin contact with the substance (Table 19.2). Activities that can result in dermal exposure include splashes when working with or near liquids, wiping with contaminated rags, contact with contaminated tools or surfaces, and immersing hands in solvents and other chemicals. These exposures can lead to chemical absorption or to ingestion if a residue remains on the workers' hands during eating, or smoking breaks. Contaminated clothing that is worn through the day or over several days can also be a source of exposure.

Dermal exposure evaluations should consider the impact of all relevant factors:

- Degree of contact with skin and the time duration of exposure are important considerations.[1]
- Some chemicals will evaporate from the skin before they are completely absorbed. This will reduce the dermal dose but may cause added inhalation exposure.
- Sweating, abrasion, or irritation can enhance chemical penetration through the skin layer.
- Occlusion (i.e., holding the chemical against the skin) as can happen if the chemical gets under protective clothing or protective barrier creams can cause irritation as well as increase its penetration.
- Different parts of the body have skin with differing ability to resist chemical penetration.
- Characteristics of the chemical such as solubility, oil/water partition coefficient, and molecular size affect speed and degree of penetration through intact skin.
- Protective glove and clothing materials vary in their ability to withstand different chemicals. Protective garments can fail either through being damaged by the chemical or allowing the chemical to penetrate the intact material. There is extensive literature and suppliers' recommendation for protective equipment selection against different chemicals. In some cases an impermeable glove's effectiveness can be evaluated by wiping the interior of the glove to determine contamination; this technique is only valid when there has been

TABLE 19.2. Chemicals that Have an ACGIH TLV® Skin Notation

Acetone cyanohydrin	1,4-Dichloro-2-butene
Acetonitrile	Dichloroethyl ether
Acrolein	1,3-Dichloropropene
Acrylamide	Dichlorvos
Acrylic acid	Dicrotophos
Acrylonitrile	Dieldrin
Adiponitrile	Diesel fuel
Aldrin	Diethanolamine
Allyl alcohol	Diethylamine
4-Aminodiphenyl	2-Diethylaminoethanol
Ammonium perfluorooctanoate	Diethylene triamine
Aniline	Diisopropylamine
Anisidine	N,N-Dimethylacetamide
Azinphos-methyl	Dimethylaniline
Benzene	Dimethylformamide
Benzidine	1,1-Dimethylhydrazine
Benzotrichloride	Dimethyl sulfate
Bis(2-dimethylaminoethyl)ether	Dinitrobenzene
Bromoform	Dinitrol-O-cresol
n-Butylamine	Dinitrotoluene
tert-Butyl chromate	1,4-Dioxane
o-sec-Butylphenol	Dioxathion
Captafol	Diquat
Carbon disulfide	Disulfoton
Carbon tetrachloride	Endosulfan
Catechol	Endrin
Chlordane	Epichlorohydrin
Chlorinated camphene	EPN
Chloroacetone	Ethion
Chloroacetyl chloride	2-Ethoxyethanol
o-Chlorobenzylidene malononitrile	2-Ethoxyethyl acetate
Chlorodiphenyl (PCBs)-42% and 54% chlorine	Ethlyamine
1-Chloro-2-propanol	Ethly bromide
2-Chloro-1-propanol	Ethly chloride
β-Chloroprene	Ethylene chlorohydrin
2-Chloropropionic acid	Ethylenediamine
Chlorpyrifos	Ethylene dibromide
Cresol	Ethylene glycol dinitrate
Crotonaldehyde	Ethylenimine
Cyclohexanol	n-Ethylmorpholine
Cyclohexanone	Fenamiphos
Cyclonite	Fenthion
Decaborane	Fonofos
Demeton	Formamide
Demoton-S-methyl	Furfural
Diazinon	Furfuryl alcohol
2-N-Dibutylaminoethanol	Heptachlor
Dibutyl phenyl phosphate	Hexachlorobenzene
3,3'-Dichlorobenzidine	Hexachlorobutadiene
	Hexachloroethane

TABLE 19.2. *Continued*

Hexachloronaphthalene	*p*-Nitrochlorobenzene
Hexafluoroacetone	4-Nitrodiphenyl
Hexamethyl phosphoramide	Nitroglycerin
n-Hexane	*n*-Nitrosodimethylamine
Hydrazine	Nitrotoluene
Hydrogen cyanide and cyanide salts	Octachloronaphthalene
2-Hydroxypropyl acrylate	Parathion
Isooctyl alcohol	Pentachloronaphthalene
2-Isopropoxyethanol	Pentachlorophenol
n-Isopropylaniline	Phenol
Lindane	Phenothiazine
Malathion	Phenyl glycidyl ether
Manganese cyclopentadienyl tricarbonyl	Phenylhydrazine
Mercury vapor, aryl and inorganic compounds	Phorate
Mercury, alkyl compounds	Propargyl alcohol
Methanol	Propyl alcohol
2-Methoxyethanol	Propylene glycol dinitrate
2-Methoxyethyl acetate	Propylene imine
bis-(2-Methoxypropyl)ether	Sodium fluoroacetate
Methyl acrylate	Sulfotep
Methyl acrylonitrile	Terbufos
n-Methyl aniline	1,1,2,2,-Tetrachloroethane
Methyl bromide	Tetraethyl lead
Methyl *n*-butyl ketone	Tetraethyl pyrophosphate
Methyl chloride	Tetramethyl lead
O-Methylcyclohexanone	Tetramethyl succinonitrile
2-Methylcyclopentadienyl manganese tricarbonyl	Thallium, soluble compounds
	Thioglycolic acid
Methyl demeton	Tin, organic compounds
4,4′-Methylene *bis*(2-chloroaniline)	Toluene
4,4′-Methylene dianiline	*o*-Tolidine
Methyl hydrazine	*o*-Toluidine
Methyl iodide	*m*-Toluidine
Methyl isobutyl carbinol	*o*-Toluidine
Methyl isocyanate	*p*-Toluidine
Methyl parathion	1,1,2-Trichloroethane
Methyl vinyl ketone	Trichloronaphthalene
Mevinphos	1,2,3-Trichloropropane
Menocrotophos	Triethanolamine
Morpholine	2,4,6-Trinitrotoluene (TNT)
Naled	Triorthocresyl phosphate
Naphthalene	Vinyl cyclohexene dioxide
Nicotine	*m*-Xylene, *a*, *a*′-diamine
p-Nitroaniline	Xylidine
Nitrobenzene	

no opportunity for chemicals to enter the glove through the wrist opening.

• Carrier solvents or other relatively nontoxic materials can enhance the absorption of toxic chemicals or cause them to be retained on the skin for long periods.

There are two methods of surface sampling to estimate an individual's exposure: (a) indirect methods that measure the degree of contamination on surfaces and (b) those that assess the degree of contamination directly by measurements on the individual's skin. Indirect methods involve using filters, gauze pads or swabs to wipe surface, instruments to "sniff" surfaces for volatile compounds, adhesive tape to lift dust from surfaces, and "micro-vacuum cleaners" to collect dusts in air sampling filter cassettes. Direct methods are typically less standardized than indirect methods and include gauze pads or charcoal pads attached to the body, direct washing of the chemical from body surfaces, and detection via fluorescence for certain chemicals. Extreme care is required when designing direct dermal evaluations since some techniques can expose the individual to additional stressors such as the possibility of occluding contaminants under a skin patch or exposure to solvents.

WIPE SAMPLING

Chemical Selection

Chemicals that can penetrate or injure the skin are the primary types of compounds for which surface contamination is a concern. Compounds such as the heavy metals that do not penetrate the skin but can be transferred to the mouth if an eating area is contaminated are also good candidates for surface sampling. Examples of heavy metals are lead, arsenic, and cadmium.

Chemicals for which the American Conference of Governmental Industrial Hygienists (ACGIH) has a "skin" notation (Table 19.2), or substance that has a skin LD_{50} (to rabbits) of 200 mg/kg or less are considered to have significant skin penetration properties and should be evaluated.

A class of irritants for which wipe sampling is commonly done is amines, since they can stick to surfaces. Some amines can also be sensitizers. The mode of action and degree of irritation can vary between chemicals. For example, hydrogen fluoride molecules actually "burrow" through the skin to the bone. In some cases, the impact of a corrosive chemical such as concentrated sodium hydroxide can be very severe and can cause significant burns to the skin's surface. The following is a list of selected chemicals that are skin irritants:

Acrolein
Allyl alcohol
Allyl glycidyl ether
Ammonia
Ammonium chloride fume
n-Butyl acetate
Caprolactam, dust and vapor
Chlorine
Chloroacetyl chloride
o-Chlorobenzylidene malononitrile
Diethylamine
Ethyl benzene
Glutaraldehyde
2-Hydroxypropyl acrylate
Methyl 2-cyanoacrylate
Phosphoric acid
Potassium hydroxide
Propylene glycol monomethyl ether
Sodium bisulfite
Sodium hydroxide
Thioglycolic acid
1,2,4-Trichlorobenzene
Triethylamine
Tetrasodium borate salts

A special case is allergic sensitization of the skin, such as that caused by isocyanates and cobalt, where exposure can result in an individual becoming unable to work around the material and requiring the worker to change jobs or leave a facility entirely. Another type of effect is photosensitization (increased sensitivity to sunlight) that results from exposure to certain compounds such as petroleum asphalt fumes and ultraviolet (UV) light, usually from the sun. The following chemicals are examples of skin sensitizers:

Captafol

Cobalt metal, fume and dust

Isophorone diisocyanate

Phenothiazine

Phenyl glycidyl ether

Picric acid

Subtilisins

Toluene-2,4-diisocyanate

In selecting chemicals to sample for ingestion potential, their oral toxicity is important. In general, compounds with oral LD_{50}'s (to rats) of 50 mg/kg and less are high candidates, but those with higher oral LD_{50}'s (meaning less toxic) must be evaluated on a case-by-case basis.

Chemicals for which wipe sampling is ineffective even though they can penetrate the skin include volatile solvents such as benzene and n-butyl alcohol, because their evaporation from surfaces can be rapid. The rapid evaporation reduces the skin adsorption hazard. Most gases (e.g., bromine and boron trifluoride) do not redeposit on surfaces, so sampling for them is ineffective; however, some gases (e.g., arsine and stibine) may revert to their metallic form after contact with surfaces and remain. In some cases, biological monitoring such as collecting urine samples from benzene-exposed workers for phenol analysis can be an effective means of determining whether skin absorption is a signif-icant concern for contaminants that are difficult to sample.

Sampling Materials

Generally there are two types of filters recommended for taking wipe samples[2]:

1. Glass fiber filters (37 mm) are usually used for materials analyzed by high-performance liquid chromatography (HPLC) and often for substances analyzed by gas chromatography (GC).
2. Cellulose (paper) filters are generally used for metals, and may be used for anything not analyzed by HPLC. For convenience, the Whatman smear tab may be used.

Gauze is sometimes used rather than filter papers. An example is for PCB sampling. Gauze is generally extracted prior to use using the Soxhlet technique to ensure its purity.

Acids and bases (or alkalies) can be detected by their reaction with pH paper. For example, ammonia and amines are very basic and can be detected by their characteristic high pH. This method can also be used as a first identification if an unknown spill is detected. Acids turn litmus (pH) paper blue, and bases turn it pink.

Purity of solvents used to "wet" the wipe sample filter is important. Although they need not be spectroscopic grade, they should be of sufficient purity not to contaminate the sample. For example, distilled water should be used rather than tap water. Solvents are not required, but they can enhance collection if appropriate for the contaminant under investigation (Table 19.3). Solvents other than distilled water should not be used to sample direct skin. Filters are fragile and when using solvents extreme care must be used or they may be damaged.

TABLE 19.3. Solvents Used in Wipe Sampling

Compound	Solvent
PCBs	Hexane
Aromatic amines	Methanol
4-Aminodiphenyl	
Azinphos methyl	
Toxaphene	
DDVP	
Diazinon	Ethylene glycol
Dieldrin	
Dinitrotoluenes	
Lindane	
Malathion	
Parathion	
4-Aminopyridine	
Aniline	
Anisidine	
Benzidine	
Heptachlor	
Nitroglycerin	Isopropanol
Pentachloronaphthalene	
TEDP	
Tetrachloronaphthalene	
o-Toluidines	
o-Toluidene	
Trinitrotoluene (TNT)	
General	
Metals and salts	Distilled water
Low-chain hydrocarbons	Distilled water
Bases	Dilute acids
Amines	Dilute acids
Hydrazines	Dilute acids
Acids	Dilute bases (detergents)
Phenols	Dilute bases (detergents)
Thiols	Dilute bases (detergents)
Nonpolar hydrocarbons	Organic solvents

Sampling Methods

Direct evaluation of dermal exposure usually is far more complicated than that of airborne exposures due to variability in deposition rates onto the body, the effect of clothing, the duration of actual skin contact with the chemical, and the importance of time in the retention and permeation of the chemical through the skin. A number of methods for assessing surface and personal contamination have been developed, including wet and dry wipe samples, adhesive tape sampling, and skin-washing techniques. Wipe sampling variables include the degree of pressure applied, accuracy of selecting the area to sample, types of wipe media used, and the physical nature of the contaminated surface (porous versus nonpermeable or smooth) since particles will be deposited in surface cracks and crevices where they are not removed by the wiping.[3]

For some types of surface contamination that do not respond well to manual collection, such as mercury, direct reading instruments like a mercury sniffer may be used. Other examples are general survey instruments, such as an organic vapor analyzer for organic vapor contamination. In this case, it is important to remember that unless the instrument is calibrated with the chemical in question, the results are all relative to the material with which it was calibrated.

It can be difficult to determine whether equipment or instruments are clean after they have been used in a contaminated area, such as on a hazardous waste site or for sampling of concentrated materials. Wipe sampling can be useful in determining whether these procedures are effective.

Wipe sampling is inappropriate for porous surfaces that would absorb the compound of interest, such as PCBs. These include wood and asphalt. Instead, a bulk sample should be collected in these situations, that includes collection of the surface (1-cm deep) material.

The patch technique (Figure 19.1) has been the most widely used method for estimating dermal exposure, but there are concerns that it has never been properly validated. Its primary limitation is the assumption that exposures are uniform

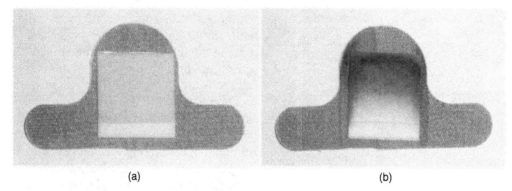

| (a) | (b) |

Figure 19.1. SKC Permea-Tec™ aromatic amine sensors to evaluate the effectiveness of protective equipment: (a) Unused sensor; (b) exposed sensor. (Courtesy of SKC, Inc.)

over various body parts. Since patches generally cover 6% or less of the body, exposures could be grossly overestimated or underestimated if droplets hit or miss the patch in situations where spray applications are being monitored.

General Procedure[2]

1. Preload a group of vials with appropriate filters. Make sure the vials have labels.

2. Always wear clean impervious gloves when doing wipes. Disposable gloves are preferable, since a clean set of gloves should be used with each individual sample. The selection of gloves will depend on the contaminants being sampled and the types of solvents in use.

3. Prepare a diagram of the area or room(s) to be wipe sampled along with locations of key surfaces.

4. Label the sample vial with the place where the sample is being collected and a unique identification number. Withdraw the filter or other media from the vial. If a damp wipe sample is needed, moisten the filter with distilled water or other solvents as

recommended for the contaminant being sampled.

5. Wipe approximately $100\,cm^2$ of the surface to be sampled. The purpose is to have consistent areas to compare among samples. Even if standards are not available, samples should be comparable for determining which areas are the most contaminated.

6. Without allowing the filter to contact any other surface, fold the filter with exposed sides against each other and then fold it again. Put the filter in its sample vial, cap the vial, and place a corresponding number at the sample location on the diagram. Some substances, such as benzidine, must have solvent added to the vial as soon as the wipe is placed inside.

7. Take notes as well as including any further descriptions that may later prove useful when evaluating sample results, for example, employees' names if personal protective equipment is being wiped.

8. At least one blank filter should be folded and put into a vial. Be sure to use clean gloves and remember that no contact with surfaces is permitted

for the blank filter. Also provide a sample of the solvent used as a blank in case it is needed.

9. Samples that can evaporate must be contained within airtight bottles or samples will be lost. Shipping wipe samples containing solvents may be a problem due to shipping restrictions and may delay receipt at the laboratory from normal air transport times. For example, hexane, which is used to sample PCBs, is highly flammable and cannot be shipped by air.

Wet Wipe Test for Arsenic[2]. This method is appropriate for arsenic and arsenic-containing inorganic compounds.

Procedure

1. Using a clean, disposable glove, remove a 7-cm (2¾-in.) diameter Whatman 41 filter from the box.
2. Moisten the filter with distilled water. Use a dry filter if sampling for a liquid residue of arsenic trichloride or arsenic trifluoride.
3. Select a sampling area that is at least $100 \, cm^2$.
4. With the leading edge of the filter slightly raised, wipe the surface in a back and forth and up and down motion. A 10-cm by 10-cm wire frame can be used as a guide.
5. Pick up the filter paper, place it on a clean sheet of paper, fold the contaminated side inward, and then make one more fold to form a 90° angle in the center of the filter.
6. Place the filter, angle first, into a glass vial and close the lid tightly.

Modified Wet Wipe for Aromatic Amines. MOCA (4,4′-methylene bis(2-chloroaniline)) is an example of an aromatic amine that is also a potent animal carcinogen used as a curing agent for some isocyanate-containing polymers. It is regulated by OSHA in a substance-specific standard. Surface contamination can occur while mixing the urethane with this catalyst. Residual material could be contacted by workers brushing against bench tops, opening doors, or handling containers. Wipe samples would include all surfaces in the area where MOCA is stored, used, or handled in its pure form, such as mixing benches, door knobs, personal protective equipment like gloves and aprons.[2] Also, if it is suspected that the employees are not always washing after using the material, or that other modes of contamination may be present, then wipes may also have to be collected on lunchroom surfaces or other surfaces where other employees may be exposed. It is not uncommon for the mixing area to be separate from the molding area. In the OSHA standard it is required that areas where employees may be exposed to MOCA be designated as "regulated areas," that is, areas where entry is restricted and where certain controls, such as personal protective equipment and training, are required. Once the material is mixed into the urethane, curing should begin, and the exposure to residual isocyanate is a greater concern than the small amounts of catalyst that are generally used in these mixtures.

Procedure

1. Follow steps 1 through 3 of the General Procedure.
2. Wipe approximately $100 \, cm^3$ with a Whatman 42 filter that is 7 cm (2.8 in.) in diameter after moistening its center with 5 drops of methanol.
3. After wiping the sample area, apply 3 drops of fluoroescamine (a visualization reagent) to the contaminated area of the filter. Also place a drop of this reagent on an area of the filter that has not contacted the wiped surface. This area becomes a blank,

and using the same filter allows for better comparison.

4. Allow 6 minutes for the reaction, then irradiate the filter with a 366-nm ultraviolet (UV) light. Compare the color development of the contaminated area with the area designated as the blank. A color change to yellow is a positive for contamination with the following amines: MOCA, benzidine, 3,3′-dichlorobenzidine, alpha-naphthylamine, beta-naphthylamine, and 4-aminodiphenyl.

5. Additional samples should be collected and sent to a laboratory for confirmation. Send a vial containing a blank filter and a small sample of the methanol along with the wipe samples.

Wipe Test Using Gauze for Polychlorinated Biphenyls. Other situations where PCB wipe sampling may be needed include testing decontamination procedures on waste sites where PCB cleanup is in progress and in any areas where leaks of PCB-containing equipment may have occurred. Another situation is to determine whether permeation has occurred through personal protective equipment. If the surface to be sampled is smooth and impervious, a wipe sample will be effective in detecting PCBs. It should be noted that a bulk sample of any cleaning solutions should be analyzed prior to disposal for PCB levels in excess of 50 ppm in order to determine whether they must be disposed of as a hazardous waste. For more information on collection bulk samples of water and sending them in for analyses, see the Chapter 20.

Procedure

1. Use 3-inch by 3-inch gauze pads that have been Soxhlet extracted with hexane. Prior to sampling, wet each pad with 8 ml of pesticide grade hexane.

2. Put on a phthalate-free glove prior to each sample.

3. Mark off a surface area of $100 \, cm^2$ and wipe the surface with the hexane-saturated pad initially in a horizontal direction using a forward and backward motion. Do a second wiping of the surface using a clean portion of the same gauze pad in the vertical direction with the same forward and backward motion.

4. Place the gauze pad in a brown glass sample container equipped with a Teflon-lined lid.

Dry Wipe Test

Procedure

1. Using the tip of the thumb, wipe a 2.4-cm-diameter filter paper disk in a "Z" or "S" pattern over a representative portion of the surface to be sampled. The length of the wipe should be 50 cm. The pressure-bearing portion of the filter paper disk will be about 2 cm wide; therefore, the area of the surface that is sampled will be approximately 100 $cm.^2$

2. Avoid contacting excess dirt when wiping an area.

Swab Test. Analysis of these samples is qualitative, but will reflect the general degree of surface contamination.

Procedure[4]

1. Assemble the following materials:
   Cotton swab with wooden stem
   Acetone, "distilled-in-glass" Nanograde, or other proper solvent
   Hexane, pesticide grade
   Isooctane, pesticide grade
   Metal clamp

Containers

Glass-stoppered glass jar

10-mL cone-shaped bottom vial with Teflon-lined screw cap

2-dram glass vial with Teflon-lined cap

Amber glass bottle, 1 pint

Plastic Nalgene bottle, 1 quart

Protective Equipment

Butyl rubber gloves

Plastic disposal bag

2. Wipe off a square area of approximately $0.24 M^2$ on the surface to be sampled. Alternatively, mark off five 2-inch-diameter circles distributed at the four corners and center of a $1\text{-}M^2$ area for building surfaces, or one 2-inch-diameter circle for vents and other surfaces. If sampling an area of known contamination, select an area of $4 in.^2$ in the center of the contamination.

3. While holding a swab in a clean metal clamp, saturate it with $20 mL$ to $30 mL$ of a $1:4$ acetone/hexane mixture. Continue holding the swab in the clamp while wiping the sampling area back and forth in a vertical direction, applying moderate pressure. Wipe several times. Turn over the swab and wipe back and forth in the horizontal direction.

4. Alternatively, dip a swab in a 2-dram vial containing $1.5 mL$ of acetone or other solvent, and swab one circle at a time, dipping the swab in the solvent before and after each circle is swabbed.

5. Wrap each swab in aluminum foil and place the wrapped swab in a clean, labeled glass container with a Teflon-lined lid, and close the cap until extraction and analysis can be performed. When all circles have been swabbed, dip a swab into the solvent, wrap it in foil, and put it into a jar labeled "blank." Tightly seal the

acetone-containing vial with a Teflon-lined cap.

6. Preserve the collected samples and blank at $4°C$ in a refrigerated box.

7. When resampling a surface after decontamination, position the sampling grid 6 inches to the right of the initial sampling points, or if movement to the right is restricted, position 6 inches downward.

OTHER SURFACE SAMPLING METHODS

"Sniff" Test

For volatile contaminants, a general survey monitor with a photoionization detector (PID) or a flame ionization detector (FID) can be used. The probe is moved around the article to see if there are any increases in levels over background. For a discussion on how to use these instruments, see the chapter on General Survey Instruments for Gases and Vapors. The following is an example of how this test can be used to identify mercury contamination.[5]

Procedure

1. Enclose the suspect material in a polyethylene bag or close-fitting airtight container for 8 hours at room temperature ($76–80°F[24–30°C]$), or place the bag in an oven set at $125°F \pm 5°F$ ($52°C \pm 2°C$), for 1 hour.

2. Make a small slit in the bag and sample the air with a mercury vapor analyzer.

3. If the mercury vapor concentration is greater than $0.01 mg/M^3$, the material is contaminated.

Surface Dust Contamination. This procedure is particularly useful for collecting samples from rafters and beams. One application is in asbestos abatement work where residual materials on areas such as window sills, pipes, and floor cracks are suspected.[6] This method could also be used to determine if residual dust, such as lead, is present on a worker's clothing, and therefore could be contributing to the worker's exposure.

Tape samples are used primarily for surface asbestos contamination, but can be used for other dusts as well. It has been suggested that adhesive tape may be more efficient in taking samples from rough surfaces than filter paper.[3]

Procedure

1. Calibrate a personal air sampling pump at 2 Lpm. Attach a filter in a cassette to the pump with a piece of flexible tubing. Depending on the analysis, the type of filter will vary. Another short length of tubing is attached to the inlet of the cassette (Figure 19.2).

2. Start the pump and use the exposed end of the second piece of tubing like a vacuum cleaner.

3. Move the sample collector through the work area, stopping at 10 sites or more that represent the most likely places where dust might collect. Each sampled spot should be marked with an "X" on a diagram, and the specific area should be described on a list, such as window still, left sleeve, cuff, top of pipe.

4. The amount of time spent sampling each site will be determined by filter loading. If the dust is readily visible, the filter will rapidly become loaded.

METHODS THAT DIRECTLY ASSESS WORKER EXPOSURE

A number of techniques to directly monitor for dermal exposure in addition to chemical-specific patches (Figure 19-1)

Figure 19.2. Special sampling kit is designed to collect fungal spores and other particulates from a fixed area on carpets. (Courtesy of SKC, Inc.)

TABLE 19.4. Surface Areas of the Body

Body Part	Surface Area (%)
Whole body	100.0
Face	3.5
Hands	4.4
Forearms	6.5
Back of neck	0.6
Front of neck and "V" of chest	0.8

Source: From Durham, W. F., and H. R. Wolfe. Measurement of the exposure of workers to pesticides. *WHO Bull.* **26**:75–91, 1962.

have also been tried. These include a patch type of dermal dosimeter made of gauze or charcoal cloth, a skin wash technique (most appropriate for chemicals with low rates of dermal absorption), urinary excretion of chemicals readily absorbed, and fluorescence of selective chemicals.[7] Common areas to sample are forearms, hands, wrists, feet, ankles, and neck (Table 19.4 and Figure 19.3). The neck can be a good site for sampling since it is generally exposed and is usually not washed during the course of the day as the hands are.

By placing the pad under the clothing, gauze and charcoal pads can be used to monitor exposure of exposed skin exposed through clothing permeation as well as

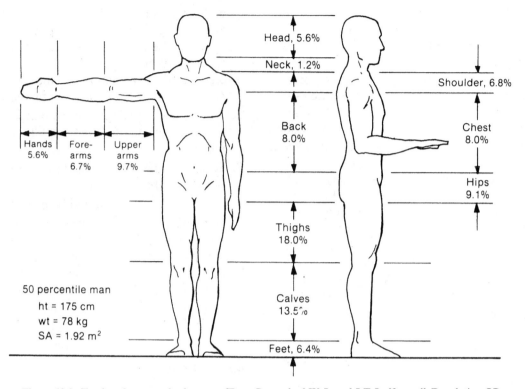

Figure 19.3. Total surface area for humans. (From Popendorf, W. J., and J. T. Leffingwell. Regulating OP pesticide residues for farmworker protection. *Res. Rev.* **82**:125–201, 1982.)

exposed skin. The pad is attached to an elastic band, and slid onto an arm or leg or pinned to underwear. Hand or face exposure can be monitored by attaching these pads, often called dosimeters, to thin cotton gloves or a hat. Outside clothing exposure can be monitored by attaching the pads using strips of Velcro.[8]

Any direct dermal sampling programs should be designed in conjunction with the health professionals who service the organization. They can advise on sampling methodologies that will evaluate dermal dose yet not pose a risk of irritation or injury to workers. Having the health professionals involved from the outset will also put them in the best position to answer any questions or concerns raised by the employees or that result from the monitoring data.

Gauze Patches or Pads

Patch or pad samplers have been primarily used to estimate exposures to pesticides, because of the significant potential for skin contact during reentry after spraying crops. These absorbent gauze pads can be attached to the hands, face, and other parts of the body that are normally unclothed. When placed underneath personal protective equipment, they can be used to estimate the degree of protection afforded by this clothing. Crop harvesters have been sampled extensively using this method, because of their exposure to dust on the crops which may contain agricultural chemical residues.

A limitation of this technique is the necessity of assuming that the area covered by the pad is representative of the entire body being sampled. The pads have several advantages in that they are small, portable, inexpensive, and passive. They tend to absorb and retain materials like spray mists or particulates that have low vapor pressures. One limitation to accurate dose estimates using these pads is that exposures must be relatively uniform between monitored and unmonitored points of the body.

Another disadvantage to gauze pads is that because they do not physically resemble the skin, adsorption, absorption, and retention of volatile or even nonvolatile materials by the cotton gauze may not be similar to the skin. There are a wide range of skin variations both between and within individuals due to such factors as hairiness, ease of sweating, wrinkling, callouses, and smoothness.[1]

The composition of a gauze pad for dermal sampling is several (often 16) layers of gauze or alpha-cellulose material, an impervious backing to prevent loss of material, and tape or other means to hold the pad together and attach it to the wearer. Both polyethylene and aluminum foil have been used as backings. Surgical tape, safety pins, or the use of an overgarment, such as a vest containing special pockets, have been used for attachment. However, ideally to measure dermal dose the pads should be worn under the clothing against the skin and on exposed skin.[9]

Procedure
1. Pads of alpha-cellulose have been used for measuring sprays, and pads made from 32 layers of surgical gauze and backed by filter paper have been used for measuring dusts. The gauze pads should be taped around the entire circumference. Whatever material is selected must be pre-extracted (Soxhlet) with the same solvent be used during analysis. Pads and gauze should be free of ink, oils, and other markings. The size of the pads is optional, although 5-cm by 5-cm gauze (exposed area) patches appear to be a useful size. The initial size has to be larger than 5-cm by 5-cm to allow a margin around the perimeter to tape the patch in place.
2. For transport, pads should be placed between folds of Whatman filter paper.
3. Secure the pad with surgical tape to the area to be monitored.
4. After use, each pad should be placed in an airtight, labeled glass container.
5. For analysis, the tape is cut away from the pad, leaving a 5-cm by 5-cm square. This 25-cm^2 piece is put through a Soxhlet extraction, and the extractant is analyzed. Results are reported in $\mu g/cm^2$.

Charcoal Pads

Charcoal pads have been used to monitor dermal exposure to volatile compounds, such as organic solvents and fumigants. The pads consist of a commercially available charcoal cloth[10] cut into 8-cm by 5-cm squares. Four layers are then sewn together and a protective covering of a thin polyester fabric ("Horizon East" Japan) is applied, and then a backing of a strong polyethylene-coated belting material provided.

These dosimeters have been recommended for monitoring highly toxic materials that are either harmful to the skin or can permeate the skin readily.[11] Materials with vapor pressures of 10 mm Hg and greater should be monitored only where ventilation effectively reduces vapor concentration. Chemical volatility also affects the usefulness of these dosimeters in that a liquid deposit on the skin is subject to air evaporation as well as absorption by the skin. Therefore, the more volatile a compound, the shorter will be the contact period on the skin, thus decreasing dermal absorption.

TABLE 19.5. Retention Efficiencies for Charcoal Cloth

VP (mm Hg):	0.3				1.0				4.0				10				15			
RL (µL):	10	25	60	100	10	25	60	100	10	25	60	100	10	25	60	100	10	25	60	100
RH (%)																				
30	94	93	92	92	92	90	89	88	87	85	83	82	83	81	77	75	81	78	75	72
50	91	90	89	88	88	86	84	83	81	78	75	73	75	71	66	63	72	67	61	58
70	89	87	85	84	84	82	79	77	75	72	67	64	67	62	55	51	62	56	49	43
90	87	84	82	81	81	78	74	72	70	75	60	56	60	52	44	38	53	45	34	26

Collection techniques will be similar to those used for gauze pads; however, sample handling following exposure is different. The units must be stored at −20°C or less until analysis to minimize evaporation.

Following collection, a correction must be made, because charcoal has a much higher adsorptive surface area than human skin, and the amount of vapors adsorbed onto the charcoal pad will be several times higher than the amount that could be adsorbed onto a similar area of human skin. Also, the total amount of chemical collected on the dosimeter would include depositions from both the liquid phase and the vapor phase in the case of materials that are being sprayed. The following calculation can be used to determine the initial liquid deposit to the charcoal pad.

$$\text{Initial liquid deposit} = \frac{\text{Total amount} - \text{Adsorbed vapors}}{RE/100}$$

where the total amount is obtained from the desorption and analysis of the cloth in the dosimeter; RE is the retention efficiency (see Table 19.5); and the amount of adsorbed vapors is unknown.

Table 19.5 lists estimated retention efficiencies for charcoal cloth patches as a function of vapor pressure (VP) of the chemical, retained liquid (RL), and relative humidity (RH).[11]

Skin Wash

Another technique involves wiping the skin directly. Skin washing has been used to estimate dermal exposures to organophosphates by using alcohol to wash unabsorbed residues from workers' skin and clothing. This technique may be more representative of actual exposure because the skin is subjected to more variables (e.g., sweating and changes in body temperature) than are external collectors such as pads. However, the effectiveness of this method is severely limited by the rapid rate with which the skin absorbs and retains certain compounds, such as organophosphates. Therefore, skin rinsing may underestimate the potential dose, because only the unabsorbed or slowly absorbed fractions of a pesticide formulation may be collected.[1] Solvents should be selected so that they are not irritating or readily absorbed by the skin. In addition, they should not facilitate skin penetration of the material being sampled.

Procedure. Prior to exposure, the area of the skin to be monitored must be cleansed.[11]

1. Swabs consist of two 8-ply 10-cm by 10-cm surgical gauze squares that have been extracted with the same solvent to be used for analyses. The squares are folded twice to form 5-cm by 5-cm squares and then stapled shut.

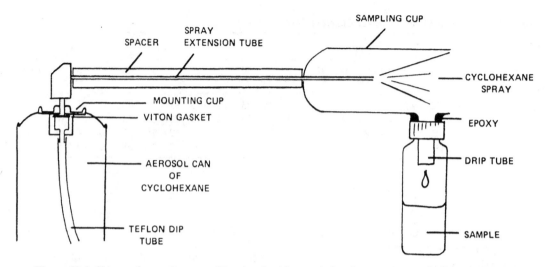

Figure 19.4. Skin wash sampling cup. (Reprinted with permission from *American Industrial Hygiene Association Journal* **43**:474, 1982.

2. Place the swabs in a glass jar saturated with 95% ethanol. The lid of the jar should be lined with aluminum foil or Teflon to prevent contamination of the gauze with residue on the lid. The jar should be kept shut until use.

3. In the field, open the jar and grasp a swab at the stapled edge using ring forceps or another similar device. Allow the excess ethanol to be squeezed off by rubbing the pad against the inside of the jar.

4. Continuing to hold the pad using forceps, the skin area selected is "washed" by rubbing the cloth back and forth over the surface with light pressure. Each area selected should be swabbed up to four times.

5. Pads for the same area can be combined into one labeled jar as a single sample.

Alternate Procedure
1. A prelabeled 12-mL vial is affixed to a skin wash sampling cup and the cup is held against the surface of the skin to be sampled (Figure 19.4).

2. Cyclohexane is sprayed through the cup onto the skin from an aerosol can. The cup will drain the skin washings into the sample bottle.

3. Sampling is terminated when the level of cyclohexane in the 12-mL vial reaches the sampling cup drip tube. This process happens very fast, in approximately 10 seconds.

4. The vial is then removed and sealed with a Teflon-lined screw cap and labeled. The cup is cleaned between samples by rinsing with a small amount of solvent from the can, and the rinse is discarded into a waste container.

5. A blank is also created by spraying 10 mL of solvent directly into a vial.

Fluorescence Monitoring

Fluorescence has been used to identify dermal contamination in two situations: when compounds such as polynuclear aromatics (PNAs) possess a natural fluorescence and when fluorescent compounds have been deliberately added to contaminants. Interferences include deodorants, coal-tar shampoos, and other cosmetic preparations that can also fluoresce. Therefore, when samples are obtained, the sampling practitioner should ask what cosmetics or shampoos have been used in order to determine potential interferences.

One fluorescence technique is to wash the skin with a solvent such as cyclohexane and then analyze the solvent with a UV fluorimeter.[12] The best use of this technique is in qualitative detection of contamination or to verify that decontamination procedures are adequate.

The addition of fluorescent compounds to mixtures has been used during pesticide application to "trace" leaf coverage, equipment efficiency, droplet size, and "drift" of spray clouds. The EPA has used fluorescent compounds to investigate visual warning methods for farmworker exposure to pesticide residues.[13]

The gauze pad technique described earlier can also be used to collect fluorescent samples. The basis is the mixing of the fluorescent material with the pesticide to produce a "tracer" effect, assuming quantitatively similar behavior of the two compounds from mixing to deposition on the skin surface. A concern is the ability of the tracer and the formulation being applied to be compatible due to solubility and mixing characteristics. This will determine the likelihood of partitioning of the tracer from the rest of the formulation during the application process. Problems will most likely be the result of a combination of solubility differences and incomplete mixing by the spray apparatus. This procedure has the potential to quantify worker exposure as well as to evaluate the effectiveness of protective clothing under actual field conditions. It also provides a measure of the effect of personal hygiene and work practices upon exposure. It can also be used for worker training in order to enhance awareness of the ramifications of not using the correct handling procedures.[14]

Fluorescent tracers were used in a study designed to see if workers could be moti-vated to use good work practices that minimize the potential for skin contamination if they could actually see the contamination on their skin. Sawmill employees involved in treating lumber with chlorophenols or handling the treated wood were monitored. Diisodium fluorescein was added daily to the mixing tanks supplying Permatox, a chlorophenol-containing wood preservative. Daily visualization of skin and clothing was done using long wavelength UV lamps. As a result, it was concluded that employees were motivated to practice better hygiene and work practices while on the job: Their levels of exposure, as measured during urine monitoring, decreased over the period of the study.[15]

Fluorescent tracers are not yet widely used. One problem is a lack of nontoxic fluorescent tracer compounds that would be compatible with a wide variety of contaminants.[16] The major disadvantage of this technique when compared to the patch method is its relative complexity in the field along with high cost compared to that of the patch tests. The stability of the UV illumination and the instrument's performance must be monitored. The ratio of pesticide to tracer deposition on the skin's surface must be determined through field sampling in each survey. In some cases, clothing penetration may be different for the two compounds, requiring a correction factor. The problem of quenching is inherent to any method measuring dermal fluorescence. If deposition of fluorescent material is excessive, the instrument response is no longer proportional to deposition due to a phenomenon called "quenching." Therefore, areas of high concentration may be underestimated.[17] Finally, discretionary exposure of employees to UV light should not be undertaken without qualified medical review and employee agreement.

EVALUATING SAMPLE RESULTS

As discussed before, there are no standards related to skin exposure; each situation must be handled on a case by case basis. The type of health effects caused, the contribution of skin absorption or ingestion to the total dose, and air sample results must be considered when evaluating the results of wipe samples. For example, detection of lead on the surfaces of an office associated with a radiator shop or lead battery shop means that individuals other than production workers are potentially exposed. Because of the toxicity of lead, it is prudent to limit exposure when possible. For wipe samples, it must be remembered that quantitation is related to the specific area that was sampled. Since it is generally impossible to wipe all areas of a surface, other portions could have higher levels of contamination.

In general, OSHA uses wipe sampling to establish the presence of a toxic material posing a potential absorption or ingestion hazard. A citation for an ingestion hazard can be issued when there is reasonable probability that in areas where employees consume food or beverages (including drinking fountains) a toxic material may be ingested and subsequently absorbed. A citation for exposure to materials that can be absorbed through the skin or can cause a skin effect such as dermatitis may be issued where appropriate personal protective equipment is necessary but not worn. Neither of these citations require any air sampling in addition to wipe sampling. There are two primary considerations when OSHA issues a citation for an ingestion or absorption hazard, such as a citation for lack of protective clothing: The first consideration is whether a health risk exists as demonstrated by a potential for an illness, such as dermatitis, and/or the presence of a toxic material that can be ingested or absorbed through the skin or in some other manner. The second is if there is a potential that the toxic material can be ingested or absorbed, meaning that it can be present on the skin of the employee, and can be established by evaluating the conditions of use and determining the possibility that a health hazard exists.

There is always the possibility that false negative results, that is, nondetection of existing contamination, will occur because the surface contamination is not removed by a wipe sample. This can be due to the selection of the wrong solvent or permeation of the contamination into a porous material. Dirt and grease on surfaces will adhere to wipes, obscuring sample results and making analyses impossible or difficult.

SUMMARY

Surface sampling can be a valuable tool in evaluating surface contamination and possible dermal dose.

Exposure is dependent on the total amount absorbed, the amount available, and the rate of absorption. There are a number of mechanisms that affect absorption of chemicals through the skin. In some cases, a portion of the initial deposition volatilizes before complete absorption can take place. Other factors include skin hydration from sweating, preexisting dermatitis, abrasions, and wearing contaminated work clothing for extended periods of time. In addition, the absence or presence of adequate warning properties, such as visual changes to the skin, irritation, or corrosive properties that would make a worker more prone to wash the skin rapidly, thus limiting the exposure period are important.

Skin contact can be minimized by the use of PPE such as gloves, aprons, and boots; replacing PPE when worn or damaged and decontaminating it each day; practicing good hygiene including washing hands at lunch and breaks; wearing fresh work clothing each day; not smoking, eating, or drinking liquids in work areas; and washing contaminated skin immediately.[15]

REFERENCES

1. Webster, R. C., and H. I. Maibach. Cutaneous pharmacokinetics: Ten steps to percutaneous absorption. *Drug Metab. Rev.* **14**: 169–205, 1983.

2. OSHA. *OSHA Industrial Hygiene Technical Manual.* 1984.

3. Chavalitnitikul, C., and L. Levin. A laboratory evaluation of wipe testing based on lead oxide surface contamination. In *Fundamentals of Analytical Procedures in Industrial Hygiene.* Akron, OH: AIHA, 1987.

4. Rosbury, K. D. Handbook: *Dust Control at Hazardous Waste Sites.* EPA-540/2-85/003.

5. Arizona Instrument Co., Tempe, AZ.

6. Natale, A., and H. Levins. *Asbestos Removal and Control: An Insider's Guide to the Business.* Cherry Hill, NJ: Source Finders, 1984.

7. Popendorf, W. J. Workshop: Predicting workplace exposure to new chemical. *Appl. Ind. Hyg.* **1**(3):R-11–R-13, 1986.

8. Cohen, B. M., and W. Popendorf. A method for monitoring dermal exposure to volatile chemicals. *AIHA J.* **50**(4):216–223, 1989.

9. Durham, W. F., and H. R. Wolfe. Measurement of workers to pesticides. *World Health Organ. Bull.* **26**:75, 1962.

10. MDA Scientific, Glenview IL.

11. Cohen, B. M., and W. Popendorf. A method for monitoring dermal exposure to volatile chemicals. *AIHA J.* **50**(4):216–223, 1989.

12. Keenan, R. R., and S. B. Cole. A sampling and analytical procedure for skin contamination evaluation. *AIHA J.* **43**:473–476, 1982.

13. Johnson, D. E., L. M. Adams, and J. D. Millar. *Sensory Chemical Pesticide Warning System, Part I. Experimental, Summary and Recommendations.* San Antonio, TX: Southwest Research Institute, 1975. EPA-540/9-75-209.

14. Fenske, R. A., et al. A video imaging technique for assessing dermal exposure. II. Fluorescent tracer testing. *AIHA J.* **47**(12):771–775, 1986.

15. Bentley, R. K., S. W. Horstman, and M. S. Morgan. Reduction of sawmill worker exposure to chlorophenols. *Appl. Ind. Hyg.* **4**(3): 69–74, 1989.

16. Dubelman, S., and J. E. Cowell. Biological monitoring technology for measurement of applicator exposure. In *Biological Monitoring for Pesticide Exposure.* Washington, D.C.: American Chemical Society, 1989.

17. Fenske, R. A. Validation of environmental monitoring by biological monitoring. In *Biological Monitoring for Pesticide Exposure.* Washington D.C.: American Chemical Society, 1989.

CHAPTER 20

BULK SAMPLING METHODS

Including bulk samples in a sampling strategy can often make the difference between a successful or unsuccessful air sampling effort. Bulk sample results can assist in making air sampling decisions, because laboratory methods can identify their constituents. Bulk samples can be collected from air, soil, water, chemicals, and many other media, such as carpet and filters from heating, ventilating, and air-conditioning (HVAC) systems. Chemicals, including chemical wastes, are the most common materials from which a bulk sample is collected.

While the technical aspects of bulk sampling are straightforward, regulatory standards may contained detailed requirements for how bulk samples are to be collected. This is especially true for environmental samples under Environmental Protection Agency (EPA) regulations.

PURPOSE

Bulk samples are collected for both occupational and environmental purposes. When used to supplement air sampling for occupational exposures, samples are most often of raw or process materials. Reasons for collecting occupational bulk samples include providing support for certain types of air sampling, such as silica sampling. An employer may have recently changed suppliers for a solvent and need composition information to aid in personal protective equipment selection. Occupational samples are also collected if interferences are suspected or for use as an analytical reference. Environmental samples are generally of chemical wastes, soil, sludge, or water. The EPA has specific definitions for materials in which environmental samples are collected. The term *solids* is applied to soils, sludges, sediments, liquids, and other bulk materials. Sludges are further defined as semidry materials ranging from dewatered solids to high-viscosity liquids and can be found on the bottom of creeks, ponds, and tanks or in any other place where solids can settle out of a body of liquid.

Air Monitoring for Toxic Exposures, Second Edition. By Henry J. McDermott
ISBN 0-471-45435-4   © 2004 John Wiley & Sons, Inc.

Water and soil are sampled on hazardous waste sites or during environmental audits for real estate transfers. Bulk soil and water samples are useful for identifying contaminants for which air sampling must be conducted during remediation work. Air samples are often collected along the site boundaries to determine if there is a potential for community exposure and on site for occupational exposures. These are performed using integrated and real time techniques as discussed in previous chapters. It is important to note that any situation where extensive samples of soil and/or surface or groundwater are collected requires the involvement of a geologist, hydrogeologist or engineer since an understanding of soil and groundwater characteristics and chemistry is required. Discharge water sampling to meet EPA requirements requires special techniques and knowledge as well.

Sampling of tap water is often performed in conjunction with evaluations for contamination of potable water sources as required by the Occupational Safety and Health Administration (OSHA). The most common compounds for which analysis is performed on water are lead, volatile organic compounds, and radon.

Bulk samples of air can be collected in bags, evacuated containers, or sorbent tubes and then analyzed to identify hazardous constituents, thus allowing an air sampling strategy to be developed for quantitation of specific compounds. Bags are used to collect gases and vapors for high-resolution analysis of trace constituents when sampling in open fields or vapor wells.[1] Drums and tanks are the most common types of chemical containers that are sampled. Both concentrated chemicals and wastes are stored in drums and tanks. Samples may be collected to identify the composition of the material or to determine if a waste meets any of the EPA criteria set for characterization of hazardous wastes under RCRA regulations. A tank may be scheduled for cleaning and the composition of the sludge inside may be uncertain. In this situation a bulk sample will assist in selecting adequate personal protective equipment and other controls to make sure workers who enter the tanks to remove the sludge are protected.

Analytical methods differ depending on the composition of the sample. For soil and water samples, EPA analytical methods are generally used. For bulk chemical samples the techniques may be from OSHA, NIOSH, EPA, or other source. If the material is classified as a waste, EPA methods are generally used. Samples of bulk air are most commonly analyzed using gas chromatography for separation and a mass spectrograph for detection. If collected on a charcoal tube, a NIOSH analytical method will be used whereas for other situations EPA methods are more likely. When interpreting the results of analysis of a bulk sample containing volatile components, the sampling practitioner should be aware that the percent composition of the mixture released to the air may be different from that remaining in the container. The airborne mixture will generally reflect a larger percent of the volatiles than were present in the bulk sample.

SAMPLE COLLECTION STRATEGIES

Most commercially available solid sampling devices are steel, brass, or plastic. Stainless steel is considered one of the most practical materials. Some devices are plated with chrome or nickel. They are not advisable to use, since scratches and flaking of the plating can drastically alter the results of analysis. Sample containers used to collect chemicals should be compatible with the material to be sampled. Polyvinyl chloride sample bottles can be used for acids and bases and other water-soluble materials. Glass, preferably with a safety plastic coat, should be used for hydrocar-

TABLE 20.1. Sampling Points Recommended for Most Waste Containers

Container Type	Sampling Point
Drum, bung on one end	Withdraw sample through bung opening.
Drum, bung on side	Lay drum on side with bung up. Withdraw sample through the bung opening.
Barrel, fiberdrum, buckets, sacks, bags	Withdraw samples through the top of barrels, fiberdrums, buckets, and similar containers. Withdraw samples through fill openings of bags and sacks. Withdraw samples through the center of the containers and to different points diagonally opposite the point of entry.
Vacuum truck	Withdraw sample through open hatch. Sample all other hatches.
Waste pile	Withdraw samples through at least three different points near the top of the pile to points diagonally opposite the point of entry.
Storage tank	Sample from the top of the sampling hole.
Soil	Divide the surface area into an imaginary grid. Sample each grid.

bons and solvents. Bakelite tops with Teflon seals should be used with glass bottles. As a general rule, equipment used to sample hazardous wastes should be disposable. If not, it must be carefully decontaminated between each sample to prevent contaminating subsequent samples. In general, metal sample containers should not be used to collect samples of liquid chemicals or wastes.[2]

Strategies for bulk sampling rarely involve collection of a single bulk sample. At a minimum, several samples are most likely to be collected and mixed together, thus creating a composited sample. If only surface samples are to be collected, then a design must be developed to maximize collection of a representative sample. Generally this involves the use of a grid pattern over the area of interest. The number of samples and distance between samples depends on the surface area to be sampled and other factors such as the cost of analyses. When depth of sampling is involved for samples that are homogenized, as are many liquids, three samples are often sufficient, each one at a different depth and then combined. In other cases a grid is developed for each of several levels. Table 20.1 describes sampling points recommended for waste containers.

Two general types of design are possible for most types of bulk sampling situations where multiple samples are required: grid designs and random designs. A grid system uses a regular pattern, either rectangular or triangular, to determine regular or random sampling points. A circular pattern of sampling around a central point may also be used. Random designs have some disadvantages compared to grid designs in that random designs are more difficult to implement in the field, since the sampling practitioner must be specifically trained to generate random patterns on-site, and since the resulting pattern is irregular. Grid designs (Figure 20.1) are more efficient in that they are certain to detect a sufficiently large contaminated area, whereas many random designs are not. For grid sampling, there are equations that may be used to determine grid intervals and the number of samples in a given area.[3] For sites larger than 3 acres,

$$GI = (A\pi/GL)^{0.5}$$

For sites smaller than 3 acres,

$$GI = \frac{(A/\pi)^{0.5}}{2}$$

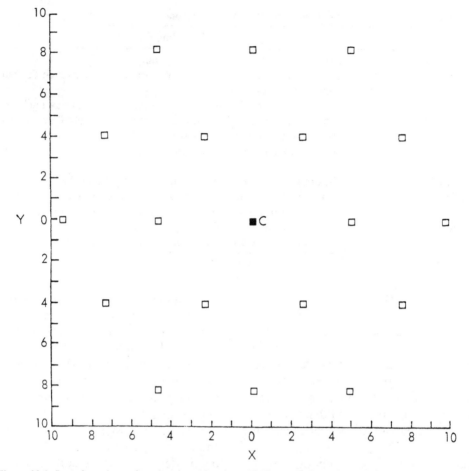

Figure 20.1. Location of sample points in a 19-point grid. The outer boundary of the contaminated area is assumed to be 10 feet from the center (C) of the spill site. (From EPA. *Verification of PCB Spill Cleanup by Sampling and Analysis.* EPA-560/5-85-026, August 1985.)

where GI = grid interval, A = area to sampled, and GL = grid length.

Compositing is often done to save on analytical costs. Several samples are collected and mixed together to form one sample (Figure 20.2). Situations where compositing is often done are tank, drum, waste pile, soil, and water sample collection. The disadvantage of sample compositing is the loss of specific data for specific levels or areas, but this loss is usually offset by having a more representative estimate of concentration. Samples can be composited in the field or the labo-

ratory, although compositing is most often done in the laboratory so individual analysis of specific samples can be performed if any of the composite samples show unusual results. The actual compositing of samples requires that they be homogenized. This procedure varies according to the type of waste being composited and the parameters to be measured.

The applicability of compositing is dependent on whether individual specimens contain sufficient material to form a composite and leave enough material for individual analyses if needed. Other limit-

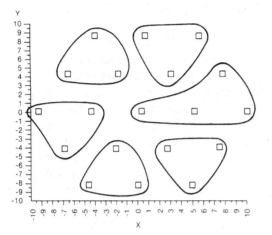

Figure 20.2. Plan showing a six-group compositing design. (From EPA. *Verification of PCB Spill Cleanup by Sampling and Analysis.* EPA-560/5-85-026, August 1985.)

ing factors are the analytical method and cleanup criteria. As a general rule, the EPA cleanup level divided by the number of composited samples should not exceed the minimum detection level of the method. Cleanup levels are generally set by individual states or the EPA and information on specific chemicals can be obtained from the state environmental agencies or the EPA.

CONTAINERS AND SHIPPING

The most important factors to consider when choosing containers for bulk samples are compatibility with the material sampled, cost, resistance to breakage, and volume. Containers must not distort, rupture, or leak as a result of chemical reactions with constituents of waste samples. Therefore, it is important to have some idea of the properties and composition of the material sampled.

All samples should be labeled with a unique sampling number or code and identified as such on the laboratory request form. It is best to label the bottle or bag first and then collect the sample. If the samples are related to air samples, then an exact ref-

erence including their number, should be made to them on the analytical request sheet.

In general, leakproof glass containers are best since they will not react with most chemicals. Brown glass is needed for photosensitive waste. However, polyethylene containers can be used for most dusts (but not soils). Bulk samples are generally collected in vials of 20 mL to 50 mL in volume with Teflon liners in screw lids. A recommended container is a 20-mL scintillation vial with a PTFE-lined cap. Specific chemicals for which glass vials should be used include aromatic compounds, chlorinated hydrocarbons, and strong acids. Generally, large quantities of material are not necessary for analyses. The shorter the time between the collection of a sample and its analysis, the more reliable will be the analytical results. Before the samples are removed from the sampling site, a check should be made to ensure that the tops are correctly and securely fastened. Decontaminate the outside of each sample container thoroughly before packing for transit.

Care should be taken when storing or shipping samples in plastic bottles or bags, since gases and vapors can diffuse in and out of these materials. Exhaust components from transportation vehicles, including jet engine exhaust and gasoline vapors, may contaminate the samples during shipment. Containers should be airtight.

The most important rule in shipping bulk samples is to ship them by themselves, although similar types of bulk samples may be sent together. Bulk samples should never be sent in the same package as air samples. If a material safety data sheet is available, it is a good practice to send it along with the bulk sample to the laboratory.

All applicable shipping regulations must be followed. If a sample is flammable, combustible, corrosive, or poisonous, the shipping options are usually limited to a land carrier. For example, under Department of Transportation (DOT) labeling requirements certain chemicals are not allowed to be mailed as bulk samples: nitric acid, gasoline, perchloric acid, benzoyl peroxide, class A poisons, aniline, chloropicrin, or organophosphate pesticides.[4] Shipping containers for liquids should be cushioned within another sealed waterproof container with absorbent material sufficient to take up all leakage or breakage of the sample container. It is generally a good practice to ship containers of liquids by surface mail.

Sample preservation is sometimes required, especially for water samples. Methods of preservation are relatively limited, and are generally intended to retard biological action, to retard hydrolysis of chemical compounds and complexes, and to reduce volatility of constituents. Preservation methods are limited to pH control, chemical addition, refrigeration, and freezing. Storage at a low temperature, usually 4°C, is perhaps the best way to preserve most samples until the next day. Use chemical preservatives only when they are shown not to interfere with the analysis being made. When they are used, add them to the sample bottle initially so that all sample portions are preserved as soon as collected.

In the case of volatile bulk samples, consideration should be given to shipping the samples on dry ice or with a bagged refrigerant inside an ice chest. The carrier should be consulted about possible restrictions on the amount of dry ice accepted. Specific package labels are usually required when dry ice is used. Drawbacks include obtaining dry ice and the need for adequate refrigeration in order to cool bagged refrigerants such as "Blue Ice" enough to get it thoroughly frozen. Occasionally a sample may require a preservative other than cooling. Usually this requirement is limited to environmental samples undergoing a certain type of analysis.

PERSONAL PROTECTION

Many materials from which bulk samples are collected are in a concentrated form; therefore, when collecting samples it is important for all personnel to wear personal protective equipment. The type of equipment needed will vary depending on whether liquids that might splash the eyes or skin are being handled, or whether toxic dusts such as asbestos are being collected. The selection of proper gloves will vary depending on the characteristics of the material, for example, whether the material is corrosive or is a solvent. Respirators should have the proper type of chemical cartridge or filter for the contaminant(s) being sampled. Any use of respirators must follow applicable OSHA or other requirements for an adequate respiratory protection program. When sampling unknown materials, it is especially important to have proper personal protective equipment available and to use it.

Sampling of drums and tanks may require the use of self-contained breathing apparatus (SCBA) or other supplied-air respiratory protective equipment, such as pressure-demand type airline equipment. These drums and tanks may hold hazardous materials and special procedures are often required.

It is *not* considered good practice to sniff samples to determine if any chemicals are present, since some chemicals can be hazardous at very low levels. For example, benzene, a carcinogen, has an odor threshold of 12 ppm,[5] which is much higher than the exposure levels considered safe. Also some compounds at levels of environmental concern do not have a significant odor, and thus the lack of odor is an invalid criterion for not analyzing the sample.

BULK AIR SAMPLES

Bulk air samples can be collected in bags, evacuated containers, or on sorbent tubes.

Another variation of sampling bulk air involves using a direct reading instrument to identify levels of volatiles above a bulk sample of a solid or liquid.

Generally, bulk air samples are collected for the purpose of qualitative analysis, that is, to see what contaminants might be present. An example is a situation where occupants are concerned about odors in a building where a carpet was recently installed and the exact identity of suspected airborne hydrocarbons is unknown. Generally reactive gases, such as nitrogen oxides, hydrogen sulfide, and isocyanates, cannot be collected in this manner unless analysis to be done in the field immediately after sampling, because these gases can react with dust particles, moisture, sealing compounds, glass, or metal, thus altering the sample's composition.

During hot processes, decomposition or off-gassing of products that are part of the formulation can occur. In addition, process intermediates, which are temporary products that are either extracted during production or react into other compounds during the manufacturing process, are often of interest. Sometimes these are given off by an interim step, for example, in a reaction vessel that has local exhaust.

The use of bags to collect air samples for personal exposures was discussed in Chapter 6. Bulk air samples can also be used to detect concentrations of gases and vapors in exhaled breath. For more information on this sampling technique, see Chapter 18.

Headspace Vapor Samples

Headspace vapor sampling is often done on samples of soil, water, and unknown liquids as a screening technique. It is done by putting the material into a container, capping it, and using a general survey monitor such as photoionization detector (PID) or flame ionization detector (FID) to "sniff" the sample. For example, the

Figure 20.3. The ppbRAE is a Photoionization Detector device that provides parts per billion detection of volatile organic compounds at hazardous waste sites. (Courtesy of REA Systems.)

ppbRAE from RAE systems (Figure 20.3) is a photoionization detector (PID) instrument that provides parts-per-billion (ppb) detection of volatile organic compounds (VOCs) with more than 300 correction factors for use at hazardous waste sites and similar applications. For soil and groundwater samples, it is sometimes used as a way of screening samples to select the most contaminated samples to send to the laboratory, thus minimizing laboratory fees.

Headspeace gases are the accumulated gaseous components found above solid or liquid layers in closed vessels. These gases may be the result of volatilization, degradation, or chemical reaction and concentrations are generally high. Since higher concentration ranges are to be expected, real time monitoring instruments with higher-range scales or a sample dilution system should be used in order to prevent saturation or deterioration of the detector.

Any sign of bulging or distortion in a drum, tank, or similar container may indicate that decomposition or other chemical reaction has occurred, and so precautions must be taken when sampling these containers. A method for taking a headspace vapor sample over a bulk soil or water sample is to put the material in a cup and cover it with cellophane. Puncture the cellophane with a probe and take a reading with a PID or FID analyzer. When monitoring soils for organic vapor content, conditions such as volatility of the sample, porosity of the soil, general weather conditions, and the ambient background hydrocarbon concentration must be taken into consideration.

Most vessels can be sampled though small hatches or openings. Fully sealed vessels require special techniques, such as the use of remote control units to drill openings, use of safety screens for explosion protection, and vessel pressure monitoring, and these vessels should not be disturbed unless the sampling practitioner has received special training in this area. In some cases, such as headspace vapor sampling on large tanks, an extension tube will be useful for collecting samples at various distances above the liquid surface. However, if the depth of the liquid is unknown, this not advisable since liquid may be drawn up the tubing into the instrument. Where there is no conveniently accessible sampling port, a preliminary scan of the external seams, edges, or any corroded areas of the vessel with a PID or FID may indicate if any vapors are present.

Sorbent Tubes

Sometimes methods using sorbent tubes are used to collect high-volume air samples for qualitative analysis. For most volatile and semivolatile organic compounds, a sample can be collected using a charcoal or Tentax tube connected to a personal sampling pump and collecting air at 1 L/min for several hours. Although the sample is likely to exhibit breakthrough, it does not matter since the purpose of the sample is to determine what substances are present rather than their exact concentrations. In some situations where the concentrations remain constant and the air is stagnant, a grab sample might be sufficiently representative of the actual concentrations present. However, the sampling practitioner should always exercise caution when interpreting the results of this type of sampling. For more information on collection techniques using sorbent tubes, see Chapter 6.

Bag Sampling for High-Resolution Analysis

Bags are often used to collect gases and vapors for high-resolution analysis of trace levels. This method is commonly used for environmental sampling in open fields or vapor wells. The sampling apparatus consists of a 5-gallon open-top container with a sealable lid, a Teflon stop-cock, Teflon bulkheads, Teflon tubing, and a sampling pump[1] (Figure 20.4). With this apparatus an ambient air sample is drawn into the bag as air is pumped out of the 5-gallon container.

Sample collection bags can also be constructed of a number of synthetic materials, including polyethylene, Saran, Mylar, Teflon, and Tedlar. Tedlar is produced without plasticizers, thereby eliminating

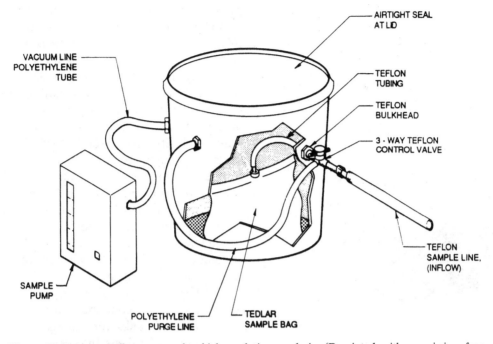

Figure 20.4. Air sampling system for high-resolution analysis. (Reprinted with permission from *American Industrial Hygiene Association Journal* **50**:A-591, 1989.)

the outgassing problem common to some other materials. For a discussion of bag materials and sampling issues, see Chapter 6.

Procedure[1]

1. A 5-L or 10-L Tedlar sampling bag is placed inside the container and connected to the small section of Teflon tubing attached to the bulkhead. The lid, which requires a gasket to ensure an airtight seal, is place tightly on the container.

2. Prior to collection an air sample, the stopcock should be positioned to close off the line to the sampling bag. The pump is then turned on and the air in the sample line is purged. If necessary, the air between the stop-cock and the sample bag can also be purged by putting a small amount of ultra-high pure nitrogen in the bag and pulling it out through the sample line.

3. The air sample is collected by opening the stop-cock to allow air to be pulled through both lines into the plastic container. In this arrangement, air is drawn through the sample line into the bag, due to the vacuum that is created by the sampling pump onto the sampling apparatus.

4. After the sampling is completed, the stopcock is returned to the original position. In this position, the air in the sample bag is closed off and simply redirected through the sampling apparatus. The pump is then shut off, the lid is opened, and the sample bag is removed. If the contaminants of interest are photochemically reactive, the sample bag immediately must be placed into a light-occlusive container.

Evacuated Containers

In many cases, the sampling practitioner simply wants to collect a quick sample of

Figure 20.5. Evacuated cylinders are available in several sizes for bulk atmospheric sampling (Courtesy of SKC, Inc.)

air from a given area. It might be during a short-term operation or from a location with homogeneous airborne composition. Evacuated flasks and cans can be used for this collection. Evacuation-type devices are sealed vessels with only one entry in which a vacuum has been created. These types of devices can be used to collect gross gas contamination, such as methane or sewer gas. There are many different sized containers for this purpose-container volume is chosen to detect the contaminants at their concentration (Figure 20.5). See Chapters 5 and 6 for more information on these samples.

BULK SAMPLES OF SOLID OR LIQUID CHEMICALS

When choosing to sample bulk chemicals there can be a number of different goals, but most commonly the goals are to provide supplementary information for air samples, to identify the constituents of a mixture, to identify a suspected contaminant, or to characterize a hazardous waste. This is generally done when all other means of obtaining information regarding the chemical composition of a process materials have been exhausted, for example, the material safety data sheet (MSDS) is not specific enough, the manufacturer has claimed a trade secret exemption, or a source of contamination is suspected. Mixtures in which the constituents are unknown are often wastes and can be found in any type of container as well as in soil and water. The composition of the material as well as its container will determine the method of collection. The most common chemical containers sampled are (a) drums and tanks for liquids and (b) sacks and bags for dusts. In each case, a different type of sampler is required. The collection techniques are discussed further in each of the specific sections for sampling these containers.

In situations where air sampling is being conducted for certain compounds used in specialized processes, laboratories may request a bulk sample to use as a standard or to be certain the bulk material has the same composition as their standard. Bulk samples are used to confirm the presence of free silica in respirable air samples and to assess the presence of other substances that may interfere in their analysis. The following are examples of chemicals with variable composition for which bulk samples should always be collected along with air samples:

Chlorinated camphene
Chlorinated diphenyl oxide
Chlorodiphenyl
Gasoline
Hydrogenated terphenyls
Kerosene
Mineral oils

Naphtha
PCBs
Petroleum distillates
Stoddard solvent
Turpentine

Never include bulk samples in the same shipping container as air samples to prevent possible contamination of the air samples.

Fugitive and Other Dusts

The term *fugitive dusts* refer to dust that has been released from unidentified sources. They may be on the rafters or other surfaces in a manufacturing operation or warehouse. Settled dust samples collected from rafters or near a worker's job site are sometimes considered representative of the airborne dust to which workers are being exposed. Usually rafter samples are used to determine the average size distribution and composition of settled airborne dusts.[6] Of course these samples may not reflect the concentration of very small particles that represent the greatest inhalation hazard. Common contaminants found in dusts include lead in processes where it is heated, such as battery manufacturing and radiator repair, because of its tendency to settle over a widespread area, and tremolite, a type of asbestos often present in talc, because it tends to form in conjunction with talc deposits. A process material might also be sampled to determine the composition of the material before it is airborne. Sampling dusts and other dry materials in containers are discussed in the next section. Samples of suspect asbestos-containing materials are collected to determine asbestos content.

When used in conjunction with air samples, the method of collection as well as the source of dust samples must be selected carefully in order to be representative of the situation in which the material is collected. If a bulk sample of the raw material

used in the manufacturing process is sampled, it may not represent the particle size ranges likely to be present following processing, thus not providing a good estimate of the respirable hazard. The next choice would be to fill a bottle with settled dust from a rafter near the operation of interest. This would be representative of the changes in particle size due to the processing of the raw material. An even better choice might be to attach a cyclone containing a filter cassette to a personal air sampling pump and "vacuum" up the dust from the rafter or near the process. This would collect the particles that were $10\,\mu m$ and less on the filter. If the material collected in the bottom of the cyclone is weighed, as well as that on the filter, an estimation of the percent of respirable dust present can be made. For more information on the use of cyclone collectors, see Chapter 7.

Another concern is exterior paint on tanks, piping, and other metal surfaces that may be sandblasted or cut using oxyacetylene torches. At one time, most paints used for these applications contained varying amounts of lead that will become airborne during the aforementioned procedures. A sample should be collected by scraping the surface down to bare metal. If any detectable levels of lead are present in the paint, precautions should be set up for workers. At a minimum, air sampling should be done during the work to determine what levels of lead are present.

Bags, Sacks, Fiberdrums, and Waste Piles Containing Dry Materials

Laboratory scoops or triers, also called sampling probes or tubes, can be used to sample dry bulk material in containers. A typical sampling trier (Figure 20.6) is a long tube usually made of stainless steel cut in half lengthwise with a sharpened tip. In some models an ejector is incorporated into the sampler for easier removal of the

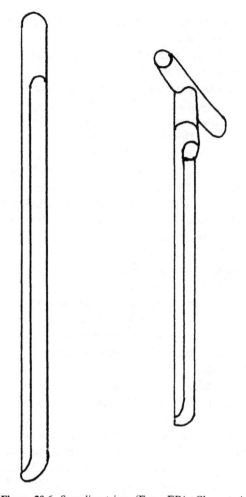

Figure 20.6. Sampling triers. (From EPA. *Characterization of Hazardous Waste Sites—A Methods Manual: Vol. II. Available Sampling Methods.* EPA-600/4-83-040. September 1983.)

depth is determined by the hardness and types of solids being collected.[2] Scoops are used to sample dry powdered or granular waste in containers such as barrels, fiberdrums, sacks, or bags. Since these wastes tend to generate airborne particles when the containers are disturbed, containers must be opened slowly.

Trier samplers are generally used to sample waste piles. A waste pile can range from a small heap to a large aggregate of wastes. The wastes are predominantly solid and can be mixtures of powders, granules, and large chunks. Hazardous materials stored in waste piles are usually of a small granular size, such as sand and dust. Waste piles should be approached from upwind due to the potential for dust emissions. Sampling should be designed to disturb the pile as little as possible, minimizing the amount of dust released.

In waste piles, the accessibility of the waste for sampling is usually a function of pile size. Ideally, piles containing unknown wastes should be sampled using a three-dimensional simple random sampling strategy. This strategy can be employed only if all points within the pile can be accessed. In such cases, the pile should be divided into a three-dimensional grid system, the grid sections are assigned numbers, and the sampling points are then chosen using random-number tables. If sampling is limited to certain portions of the pile, the collected sample will be representative only of those portions, unless the waste is known to be homogeneous.[2]

When using a trier to sample waste piles, the sampler must be decontaminated in between piles or disposed of after each set of samples. A number of core samples, depending on the size of the waste pile, must be taken at different angles and composited in order to obtain a sample that upon analysis[7] will give average values for the hazardous components in the waste pile.

sample. The ejector snaps into the probe's slot, sliding up the probe as it is inserted into the material. When the trier is pulled up, the ejector slides down the probe and the sample is pushed out. Triers are preferred when the granular material to be sampled is moist or sticky and are also used to sample surface soil. Scoops are preferred when the material to be sampled is dry, granular, or powdered. The scoop or trier can be used to collect samples of solids at depths greater than 3 inches. The sampling

Procedure

1. Barrels, fiberdrums, and cans must be positioned upright for sampling. If possible, sample sacks or bags in the position in which they are found, since standing them upright might rupture them. For waste piles, determine the number and locations of sampling points.

2. Collect a composite sample from the container with a trier or stainless steel scoop. Insert the trier into the material at 0° to 45°C angle from the horizontal. This orientation minimizes the spillage of sample from the trier. Rotate the trier two or three times to cut a core of material.

3. Slowly withdraw the trier, making sure the slot is facing upward. It might require tilting of the container. Withdraw samples through the top of barrels, fiberdrums, buckets, and similar receptacles. Withdraw samples through fill openings of bags and sacks. Withdraw samples through the center of the receptacles and to different points diagonally opposite the point of entry. Transfer the sample into a suitable container with the aid of a spatula and/or brush.

4. If composite sampling is desired, repeat the sampling at different points two or more times, and combine the samples in the same container with the sample from step 3 and cap it. For waste piles, sample through at least three different points near the top of the pile to points diagonally opposite the point of entry, and composite these in the same sample container.

5. Wipe the sampler clean and store it in a plastic bag for subsequent cleaning.

6. Where there is more than one container of wastes on a site, segregate the containers by waste type and sample according to a table of random numbers. For more information on random numbers, see Chapter 3.

Drum Sampling

The 55-gallon drum is the most frequently used container for storage and disposal of hazardous wastes. It is also one of the most common fixtures that are sampled and analyzed.[8] Because of the drum's frequent appearance at hazardous waste dump sites, and the number of problems associated with sampling it, special precautions and techniques must be followed. Structurally sound drums present the least amount of risk of rupture during mechanical handling. A responsible, experienced member of the sampling party should determine whether the drums can be opened or punctured and sampled in place. This decision will be based on the extent of cleanup in case of rupture, the danger involved in a rupture, and other factors such as the size of the site and drum spacing. Drum sampling can be one of the most hazardous activities to worker safety and health because it often involves direct contact with unidentified wastes and concentrated materials.

In some situations the general composition of a waste may be known yet it still requires a sample. For example, drums of mineral oil waste generated by a process such as metalworking very likely contain dissolved amounts of the metals being tooled. In the case of carbide tools, this might include cobalt and tungsten. Therefore, drums used to store waste material will most likely require sampling before a determination of the proper category for this waste can be achieved. Printing processes often end up with waste mixtures of solvents and inks. A representative sample of the waste has to be collected to determine the ranges of the components in order to submit it to a disposal facility.

The following situations require extraordinary procedures and are beyond the scope of this book. Contact specialists whenever these are present: drums that may contain shock-sensitive wastes, drums containing radioactive wastes, drums containing laboratory packs (laboratory wastes), and buried drums. Identifiers of high-risk drums include drums with bulged heads most likely due to internal gas formation, drums with bulges on the side or bottom most likely due to freezing and expansion of the contents, and drums that have been deformed due to mishandling.

All tools used for drum opening should be of the nonsparking variety. Often these tools are made of brass. For drums that may contain unknown hazardous wastes, remote-controlled devices such as pneumatically operated impact wrenches, hydraulically or pneumatically operated drum piercers, or a backhoe equipped with a bronze spike to penetrate the drum top are preferable. Do not use picks, chisels, and firearms to open drums. Reseal openings as soon as possible to minimize vapor release. Decontaminate equipment after each use to avoid mixing incompatible wastes and contaminating future samples.

Drum sampling is generally done using a containerized liquid waste sampler (COLIWASA) (Figure 20.7) readily available from scientific and hazardous waste supply houses. It is especially useful for sampling wastes that consist of several immiscible liquid phases. It is made of glass or plastic and can be either disposable or reusable. One reusable model is made of borosilicate glass and has a sample capacity of 200 mL. A general-purpose disposable model has a slightly tapered bottom that accepts a glass inner tube with a blown ball at the end. The ball can be transferred to a sample bottle with minimal loss. Tubes come prescored to permit components to be snapped in half for easy disposal. The glass COLIWASA is used to sample hydrocarbons such as solvents and all other con-

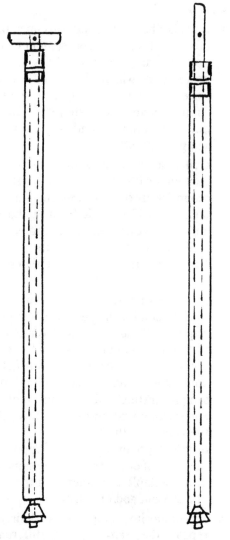

Figure 20.7. COLIWASA. (From EPA. *Characterization of Hazardous Waste Sites—A Methods Manual: Vol. II. Available Sampling Methods*. EPA-600/4-83-040, September 1983.)

taminated liquids that cannot be sampled with the plastic COLIWASA except for strong alkali and hydrofluoric acid solutions. PVC COLIWASAs are best for acids, bases, or other water-based substances. The COLIWASA can also be used to sample free-flowing liquids and slurries contained in shallow tanks, pits, and similar containers.

Make sure COLIWASAs are clean between successive samples or use a different tube for each sample. To clean, use a long-handled brush, rags, and a solvent and clean the tube inside and outside. Proper PPE must be worn during decontamination of the sampling device. Using a separate brush and rags, wash with soap and water and then rinse with clean water. Before using the tube to sample again, inspect it for signs of deterioration. All wash materials must be decontaminated or disposed of properly. Any solids should be removed and the sampler drained before using it for additional sampling. Check to be sure that the sampler is functioning properly.

Procedure

1. Develop a sampling plan. Research background information about the wastes. Determine which drums should be sampled. Select the appropriate sampling devices and containers. Decide the number, volume, and locations where samples are to be taken. When there is more than one drum of wastes to be sampled at a site, segregate the drums according to waste types and use a table of random numbers to determine which drums to sample. For sampling of known wastes a composite sample should be collected by pulling samples from at least 10 percent of the containers. Drums should be selected at random, taking care not to sample only easier-to-reach drums on the perimeter.[9]

2. Develop standard operating procedures for opening drums, sampling, and sample packaging and transportation Based on available information about the wastes and site conditions, have a trained health and safety professional determine the appropriate personal protection to be used during sampling, decontamination, and packaging of the samples.

3. Position the drum to be sampled so the bung is facing up. Drums with bungs on the ends should be positioned upright. Drums with bungs on the side should be lying on the side with the bungs upright. Use a manual nonsparking hand wrench and slowly loosen the large bung to allow any gas pressure to escape. If the bung cannot be removed, the drum should be punctured.

4. Select a clean plastic or glass COLIWASA for the liquid waste to be sampled, and assemble the sampler. Check the sampler to make sure it is functioning properly. Adjust the locking mechanism if necessary to make sure the neoprene rubber stopper provides a tight closure. Put the sampler in the open position by placing the stopper rod handle in the "T" position and pushing the rod down until the handle site against the sampler's locking block. While wearing appropriate PPE, slowly lower the sampler into the liquid waste at a rate that permits the levels of the liquid inside and outside the sampler tube to be about the same. If the level of the liquid in the sampler tube is lower than outside the sampler, the sampling rate is too fast and will result in a nonrepresentative sample.

5. When the sampler stopper hits the bottom of the waste container, push the sampler tube downward against the stopper to close the sampler. It should be noted that this sampler will not sample the bottom 1 to 2 inches of the drummed material, nor will it sample solids. Lock the sampler in the close position by turning the "T" handle until it is upright and one end rests tightly on the locking block. Slowly withdraw the sampler from the waste container with one hand

while wiping the sampler tube with a disposable cloth or rag with the other hand. Dispose of the cloth into an appropriate container.

6. Secure the container to keep it from tipping over before transferring the sample. The container to receive the sample should have a mouth wide enough for the COLIWASA to fit into it and large enough to hold the volume contained in the COLIWASA. Carefully discharge the sample into a labeled sample container by slowly pulling the lower end of "T" handle away from the locking block while the lower end of the sampler is positioned in the sample container. Continue to add other samples in the same way if this is to be a composite sample. Mix the contents of the random samples together after each addition. Cap the container in between samples. Most laboratories require only a pint of liquid to conduct analytical tests. Return any excess material to the drums.

7. After the last sample has been added to the bottle, making sure it is securely closed, invert the sample bottle a few times to check for leaks. Regardless of whether visible leaks are detectable or not, wipe the bottles with rags to remove any wastes on the outside.

8. Unscrew the "T" handle of the COLIWASA and disengage the locking block. Clean the sampler on-site or store the contaminated parts of the sampler in a plastic storage tube for subsequent cleaning.

9. Mark the drum with paint or other indelible marking compound so that it can be identified later. Put a plastic drum cover over the drum, or reinstall the bung, to prevent any liquid such as rain from entering. Do not use the drum cover as the drum marker as it may be blown off by the wind.

Procedure Using Glass Tubing. The simplest method for sampling drums is to use a length of glass tubing to collect samples.[7] The tube is normally 122 cm long and ranges in size from 6 mm to 16 mm inside diameter with larger tubes used for more viscous fluids. The tubing is broken and discarded in the container after the sample has been collected. This is a quick and relatively inexpensive means of collecting concentrated samples of containerized materials. Nonviscous materials are more difficult to retain in the tubing. The sampling practitioner should be protected against splashes, wearing at a minimum neoprene gloves, butyl rubber apron, and a faceshield.

1. Remove cover from container to be sampled.

2. Insert glass tubing almost to the bottom of the container, keeping at least 30 cm of tubing above the top of the container.

3. Allow the liquid in the drum to reach natural level in the tube. Hold a thumb (must have gloves on) over the exposed end of the tube or put a rubber stopper in it.

4. Carefully remove the tube from the drum and insert the uncapped end into the sample container. Remove thumb or stopper to allow the sample to flow into the bottle. Cap the sample container and make sure it is labeled with a unique sample number.

5. Repeat sample collection until sufficient volume is collected. Discard the glass tube in the drum, breaking it in such a way that all of it is inside the drum.

6. Replace the cover on the container.

Tank Sampling

For tanks, procedures similar to drum sampling are followed. Sampling above-ground storage tanks can require great dexterity.

Usually it requires climbing to the top of the tank through a narrow vertical or spiral stairway while wearing protective equipment and carrying sampling paraphernalia. At least two persons must always perform the sampling: one to collect the actual samples; the other to stand back, usually at the head of the stairway, and observe, ready to assist or call for help in case of problems.

When opening a tank hatch, make sure that excess pressure from stored volatiles has been vented. Often underground storage tanks have vent pipes. If there is any doubt, call in experts prior to sampling. Guard manholes to prevent personnel from falling into the tank.

A *bomb sampler* is a composite sampling device for storage tanks. Bomb samplers are available in stainless steel and consist of a cylindrical reservoir chamber, a weighted plunger that seals the chamber at the bottom, and a cable attachment for suspending the apparatus and activating the sampling device. These samplers typically hold 500 mL of sample.

Procedure

1. Collect one sample each from the upper, middle, and lower sections of the tank contents with a weighted bottle sampler.
2. Combine the samples in one container and submit the container as a composite sample for analysis.[8]

SOIL SAMPLING

Soil samples can provide useful information about the exposures to personnel who might have skin contact with the soil, or about the potential for dust clouds to occur (fugitive emissions) due to activities such as grading or excavating or the wind. For volatile materials trapped in soil, airborne vapors will be a concern as well. Generally these tend to increase during excavations.

In general, soil samples are taken in a grid pattern over the entire site to ensure a uniform coverage of the site. As noted in the section on grid sampling, there are elaborate statistically designed patterns for sampling soils. Soils sampling increments are often 6 inches for the upper 2 feet of soil and every 12 inches below a depth of 2 feet.[3] The depth will depend on the degree of contamination and it is usually the depth at which the cleanup criterion (concentration) is met.

Background samples also must be collected to allow for statistical comparison of the natural condition to the potentially contaminated area. They can be used to differentiate between contaminants and materials naturally occurring in the soil, such as heavy metals. A few contaminants such as formaldehyde and phenol may be naturally produced, but their concentrations in soils are typically very low and near or below detection limits. Background samples should be taken in areas not affected by the hazardous waste units or by the site itself. They may be collected away from the site, but the sampling location should be as close as possible to the site. Background samples should be taken from soil depths and soil horizon materials similar to those of the potentially contaminated area. The location and depth of background samples must be indicated. As many as 10 background samples from each soil strata may be needed.[3]

Soil sampling intervals and total depth may be dependent on several factors, including soil type and hydraulic conductivity; suspected magnitude of surface contamination; physical state of the waste and its mobility; height of liquid head at the ground surface; length of time that the waste was present at the site; relative toxicity of the waste; and location of the waste management unit (indoors or outdoors).[1] Many of these decisions require the expertise of geologists and engineers. Soil types vary considerably; therefore, it is important to maintain a detailed record during sampling, especially listing locations and depth sampled, and characteristics such as soil grain size, color, and the presence of an odor.

Soil samples for underground storage tank (UST) closures should be taken beneath the tank invert and as close to the tank as possible. Excavations should only be entered after the sides have been properly sloped, shored, or braced. Near-surface soil samples should also be taken in loading/unloading areas adjacent to all storage tanks to determine whether soil contamination from spills has occurred.[3]

Surface Soil Samples

When samples of surface soil, sand, or sediment are to be collected, scrape samples from the surface can be collected. Using a 10-cm by 10-cm template to mark the area to be sampled, the surface should be scraped to a depth of 1 cm with a stainless steel trowel or similar implement. The yield should be at least 100 g soil. If more sample is required, expand the area but do not sample deeper. Use a disposable template or thoroughly clean the template between samples to prevent contamination of subsequent samples. The sample should be scraped directly into a precleaned glass bottle. If it is free-flowing, the sample should be thoroughly homogenized by tumbling. If not, successive subdivision in a stainless steel bowl should be used to create a representative subsample.

In some cases, such as sod, scrape samples may not be appropriate. For these cases, core samples not more than 5 cm deep should be taken using a soil coring device such as the corers discussed in the section on Sludge Sampling. These core samples should be well homogenized in a stainless steel bowl by successive subdivision. A portion of each sample should then be removed, weighed, and analyzed.

The simplest method of collecting surface soil samples for analysis is with the use of a trowel or scoop. The laboratory

Figure 20.8. Sampling scoop for soil and bulk granular material. (Courtesy of Lab Safety Supply, Inc.)

scoop has a curved blade and a closed upper end to permit the containment of material. Scoops come in different sizes and makes and the size will depend on the amount of material that needs to be collected. The stainless steel laboratory scoop is the preferable choice for sampling. Identical sample amounts for a composite sample are difficult to collect with this sampler. If undisturbed sections of soil are needed, a flat, pointed trowel can be used to cut a block of the desired soil rather than the scoop (Figure 20.8).[7]

Procedure

1. Divide the area to be sampled into an imaginary grid on the site map with number codes in two perpendicular directions. Then mark the grid on the ground with numbered stakes and flags.

2. If it is not known whether contamination is present, use a calculator or random-number table to choose random numbers; two numbers are required to choose a sampling site, and these can be used to number the sample.

3. If discolored areas of soil exist, which strongly suggest chemical contamination, plan to sample each area as well as the grid.

4. Do not kneel, squat, or otherwise touch the ground, because contaminants could be transferred from boots to clothing. Use a ground cover such as plastic for kneeling or placement of equipment.

5. Carefully remove the top layer of soil to the desired sample depth with a spade.

6. To sample up to 8 cm deep, collect samples with a scoop or spoon. Dig down to the desired depth; then use a spatula to take a "channel" sample composited over the vertical interval of interest.

7. Sample each section of the grid, and combine appropriate samples into composite samples. Do not combine if sampling in highly contaminated areas.

8. Generally a stainless steel sampler is used to sample volatile organic compounds (VOCs) in soil. If sampling for VOCs, do not composite, mix, or aerate the soil because volatilization may result, causing loss of sample. Instead quickly fill a 40-mL vial with soil, wipe its rim, screw the septum cap tightly, and keep the vial on ice. Fill the sample container completely to eliminate any headspace.

9. Decontaminate the auger and bucket before moving on to the next sample site. Following collection, transport the samples to the laboratory as soon as possible, generally within 24 hours of sampling.

Subsurface Soil Sampling with Augers and Thin-Tube Samplers

The auger boring is the most common method of soil investigation and sampling.

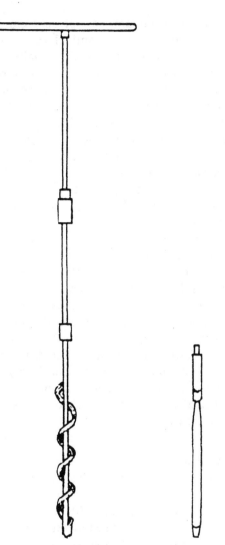

Figure 20.9. Screw auger and thin tube samplers. (From EPA. *Characterization of Hazardous Waste Sites—A Methods Manual: Vol. II. Available Sampling Methods.* EPA-600/4-83-040, September 1983.)

The soil auger can be used for both boring the hole and for bringing up samples of the soil.

The auger (Figure 20.9) consists of an auger bit at the end of a cylinder, a series of drill rods, and a "T" handle. A threaded coupling at the top of the auger allows different size of extension tubing to be attached. Bits are generally made of high-carbon alloy steel and are available in a variety of configurations for different soil types such as sandy and clayey. Screw augers are used for collecting smaller soil samples. Augers are available in different size diameters. In order to push the auger through the soil, techniques that are easier than sheer physical strength are available. The simplest modification is using a mallet with a rubber or urethane head to pound the auger into the ground. Other modification include foot-operated "jacks" on the extension tubing that take the stress off the arms, back, and shoulders.

This system can be used in a variety of soil conditions. It can be used to sample both from the surface or to a depth of 6 meters. The presence of rock layers and the collapse of the borehole usually prohibit sampling at depths in excess of 2 meters. Since this sampler destroys the structure of cohesive soil and does not distinguish between samples collected near the surface or toward the bottom, it is not recommended when an undisturbed soil sample is desired.[7]

Procedure

1. Wearing appropriate PPE, select the necessary sampling points and remove unnecessary rocks, twigs, and other nonsoil materials.

2. Attach the auger bit to a drill rod extension and further attach the "T" handle to the drill rod.

3. Bore a hole through the middle of an aluminum pie pan large enough to allow the blades of the auger to pass through. The pan will be used to catch the sample brought to the surface by the auger.

4. Spot the pan against the sampling point. When using the auger, twist it down to the requisite depth, collecting all auger scrapings in a bucket for a composite. Note that augers tend to cross contaminate between depths.

5. The auger bit is used to bore through the hole in the middle of the aluminum pie pan to the desired sampling depth and then it is withdrawn. Transfer the sample collected in the pie pan and the sample adhering to the auger to a labeled container. Spoon out the rest of the loosened sample to the same container.

6. The auger tip is then replaced with thin-wall tube sampler and the proper cutting tip is installed.

7. Carefully lower the tube into the borehole and gradually force the tube into the soil. Care should be taken to avoid scraping the bore hole sides. Hammering of the drill rods to facilitate coring should be avoided as the vibrations may cause the boring walls to collapse.

8. The tube is then withdrawn and the drill rods are unscrewed. The cutting tip and tube are removed from the device.

9. Discard the top of the tube sample, which represents any material collected by the tube before penetration of the layer in question. Place the remaining sample into a labeled container and cap the container.

10. Brush off and wipe the sampler clean or store it in a plastic bag for subsequent cleaning. Using a stainless steel scoop, collect the desired quantity of soil, transfer the sample to a labeled container, and cap the container.

Sampling of Sludges and Sediments

Sludges are defined as semidry materials ranging from dewatered solids to high-viscosity liquids. Frequently sludges form as a result of settling of the higher-density components of a liquid. In this instance, the sludge may still have a liquid layer above

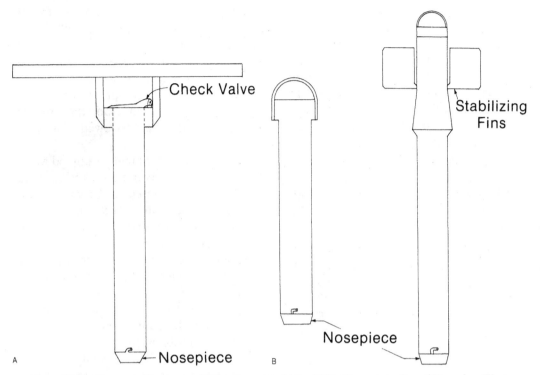

Check Valve

Stabilizing
Fins

Nosepiece

Nosepiece

A   B

Figure 20.10. Corers. A: Hand corer. B: Gravity corers. (After EPA. *Characterization of Hazardous Waste Sites—A Methods Manual: Vol. II. Available Sampling Methods.* EPA-600/4-83-040, September 1983.)

it. The primary characteristic of a sludge is that the material is completely saturated with liquid. Sediments are the deposited material underlying a body of water.[2]

Corers (Figure 20.10) are used for collecting samples of most sludges and sediments. They collect essentially undisturbed samples that represent the profile of strata that may develop in sediments and sludges during variations in the deposition process. Depending on the density of the substrate and the weight of the corer, penetration of depth of 30 inches can be attained. Improvements on the basic core sampler are available. One model features liners so that many samples can be collected with the same corer. The ends of the liner are capped following collection and a new liner is inserted. Liners are available made of plastic, stainless steel, aluminum, and Teflon. A butterfly valve has been incorporated into one model. When the core

sampler is pushed into the sludge, the valve opens, allowing the material to enter the liner, and upon removal the valve closes to prevent loss of sample. Another model features an auger bit for easier penetration of denser sludges. Some hand corers can be fitted with extension handles that will allow the collection of samples underlying a shallow layer of liquid.

A gravity corer is a metal tube with a replaceable tapered nosepiece on the bottom and a ball or other type of check valve on the top. The check valve allows water to pass through the corer upon descent but prevents washout during recovery. The tapered nosepiece facilitates cutting and reduces core disturbance during penetration.

Procedure[8]

1. Inspect the corer for proper cleanliness.

2. Attach a precleaned corer to a length of sample line. Solid-braided $\frac{3}{16}$-inch nylon line is sufficient, although $\frac{3}{4}$-inch nylon line is easier to grasp during hand hoisting. Secure the free end of the line to a fixed support to prevent accidental loss of the corer. Allow the corer to free fall through the liquid to the bottom.

3. Force in the corer with a smooth continuous motion. Twist the corer and then withdraw it in a single smooth motion. An alternative is to use the gravity corer.

4. Retrieve the corer with a smooth, continuous lifting motion. Do not bump the corer since sample loss may result. Remove the liner from the corer, cap the ends, label, and wrap in foil. If a liner is not being used, transfer the sample from the corer into a labeled sample bottle and cap tightly. Refrigerate the sample in an ice chest until it reaches the laboratory.

WATER SAMPLING

Water is commonly sampled as a result of environmental audits for real estate transfer, to identify whether contaminants are being transferred to decontamination water, to identify potential contaminants in indoor air investigations, and for the OSHA requirement that potable water be used by workers for washing and cleaning utensils. This may involves sampling drinking water from a tap, chiller, drinking fountain or private well. Sumps in basements and for decontamination systems are also frequently sampled.[10] Some situations such as stagnant water in a basement or HVAC ducts and requirements for tap water may involve sampling for microorganisms. This type of sampling is discussed further in the chapter on Sampling for Bioaerosols. Likewise, radon in water is discussed in Chapter 15.

Lead is the most common metal of interest in drinking water samples, and procedures for collection of these samples are well established.[11] Other metals can be more difficult to sample and often require preservatives. Therefore, prior to sampling for metals other than lead, the specific method should be reviewed for that element.[12]

VOCs are also frequently sampled for in water. The most common VOCs found in water are chlorinated and aromatic compounds. These might be encountered in situations associated with leaking underground storage tanks. High levels of trihalomethanes are often the result of organic compounds reacting with chlorine in drinking water systems. Vinyl chloride and methane have been found in water contaminated by nearby sanitary landfills. The biggest problem associated with sampling for VOCs is their volatility, resulting in loss of sample due to evaporation during collection. Sampling procedures for these compounds are designed to minimize agitation and contact of the sample with air. Samples must be kept cool ($4°C$) to avoid degradation of organics.

Bailers are one of the most common samplers for water and can be constructed from a wide variety of materials in various designs. The conventional bailer consists of a weighted bottle or basally capped length of pipe attached to a cable or cord that fills from the top as it is lowered into the well or sump to retrieve a sample. Check-valve bailers (Figure 20.11) have a valve located at the base that allows them to fill from the bottom. When the bailer is pulled, the valve closes, retaining the sample as the bailer is raised to the surface. Bailers can be constructed of plastics, Teflon, or borosilicate glass: the need dictates the material, although plastic is the most common. Glass bailers resist chemicals and scratching better than most plastics and therefore are useful for collecting samples for analysis for organics. Teflon bailers are often used

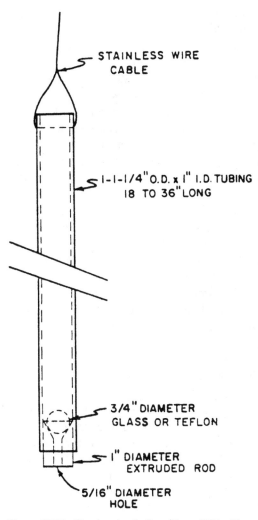

STAINLESS WIRE CABLE

I-I-1/4" O.D. x I" I.D. TUBING 18 TO 36" LONG

3/4" DIAMETER GLASS OR TEFLON

I" DIAMETER EXTRUDED ROD

5/16" DIAMETER HOLE

Figure 20.11. Check valve bailer. (From EPA. *Characterization of Hazardous Waste Sites—A Methods Manual: Vol. II. Available Sampling Methods.* EPA-600/4-83-040, September 1983.)

for groundwater monitoring. Since plastic bailers are relatively inexpensive, they can be disposed of after sampling a source. Bailers allow samples to be recovered with a minimum of aeration if care is taken when lowering the bailer. The primary disadvantages of bailers are limited sample volume and the inability to collect discrete samples from various depths.[7]

The following is a general procedure for collecting samples of water from standing bodies of water, sumps, and private wells using a bailer.[7] It is not designed for groundwater sampling situations, since these require more sophisticated techniques[13] as well as the involvement of geologists, hydrogeologists, and engineers. There are also specific EPA methods for collection and analyses of water from standing sources for chemicals that may prove useful to review.[12,14]

General Procedure

1. Gather all necessary equipment, making sure it has been decontaminated from prior uses.

2. Record observations including type of water being sampled, location, incongruities about the surface such as a sheen or scum, and presence or lack of an odor.

3. Evaluate the area around the sampling point for possible contamination from external sources such as a loosely sealed gasoline can or solvent drum, automobile, or exhaust from nearby processes. Make sure the rinse water for the samplers is not near any of these sources.

4. Place a plastic tarp on the ground next to the sampling site to prevent equipment from touching the ground. Gloves and other PPE should be donned prior to collecting samples.

5. Prior to sampling each point, label sample bottle with unique sampling number and record it on the sampling data sheet.

6. Attach a bailer to a cable or line. Gradually lower the bailer until it contacts the source. Allow the bailer to sink and fill with a minimum of surface disturbance.

7. Slowly raise the bailer. Do not allow the bailer line to contact the ground.

8. Tip the bailer to allow slow discharge from the top to flow gently down the inside of the sample bottle with a minimum of entry turbulence. Repeat collection until a sufficient volume is obtained.

9. Filter samples if the procedure requires it. Wipe the exterior of the container, label it with a unique sample number, and record this on the sampling data sheet.

10. Put labeled sample into a plastic bag and seal. Put bagged sample into clean metal can with top. Fill can with vermiculite and attach lid. Put container into cooler. Change outer gloves to avoid contaminating the next sample or other equipment during the next phase of the work.

11. Generally field blanks are collected after the samples have been collected. Other quality control samples may be collected as well, such as duplicate samples or splitting a single sample for separate analyses by different laboratories.

12. Store samples and any blanks in a cooler with a thermometer to make sure the temperature is maintained at 4°C.

13. If possible, deliver the samples to the laboratory on the same day that they were collected. If this is not possible, store samples in a refrigerator overnight and deliver the next day. Shipping water samples is undesirable due to the potential for delays, which can result in warming of the samples and rough treatment that can cause containers to leak.

Tap Water Samples

The most common purpose for sampling tap water for chemical analysis is for lead detection. The primary sources of lead contamination are lead-containing pipes and plumbing system components and lead-containing solder and flux used in drinking water distribution lines and home plumbing. These sources can release lead into the water supply, especially in areas with particularly corrosive water and where lead solder is less than 5 years old. Water samples for lead are rarely collected as part of an air sampling project. Instead, sampling is performed to evaluate a potential hazard from ingesting toxic lead in drinking water.

Some municipalities have water supply systems made of copper pipe with lead solder. Older cities, such as Chicago, specified the use of lead pipe in their building codes at certain periods. Generally, it was used in smaller services, such as the 2-inch pipe used in residences. Lead levels are usually highest in water that has the longest contact time with the plumbing. The highest levels are usually found in faucets that are seldom used, or from first-draw samples in the morning.

Through the early 1900s, it was a common practice in some areas of the country to use lead pipes for interior plumbing. Also, lead piping was often used for the service connections that join building piping to the public water system mains. Plumbing installed before 1930 is most likely to contain lead.

Since the 1930s, interior plumbing has usually been constructed from galvanized iron or copper pipe. Copper pipe was usually soldered with 50% tin and 50% lead solder until the 1970s or 1980s when many city building codes changed to require lead-free solder. Although relatively little solder is exposed to the interior of the pipe on a well-made joint, significant amounts of lead can leach into the water from lead solder joints. The degree of lead released from soldered joints depends on the pH of the water; the lower the pH, the more likely it is that lead will be released. Newer lead-containing plumbing is affected most.

The configuration of the interior plumbing can vary depending on the layout of the building. Therefore, it is useful prior to sampling to study the blueprint(s) of the building's plumbing system. Locate service intakes, headers, laterals, fixture supply pipes, drinking water fountains, central chiller units, storage tanks, riser pipes, and different drinking water loops. In multistory buildings water is elevated to the floors by one or more riser pipes. Water from the riser pipes is usually distributed through several different drinking water loops and in some buildings may be stored in a tank prior to distribution. In single-story buildings water comes from the service connection via main plumbing branches, often called headers. These in turn supply water to laterals. Smaller plumbing connections from the laterals and loops supply water to the faucets, drinking fountains, and other outlets. Water within a plumbing system moves downstream from the source, for example, from the distribution main in the street. Figure 20.12 shows suggested sample sites in single-level and multistory buildings.[11]

There are different techniques for sampling tap water. One of the most common is a first-draw sample collected after overnight stagnation; others involve collecting random daytime samples and flushed samples. First-draw samples generally have the highest lead concentrations, but are the least convenient to collect. Samples collected after flushing a system result in the most consistent values, but the data reflect the *minimum* exposure of the water to lead. Random daytime samples are most representative of lead levels in the water being consumed, but are the most variable, so a great number of samples are required. Sampling variables include the flow rate from the tap and the volume collected. If the flow rates vary from tap to tap, it may be difficult to compare samples from different taps. The volume must be consistent from sample to sample or changes in water quality may not be detected. Analy-

sis of drinking water samples for lead should be done by a state-certified laboratory using EPA methods.[14]

Sample Collection for VOCs in Drinking Water

Procedure

1. Prior to sampling each point, label two vials with a unique sampling number and record it on the sampling data sheet. For duplicate samples an "A" and "B" designation added to the unique number can be used.

2. Turn on the tap and minimize air contact with sampled water while filling 2 vials to overflowing. Make sure the cap liner is in the correct position with the Teflon liner, which is usually white, on the outside; when the lid is screwed on, it should point toward the sample. Cap the vials immediately, screwing on the caps tightly.

3. Fill two vials from each sampling point.

4. There should be no air space at the top of the vials. To test this, invert each vial and tap it on a hard surface to determine if any air bubbles remain inside. If there are any, refill it and again inspect for air bubbles. Continue to work with the sample until no bubbles are present.

5. If samples are being collected for compliance, split samples are usually taken, meaning each sample is usually divided between the facility owner and the regulatory agency. In the case of VOCs, it is best to fill all vials from the same tap.

Sample Collection of Lead in Drinking Water. Sampling sites most likely to show lead contamination include areas where the plumbing is used to ground electrical circuits; areas where corrosive water having

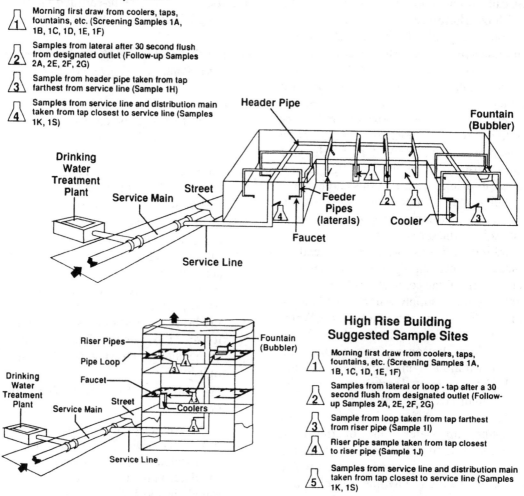

Figure 20.12. Suggested sampling sites for lead in water in a single-level building (*top*) and in a multi-story building (*bottom*). (From EPA. *Lead in School's Drinking Water*. EPA-570/9-88-001, January 1989.)

low pH and alkalinity is distributed; areas of low flow and/or infrequent use, such as where water is in contact for a long time with sediments or plumbing containing lead; areas containing lead pipes or areas in which lead solder or materials containing lead were used; and water coolers identified by the EPA as having lead-lined storage tanks or other parts containing lead.

The procedure for collecting samples involves collecting screening samples from outlets that provide drinking water, and doing follow-up sampling of outlets that showed significant lead levels in the initial screening in order to identify the sources of lead in the drinking water.[11]

Procedure

1. Determine the source of the water and the type of piping used to carry the water both for the municipality and the building being sampled. Individual properties as well as townships can rely on wells for water. Pipelines

may be polyvinyl chloride (PVC), copper with lead–tin solder, cast iron, asbestos concrete (A/C), and galvanized pipe.

2. The faucet chosen for sampling must be a supply without filters, softeners, or heaters. This sample consists of water that has been in contact with the fixture and the plumbing connecting the faucet to the lateral. It is representative of the water that may be consumed at the beginning of the day or after infrequent use. It should be collected before any water is used for the day.

3. Prior to collecting samples, lines must be flushed sufficiently to ensure that the sample is representative of water supply, taking into account the diameter and length of the pipe to be flushed and the velocity of flow. For volatile organics, let the water just dribble into the vial to avoid loss of volatiles. Do not use plastic containers for volatiles. For bacteria, be aware that contamination may be in any part of the system: the well or other source, pipes, filter medium, or faucet. These possibilities can impact the sampling strategy and frequently call for multiple samples. For more information on sampling for bacteria, see the chapter on Sampling for Bioaerosols.

4. Remove the aerator if there is one. Aerators are usually found on the kitchen and bathroom faucets. Collect the water immediately after opening the faucet without allowing any water to run into the sink. Fill the container with 250 mL of water.

5. Following-up samples should be taken from those water faucets where test results indicate lead levels over 20 ppb. The follow-up sample is representative of the water that is in the plumbing upstream from the faucet.

Take this sample before any water is used in the morning. Let the water from the faucet run for 30 seconds before collecting the sample. Collect 250 mL of water.

Interpreting the Results

1. If the lead level in the first sample is higher than in the second, the source of lead is the water faucet and/or the plumbing upstream from the faucet.

2. If the lead level in the second sample is very low—close to 5 ppb—very little lead is coming from the plumbing upstream from the faucet. The majority or all of the lead in the water is from the faucet and/or the plumbing connecting the faucet to the lateral.

3. If the lead level in the second sample significantly exceeds 5 ppb (e.g., 10 ppb), lead may be contributed from the plumbing upstream from the faucet.

Service Connections and Distribution Mains. The most common sample collected at these points is for lead analysis.[11]

Procedure

1. A first-morning sample should collected.

2. Open the tap closest to the service connection.

3. Let the water run and feel the temperature of the water. As soon as the water changes from warm to cold, collect the sample. Because water usually warms slightly after standing in the interior plumbing, this colder water sample represents the water that had been standing just outside of the building and in contact with the service connector. Fill the sample container with 250 mL of water.

4. Allow the water to run for an additional 3 minutes and then collect a

second sample. Fill this container with another 250 mL of water. This sample is representative of the water that has been standing in the distribution main.

Interpreting the Results

1. If the lead level of the service connection sample significantly exceeds 5 ppb and is higher than in the second sample of the distribution main, lead is being contributed from the service connector. Check for the presence of a lead service line. In the absence of a lead service connector, lead goosenecks or other appurtenances containing lead in line with the service connection may be the sources of contamination. The distribution main is rarely the source of contamination.

2. If the lead level of the service main sample significantly exceeds 5 ppb, lead in the water may be attributed to the source water, sediments in the main, or possibly lead joints used in the installation or repair of cast iron pipes. If the water supplied is from a well, a lead packer in the well may also contribute lead to the water.

3. If the lead level of both samples is low, close to 5 ppb, very little lead is picked up from the service line or the distribution main.

Interior Plumbing. Sampling interior plumbing should proceed systematically upstream from the initial follow-up sample sites. The goal is to isolate sections of interior plumbing that contribute lead to the water by comparing the results of these samples with results of previous samples.

Procedure for Laterals[11]. Laterals are the plumbing branches between a fixture or group of fixtures, such as taps and water fountains.

1. Open the tap that has been designated as the sample site for the lateral pipe.

2. Let the water run for 30 seconds before collecting the sample. The purpose of flushing the water is to clear the plumbing between the sample site and the lateral pipe, which will ensure collection of a representative sample. Fill the container with 250 mL of water.

Interpreting the Results

1. If the lead level exceeds 20 ppb, collect additional samples from the plumbing upstream (the service line, the riser pipe, the loop or header supplying water to the lateral). High lead levels may also be caused by recent repairs and additions using lead solder, or by sediments and debris in the pipe. Debris in the plumbing is most often found in areas of infrequent use, and a sample should be sent to the laboratory for analysis.

2. If the lead level of this sample is the same as the lead level in a sample taken downstream from the sample site, lead is contributed from the lateral or from interior plumbing upstream from the lateral. Possible sources of lead may be the loop, header, riser pipe, or service connection.

3. If the lead level is very low—close to 5 ppb—the portion of the lateral upstream from the sample site and the interior plumbing supplying water to the lateral are not contributing lead to the water.

4. If the lead level significantly exceeds 5 ppb, and is less than the lead level in a sample taken downstream from the sample site, a portion of the lead is contributed downstream from the sample site.

Procedure for Loops and/or Headers[11]. A loop is a closed circuit of a plumbing branch that supplies water from the riser to a fixture or a group of fixtures. A header is the main pipe in the internal plumbing system of a building. The header supplies water to lateral pipes.

1. Locate the sampling point farthest from the service connection or riser pipe on a floor.
2. Open the faucet and let it run for 30 seconds before collecting the sample. The purpose of flushing the water is to clear the faucet and plumbing between the sample site and the loop and/or header pipe, thus assuring collection of a representative sample. Fill the container with 250 mL of water.

Interpreting the Results

1. If the lead level is over 20 ppb, collect additional samples from the plumbing upstream supplying water to the loop or header. Compare the sample results with those taken from the service line or the riser pipe that supplies water to the loop and/or header. High lead levels may also be caused by recent repairs and additions using lead solders, or by sediment and debris in the pipe. Debris in the plumbing is most often found in areas of infrequent use, and a sample should be sent to the laboratory for analysis.
2. If the lead level in either sample is equal to the lead level in a sample taken downstream from the sample site, the lead is contributed from the header or the loop and from the interior plumbing upstream from the header or loop. Possible sources of lead may be the loop, header, riser pipe, or service connection.

3. If the lead in either sample is close to or equal to 5 ppb, the portion of the header or loop upstream from the sample site and the interior plumbing supplying water to the loop or header are not contributing lead to the drinking water. The source of lead is downstream from the site.
4. If the lead level in either sample significantly exceeds 5 ppb, and is less than the lead level in a sample taken downstream from the sample site, a portion of the lead is contributed downstream from the sample site.

Procedure for Riser Pipes[11]. A riser is the vertical pipe that carries the water from one floor to another.

1. Open the tap closest to the riser pipe.
2. Let the water run for 30 seconds before collecting the sample. The purpose of flushing the water is to clear the faucet and plumbing between the sample site and the riser pipe to assure collection of a representative sample. Collect 250 mL in the sample container.

Interpreting the Results

1. If lead levels exceed 20 ppb, collect additional samples from the plumbing upstream from the riser. High lead levels in the riser pipes may also be caused by repairs and additions using lead solder.
2. If the lead level in the sample is equal to the lead level in a sample taken downstream from this site, the source of lead is the riser pipe or the plumbing and service connection upstream from the riser pipe.
3. If the lead level in the sample is close to or equal to 5 ppb, the portion of the riser pipe and plumbing upstream from the sample site and the service

Suggested Sample Sites

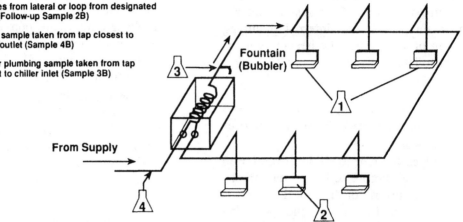

/1\ Morning first draw from coolers, taps, fountains, etc. (Screening Samples 1B)

/2\ Samples from lateral or loop from designated outlet (Follow-up Sample 2B)

/3\ Chiller sample taken from tap closest to chiller outlet (Sample 4B)

/4\ Interior plumbing sample taken from tap closest to chiller inlet (Sample 3B)

Fountain (Bubbler)

From Supply

Figure 20.13. Suggested sampling sites for water supplied from a central chiller to water fountains. (From EPA. *Lead in School's Drinking Water.* EPA-570/9-88-001, January 1989.)

connection are not contributing lead to the water. The source of lead is downstream from the sample site.

4. If the lead level in the sample significantly exceeds 5 ppb, and is less than the lead level in a sample taken downstream from the sample site, a portion of the lead is contributed downstream from the sample site.

Drinking Water Fountains. Drinking water fountains have three basic configurations: (1) a self-contained water cooler equipped with its own cooling system that receives water from the building's supply; (2) a system of drinking fountains that receive water directly from the building's plumbing system via a central chiller (Figure 20.13); and (3) a drinking fountain that receives water directly from the building's plumbing system.

For sampling, valves to the water fountains should not be closed to prevent their use, since minute amounts of scrapings from the valves will be collected, resulting in higher than actual lead levels. All

samples should be collected with the taps fully open.[11] A drinking water cooler is considered lead-free when no component that comes into contact with drinking water contains more than 8% lead, and no solder, flux, or storage tank interior that may come into contact with drinking water contains more than 0.2% lead.[11]

Sampling Bubblers with or without a Central Chiller. This sample is representative of the water that may be consumed at the beginning of the day or after infrequent use. It consists of water that has been in contact with the bubbler valve and fittings and the section of plumbing closest to the outlet of the unit.

Procedure

1. Take this sample before any water is used. Collect the water immediately after opening the faucet without allowing any water to run into the sink. Fill the container with 250 mL of water.

2. Follow-up samples should be taken from water fountains where test results indicate lead levels over 20 ppb. These consist of samples that are representative of the water in the plumbing upstream from the bubbler. Take them before any water is used. Let the water from the fountain run for 30 seconds before collecting the sample. Fill the container with 250 mL.

Interpreting the Results

1. If the lead level in the first sample is higher than in the follow-up, a portion of lead in the drinking water is contributed from the bubbler.

2. If the lead level in the follow-up sample is very low—close to 5 ppb—very little lead is picked up from the plumbing upstream from the outlet. The majority or all of the lead in the water is contributed from the bubbler.

3. For a bubbler without a central chiller, if the lead level in the follow-up sample exceeds 5 ppb, lead in the drinking water is also contributed from the plumbing upstream from the bubbler. If the lead level in the follow-up sample exceeds 20 ppb, sampling from the header or loop supplying water to the lateral should be done to locate the source of contamination. For a bubbler with a central chiller, if the lead level in the follow-up sample significantly exceeds 5 ppb, the lead in the drinking water may be contributed from the plumbing supplying the water from the chiller to the bubbler, from the chiller, or from the plumbing supplying water to the chiller.

4. If the lead level in the follow-up sample exceeds 20 ppb, sample from the chiller unit supplying water to the lateral to locate the source of contamination.

Sampling Water Fountains with Coolers.

There are two types of water coolers: wall-mounted coolers and free-standing coolers. Water in the cooler in stored in a pipe coil or in a reservoir. Refrigerant coils in contact with either of these storage units cool the water. Sources of lead in the water may be the internal components of the cooler, including a lead-lined storage unit; the section of the pipe connecting the cooler to the lateral; and/or the interior plumbing. Some water coolers have storage tanks linked with materials containing lead. Sediments and debris containing lead on screens or in the plumbing frequently produce significant lead levels. Lead solder in the plumbing can also contribute to the problem.[11]

Collect a sample that is representative of the water that may be consumed at the beginning of the day or after infrequent use. It consists of water that has been in contact with the valve and fittings, the storage unit, and the section of plumbing closest to the outlet of the unit.

Procedure

1. Take the sample before any water is used. Collect the water immediately after opening the faucet, without allowing any water to be wasted. Fill the sample container with 250 mL of water.

2. Follow-up samples should be taken from water coolers where test results indicate lead levels over 20 ppb. These water samples are representative of the water that is in contact with the plumbing upstream from the cooler. Take these samples at the end of the day, letting the water run from the fountain for 15 minutes before collecting them. Flushing for 15 minutes is necessary to ensure that no stag-

nant water is left in the storage unit. Fill the containers with 250 mL of water.

3. This sample must be taken the morning following the sample collected in step 2. Because the water in the cooler should be flushed the afternoon prior to collecting this sample, it is representative of the water that was in contact with the cooler overnight, not in extended contact with the plumbing upstream. Take this sample before any water is used. Collect the water immediately after opening the faucet, without allowing any water to be wasted.

Interpreting the Results

1. If the lead level in the sample collected in step 3 is higher than the level in the sample collected in step 2, the water cooler is contributing lead to the water.

2. If the lead level in the sample collected in step 3 is higher than the level in the sample collected in step 2, and the lead level in the sample collected in step 1 is higher than the level in the sample collected in step 2, the plumbing upstream from the water cooler may also be contributing lead to the water.

3. If the lead level in the sample collected in step 3 is identical or close to the level of the sample collected in step 2, the water cooler probably is not contributing lead to the water.

4. If the lead level in the sample collected in step 1 is higher than the level in the sample collected in step 3, and if the lead levels in the samples collected in steps 2 and 3 are close or identical, the plumbing upstream from the cooler and/or the plumbing connection leading to the cooler are contributing lead to the water.

5. If the lead level in the sample collected in step 2 is in excess of 10 ppb, and is equal to or greater than the lead levels in the samples collected in steps 1 and 3, the source of the lead may be sediments contained in the cooler storage tank, screens, or the plumbing upstream from the cooler.

Verifying the Source of Lead

1. Take a 30-second flushed sample from a tap upstream from the cooler. If a low lead level is found in this sample, the source of lead may be sediments in the cooler or the plumbing connecting the cooler to the lateral, or lead solder in the plumbing between the taps.

2. If the flushed samples from the upstream outlets have lead levels in excess of 5 ppb, then the cooler and the upstream plumbing may both contribute lead to the water.

Confirming Whether the Cooler Is a Source of Lead

1. Turn off the valve leading to the cooler. Disconnect the cooler from the plumbing and look for a screen at the inlet.

2. Remove the screen. If there is debris present, check for the presence of lead solder by sending a sample of the debris to the laboratory for analysis.

3. Some coolers also have a screen installed at their bubbler outlet. Carefully remove the bubbler outlet by unscrewing it. Check for a screen and debris, and have a sample of any debris analyzed.

4. Some coolers are equipped with a drain valve at the bottom of the water reservoir. Water from the bottom of the water reservoir should be sampled, and any debris should

be analyzed. Collect 250 mL in a container.

5. Collect a sample from the disconnected plumbing outlet. Take the sample before any water is used. Collect this fraction as soon as possible after disconnecting the plumbing from the cooler.

Interpreting the Results

1. If the lead level in the last sample collected from the disconnected plumbing outlet is less than 5 ppb, lead is coming from the debris in the cooler or the screen.

2. If the lead level in the last sample is significantly higher than 5 ppb, the source of lead is the plumbing upstream from the cooler.

SUMMARY

Bulk samples of air, solids, and liquids are collected to determine the composition of these materials and especially to identify the presence of any hazardous chemicals. Bulk materials include raw materials, wastes, and even potable water for the presence of toxic lead. Grab air samples are considered as bulk samples when they are used to identify composition rather than to define exposures. Information from bulk samples is critical in designing air sampling programs, and evaluating and controlling potential hazards to employees and members of the public.

Often bulk sampling techniques are straightforward. They focus on the steps required to obtain valid representative samples of the bulk material that are suitable for laboratory analysis. However, when samples are collected to meet U.S. EPA or other regulatory requirements, the sampling protocol may be spelled out in detail in agency regulations. In this case, compliance can only be achieved by scrupulously following the required sampling methodology.

REFERENCES

1. Hamaan, M. E. An air sampling collection apparatus for high resolution air analysis. *AIHA J.* **50**(8):A-591–A592, 1989.

2. U.S. Environmental Protection Agency. *RCRA Inspection Manual.* Washington, D.C.: EPA, 1993.

3. State of Illinois Environmental Protection Agency. *Instructions for the Preparation of Closure Plans for Interim Status RCRA Hazardous Waste Facilities.* Springfield, IL, March 2, 1989.

4. U.S. Code of Federal Regulations, Title 49, Parts 100–199.

5. Notice of intended changes for benzene TLV. *Appl. Ind. Hyg.* **5**(7):453–463, 1990.

6. First, M. W. Air sampling: *Appl. Ind. Hyg.* **3**(12):F-20, 1988.

7. Environmental Protection Agency. *Characterization of Hazardous Waste Sites—A Methods Manual: Vol. II. Available Sampling Methods.* EPA 600/4-83-040. Washington, D.C.: EPA, 1983.

8. USEPA. *Test Methods for Evaluating Solid Waste,* SW 846, 1998.

9. Chichowicz, J. Hazmat samples: Proper packaging, labeling and shipping are critical. *Hazmat World,* June 1989, pp. 60–61.

10. Rosbury, K. D. *Handbook: Dust Control at Hazardous Waste Sites.* EPA-540/2-85/003.

11. Environmental Protection Agency. *Lead in School's Drinking Water.* EPA 570/9-89-001. Washington, D.C.: EPA, 1989.

12. American Public Health Association. *Standard Methods for the Examination of Water and Wastewater,* 20th ed. APHA. 1999

13. Environmental Protection Agency. *Practical Guide for Ground-Water Sampling.* EPA/600/2-85/104. Washington, D.C.: EPA, 1985.

14. Environmental Protection Agency. *Methods for Chemical Analysis of Water and Wastes.* EPA-600/4-79-020. Cincinnati: US EPA, 1983.

APPENDICES

APPENDIX A

AIR SAMPLING PROCEDURES

This Appendix contains separate sampling procedures that describe how to collect samples for the following airborne contaminants:

- Dusts, Mists, and Fumes
- Asbestos Fibers
- Active Sampling for Organic Vapors: Adsorption Tubes
- Gases and Vapors: Bubblers and Impingers
- Passive Sampling for Organic Vapors: Badges or Dosimeters
- Respirable Dust Using a Cyclone
- Silica
- Total Dust
- Gasoline and Light Hydrocarbons
- Welding Fumes
- Benzene

DUSTS, MISTS, AND FUMES

Purpose/Scope. To measure the airborne concentration of particulate contaminants where a chemical or physical analytical technique for the substance is available.

Airborne particulates are collected on a filter of suitable material in a 37-mm cassette attached to a high-flow personal sampling pump. The filter cassette is returned to the laboratory to determine the concentration of specific contaminants in the air.

1. For most metal analyses (welding fumes, etc.), asbestos, and other fibers, a mixed cellulose ester (MCE) membrane filter is used.
2. For silica, hexavalent chromium, and respirable or total dust, a polyvinyl chloride (PVC) filter must be used.
3. For oil mists or polynuclear aromatic compounds in high-boiling-point petroleum fractions, a Teflon membrane filter or glass fiber filter is used in the cassette.
4. For recommendations for other substances, consult the analytical laboratory for advice.

Air Monitoring for Toxic Exposures, Second Edition. By Henry J. McDermott
ISBN 0-471-45435-4   © 2004 John Wiley & Sons, Inc.

Equipment

1. Battery-operated high flow pump, with 1- to 3-L/min flowrate.
2. 37-mm filter cassettes containing the proper filter medium. If a total airborne dust measurement is required, matched-weight filters must be used. See Procedure entitled "Total Dust."
3. Flowmeter.
4. Sampling hose.
5. Filter cassette coupler.

Calibration. Calibrate the pump with cassette inline as described in "Calibration with a Rotameter" in Chapter 5.

Sampling Procedure

1. For a breathing zone sample, attach the sampling pump to the worker's belt and attach the filter cassette to the collar with a 3-foot section of sampling hose. Check that the inlet pin has been removed from the cassette, the worker's clothing does not obstruct the inlet to the cassette, and the sampling hose does not interfere with the worker.
2. Start the pump and check the flow reading. Record the time, the flow reading on the pump, and other pertinent information on the sampling data sheet.
3. Check the flowrate periodically throughout the sampling period (every 60 minutes) using the flowmeter on the pump. If the flowrate varies more than 10%, terminate the sample, change the filter, and start a new sample using a freshly charged pump (if needed) to cover the remainder of the exposure period.
4. At the end of the sampling period, check the flowrate to verify that the rate is unchanged. Then stop the pump and record the time. Remove the filter cassette and replace inlet and outlet plugs. Make sure the cassette is properly labeled with the sample number that is indicated on the corresponding sampling data form.
5. When the samples are returned to the laboratory for analysis, submit a field blank with each batch of samples. The field blank should be a filter cassette from the same batch of filters used for sampling. The blank should be handled the same as sample filters, but no air is drawn through it.

Sample Analysis

1. If you want the laboratory to calculate the final airborne concentration, inform the laboratory as to the air volume sampled and request that the filters be analyzed for the contaminants desired in mg/sample and mg/m^3. Also, request the laboratory to identify the analytical method used (atomic absorption, etc.) for each contaminant by NIOSH Method or other reference number.

ASBESTOS FIBERS

Purpose/Scope. To measure the airborne concentration of asbestos fibers greater than 5 μm in length.

Airborne asbestos fibers are collected "open face" on a mixed cellulose ester (MCE)—0.8-μm pore size membrane filter in a 3-piece 25-mm cassette, with 2-inch extension cowl, attached to a high flow rate personal sampling pump. The filter cassette is returned to the laboratory where the filter is cleared (i.e., made transparent) and the fibers counted using a phase contrast optical microscope. Other analysis (metals, etc.) cannot be performed on the same filter.

Equipment

1. Battery-operated, high flow pump with 1- to 3-L/min flowrate.
2. Three-piece 25-mm filter cassettes with 2-inch extension cowl and 0.8-μm pore size, mixed cellulose ester (MCE) membrane filter.
   Note: Use electrically conductive cassettes.
3. Flowmeter.
4. Sampling hose.

Calibration. Calibrate the pump with a sampling cassette in line at 0.5–2.5 L/min, as described in "Calibration with a Rotameter" in Chapter 5. OSHA regulations specify calibrating before and after monitoring.

Sampling Procedure

1. Assemble 25-mm filter cassette, hose, and pump.
2. Attach the pump to the worker's belt and attach the cassette to the collar. Make certain that the cassette is facing down, to avoid large particles and fiber bundles from falling onto the filter. Remove the top section of the cassette so the fibers will collect over the entire surface of the filter (open face). Note that the sample must be collected "open face." If the sample is collected "closed face," the particles and fibers accumulate directly under the entry port and cannot be analyzed. Ensure that the worker's clothing does not obstruct the filter inlet, and also ensure that the sampling hose does not interfere with the worker's normal movement.
3. Start the pump and record all pertinent sampling data on the sampling data form.
4. Sampling time is dependent on the total airborne dust concentration and

sampling rate. It can range from 10 minutes in very dusty areas up to 4–8 hours in clean areas. Periodically check the filter. If a visible color starts to show or loose dust is present, stop the sampling and change filter cassettes. Heavy loading of dust on the filter may make fiber counting with the microscope impossible.

5. Upon completion of sampling period, stop the pump and immediately replace top section of cassette, being careful not to touch the surface of the filter.
6. Before removing the filter from the sampling line, recheck the flowrate using a flowmeter. If there is any change from the flowrate at the start of the sampling period, use the average of the initial and final readings to calculate the sample volume.
7. Replace inlet and outlet plugs, making sure that the cassette is properly labeled with the sample number that is indicated on the corresponding sample data form. Submit a blank filter cassette for every ten samples in a batch. The blank should be from the same batch of filters as the samples. A blank should be handled the same as the sample filters, but no air is drawn through it.
8. For shipment to analytical laboratory, do not use polystyrene packing materials, because they have high static charges and may cause fibers from filter to be transferred to walls of the plastic cassette.

ACTIVE SAMPLING FOR ORGANIC VAPORS: ADSORPTION TUBES

Purpose/Scope. To measure airborne concentration of various organic vapors or gases.

1. Organic vapors are adsorbed on activated charcoal as air is drawn through the tube using a battery-operated pump. The compounds are desorbed from the charcoal in the analytical laboratory and analyzed by gas chromatography, or another analytical method. Due to differences in analytical procedures, some organic compounds are not compatible for analysis with other compounds on the same charcoal tube. If you have any questions, consult with the analytical laboratory.

2. For materials that do not collect on charcoal, such as methanol and some amine compounds, tubes containing silica gel or other adsorbent may be specified. Follow the same procedure to sample with these tubes.

3. Alternate procedure: Organic Vapor Badge. See the Monitoring Procedure for Passive Dosimeters.

Equipment

1. Battery-operated sampling pump (either high-flow or low-flow, depending on the sampling criteria).

2. Standard charcoal tubes. (For some compounds, the analytical laboratory or industrial hygienist may specify large charcoal tubes, chemically treated charcoal tubes, or tubes containing silica gel.)

3. Tube holder with sampling hose.

4. 2.5 × 4-inch manila envelope.

Calibration. Calibrate the pump and sampling train as described in "Calibration with a Burette" in Chapter 5.

The flowrate is dependent on the compounds to be sampled and the type of sample (i.e., ceiling or 8-hour TWA), typically 20 mL/min to 1.0 L/min. See Table A.1 for guidelines.

Pumps must be calibrated before and after sampling to accurately determine the volume of air sampled.

Sampling Procedure

1. Immediately before sampling, break the ends of the tube to provide an opening of at least one-half the inside diameter of the tube (approximately 2 mm). There is a special breaking tool that can be used. Do not use the pump's charging jack on the pump. **Take care to avoid injury when handling broken glass!**

2. Insert the charcoal tube into the tube holder, making sure that the arrow on the tube is pointing in the direction of airflow (toward the pump). If there is no arrow to indicate air flow, insert the tube so that the end with the larger section of charcoal is open to the air, and the end with the smaller section of charcoal is attached to the tubing.

3. Attach the pump to the worker's belt, and attach the charcoal tube holder with the sampling hose to the lapel. The charcoal tube should be vertical during sampling to reduce channeling through the charcoal tube.

4. Air being sampled should not pass through any hose or tubing before entering the charcoal tube.

5. Note the time, and start the pump.

6. Record all pertinent data on the sample data form.

7. Turn off the pump at completion of the sampling period. Record the time and the final counter reading.

8. Remove the charcoal tube from the holder, and immediately seal both ends with the plastic caps provided. Place the tube in a small envelope and write the sample identifying information on the envelope.

9. One blank charcoal tube should be

TABLE A.1. Charcoal Tube Monitoring—Suggested Sampling Times[a]

Substance	Expected Concentration Range (ppm)	Flowrate (mL/min)			
		Fifteen-Minute Sample	One-Hour Sample	Four-Hour Sample	Eight-Hour Sample
Acetone	5–200	200	100	50	NR[b]
	200–2000	100	10	NR	NR
Benzene	0.1–20	200	100	50	20
Diesel fuel	20–1000	200	100	50	NR
Freon™	100–2000	1001	0	NR	NR
Gasoline	0–50	200	50	20	NR
	0–300	100	20	NR	NR
Heptane	5–100	200	100	50	NR
	100–1000	50	10	NR	NR
Hexane	5–100	200	100	50	NR
	100–1000	50	10	NR	NR
Methyl ethyl ketone	2–50	2–50	100	100	50
(2-butanone)	50–400	50–400	50	50	NR
Methyl isobutyl ketone	1–15	200	200	100	50
	15–200	100	50	NR	NR
Mineral spirits	5–75	200	200	100	50
	75–1000	50	NR	NR	NR
Octane	5–75	200	200	100	50
	75–1000	50	NR	NR	NR
Toluene	1–15	500	200	100	50
	15–200	200	100	50	NR
Xylene	1–15	300	200	100	50
	15–200	200	100	50	NR

[a] Sampling times and rates are selected to permit detection of 0.1 to 2 times the TLV for the substance.
[b] NR, not recommended. Contact the analytical laboratory for assistance.

handled in the same manner as the sample tubes (break ends and seal with caps in field), except that no air is drawn through this tube. Submit at least one blank for every 10 samples.

10. Ship the samples to an analytical laboratory as soon as possible. Store the samples in a freezer if possible, until they are shipped to analytical laboratory, especially if delay exceeds 2–3 days. Package the samples to avoid breakage (use a mailing tube or sturdy carton with packing material).

Sample Analysis

1. If you want the laboratory to calculate the airborne concentration, inform the laboratory of the air volume sampled with each tube and a list of compounds to be analyzed. Instruct them to report separately the number of micrograms in both the front and back sections of the tube and also the total ppm.

2. Request the laboratory to provide the method (NIOSH or other) used for each contaminant.

GASES AND VAPORS: BUBBLERS AND IMPINGERS

Purpose/Scope. To measure the airborne concentrations of various contaminants.

1. The contaminant being sampled is adsorbed into a liquid absorber solution that is compatible with the analytical procedures for the contaminant. For gases and vapors in the absence of particulates, a fritted bubbler is preferred to break the air stream into smaller bubbles than an impinger, which improves the absorption efficiency. The absorber solution is returned to the laboratory for analysis.

2. Due to the difficulty of handling liquids in the field, this technique should be used only if there is no other collection method available for the contaminant.

Equipment
1. Battery operated high-flow personal sampling pump capable of 1–3 L/min.
2. Midget impinger or bubbler.
3. Absorber solution for specific compound of interest.
4. Flowmeter.
5. Holder for impinger or bubbler.
6. 30-mL glass vials.
7. Sampling hose.

Calibration. Calibrate the pump and impinger/bubbler (as described in "Calibration with a Rotameter" in Chapter 5) at 1.0 L/min. Higher rates may cause carryover of solution into the pump.

Sampling Procedure
1. Impingers and bubblers should be cleaned before use. The cleaning procedure varies with the contaminants being sampled; instructions can be obtained from the analytical laboratory.

2. Remove the top section of the impinger and add the volume of absorber solution specified by the laboratory, generally 10–50 mL, depending on size and type of device.

3. Attach the hose to the side arm of the impinger/bubbler and pump inlet. Attach the pump to the worker's belt, and attach the device to the collar using a holder. (*Caution*: Avoid tipping the device during sampling as liquid may be drawn into pump.)

4. If using a glass impinger or bubbler, use a protective holder or else protect the glass by taping if needed to avoid breakage and possible injury to the worker.

5. Start the pump and record all pertinent information on the sample data form.

6. Check the absorber volume periodically. Replenish the absorber if needed.

7. Upon completion of the sampling period, transfer the absorber solution from the impinger or bubbler to a clean vial. Rinse the device with a small amount of fresh absorber solution, which is added to the vial. Label the vial with the appropriate sample designation.

8. Prepare a field blank by adding the required volume of absorber solution to a clean impinger and transfer it to a vial. Ship the blank with samples to the laboratory that provided the solution. Package samples to prevent breakage. Comply with applicable shipping regulations if the solution is flammable or toxic.

Sample Analysis
1. If you want the laboratory to calculate the airborne concentration,

inform the laboratory as to the air volume sampled and instruct them as to the contaminant to be analyzed for and any interferences which may have been present. Request the results be reported as µg per sample and ppm (parts per million).

2. For assistance in arranging for laboratory analysis, shipping requirements, or interpreting results, contact the analytical laboratory.

Other

1. Impinger and bubbler sampling methods should only be used when no other sampling methods are available for the contaminant of interest. Some disadvantages are as follows:

   a. If glass bubblers and impingers are used, they are easily broken, causing loss of sample and possible injury to worker. Devices made of Teflon™ and other break-resistant materials are available.

   b. If the device is tipped, the liquid may be drawn into the sampling pump, resulting in sample loss and a contaminated pump.

   c. High flowrates may cause bubbling to the point where some liquid is carried over into the sample hose and pump.

   d. Equipment must be cleaned properly before use and between samples to avoid contamination.

PASSIVE SAMPLING FOR ORGANIC VAPORS: BADGES OR DOSIMETERS

Purpose. To measure the airborne concentration of various organic vapors.

1. Passive, diffusion monitors work on the principle of gas diffusion. Gas molecules penetrate a porous protec-

tive membrane and migrate to the charcoal element in the bottom of the holder, where they are trapped until analysis is performed at the laboratory.

2. They do not require the use of sampling pumps, nor is it necessary to perform any calibration.

3. There are some physical parameters that present certain considerations in the proper use of organic vapor badges.

The following guidelines must be followed to ensure that a representative and reliable sample is obtained.

Stagnant Atmospheres. Organic vapor badges require a minimum face velocity for accurate readings. Refer to manufacturer's guidelines for the appropriate minimum velocity. Therefore, area samples in confined areas or offices with minimal airflow should not be evaluated using an organic vapor monitor.

High Relative Humidity. Water vapor molecules compete with organic hydrocarbon molecules for active adsorption sites on the charcoal. In atmospheres with very high relative humidity levels, a significant portion of the adsorption sites can be occupied by water molecules, thereby reducing the amount of organic hydrocarbon the charcoal will adsorb. Consult with the manufacturer to determine if your operation will exceed the recommended relative humidity levels.

Breakthrough Concentrations of Hydrocarbons. Very high concentrations of hydrocarbons may saturate the charcoal element. This would prevent the adsorption of any additional hydrocarbon. If concentrations of hydrocarbons could be high enough to cause saturation, a charcoal tube should be used. The charcoal tubes have a backup section that can be analyzed separately, this will show if the front section became saturated.

Short-Duration Samples. The minimum sampling time when using badges is 15 minutes. Sampling for short term-exposure limit (STEL) compliance should be given careful consideration. For some compounds the sampling rate and the analytical detection limit may not be sufficiently low.

Suitable Compounds for Sampling. Badges are not intended for all compounds. Their use should be restricted to those compounds suggested by the manufacturer or laboratory. Also, be aware of the sampling time restrictions, because they relate to the expected organic vapor concentrations and the percent relative humidity.

Calibration. None required. The equivalent sampling rate for each compound is determined by the manufacturer. These data will be used by the laboratory to calculate the concentration of a compound based on the sampling time.

Sampling Procedure

1. Just prior to use, open the container and remove the badge from the packaging material. Retain any additional parts for future use.
2. Note the date and start time.
3. Clip the badge to the worker's lapel so it is near the breathing zone. The sampling surface should be facing away from the worker's body. Check to ensure that no clothing will cover the badge during normal movements.
4. Fill out the relevant information on the sample data form.
5. The badge should remain on the worker for the entire shift or all of the exposure period.

6. At the end of the shift or end of the exposure period, remove the badge from the worker and record the end time on the back of the badge and the sample data form.
7. Snap the provided seal in place. This stops the collection of any more material on the badge. Note that the exposed charcoal element will continue to capture any hydrocarbons present until it is capped.
8. Replace the badge in its original container or packing for shipment to the laboratory.

Blanks

1. At least one badge from the same lot as the field samples should be submitted with each group of samples. The blank should be opened and resealed in the same area the field samples were recapped. This badge should be labeled as a "BLANK."
2. If a large number of samples are being submitted to the laboratory for analysis as a single batch, include one blank for every 10 samples.

Shipping. Package the samples in a sturdy container with adequate padding to prevent breakage during shipment to an AIHA accredited laboratory.

Sample Analysis

1. Request the laboratory to provide the method (NIOSH or other) used for each contaminant.

RESPIRABLE DUST USING A CYCLONE

Purpose/Scope. To measure the airborne concentration of respirable dust.

1. The respirable portion of the dust is collected on matched-weight 5-μm pore-size PVC filters in a 37-mm cassette attached to a cyclone assembly and high flow pump. The cyclone separates particulates greater than about 10 μm; particles passing through the cyclone collect on the filter. The filter cassette is returned to the laboratory to determine the weight of respirable dust. Other analyses (metals, etc.) may be performed on the same filter if the analysis requested is compatible with the PVC filter.

Equipment

1. Battery operated high flow-pump capable of 1–3 L/min flowrate.
2. Size-selective cyclone.
3. 37-mm filter cassettes containing matched-weight 5-μm pore-size PVC filters. The samples must be returned to a suitable laboratory for gravimetric determination of collected dust.
4. Tygon tubing or other sampling hose.
5. "Matched-weight" refers to two filters that are matched in weight and loaded into a single cassette in a controlled laboratory environment. The top filter collects the contaminant, and the bottom filter serves as a control. The laboratory removes and weighs both filters. The difference between the two is the weight of the sample collected.

Calibration. Calibrate the pump and sampling train at the suggested flowrate. Refer to cyclone manufacturer and/or NIOSH sampling method for correct flow. See "Calibration with a Rotameter" in Chapter 5.

Sampling Procedure

1. Clean cyclone prior to use. Refer to the manufacturer's instructions and disassemble cyclone, clean the parts with isopropyl alcohol and pipe cleaners or cotton swabs, and reassemble. Make sure that the "grit pot" at the bottom of the cyclone is empty.

2. Install the filter cassette in cyclone, being certain that the filter inlet is facing toward the airflow from the cyclone outlet. If the filter cassette is not installed properly, the dust will be collected on the support pad and cannot be analyzed. Make certain that the cassette ports are properly seated in the cyclone assembly to prevent leakage.

3. Connect the cyclone outlet to the calibrated pump with a 3-foot section of sampling hose.

4. Place the pump on the belt of the worker being monitored in a manner that is most comfortable to the worker.

5. Clip cyclone unit to worker's collar. Check to ensure that clothing does not obstruct the inlet to the cyclone during normal movement and that the sampling hose does not interfere with the worker. Instruct the worker to not allow the cyclone to become inverted, because the large non-respirable particles in the grit pot at the bottom of the cyclone may fall onto the filter and give an erroneous result.

6. Start pump and record the time and the float position on the pump's flowmeter. Since the cyclone functions properly only at the appropriate flowrate, it is important to check the pump approximately every 45–60 minutes during the sampling period to verify flowrate. If the proper flowrate cannot be achieved, then terminate the sample, change filter, and start a new sample.

7. Record all pertinent data on sample data form.

8. At the end of sampling period, stop the pump and record the time. Remove the filter cassette from the cyclone and replace inlet and outlet plugs. Make sure the cassette is properly labeled with the sample number that is indicated on the corresponding sample data form.

9. When samples are returned to laboratory for analysis, submit a field blank with each batch of samples. The field blank should be a filter cassette from the same lot as those used for sampling. The blank should be handled in the same manner as the sample filters, but no air is drawn through it.

Sample Analysis

1. If you want the laboratory to calculate the airborne concentration, inform the laboratory as to air volume sampled and request that the filters be weighed for dust in mg/sample and mg/m^3. Also request metals or other desired analyses, if applicable.

2. Request the laboratory to provide the method number (NIOSH or other) for each contaminant.

SILICA

Introduction. Silica (SiO_2) can exist in two basic forms: crystalline silica and amorphous silica. Crystalline silica is the form responsible for the most important adverse health effects on the lung, including silicosis and possibly lung cancer. Amorphous silica is much less toxic. This procedure will address the methods for monitoring both crystalline and amorphous silica.

Crystalline Silica

Purpose/Scope. There are two types of crystalline silica normally encountered: quartz and cristobalite. A third type of crystalline silica is not as common: tridymite. The basic steps for monitoring all types of crystalline silica are the same:

1. Collect respirable dust samples using a filter and cyclone. Use a 5-μm PVC filter in a 37-mm cassette attached to a size-selective cyclone assembly and a high-flow pump.

2. If only the crystalline silica concentration is desired, a non-matched-weight filter can be used. If both the respirable dust concentration and the crystalline silica concentration are desired, a matched-weight filter must be used.

3. Return the filter cassette to the laboratory for analysis. Other types of analyses (e.g., metals) cannot be performed on the same filter. A separate sample must be collected to perform other analyses.

4. It is no longer necessary to collect total crystalline silica dust samples.

Equipment
• Same as **Respirable Dust**

Calibration. Calibrate the pump and sampling train at the manufacturer's recommended flowrate. See "Calibration with a Rotameter" in Chapter 5. The cyclone will not function properly at an incorrect flowrate.

Sampling Procedure. Whenever possible, the sample should be collected for 6–8 hours. If the pump runs for short time periods (e.g., 1 hour or less), the sample result may be reported at the limit of detection (LOQ), which may be greater than the standard.

1. Clean the cyclone prior to use. Refer to the manufacturer's instructions and disassemble the cyclone, clean the parts with isopropyl alcohol and pipe cleaners or cotton swabs, and reassemble. Make sure that the "grit pot" is empty.

2. Attach the filter cassette to the cyclone, being certain the filter inlet is facing toward the cyclone outlet. If the filter cassette is not installed properly, the dust will be collected on the support pad and cannot be analyzed. Make certain that the cassette ports are seated in the cyclone assembly.

3. Connect the cyclone outlet to the calibrated pump with a 3-foot section of sampling hose.

4. Place the pump on the belt of the worker being monitored in a manner that is most comfortable to the worker.

5. Clip the cyclone to the worker's collar. Check to ensure that clothing does not obstruct the inlet to the cyclone with normal movement and that the sampling hose does not interfere with the worker. Instruct the worker to not allow the cyclone to become inverted, because the large nonrespirable particles in the bottom of the cyclone may fall onto the filter and give an erroneous result.

6. Start the pump and record the time and note the float position on the pump flowmeter. It is important to check the flowrate approximately every 45–60 minutes during the sampling period to ensure that the flowrate is constant. If the proper flow cannot be maintained, start a new sample. Otherwise, sample for at least 6 hours to collect enough dust for adequate analytical sensitivity.

7. Record all pertinent data on the sample data form.

8. At the end of the sampling period, stop the pump and record the time. Remove the filter cassette from the cyclone and replace the inlet and outlet plugs. Make sure that the cassette is properly labeled with the sample number that is indicated on the corresponding sample data form.

9. When the samples are returned to the laboratory for analysis, submit a field blank with each batch of samples. The field blank should be a matched-weight filter cassette from the same batch of filters used for sampling. The blank should be handled the same as the sample filters, but no air is drawn through it.

Sample Analysis

1. Send the sample to an approved laboratory for analysis. Request that they determine the respirable dust concentration using gravimetric analysis and determine the crystalline silica concentration using X-ray diffraction. Report both results in units of mg/m^3.

2. For assistance in arranging for laboratory analysis or interpreting results, contact the analytical laboratory.

Amorphous Silica

The monitoring procedure for amorphous silica is the same as the procedure for either "Total Dust" or "Respirable Dust". The current ACGIH TLV® for amorphous silica is 10 mg/m^3 (inhalable) and 3 mg/m^3 (respirable). Compare the results with the TLV, or with the OSHA PEL if it is lower.

TOTAL DUST

Purpose/Scope. To measure the airborne concentration of total dust.

Total dust is collected on a matched-weight 5-μm pore-size PVC filter in a 37-mm filter cassette, attached to a high-flow pump. After collection, the filter cassette is returned to the laboratory to determine the weight of dust collected. Other analyses (metals, etc.) may be performed on the sample if the analyses requested are compatible with the PVC filter.

Equipment

1. Battery-operated high-flow pump.
2. 37-mm filter cassette, containing matched-weight 5-μm pore-size PVC filter.
3. Flowmeter.
4. Sampling hose.
5. Filter cassette coupler.

Calibration. Calibrate the high-flow pump with cassette inline as described in "Calibration with a Rotameter" in Chapter 5, at 1.0–2.0 L/min.

Sampling Procedure

1. For a breathing zone sample, attach sampling pump to worker's belt and the filter cassette to the collar with a 3-foot section of sampling hose. Check that the "inlet" pin has been removed from the cassette, the worker's clothing does not obstruct the inlet to the cassette, and the sampling hose does not interfere with the worker.
2. Start the pump and record the time, flowrate, and other pertinent information on the sample data form.
3. Check the flowrate periodically throughout the sampling period (every 30–60 minutes) using the flowmeter on the pump. If the flowrate varies by more than 10%, terminate the sample. Change the filter and start a new sample using a freshly charged pump (if needed).

Otherwise, sample for at least 6 hours to collect enough dust for adequate analytical sensitivity.

4. At the end of sampling period, stop the pump and record the time. Remove the filter cassette and replace inlet and outlet plugs. Make sure that the cassette is properly labeled with the sample number that is indicated on the corresponding sample data form.
5. When samples are returned to the laboratory for analysis, submit a field blank with each batch of samples. The field blank should be a filter cassette from the same batch of filters used for sampling. The blank should be handled the same as sample filters, but no air is drawn through it.

Sample Analysis

1. If you want the laboratory to calculate the airborne concentration, inform the laboratory as to the air volume sampled and request that the filters be weighed for total dust and reported in mg/sample and mg/m^3.
2. For assistance in arranging for laboratory analysis or interpreting results, contact the analytical laboratory.

GASOLINE AND LIGHT HYDROCARBONS

Purpose/Scope. This procedure describes how to conduct exposure monitoring for gasoline and gasoline-like light hydrocarbon vapors.

The airborne vapors are collected on activated charcoal tubes, and then they are desorbed and analyzed by a qualified analytical laboratory using gas chromatography.

The following protocol is applicable only to charcoal tube air monitoring. The use of passive monitors is presented in another procedure.

Equipment

1. Battery-operated low-flow pump
2. Charcoal tubes
3. Tube holder
4. Manila sample envelope

Calibration. Calibrate the pump and sampling train for a flowrate of 50 mL/min ± 10%. This flowrate is recommended to prevent the hydrocarbons from being carried through the tube. See "Pumps" in Chapter 5 for instructions.

Sampling Procedure

1. Calibrate the pump as described above. A 750-mL (15-min) sample is recommended as a minimum. The maximum volume should not exceed 40 L.
2. Immediately before sampling, break the ends of the charcoal tube to provide an opening that is at least half the inside diameter of the tube.
3. Insert the charcoal tube in the holder, making sure that the flow indicator arrow is pointing toward the pump (if there is no arrow, the larger section faces out, and the smaller section connects to the tube). Replace the cover on the tube holder.
4. Attach the pump to the workers belt or pocket and the tube holder to the shirt lapel. The tube should be vertical to reduce channeling through the tube.
5. Record the time and start the pump.
6. At the completion of the sampling period, stop the pump and record the ending time.
7. Remove the charcoal tube from the holder, and seal with plastic caps. Take special care when handling the broken glass. Place tube in manila envelope and write sample identification information on the envelope. Record all pertinent data on the sample data form.
8. Submit a blank with each batch of samples. The blank should be a tube from the same lot as the sample tubes; it is handled in the same manner as a sample, except that no air is drawn through it. Label this as "Blank".
9. Store samples in freezer until they are ready to be shipped to the laboratory for analysis.

Shipping. Package samples to avoid breakage. Use a sturdy carton or mailing tube with packing material. Samples should be shipped via an air carrier and received by the laboratory preferably within one, and no later than two, days from shipping date.

WELDING FUMES

Purpose/Scope. To measure airborne concentrations of particulates generated from welding, and arc or torch cutting operations.

1. Welding operations can produce two types of potential exposures; particles (primarily metal fumes), and in some cases various gases. Generally, the most important exposures of concern are to the welding fume particles. The composition of welding fume can vary greatly depending on a number of factors—for example, the composition of the rods or electrodes used, the welding process, any fluxes, and the base metal. Another factor which can affect the composition of the fume is the presence of any coating or plating materials on the base metal—for example, paints, galvanized coating

(zinc), cleaning materials, or chemical contamination on recycled materials.

2. In addition to the welding fume, exposures to hazardous gases can also occur during some welding operations. The most important gases include carbon monoxide, nitrogen oxides, ozone, phosgene, and fluorides. The gases generated will depend on the welding process used. For information on exposure monitoring of these gases, contact a qualified industrial hygienist or equivalent specialist.

Equipment

1. Battery-operated high- or low-flow pump.
2. Two-piece or three-piece 37-mm filter cassette.
   a. Mixed cellulose ester membrane filter with 0.8-μm pore size for monitoring for individual metals; analysis by atomic absorption.
   b. PVC filter with 5.0-μm pore size for monitoring the total metal fume present (analysis by gravimetric weight) and hexavalent chromium [Cr^{6+}] (analysis by colorimetric method).
3. Flowmeter.
4. Sampling hose.

Pump Calibration. Calibrate the sampling pump with the filter cassette in line, as described in "Pump Calibration" in Chapter 5. Calibrate before and after monitoring. Use the average of the initial and final flowrate to calculate total volume sampled.

Monitoring Procedure

1. Assemble the sampling train consisting of the filter cassette, sampling hose, and pump.

2. If possible, attach the filter cassette to the inside of the welder's hood with duct tape. Position the cassette so it will be near the cheek when the hood is lowered. This will prevent moisture from the welder's exhaled breath from being drawn into the cassette, plugging up the filter, and reducing the air flow. Tape the sampling hose to the inside of the hood and exit near the pivot point of the hood. The sampling hose should be attached to the pump on the welder's belt. Allow enough slack in the sampling hose to allow the hood to be raised and lowered without interference.

   If the welding hood design is such that the 37-mm-diameter filter cassette rests against the welder's face when the hood is lowered, a 25-mm-diameter cassette may be used. Ad a last option, the filter can be mounted outside the helmet on the welder's shoulder. Remove the pin from the cassette inlet.

3. Start the pump and record the starting time and other pertinent data required on the sample data form.

4. Periodically throughout the sampling period, check the flowrate of the pump by using the pump flowmeter. If the flowrate decreases by more than 10%, check the pump's flowrate with a calibrated rotameter or other flowmeter, and then terminate that sample. Start a second sample with a new filter cassette and freshly charged pump (if needed) to sample the remaining exposure time. Combine the results of the two samples to determine the TWA.

5. At the end of the sampling period, recheck the flowrate of the pump and record the ending flowrate and stop time on the sample data form. Use the average of the initial and final flowrates to calculate sample volume.

6. Remove the filter cassette from the sampling train and replace the pins in the inlet and outlet openings of the cassette.

7. Make sure that the cassette is properly labeled with the same sample number as is indicated on the corresponding sample data form.

8. Send the samples to an approved laboratory for analysis. Submit at least one blank filter cassette from the same batch of filters with each group of samples submitted for analysis. Blanks should be handled in the same manner as field samples except that no air is drawn through them.

Sample Analysis. For assistance in interpreting results, contact the analytical laboratory.

Selection of Analyses and Sampling Devices

1. The best method for determining which contaminants can be present in the welding fume is to review the "Material Safety Data Sheets" for both the base metal and filler materials (i.e., the rods or electrodes) being used in the welding operation being monitored. This information should be used to identify the major metal constituents to be monitored. Sampling for every contaminant known to be present would be difficult and time consuming. By selecting only the contaminants of most concern—that is, those contaminants with low TLVs® or a high percent composition in the base metal or filler material—the number of contaminants can be reduced to a manageable level.

2. As an illustration, Table A.2 presents a list of principal contaminants of concern, sampling devices, and analytical methods for typical welding operations. The table does not cover all possible welding situations and materials. If you encounter welding materials not covered in the table, contact the analytical laboratory or a qualified industrial hygienist.

To use the table, find in the first column the applicable base metal and filler materials that are being used. The table also indicates in parentheses those alloys that contain varying amounts of important metals. The percent composition of these metals will determine how they are to be monitored and analyzed. For example, some carbon steels contain manganese in varying amounts. If the composition of manganese is less than 10%, monitor and analyze for the "total fume." If the composition of manganese is greater than 10%, then monitor for iron and manganese. Column 2 indicates the individual metals that should be analyzed for by the laboratory. Column 3 indicates the type of filter to be used for monitoring.

3. When monitoring for chromium from welding on stainless steel, or other metal alloys containing more than 1% chromium composition, there are several forms of chromium that are present in the welding fume. The most hazardous form of chromium is the "hexavalent" form (CR^{6+}). Although chromium occurs in other forms in the welding fume, hexavalent chromium is the most toxic. As shown in Table A.2, the monitoring procedure for hexavalent chromium is different from the procedure for total chromium (i.e., all forms including hexavalent). The procedure for total chromium (i.e., MCE filter with atomic absorption analysis) can be used to determine the exposure to all forms of chromium combined

TABLE A.2. Recommended Analytes for Welding Fumes

Base Metals/Filler Materials (Rods)	Contaminants of Concern	Filter Type	Analysis Method
• Carbon steel (mild steel)			
(Manganese < 10%)[a]	Total fume	PVC	GR
(Manganese > 10%)	Fe, Mn	MCE	AA
• Low alloy steels			
(Chromium < 1%, nickel < 2%)	Total fume	PVC	GR
(Chromium > 1%, nickel < 2%)	Fe, Cr (total), Ni, hexavalent Cr	MCE	AA
		PVC	CM
• Aluminum	Total fume	PVC	GR
• Aluminum alloys			
(Copper > 10%)	Al, Cu	MCE	AA
• Copper/copper alloys	Cu	MCE	AA
• Copper/molybdenum alloys	Cr (total), Mo, hexavalent Cr	MCE	AA
		PVC	CM
• Stainless steels	Fe, Cr (total), Ni, Mn	MCE	AA
	Hexavalent Cr	PVC	CM
• Galvanized metals			
Iron, zinc	Fe, Zn	MCE	AA
Tin, zinc	Sn, Zn	MCE	AA

Definitions of Terms:
MCE, mixed cellulose ester membrane filter, 0.8-μm pore size; PVC, polyvinyl chloride membrane filter, 5-μm pore size; GR, gravimetric analysis, not specific for metals; AA, Atomic absorption analysis, for specific metals; CM, colorimetric analysis.

Chemical Symbols/Names:
Al, aluminum (welding fume); Cr, chromium (either total or hexavalent); Cu, copper (fume); Fe, iron (iron oxide fume); Mn, manganese (fume); Mo, molybdenum (soluble); Ni, nickel (metal); Sn, tin (oxide); Zn, zinc (zinc oxide fume).

[a] Based on TLV® for Mn of $0.2\,mg/m^3$. This guidance may not apply for a lower Mn TLV®.

(including hexavalent) as well as to the other metals present. The procedure for hexavalent chromium provides only the hexavalent chromium and no other metals or forms of chromium. Both procedures are needed if exposures to the hexavalent chromium and total chromium (plus other metals) are to be determined.

BENZENE

Purpose/Scope. To measure airborne concentrations of benzene vapor in the workplace.

The current OSHA PEL for benzene is 1 ppm (part per million), with an action level of 0.5 ppm along with a short-term exposure limit (STEL) of 5 ppm over 15 minutes. The standard contains specific monitoring requirements.

There are two sampling methods that can be used to measure the concentration of benzene in an employee's breathing zone: the charcoal tube method and the organic vapor badge method. Both of these methods are addressed in this procedure.

Methods

A. Charcoal Tube Method
1. PRINCIPLE OF OPERATION. Air is drawn through a tube containing activated charcoal, using a battery powered

pump. Any hydrocarbon molecules including benzene in the air stream will adsorb onto the charcoal. After sampling, the charcoal tube is analyzed by gas chromatography in an accredited laboratory.

2. EQUIPMENT
   - Battery operated personal sampling pump, capable of drawing a flowrate of 0.05–0.2 Lpm (50–200 mL/min).
   - Standard charcoal tube 100-mg front section, 50-mg back section.
   - Standard tube holder, for charcoal tube, with connective tubing.
   - Coin envelopes.

3. CALIBRATION. Calibrate the pump and sampling train as indicated in "Pumps" in Chapter 5.

   The flowrate should be selected based upon the length of sampling time as indicated below.

RECOMMENDED FLOWRATES

Sampling Time	Flowrate
15 minutes–2 hours	200 mL/min
2 hours–4 hours	100 mL/min
4 hours–8 hours	50 mL/min

4. BREAKTHROUGH. It should be noted that breakthrough may occur if there are high concentrations of other hydrocarbons present or if the relative humidity is greater than 70%. Other hydrocarbons or water vapor may saturate the front section of the charcoal tube, allowing benzene to breakthrough to the back section. Breakthrough can be detected by analyzing the front and back sections of the tube separately. It can be prevented by using a large charcoal tube (400-mg front, 200-mg back) or by reducing the volume of air sampled.

5. SAMPLING PROCEDURE
   - Immediately before sampling, break off the ends of the charcoal tube and place it in the tube holder. Make sure that the arrow printed on the tube is pointing in the direction of airflow (toward the pump). *Note*: If there is no arrow on the tube, place the end with the large section "out" and place the end with the smaller section toward the tubing.
   - Attach the sampling pump to the worker's belt and attach the tube holder to the collar, so the charcoal tube is in the worker's breathing zone. The charcoal tube should be in a vertical position to avoid channeling.
   - Note the time, and start the pump.
   - Record all pertinent data on the sample data form.
   - Turn off the pump upon completion of the sampling period. Record the time on the sample data form.
   - Remove the charcoal tube from the holder and immediately seal both ends with the caps provided. Place the sealed tube in a small coin envelope and write the sample identifying information on the envelope (e.g., sample number).
   - Calculate sampling time and volume sampled, and record this information on the sample data form.

6. BLANKS
   - At least one charcoal tube from the same lot as those used to collect the samples should be submitted as a "field blank." The tips of the tube should be broken off in the field and sealed with the caps without drawing any air through the tube.
   - For large numbers of samples (greater than 10) in a batch sub-

mitted for analysis, 10% should be field blanks.

B. Organic Vapor Badge Method

1. PRINCIPLE OF OPERATION. Organic vapor badges work on the gas diffusion principle. They do not require a sampling pump to operate, and there is no calibration to perform.

   The following guidelines must be followed to ensure a representative and reliable sample is obtained.

   - Stagnant Atmospheres: Organic vapor badges require a minimum face velocity to operate properly. Consult with the badge manufacturer to obtain the required conditions. Therefore, area samples in confined areas or offices with minimal air flow should not be evaluated with a badge. This restriction should not affect most personal samples.
   - Short-Duration Samples: Badges have a fixed "equivalent sampling rate" that might be too low to give adequate sensitivity with low airborne levels. For example, a 15-minute sample might have a lower limit of detection of 2 ppm. In order to achieve a lower detection limit of 0.5 ppm, which is the action level for the OSHA Benzene Standard, a minimum of 1 hour of sampling time is required.
   - High Relative Humidity/Other Hydrocarbons Present: Water vapor molecules and other organic vapor molecules compete with benzene for the available charcoal adsorption sites. Therefore, some badge manufacturers offer monitors with a backup section for use in atmospheres where the relative humidity is greater than 70%, or where the concentration of total organic vapor exceeds the TLV.

Consult with the badge manufacturer to see if a dual stage monitor is necessary.

2. EQUIPMENT. Organic vapor badges.

3. CALIBRATION. None required.

4. BREAKTHROUGH. High relative humidity or high concentrations of other hydrocarbons may cause breakthrough to occur. This cannot be detected during laboratory analysis of the samples if a single stage badge is used. It will be detected by the analysis procedure when using a dual stage badge.

5. SAMPLING PROCEDURE

   - Just prior to use, open the container and remove the badge from the packaging material. Retain the additional parts for future use.
   - Note the date and start time.
   - Clip the badge to the worker's lapel so it is near the breathing zone. The sampling face should be facing away from the worker's body. Check to ensure that no clothing will cover the badge face during normal movements.
   - Fill out the pertinent information on the sample data form.
   - The badge should remain on the worker for the entire shift or all of the exposure period.
   - At the end of the shift or end of the exposure period, remove the badge from the worker and record the ending time on the back of the badge and the sample data form.
   - Snap the provided seal in place. This stops the collection of any more material on the badge. The exposed charcoal element will continue to capture any hydrocarbons present until it is capped.
   - Replace the badge in the container that it came in.

6. BLANKS

- At least one badge from the same lot as the field samples should be submitted with each group of samples. The blank should be opened and resealed in the same area the field samples were recapped. This badge should be labeled as a "BLANK."

- If a large number of samples are being submitted to the laboratory for analysis as a single batch, include one blank for every 10 samples.

Shipping and Analysis

1. Package the samples in a sturdy container with adequate padding to prevent breakage during shipping. If shipping of samples is delayed for more than one day, store in a freezer until shipped.

2. Ship samples along with sampling time to the laboratory for analysis. Request that the laboratory analyze the sample in accordance with NIOSH Method 1501, *Hydrocarbons, Aromatic.*

Results. The results should be recorded on the sample data form. According to the "Notification of Monitoring Results" requirement in the OSHA standard, the results are to be reported to the employees who were monitored and other employees who have the same job.

APPENDIX B

GAS AND VAPOR CALIBRATIONS

This Appendix describes the theory and procedures for generating known concentrations of gases and vapors in air for the purpose of calibrating direct-reading instruments or evaluating sample collection device methods. Appendix C is a stream lined version of calibration information for direct-reading sensors. Appendix C is intended for sampling practitioners with a more focused need for calibration information, who do not need the extra detail in this Appendix.

For calibration of direct reading and real time instruments that measure gases and vapors, a known concentration of a specific contaminant is the best way to ensure accuracy. Known mixtures of gases and vapors in air are used to set the sensitivity, range, or span of an instrument's response. They are often called span gases. They are limited to gases and vapors that will remain in the gaseous phase over the period of time needed for calibration.

Sample collection device methods for gases and vapors often use similar "calibration" mixtures to evaluate performance of the collection devices or to evaluate the recovery and accuracy of the analytical method.

One of the major problems associated with toxic gas monitoring is the availability of reliable calibration gases due to the varying physical properties of chemicals that can make storage difficult. For example, carbon monoxide is readily available in cylinders, whereas nitrogen oxides are reactive and must be generated in a laboratory using permeation tubes and special instrumentation. Depending on the specific characteristics of the compound, it could be packaged in a cylinder, made up in a bag using a syringe, mixed in a bottle using premeasured ampules, or more complex techniques may be required such as permeation tubes or dynamic gas generating systems. For materials such as mercury vapor. Table B.1 shows typical test atmospheres of commonly sampled chemicals.[1]

There are several methods for generating calibration gases. These include both

Air Monitoring for Toxic Exposures, Second Edition. By Henry J. McDermott
ISBN 0-471-45435-4 © 2004 John Wiley & Sons, Inc.

TABLE B.1. Typical Test Atmospheres of Commonly Sampled Chemicals

Compound	Method of Generation
Ammonia	Permeation device or ampule
Carbon dioxide	Gas cylinder with zero air or nitrogen containing carbon dioxide in the required concentration
Carbon monoxide	Gas cylinder with zero air or nitrogen containing carbon monoxide in the required concentration
Ethane	Gas cylinder of prepurified nitrogen containing ethane as required
Hydrogen chloride	Cylinder of prepurified nitrogen containing hydrogen chloride as required
Hydrogen sulfide	Permeation device, cylinder, or ampule
Methane	Gas cylinder of zero air containing methane as required
Nitrogen dioxide	Permeation device
Oxygen	Use outdoor air at sea level
Ozone	Calibrated ozone generator
Sulfur dioxide	Permeation device
Xylene	Cylinder of prepurified nitrogen containing xylene as required or gas bag with appropriate amount of solvent to make up desired concentration

Source: 40 CFR 53 EPA.[1]

static and dynamic systems. A static system is a mixture in a fixed volume such as a Tedlar bag or Teflon bottle. Preparation of a batch mixture in a bottle or a bag has the advantage of simplicity and convenience in some cases. Dynamic systems generate a continuous flow from a gas cylinder or from bubbling air through a liquid solvent. Flow dilution systems are capable of providing theoretically unlimited volumes at known low concentrations that can be rapidly changed if desired. The disadvantages of flow dilution systems are the skill, exactness, and complex laboratory setups required. Although these are beyond the scope of this book, information is available from other sources.[2] Permeation tubes also fall into this category, although recent advances in simplified calibration apparatus that provide sufficient accuracy for field use have made them somewhat simpler to use than in the past.

Many gases and vapors in commonly used concentrations are available from several manufacturers. Usually they are contained in compressed gas cylinders,

ampules, or aerosol cans. Filling high-pressure gas cylinders is better left to the experts. When purchasing calibration or span gases, one assurance that the source is reliable is certification of the concentration and contents of the mixture from the manufacturer. The gas supplier should be asked to indicate on the certificate of analysis the analytical method used and to estimate the means used to compute the accuracy, including the calibration technique. Some direct reading instruments can accommodate a compound mixed in nitrogen; however, many instruments require oxygen to operate and so for these calibration mixtures must be made up using air.

Not every gas mixture can be pressurized. For example, the vapor pressure of toluene at 20°C is only high enough to produce a compressed gas mixture of 1200 ppm at 300 psi; therefore, it is not possible to purchase a cylinder of toluene gas for calibration in the lower explosive level (LEL) range. For this reason, as well as others, sometimes the gas selected for calibration is not the one that will actually be

TABLE B.2. Overview of Selected Calibration Methods

Method	Typical Accuracy (%)	Precision	Range	Cost	Ease of Use
Permeation tubes	97–99[a]	Good	ppb to low ppm	Moderate	Difficult
Gas cylinder[b]	95–99	±1% to 2%	ppm to %	Moderate	Easy
Dynamic dilution of pure gas or higher concentration	93–97[c]	±3%	% to ppm	Moderate	Moderately easy
Static dilution	90–97[d]	±3% to 10%	ppm to %	Inexpensive	Easy

Source: Becker, J. H., et al. Instrument calibration with toxic and hazardous materials. *Ind. Hyg. News*, July 1983.

[a] Depends on good temperature control, balance accuracy, and flow rate.
[b] Data from manufacturer's literature; refers to analyzed standards only.
[c] Includes inaccuracies in calibration gas plus flow rate.
[d] Depends on analyst's technique.

monitored, but instead one for which the response of the instrument can be compared to the gas of interest. An example is the use of isobutylene for calibrating instruments to detect organic vapors similar to benzene. If a standard is to be used to calibrate a combustible gas indicator, do not prepare concentrations greater than 25% of the LEL for that compound.

Making up standards of calibration gases can be difficult unless the sampler is familiar with the characteristics associated with each compound and the system to be used (Table B.2). For example, isocyanates are very reactive. Other effects, such as container wall absorption, humidity and volumetric errors, must be recognized. Appropriate precautions must be taken to avoid concentration errors derived from these seemingly extraneous sources when attempts are made to produce homogeneous, known concentrations for calibration purposes.

When calibrating any instrument, several known concentrations should be used. These should encompass the range of the instrument's measuring capability. Periodically, a calibration curve should be prepared. For field use, a calibration at one point, such as the threshold limit value

(TLV), can suffice as long as the instrument is periodically calibrated at several points.

The environmental conditions, including temperature, relative humidity, and barometric pressure, of the instrument at the time of calibration should be as near as possible to what will be encountered in the field. Temperature is the most important, because changes in temperature are most often encountered in the field and can cause bias in the readings obtained. If it is not possible to calibrate at the working temperature, the user must allow sufficient time for field equilibration of temperature. Since many of these units are battery operated, it should be noted that temperature extremes can also adversely affect the performance or life of the batteries used in these devices.

Whenever possible do calibrations in a laboratory hood. Always work with the smallest quantity of material possible, including using the lowest practical concentration in gas cylinders. Use appropriate personal protection for the type of contaminant. Leak test lines prior to using. If the instrument being calibrated is a nondestructive analyzer, either use it in a hood or make sure a window or other source of air is available and "vent" the instrument in this direction. If calibrations will be performed in one given area, consider installing a local exhaust vent. For instruments like the flame ionization detector (FID), the exhaust gases should be vented as well. If the instrument's response is very rapid, large volumes of calibration mixtures are not needed.

Make sure that the calibration procedure itself does not pressurize the sensing cell. It is especially important to observe when pressurized cylinders of standard gases are used for calibration. Overpressurization of the sensor can be avoided by using a pressure regulator on the calibration gas cylinder, or by installing a "T" fitting in the line to reduce the stream to atmospheric pressure. Another method is to fill a bag with gas from the cylinder so that it can be presented to the sensor at atmospheric pressure from the bag.

It is important that appropriate accessories be used depending on the chemical in use. For example, substituting tygon tubing for Teflon tubing may cause some sample to be lost by absorption loss in the tubing walls for some compounds. Some instruments require a humidifier to compensate for the humidity effects caused by using a dry compressed gas mixed with nitrogen. Dispose of unneeded gases from bags and bottles properly. Keep bags and bottles clamped shut until ready to dispose of the gases. Decide where to release the unused contents of the calibration mixture once calibration is complete. It should be outdoors, or if the gas is soluble in water (sulfur dioxide, hydrogen chloride, chlorine) hooking up the bag to a jar of water via tubing and allowing the bag to slowly release its contents.

PREMIXED GASES AND VAPORS IN CYLINDERS

Most scientific gas suppliers rate their mixture accuracy at 95% to 99%, depending on the type of gas and the concentration level. For some purchased calibration mixtures, the specified concentration may depend on the analytical method used by the supplier. The worst case could involve calibration values from different manufacturers determined by two different analytical techniques, such as hydrocarbons by FID or an infrared detector. In this situation, the difference between the analytical techniques could cause a 10% to 20% discrepancy.[3] If using premixed gases and two different batches of gas, and the resulting measurements do not agree, a third batch of gas should be obtained and used. If it too is contradictory, change sources. For many gases, the labeled ppm value defines the analysis at the time of manufacture only. For certain gases and certain manufacturers, labeled ppm values are applicable for a year or longer. Unfortunately, not all commercially available cylinder standard gases have such stability, and some may become unreliable in 2 to 3 weeks or less.[2] As a rule of thumb, at least 200 psi of gas is required in a calibration cylinder for accurate calibration.

Premixed gas standards purchased commercially can be diluted, if purchased in a high enough concentration to provide a wide-range calibration curve. However, a possible source of error with dilution calibration system is working at the lower ranges of rotameters, where they are less accurate.

Compressed gas cylinders present a number of hazards. They are often under extremely high pressure when full, the size of the tank frequently determining how much pressure it can hold. Striking the valve at the top of the tank may shear off the valve assembly, venting the pressurized gas. In addition to the potential for fire, explosion, or toxic atmosphere the velocity of the existing gas may propel the cylinder at hazardous speeds. Check the hydrostatic or other test expiration dates for compressed gas cylinders. Also always check tanks for pitting and rusting. Any sign of deterioration should be reported immediately, and the tank should be removed from service. Never assume that the color of the tank indicates the contents. Never add adaptors or other gear to a regulator to make equipment fit.

Often special threads and sizes are used for regulators to forewarn or prevent certain types of equipment from being used or attached to the tanks. Regulators are specific for different types of tanks and should not be switched between tanks of different manufacturers. The direction of the threads on tanks may be reversed from the normal directions used in other equipment. Never attempt to force threads or nuts. Never store tanks in direct sunlight or near excessive heat. Nonflammable gases, such as carbon dioxide, may rupture with a force equal to or greater than that of flammable gases. Do not store calibration gas tanks in rooms where individuals work. Calibration gas tank contents are under pressure; therefore, do not use oil, grease, or flammable solvents on the flow control or the calibration gas tank. Do not throw cylinders into a fire, incinerate, or puncture. It is illegal in some cases to refill certain types of tanks.

Procedure for Using Gas Cylinders

1. Attach the piece of tygon tubing connected to the bag to the beveled connector attached to the pressure gauge on the cylinder. Never attempt to use a high-pressure cylinder without a pressure regulator as a source of calibration mixture. If the gas in the cylinder is sufficiently toxic, this operation should be done inside a laboratory hood or outdoors.

2. Turn on the cylinder slowly and allow the bag to fill. When full, turn off the cylinder, clamp the bag shut, and remove the tubing.

3. Attach the tubing to the instrument. Open the clamp, and turn on the instrument, which should have been zeroed earlier in a noncontaminated area.

STATIC CALIBRATION MIXTURES

Static or batch calibration mixtures allow the users to make up their own concentrations, thus providing more flexibility and offering a wide variety of concentrations. Static calibration mixtures are prepared by introducing a known volume of contaminant into a specific volume of air inside a container. Usually the container is either a bag or a bottle (rigid chamber). Two options for rigid vessels are a 1-L Teflon bottle whose cap has been modified to accept a septum and a 1-L cylindrical glass vessel with Teflon stop-cocks in each end and a septum port projecting from the side wall (Figure B.1). Thin-walled Teflon bottles may be susceptible to loss via diffusion after extended heating to "bakeout" contaminants.[4] The glass vessel is purged by passing a stream of clean air through the two stopcocks for 2 to 5 minutes. A way of determining if contamination is present is to inject a sample of inside air into a gas chromatograph. Bottles should also be calibrated, which can be done by filling them with water using a graduated cylinder. The most commonly used bottle is a 5-gallon narrow-neck carboy. It is large enough to allow sufficient volume for the liquid to evaporate. The sampler can utilize approximately one-tenth of the volume of this container before the mixture is diluted appreciably by room air. Placing two or three carboys in series and preparing the same concentration in each reduces the dilution effects even further and allows more of the original container to be used. In general, the lower the concentration of a static standard, the more frequently it should be replaced.

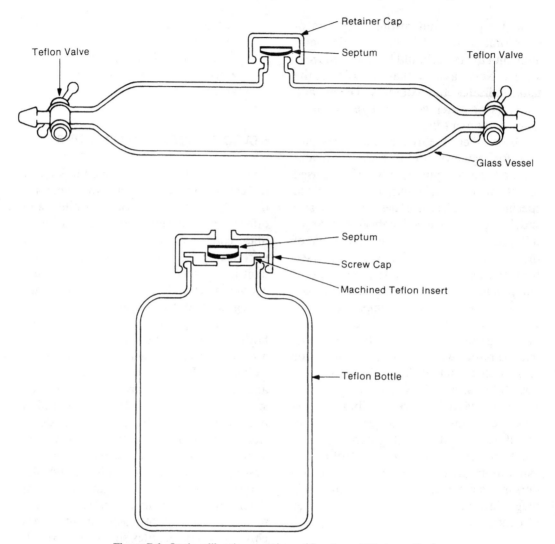

Figure B.1. Static calibration containers. (Courtesy of Photovac, Inc.)

The major advantage of using flexible bags is that no dilution from room air due to displacement of volume occurs as the sample is withdrawn. Bags should be tested frequently for pinhole leaks. Testing is done by filling bags with clean air and sealing them. If no detectable flattening occurs within 24 hours, the leakage is negligible.

Adsorption and reaction on the walls is not a great problem for relatively high concentrations of inert materials, such as carbon monoxide. However, low concentrations of reactive materials, such as sulfur dioxide, nitrogen dioxide, and ozone, are partly lost, even with prior conditioning of the bags. Prior conditioning is accomplished by putting a similar mixture in a bag and allowing it to sit for at least 24 hours in the bag, and then evacuating the bag just before use. Larger-size bags are preferable to minimize surface-to-volume ratio. Losses of 5% to 10% can occur during the first hour, so calibrations should be done immediately following the mixing. Precautions must be taken in order to avoid contamination of calibration gases by prior contents of bags: Bags should be dedicated to certain chemicals or new bags should be used to do calibrations.

One static calibration method involves the use of ampule kits. A kit generally contains ampules of a specific compound, a calibration bottle, and some device for breaking the ampules inside the bottle, such as ceramic balls. For calibration, the ampules are broken in the calibration bottle and the resulting gas mixture within the bottle equals the concentration specified on the label of the ampule. The calibration bottle is then placed over the head of the sensor for calibration. Ampules are available for hydrogen sulfide, ammonia, sulfur dioxide, hydrogen cyanide, and methane. Ampules are used primarily for field checks.

Static calibration mixtures using liquids are prepared by injecting either the liquid or head space vapor from a glass microsyringe, gas syringe, or pipet containing a calculated volume of the chemical of interest into a bag of known volume that contains a sufficient volume of purified air to produce the desired concentration. Generally, syringes are better for handling concentrated solutions of volatile materials than pipets. Microliter-sized auto pipets are often made of plastic. They are usually designed to deliver water-based solutions rather than compounds with a high vapor pressure (VP). With plastic pipets the VP over the liquid can displace some of the liquid, causing inaccurate deliveries. The use of syringes is discussed later in this section.

Contaminants that are gases at room temperature are usually introduced via a gas syringe. Gas-tight syringes are leak-free only when new and should be tested regularly or injections will be less than indicated, resulting in inaccurate mixtures. Syringes can absorb amounts of contaminants into the walls.

Calculating Concentrations for Static Calibrations

When making up a calibration mixture from a liquid, the amount to add to a specific volume can be calculated.

$$\text{ppm} = \frac{(x_g)(24.5)(10^6)}{(\text{vol.})(\text{MW})}$$

where ppm is the desired concentration, x_g represents the number of grams of compound needed for desired concentration, vol. stands for the amount of air to be metered into bag (L), MW is the molecular weight of compound, and 24.5 represents the number of liters of vapor per mole of contaminant at 25°C and 760 mm Hg. To convert from grams of compound to milliliters,

$$\frac{24.5\,\text{L}}{\text{mole}} = \frac{24,500\,\text{mL}}{\text{mole}}$$

$$\frac{(x_g)24,500\,\text{mL/mole}}{\text{MW}(\text{grams/mole})} = \text{mL of compound}$$

For compounds that are gases at room temperature,

$$\text{ppm} = \frac{(\text{Vapor volume in mL})(10^3)}{\text{Air volume in liters}}$$

The headspace vapor technique can also be used for making up smaller amounts of calibration mixtures.[4] These concentrations can be made up using the headspace VP. The temperature of the room must be measured and referenced to the VP of the compound at that temperature. A pure gas has a vapor pressure greater than 760 mm Hg; therefore, VP is not a concern, and to make up a mixture the ratio needed is calculated using the desired final concentration.

$$\text{ppm} = \frac{\mu\text{L of compound}}{\text{volume in liters}}$$
$$= \mu\text{L of compound}$$
$$\times \left(\frac{760\,\text{mm Hg}}{\text{VP of compound}}\right)$$

Syringes

Accurate techniques for using the syringe should be followed. A syringe twice as big as the amount of sample needed is generally required. For example, a 10-μL syringe would be used to deliver 5 μL of sample. The syringe should be flushed with the chemical several times before using. A small amount of the liquid (1 μL) is pulled into the syringe initially, and then a small amount of air. Pull the liquid–air interface in until all of the 1 μL of the liquid can be seen. Then depress the plunger until the liquid is just above the air, and pull in the amount of sample needed. Pull the sample into the barrel to make sure the amount is accurate; then push the sample to the bottom of the barrel. Inject the sample into the bag.

With a gas, a gas syringe can be used to pull air from a tygon tubing attached to a cylinder with a regulator on it. This operation is best done under a laboratory hood. The tubing should be pinched off at the end with a sturdy pinch clamp, and the cylinder should be turned on. The gas syringe can then be inserted into the tubing, and the sample can be pulled out. The sample is then injected into the bag. For samples suspected of containing high concentrations of volatile compounds, disposable glass syringes with stainless steel/Teflon hub needles are used.[5] One way to increase the accuracy of measurement of the amount delivered by syringe is for the sample to be weighed out.[4]

Syringes can develop problems: Teflon plunger tips become worn and leaks can develop, seals on screw-on needles can become loose, and needles can become plugged. With large syringes a blockage can be detected because there is a resistance as the plunger is pushed in. With smaller syringes blowing air through the syringe into clean water and observing whether bubbles emerge throughout the travel of the plunger down the barrel can detect blockage. Manufacturers generally provide fine wire for cleaning blocked needles or alternatively the needle can be replaced.

Preparation of Known Concentrations Using Bags

Bags can be used for compounds that are liquids or gases at room temperature. Following injection of a liquid or vapor into a bag, a sufficient amount of time to allow equilibration is needed (Figure B.2). Recommendations range from one hour for liquids to 15 to 30 minutes for gases and headspace vapors.[6]

Procedure
1. Select a 5-L to 10-L bag. The volumes of air and gas or liquid needed should be calculated ahead of time in order

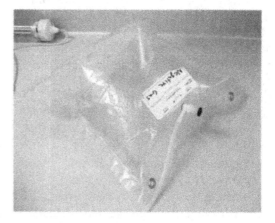

Figure B.2. Calibration gas mixture prepared in a bag. (Courtesy of International Sensor Technology.)

to make up the desired concentration and will be dependent on the size of the bag.

2. Partially fill the bag with air that has been passed through a series of traps, one of which is filled with a solid desiccant and another that is filled with silica gel or activated charcoal, or draw clean air (or oxygen or nitrogen) from a supply cylinder into the syringe for measured transfer into the bag and then completely evacuate it using either a personal air sampling pump or a 1-L to 2-L syringe. The procedure should be repeated several times in order to condition and purge the bag.

3. Following this purging cycle, a specific volume of air should be metered into the bag using a dry test meter or personal sampling pump at a flow rate appropriate for the size of the bag. For example, 5 Lpm is appropriate for a 10-L bag. When done, close the inlet to the bag. Often it involves not only pushing the valve down but also turning it a few times to lock it.

4. Add the chemical through the bag's septum (usually located behind the valve) or through tubing attached to the valve using a microliter syringe if it is a liquid, or a gas syringe if it is a gas, using proper techniques to assure accuracy.

5. The contents may be mixed by gently kneading the bag with the hands. Following injection of a liquid, inspect the inside of the bag for signs of liquid if the bag is the see-through type. Liquid is an indication that the material has not completely volatilized. Continue to knead the bag until the material evaporates.

6. Perform calibration of the desired instrument following the manufacturer's recommendations.

Preparation of Known Concentrations Using Bottles

Generally, bottles are best used for compounds that are liquids at room temperature (Table B.3).[2]

Procedure

1. Clean and dry the bottle. Any residue inside should be avoided, since it may absorb the gas or vapor added. The bottle should have a cap or plug in which two short glass tubes have been inserted. A short piece of tubing should be attached to one glass tube and a long one to the other. Both are then clamped off.

2. Cut two squares of aluminum foil approximately 10 cm by 10 cm, crumple them up, and put them inside the bottle. They will be used to provide a means to agitate the mixture inside of the bottle (Figure B.3).

3. Measure out the liquid using a microsyringe and proper technique. Unclamp one of the tubes, carefully insert the syringe, and carefully deliver the liquid into the bottle. Reclamp the tubing. Another technique is to take the top off the bottle and insert the needle, covering it with the top. Deliver the liquid, quickly remove the needle, and recap.

4. Pick up the bottle and shake it so that the foil strips rotate as they move up and down. Continue mixing until all the liquid has evaporated. The mixture is now ready to use.

GAS PERMEATION TUBES

Gas permeation tubes allow for very accurate calibration of some instruments at low gas concentrations. The primary value of these tubes is the ability to generate atmos-

TABLE B.3. Concentrations of Selected Compounds to Produce 100 ppm in a 5-gallon Bottle

Compound	Concentration (μL)
Acetaldehyde	4.5
Acetone	5.9
Acetonitrile	4.2
Acrolein	5.3
Acrylonitrile	5.3
Benzyl chloride	9.2
n-Butanol	7.3
Butyl acetate	10.5
Carbon disulfide	4.8
Carbon tetrachloride	7.7
Chloroform	6.4
Cumene	11.2
Decane	15.6
1,2-Dibromoethane	6.9
1,2-Dichloroethane	6.3
1,4-Dioxane	5.2
Ethanol	4.7
Ethyl acetate	7.8
Ethyl acrylate	8.7
Ethyl benzene	10.0
Ethyl bromide	6.1
Ethyl mercaptan	5.9
Freon 11	7.4
Freon 113	9.6
Heptane	11.7
Hexane	10.4
Hexanol	5.8
Isopropanol	6.1
Methanol	3.2
Methyl acrylate	7.2
Methylene chloride	5.1
Methyl ethyl ketone	7.2
Methyl isobutyl ketone	10.0
Methyl methacrylate	8.9
Methyl propyl ketone	8.5
Nitromethane	4.3
2-Nitropropane	7.2
Nonane	14.3
Pentane	9.2
n-Propanol	6.0
Propylene oxide	5.4
Pyridine	6.4
Styrene	9.2
Tetrachloroethylene	8.2
Tetrahydrofuran	6.4
Toluene	8.5
1,1,1-Trichloroethane	8.4
1,1,2-Trichloroethane	6.8
Trichloroethylene	7.2
Xylene	9.9

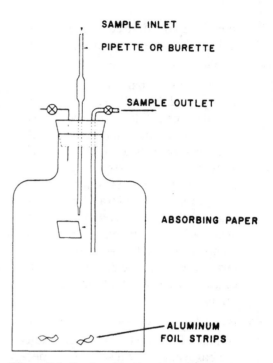

Figure B.3. Batch calibration in a 5-gallon bottle. (From NIOSH. *The Industrial Environment—Its Evaluation & Control.* Washington, D.C., 1973.)

pheres of reactive gases, such as sulfur dioxide and nitrogen dioxide, that would be difficult otherwise. The tubes require a prolonged equilibration period prior to use, and the equipment to house them in a constant-temperature environment is bulky. The principle is that any material whose critical temperature is above 20°C to 25°C can be sealed in Teflon tubing (or in some cases an impermeable tube with Teflon end caps or internal bladder). The material then permeates through the Teflon surface and will diffuse out at a rate dependent on thickness and area (fixed parameters) and temperature. At constant temperature, the rate of weight loss is constant as long as there is liquid in the tube. Permeation tubes for many compounds, including sulfur dioxide, nitrogen dioxide, hydrogen sulfide, chlorine, propane, butane, and methyl mercaptan, are available. For compounds such as nitrogen oxides and formaldehyde permeation tubes may be the only choice. Dilution gas systems can be used to generate varying concentrations from individual permeation tubes.

The liquid or gas in permeation tubes maintains a constant vapor pressure in contact with the inner wall of the tube. The compound's molecules pass through the tubing wall at a very small, extremely constant, flow rate. After an initial stabilization period of 1 to 3 weeks, the material permeates at a uniform rate through the walls of the tubing as long as it is kept at a constant temperature, and the output rate of the tube will remain essentially constant until nearly all of the liquid has permeated through the tube. In general, permeation tubes can be used to generate concentrations between 0.1 ppm and 50 ppm.

Two levels of accuracy are used in the manufacture of permeation tubes. Certified tubes are rigorously calibrated and their rates are given a factory calibration. Batch calibrated tubes are identical in construction to certified tubes, but do not go through the same rigorous calibration method. Their calibration is based on the knowledge of the rate of certified tubes from the same batch of tubing material. Accuracy for batch tubes is generally ±5%. A graph of permeation rate versus temperature is provided with permeation tubes.

Tubes can be disposable or refillable. Disposable tubes are usually short lengths of Teflon tubing with the liquid inside. The tubing used most commonly is either FEP Teflon or TFE Teflon. The tubing material is selected to match the permeation characteristics of the chemical. Refillable tubes are small stainless steel cylinders with a Teflon tubular membrane sealed inside. The cylinder serves as a large-volume reservoir for the component that is independent of the tubular membrane. The compound surrounds the membrane and permeates through the membrane wall to mix with the dilution gas flowing inside the membrane.

Refillable permeation tubes are certified by vacuum leak measurement and are supplied with graphs describing the emission rates over their entire operating temperature range. Refilling the tubes does not disturb the membrane inside, thus eliminating any need for recalibration according to manufacturers. Some tubes have two layers of polymer. In this case, the liquid or gas inside the device permeates through the first polymer layer to a gaseous form that then permeates through the second polymer layer at a controlled rate allowing for much lower concentrations. Some tubes have a vial attached that serves as a reservoir, giving the device an extended life and making lower permeation rates obtainable.

The permeation rate is dependent on temperature. The sensitivity of tubes ranges from a 1% to 15% change in permeation rate per °C. Some tables give the permeation rate per cm of length of tube. In general, as the length of the tubing doubles, the permeation rate also doubles. Permeation rate can also be controlled to some extent by the wall thickness of the permeation tube and tubes having three different wall thicknesses are available. The life of the tube depends on the volume of the tube, the weight of the material inside the tube, and the permeation rate of that particular tube. Permeation tubes are calibrated by weight loss over a known time interval. Generally, a semi-micro balance is chosen to measure these weight losses. Fingerprints or additional dirt on the outside of the tube can seriously affect the accuracy of these weights. A wire loop is supplied with many diffusion tubes to help facilitate handling of the tubes. In general, more than three weighings are obtained on a tube to determine its permeation rate.

If a higher concentration is desired, some systems can accommodate more than one tube. The permeation rate is then the sum of the rates of these tubes. Increasing the temperature of the calibrator if it has

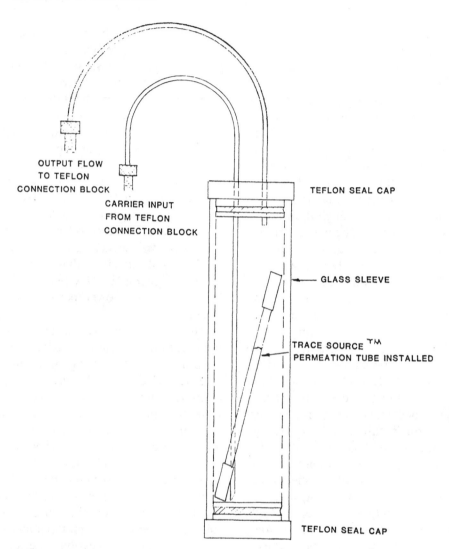

OUTPUT FLOW
TO TEFLON
CONNECTION BLOCK

CARRIER INPUT
FROM TEFLON
CONNECTION BLOCK

TEFLON SEAL CAP

GLASS SLEEVE

TRACE SOURCE ^{TM}
PERMEATION TUBE INSTALLED

TEFLON SEAL CAP

Figure B.4. Permeation tube holder for a portable permeation tube system. (Courtesy of Kin-Tek Laboratories.)

an oven will also increase permeation rates up to a certain point.

A typical calibration device using a permeation tube gives the user the option of selecting the desired concentration. After selecting the desired concentration, the temperature and other data are used to calculate the flowrate of the calibrator. The calibrator is turned on and the permeation tube is placed inside the bottle or other holder and put inside of the calibrator (Figures B.4 and B.5). The instrument is hooked up and turned on and calibration proceeds. Effective use of permeation tubes requires precise control of the tube operating temperature and the flow of dilution air. In the case of an ovenless calibrator, the temperature inside is kept constant through insulation.

In use, a time period of up to several hours is required for the calibration system to come to thermal equilibrium, thus pro-

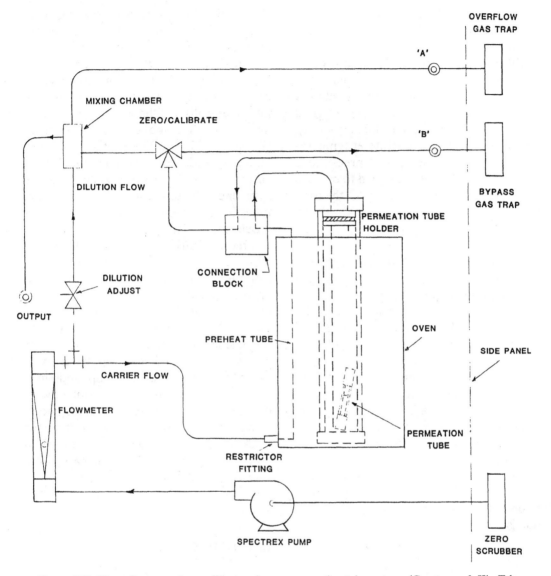

Figure B.5. Flow diagram of a calibrator for a permeation-tube setup. (Courtesy of Kin-Tek Laboratories.)

ducing a constant permeation rate. Generally, at least 1 hour is required for warm-up and stabilization of oven units. A precision rotameter is used to measure the span gas output. Airflow to dilute the concentration to the desired flowrate is adjustable and often measured with a precision rotameter. Most tubes require 30 minutes to 3 hours to reach equilibrium. Heavy wall tubes, low vapor pressure compounds, and halo-

genated compounds typically take longer. The best procedure is to set up the calibration system the day before it is needed, allowing the system to equilibrate overnight.[7] Gas flow is passed across the tube. Typically the flowrate is kept fairly low to preserve the thermal stability of the permeation tube and thus its permeation rate. The low gas flow transports all of the material that has permeated from the tube and is diluted by a high airflow commonly called the diluent airflow to the desired concentration.

Tubes can be stored and reused without affecting their permeation rate. Some tubes can be stored up to a year. Storing tubes at a lower temperature than they are designed for use at will prolong their life. Generally, for tubes designed to operate at temperatures above 60°C, no special precautions are required. For tubes designed to operate at low temperatures, such as 30°C, the total tube life can be extended significantly by storing the tube at reduced temperatures; however, storage under conditions that will freeze the liquid in the tubes is not recommended. Tubes can also be stored inside a laboratory hood. The proper procedure for storing a permeation tube depends on the type of material in the tube. It is suggested that the tube be stored in a sealed container together with packets of activated charcoal to prevent contamination to the outer surface of the tube and to prevent the tube from contaminating the room by permeation. Permeation tubes continue to emit their component during storage, even at reduced temperatures.

For calculations of concentrations generated by permeation tubes and flow rates to operate calibrators, the following equation is used:

$$Co = \frac{(K)(E)(L)}{(F_1 + F_2)}$$

where Co is the concentration, ppm; E is the permeation tube emission rate, ng/(min-cm); L is the permeation tube length, cm; K is the factor for a particular gas to convert from ng/mL to ppm at 25°C; F is the total dilution airflow rate (mL/min) $= F_1 + F_2$; F_1 is the carrier gas flowrate, mL/min; and F_2 is the dilution gas flowrate, mL/min.

Some systems use only dilution gas and then $(F_1 + F_2)$ in the equation simplifies to

F. K is usually provided by the manufacturer. In order to make additional concentrations, for each Co a different flow rate for the dilution gas, F_2, is selected.

Advantages of permeation tubes include no loss of material due to adsorption into the walls of the container such as happens with static calibration systems. By varying the airflow of the calibrator, multiple standards can be generated for linearity checks. Since toxic compounds are kept encapsulated inside the permeation tube, the risk of exposure to the operator is reduced.

Precautionary measures include not allowing material from a calibrator to vent continuously into a closed or poorly ventilated room. Permeation tubes should be opened outdoors or under a laboratory hood. Avoid overheating permeation tubes, since some tubes have a high vapor pressure inside and may rupture if overheated. Some less stable compounds may decompose, or polymerize violently, if overheated. Do not store tubes in poorly ventilated areas such as a car interior or truck cab. Care should be taken when removing permeation tubes from calibrators because they can be very hot.

REFERENCES

1. 40 CFR 53 EPA. *Ambient Air Monitoring Reference and Equivalent Methods.*

2. Nelson, G. O. *Gas Mixtures: Preparation and Control.* Lewis Publishers, Chelsea, MI, 1992.

3. Becker, J. H., et al. Instrument calibration with toxic and hazardous materials. *Ind. Hyg. News*, July 1983.

4. *Photovac 10S50 Gas Chromatograph Operating Manual.* Photovac, Inc., Thornhill, Ontario, Canada, 1991.

5. EPA. *Compendium Method TO-14: The Determination of VOCs in Ambient Air Using Summa Passivated Canister Sampling*

and *Gas Chromatographic Analyses.* Washington, D.C., 1997.

6. EPA. *Portable Instruments User's Manual for Monitoring VOC Sources* (EPA 340/1-86-015). Washington, D.C., 1986.

7. Technical Note 1001. *Generating Calibration Gas Standards with Dynacal® Permeation Devices.* Houston, TX: VICI Medtronics, Inc., 2004.

FIELD CALIBRATION OF GAS AND VAPOR SENSORS

Note: This Appendix contains easy-to-apply information on field calibration of gas and vapor sensors for direct reading instruments as a supplement to Chapter 10. Appendix B contains more complete information of calibration gas mixtures. This appendix is excerpted from *Hazardous Gas Monitors—a Practical Guide to Selection, Operation and Applications* by Jack Chou (McGraw-Hill, 2000). It is reproduced with permission of the copyright owner, International Sensor Technology, Inc. (www.intlsensor.com).

Gas sensors need to be calibrated and periodically checked to ensure sensor accuracy and system integrity. It is important to install stationary sensors in locations where the calibration can be performed easily. The intervals between calibrations can be different from sensor to sensor. Generally, the manufacturer of the sensor will recommend a time interval between calibrations. However, it is good general practice to check the sensor more closely during the first 30 days after installation. During

this period, it is possible to observe how well the sensor is adapting to its new environment.

Also, factors that were not accounted for in the design of the system might surface and can affect the sensor's performance. If the sensor functions properly for 30 continuous days, this provides a good degree of confidence about the installation. Any possible problems can be identified and corrected during this time. Experience indicates that a sensor surviving 30 days after the initial installation will have a good chance of performing its function for the duration expected. Most problems—such as an inappropriate sensor location, interference from other gases, or the loss of sensitivity—will surface during this time.

During the first 30 days, the sensor should be checked weekly. Afterward, a maintenance schedule, including calibration intervals, should be established. Normally, a monthly calibration is adequate to ensure the effectiveness and sensibility of each sensor; this monthly check will also

Air Monitoring for Toxic Exposures, Second Edition. By Henry J. McDermott
ISBN 0-471-45435-4   © 2004 John Wiley & Sons, Inc.

afford you the opportunity to maintain the system's accuracy.

The method and procedure for calibrating the sensors should be established immediately. The calibration procedure should be simple, straightforward, and easily executed by regular personnel. Calibration here is simply a safety check, unlike laboratory analyzers that require a high degree of accuracy. For area air quality and safety gas monitors, the requirements need to be simple, repeatable, and economical. The procedure should be consistent and traceable. The calibration will be performed in the field where sensors are installed so it can occur in any type environment. Calibration of the gas sensor involves two steps. First the "zero" must be set and then the "span" must be calibrated.

STEP ONE: SETTING THE "ZERO" READING

There is no established standard that defines zero air. Many analytical procedures, including some specific analyzer procedures such as EPA methods, use pure nitrogen or pure synthetic air to establish the zero point. The reason for this is that bottled nitrogen and pure synthetic air are readily available. As a result, it is popularly believed that using bottled nitrogen or synthetic air is a good method to zero a sensor.

Unfortunately, this is not correct. Normal air contains traces of different gases besides nitrogen and oxygen. Also, ambient air normally contains a small percentage of water vapor. Therefore, it is much more realistic and practical to zero the sensor using the air surrounding the sensor when the area is considered to be clean. This reference point can be difficult to establish. Therefore, a good reference point can be in the area where air is always considered clean, such as in an office area. This will give a more realistic representation of the zero point because it will be representative of the local ambient air condition. The lack of water vapor can cause the zero point setting to read lower than in ambient air making the sensor zero appear to drift. This is most noticeable in solid-state sensors and PIDs.

Calibration Methods

Taking all factors such as the type of sensor and the conditions of the application into consideration, the following are some proposed methods of calibration:

1. In applications where the ambient air is normally clean, and, based on the operator's judgment that no abnormal condition exists and the instrument is indicating a close to zero reading, the procedure to zero the sensor can be skipped. When in doubt, use a plastic bag to get a sample of what is considered to be "clean air" in the facility and expose it to the sensor for a few minutes. This is a very quick and easy procedure. It is also a very effective way to differentiate a real alarm from a false alarm.

2. Compressed air has the advantage that it is easy to regulate and can be carried around in a bottle. Also, in many facilities, shop air is available throughout the plant, making it very accessible and convenient. However, most shop air contains small concentrations of hydrocarbons, carbon monoxide, carbon dioxide, and possibly other interference gases. Also, the air is typically very low in humidity. A solution to this is that the air can be filtered through activated charcoal to remove most of the unwanted gases, and water vapor can be added into the air using a humidifier in the sampling system. After this conditioning, the air can be used to calibrate most types of sensors. However, it is impor-

tant to note that carbon monoxide is *not* removed by charcoal filters.

It is therefore imperative to make sure that the CO concentration in the shop air is the same as in the ambient air. Furthermore, a soda ash filter should be used to remove carbon dioxide. This is also a very good way to zero carbon dioxide sensors since placing a soda ash filter in-line with the sampling system will remove all carbon dioxide, thus providing an easily obtainable zero baseline.

Although synthetic air is usually very pure, it cannot be used with solid-state sensors or PID sensors because these sensors require some water vapor in the sample stream. A simple solution to this problem is to add a wet tissue paper in the sample line. This acts as a humidifier in the sample stream and provides enough water vapor for the sensor to read properly.

STEP TWO: SPAN CALIBRATION

The span calibration can be quite easy or it can be very complicated and expensive, depending on the gas type and concentration range. In principle, to achieve the best accuracy, *a mixture of the target gas balanced in the background environmental air is the best calibration gas.* However, although this can be done, it usually requires that the operators be more skilled than would usually be required. In practice, most calibration gases are purchased from commercial suppliers. The following section describes a few methods of span calibration.

A. Premixed Calibration Gas

This is the preferred and most popular way to calibrate gas sensors. Premixed gas mixtures are compressed and stored under pressure in a gas bottle. The bottles are available in many sizes, but most field cali-

brators employ smaller, lightweight bottles. These small portable bottles come in two different categories: a low-pressure and a high-pressure version. The low-pressure bottles are thin-walled, lightweight bottles that are usually nonreturnable and disposable. High-pressure bottles are designed to bottle pure hazardous chemicals. For calibration gases, these bottles are normally made of thick-walled aluminum which has a service pressure of 2000 psi.

To get this highly pressurized gas out of the bottle in order to calibrate the sensor, a regulator assembly is needed. This assembly consists of a pressure regulator, a pressure gauge, and an orifice flow restrictor. The orifice flow restrictor is a fitting with a hairline hole that allows a constant airflow at a given pressure difference. In operation, the high pressure from the bottle is reduced to a lower pressure of only a few psi, which provides a constant air flow through the orifice. Flowrates between 600–1000 cc/min are most common. Models can be fitted with an adjustable pressure regulator so that the flowrate can be adjusted accordingly.

Many gases can be premixed with air and stored under pressure, but some gases can only be mixed in inert gas backgrounds, such as nitrogen. Some mixtures can only be stored in bottles that are specially treated or conditioned. Each type of mixture will have a different amount of time before it expires or before it can no longer be used.

Detailed information about storage and shelf life can be obtained from the manufacturer. Generally, high-vapor-pressure gases with low reactivity, such as methane, carbon monoxide, and carbon dioxide, can be mixed with air and stored under high pressure. Low-vapor-pressure gases, such as liquid hydrocarbon solvents, can only be mixed with air and stored under low pressure. Most highly reactive chemicals are mixed with a nitrogen background. With certain sensors, such as solid-state sensors, whether the mixture of the gas is in the air or in the nitrogen background will dramatically affect the sensor reading. During calibration, some sensors may need moisture to get a proper reading. Moisture can be added by following the same procedure described in Step 1 for zeroing the sensor. To estimate the volume of a pressurized gas in a cylinder, take the total pressure (P) divided by the atmospheric pressure (P_a) and multiply this ratio by the volume of the cylinder:

$$V_{mix} = V(P/P_a)$$

where V_{mix} is the volume of the gas mixture, V is the volume of the cylinder, P is the pressure in the cylinder, and P_a is the atmospheric pressure. For example, let's say a given lecture bottle has a 440-cc volume (V). Assume the bottle has a 1200-psi pressure. The estimated volume of the premixed gas at atmospheric pressure is

$$V_{mix} = (440\,cc) \times (1200/14.7) = 35,918\,cc$$

If the flowrate of the calibration gas is 1000 cc/min and it takes approximately one minute per sensor to calibrate, a single cylinder can be used to calibrate a sensor approximately 30 times.

B. Cross Calibration

Cross calibration takes advantage of the fact that every sensor is subject to interference by other gases. For example, for a sensor calibrated to 100% LEL hexane, it is usually much easier to use 50% LEL *methane* gas to calibrate the sensor instead of using an actual hexane mixture. This is because hexane is a liquid at room temperature and it has a low vapor pressure. Therefore, it is more difficult to make an accurate mixture and to keep it under high pressure.

On the other hand, methane has a very high vapor pressure and is very stable. Fur-

thermore, it can be mixed with air and still be kept under high pressure. It can be used for many more calibrations than a hexane mixture in the same size bottle, and it has a long shelf life. A 50% LEL methane mixture is also readily available. Therefore, it is common practice for manufacturers of combustible gas instruments to recommend the use of methane as a substitute to calibrate for other gases.

There are two ways to accomplish this task. The first method is to calibrate the instrument to methane while other gas readings are obtained by multiplying the methane reading by response factors that are included in the operating manual. This is commonly done with catalytic sensors. Catalytic sensors have a linear output, and therefore the use of this response factor is applicable to the full-scale range. For example, pentane has an output of only half that of methane gas when the sensor is calibrated to methane. Therefore, it has a response factor of 0.5. So, if the instrument is calibrated to methane but is used to measure pentane, the reading is multiplied by 0.5 to obtain the pentane reading.

The second method is to still use methane as the calibration gas, but double the value of the reading of the calibration. For instance, use 50% LEL methane calibration gas and calibrate with this as 100% LEL pentane. After the calibration, the instrument directly indicates the pentane gas concentration although it was calibrated using methane gas.

Many low-range toxic gas sensors can be calibrated using cross gas calibration. Also, with infrared instruments, any gas within the same wavelength of absorption can be used for cross calibration. The advantage of cross calibration is that it allows the sensor to be calibrated with a gas and range that is easier to obtain and handle.

However, there are some problems with using cross calibration. One is that the response factors for each sensor can be different because it is generally impossible to make most sensors exactly alike. For example, in catalytic sensors, the heater voltage has to be as specified in the operating manual; otherwise, the response factor will not be applicable. The response characteristics will vary with different heater voltage settings. Therefore, it is a good practice to periodically check the calibration of the sensor with the actual target gas.

Mixtures of stable noncombustible and nontoxic gases with various concentrations are available from many supply sources. Check with the instrument manufacturer for more detailed information.

C. Gas Mixing

Not all calibration gases are available. Even if they are available, it is very possible that they would not be available in the right concentration or in the proper background mixture. However, many mixtures are available for some process uses which can be diluted to use in calibration of gas monitors in lower concentration ranges. For example, 50% LEL methane has a concentration of 2.5% or 25,000 ppm. To make a 20% LEL mixture having a volume of 2000 cc, the following formula can be used:

$$V_b = \frac{C}{C_b} \times V, \qquad V_a = \frac{(C - C_b)}{C} \times V,$$
$$V_a = V - V_b$$

where C_b is the concentration in the bottle, 50% in this case; C is the new concentration, 20% in this case; V is the total final volume, 2000 cc in this case; V_b is the volume of mixture; and V_a is the volume of air or other diluent.

$$V_b = 20/50 \times 2000 = 800 \text{ cc}$$
$$V_a = 2000 - 800 = 1200 \text{ cc}$$

The final mixture would be made by taking 800 cc of the calibration gas and mixing it

with 1200 cc of air to make the mixture equal to 20% LEL.

Another example is to dilute this 25,000 ppm of methane calibration gas to make a 100 ppm of mixture.

$$V_b = 100/25,000 \times 2000 = 8 \text{ cc}$$

therefore

$$V_a = 2000 - 8 = 1992 \text{ cc}$$

By mixing 8 cc of calibration gas into 1992 cc of air, 2000 cc of 100 ppm gas mixture is obtained.

SOME CALIBRATION TOOLS

To perform the above procedure, the following tools are needed:

- *Syringe and Needle:* This is the most inexpensive way to accurately measure the amount of gas. A disposable medical syringe with a large-gauge needle is most practical, but there are few syringes with more than 100-cc volume. Hence, large volume measurements can be troublesome. However, it is easy to make a syringe using any standard size pipe having about a 2-inch diameter. It provides an easy and convenient means to make a mixture on a regular basis. For very small volume measurements, there are micro-syringes that are readily available in chemical supply catalogs.
- *Calibration Bag:* Most of the materials used in food packaging or storage are quite inert; otherwise, food would be contaminated with odor. Therefore, food storage bags can be used to hold most chemicals as long as they are used for relatively short durations.

This is an important point to keep in mind since gas molecules will eventually diffuse through the many thin layers of a plastic bag. For example, potato chips can stay fresh in their original bag for long periods of time because the bag material is less permeable by gas molecules than normal food storage bags. This is demonstrated by the fact that when the potato chips are transferred into a tightly sealed food container bag, they will lose their crispness in a very short time. There are also many commercially available sampling bags on the market. One common example is a Tedlar bag. It is made from polyvinyl fluoride and has low absorption of gas molecules. However, this type of bag is still permeable, so a heavy gauge material will be needed if permeability is a major concern. Sampling bags normally come with a valve and a septum that is used as an injection port.

CALIBRATING LIQUID CHEMICAL MIXTURES

To make a calibration mixture for liquid chemicals, a known volume of liquid is vaporized in a known volume of diluent air. The ideal gas law states that one gram mole of molecules will occupy 24,500 cc of volume at 25°C and 760 mm of mercury or sea-level atmospheric pressure. This temperature and pressure is also called the *standard condition*. At standard conditions, the equation is

$$C_{ppm} = 24.5 \times 10^9 \times (V \times D)/(V_a \times M)$$

where V is the volume of liquid, D is the density of the liquid, which is the same as the specific gravity; V_a is the volume of the diluent air; and M is the molecular weight of the liquid.

Since it is easier to measure the liquid using a micro-syringe, the equation then becomes

$$V = (C_{ppm} \times V_a \times M)/(24.5 \times 10^9 \times D)$$

where all units are in milliliters, cubic centimeters, and grams.

For example, benzene has $M = 78.1$ g and $D = 0.88$ g/cc. What is the amount of benzene needed to make a 1000-ppm mixture in a 2000-cc bottle?

$$V = (1000 \times 2000 \times 78.1)/(24.5 \times 10^9 \times 0.88)$$

which yields

$$V = 0.0072\,cc = 7.2\,\mu L$$

Table B.3 (in previous Appendix) shows the quantity (in μL) of some common liquid chemicals to be added to a 5-gallon container to produce an airborne vapor concentration of 100 ppm.

In air pollution, industrial hygiene, and medical toxicology work, the commonly used unit of concentration is milligram per cubic meter. The following equation expresses this relation, again assuming standard conditions:

$$C_{ppm} = (C_{mg/m^3} \times 24.5)/M$$

In applying these concepts, it is often convenient to use readily available containers. Since many containers are sized by "gallons." it is useful to know that one gallon is equal to 3785 cc.

For the calibration of most gas and vapor sensors, accuracy is not extremely important because the instrument is not an analytical device or system. However, it is most important to keep the calibration methods standardized and easily traceable. If procedures are standardized, data can be normalized at a later date if necessary.

APPENDIX D

CHEMICAL-SPECIFIC GUIDELINES FOR AIR SAMPLING AND ANALYSIS

This Appendix contains a listing of sample collection device methods for common air contaminants. It shows the method, analytical technique, collection medium, required air volume, sampling rate, and limit of quantitation (LOQ). It is excerpted from the analytical procedures of Galson Laboratories, East Syracuse, NY. (www.galsonlabs.com).

Air Monitoring for Toxic Exposures, Second Edition. By Henry J. McDermott
ISBN 0-471-45435-4 © 2004 John Wiley & Sons, Inc.

Chemical Specific Guidelines for Air Sampling and Analysis

Substance	Method	Analytical Technique	Collection Medium	Comp Code	Air Volume (L)	Sampling Rate (LPM)	LOQ
Acetaldehyde	Modified NIOSH 2016/TO-11	HPLC	DNPH silica gel		1–15	0.2–0.5	0.2µg
	Note: Preferred method.						
	OSHA 68	GC/NPD	2HMP-treated XAD-2	B	3	0.05	0.4µg
	NIOSH 2538	GC/NPD	2HMP-treated XAD-2	B	3	0.05	0.4µg
Acetic acid	OSHA IMIS 0020	GC/FID	Charcoal		20–300	0.01–1.0	10µg
	ID-186SG	IC	Charcoal		48	0.2	5µg
Acetic anhydride	OSHA 102	GC/NPD	Treated glass fiber filter		7.5	0.05–0.5	10µg
	Note: Treated filters should be used within 30 days of preparation.						
Acetone	NIOSH 1300	GC/FID	Charcoal	A	0.5–3	0.01–0.2	3µg
	Note: 75% DE.						
	3M	GC/FID	PM	A	2 hr		4µg
	Note: 3M recommends maximum sampling time of 2 hr.						
	OSHA 69	GC/FID	Anasorb CMS		3	0.05	3µg
	Note: Preferred method. 90–100% DE.						
Acetonitrile	NIOSH 1606	GC/FID	Charcoal		3–25	0.01–0.2	3µg
	Note: Large charcoal tube recommended for sample collection.						
	3M	GC/FID	PM		2 hr		4µg
	Note: 3M recommends maximum sampling time of 2 hr.						
Acid mist HBr, HF, HCl, HNO$_3$, H$_2$SO$_4$, H$_3$PO$_4$	NIOSH 7903	IC	Washed silica gel		3–100	0.2–0.5	1µg
Acrolein	Modified NIOSH 2016/TO-11	HPLC	DNPH-treated silica gel		1–15	0.2–0.5	0.2µg
Acrylamide	OSHA 21	GC/NPD	GFF/silica gel		120	1	1µg
	Note: Place filter into 1 mL methanol after sampling.						
Acrylic acid	OSHA 28	HPLC/UV	XAD-8		24	0.1	1µg
	Note: Collect using two tubes in series.						
Acrylonitrile	NIOSH 1604	GC/FID	Charcoal, PM		4–20	0.01–0.2	3µg, 4µg
	OSHA 37	GC/NPD	Charcoal		20	0.2	3µg

Analyte	Method	Technique	Media		Volume	Flow	Limit
Alkaline dust	NIOSH 7401	Titration	37PTFE 1.0		70–1000	1.0–4.0	40 μg
	Note: Quantified as sodium hydroxide (NaOH).						
Aluminum	Modified NIOSH 7300	ICAP/ICP-MS	37MCEF 0.8		5–100	1–4	3 μg
Aluminum oxide	NIOSH 0500	Gravimetric	37PVC 5.0 preweighed		408–960	1–2	50 μg
Ammonia	NIOSH S347	ISE	Treated silica gel		30	0.1–0.2	20 μg
	Note: Preferred method.						
	OSHA ID-164	ISE	0.1N H₂SO₄		120	1	25 μg
Amyl acetate	NIOSH 1450	GC/FID	Charcoal, PM	A	1–10	0.01–0.2	3 μg
sec-Amyl acetate	NIOSH 1450	GC/FID	Charcoal	A	1–10	0.01–0.2	3 μg
Aniline	OSHA IMIS 0220	GC/FID	H₃PO₄-treated XAD		30	0.2	5 μg
Antimony	Modified NIOSH 7300	ICAP/ICP-MS	37MCEF 0.8		45	1.5	0.9 μg
Arsenic	Modified NIOSH 7300	ICAP/ICP-MS	37MCEF 0.8		5–2000	1–3	0.3 μg
	Note: Minimum 30-L air volume required to reach 5 μg/m³ action limit.						
Arsine	NIOSH 6001 modified	ICAP/ICP-MS	Charcoal		1–10	0.01–0.2	0.05 μg
Asbestos—fibers	OSHA Ref Method	PCM	25MCEF 1.2 or 0.8		1000–3000	1.0–10.0	0.1 fiber/cc
	NIOSH 7400	PCM	25MCEF 1.2 or 0.8		1000–3000	1.0–10.0	0.1 fiber/cc
Asbestos—fibers—TEM	AHERA	TEM	25MCEF 0.45			0.5–16.0	
Asbestos—bulk (friable)	EPA 600/M4-82-020	PLM dispersion staining	Double-bag sample vial				<1%
Asbestos-bulk (nonfriable)	EPA 600/M4-82-020	PLM pretreatment	Double-bag sample vial				<1%
	Note: Positive results reported as greater than 1% (>1%). Organically bound samples collected in New York State will require analysis by TEM if found to be negative by PLM.						
Asbestos-bulk (vacuum/wipe)	EPA	PLM dispersion staining	25 or 37MCEF 0.8 or 1.2				<1%
Asbestos-bulk Nonfriable organically bound sample (NOBs)		Matrix reduction only and PLM					<1%
Asbestos-bulk TEM analysis		Matrix reduction and ELAP 198.1 and 198.4					
Asphalt Fume							
Total particulate	NIOSH 5042	Gravimetric	PTFE 2.0 preweighed		960	2	50 μg
Benzene soluble fraction							
Bacteria—see Microbiological							
Barium	Modified NIOSH 7300	ICAP/ICP-MS	37MCEF 0.8		50–2000	1–4	0.15 μg

Chemical Specific Guidelines for Air Sampling and Analysis

Substance	Method	Analytical Technique	Collection Medium	Comp Code	Air Volume (L)	Sampling Rate (LPM)	LOQ
Barium, soluble compounds (as Ba)	OSHA ID-121	ICAP/ICP-MS	37MCEF 0.8		480–960	2	TBD
Benzene	NIOSH 1500/1501	GC/FID	Charcoal, PM	A	2–30	0.01–0.2	2 µg, 3 µg
	OSHA 12	GC/FID	Charcoal, PM	A	10	0.2	2 µg, 3 µg
BTEX	NIOSH 1500/1501	GC/FID	Charcoal, PM	A	2–30	0.01–0.2	2 µg, 3 µg
Benzyl Chloride	NIOSH 1003	GC/FID	Charcoal, PM	A	6–50	0.01–0.2	5 µg
Berryllium	Modified NIOSH 7300	ICAP	37MCEF 0.8		1250–2000	1–4	0.15 µg
	NIOSH 7300	ICP-MS	37MCEF 0.8		1250–2000	1–4	0.015 µg
BHT (butylated hydroxy toluene)	OSHA IMIS 2683	GC/FID	OVS		10–100	1	9 µg
Bismuth	Modified NIOSH 7300	ICAP/ICP-MS	37MCEF 0.8		1250–2000	1–4	0.75 µg
Bisphenyl A	NIOSH 333	HPLC/UV	GFF		288	1.6	0.2 µg
Biphenyl	NIOSH 2530	GC/FID	Tenax GC	D	15–30	0.01–0.5	2 µg
Boron	Modified NIOSH 7300	ICAP/ICP-MS	37MCEF 0.8		25–2000	1–4	1.5 µg
Bromine	OSHA ID-108	IC	$NaHCO_3/Na_2CO_3$		30	0.5	1 µg
	NIOSH 6011 (2)	IC	Silver membrane		8–360	0.3–1.0	2 µg
	Note: Preferred method, Protect cassettes from light; call in advance of sampling to order media.						
Bromoform	NIOSH 1003	GC/FID	Charcoal, PM	A	4–70	0.01–0.2	20 µg
1,3-Butadiene	NIOSH 1024	GC/FID	Charcoal, PM		5–25	0.01–0.5	2 µg
	Note: Ship overnight delivery. Sampler: 400 mg and 200 mg charcoal in separate tubes connected in series. 30 L of air needed to reach 1/10th PEL.						
	Sampler handling: Chill below –4°C. Separate front and back tubes. Interferences: Pentane, methyl acetylene, or vinylidene chloride may chromatographically interfere at high levels.						
	OSHA 56	GC/FID	4tBC treated charcoal		3	0.05	2 µg
2-Butanone (methyl ethyl ketone)	NIOSH 2500	GC/FID	Ambersorb, PM		0.2–12	0.01–0.2	2 µg, 3 µg
	OSHA 16	GC/FID	Silica gel		3	0.1	2 µg
	Note: Collect using 2 tubes in series. Tube sections will be combined.						
	OSHA 84	GC/FID	Anasorb CMS		3	0.05	2 µg
2-Butoxyethanol (butyl Cellosolve)	NIOSH 1403	GC/FID	Charcoal, PM	F	1–10	0.01–0.05	5 µg, 8 µg
	OSHA 83	GC/FID	Charcoal	F	48	0.1	5 µg

Substance	Method	Analysis	Media		Volume	Flow	Amount
Butoxyethyl Acetate (butyl Cellosolve acetate)	OSHA 83	GC/FID	Charcoal, PM	F	48	0.1	5 µg
n-Butyl acetate	NIOSH 1450	GC/FID	Charcoal, PM	A	1–10	0.01–0.2	3 µg, 4 µg
	Note: Ship overnight refrigerated.						
sec-Butyl acetate	NIOSH 1450	GC/FID	Charcoal, PM	A	1–10	0.01–0.2	3 µg, 4 µg
	Note: Ship overnight refrigerated.						
tert-Butyl acetate	NIOSH 1450	GC/FID	Charcoal, PM	A	1–10	0.01–0.2	3 µg, 5 µg
	Note: Ship overnight refrigerated.						
Butyl acrylate	NIOSH 1450	GC/FID	Charcoal, PM	A	1–10	0.01–0.2	2 µg, 3 µg
	Note: Ship overnight refrigerated.						
n-Butyl alcohol	NIOSH 1401	GC/FID	Charcoal, PM		2–10	0.01–0.2	3 µg
sec-Butyl alcohol	NIOSH 1401	GC/FID	Charcoal, PM		2–10	0.01–0.2	3 µg
tert-Butyl alcohol	NIOSH 1400	GC/FID	Charcoal, PM		1–10	0.01–0.2	10 µg
Butyl carbitol	OSHA 53	GC/FID	Charcoal, PM		10	0.1	10 µg
Butyl carbitol acetate	OSHA 53	GC/FID	Charcoal, PM		10	0.1	10 µg
n-Butyl glycidyl ether	NIOSH 1616	GC/FID	Charcoal, PM	A	15–30	0.01–0.2	5 µg
Cadmium	Modified NIOSH 7300	ICAP/ICP-MS	37MCEF 0.8		13–2000	1–4	0.15 µg
Calcium	Modified NIOSH 7300	ICAP/ICP-MS	37MCEF 0.8		5–200	1–4	7.5 µg
Caprolactam	OSHA IMIS 0524	HPLC	OVS		100	1	
	Note: Additional phase $40.						
Carbon black	NIOSH 5000	Gravimetric	37PVC 5.0 preweighed		30–570	1–2	50 µg
	Note: Call in advance of sampling to order media.						
	OSHA ID-196	THF extraction gravimetric	37PVC 5.0 preweighed		480–960	2	1 mg
	Note: Call in advance of sampling to order media.						
Carbon, elemental & organic	NIOSH 5040	EGA (evolved gas analysis)	37-Quartz fiber or jeweled impactor		150	2–4	3 µg
Carbon tetrachloride	NIOSH 1003	GC/FID	Charcoal, PM	A	3–150	0.01–0.2	20 µg, 30 µg
Carpet gas (4-phenylcyclohexene)	OSHA IMIS R222	GC/FID	Charcoal		10–1200	0.5–2	2 µg
Catechol	OSHA 32	HPLC/UV	XAD-7		24	0.1	1 µg
Chloramines—see Nitrogen trichloride							
Chlordane	NIOSH 5510	GC/ECD	37MCEF 0.8/CSORB		10–200	0.05–1	0.2 µg
	OSHA 67	GC/ECD	OVS		480	1	0.2 µg

Chemical Specific Guidelines for Air Sampling and Analysis

Substance	Method	Analytical Technique	Collection Medium	Comp Code	Air Volume (L)	Sampling Rate (LPM)	LOQ
Chlorine	OSHA ID-101	ISE	Sulfamic acid		15–240	1	10 µg
	Note: Ship overnight refrigerated.						
	NCASI 520	Titration	2% KI		10–100	0.5–1.0	20 µg
	NIOSH 6011	IC	Silver membrane		10–90	0.3–1.0	5 µg
	Note: Preferred method. Protect cassettes from light; call in advance of sampling to order media.						
Chlorine dioxide	NCASI 520	Titration	2% KI		40–100	0.5–1.0	10 µg
	Note: Chlorine and chlorine dioxide may be collected and analyzed together by NCASI 520.						
Chlorobenzene	NIOSH 1003	GC/FID	Charcoal, PM	A	1.5–40	0.01–0.2	2 µg, 3 µg
Chloroform	NIOSH 1003	GC/FID	Charcoal, PM	A	1–50	0.01–0.2	10 µg, 15 µg
	OSHA 05	GC/FID	Charcoal, PM	A	10	0.2	10 µg, 15 µg
Chloromethane (methyl chloride)	NIOSH 1001	GC/FID	Charcoal		0.4–3	0.01–0.1	10 µg
Chlorpyrifos (Dursban)	OSHA 62	GC/ECD	OVS	E	480	1	0.04 µg
Chromic acid	NIOSH 7600	Colorimetric	37PVC 5.0		8–400	1.0–4.0	0.1 µg
	Note: Filter must be removed from cassette and put into glass vial immediately after sampling.						
Chromium (hexavalent) Insoluble total	NIOSH 7600	Colorimetric	37PVC 5.0		8–400	1.0–4.0	0.05 µg
	Note: Filter must be removed from cassette and put into glass vial immediately after sampling.						
Chromium (hexavalent) soluble	NIOSH 7600	Colorimetric	37PVC 5.0		8–400	1.0–4.0	0.05 µg
	Note: Filter must be removed from cassette and put into glass vial immediately after sampling.						
Chromium	Modified NIOSH 7300	ICAP/ICP-MS	37MCEF 0.8		5–1000	1.0–4.0	1.5 µg
Cigarette smoke (nicotine)	NIOSH 2551	GC/NPD	XAD-4		600	1	0.5 µg
Cigarette smoke (2-ethenyl pyridine)	NIOSH 2551	GC/NPD	XAD-4		600	1	0.5 µg
Coal dust (respirable)	NIOSH 0600	Gravimetric	37PVC 5.0 preweighed		20–400	1.7 or 2.2	50 µg
	Note: Call in advance of sampling to order media.						
Coal tar pitch volatiles (also see PNAH)	OSHA 58	Gravimetric	Glass fiber filter		960	2	50 µg
	Note: Remove filter from cassette and place into vial, wrap with aluminum foil to protect from sunlight. Send samples refrigerated, overnight delivery.						
	OSHA 58	HPLC/UV-FL	Glass fiber filter		960	2	3 µg

Note: Remove filter from cassette and place into vial, wrap with aluminum foil to protect from sunlight. Send samples refrigerated, overnight delivery.

Substance	Method	Analysis	Media		Volume	Conc	Amount
Cobalt	Modified NIOSH 7300	ICAP/ICP-MS	37MCEF 0.8		25–2000	1.0–4.0	0.45 µg
Copper	Modified NIOSH 7300	ICAP/ICP-MS	37MCEF 0.8		5–1000	1.0–4.0	0.3 µg
Cotton dust	NIOSH 0500	Gravimetric	37PVC 5.0 preweighed		7–133	1–2	50 µg

Note: Call in advance of sampling to order media.

Substance	Method	Analysis	Media		Volume	Conc	Amount
Cresols (each isomer)	OSHA 32	HPLC/UV	XAD-7		24	0.1	0.4 µg

Note: Preferred method.

Substance	Method	Analysis	Media		Volume	Conc	Amount
	NIOSH 2546	GC/FID	XAD-7	A	1–24	0.01–0.1	2 µg
Cumene	NIOSH 1501	GC/FID	Charcoal, PM	A	1–30	0.01–0.2	2 µg, 3 µg
Cyanide (gaseous)	NIOSH 7904	ISE	0.1 N KOH		10–180	0.5–1.0	2.6 µg
	NIOSH 6010	Colorimetric	Soda lime tube		0.6–90	0.05–0.2	2.6 µg
Cyanide (particulate)	NIOSH 7904	ISE	37PVC 0.8		10–180	0.5–1.0	2.5 µg
Cyclohexane	NIOSH 1500	GC/FID	Charcoal	A	2.5–5	0.01–0.2	2 µg
	3M	GC/FID	PM	A	6 hr		3 µg

Note: 3M recommends maximum sampling time of 6hr.

Substance	Method	Analysis	Media		Volume	Conc	Amount
Cyclohexanol	NIOSH 1402	GC/FID	Charcoal, PM	A	1–10	0.01–0.2	3 µg
Cyclohexanone	NIOSH 1300	GC/FID	Charcoal, PM	A	1–10	0.01–0.2	3 µg
	OSHA 01	GC/FID	Chromosorb 106		10	0.05–0.2	3 µg
Cyclohexene	NIOSH 1500	GC/FID	Charcoal, PM	A	1–7	0.01–0.2	2 µg
2,4-D	NIOSH 5001	HPLC/UV	Glass fiber filter		15–200	1–3	2 µg
2,4,5-T	NIOSH 5001	HPLC/UV	Glass fiber filter		15–200	1–3	2 µg
DDVP (Dichlorvos)	OSHA 62	GC/ECD	OVS	E	480	1	0.04 µg
Diacetone alcohol	NIOSH 1402	GC/FID	Charcoal, PM		1–10	0.01–0.2	3 µg
Diazinon	OSHA 62	GC/NPD	OVS	E	480	1	0.04 µg
Dibutylphthalate	NIOSH 5020	GC/FID	37MCEF 0.8		6–200	1.0–3.0	5 µg
	OSHA 104	GC/FID	OVS		240	1	5 µg
Dichlorobenzene (each isomer)	NIOSH 1003	GC/FID	Charcoal, PM	A	1–60	0.01–0.2	2 µg, 3 µg

Note: Please specify isomer required.

Substance	Method	Analysis	Media		Volume	Conc	Amount
Dichlorobenzenes, total	NIOSH 1003	GC/FID	Charcoal, PM	A	1–60	0.01–0.2	6 µg, 9 µg
1,1-Dichloroethane	NIOSH 1003	GC/FID	Charcoal, PM	A	0.5–15	0.01–0.2	3 µg
1,2-Dichloroethane (ethylene dichloride)	NIOSH 1003	GC/FID	Charcoal, PM	A	0.5–10	0.01–0.2	3 µg, 4 µg
	OSHA 7	GC/FID	Charcoal, PM	A	10	0.2	
1,2-Dichloroethylene	NIOSH 1003	GC/FID	Charcoal, PM	A	1–50	0.01–0.2	5 µg
Dichlorvos (DDVP)	OSHA 62	GC/ECD	OVS	E	480	1	0.04 µg

Chemical Specific Guidelines for Air Sampling and Analysis

Substance	Method	Analytical Technique	Collection Medium	Comp Code	Air Volume (L)	Sampling Rate (LPM)	LOQ
Diethanolamine	OSHA IMIS D129	HPLC/UV	NITC Treated XAD-2		10	0.1	0.5 μg
Diethylene glycol	OSHA 53	GC/FID	Charcoal	F	10	0.1	10 μg
Duthy lene glycol methyl ether	Note: [2-(2-Methoxyethoxy) ethanol] or methyl carbitol.						
Diethylene triamine	OSHA 60	HPLC/UV	NITC-treated XAD-2		10	0.1	1.0 μg
	NIOSH 2540	HPLC/UV	NITC-treated XAD-2		1–20	0.01–0.1	1.0 μg
Di(2-ethylhexyl) phthalate (dioctylphthalate)	NIOSH 5020	GC/FID	37MCEF 0.8		10–200	1.0–3.0	5 μg
	OSHA 104	GC/FID	OVS		240	1	5 μg
	Note: Preferred method.						
Dimethylacetamide	NIOSH 2004	GC/NPD	Silica gel		15–80	0.01–1.0	0.5 μg
n,n-Dimethylaniline	OSHA IMIS 0931	GC/FID	H_3PO_4-treated XAD		30	0.2	5 μg
Dimethylformamide (DMF)	NIOSH 2004	GC/FID	Silica gel		15–80	0.01–1.0	5 μg
	OSHA 66	GC/NPD	Charcoal, PM		10	0.2	5 μg
1,4-Dioxane	NIOSH 1602	GC/FID	Charcoal, PM	A	0.5–15	0.01–0.2	3 μg, 4 μg
Diphenyl	NIOSH 2530	GC/FID	Tenax	D	15–30	0.01–0.5	2 μg
Diphenylamine	OSHA 78	HPLC/UV	H_2SO_4-treated filter		100	1	1 μg
Dipropylene glycol monomethyl ether	OSHA 53	GC/FID	Charcoal	F	10	0.1	20 μg
Dursban—see Chlorpyrifos	OSHA 62	GC/ECD	OVS	E	480	1	0.04 μg
Dust, alkaline	NIOSH 7401	Titration	1-μm Teflon		70–1000	1.0–4.0	40 μg
	Note: Quantified as sodium hydroxide (NaOH).						
Dust, nuisance	NIOSH 0500	Gravimetric	37PVC 5.0 preweighed		7–133	1–2	50 μg
	Note: Call in advance of sampling to order media.						
Dust, respirable	NIOSH 0600	Gravimetric	37PVC 5.0 preweighed		20–400	1.7 or 2.5	50 μg
	Note: Call in advance of sampling to order media. Flowrate cyclone-dependent						
Epichlorohydrin	NIOSH 1010	GC/FID	Charcoal, PM	A	2–30	0.01–0.2	3 μg, 4 μg
Ethanolamine	OSHA IMIS 1030	HPLC/UV	NITC-treated XAD-2		10	0.1	1 μg
2-Ethenyl pyridine	EPA IP-2A	GC/NPD	XAD-4		60–480	1	0.7 μg

Compound	Method	Analysis	Media		Sampling time/rate	Flow rate	Mass
2-Ethoxyethanol (Cellosolve)	NIOSH 1403	GC/FID	Charcoal	F	1–6	0.01–0.05	5 µg
	OSHA 53	GC/FID	Charcoal	F	10	0.1	5 µg
2-Ethoxyethyl acetate (Cellosolve acetate)	NIOSH 1450	GC/FID	Charcoal	A	1–10	0.01–0.2	5 µg
	OSHA 53	GC/FID	Charcoal	F	10	0.1	5 µg
Ethyl acetate	NIOSH 1475	GC/FID	Charcoal	A	0.1–10	0.01–0.2	3 µg
Note: Ship samples overnight refrigerated.							
	3M	GC/FID	PM	A	6 hr		4 µg
Note: 3M recommends a maximum sampling time of 6hr.							
Ethyl acrylate	NIOSH 1450	GC/FID	Charcoal, PM	A	1–10	0.01–0.2	3 µg, 4 µg
Ethyl alcohol	NIOSH 1400	GC/FID	Charcoal		0.1–1	0.01–0.05	3 µg
Note: Store samples in the freezer, Ship samples overnight refrigerated.							
	3M	GC/FID	PM		1 hr		4 µg
Note: 3M recommends a maximum sampling time of 1hr.							
	OSHA 100	GC/FID	Anasorb 747		12	0.05	TBD
Note: Collect using two single section tubes in series. Separate tubes immediately after collection.							
Ethylamine	OSHA 36	HPLC/FL	Treated XAD-7		10	0.2	1.0 µg
Ethylbenzene	NIOSH 1501	GC/FID	Charcoal, PM	A	10–24	0.01–0.2	2 µg, 3 µg
	OSHA 1002	GC/FID	Charcoal, PM	A	1–30	0.05	TBD
Ethyl-2-cyanoacrylate	OSHA 55	HPLC/UV	H_3PO_4-treated XAD-7		12	0.1	0.7 µg
Note: Refrigerate samples after collection and during shipment to the laboratory.							
Ethyl ether	NIOSH 1610	GC/FID	Charcoal		0.25–3	0.01–0.2	3 µg
	3M	GC/FID	PM		4 hr		4 µg
Note: 3M recommends a maximum sampling time of 4hr.							
Ethyl silicate	NIOSH S264	GC/FID	XAD-2		10	0.05	3 µg
Ethylenediamine	OSHA 60	HPLC/UV	NITC-treated XAD-2		10	0.1	0.2 µg
	NIOSH 2540	HPLC/UV	NITC-treated XAD-2		1–20	0.01–0.1	0.2 µg
Ethylene dibromide	NIOSH 1003	GC/FID	Charcoal		1–25	0.01–0.2	5 µg
Ethylene glycol	NIOSH 5523	GC/FID	OVS		5–60	0.5–2	5 µg
Note: Ship overnight refrigerated.							
Ethylene oxide	NIOSH 1614	GC/ECD	Treated charcoal		1–24	0.05–0.15	0.5 µg
Note: Ship overnight refrigerated.							
	OSHA 50	GC/ECD	Treated charcoal		24	0.1	0.5 µg
Note: Preferred method. Ship overnight refrigerated. 3-smaple minimum.							
	OSHA 49	GC/ECD	PM		4–6 hr		0.75 µg
Note: Ship overnight refrigerated. Call in advance to arrange analysis.							

Chemical Specific Guidelines for Air Sampling and Analysis

Substance	Method	Analytical Technique	Collection Medium	Comp Code	Air Volume (L)	Sampling Rate (LPM)	LOQ
Fibers—Asbestos							
Fiber count	OSHA Ref. Method	PCM	25MCEF 1.2 or 0.8		1000–3000	1.0–10.0	0.1 fiber/cc
	NIOSH 7400	PCM	25MCEF 1.2 or 0.8		1000–3000	1.0–10.0	0.1 fiber/cc
Fibers—TEM analysis	NIOSH 7402		25MCEF 0.45		See method	0.5–16.0	
	EPA AHERA		25MCEF 0.45		See method	0.5–16.0	
Fibrous glass	NIOSH 7400 B rules	PCM	25MCEF 1.2 or 0.8		400	0.5–16.0	0.1 fiber/cc
Fluoride, gaseous	NIOSH 7902	ISE	Treated pad		12–800	1.0–2.0	5 µg
Fluoride, particulate (soluble)	NIOSH 7902	ISE	37MCEF 0.8		12–800	1.0–2.0	5 µg
Fluoride, particulate (total)	NIOSH 7902	ISE	37MCEF 0.8		12–800	1.0–2.0	5 µg
Fluoride, gaseous and particulate (soluble) or total	NIOSH 7902	ISE	Treated pad and 37MCEF 0.8		12–800	1.0–2.0	5 µg
Formaldehyde	NIOSH 2016	HPLC	DNPH silica gel		1–15	0.1–1.5	0.1 µg
	Note: Preferred method.						
	OSHA 52	GC/NPD	2HMP Treated XAD	B	3–24	0.1–0.2	0.2 µg
	NIOSH 2541	GC/NPD	2HMP Treated XAD	B	1–36	0.01–0.1	0.2 µg
	NIOSH 3500	Colorimetric	1% NaHSO$_3$		1–100	0.2–1.0	5 µg
	Note: Transfer impinger solution to glass vial for shipping. May be modified for particulate.						
	Assay	HPLC	PM				
	OSHA ID 205	Colorimetric	PM		8 hr		3 µg
	Note: 3M—not recommended for STEL sampling.						
	SKC-UME	HPLC	PM		8–24 hr		TBD
	SKC-STEL	Colorimetric	PM		15 min		0.5 ppm
	SKC-PEL	Colorimetric	PM		8 hr		0.2 ppm
	SKC-IAQ	Colorimetric	PM		5–7 days		0.01 ppm
	Note: Freezer storage necessary for media; ship samples back to the laboratory refrigerated.						
Formic acid	OSHA ID-112	IC	0.01 N NaOH		120	1	5 µg
	NIOSH 2011	IC	PTFE/washed silica gel		1–24	0.05–0.2	5 µg
Freons (specify each type) (11, 12, 22, 113, 141B)	Various NIOSH methods GC/FID		Call for correct media		0.5–2.5	0.01–0.05	4–50 µg

668

Substance	Method	Analysis	Media	Group			Amount
Fungi—see microbials							
Furfuryl alcohol	NIOSH 2505	GC/FID	Porapak-Q tube		3–25	0.05–0.5	3 µg
Furfuryl aldehyde (furfural)	OSHA 72	GC/FID	Petroleum charcoal		180	1	10 µg
Glutaraldehyde	NIOSH 2532	HPLC/UV	DNPH silica gel		1–30	0.05–0.5	0.3 µg
Note: Preferred method.							
	OSHA 64	HPLC/UV	Treated GFF		15–120	1	0.3 µg
Note: Cover cassettes with foil during sampling. Sample open faced, preferred media storage is under refrigeration.							
Assay		HPLC/UV	Aldehyde PM				
Heptane	NIOSH 1500	GC/FID	Charcoal, PM	A	4	0.01–0.2	2 µg, 3 µg
Hexane (*n*-hexane)	NIOSH 1500	GC/FID	Charcoal	A	4	0.2	2 µg
	3M	GC/FID	PM	A	7hr		3 µg
Note: 3M recommends a maximum sampling time of 7hr.							
HDI—hexamethylene diisocyanate	OSHA 42	HPLC/UV	Treated GFF	G	15–240	1	0.3 µg
Note: Freezer storage is necessary for media. Sample open-faced.							
IsoChek	HPLC/UV/FL	IsoChek filter		15–30	1	0.03 µg
Note: Monomer (10-day TAT on IsoChek analysis); oligomer as NCO equivalent.							
Hydrobromic acid	NIOSH 7903	IC	Washed silica gel		3–100	0.2–0.5	1 µg
Hydrocarbons, total	NIOSH 1500/1501	GC/FID	Charcoal, PM	A	4	0.01–0.2	10 µg, 15 µg
Note: Please specify aliphatic or aromatic. Aliphatic quantified using the response factor and molecular weight of *n*-hexane, aromatic uses toluene.							
Hydrochloric acid	NIOSH 7903	IC	Washed silica gel		3–100	0.2–0.5	1 µg
Hydrofluoric acid	NIOSH 7903	IC	Washed silica gel		3–100	0.2–0.3	1 µg
Note: Do not exceed 0.3 LPM sampling rate.							
Hydrogen peroxide	OSHA IMIS 1470	Colorimetric	TiOSO4		100	0.5	14 µg
Note: Transfer impinger solution to glass vial for shipping.							
Hydrogen sulfide	NIOSH 6013	IC	Treated charcoal		1.2–40	0.1–1.5	3 µg
Note: Method recommended flow rate is 0.2 LPM.							
Hydroquinone	OSHA IMIS 1490	HPLC/UV	Treated XAD-7		20	0.2	0.2 µg
Iodine	OSHA ID-212	ISE	Treated charcoal (KOH)		2.5–7.5	0.5–1.0	2.5 µg
Iron	Modified NIOSH 7300	ICAP/ICP-MS	37MCEF 0.8		5–100	1.0–4.0	7.5 µg

Chemical Specific Guidelines for Air Sampling and Analysis

Substance	Method	Analytical Technique	Collection Medium	Comp Code	Air Volume (L)	Sampling Rate (LPM)	LOQ
Iron (oxide fume)	Modified NIOSH 7300	ICAP/ICP-MS	37MCEF 0.8		5–100	1.0–4.0	11 µg
Isoamyl acetate	NIOSH 1450	GC/FID	Charcoal, PM	A	1–10	0.01–0.2	3 µg, 4 µg
Isoamyl alcohol	NIOSH 1402	GC/FID	Charcoal		1–10	0.01–0.2	2 µg
Isobutyl acetate	NIOSH 1450	GC/FID	Charcoal, PM	A	1–10	0.01–0.2	3 µg, 4 µg
Isobutyl alcohol	NIOSH 1401	GC/FID	Charcoal		2–10	0.01–0.2	3 µg
Isophorone diisocyanate (IPDI)	OSHA 42	HPLC/UV	Treated GFF	G	15	1	0.3 µg
	Note: Freezer storage is necessary for media. Sample open-faced.						
	IsoChek	HPLC/UV/FL	IsoChek filter		15–30	1	0.03 µg
	Note: Monomer (10-day TAT on IsoChek analysis).						
Isopropyl acetate	NIOSH 1454	GC/FID	Charcoal	A	0.1–9	0.02–0.2	3 µg
	3M	GC/FID	PM	A	7 hr		4 µg
	Note: 3M recommends a maximum sampling time of 7 hr.						
Isopropyl alcohol	NIOSH 1400	GC/FID	Charcoal, PM		0.2–3	0.01–0.2	3 µg, 4 µg
	Note: Refrigerate samples after collection and during shipment to the laboratory.						
	OSHA 109	GC/FID	Anasorb 747		18	0.05–0.2	TBD
	Note: Collect using two single-section tubes in series. Separate tubes immediately after collection.						
Isooctane	NIOSH 1500	GC/FID	Charcoal, PM	A	4	0.01–0.2	2 µg, 3 µg
Kerosene	NIOSH 1550	GC/FID	Charcoal	A	1–20	0.01–0.2	100 µg
	Note: Please send, under separate cover, a sample (bulk) of your particular solvent for use as a reference standard.						
Lead (air)	Modified NIOSH 7300	ICAP/ICP-MS	37MCEF 0.8		50–2000	1.0–4.0	0.38 µg
Lead (wipe)	Modified NIOSH 9100	ICAP/ICP-MS	2 × 2 Gauze or Whatman filter, ghost wipe				0.38 µg
	Note: Moisten gauze with de-ionized water.						
	Modified NIOSH 9100	ICAP/ICP-MS	Pace wipe or large gauze				1.2 µg
	Note: Moisten gauze with de-ionized water.						
Lead (paint)	EPA Pb92-114172 modified	ICAP/ICP-MS	ICAP/ICP-MS	3 Grams			50 mg/kg
	Note: Minimum of 1 g required.						
Magnesium (oxide fume)	Modified NIOSH 7300	ICAP/ICP-MS	37MCEF 0.8		5–67	1.0–4.0	3 µg
Malathion	NIOSH 5600 modified	GC/NPD	OVS	E	12–60	0.2–1	0.04 µg
	OSHA 62	GC/NPD	OVS	E	60	1	0.04 µg

Compound	Method	Analysis	Media		Volume (L)	Flow (L/min)	Detection limit
Maleic anhydride	OSHA 86	HPLC/UV	Treated filters		60	0.5	0.5 μg
	Note: Store samples in freezer; ship overnight, refrigerated.						
Manganese	Modified NIOSH 7300	ICAP/ICP-MS	37MCEF 0.8		5–200	1.0–4.0	0.15 μg
MDI—Methylene bisphenyl isocyanate	OSHA 47	HPLC/UV	Treated GFF	G	15–240	1	0.3 μg
	Note: Freezer storage is necessary for media. Sample open-faced.						
	IsoChek	HPLC/UV/FL	IsoChek filter		15–80	1	0.04 μg
	Note: Monomer (10-day TAT on IsoChek analysis); oligomer as NCO equivalent.						
Mercury (vapor/particulate)	NIOSH 6009	CVAA	Hydrar or filter		2–100	0.15–0.25	0.06 μg
	NIOSH 6009	CVAA	Filter/Hydrar (both)		2–100	0.15–0.25	0.06 μg
	OSHA 140	CVAA	Badge				0.1 μg
Mercury (wipe)	NIOSH 6009 modified	CVAA	MCEF or Whatman filter				0.04 μg
	Note: Moisten filter with de-ionized water.						
Metal removal fluid aerosol							
Total	ASTM PS 42-97	Gravimetric	Preweighed Teflon		500–960	2	250 μg
	Note: Call in advance to order media.						
Extractable	ASTM PS 42-97	Gravimetric	Preweighed Teflon		500–960	2	250 μg
	Note: Call in advance to order media. Includes "total."						
2-Methoxyethanol (methyl Cellosolve)	OSHA 53	GC/FID	Charcoal	F	10	0.1	3 μg
	NIOSH 1403	GC/FID	Charcoal	F	6–50	0.01–0.05	3 μg
2-Methoxyethyl acetate (methyl Cellosolve acetate)	OSHA 53	GC/FID	Charcoal	F	10	0.1	5 μg
	NIOSH 1451	GC/FID	Charcoal	A	0.2–20	0.01–0.2	5 μg
1-Methoxy-2-propanol (propylene glycol methyl ether)	OSHA 99	GC/FID	Charcoal		10	0.1	5 μg
1-Methoxy-2-propyl acetate (propylene glycol methyl ether acetate)	OSHA 99	GC/FID	Charcoal		10	0.1	5 μg
Methyl acetate	NIOSH 1458	GC/FID	Charcoal	A	0.2–10	0.01–0.2	3 μg
	3M	GC/FID	PM		2 hr		4 μg
Methyl acrylate	NIOSH 1459	GC/FID	Charcoal	A	1–5	0.01–0.2	3 μg

671

Chemical Specific Guidelines for Air Sampling and Analysis

Substance	Method	Analytical Technique	Collection Medium	Comp Code	Air Volume (L)	Sampling Rate (LPM)	LOQ
Methyl alcohol	NIOSH 2000	GC/FID	Silica gel		1–5	0.02–0.2	5 µg
	Note: Refrigerate samples after collection and during shipment to the laboratory.						
	OSHA 91	GC/FID	Anasorb 747		3–5	0.05	5 µg
	Note: Samples collected in series. The tubes must be separated as soon as possible after sampling. Do not exceed maximum recommended volume.						
Methyl amyl ketone	NIOSH 1300	GC/FID	Charcoal, PM	A	1–25	0.01–0.2	3 µg, 4 µg
	NIOSH 1301	GC/FID	Charcoal		1–25	0.01–0.2	3 µg
Methyl bromide	OSHA IMIS 1680	GC/FID	Anasorb 747		2.5–11	0.01–1	4 µg
Methyl chloride	NIOSH 1001	GC/FID	Charcoal		0.4–3	0.01–0.1	3 µg
Methyl chloroform	NIOSH 1003	GC/FID	Charcoal, PM	A	0.1–8	0.01–0.2	5 µg, 8 µg
(1,1,1-trichloroethane)	OSHA 14	GC/FID	Charcoal, PM	A	3	0.2	5 µg, 8 µg
Methyl-2-cyanoacrylate	OSHA 55	HPLC/UV	H_3PO_4-treated XAD-7		12	0.1	2 µg
	Note: Refrigerate samples after collection and during shipment to the laboratory.						
Methylene bisphenyl isocyanate— see MDI							
Methylene chloride	OSHA 80	GC/FID	Anasorb CMS		3	0.05	5 µg
	Note: Preferred method.						
	NIOSH 1005	GC/FID	Two-in-series charcoal, PM		0.5–2.5	0.01–0.2	5 µg, 8 µg
	Note: Tube sections will be combined.						
	OSHA 59	GC/FID	Three-section charcoal tube		10	0.05	5 µg
4,4-Methylene dianiline (MDA)	NIOSH 5029	HPLC	Treated GFF		10–1000	1–2	0.1 µg
	Note: Filter must be transferred into vial containing 4ml methanolic KOH within 4hr of sampling. Sample open faced.						
Methyl ethyl ketone (2-butanone)	NIOSH 2500	GC/FID	Ambersorb, PM		0.2–12	0.01–0.2	2 µg, 3 µg
	OSHA 16	GC/FID	Silica gel		3	0.1	2 µg
	Note: Collect using two tubes in series. Tube sections will be combined.						
Methyl isobutyl ketone (hexone)	OSHA 84	GC/FID	Anasorb CMS		3	0.05	TBD
	NIOSH 1300	GC/FID	Charcoal, PM	A	1–10	0.01–0.2	2 µg, 3 µg
	Note: Refrigerate samples after collection and during shipment to the laboratory.						

Analyte	Method	Analysis	Media				
Methyl mercaptan	OSHA 26 modified	GC/PID	Treated GFF		20	0.2	0.4 µg
Methyl methacrylate	NIOSH 2537	GC/FID	XAD-2		1–8	0.01–0.05	3 µg
	Note: Ship with dry ice overnight.						
	OSHA 94	GC/FID	4tBC-treated charcoal, PM		3	0.05	TBD
	Note: Send samples refrigerated, overnight delivery.						
Methyl-*n*-butyl ketone (2-hexanone)	NIOSH 1300	GC/FID	Charcoal, PM	A	1–10	0.01–0.2	2 µg, 3 µg
1-Methyl-2-pyrrolidinone	NIOSH 1302	GC/FID	Charcoal		0.5–125	0.05–0.2	5 µg
	Note: Ship samples overnight refrigerated. Protect samples from light.						
Methyl styrene (each isomer)	NIOSH 1501	GC/FID	Charcoal	A	3–30	0.02–0.2	3 µg
Methyl *tert*-butyl ether	NIOSH 1615	GC/FID	Charcoal		2–96	0.1–0.2	20 µg
Microbials							
Spores, mycelial fragments, pollen	In-house	Microscopy	Zefon 37-mm Air-O-Cell		15–150 L	15	1 spore
Spores, mycelial fragments, pollen fibers, skin cells	In-house	Microscopy	Zefon 37-mm Air-O-Cell		15–150 L	15	1 spore
Viable mold	In-house	Microscopy	PDA agar plate		28–280 L	28	1 CFU
Viable bacteria	In-house	Microscopy	TSA agar plate		28–280 L	28	1 CFU
Viable thermophilic mold	In-house	Microscopy	PDA agar plate		28–280 L	28	1 CFU
Viable thermophilic bacteria	In-house	Microscopy	TSA agar plate		28–280 L	28	1 CFU
Spores, mycelial fragments	In-house	Microscopy	Bio tape		1–100 cm^2		1 LOC
Viable mold	In-house	Microscopy	Swab		1–100 cm^2		20 CFU
Viable bacteria	In-house	Microscopy	Swab		1–100 cm^2		20 CFU
Viable thermophilic mold	In-house	Microscopy	Swab		1–100 cm^2		20 CFU
Viable thermophilic bacteria	In-house	Microscopy	Swab		1–100 cm^2		20 CFU

Chemical Specific Guidelines for Air Sampling and Analysis

Substance	Method	Analytical Technique	Collection Medium	Comp Code	Air Volume (L)	Sampling Rate (LPM)	LOQ
Spores, mycelial fragments	In-house	Microscopy	Bulk		0.5–100 g		1 LOC
Viable mold	In-house	Microscopy	Bulk		0.5–100 g		20 CFU
Viable bacteria	In-house	Microscopy	Bulk		0.5–100 g		20 CFU
Viable thermophilic mold	In-house	Microscopy	Bulk		0.5–100 g		20 CFU
Viable thermophilic bacteria	In-house	Microscopy	Bulk		0.5–100 g		20 CFU
Spores, mycelial fragments	In-house	Microscopy	Condensate		1–10 mL		1 LOC
Viable mold	In-house	Microscopy	Condensate		1–10 mL		1 CFU/mL
Viable bacteria	In-house	Microscopy	Condensate		1–10 mL		1 CFU/mL
Viable thermophilic mold	In-house	Microscopy	Condensate		1–10 mL		1 CFU/mL
Viable thermophilic bacteria	In-house	Microscopy	Condensate		1–10 mL		1 CFU/mL
Identification of predominant bacterium	API	Biochemical	PDA agar plate		28–280 L	28	1 CFU
Identification of predominant bacterium	API	Biochemical	Swab		1–100 cm^2		20 CFU
Identification of predominant bacterium	API	Biochemical	Bulk		0.5–100 g		20 CFU
Identification of predominant bacterium	API	Biochemical	Condensate		1–10 mL		1 CFU/mL
Mold by PCR (air)	EPA patent no. 6,387,652	Real-time PCR	Sterilized 37-mm polycarbonate		200–1000	5–10	Species-dependent (1–71 spores/filter)

Analyte	Method	Analysis	Media				
Mold by PCR (bulk)	6,387,652	Real-time PCR	10mg				
Molybdenum	Modified NIOSH 7300	ICAP/ICP-MS	37MCEF 0.8	5–67		1.0–4.0	0.15 µg
Molybdenum, soluble compounds (as Mo)	OSHA ID-121	ICAP-ICP-MS	37MCEF 0.8	480–960		2	TBD
Naphtha (VM&P)	NIOSH 1550	GC/FID	Charcoal, PM	1.3–20	A	0.01–0.2	35 µg, 52 µg

Note: Please send, under separate cover, a sample (bulk) of your particular solvent for use as a reference standard. Additional analyses may be required to determine the DE for results that do not match a common standard.

Analyte	Method	Analysis	Media				
Naphthalene	NIOSH 1501	GC/FID	Charcoal, PM	100–200	A	0.01–1.0	2 µg, 3 µg
	OSHA 35	GC/FID	Chromosorb 106	10		0.2	2 µg

Note: Preferred method.

Analyte	Method	Analysis	Media				
Nickel	Modified NIOSH 7300	ICAP/ICP-MS	37MCEF 0.8	25–1000		1.0–4.0	0.3 µg
Nickel, soluble compounds (as Ni)	OSHA ID-121	ICAP/ICP/MS	37MCEF 0.8	480–960		2	TBD
Nicotine	NIOSH 2551	GC/NPD	XAD-4	600		1	0.5 µg
Nitric acid	NIOSH 7903	IC	Washed silica gel	3–100		0.2–0.5	5 µg
Nitric oxide	NIOSH 6014	Colorimetric	TEA tube/oxidizer	1.5–6		0.025	1.3 µg
	OSHA ID-190	IC	TEA tube/oxidizer	6		0.025	0.5 µg

Note: Freezer storage of media preferred.

Analyte	Method	Analysis	Media				
Nitrogen dioxide	NIOSH 6014	Colorimetric	TEA tube	1.5–6		0.025–0.2	2 µg
	OSHA ID-182	IC	TEA tube	3		0.2	0.8 µg
	EPA IP-5B	Colorimetric	Drager tube—passive	8hr–14 day			0.15 µg
Nitrogen trichloride	In-house	IC	Treated filter	160		1	5 µg

Note: Call in advance of sampling to order media.

Analyte	Method	Analysis	Media				
Nitroglycerin	NIOSH 2507	GC/ECD	Tenax	3–100		0.2–1.0	0.05 µg
1-Nitropropane	OSHA 46	GC/FID	XAD-4	4		0.1	3 µg
2-Nitropropane	OSHA 46	GC/FID	XAD-4	4		0.1	3 µg

Note: Preferred method.

Analyte	Method	Analysis	Media				
Nitrosamines (each isomer)	NIOSH 2528	GC/FID	Chromosorb 106	0.1–2		0.01–0.05	3 µg
	NIOSH 2522 modified	GC/NPD	Thermosorb/N	15–1000		0.2–2.0	0.04 µg
	OSHA 27 modified	GC/NPD	Thermosorb/N	75		0.2–2.0	0.04 µg
Octane	NIOSH 1500	GC/FID	Charcoal, PM	4	A	0.01–0.2	2 µg, 3 µg
Oil mist Total	ASTM PS 42-97	Gravimetric	Preweighed Teflon	960		2	50 µg

Note: Call in advance to order media.

Analyte	Method	Analysis	Media				
Extractable	ASTM PS 42-97	Gravimetric	Preweighed Teflon	960		2	50 µg

Chemical Specific Guidelines for Air Sampling and Analysis

Substance	Method	Analytical Technique	Collection Medium	Comp Code	Air Volume (L)	Sampling Rate (LPM)	LOQ
Ozone	*Note:* Call in advance to order media. Includes "total."						
	OSHA ID-214	IC	Treated filter		90–120	0.25–0.5	5 µg
	Note: Treated filters must be used within four weeks of preparation.						
Parathion	OSHA 62	GC/NPD	OVS	E	480	1	0.04 µg
Parrafin wax fume	OSHA IMIS 2000	GC/FID	Glass fiber filter		100	1	50 µg
PCB air	NIOSH 5503	GC/ECD	GFF & Florisil		1–50	0.05–0.2	0.05 µg
(polychlorinated biphenyls)	*Note:* Method allows sampling—1 LPM for 24 hr.						
	OSHA STOPGAP	GC/ECD	OVS		60	1	0.05 µg
PCB wipe	40 CFR 761	GC/ECD	Gauze				0.5–1.0 µg
(polychlorinated biphenyls)	*Note:* Moisten wipe with hexane prior to sampling.						
PCB oil	600/4-81-045	GC/ECD	Glass vial				5 mg/kg
(polychlorinated biphenyls)							
n-Pentane	NIOSH 1500	GC/FID	Charcoal	A	2	0.01–0.05	2 µg
	3M	GC/FID	PM		3 hr		3 µg
Perchloroethylene	NIOSH 1003	GC/FID	Charcoal, PM	A	0.2–40	0.01–0.2	5 µg, 8 µg
(tetrachloroethylene)	OSHA 1001	GC/FID	Charcoal, PM		1–23	0.05	TBD
	NYS-DOH 311-9	GC-FID	PM				0.03 µg
Phenol	OSHA 32	HPLC/UV	XAD-7		24	0.1	1 µg
	Note: Preferred method.						
	NIOSH 2546	GC/FID	XAD-7		1–24	0.01–0.1	TBD
4-Phenylcyclohexene	OSHA IMIS R222	GC/FID	Charcoal		10–1200	0.05–2	5 µg
Phenyl ether	NIOSH 1617	GC/FID	Charcoal		1–50	0.01–0.2	3 µg
Phenyl ether–diphenyl mixture	NIOSH 2013	GC/FID	Charcoal		1–40	0.01–0.2	TBD
Phenyl hydrazine	Assay	HPLC	PM				2 µg
Phosphoric acid	NIOSH 7903	IC	Washed silica gel		3–100	0.2–0.5	1 µg
PNAH (polynuclear aromatic hydrocarbons)							

Analyte	Method	Analysis	Media		Volume	Flow	Amount
5-Compound profile	OSHA 58	HPLC/UV-FL	Glass fiber filter		960	2	0.3 μg
	Note: Remove filter from cassette and place into vial, wrap with aluminum foil to protect from sunlight. Send samples refrigerated, overnight delivery.						
17-Compound profile	NIOSH 5506	HPLC/UV-FL	37PTFE 1.0/XAD		200–100	2	0.3 μg
	Note: Remove filter from cassette and place into vial, wrap with aluminum foil to protect from sunlight. Send samples refrigerated, overnight delivery.						
Coal tar pitch volatiles	OSHA 58	Gravimetric	Glass fiber filter		480–960	1.5–2	50 μg
	Note: Remove filter from cassette and place into vial, wrap with aluminum foil to protect from sunlight. Send samples refrigerated, overnight delivery.						
Potassium	Modified NIOSH 7300	ICAP/ICP-MS	37MCEF 0.8		120–240	1.0–2.0	15 μg
Potassium hydroxide	NIOSH 7401	Titration	37PTFE 1.0		70–1000	1.0–4.0	60 μg
n-Propyl acetate	NIOSH 1450	GC/FID	Charcoal, PM	A	1–10	0.01–0.2	3 μg, 4 μg
n-Propyl alcohol	NIOSH 1401	GC/FID	Charcoal		1–10	0.01–0.2	3 μg
	3M	GC/FID	PM		6 hr		4 μg
	Note: 3M recommends a maximum sampling time of 6hr.						
Propylene oxide	OXHA 88	GC/FID	Anasorb 747		5	0.1	TBD
	Note: Store collected samples in freezer; however, overnight refrigerated shipment is not necessary.						
Pyrethrum	NIOSH 5008	HPLC/UV	Glass Fiber Filter		20–400	1–4	0.5 μg
	Note: Preferred method.						
Pyridine	OSHA 70	GC/ECD	XAD-2		60	1	0.5 μg
	NIOSH 1613	GC/FID	Charcoal		18–150	0.01–1.0	2 μg
Rotenone	NIOSH 5007	HPLC/UV	37PTFE 1.0		8–400	1–4	1.0 μg
Selenium	Modified NIOSH 7300	ICAP/ICP-MS	37MCEF 0.8		5–2000	1–4	0.3 μg
Silica, crystalline	NIOSH 7603	IR	37PVC 5.0		300–1000	1.7 or 2.2	30 μg
	Note: Sample with cyclone to collect respirable fraction. Flow rate cyclone dependent.						
Quartz (includes dust)	NIOSH 7500	XRD	37PVC 5.0		400–1000	1.7 or 2.2 or 2.5	10 μg
		Flow rate cyclone dependent.					
Quartz (without dust)	NIOSH 7500	XRD	37PVC 5.0		400–1000	1.7 or 2.2 or 2.5	10 μg
		Flow rate cyclone dependent.					
Silica, crystalline (bulk)	NIOSH 7500	XRD	Bulk				1%
	Note: Includes quartz, cristobalite, and tridymite.						
Silicon	Modified NIOSH 7300	ICAP	37MCEF 0.8		120–240	1.0–2.0	15 μg
Silver	Modified NIOSH 7300	ICAP/ICP-MS	37MCEF 0.8		250–2000	1.0–4.0	0.45 μg
Sodium	Modified NIOSH 7300	ICAP/ICP-MS	37MCEF 0.8		13–2000	1.0–4.0	30 μg

677

Chemical Specific Guidelines for Air Sampling and Analysis

Substance	Method	Analytical Technique	Collection Medium	Comp Code	Air Volume (L)	Sampling Rate (LPM)	LOQ
Sodium hydroxide	NIOSH 7401	Titration	37PTFE 1.0		70–1000	1.0–4.0	40 µg
Stoddard solvent	NIOSH 1550	GC/FID	Charcoal, PM	A	1.3–20	0.01–0.2	50 µg, 75 µg
	Note: Please send, under separate cover, a sample (bulk) of your particular solvent for use as a reference standard. Additional analyses may be required to determine the DE for results that do not match a common standard.						
Strontium	Modified NIOSH 7300	ICAP/ICP-MS	37MCEF 0.8		25–2000	1–4	0.75 µg
Styrene	OSHA 89	GC/FID	Treated charcoal		0.75–12	0.01–0.05	2 µg
	Note: Preferred method.						
	NIOSH 1501	GC/FID	Charcoal, PM	A	1–14	0.01–1.0	2 µg, 3 µg
	OSHA 09	GC/FID	Charcoal, PM	A	10–15	0.2–1.0	2 µg, 3 µg
Sulfates	NIOSH 6004	IC	37MCEF 0.8		200	1.5	1 µg
	Note: Remove filter and place in vial prior to shipping.						
Sulfite	NIOSH 6004	IC	37MCEF 0.8		200	1.5	1 µg
	Note: Remove filter and place in vial prior to shipping.						
Sulfuric acid	NIOSH 7903	IC	Washed silica gel		3–100	0.2–0.5	1 µg
	P&CAM S174	Titration	37MCEF 0.8		180	1.5	50 µg
	Note: Remove filter from cassette and place into glass vial within 1 hr of sampling. Discard backup pad.						
Sulfur dioxide	NIOSH 6004	IC	Treated filter		4–200	0.5–1.5	10 µg
	OSHA ID-200	IC	Treated charcoal bead		12	0.1	10 µg
TDI—toluene diisocyanate (each isomer)	OSHA 42	HPLC/UV	Treated GFF	G	15–240	1	0.4 µg
	Note: Freezer storage necessary for media, sample open-faced. Specify 2,4-TDI or 2,6-TDI or both.						
	IsoChek	HPLC/UV/FL	IsoChek filter		15–30	1	0.03 µg
	Note: Specify 2,4-TDI or 2,6-TDI or both. (10-day TAT on IsoChek analysis).						
1,1,2,2-Tetrachloroethane	NIOSH 1019	GC/FID	Petroleum charcoal		3–30	0.01–0.2	10 µg
Tetrachloroethylene—see Perchloroethylene							
Tetraethyl lead	NIOSH 2533	GC/PID	XAD-2		100–200	0.1–1	0.5 µg
Tetrahydrofuran	NIOSH 1609	GC/FID	Charcoal, PM	A	1–9	0.01–0.2	3 µg, 4 µg
Thallium	Modified NIOSH 7300	ICAP/ICP-MS	37MCEF 0.8		25–2000	1–4	0.15 µg
Thallium, soluble compounds (as Tl)	OSHA ID-121	ICAP/ICP-MS	37MCEF 0.8		480–960	1–4	TBD
Thiourea	OSHA IMIS T109	HPLC/UV	Glass fiber filter		50	1.5	2.5 µg

Compound	Method	Technique	Media		Range		Amount
Thiram	NIOSH 5005	HPLC/UV	37PTFE 1.0		10–400	1.0–3.0	100 µg
Tin, Inorganic	OSHA ID-121	ICAP/ICAP-MS	37MCEF 0.8		5–500	1.0–4.0	3 µg
	Note: Sample separately from other metals.						
Titanium	Modified NIOSH 7300	ICAP/ICP-MS	37MCEF 0.8		5–100	1.0–4.0	0.15 µg
Tobacco smoke (nicotine)	NIOSH 2551	GC/NPD	XAD-4		60–480	1	1 µg
Tobacco smoke (2-ethenyl pyridine)	NIOSH 2551	GC/NPD	XAD-4		60–480	1	1 µg
Toluene	NIOSH 1500/1501	GC/FID	Charcoal, PM	A	2–8	0.01–0.2	2 µg, 3 µg
	NIOSH 4000	GC/FID	PM				3 µg
	OSHA 111	GC/FID	Charcoal, PM, Anasorb 747		1–12	0.05	TBD
Toluidine (each isomer) (*o*, *m*, and *p*)	NIOSH 2002	GC/FID	Silica gel	C	30	0.01–0.2	5 µg
Triethylene tetramine	NIOSH 2540	HPLC/UV	NITC-treated XAD-2		1–20	0.01–0.1	TBD
	OSHA 60	HPLC/UV	NITC-treated XAD-2		10	0.01	1.0 µg
1,1,2-Trichloroethane	NIOSH 1003	GC/FID	Charcoal	A	2–60	0.01–0.2	10 µg
	OSHA 11	GC/FID	Charcoal	A	10	0.2	10 µg
Trichloroethylene	NIOSH 1022	GC/FID	Charcoal, PM	A	1–30	0.01–0.2	5 µg, 8 µg
	OSHA 1001	GC/FID	Charcoal, PM		1–23	0.05	TBD
Triethanolamine	OSHA IMIS T185	GC/FID	GFF		100	1	250 µg
Trimethylbenzenes, total	OSHA IMIS 2505	GC/FID	Charcoal	A	10	0.1	
	Note: Each isomer (1,2,3-trimethylbenzene; 1,2,4-trimethylbenzene; 1,3,5-trimethylbenzene) can be reported individually.						
Triphenyl phosphate	NIOSH 5038	GC/NPD	37MCEF 0.8		10–400	1–3	0.6 µg
Tungsten	Modified NIOSH 7300	ICAP/ICP-MS	37MCEF 0.8		5–200	1.0–4.0	0.75 µg
Turpentine	NIOSH 1550	GC/FID	Charcoal	A	1–10	0.01–0.2	10 µg
	Note: Please send, under separate cover, a sample (bulk) of your particular solvent for use as a reference standard.						
Vanadium	Modified NIOSH 7300	ICAP/ICP-MS	37MCEF 0.8		5–2000	1.0–4.0	0.45 µg
Vinyl acetate	NIOSH 1453	GC/FID	Carbon molecular sieve		0.72–24	0.1–0.2	2 µg
Vinyl chloride	NIOSH 1007	GC/FID	Charcoal, PM		0.7–5	0.05	3 µg, 4 µg
	Note: Samples collected using two tubes in series.						
	OSHA 4	GC/FID	Charcoal, PM				

Chemical Specific Guidelines for Air Sampling and Analysis

Substance	Method	Analytical Technique	Collection Medium	Comp Code	Air Volume (L)	Sampling Rate (LPM)	LOQ
	OSHA 75	GC/FID	Carbosieve S-III		3	0.05	3 µg
	Note: Preferred method.						
Vinylidene chloride (1,1-Dichloroethene)	NIOSH 1015	GC/FID	Charcoal, PM	A	2.5–7	0.01–0.2	3 µg
	OSHA 19	GC/FID	Charcoal	A	3	0.2	3 µg
Vinyl toluene	NIOSH 1501	GC/FID	Charcoal, PM	A	2–30	0.01–0.2	2 µg, 3 µg
Volatile organics	EPA TO15	GC/MS	Mini-can		0.4–1.0	Varies	5 ppb
	Multiple NIOSH	GC/FID	Charcoal		2–8	0.01–0.2	2–20 µg
Welding fumes, total particulate	NIOSH 0500	Gravimetric	37PVC 5.0 preweighed		7–133	1–2	50 µg
	Note: Call in advance of sampling to order media.						
Wood dust	NIOSH 0500	Gravimetric	37PVC 5.0 preweighed		7–133	1–2	50 µg
	Note: Call in advance of sampling to order media.						
Xylenes, total	NIOSH 1501	GC/FID	Charcoal, PM	A	2–23	0.01–0.2	4 µg, 6 µg
	OSHA 1002	GC/FID	Charcoal, PM		1–30	0.05	TBD
Zinc	Modified NIOSH 7300	ICAP/ICP-MS	37MCEF 0.8		5–200	1.0–4.0	1.5 µg
Zirconium	Modified NIOSH 7300	ICAP/ICP-MS	37MCEF 0.8		5–200	1.0–4.0	0.75 µg

Abbreviations

CFU, Colony-forming unit; CVAA, Cold vapor atomic absorption; DE, Desorption efficiency; DNPH, Dinitrophenylhydrazine; ECD, Electron capture detector; ELCD, Electrolytic conductivity detector; FID, Flame ionization detector; FL, Fluorescence detector; GC, Gas chromatography; GC/MS, Gas chromatography/mass spectrometry; GFF, Glass fiber filter; 2HMP, 2-(Hydroxymethyl) piperidine; HPLC, High-performance liquid chromatography; IC, Ion chromatography; ICAP, Inductively coupled argon plasma emission spectroscopy; IMIS, OSHA Salt Lake City Laboratory in-house file; IOM, Institute of Occupational Medicine; IR, Infrared spectrophotometry; ISE, Ion-specific electrode; LOC, Level of contamination; LOQ, Limit of quantitation; MCEF, Mixed cellulose ester filter; MF, Membrane filter; NBDC, 7-Chloro-4-nitrobenzo-2-oxa-1,3-diazole; NCO, Isocyanate group; NITC, 1-Naphthylisothiocyanate; NOB, Non-organically bound; NPD, Nitrogen phosphorus detector; OVS, OSHA versatile sampler; PCB, Polychlorinated biphenyl compounds; PCM, Phase contract microscopy; PID, Photoionization detector; PLM, Polarized light microscopy; PM, Passive monitor; PNAH, Polynuclear aromatic hydrocarbons; PTFE, Teflon membrane filter; PUF, Polyurethane foam; PVC, Polyvinyl chloride filter; STEL, Short-term exposure limit; 4tBC, 4-tert-Butylcatechol; TEM, Transmission electron microscopy; UV, Ultraviolet detector; XRD, X-ray diffraction

Method References

NIOSH, Manual of Analytical Methods; OSHA, Analytical Methods Manual

Compatibility Codes

A, Charcoal and carbon disulfide desorption; B, Treated XAD-2 (2-(hydroxymethyl) piperidine and toluene desorption; C, Silica gel and 95% ethanol desorption; D, Tenax and carbon tetrachloride desorption; E, OVS sampler and toluene desorption; F, Charcoal and 5% methanol in methylene chloride; G, Treated GFF (1-(2 pyridyl) piperazine) and acetonitrile/dimethyl sulfoxide desorption

INDEX

Air Monitoring for Toxic Exposures, Second Edition. By Henry J. McDermott
ISBN 0-471-45435-4   © 2004 John Wiley & Sons, Inc.